PUBLISHED PAPERS OF E.A. CORNISH

PUBLISHED PAPERS

of

E.A. CORNISH

A Memorial Volume

E.A. Cornish Memorial Appeal
Adelaide
1974

Published by
E.A. Cornish Memorial Appeal
Adelaide
1974

Printed by
Coudrey Offset Press Pty. Ltd.
Adelaide

National Library of Australia Card Number
and I.S.B.N. 0 9598601 0 X

PREFACE

Dr. Edmund Alfred Cornish (1909 - 1973)*

As homage to the late Dr. Edmund Alfred Cornish his widow, Mrs. Gene May Cornish, and his colleagues in the CSIRO Divison of Mathematical Statistics of which he was Chief have decided to bring out this commemorative volume of his collected works.

Cornish was an esteemed chief, a sincere colleague, an inspired scientist and a fine person to know. His interests outside his work ranged from gardening and using his workshop, to cricket and ballroom dancing. He had a casual wit and a flair for writing lucid crisp English. His stern expression his minimal response to small talk at casual meetings, his conversation interspersed with gaps of silence made him seem aloof to some, inhibiting to others. But his friends fondly remember him for the warmth he conveyed, and reminisce with sadness about the relaxed thoughtful discussions in which short crackling laughs accompanied his accounts of humorous situations. His passionate sense of fair play and his loyalty made him a never-to-be-forgotten standby in any crisis which affected his family, friends or colleagues of whatever rank. Cornish's success in guiding his Division over the last thirty years, the acceptance of his promptings for the establishment of a Chair of Mathematical Statistics at the University of Adelaide and the adoption of his proposals for large-scale computing facilities in CSIRO bear witness to the high regard held for his counsel.

By introducing effective mathematical and statistical underpinning for scientific and industrial research in CSIRO since 1941, Cornish was responsible for the high standards in experimental design and data analysis which are now taken for granted. However, staff losses from senior positions in his Division to university chairs, both local and overseas, in the post-sputnik era continued to frustrate his plans to have sufficient high calibre staff to give all of CSIRO easy access to the impetus and power that mathematics and statistics are able to provide.

Cornish anticipated correctly that efficient use of the modern digital computer would greatly increase the scope and need for mathematics and statistics in research. In spite of having to grapple with the extra load on his Division and himself that this brought, he nevertheless found time to continue his work as an active scientist and to maintain a keen but non-interfering interest in the research projects of his staff, of whose achievements he was intensely proud.

An Agricultural Science degree from the University of Melbourne gave Cornish a broad education in plant, animal and physical sciences. This, together with his subsequent mathematics degree from the University of Adelaide, and the stimulating influence of his friend Sir Ronald Fisher, account for the variety of his research projects, his choice of staff and his interest in providing an interface between his and other divisions of CSIRO. The obvious links between his own theoretical work and

* A detailed curriculum vitae may be obtained from the 'Records of the Australian Academy of Science' (Vol. II, No. 4, 1973) in a Biographical Memoir by P. A. P. Moran, or from 'Biometrics', September 1973, in an Obituary by E. J. Williams.

the practical needs which gave rise to it, strengthened his conviction that mathematics is there to be used and developed as a tool and not as a toy. He accordingly envisaged that each of the diverse activities of CSIRO would generate its own need for mathematical and statistical know-how, and that staff recruited to meet this need would enter a temporary partnership with the host researcher. In this way he felt that his staff, while satisfying the immediate requirements of the work in hand would 'grow' in experience and knowledge and would proceed to make contributions to mathematics and statistics where they were most needed.

Cornish's contribution to multivariate distributions, data analysis and joint fiducial distributions made him a key-figure in the theory and practice of statistical inference. He knew from experience that, in statistical practice, the application of Bayes theorem to experimental data, though valid, can seldom be applied to experimental data, because often there is no prior distribution and, when there is, it is usually not known. In analysing such data he would not accept prior distributions which were made to 'materialise' by invoking Laplace's dictum that a parameter, the value of which is unknown, may be taken to be a sample from a (uniform) prior distribution. Instead he invariably used statistical methodology based on Fisher's notion of fiducial probability, where the truth of an assertion regarding a parameter is brought into one-one correspondence with the occurrence of an event with known frequency probability; the probability of the event is taken as a measure of uncertainty of the assertion and called fiducial probability. In the early sixties it became clear from the literature that, of the alleged paradoxes of fiducial probability, all except one, namely non-uniqueness, stemmed from misapprehensions regarding the notion itself or its logical manipulations. With respect to this exception Cornish showed that non-unique fiducial distributions can be obtained using methods advocated by Fisher himself.

Cornish's plan for retirement included a study of claims in the current literature to solutions of this non-uniqueness difficulty. His illness and death prevented this, interrupted the completion of his work on applications of the Cornish-Fisher expansion (presented as the third Fisher Memorial Lecture) and left unfinished substantial extensions of his D.Sc. work on yield trends and the influence of rainfall on wheat yields in South Australia. It is hoped that the publication of the present volume will stimulate further work in the areas so competently pioneered by Cornish.

A.M.W. Verhagen

Chairman, E.A. CORNISH
MEMORIAL APPEAL
Commonwealth Scientific And
Industrial Research Organization
Division of Mathematical
Statistics

November 1973

Grateful acknowledgment is made to the publishers and others concerned for their cooperation in permitting reproduction of the papers contained in this volume. Each paper, as well as the Contents, states the source of publication.

CONTENTS

[*Reprint from the Journal of the Council for Scientific and Industrial Research, Vol. 9. No. 1, February, 1936.*]

The Influence of Rainfall on the Yield of a Natural Pasture.

By H. C. Trumble, M.Agr.Sc., and E. A. Cornish, B.Agr.Sc.**

(*From the Waite Agricultural Research Institute*).

Summary.

The influence of annual, half-yearly, tri-monthly, bi-monthly, and monthly rainfall on the yield of unfertilized and top-dressed natural pasture, for a period of ten seasons at the Waite Institute, has been investigated by correlation and regression.

High significant positive correlations were obtained between the yield of each pasture and the rainfall for January-December, January-June, the tri-monthly and bi-monthly periods between February and July, and the single month of April. The correlation was strongest in the period April to June inclusive, which coincides with the early stages of seasonal growth.

High significant negative correlations were found between the yield of each pasture and November rainfall. There was, however, a significant negative relationship, for the ten years in question, between April-June rainfall and November rainfall. Elimination of this effect by partial correlation reduced the association between November rainfall and yield to insignificance.

The effectiveness of autumnal and early winter rainfall and the relative ineffectiveness of late winter and spring rainfall in determining yield appear to be associated with the type of pasture investigated, which is composed of species with a restricted growing period and a limited capacity for production.

The replacement of these species by cultivated herbage plants with a perennial deep-rooted habit and a capacity for growth over all or most of the year enables the rainfall to be used more effectively, and constitutes an important application of the results in practice.

1. Introduction.

Among the numerous factors which influence agricultural production, rainfall is largely the major determinant of yield, especially where the reliability of effective rains is not high. The amount of annual rainfall, however, is not usually so important as the quantities falling at critical periods of crop and pasture growth. Unfortunately, there is a paucity of data concerning the question under Australian conditions. Despite this, however, the concept of an "ideal" type of rainfall for given requirements is generally admitted. Apart from rain *per se,* the associated atmospheric factors of temperature, humidity, and wind velocity, by governing the rate at which moisture is dissipated through evaporation and transpiration, influence the effectiveness of rainfall to a marked degree.

The influence of weather on crop yields has received detailed attention in England, (3), (4), (5), (7), and much work on this aspect has been carried out in other countries.

In Australia, the problem has been studied to a limited extent, chiefly by Barkley, who concluded (1) that the rain falling immediately prior to the flowering of the wheat plant—that is, the rainfall for August

* Officers of the Waite Agricultural Research Institute of the University of Adelaide.

C.2679.

and September, in northern Victoria—exercised a powerful influence on the resulting harvest. The work of Richardson (6) on the water requirements of farm crops under Victorian conditions stressed the value of September-October rains in increasing wheat yield; it was shown that maximum losses of water by transpiration occurred at that particular time of the year. Apart from moisture actually gained from rain falling at, or immediately before, flowering, however, soil and sub-soil reserves of moisture accumulated previously must play a substantial part in supplying water during the spring period, and furthermore, the conservation of soil moisture by lessened transpiration following increased humidity and lowered temperature is probably an important associated effect of rainfall.

In connexion with pastoral production, Barkley (2) also investigated the influence of seasonal rainfall on the yearly wool clip of a station situated in the Western District of Victoria, and concluded that the major influence of rainfall in this case was restricted to the two-month period January and February, eight months prior to shearing early in November.

The effect of seasonal rainfall on pasture yields appears to have been little investigated. Few instances are available in which yields of pasture grass or herbage have been taken over a sufficient period to enable statistical treatment to be applied.

2. Scope of Investigation.

In the present case, advantage has been taken of a ten years' sequence of yields from plots established on an area of natural pasture at the Waite Institute in 1925. This area is typical of much grazing land in the better rainfall country of South Australia, and is dominated in its unfertilized condition by *Danthonia,* with smaller quantities of exotic annuals such as *Festuca myuros, Trifolium arvense, T. procumbens, Erodium botrys,* and *Echium plantagineum.* Under top-dressing with superphosphate, the pasture has become dominated by these annuals, with considerable reduction in the amount of *Danthonia* present.

The yields of pasture were taken from two plots of 0.8 acre each, one unfertilized and the other top-dressed with superphosphate (40 lb. P_2O_5) for a period of ten consecutive seasons at the Waite Institute. The plots were grazed by sheep, and the yields obtained by sampling from eight randomized quadrats of 2½ square metres each per plot; the quadrat frames were removed to grazed pasture at the commencement of each season. The rainfall figures are those taken from the Waite Institute rain gauge, situated approximately 10 chains from the plots.

3. Experimental Data.

The yields of natural pasture from the unfertilized and the top-dressed plots, together with the monthly rainfall for each of the years 1925-1934, are given in Table 1. The pasture yields are shown graphically in conjunction with tri-monthly rainfall for the periods January-March, April-June, July-September, and October-December, in Fig. 1.

TABLE 1.—SHOWING YIELDS OF NATURAL PASTURE AND MONTHLY RAINFALL FOR EACH OF THE SEASONS 1925-34, WAITE INSTITUTE.

—	1925.	1926.	1927.	1928.	1929.	1930.	1931.	1932.	1933.	1934.	Mean.	Standard. Error
1. Yield of Natural Pasture (cwt. per acre)—												
(*a*) Unfertilized ..	21·76	28·35	18·00	24·31	17·69	20·67	41·43	56·92	51·43	20·23	**30·08**	±4·22
(*b*) Top dressed with superphosphate	27·74	43·20	28·54	39·63	33·70	30·32	47·50	75·66	64·70	38·00	**42·90**	±4·37
(*c*) Increase ..	5·98	14·85	10·54	15·32	16·01	9·65	6·07	18·74	13·27	17·77	**12·82**	±1·44
						Monthly Total.						
2. Rainfall (inches)—												
January	0·57	..	0·24	1·09	0·67	0·03	0·76	0·23	1·22	0·58	**0·54**	
February ..	3·16	1·08	1·09	2·63	0·03	0·36	0·27	1·60	0·40	0·14	**1·08**	
March	0·37	0·01	1·15	1·37	0·25	0·05	1·72	1·64	1·42	0·72	**0·87**	
April	1·16	1·83	0·42	1·36	0·65	1·02	1·06	5·51	2·09	1·48	**1·66**	
May	5·11	6·27	3·31	2·65	1·73	0·97	3·25	2·11	6·85	0·06	**3·23**	
June	2·20	2·14	2·53	4·58	4·26	1·52	5·97	5·27	1·80	1·14	**3·14**	
July	3·14	2·29	3·51	3·33	2·65	4·90	4·17	3·53	2·19	1·18	**3·09**	
August	1·84	4·71	4·12	0·84	2·37	3·73	3·26	3·61	3·85	4·90	**3·32**	
September ..	3·76	2·76	1·17	2·04	2·72	3·21	4·51	1·94	3·71	4·05	**2·99**	
October ..	1·19	1·91	0·41	2·97	1·17	3·35	0·63	2·43	0·68	2·50	**1·72**	
November ..	0·47	0·79	1·42	0·53	1·26	1·12	1·07	0·56	0·26	3·79	**1·13**	
December ..	0·14	1·97	1·62	0·15	3·68	0·92	0·21	0·40	0·42	1·53	**1·10**	
Total ..	23·12	25·77	21·00	23·54	21·44	21·18	26·90	28·83	24·90	22·07	**23·88**	

(i) *Statistical method.*

The comparatively short sequence of seasons available precludes a detailed analysis according to the methods developed by Fisher for investigating the influence of rainfall on the yield of wheat, and which were subsequently used in similar investigations with barley and mangolds.

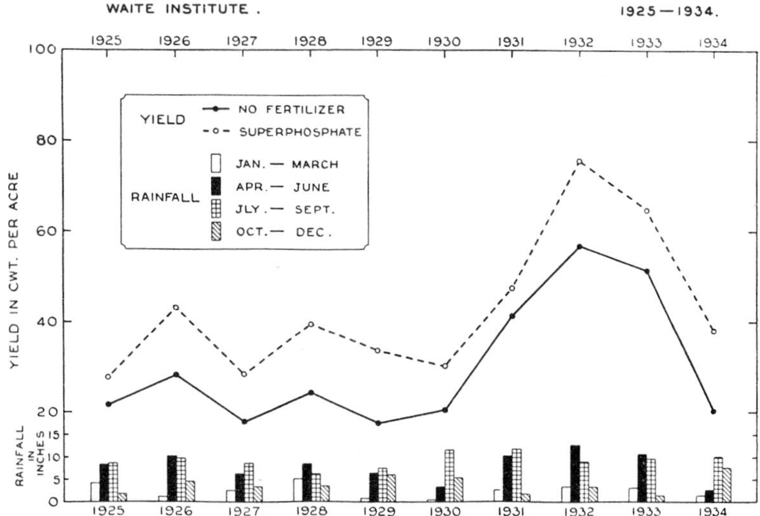

FIG. 1.—Graph showing the annual yield of natural pasture at the Waite Institute, 1925–34, (*a*) unfertilized, (*b*) top-dressed with 185 lb. superphosphate per acre, together with the rainfall in quarterly periods for each year.

A preliminary examination was commenced prior to the completion of the tenth season's harvest, and, in view of the marked increases in yield during the seventh, eighth, and ninth seasons, it was deemed advisable to take account of the yield trend by fitting a curve of the first degree and similarly treating the rainfall figures before correlating the variables. The inclusion of the tenth year considerably reduced the regression of yield on time, but the co-efficient, though insignificant, still remained sufficiently large to reduce the standard error of the mean yield; for this reason it was retained.

The yield residuals were correlated with the rainfall residuals of the following periods:—(i) annual, (ii) half-yearly, (iii) tri-monthly, (iv) bi-monthly, (v) monthly.

(ii) *Results of analysis by correlation and regression.*

The correlation coefficients for the yield of pasture and the annual, half-yearly, quarterly, bi-monthly, and monthly rainfall are given in Table 2. For brevity, only values greater than + 0.49 and less than — 0.49 have been included.

TABLE 2.—RAINFALL PERIOD.

Annual.	Half-yearly.	Quarterly.	Bi-monthly.	Monthly.
		(a) No fertilizer.		
Jan.-Dec. + 0·90*	Jan.-June + 0·91*	Feb.-Apl. + 0·73‡	Feb.-Mar. + 0·53	March + 0·57
	July-Dec. — 0·62	Mar.-May + 0·93*	Mar.-Apl. + 0·82*	April + 0·73‡
		Apl.-June + 0·96*	Apl.-May + 0·89*	May + 0·67‡
		May-July + 0·85*	May-June + 0·85*	Nov. — 0·81*
		Sept.-Nov. — 0·63	Oct.-Nov. — 0·68‡	Dec. — 0·50
		Oct.-Dec. — 0·78†	Nov.-Dec. — 0·76†	
		(b) Superphosphate.		
Jan. Dec. + 0·88*	Jan. June + 0·89*	Feb.-Apl. + 0·78†	Feb.-Mar. + 0·50	March + 0·50
	July Dec. — 0·60	Mar.-May + 0·91*	Mar.-Apl. + 0·89*	April + 0·84*
		Apl.-June + 0·95*	Apl.-May + 0·88*	May + 0·59
		May-July + 0·72‡	May-June + 0·78†	Nov. — 0·74‡
		Sept.-Nov. — 0·62	Oct.-Nov. — 0·54	
		Oct.-Dec. — 0·58	Nov.-Dec. — 0·61	

Significance of the correlation coefficient.

* $P < 0{\cdot}01$. † $P < 0{\cdot}02$. ‡ $P < 0{\cdot}05$.

In Table 3, the complete series of monthly and bi-monthly values for the correlation coefficient (r) and the regression coefficient (b) in cwts. per acre per inch of rain are given. The seasonal trend of the correlation coefficient for monthly and bi-monthly periods of rainfall is shown graphically in Figs. 2 and 3 respectively.

The values for the correlation coefficient given in Tables 2 and 3 indicate a relatively high degree of association between yield and rain falling within the March to June period, with the relationship strongest in the month of April. From June to August, the correlation coefficient falls towards zero, and from August onwards becomes negative, reaching a high and significant negative value in the month of November. From November to April, there is a continuous change in the reverse direction, zero being approached again in January. The bi-monthly values show a similar seasonal trend, but the positive figures for March-April and April-May are both higher than for the single

TABLE 3.

(i) *Monthly Intervals.*

—	Jan.	Feb.	Mar.	Apl.	May.	June.	July.	Aug.	Sept.	Oct.	Nov.	Dec.
						1. *No Fertilizer.*						
r ..	+0·09	+0·38	+ 0·57	+0·73‡	+0·67‡	+0·45	+0·19	−0·06	−0·08	−0·25	− 0·81*	−0·50
b ..	+2·97	+5·49	+12·09	+6·90‡	+4·06‡	+3·26	+2·36	−0·60	−0·97	−3·00	−11·17*	−5·64
						2. *Superphosphate.*						
r ..	+0·04	+0·39	+ 0·50	+0·84*	+0·59	+0·43	+0·02	−0·04	−0·28	−0·11	−0·74‡	−0·34
b ..	+1·46	+5·98	+10·88	+8·25*	+3·74	+3·26	+0·29	−0·42	−3·60	−1·41	−10·55‡	−3·92

(ii) *Bi-monthly Intervals.*

—	Dec.-Jan.	Jan.-Feb.	Feb.-Mar.	Mar.-Apl.	Apl.-May.	May-June.	June-July.	July-Aug.	Aug.-Sept.	Sept.-Oct.	Oct.-Nov.	Nov.-Dec.
						1. *No Fertilizer.*						
r ..	−0·34	+0·34	+0·53	+0·82*	+0·89*	+0·85*	+0·41	+0·10	−0·09	−0·26	−0·68‡	−0·76†
b ..	−3·84	+3·98	+5·24	+6·43*	+4·19*	+4·27*	+2·22	+0·89	−0·79	−2·46	−5·85‡	−5·48†
						2. *Superphosphate.*						
r ..	−0·46	+0·33	+0·50	+0·89*	+0·88*	+0·78†	+0·33	−0·02	−0·21	−0·29	−0·54	−0·61
b ..	−5·36	+4·08	+5·20	+7·20*	+4·33*	+4·06†	+1·82	−0·15	−1·79	−2·95	−4·81	−4·59

Significance of correlation or regression.
* $P < 0 \cdot 01$. † $P < 0 \cdot 02$. ‡ $P < 0 \cdot 05$.

month of April, whereas the negative figures for October-November and November-December are both lower than for the single month of November.

While the highest positive correlation coefficients for monthly rainfall occur in April, the regression coefficients reach their highest positive values in March, indicating that March rainfall is more effective in increasing the yield of pasture than rain falling in any other month. The March values for each treatment, however, do not differ significantly from zero. Negative regression, as in the case of negative correlation, is highest in November.

The points of special interest arising from the data given in Tables 2 and 3 are, firstly, the apparent dependence of the pasture yield under these conditions on autumnal rainfall; secondly, the absence of relationship between yield and the amount of rain falling between July and October; and, finally, significant negative values for correlation between yield and rain falling in November or in the October-December period.

The effectiveness of autumn rainfall and the comparative ineffectiveness of winter rainfall in determining yield are dealt with in the discussion (q.v.). Due consideration of the pasture species concerned, in relation to varying factors of their environment, indicates that the values obtained for the period March to September at least are cogent, and have an important practical bearing in connexion with the replacement of natural grassland by seeded pastures.

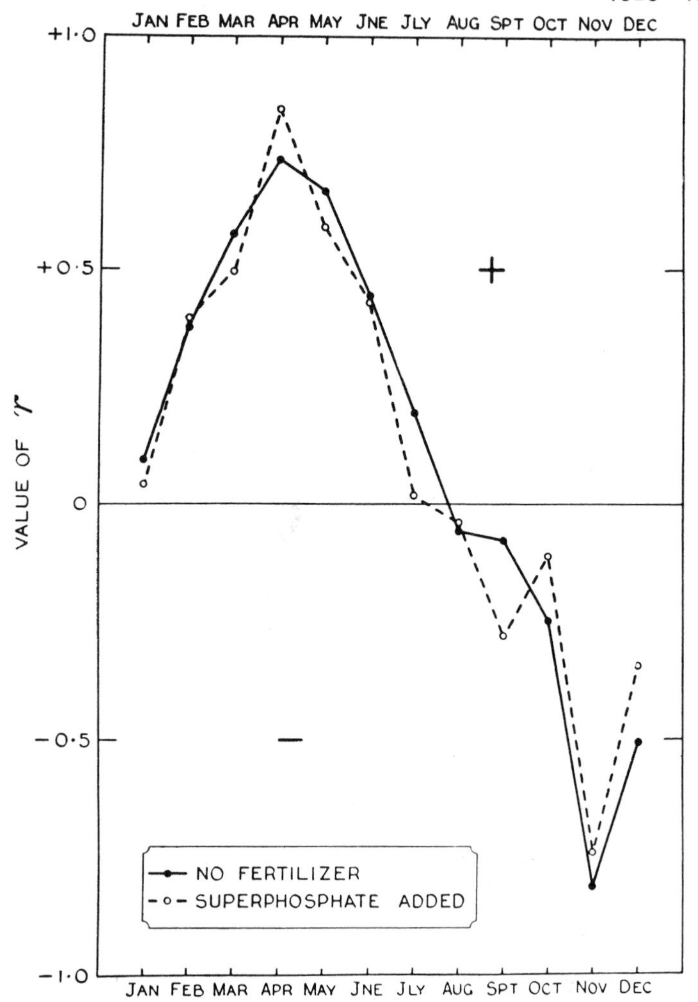

FIG. 2.—Graph showing correlation coefficient (r) for the annual yield of natural pasture, (a) unfertilized, (b) top-dressed with superphosphate, and the rainfall for each month, Waite Institute, 1925–34.

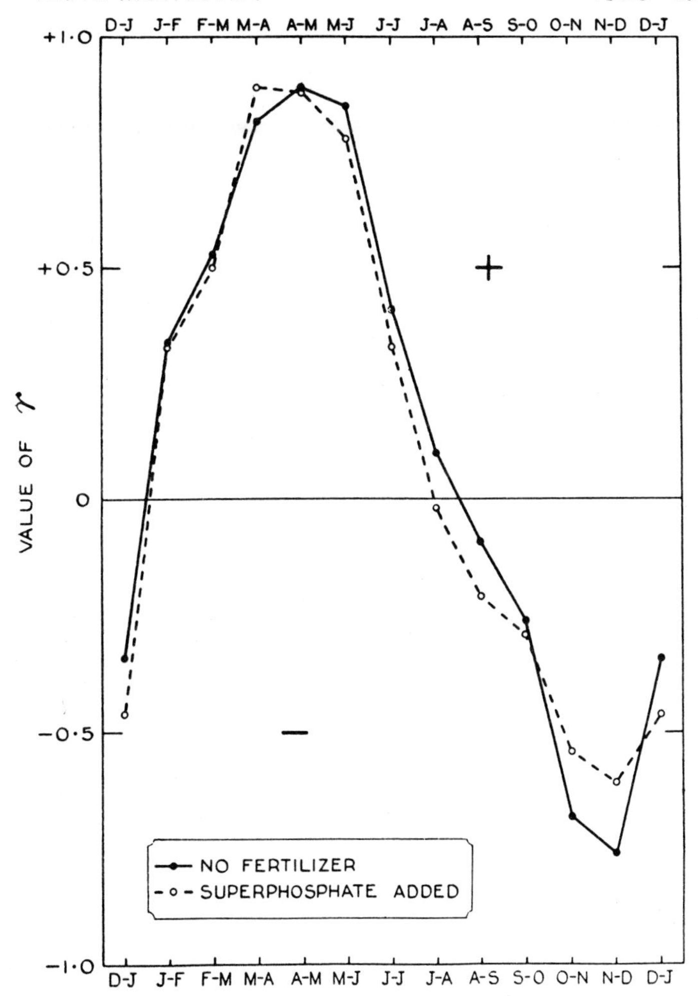

Fig. 3.—Graph showing correlation coefficient (r) for the annual yield of natural pasture, (a) unfertilized, (b) top-dressed with superphosphate, and the rainfall in bi-monthly periods, Waite Institute, 1925–34.

The high negative correlation between yield and November rainfall is not by itself acceptable, however, because in each season the harvesting of the pasture samples was completed early in November, growth normally ceasing at this time. It is manifestly unlikely that rain falling at or after harvest could significantly affect the total yield of pasture in either an adverse or beneficial manner. It is reasonable, therefore, to suppose that a large proportion of the apparent negative association between yield and November rain may be due to the fact that November rainfall is negatively correlated with rain in other parts of the season. This actually proved to be the case, as will be seen from Table 4, wherein are given the values of the correlation between yield and spring-early summer rainfall after allowance has been made for April-June rainfall.

TABLE 4.—PARTIAL CORRELATION COEFFICIENTS (APRIL–JUNE RAINFALL ELIMINATED).

(a) Monthly Intervals.

—	Aug.	Sept.	Oct.	Nov.	Dec.
Nil	+0·36	+0·05	+0·14	−0·43	−0·54
Super. ..	+0·37	−0·60	+0·60	−0·05	+0·10

(b) Bi-monthly Intervals.

—	Oct.-Nov.	Nov.-Dec.
Nil	−0·10	−0·62

* None of the above values is significant.

On removing the correlation between April-June rainfall and August-September, &c., rainfall, the value of r for yield correlated with rainfall for the months following July becomes irregular and in all cases insignificant.

The correlation coefficient for April-June rainfall with November rainfall is −0.77 and is significant at the 2 per cent. ~~point.~~ level Allowing for this relation, the value of r for yield with November rainfall becomes reduced from −0.81 (significant at the 1 per cent. point) to −0.43 (insignificant) in the case of the unfertilized pasture and from −0.74 (significant at the 5 per cent. point) to −0.05 (insignificant) with phosphated pasture.

Thus, for the period of ten seasons under review, there appears to be a significant negative relationship between November rainfall and April-June rainfall, and the recorded negative correlations between yield and November rainfall are probably not due to any influence of November rainfall on yield.

4. Discussion.

It seems apparent that the total seasonal yield of natural pasture at the Waite Institute over the period 1925-34 has been largely determined by the amount of rain received during March, April, May, and June. Furthermore, rain falling after June or July appears to have had little significant influence on the total pasture yield, despite the fact that growth each year proceeded until November. In the event of negligible rain prior to July, succeeding rains would, of course, be important, but a low final yield would be expected.

The importance of autumnal rains in a winter rainfall environment may be accounted for by the fact that at this time of the year soil temperatures are rapidly falling to values which in June, July, and August render the germination and vegetative growth of herbage plants particularly slow. At the Waite Institute, the mean monthly soil temperature at a depth of 1 inch is approximately 70°F. in April, compared with 55°F. in June, July, and August, with a mean monthly minimum value of 46.4° F. in July. The increased length of day in early autumn, compared with late autumn and winter, may also be important.

It is obvious, then, that given suitable soil conditions, rain falling in early autumn will tend to produce considerably more growth than similar quantities received in late autumn or winter, provided rainfall suitably exceeds evaporation for a sufficient interval of time. In the case of the ten seasons under review, April-June rainfall exceeded April-June evaporation in the four years giving the four highest yields. In each of the four years with the four lowest yields there was a deficiency of more than 3 inches for this period. The remaining two yields were also on the low side. In one case (1928), there was a deficiency of 1.57 inches; in the other (1925), rainfall exceeded evaporation by 0.49 inches over April-June, with a deficiency, however, of 0.21 inches over April-May.

The major influences of this early development on ultimate production are at least threefold. Firstly, there is the direct effect on growth, which in the early stages is greatly dependent on previous accumulation. Secondly, the earlier the commencement of growth the longer the interval of time over which growth can proceed, as the end point with these species has been remarkably constant, depending for the most part on the more or less constant length of day factor. Finally, early development of the root system becomes important in winter when the increased depth of penetration attained secures the advantages of warmer soil and greater capacity both for root respiration and the picking up of partially leached nitrates.

The fact that growth and transpiration accelerate from July onwards, reaching a maximum in September and October, tends to suggest that spring rains also should be highly effective in determining the ultimate pasture yield. The absence of relationship between yield and rain falling after the end of June, therefore, may at first sight seem difficult to comprehend, and was at least unexpected.

A consideration of the particular species concerned, however, provides an explanation for the relative ineffectiveness of late winter and spring rains compared with autumnal rains. These species are adapted to exigent soil conditions and to a comparatively short rainy period; their potentialities for increased production are not high.

It is of interest to note that the mean rainfall for the April-June period, over this sequence of ten years, is 8.03 inches. Transpiration experiments carried out at the Waite Institute during the same period suggest that the transpiration requirements of the mean yield from the higher yielding plot would be approximately this amount.

It would thus be seen that the ultimate development of the inherently early-maturing and short-lived species which comprise these natural pastures is largely predetermined by conditions during the March-June period, and that rains falling later, whilst by no means negligible, are by comparison much less important for this type of pasture, and in certain seasons may be quite ineffective.

The above results, whilst admittedly tentative, provide a new interpretation for differences in yield due to seasonal variation and also indicate an important application in practice. This is the establishment of herbage species with the combined capacities of drought-resistance and sustained vegetative growth, enabling full advantage to be taken of late winter and spring rains as well as rains falling earlier in the season. Two species which fulfil these requirements in particular are lucerne and *Phalaris tuberosa,* which, by virtue of their perennial deep-rooting habit, drought-resistance, and capacity for growth over all or most of the year, are able to utilize the rainfall to a maximum degree and frequently to provide green herbage at those times of the year when it is normally absent.

5. Acknowledgment.

We desire to thank Mr. C. A. N. Smith, B.Agr.Sc., for carrying out much of the arithmetical work involved in the calculation of the correlation and regression values.

6. References.

1. Barkley, H.—*Economic Record* (*Sydney*) **2**: 161-173, 1926.
2. Barkley, H.—*Pastoral Review* (Melbourne) **37**: 759-761, 1927.
3. Fisher, R. A.—*Phil. Trans. B.* **213**: 89-142, 1924.
4. Hooker, R. H.—*Quart. J. Roy. Met. Soc.* **47**: 75-100, 1921
5. Kalamkar, R. J.—*J. Agr.. Sc.* **23**: 571, 1933.
6. Richardson, A. E. V.—*J. Vic. Dept. Agr.*, 1923.
7. Wishart, J., and Mackenzie (Tyrrell), W.A.—*J. Agr. Sc.* **20**: 417, 1930.

H. J. GREEN, Government Printer, Melbourne.

Reprinted from *The Journal of the Australian Institute of Agricultural Science*, Vol. 2, No. 2, pp. 79-82, June, 1936.

NON-REPLICATED FACTORIAL EXPERIMENTS.

Owing to the stimulus given by the Statistical Department at Rothamsted, the development of agricultural research, particularly in the branch of field experimentation, has made rapid progress during the last decade. Outstanding among the various accomplishments of this department have been the researches on the factorial design of field experiments and the artifice of confounding as a means of controlling soil heterogeneity. These experimental devices have been the subject of a full and able treatment in a number of well-known papers by R. A. Fisher, F. Yates and others; it would be redundant therefore to attempt in this note any discussion of the advantages which accrue from their use.

It is in the large factorial experiment that the confounding of unimportant high-order interactions is most usefully employed, since the sacrifice of information on the comparisons involved is of little consequence. The evidence, accumulated from the analyses of experiments conducted at the Waite Institute, has shown that—as in the Rothamsted work—the high-order interactions have been, in general, of the same magnitude as experimental error. This fact suggested that these interactions might be exploited, not only for the purpose of confounding, but also as a means of providing an estimate of error in circumstances where it may be desirable to dispense with replication. In a factorial experiment of the type considered a single set of all combinations of factors will usually provide sufficient replication *within the set* to give an accurate determination of the main effects. For example, if it were desired to test three varieties of wheat at three levels of nitrogen and three levels of phosphorus, one replication of all combinations of these factors would require 27 plots. All plots contribute to the answering of every question and, therefore, the main effects will be measured with the accuracy appropriate to means of 9 plots.

If only one complete replication is employed all the independent comparisons, which the experiment provides, will be assigned to the effects of variety, phosphorus and nitrogen, together with their interactions, there being none attributable to pure error. Under these conditions the allocation of the degrees of freedom will be as follows:

	Degrees freedom
V	2
P	2
N	2
N × P	4
N × V	4
P × V	4
N × P × V	8
	26

The 8 degrees of freedom for the interaction N × P × V may be used as a basis for testing the significance of the main effects and of the first-order interactions, and if the latter proved to be insignificant they may legitimately be pooled with the second-order interaction to provide the estimate of error.

An improvement in the design may be effected by dividing the 27 plots into 3 blocks of 9 plots and making a suitable selection of the treatment comparisons, corresponding to one pair of degrees of freedom for the interaction N × P × V, to occupy the blocks. The experiment thereby gains the advantage of the greater precision of the smaller blocks through the sacrifice of an unimportant differential reaction. This results in the following partitioning of the degrees of freedom.

	Degrees freedom
Blocks	2
V	2
N	2
P	2
N × P	4
N × V	4
P × V	4
N × P × V	6
	26

As before, the error is estimated from the unconfounded interactions.

An experiment of this type was conducted at the Waite Institute during 1935; 3 varieties of wheat were used with 3 levels of nitrogen and 3 levels of phosphorus. The experiment consisted of two replications of the 27 treatments, each set of 27 plots being divided into 3 randomised sub-blocks to confound the same pair of degrees of freedom for second-order interaction in each main block. The objective was to observe the agreement of the results of each set of 27 plots, and to provide a check on the magnitude of the interactions from an analysis of all 54 plots, in which pure error degrees of freedom were available. Table I gives the analysis of variance of plots 28-54.

Since the fertilisers were applied in equal increments, the several effects have been partitioned, by utilising the orthogonal polynomials devised by Fisher (1), into their orthogonal components which correspond to the physical characters of the response curves.

In Table II are shown the standard deviations of a single plot calculated by the various methods and expressed as percentages of the corresponding mean yields. It will be observed that the agreement is good and that the interactions are, in fact, negligible in comparison with pure error.

The standard deviation of a mean of 9 plots, derived from the analyses of the sets of 27 plots is, in both cases, approximately 1 bush./acre and therefore the experiment is very accurate when consideration is taken of its size.

TABLE I.

Analysis of Variance (lbs. grain per plot)

Variation due to	Deg. freedom		Sum. Sq.		Mean Sq.	S.D.
Blocks	2		84.57			
N lin. regr.†	1		99.17			
N quad. regr.*	1		6.86			
P lin. regr.	1		595.13			
P quad. regr.	1		9.80			
V	2		799.23			
N lin. regr. × P lin. regr.	1		5.67			
N lin. regr. × P quad. regr.	1		8.27			
N quad. regr. × P lin. regr.	1		45.00			
N quad. regr. × P quad. regr.	1		0.36			
V × N lin. regr.	2	18	17.44	219.36	12.1875	3.4911
V × N quad. regr.	2		16.59			
V × P lin. regr.	2		20.31			
V × P quad. regr.	2		29.04			
N × P × V	6		76.68		12.7801	3.5749
Total	26		1814.12			

†linear regression *quadratic regression

TABLE II.

Estimates of Error

	General mean yield bush. per acr	Error estimated from— 2nd order interaction 6 deg. freed.		Pooled interaction 18 deg. freed.		Pure Error 24 deg. freed.	
		S.D. of a single plot bush. per acre	% of general mean	S.D. of a single plot bush. per acre	% of general mean	S.D. of a single plot bush. per acre.	% of general mean
Plots 1-27	46.67	2.84	6.09	3.02	6.46	—	—
„ 28-54	49.54	2.98	6.01	2.91	5.87	—	—
„ 1-54	48.11	—	—	—	—	3.14	6.5

It should be borne in mind, however, that such an experiment is liable to failure in the sense that the error of the comparisons may be overestimated, since it may happen that the interactions employed as error represent real effects, whereas the true error will be smaller. The experiment should be tried only in circumstances where there is evidence that the interactions would be too small to be statistically significant, due consideration being given to the size of the experiment proposed. It is worth noting that an indication of the presence of large high-order interaction effects may be obtained from the magnitudes of the interactions between pairs of factors; if any one of these were large, it would be worth while examining the three factor interactions involving this pair. Those which prove to be of considerable magnitude should be isolated from error.

The following elaboration of the design illustrated is of especial value in an experimental programme designed to investigate the manurial requirements of a representative selection of varieties of a particular crop, as influenced by the simultaneous action of soil and climatic conditions. If four sites of the same general soil type are chosen within a district, and 27 plots used at each centre, it would be possible, by confounding a different pair of degrees of freedom at each place, to provide a wide inductive basis to any conclusions which may be drawn from the data.

The procedure illustrated is applicable with appropriate modifications, to other treatment systems, e.g., a 2 x 2 x 2 x 2 x 2. In certain circumstances, for example at various farms associated with a central experimental station, facilities for the convenient handling of more than approximately 30 plots are not usually available. Under such limiting conditions a single replication of a factorial system of treatments makes the most efficient use of the means at the disposal of the experimenter.

It should be emphasised, however, that the type of experiment described does not provide a means of dispensing with absolute replication on all occasions.

In concluding, it should be stated that the utility of non-replicated factorial designs has been known at Rothamsted for some time (F. Yates (2)), but was discovered independently at the Waite Institute in 1935.

E. A. CORNISH.

Waite Agricultural Research Institute.
South Australia.

REFERENCES

(1) Fisher, R. A.: Statistical Methods for Research Workers, 4th ed., 1932.
(2) Yates, F.: Complex Experiments. Supp. Journ. Roy. Stat. Soc. 2: 181, 1935.

Reprinted from The Australian Journal of Experimental Biology and Medical Science, Volume XIV (1936).

PHYSIOLOGICAL ONTOGENY IN PLANTS AND ITS RELATION TO NUTRITION

I. THE EFFECT OF NITROGEN SUPPLY ON THE GROWTH OF THE PLANT AND ITS PARTS

by

L. A. T. BALLARD AND A. H. K. PETRIE

With a Statistical Appendix by

E. A. CORNISH.

(From the Waite Agricultural Research Institute, University of Adelaide).[1]

(Submitted for publication 7th July, 1936.)

INTRODUCTION.

The growth and development of an organism is the result of a large number of directed and integrated metabolic reactions. The integration is such that there are temporal drifts in metabolism which underlie drifts in size, structure, chemical composition, and in the rates of change of these attributes; these drifts can be recognized as characteristic features of ontogeny. Integration is not confined to direct relations among chemical or metabolic reactions, but also involves the changes in structure and composition, which themselves influence the reactions that caused them. The elucidation of this complex integration represents the ultimate task in the study of growth; it can be achieved only after intensive investigation of the nature of ontogenetic drifts of various kinds and of their interrelations. To this investigation the present work forms a contribution.

It is known that the chemical composition of the plant not only drifts with time, but varies with the nutrient supply. It has therefore been considered that the observation of ontogenetic drifts under different initial supplies of nutrients should be a valuable method of approach to the study of physiological ontogeny. This paper, apart from the presentation of certain introductory conceptions, concerns the definition of the effect of nitrogen supply on the drifting size and growth rate. The observations have been confined mainly to annual or annually renascent gramineous plants; in these the drifts with which we are concerned are shown in comparatively simple form.

[1] This represents one of a series of investigations carried out at the Waite Institute and financed co-operatively by the Carnegie Corporation of New York, the Council for Scientific and Industrial Research, and the University of Adelaide.

GROWTH STAGES.

The life-history of a single vegetative organ such as a leaf is marked by three phases: (*a*) an initial period of growth which we shall term adolescence; (*b*) a period of constant weight after the cessation of growth, known as maturity; and (*c*) a period of decline in weight and internal disorganization known as senescence. By analogy, these terms may be applied to the totality of one type of organ on the plant, such as the leaves. The terms then become only average expressions of the ontogenetic state of the organs; thus, when the leaves as a whole cease to increase in weight, some will be declining in weight and others increasing. The analogy, however, is justified since the general ontogeny of the totality of organs of one type reflects that of a single organ of the same type.

RELATIVE GROWTH RATE AND ITS COMPONENTS.

The primary measure of the rate of change in size that has been adopted is the relative growth rate, R, given by the equation

$$R = \frac{1}{W}\frac{dW}{dt}, \qquad (1)$$

where W is the dry weight of the whole plant and t the time. The characteristic time curve of this rate has been demonstrated by Briggs, Kidd, and West (1920). The relative growth rate is a convenient measure for comparing rate of increase in dry weight at different times and in different plants, but on account of the heterogeneity of the plant it has no direct physiological significance. It has been resolved, however, into two components, the fraction of the plant that is producing new material and the rate of production by this fraction.

The new material is produced mainly by the leaves, and the fraction of the plant that they represent can be measured as leaf weight ratio, F_W, given by the equation

$$F_W = W_L/W, \qquad (2)$$

where W_L is the total leaf weight. It can also be measured as leaf area ratio (F_A), substituting the total leaf area (A_L) for W_L. The time curve of F_A has been described by Briggs, Kidd, and West (1920) and by Inamdar, Singh, and Pande (1925). The latter authors showed that under conditions of bright sunlight, W_L/A_L tends to be constant; on the other hand, a complex drift of this ratio was recorded for single successive mature leaves of barley by Gregory and Richards (1929); Smirnow (1928) found in *Helianthus* and *Nicotiana* that the average value of the ratio for all but the lower leaves increased with time during the growing period; and for the total leaves of *Fagopyron* Maiwald (1930) found it to fall. In the present work we have measured leaf weight, since the measurement of leaf area was impracticable.

The second component of the relative growth rate is the increase in dry weight of the plant per unit of leaf, the unit leaf rate (E) of Briggs, Kidd, and West, which is given by the equation

$$E_A = \frac{1}{A_L} \frac{dW}{dt}. \quad (3)$$

We have used the analogous quantity

$$E_W = \frac{1}{W_L} \frac{dW}{dt}. \quad (4)$$

Since by definition $R = FE$, we may expect from the observations of Inamdar, Singh, and Pande that E will rise and then fall, although this expectation might apply only to their environmental conditions. In any case, since the assimilatory activity of the seedling leaves of those plants in which these do not act as storage organs increases at first, an initial rise in E would in such plants be expected, as Briggs, Kidd, and West point out. A subsequent fall would be caused by a decrease in assimilatory activity of the individual leaves as they grow older, and the increasing mean age of the leaves as a whole. Evidence[2] for such a decrease has been obtained by Henckel and Litvinov (1930) and by Gassner and Goeze (1934*b*). Respiration also, as Briggs, Kidd, and West point out, will make the unit leaf rate curve still more concave to the time axis.

Briggs, Kidd, and West calculated values for E_A from some data for maize, and demonstrated an initial rise. During the middle portion of the growing period, the values showed marked fluctuations about a mean, these being attributable mainly to temperature. The subsequent values they regarded as inaccurate. Wagner (1932) gives a set of data for oats from which E_W can be calculated over the later part of the life-period; a general decline with time is shown. There is a similar decline in E_A (up to leaf maturity) in some experiments of Gregory (1926). Both Wagner and Gregory, however, were concerned with plants senescing under conditions of falling temperature, and Gregory attributes the fall in unit leaf rate in his case solely to falling temperature[3].

It has thus been shown that relative growth rate can be represented as the product of two factors: the one is the net rate of increase of dry matter resulting from a number of metabolic processes, the rates of all of which except carbon assimilation are generally relatively small; the other is an easily measurable aspect of morphological structure. Both factors exhibit characteristic ontogenetic drifts. In the present paper these conceptions will be used in describing the effect of nitrogen supply on growth. The description will be based largely on two experiments performed in this laboratory, the details of which will be given in the following section.

[2] Singh and Lal (1935) have studied age-assimilation relations, but they do not give data for the drift in assimilation rate of all the leaves or a single leaf throughout the life-period under normal environmental conditions.

[3] Evidence will be given subsequently that E_W falls independently of temperature.

PRESENTATION OF NEW DATA.

Experiment 1.

On June 23, 1933, seeds of a pure line of late Gluyas wheat (*Triticum vulgare* Vill.) were planted in glazed earthenware pots containing 25 kg. water-washed sand. Fifteen seeds were planted in each of 120 pots, and about 25 seeds in each of a further 12 pots. On July 4, before the second leaf had appeared, the seedlings of the 120 pots were thinned to 6 uniform plants per pot, and those in the remaining pots were thinned to a similar standard of uniformity. At the same time, nutrient solutions were added so that each pot received the following amounts:

KH_2PO_4	0·80 gm.
K_2SO_4	2·00 „
$MgSO_4.7H_2O$	0·50 „
$CaCl_2.6H_2O$	2·00 „
$FeCl_3$	0·08 „
$MnSO_4.4H_2O$	0·0021 „

The pots were divided into 4 groups each receiving a different amount of $NaNO_3$. The pots of these groups will be referred to respectively as those of treatments I–IV. The amounts of $NaNO_3$ were as follows:

treatment I	1 gm.
II	3 „
III	6 „
IV	10 „

Distilled water was added frequently in such amount as to restore the moisture content of the sand to 50 per cent. of its maximum water-holding capacity. Evaporation loss was restricted by spreading 2·75 kg. of fine gravel on the surface of each pot, and the depletion of water never exceeded half the amount originally present.

Harvests were made on July 4, 10, 13, 18, and 31, August 28, September 19, October 13, 31, and November 9. The first harvest consisted of 5 replicates, each consisting of the shoots of 24 plants similar to those left in the thinning operation. The second harvest was made from the closely planted pots, and consisted of 5 replicates of 5 plants each per treatment. The third and fourth harvests were made by thinning of the 6 plants in each pot to 4 evenly spaced ones. The 5 replicates of treatments I, II, and III each contained 5 plants, and treatment IV had 3 replicates of 4 plants each; some of the pots of treatment IV had been rejected owing to the plants being injured at the time of adding the solutions. Harvests 5–10 consisted of the 4 remaining plants of 5 pots for each treatment except treatment IV, for which at harvest 5 no pots, and at harvest 6 only 3 pots, were available.

Previous experiments had led to the conclusion that, at the time of harvesting from pots with more than 4 plants each the fraction of nutrients removed from the sand was negligible, so that no complication was anticipated from the procedure adopted, which was necessitated by the number of pots available.

Various portions of the plants were separated at harvesting. First leaves were separated for harvests 1–4, second leaves from their appearance till harvest 5, and third leaves till harvest 7. Leaf separations were made by cutting at the ligule but, in the case of the youngest leaf of each axis, only that portion was taken which had protruded from the sheath of its predecessor. Tillers appeared at harvest 5, and from this harvest onwards the primary shoot was kept separate. The stem fraction included true stems, the leaf sheaths, portions of immature leaves and also, although these would be of negligible weight, coleoptiles and rudimentary tillers. Inflorescences, which first appeared on September 27, were separated at harvests 8–10 by cutting immediately beneath the lowest floret. From harvest 6 onwards, roots were obtained by sieving under water until practically free from sand.

Immediately after separation, the various portions of the plants were weighed, and dried at 100°C. for one night, and then at 80°C. to constant weight. The material was then ground for analytical purposes. Estimations of sand content were made on the root material, and the dry weight data corrected for the small quantities present.

In all cases the results have been expressed on a basis of one plant per replicate (see Table 2).

TABLE 1.

Harvest Details of Experiment 2.

Harvest.	Date.	No. of plants per pot.	No. of replicates.
2	Feb. 6	12	6
3	,, 18	7	6
4	,, 25	4	5
5	Mar. 12	3	5
6	April 2	3	5
7	,, 29	3	5

The date of appearance of the first inflorescence and of the exsertion of the first anthers was noted for each plant. From these was obtained for both phenomena for each treatment the date for the $\frac{1}{2}n$th plant, n being the number of plants still unharvested at that date. These dates are indicated in the graphs.

Tiller counts were made on each plant at harvesting from harvest 6 onwards, and the average number of tillers per plant recorded (see Table 6).

Experiment 2.

On January 16, 1935, seeds of Sudan grass (*Andropogon sudanensis* Leppan and Bosman) were planted in glazed earthenware pots containing 14 kg. water-washed sand, and also in seed-pans containing similar sand. Harvest 1 was taken on January 25, and comprised 15 seed-pans each containing 48 seedlings. Six of these sets of 48 were taken as replicates for dry weight determination. The remaining harvests were made from the pots, all of which were thinned to their final number of plants on January 25. On January 27, nutrient solutions were added so that each pot received the following amounts:

KH_2PO_4	0·60 gm.
K_2SO_4	2·00 ,,
$MgSO_4.7H_2O$	0·40 ,,
$CaCl_2$	0·74 ,,
$FeCl_3$	0·06 ,,
$MnSO_4.4H_2O$	0·0016 ,,

The pots were divided into three groups each receiving a different amount of $NaNO_3$ per pot. The amounts will be referred to as treatments I–III; they were as follows:

I	0·75 gm.
II	2·25 ,,
III	4·50 ,,

This was almost identical with the treatment used in a previous experiment with the same species (Ballard 1933). The moisture content of the sand was maintained at 57 per cent. of its maximum water-holding capacity. Harvest details are given in Table 1.

The plant portions were weighed and dried to constant weight at 85°C. under forced draught. The root weights were again corrected for sand content. In all cases the results are expressed on a basis of one plant per replicate (see Table 3).

At each harvest measurements were made of the respiration rate of leaves from pots not used for dry weight determinations.

The inflorescences and flowering data were obtained as in Experiment 1, except that all the plants in a pot instead of the single plant were takes as the unit.

Calculation of Derived Quantities.

Relative growth rates, leaf weight ratios, and unit leaf rates have been calculated from the dry weight data (see Tables 4 and 5), the first and third of these quantities being calculated from the equations

$$R = \frac{\log_e W_2 - \log_e W_1}{t_2 - t_1},$$

and

$$E = \frac{\log_e w_2 - \log_e w_1}{t_2 - t_1} \times \frac{W_2 - W_1}{w_2 - w_1},$$

where the subscripts 1 and 2 refer to the harvests between which the average value of the quantity is being calculated, and $w \equiv (W_L)$.

Statistical Treatment.

In the statistical treatment of the data we have had the valued advice of Mr. E. A. Cornish, who has described the methods adopted in an appendix. The results of the analyses of variance are given in the tables.

DISCUSSION.

The Growth Curves.

Treatment Effects.

The dry weight curves of the whole plants in Experiment 2 are presented in fig. 2, and those of the parts in both experiments in figs. 3 and 4. Certain characteristic effects of nitrogen treatment are obvious from these figures. In the following paragraphs are presented systematically some further conclusions drawn from the statistical examination of the data: I-IV signify plants of the various treatments, and the arabic figures the harvests.

Wheat.

First leaf. 2: IV almost significantly below II and significantly below I; thereafter insignificantly different from both II and III; 4: III significantly greater than I.

Third leaf. 3: depression insignificant; 4, 5, and 6 taken together: II and III each significantly greater than I; 7: no significant differences.

Total leaves. 2: I–III insignificantly different, IV significantly depressed; 3 and 4: IV insignificantly different from III; 6: depression in IV again significant; 8, 9, and 10 taken together: IV significantly greater than III.

Roots. 6: II and III significantly below I, IV significantly below the others; 7–10: IV insignificantly different from III.

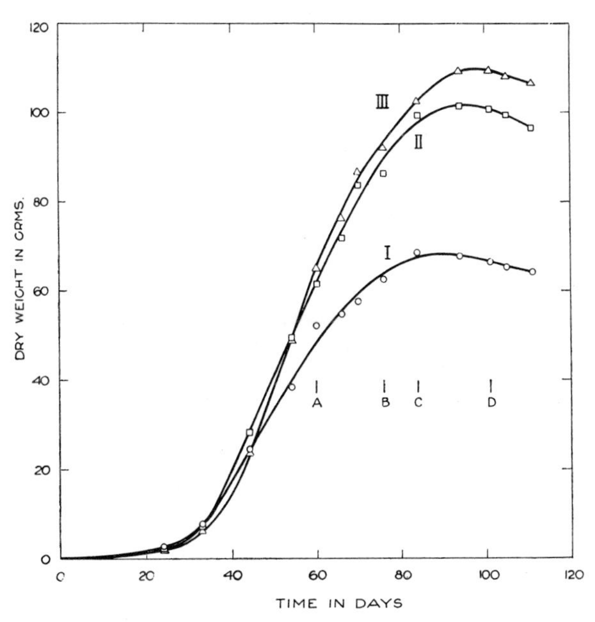

Fig. 1. Growth curves of oats with I, 0·4 gm. nitrogen; II, 0·8 gm.; and III, 1·2 gm. per pot. A, exsertion of inflorescences; B, flowering; C, grain in the pre-resting stage; D, grain in the resting stage. (From data of Blanck and Giesecke, 1934).

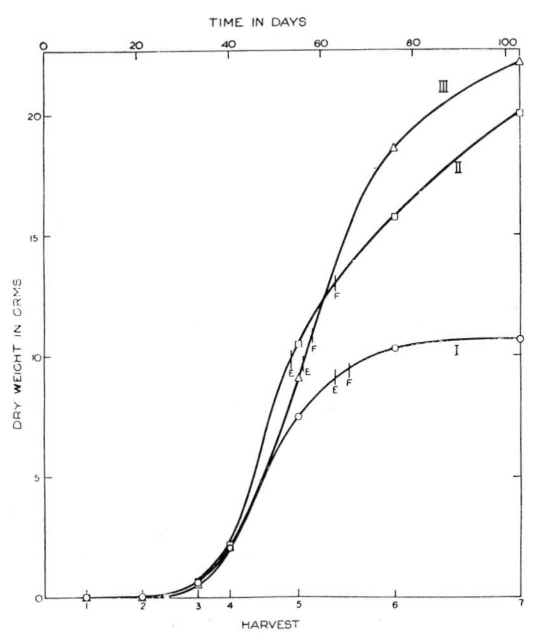

Fig. 2. Growth curves of Sudan grass with three nitrogen treatments, I, II, and III, Experiment 2. In this and subsequent figures, E indicates time of exsertion of inflorescences, and F time of flowering.

Stems. 2: I–III insignificantly different. IV significantly depressed; 3 and 4 together: IV insignificantly different from III; 6 and 7–9 together: IV again significantly depressed; 10: III and IV insignificantly different.

Inflorescences. IV significantly greater than III, taking the data as a whole.

Total shoots. 2: IV significantly below III; 3 and 4: IV insignificantly different from III; 6: IV depressed again; thereafter insignificantly different from III.

Total plant. 6: IV significantly below II; IV never significantly greater than III.

Sudan grass.

Leaves. 2 and 3: III significantly depressed, I and II insignificantly different; 6 and 7: III significantly greater than II.

Roots. 2–5 taken together: III significantly depressed below II; III never becomes significantly greater than II.

Stems. 2–5 taken together: depression in III barely significant; 6–7 taken together: IV insignificantly greater than II.

Inflorescences. Treatment effect between each pair significant on the whole.

Total plant. 2–5 taken together: III significantly depressed below II; III never significantly greater than II.

With the highest treatment, vegetative organs and whole plants are depressed in size during adolescence, the roots more markedly than the leaves. Although the depression passes off, only the leaves and inflorescences become significantly greater in weight than those of the preceding treatment; and their excess weight is insufficient to affect significantly the weight of the whole plant. There is a suggestion in the wheat experiment that the depression wears off after harvest 2 and appears again at harvest 6: such a phenomenon could be produced by interaction of treatment and environment.

Two factors may contribute to the ultimate increase in leaf weight resulting from nitrogen treatment: individual leaves may attain a greater size, or a greater number of leaves may be formed. Probably both factors operate. Increased nitrogen supply tends slightly to increase the mature weight of the early individual leaves[4]; it also causes increased development of leaf primordia (cf. Crowther, 1934) and of tillers (see Table 6).

Drifts with Time.

In the wheat experiment, the roots show maximum weight at harvest 8, the leaves at harvest 8 with higher treatments and a little earlier with lower treatments. These maxima are approximately coincident with flowering. Thereafter a significant decrease in leaf weight occurs with all treatments[5], probably in part owing to breakdown of tissues consequent upon translocation of materials to the inflorescence: the absolute decrease is greater, the greater the inflorescence weight.

[4] Increase in size of leaves with high nitrogen treatment is a generally observed phenomenon (Mothes, 1932; Garner et al., 1934).

[5] A rather similar picture is given by the data of Blanck and Giesecke (1934), except that there the root maximum is delayed with increasing treatment.

The loss from the roots (insignificant in plants of treatment IV) may be due partly to decomposition.

In the Sudan grass the roots increase in weight to the end of the experiment,

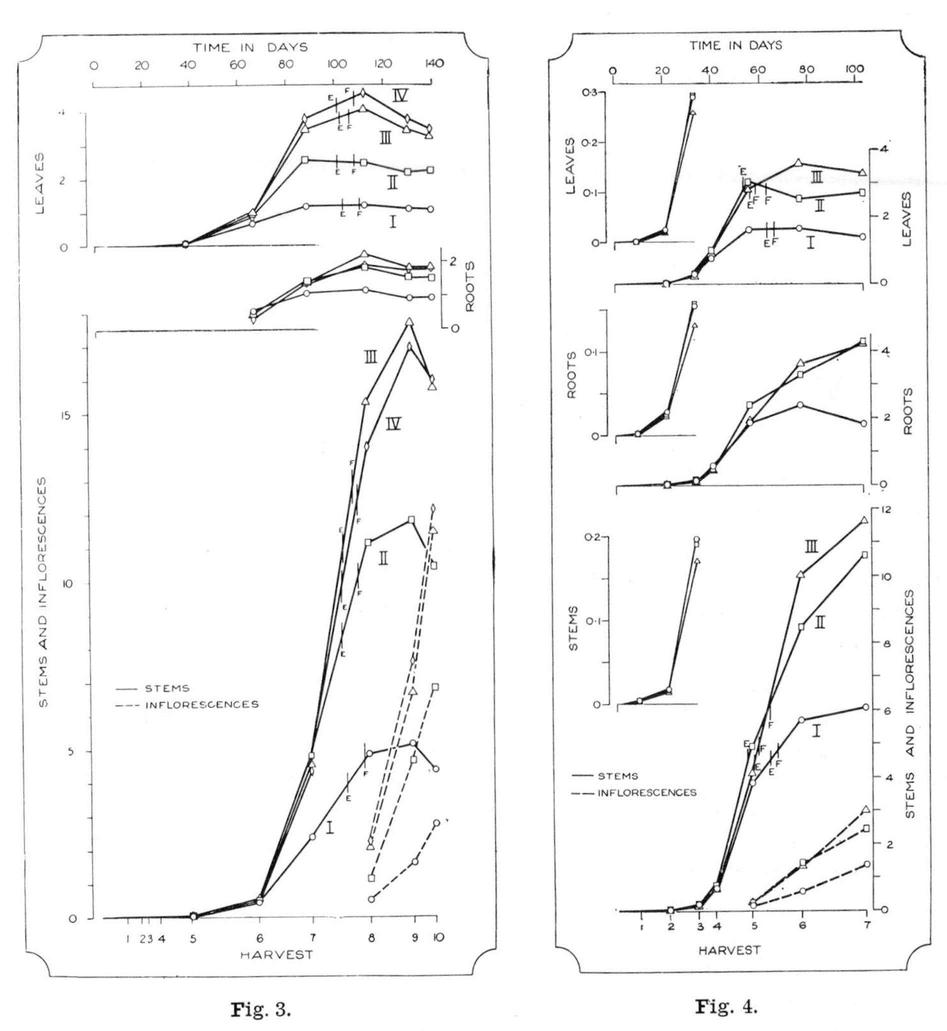

Fig. 3. **Fig. 4.**

Fig. 3. Growth curves (dry weight in gm.) of parts in wheat, Experiment 1.

Fig. 4. Growth curves of parts in Sudan grass, Experiment 2. The early portions of the curves are repeated with enlarged scale of abscissae.

except those of treatment I, whose weight falls significantly. The leaves attain maximum weight approximately at flowering (later with treatment III), but there is no appreciable loss thereafter.

The final harvest in this experiment was made before the seeds had passed into the resting stage, and loss in leaf weight might have occurred later. Nevertheless the post-flowering period of growth is clearly expanded in Sudan grass as compared with wheat; this is probably due partly to the smaller inflorescences of Sudan grass and partly to the temperature, which fell during the senescence of the Sudan grass, and rose during that of the wheat.

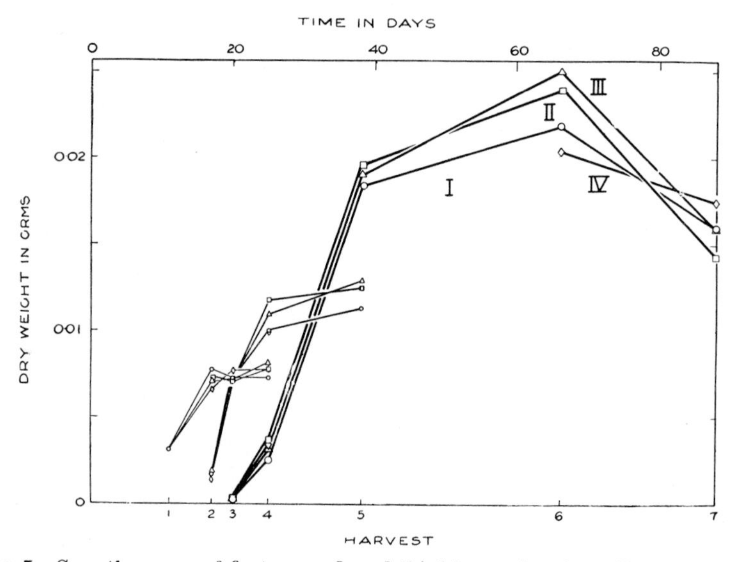

Fig. 5. Growth curves of first, second, and third leaves in wheat, Experiment 1.

The curves for the individual leaves of wheat are shown in fig. 5. It is seen that the early growth of each leaf is coincident with cessation of growth in the previous leaf, as was shown for the potato plant by Stone (1933). The data for the third leaf suggest that senescence may be retarded by increasing treatment.

Data of Other Investigators.

The effect of nitrogen on the growth of *Helianthus* has been observed by Rippel and Ludwig (1926), using ammonium nitrate; on that of oats by Meyer (1927) and Rippel and Meyer (1933), using ammonium sulphate, and by Blanck et al. (1933, 1934), using ammonium nitrate; on that of barley and wheat by Brenchley and Jackson (1921), using sodium nitrate; and on that of the sub-aerial portions of Sudan grass and barley by Ballard (1933), using sodium nitrate. The data of Blanck and Giesecke (1934) have been plotted in fig. 1.

All these investigations (except that of Brenchley and Jackson, where replication was poor) show that, after a certain point, increase of the initial nitrogen supply depresses the growth in the early stages. The depression disappears with time and, up to some optimum value of nitrogen supply, determined by the conditions and the species, plants with the greatest supply attain the greatest size. A number of the features exhibited by the present data are also to be seen in the data of Crowther (1934) for the cotton plant.

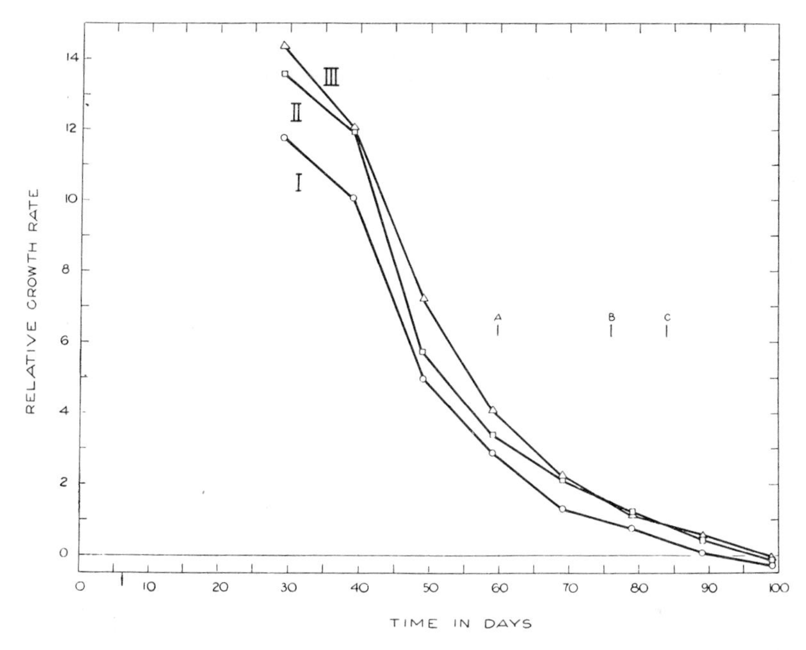

Fig. 6. Curves of relative growth rate (gm. per 100 gm. per day, i.e. $R \times 100$) of oats with three different nitrogen treatments. Calculated by interpolation in the curves of fig. 1. A, B, C, as in fig. 1.

Effect of Initial Depression on Final Weight of Plant.

The available data suggest that the initial depression prevents the full potentiality of a high nitrogen supply being manifested within the growing period. The depression appears to be due to the addition of a large amount of nitrogen to the external medium when the plant is very small. If the addition is distributed over a period during early growth, or if the main addition is made some time after the commencement of growth, the final size attained by the plant for a given amount of nitrogen supplied is greater, and the depression appears not to be manifested. Neidig and Snyder (1922) found that successive additions of sodium nitrate throughout the growing period of wheat gave yields in excess of those obtained when the total nitrate was applied at planting time. Gericke (1920, 1922) found that a greater yield of wheat was obtained when the main addition of nitrogen was made from 33 to 110 days after planting than when it was made at the time of planting. This effect of delayed application, however, holds only up to a certain point in the growing period: Davidson and LeClerc (1917, 1918, 1923) found that application at the time of exertion of inflorescences produced no increase in yield.

TABLE 2.

Mean Dry Weights in gm. per Plant for Experiment 1 (Wheat).

Treatment.

Harvest	I		II		III		IV	
First Leaf								
1	0·00306		0·00306		0·00306		0·00306	
2	0·00778		0·00727		0·00711		0·00656	
3	0·00732		0·00716		0·00731		0·00765	
4	0·00729		0·00786		0·00823		0·00778	

Analysis of variance performed on unchanged dry weight data, excluding harvest 1. Treatment as a whole insignificant, but interaction and time significant.

S.E. for means of 5 = ±0·000248.
S.E. for means of 6 = ±0·000320.

Harvest	I		II		III		IV	
Second Leaf								
2	0·00172		0·00164		0·00189		0·00138	
3	0·00715		0·00656		0·00655		0·00412	
4	0·0100		0·0117		0·0111		0·00987	
5	0·0114		0·0125		0·0129		—	
Third Leaf								
3	0·00025	±0·00008	0·00030	±0·00004	0·00027	±0·00005	0·00021	±0·00007
4	0·00247	±0·00019	0·00353	±0·00026	0·00318	±0·00020	0·00345	±0·00045
5	0·0197	±0·0008	0·0216	±0·0012	0·0206	±0·0013	—	—
6	0·0219	±0·0019	0·0245	±0·0016	0·0251	±0·0008	0·0204	±0·0023
7	0·0159	±0·0008	0·0143	±0·0008	0·0159	±0·0005	0·0174	±0·0017

Interaction insignificant. Standard errors of means given.

Harvest	I		II		III		IV	
Total Leaves								
1	0·00306		0·00306		0·00306		0·00306	
2	0·00950	*0·977*	0·00891	*0·949*	0·00899	*0·953*	0·00793	*0·894*
3	0·0147	*1·166*	0·0140	*1·145*	0·0141	*1·150*	0·0147	*1·166*
4	0·0198	*1·297*	0·0231	*1·362*	0·0226	*1·354*	0·0211	*1·323*
5	0·0640	—	0·0692	—	0·0811	—	—	—
6	0·664	*2·822*	0·936	*2·971*	0·989	*2·994*	0·857	*2·932*
7	1·17	*3·066*	2·53	*3·403*	3·43	*3·535*	3·75	*3·573*
8	1·19	*3·075*	2·45	*3·388*	4·01	*3·603*	4·47	*3·649*
9	1·08	*3·032*	2·15	*3·333*	3·40	*3·531*	3·73	*3·572*
10	1·07	*3·030*	2·20	*3·343*	3·20	*3·504*	3·42	*3·532*

Harvest 5 omitted from analysis of variance.

S.E. for means of 5 = ±*0·0140*.
S.E. for means of 3 = ±*0·0181*.

Harvest	I		II		III		IV	
Roots								
6	0·527	*2·706*	0·425	*2·619*	0·415	*2·616*	0·289	*2·453*
7	1·04	*3·018*	1·41	*3·148*	1·33	*3·122*	1·32	*3·123*
8	1·15	*3·057*	1·77	*3·247*	2·18	*3·337*	1·86	*3·266*
9	0·885	*2·946*	1·50	*3·176*	1·82	*3·258*	1·73	*3·235*
10	0·889	*2·947*	1·50	*3·171*	1·82	*3·258*	1·74	*3·240*

S.E. for means of 6 = ±*0·0260*.
S.E. for mean of 5 = ±*0·0335*.

TABLE 2 (cont.).

Harvest	Treatment. I		II		III		IV	
Stems								
1	0·00182		0·00182		0·00182		0·00182	
2	0·00321	*0·505*	0·00305	*0·484*	0·00322	*0·505*	0·00252	*0·396*
3	0·00309	*0·489*	0·00318	*0·501*	0·00343	*0·533*	0·00333	*0·522*
4	0·00556	*0·743*	0·00679	*0·829*	0·00643	*0·808*	0·00351	*0·765*
5	0·0271	—	0·0298	—	0·0338	—	—	—
6	0·425	*2·623*	0·529	*2·722*	0·570	*2·753*	0·486	*2·685*
7	2·38	*3·376*	4·80	*3·689*	4·52	*3·655*	4·32	*3·634*
8	4·85	*3·685*	11·13	*4·046*	15·33	*4·184*	13·97	*4·141*
9	5·14	*3·710*	11·81	*4·072*	17·71	*4·248*	16·96	*4·229*
10	4·36	*3·640*	10·44	*4·018*	15·76	*4·197*	15·97	*4·202*
Harvest 5 omitted from analysis of variance. S.E. for means of 5 = ±*0·0162*. S.E. for means of 3 = ±*0·0209*.								
Inflorescences								
8	0·530	±0·020	1·41	±0·05	2·03	±0·06	2·23	±0·08
9	1·62	±0·11	4·63	±0·24	6·67	±0·39	7·61	±0·38
10	2·77	±0·05	6·80	±0·13	11·47	±0·35	12·13	±0·27
Mean for all harvests		*3·123*		*3·548*		*3·729*		*3·770*
Interaction insignificant; treatment significant. S.E. for treatment means (of 15) = ±*0·0104*. S.E.'s of individual dry weight means given.								
Primary Shoot								
1	0·00488		0·00488		0·00488		0·00488	
2	0·0127		0·0120		0·0122		0·0104	
3	0·0177		0·0172		0·0176		0·0181	
4	0·0254		0·0299		0·0289		0·0270	
5	0·0738		0·0830		0·0888		—	
6	0·466		0·554		0·556		0·516	
7	1·76		2·40		2·15		2·09	
8	3·36		4·73		5·10		4·88	
9	4·39		5·87		6·33		6·03	
10	4·30		6·33		6·23		6·41	
Total Shoots								
1	0·00488		0·00488		0·00488		0·00488	
2	0·0127	*1·103*	0·0120	*1·095*	0·0122	*1·086*	0·0104	*1·013*
3	0·0177	*1·248*	0·0172	*1·234*	0·0176	*1·244*	0·0181	*1·255*
4	0·0254	*1·403*	0·0299	*1·474*	0·0289	*1·462*	0·0270	*1·430*
5	0·0898	—	0·0990	—	0·1147	—	—	—
6	1·08	*3·035*	1·47	*3·165*	1·55	*3·191*	1·34	*3·127*
7	3·55	*3·549*	7·33	*3·865*	7·94	*3·900*	8·07	*3·906*
8	6·57	*3·817*	14·99	*4·175*	21·37	*4·329*	20·66	*4·313*
9	7·83	*3·894*	18·59	*4·269*	27·78	*4·444*	28·31	*4·451*
10	8·20	*3·914*	19·44	*4·289*	30·42	*4·483*	31·53	*4·498*
Harvest 5 omitted from analysis of variance. S.E. for means of 5 = ±*0·0143*. S.E. for means of 3 = ±*0·0185*.								
Total Plant								
6	1·62	*3·206*	1·89	*3·275*	1·97	*3·295*	1·63	*3·210*
7	4·59	*3·662*	8·74	*3·941*	9·28	*3·967*	9·40	*3·972*
8	7·71	*3·887*	16·76	*4·224*	23·54	*4·371*	22·52	*4·350*
9	8·72	*3·940*	20·09	*4·303*	29·60	*4·471*	30·04	*4·479*
10	9·09	*3·958*	20·93	*4·321*	32·24	*4·508*	33·26	*4·522*
S.E. for means of 5 = ±*0·0132*. S.E. for mean of 3 = ±*0·0170*.								

NOTES.

The means of the logarithms of the dry weight data and their standard errors are in italic figures. All values are means of 5, except treatment IV, harvests 3, 4, and 6, which are means of 3.

Also the increase in yield obtained by delayed application may be dependent upon no other nutrients being limiting.

Garner et al. (1934) found with *Nicotiana tabacum* that 80 lb. nitrogen per acre applied initially in one dose gave an early retardation in growth and no greater yield than a dose of 40 lb.; but if half the quantity was applied at the commencement of growth and the other half 30 days later, a greater yield was obtained, and retardation was not evident. Somewhat similar effects were found in oats by Rackmann (1935), except that in this case the doses comprised nitrogen, phosphorus, and potassium.

The Relative Growth Rate.

Relative growth rates for the two experiments are presented in Tables 4 and 5 and in figs. 7 and 8. Values have been calculated from the data of Blanck and Giesecke also, by interpolating in the smoothed growth curves[6] of fig. 1; these are presented graphically in fig. 6. The following conclusions can be drawn from statistical examination of the data of the present experiments.

Wheat.

6–7: II and III significantly greater than I; IV significantly greater than III; 6–7 to 8–9 taken together: IV insignificantly greater than III; otherwise significant increases with treatment from 6–7 to 8–9; 9–10: no significant differences.

Sudan grass.

1–2: III significantly less than I; 2–3 to 4–5: II and III insignificantly different; 2–3 to 6–7: I and II insignificantly different, even if 3–4 omitted; 5–6: III significantly greater than II.

Evidently, owing to the insensitivity of the statistical method, in some places it has not been possible to show significant differences between relative growth rates where such differences must have occurred. Thus, at some time, R for the Sudan grass plants of treatment II must have been greater than R for those of treatment I, although we cannot conclude this from the R data.

All the curves presented, however, conform to the same general picture. In the early stages, R is less the greater the nitrogen supply; this holds presumably only beyond the level of supply at which depression becomes evident. After the maximum is reached, R is higher with the higher treatments than with the lower treatments. The curves tend to converge at the end of the growing period.

Relative growth rate curves calculated from the other data referred to in the previous section tend to conform to the general picture, although certain of them exhibit irregularities probably due to the variability of the material.

[6] Briggs, Kidd, and West found a depression in R immediately before flowering. This may have been eliminated by the smoothing method adopted in the present case, but the method enables us to depict the general trend of the effect of treatment. This depression would not be exhibited in the experimental data of the present paper owing to the large intervals between harvests.

LEAF WEIGHT RATIO.

Among the data so far considered, leaf weight ratios can be calculated only from those presented in this paper. The values are given in Tables 4 and 5, and are plotted in figs. 9 and 10. The Sudan grass curves show no significant effect of

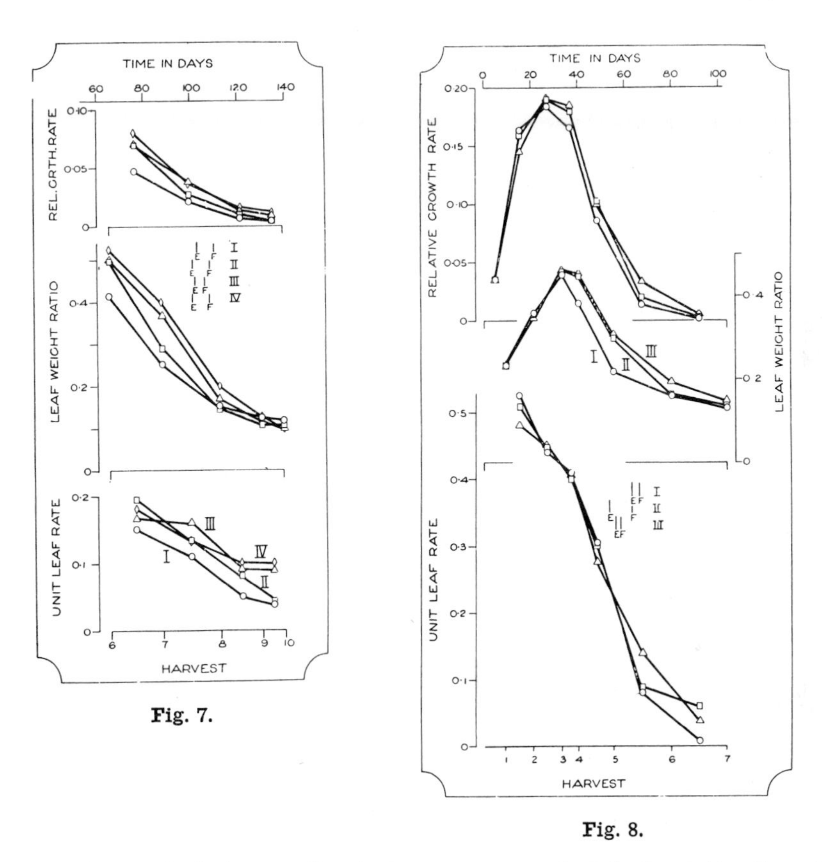

Fig. 7.

Fig. 8.

Fig. 7. Curves of relative growth rate (gm. per gm. per day), leaf weight ratio and unit leaf rate (gm. per gm. per day) in wheat, Experiment 1.

Fig. 8. Curves of relative growth rate (gm. per gm. per day) leaf weight ratio and unit leaf rate (gm. per gm. per day) in Sudan grass, Experiment 2.

treatment until after the maxima have been attained, when the ratio becomes an increasing function of treatment, and this is the state exhibited in the curves for wheat[7]. At the end of the growing period, the curves converge, becoming insignificantly different at the final harvest.

[7] The results of Müller and Larsen (1935) for *Sinapis alba* also show that increased nitrogen content of leaves is associated with increase in F_A.

Table 3.

Mean Dry Weights in gm. per Plant for Experiment 2 (Sudan Grass).

Harvest	Treatment I		Treatment II		Treatment III	
Leaves						
1	0·0023		0·0023		0·0023	
2	0·0251	*1·397*	0·0234	*1·369*	0·0199	*1·294*
3	0·288	*2·455*	0·293	*2·465*	0·258	*2·407*
4	0·773	*2·877*	1·000	*2·988*	0·928	*2·965*
5	1·63	*3·210*	3·05	*3·471*	2·82	*3·446*
6	1·65	*3·218*	2·56	*3·406*	3·61	*3·557*
7	1·39	*3·140*	2·72	*3·430*	3·30	*3·514*
	S.E. for means of 6 = ±*0·0149*.					
	S.E. for means of 5 = ±*0·0164*.					
Roots						
1	0·0032		0·0032		0·0032	
2	0·0282	*1·447*	0·0258	*1·406*	0·0225	*1·350*
3	0·154	*2·182*	0·157	*2·193*	0·131	*2·116*
4	0·597	*2·768*	0·507	*2·693*	0·465	*2·661*
5	1·87	*3·269*	2·40	*3·369*	1·93	*3·283*
6	2·37	*3·375*	3·29	*3·513*	3·63	*3·558*
7	1·81	*3·230*	4·28	*3·618*	4·24	*3·618*
	S.E. for means of 6 = ±*0·0401*.					
	S.E. for means of 5 = ±*0·0440*.					
Stems						
1	0·0044		0·0044		0·0044	
2	0·0169	*1·227*	0·0168	*1·224*	0·0148	*1·168*
2	0·197	*2·288*	0·190	*2·278*	0·171	*2·228*
4	0·682	*2·820*	0·738	*2·846*	0·660	*2·816*
5	3·82	*3·582*	4·90	*3·663*	4·11	*3·613*
6	5·68	*3·753*	8·48	*3·928*	10·01	*4·000*
7	6·07	*3·778*	10·61	*4·017*	11·62	*4·062*
	S.E. for means of 6 = ±*0·0338*.					
	S.E. for means of 5 = ±*0·0371*.					
Inflorescences						
5	0·158		0·196		0·246	
6	0·580		1·42		1·35	
7	1·37		2·44		3·00	
Mean for all harvests		*2·137*		*3·011*		*3·321*
	Interaction insignificant. Treatment significant.					
	S.E. for treatment means (of 15) = ±*0·0869*.					
Total Plant						
Seed	0·00727		0·00727		0·00727	
1	0·0099		0·0099		0·0099	
2	0·0702	*1·845*	0·0660	*1·819*	0·0572	*1·755*
3	0·639	*2·801*	0·641	*2·805*	0·560	*2·745*
4	2·05	*3·305*	2·25	*3·331*	2·05	*3·309*
5	7·49	*3·873*	10·52	*4·006*	9·11	*3·958*
6	10·29	*4·012*	15·75	*4·197*	18·61	*4·269*
7	10·64	*4·022*	20·04	*4·298*	22·16	*4·343*
	S.E. for means of 6 = ±*0·0279*.					
	S.E. for means of 5 = ±*0·0306*.					

NOTES.

The means of the logarithms of the dry weight data and their standard errors are in italic figures. Values for harvests 1–4 are means of 6; and for harvests 5–7, means of 5.

Figs. 11 and 12 are of interest in connection with these drifts. They show that, during adolescence, roots and leaves constitute most of the plant; later the stems, and finally the inflorescences, become fractions of increasing relative magnitude. Three successive phases of ontogeny are thus manifested, in which growth of leaves and roots, stems, and inflorescences respectively preponderate.

TABLE 4.

Relative Growth Rate, Unit Leaf Rate, and Leaf Weight Ratio in Experiment 1 (Wheat).

Harvest	Treatment I	II	III	IV
Relative Growth Rate (100R)				
6–7	4·73 ± 0·280	6·96 ± 0·225	7·06 ± 0·196	7·96 ± 0·357
7–8	2·16 ± 0·100	2·71 ± 0·125	3·88 ± 0·167	3·61 ± 0·261
8–9	0·68 ± 0·050	1·01 ± 0·130	1·28 ± 0·212	1·64 ± 0·327
9–10	0·46 ± 0·109	0·46 ± 0·230	0·95 ± 0·267	1·13 ± 0·409
Leaf Weight Ratio (100F_W)				
6	41·5	49·7	50·1	52·6
7	25·4	29·0	37·0	40·0
8	15·4	14·6	17·1	20·0
9	12·4	10·7	11·5	12·5
10	11·8	10·5	9·9	10·3

All values are means of 5, except that for harvest 6, treatment IV, which is a mean of 3.
S.E. for means of 5 = ±0·66.
S.E. for mean of 3 = ±0·85.

Harvest	I	II	III	IV
Unit Leaf Rate (100E_W)				
6–7	15·11 ± 0·751	19·42 ± 0·706	16·93 ± 0·546	18·02 ± 0·945
7–8	11·05 ± 0·480	13·41 ± 0·698	16·00 ± 0·973	13·33 ± 1·263
8–9	4·95 ± 0·358	8·05 ± 0·992	9·11 ± 1·485	10·22 ± 1·974
9–10	3·82 ± 0·922	4·29 ± 2·192	8·89 ± 2·543	10·02 ± 3·599

Since it will be shown in the next section that treatment does not increase E_W during early adolescence, its effect on F_W is evidently due firstly to one on the distribution of synthesized materials among the parts. The initial depression, possibly a toxic effect, is more prolonged in roots than in leaves; this presumably leads to the first increase with treatment in F_W. Immediately F_W is increased, the relative rate of dry matter production is increased, and a greater proportion of the relative excess dry matter produced with high treatment is used for leaf growth than for the other parts. This is responsible for further increase of leaf weight ratio and decrease of root weight ratio with treatment[8]. The effect on F_W becomes apparent in the data after the third harvest, when the growth curves of the leaves

[8] Decrease in root weight ratio with increasing nitrogen supply was also shown by Turner (1922), Mothes (1932), and Rippel and Meyer (1933), and can be seen in the data of Brenchley and Jackson (1921).

begin to diverge. The effect on the stem weight ratio is intermediate between those on root and leaf weight ratios; and inflorescence weight ratio, although increased in wheat, does not appear to be significantly affected in the Sudan grass experiment up to the time of the final harvest.

TABLE 5.

Relative Growth Rate, Unit Leaf Rate, and Leaf Weight Ratio in Experiment 2 (Sudan Grass).

Harvest	Treatment I	Treatment II	Treatment III
Relative Growth Rate (100*R*)			
Sowing to harvest 1	3·45 ± 0·985	3·45 ± 0·985	3·45 ± 0·985
1–2	16·31 ± 0·315	15·79 ± 0·246	14·60 ± 0·444
2–3	18·40 ± 0·676	18·94 ± 0·424	19·01 ± 0·602
3–4	16·65 ± 1·520	17·94 ± 1·558	18·54 ± 1·061
4–5	8·64 ± 0·569	10·28 ± 0·753	9·94 ± 0·487
5–6	1·51 ± 0·192	1·92 ± 0·261	3·40 ± 0·243
6–7	0·12 ± 0·254	0·89 ± 0·255	0·65 ± 0·211
Leaf Weight Ratio ($100F_W$)			
1	23·4	23·4	23·4
2	35·8	35·6	34·6
3	45·1	45·7	46·0
4	38·1	44·7	45·3
5	21·9	29·6	30·9
6	16·1	16·2	19·4
7	13·1	13·6	14·9

Values for harvests 2–4 are means of 6; of harvests 5–7, means of 5.

S.E. for means of 6 = ±0·87.
S.E. for means of 5 = ±0·96.

Harvest	Treatment I	Treatment II	Treatment III
Unit Leaf Rate ($100E_W$)			
1–2	52·47 ± 2·625	51·24 ± 7·918	48·14 ± 20·555
2–3	43·99 ± 3·958	44·85 ± 2·785	45·12 ± 9·302
3–4	41·08 ± 5·134	39·8 ± 0·614	40·67 ± 3·520
4–5	31·46 ± 2·216	29·99 ± 2·494	27·67 ± 2·169
5–6	8·13 ± 1·054	8·90 ± 1·069	14·14 ± 0·960
6–7	0·85 ± 2·359	6·02 ± 1·922	3·81 ± 1·341

Various factors contribute to the convergence during senescence. Leaf growth ceases at about the time of flowering, and in wheat the dry weights of leaves actually converge in senescence. Root growth continues later in Sudan grass in the higher treatments, causing crossing in the root weight ratio curves, and this obviously contributes to convergence in the F_W curves. Finally in wheat the inflorescence weight ratio increases with treatment. The loss in weight of the leaves is probably associated with the high inflorescence weight ratio of this plant, since in the Sudan grass the loss is inappreciable.

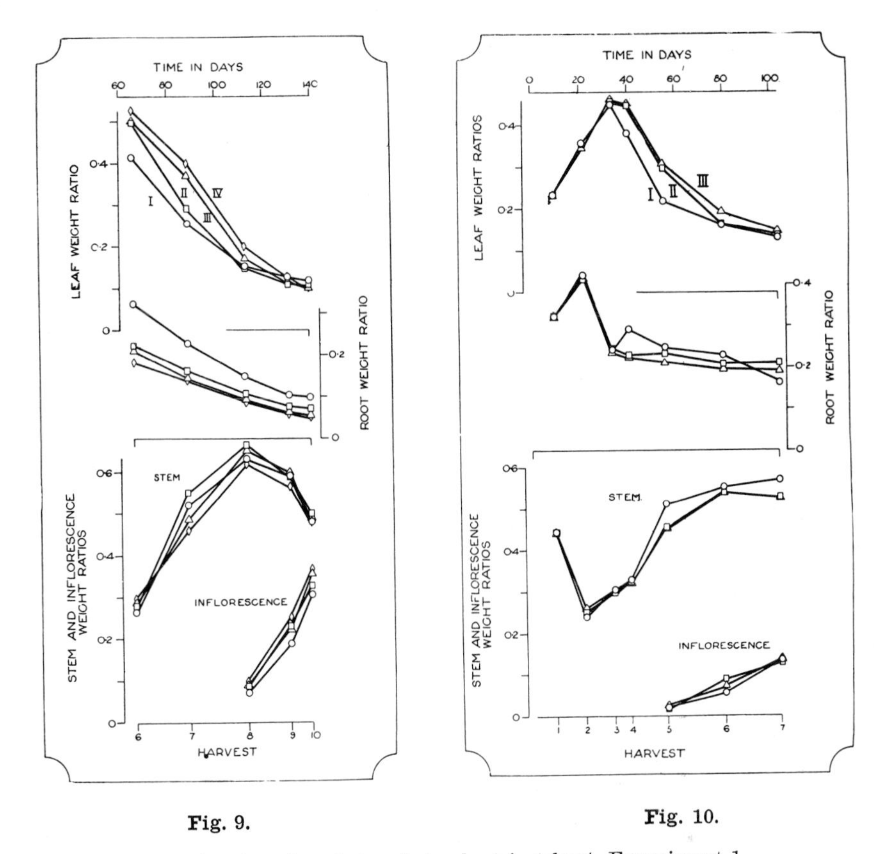

Fig. 9. Curves of ratios of parts to whole plant in wheat, Experiment 1.

Fig. 10. Curves of ratios of parts to whole plant in Sudan grass, Experiment 2.

The Unit Leaf Rate.

Unit leaf rates are presented in Tables 4 and 5 and figs. 7 and 8. The values are slightly in error on account of the exclusion of the leaf sheaths, which perform a certain amount of carbon assimilation; the error is less the younger the plant. The following conclusions have been drawn from statistical examination of the data.

Wheat.

6–7: III significantly below II; III and IV insignificantly different; 6–7 to 8–9: I significantly below II at each point; 7–8 and 8–9: II, III, and IV insignificantly different; 9–10: no significant differences.

Sudan grass.

5–6: III significantly greater than II; no other differences demonstrably significant.

The effect of treatment on E_W has clearly differed in the two experiments. In the wheat the effect is evident between plants of treatment I and the others, at least from a point very soon after harvest 6, at which time the leaves as a whole are still adolescent. In the Sudan grass there is no treatment effect except for a brief period after flowering. Depression is shown with treatment III between harvests 6 and 7 in the wheat experiment; it is not shown with treatment IV, but the value of E_W is here somewhat unreliable on account of fewness of replicates. There is no significant depression in the Sudan grass experiment.

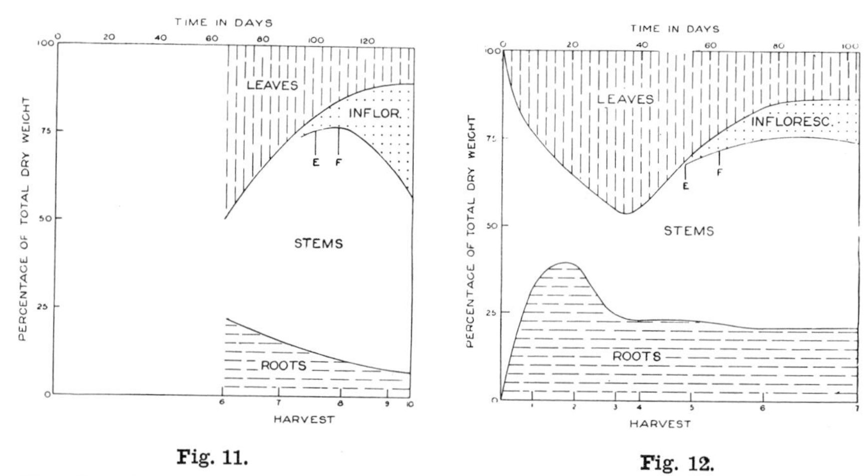

Fig. 11. **Fig. 12.**

Fig. 11. Diagram showing the relative proportions of parts in wheat during the latter half of the growing period, Experiment 1, treatment II. The conventions of separation in this and the next figure are those described in the text.

Fig. 12. Diagram showing the relative proportions of parts in Sudan grass during the growing period, Experiment 2, treatment II. The ungerminated seed has been arbitrarily classed as stem.

Gregory (1926) has presented graphically values for E_A for plants with two different nitrogen supplies. The observations ceased with attainment of maximum leaf area, which approximately coincides with flowering. Gregory concludes that difference in nitrogen supply has had no effect on E_A up to that time. Crowther (1934) also finds that nitrogen treatment has no effect on E_W during the main growth phase of the cotton plant[9]. These conclusions agree essentially with the results of the Sudan grass experiment, but not with those of the wheat.

Gregory also concludes that, up to the time of attainment of maximum leaf area, E_A is independent of the age of the plant, and that the fall with time is due to environmental changes. Observations on the environment during the course of the present experiments render it very improbable that the marked decline in

[9] Crowther omitted the root weights in the calculation of his values of E_W.

E_W was due to environmental factors. The Sudan grass senesced during conditions of falling temperature, and the wheat during conditions of rising temperature.

Since the data presented in this paper show that during part of the life-period E_W can be increased by nitrogen treatment, further analysis of this phenomenon is desirable. There are two possibilities: treatment may increase the rate of carbon assimilation of the individual leaves at a given growth stage, or it may decrease the average rate of ontogeny of the leaves as a whole, and so prolong their effective life. The latter could be accomplished either by inducing a longer life of individual leaves, or by causing a prolonged formation of young leaves.

TABLE 6.

Mean Number of Tillers per Plant for Experiment 1 (Wheat).

Harvest	Treatment I	II	III	IV
Total Tillers				
6	4·4	6·5	6·9	6·3
7	4·1	6·5	7·9	8·9
8	3·9	6·9	7·9	7·5
9	4·3	6·2	7·4	8·4
10	4·0	6·5	8·6	8·2

All values are means of 5 except that for harvest 6, treatment IV, which is a mean of 3.

S.E. for means of 5 = ±0·06.
S.E. for mean of 3 = ±0·08.

Harvest	Treatment I	II	III	IV
Inflorescence-bearing Tillers				
8	2·3	3·7	5·7	5·9
9	2·1	3·8	5·8	6·5
10	2·2	3·7	6·2	6·4

Interaction insignificant.

With regard to the first possibility, Gregory and Richards (1929) found that nitrogen deficiency lowered the rate of carbon assimilation of mature leaves at high light intensity, but the decrease was insignificant. This result, however, was obtained with high carbon dioxide concentration. Müller (1932) and Müller and Larsen (1935) found that the assimilation rate of *Sinapis alba* per unit leaf area increased with nitrogen supply and nitrogen content of the leaf; the same was found by Gassner and Goeze (1934 *a* and *b*) in wheat and rye, provided the supply of potassium was not low. The first possibility suggested is therefore probably a contributing factor.

To obtain information about the second possibility, the separation of the primary shoot and early leaves was performed in the wheat experiment. The data do not show that increasing nitrogen supply increases the longevity of the primary shoot. Fairfield Smith (1933) and Bartel (1935) have shown that tillers are not individual units, but that competition between them for elaborated materials may occur, and this may account for the result obtained.

The data of the wheat experiment suggest that the third leaves decline in weight less rapidly with treatment IV than with the other treatments. Gassner and Goeze (1934*b*) also showed that the decline with age in rate of assimilation occurs later in leaves with high nitrogen content. Such an effect on rate of ontogeny of the individual leaf could be a factor contributing to the effect of treatment on E. It is, however, unjustifiable to extend this conclusion to the leaves at the end of the growing period of the plant: other elements such as phosphorus may become limiting in the larger plants earlier than in the smaller ones, and this might even hasten senescence in the late-formed leaves.

From Table 6 it can be seen that, whereas plants of treatments I and II had ceased to tiller before harvest 6, those of the higher treatments had not ceased till some weeks later; and it is most likely that the increased unit leaf rates of treatments III and IV towards the end of the life-period are due partly to the active assimilation of the young leaves of these later-formed tillers.

F_W AND E_W AS DETERMINANTS OF R.

Reviewing the conclusions drawn from this discussion, we see that the data do not permit us to state definitely whether both these factors contribute to the early depression in R with increasing nitrogen treatment, although insignificant depressions are shown in both. In the wheat experiment, both factors contribute to an increase in R with treatment throughout a considerable part of the growing period, although the effect of E_W is shown only between treatments I and II; in the Sudan grass experiment, the effect of F_W is the main one, and that of E_W is negligible until after the exsertion of inflorescences.

ONTOGENETIC RATE.

Needham (1933) has shown that in the development of an animal embryo the ontogenetic processes can be dissociated experimentally. The dissociation may, however, be only partial, so that a certain factor may affect the rate of different processes in different ways.

In the present experiments low nitrogen supply caused late exsertion of inflorescences and flowering, although in the Sudan grass experiment this conclusion is based on too few data to be reliable. Garner et al. (1934) showed that low and also very high nitrogen supply retarded flowering in the tobacco plant. On the other hand, Borodin (1931) found that in extreme nitrogen deficiency the appearance of inflorescences was earlier in barley and unaffected in time in millet.

Nitrogen treatment in the present experiments affected other processes differently: it delayed the decline of leaf weight ratio and hence of relative growth rate, although it hastened flowering. Retardation in the development of vegetative parts was also shown by Gregory (1926) as a result of increasing nitrogen treatment.

SUMMARY.

This paper represents the first of a series devoted to the study of physiological ontogeny, and defines the effect of increasing the initial nitrogen supply to certain plants over a certain range.

It is found in general that, in the whole plant and its vegetative parts, increasing supply causes at first a depression in dry weight; later, this passes off and, provided the supply does not exceed a certain limit, gives place to an increase. In the early stages of growth, the relative growth rate is less the greater the nitrogen supply. After the maximum is attained the rate becomes greater with higher treatment. Similar phenomena are exhibited in the leaf weight ratio, only the depression could not be shown to be significant. The values of the ratio converge during senescence. The unit leaf rate was found to decline throughout practically the whole of the growing period, and no significant depression with treatment could be demonstrated. Increase with treatment was found in the later part of the growing period, and may be quite absent in the early part.

Other aspects of the change in weight of the plant and its parts are discussed.

STATISTICAL APPENDIX

by E. A. Cornish.

Before proceeding to the details of the methods employed in the statistical reduction of the data, it is necessary to remark upon the choice of a suitable variable for use in the analysis.

The analysis of variance and the z test are required here for the purpose of testing jointly the significance of the differences between several means, but the test is, strictly speaking, only applicable if the variance of a set of replicates is independent of the mean of these replicates.

An examination of the observational data indicated, in the majority of cases, the presence of a positive correlation between the treatment-harvest class (hereafter called class) mean and the class variance; and the following figures, which have been derived from an analysis of total shoots, Experiment 1, are fairly representative of the existing conditions. There was a significant positive correlation of $0 \cdot 75$ between the mean and the variance, while the correlation between the variance and the square of the class mean was $+0 \cdot 32$, a value which just failed to reach the 5 p.c. level of significance. It thus appeared that the square roots of the observations would provide a suitable variable. In this case, however, the coefficient had the value $+ 0 \cdot 62$, which, though indicating a weaker relationship, was nevertheless significant. The correlation, therefore, had not been eliminated and, in a further endeavour to do this, the logarithms of the observations were tried. On evaluation this correlation was found to be —0·44, a significant value, showing that an over-correction had now been introduced. The regression of variance on mean was, however, relatively small, and in consequence it was decided to employ the logarithms where a change of variable was necessary. The correct unit, which may be found by a process of trial and error, would be some positive power of the observations less than ½. In order to avoid negative values the dry weights were multiplied by 10^3 before transformation to logarithms.

TABLE 7.

Analysis of Variance. Total Shoots, Experiment 1 (grams).

Variation due to	Deg. freed.	Sum. sq.	Mean sq.
Between classes	34	17906·47	
Within classes	134	33·70	0·25152
Total	168	17940·17	

The following alternative test demonstrates, perhaps more forcibly, the differential variability of the classes. Working with the original weights in grams, the analysis of the total variability gave the results shown in Table 7.

If v_r $(r = 1, 2, \ldots\ldots, 35)$ stand for the variance of class r, then the mean variance, $\bar{v}$, is given by

$$\bar{v} = S(v)/35 = 2\cdot2577,$$

and the variance of variance, V, by

$$V = S(v - \bar{v})^2/34 = 891\cdot62 \text{ (34 deg. freed).}$$

If there are no significant differences in variability among the several classes, the observed variances will be distributed about the weighted mean variance, 0·25152, with variance, v_0, given by

$$v_0 = 2(\cdot25152^2)/(n-1) = 2(\cdot25152^2)/3\cdot7488 = 0\cdot033751 \text{ (134 deg. freed.),}$$

where n is the number of replicates per class. This number varied from class to class, and since the variance is inversely proportional to the number of observations, n has been taken as the harmonic mean of the class numbers. V and v_0 were compared by means of the z test, which demonstrated a significant differential variability among the classes.

METHODS OF ANALYSIS.

In the majority of cases, the experimental observations are non-orthogonal, and this fact necessitates a modification of the ordinary analytic procedure of the analysis of variance in their treatment. Several methods are available for the analysis of non-orthogonal data, of which the most general in its range of application is the fitting of constants by the method of least squares, the remaining methods being applicable only under certain conditions. For a full discussion of the subject, recourse should be made to an excellent paper by Yates (1933), who devised the methods.

TABLE 8.

Class Totals for Third Leaf, Experiment 1.

Harvest	Treatment I	II	III	IV	Total.
3	1·5977 (5)	2·3051 (5)	2·0053 (5)	0·8195 (3)	6·7276 (18)
4	6·9344 (5)	7·7157 (5)	7·4977 (5)	4·5950 (3)	26·7428 (18)
5	11·4611 (5)	11·6625 (5)	11·5476 (5)	—	34·6712 (15)
6	11·6635 (5)	11·9187 (5)	11·9908 (5)	6·9151 (3)	42·4881 (18)
7	10·9983 (5)	10·7653 (5)	11·0040 (5)	11·1652 (5)	43·9328 (20)
Total	42·6550 (25)	44·3673 (25)	44·0454 (25)	23·4948 (14)	154·5625 (89)

Figures in brackets indicate the number of observations contributing to the corresponding total.

TABLE 9.

Analysis of Variance.

Variation due to	Deg. freed.	Sum. sq.	Mean sq.
Constants t and h	7	50·8277	
Residual (interaction)	11	0·1345	0·01223
Between classes	18	50·9622	
Within classes	70	0·8579	0·01226
Total	88	51·8201	

TABLE 10.

Analysis of Variance.

Variation due to	Deg. freed.	Sum. sq.	Mean sq.
Harvest	4	3·28052	0·82013
Treatment	3	0·49413	0·16471
Residual (interaction)	12	0·12673	0·01056
Total	19	3·90138	

(1) *Fitting Constants.*

The observations on the third leaf, Experiment 1, are given in Table 8. As one class is entirely missing, these data must be analysed by fitting constants. The main purpose of the analysis is to establish the existence or otherwise of the interaction of treatment and harvest. On the hypothesis of negligible interaction, constants representing treatment and harvest are fitted, and the residual variance between classes, after fitting, is compared with the intraclass variance by means of the z test.

The preliminary analysis of variance gives the following results:

Total sum sq. $= 51 \cdot 8201$ (88 deg. freed.)
Between classes $= 50 \cdot 9622$ (18 deg. freed.)
Within classes $= 0 \cdot 8579$ (70 deg. freed.).

The constants to be determined are:

a general mean	m;
four treatment constants	t_1, t_2, t_3, t_4;
five harvest constants	h_1, h_2, h_3, h_4, h_5.

The solution of the normal equations results in the following values for the constants:

$$m = +1 \cdot 744277 \qquad h_1 = -1 \cdot 371619$$
$$t_1 = -0 \cdot 038075 \qquad h_2 = -0 \cdot 259664$$
$$t_2 = +0 \cdot 030415 \qquad h_3 = +0 \cdot 563843$$
$$t_3 = +0 \cdot 017539 \qquad h_4 = +0 \cdot 615075$$
$$t_4 = -0 \cdot 009879 \qquad h_5 = +0 \cdot 452363,$$

whence, the reduction in the sum of squares due to fitting constants, excluding the mean, is

$$1 \cdot 744277 \times 154 \cdot 5625 + (-0 \cdot 038075 \times 42 \cdot 6550) + \ldots\ldots + (-1 \cdot 371619 \times 6 \cdot 7276) + \ldots\ldots - 154 \cdot 5625^2/89 = 50 \cdot 8277.$$

The complete analysis of variance is given in Table 9.

It will be observed that all the variation between class means, except for a very small fraction, may be accounted for as the additive effect of treatment and harvest.

(2) *Approximate Analysis of Variance.*

In the remaining non-orthogonal data of Experiment 1, the class numbers do not depart appreciably from equality, so that their reduction has been effected by the method of unweighted means, an approximate analysis of variance, due to Yates. The details of this method are illustrated in the following analysis of total plant, Experiment 1. The class means are given in Table 2. The preliminary analysis results in the following partitioning of the sum of squares:

Between classes	$18 \cdot 2465$	(19 deg. freed.)
Within classes	$0 \cdot 0678$	(78 deg. freed.)
Total	$18 \cdot 3143$	

An ordinary analysis of variance is now conducted on the means of Table 2, the results of which are entered in Table 10.

In order to test the significance of the several items of Table 10, an appropriate error variance is calculated by dividing the intraclass variance, 0·00087, by the harmonic mean of the class numbers, 4·3548, a process which yields the number 0·00020. The analysis of variance then reduces to the form given in Table 11.

It should be noted that the appropriate figure upon which to base the errors of the class means is the intraclass variance 0·00087.

TABLE 11.

Analysis of Variance.

Variation due to	Deg. freed.	Sum. sq.	Mean sq.
Harvest	4	3·28052	0·82013
Treatment	3	0·49413	0·16471
Interaction	12	0·12673	0·01056
Intraclass	78	—	0·00020

TABLE 12.

Class Totals for Total Plant, Experiment 2.

Harvest	Treatment I	Treatment II	Treatment III	Total.
2	11·0672 (6)	10·9108 (6)	10·5274 (6)	32·5054 (18)
3	16·8029 (6)	16·8283 (6)	16·4694 (6)	50·1006 (18)
4	19·8320 (6)	19·9885 (6)	19·8528 (6)	59·6733 (18)
5	19·3667 (5)	20·0322 (5)	19·7881 (5)	59·1870 (15)
6	20·0593 (5)	20·9849 (5)	21·3449 (5)	62·3891 (15)
7	20·1146 (5)	21·4889 (5)	21·7160 (5)	63·3195 (15)
Total	107·2427 (33)	110·2336 (33)	109·6986 (33)	327·1749 (99)

Figures in brackets indicate the number of observations contributing to the corresponding total.

TABLE 13.

Analysis of Variance.

Variation due to	Deg. freed.	Sum. sq.	Mean sq.
Harvest	5	75·0537	15·0107
Treatment	2	0·1542	0·0771
Interaction	10	0·4089	0·0409
Within classes	81	0·3781	0·0047
Total	98	75·9949	

(3) *Proportionate Class Numbers.*

In the data of Experiment 2 (Sudan grass), class numbers which were unequal possessed, however, the property of being proportionate, i.e.

$$n_{pt}/n_{pr} = n_{qt}/n_{qr}.$$

Under these conditions, component sums of squares of any complex item in the table of analysis of variance, such as that between classes, are additive, and consequently the sums of squares corresponding to these items may be evaluated by the method given by Fisher (1934) for a single classification with unequal numbers in the classes. The following analysis of the observations on total-plant, Experiment 2 (Table 12), illustrates the method.

The total sum of squares is 75·9949, and the fractions, between and within classes, are respectively 75·6168 and 0·3781. The correction factor is $327{\cdot}1749^2/99 = 1081{\cdot}2466$.

Treatment:	$(107{\cdot}2427^2 + 110{\cdot}2336^2 + 109{\cdot}6986^2)/33 =$	1081·4008
	Deduct	1081·2466
	Treatment sum sq. =	0·1542
Harvest:	$(32{\cdot}5054^2 + 50{\cdot}1006^2 + 59{\cdot}6733^2)/18 =$	395·9763
	$(59{\cdot}1870^2 + 62{\cdot}3891^2 + 63{\cdot}3195^2)/15 =$	760·3240
	Total	1156·3003
	Deduct	1081·2466
	Harvest sum sq. =	75·0537
	Interaction sum sq. = 75·6168 — (75·0537 + 0·1542) =	0·4089

The complete analysis is given in Table 13.

VARIANCE OF R AND E_W.

Provided that the errors e_1 and e_2, in W_1 and W_2, respectively, are small compared with W_1 and W_2, an approximation to the error, A, of R, is given by

$$A = e_1 \frac{\partial f}{\partial W_1} + e_2 \frac{\partial f}{\partial W_2},$$

so that the variance of R is

$$v(R) = v(W_1)\left(\frac{\partial f}{\partial W_1}\right)^2 + v(W_2)\left(\frac{\partial f}{\partial W_2}\right)^2$$

$$= \frac{1}{(t_2 - t_1)^2}\left\{\frac{v(W_1)W_2^2 + v(W_2)W_1^2}{W_1^2 W_2^2}\right\},$$

where $v(W_1)$ and $v(W_2)$ are the variances of W_1 and W_2.

In a similar manner the unit leaf rate, E_W, has a variance given by

$$v(E_W) = v(w_1)\left(\frac{\partial f}{\partial w_1}\right)^2 + v(w_2)\left(\frac{\partial f}{\partial w_2}\right)^2 + v(W_1)\left(\frac{\partial f}{\partial W_1}\right)^2 + v(W_2)\left(\frac{\partial f}{\partial W_2}\right)^2$$

$$= \frac{1}{(t_2 - t_1)^2 (w_2 - w_1)^2}\{v(w_1)A + v(w_2)B\}$$

$$+ \frac{E_W^2}{(W_2 - W_1)^2}\{v(W_1) + v(W_2)\},$$

where $A = \{w_1 E_W(t_2 - t_1) - (W_2 - W_1)\}^2/w_1^2$,

and $B = \{w_2 E_W(t_2 - t_1) - (W_2 - W_1)\}^2/w_2^2$.

Owing to the fact that the variances of the quantities W_1, W_2, etc., have been based on comparatively few degrees of freedom, the sensitivity of tests of significance involving the variances given above is rather low. This is, however, unavoidable owing to the nature of the data.

REFERENCES.

Ballard, L. A. T. (1933) : Austral. J. exp. Biol., 11, p. 161.

Bartel, A. T., Martin, J. H. and Hawkins, R. S. (1935) : J. Amer. Soc. Agron., 27, p. 707.

Blanck, E. and Giesecke, F. (1934) : J. f. Landw., 82, p. 33.

Blanck, E., Giesecke, F. and Heukeshoven, W. (1933) : *Ibid.*, 81, p. 91.

Borodin, I. (1931) : Bull. Appl. Bot., Genetics and Plant Breeding, 27, p. 171.

Brenchley, W. and Jackson, V. G. (1921) : Ann. Bot., 35, p. 533.

Briggs, G. E., Kidd, F. and West, C. (1920) : Ann. Appl. Biol., 7, pp. 103 and 202.

Crowther, F. (1934) : Ann. Bot., 48, p. 877.

Davidson, J. and LeClerc, J. A. (1917) : J. Amer. Soc. Agron., 9, p. 145.

Davidson, J. and LeClerc, J. A. (1918) : *Ibid.*, 10, p. 193.

Davidson, J. and LeClerc, J. A. (1923) : J. Agric. Res., 23, p. 55.

Fairfield Smith, H. (1933) : J. Austral. Counc. Sci.. and Industr. Res., 6, p. 32.

Fisher, R. A. (1934) : Statistical Methods for Research Workers, 5th Edition, Edinburgh.

Garner, W. W., Bacon, C. W., Bowling, J. D. and Brown, D. E. (1934) : U.S.A. Dept. Agric. Tech. Bull. No. 414.

Gassner, G. and Goeze, G. (1934*a*) : Z. Pflanzenernähr., A, 36, p. 61.

Gassner, G. and Goeze, G. (1934*b*) : Z. f. Bot., 27, p. 257.

Gericke, W. F. (1920) : Science, 52, p. 446.

Gericke, W. F. (1922) : Soil Sci., 13, p. 135; 14, p. 103.

Gregory, F. G. (1926) : Ann. Bot., 40, p. 1.

Gregory, F. G. and Richards, F. J. (1929) : *Ibid.*, 43, p. 119.

Henckel, P. A. and Litvinov, L. S. (1930) : Bull. Inst. Rech. biol. et Sta. biol., Univ. Perm., 7, p. 133.

Inamdar, R. S., Singh, S. B. and Pande, T. D. (1925) : Ann. Bot., 39, p. 281.

Maiwald, K. (1930) : Z. Pflanzenernähr., 17, p. 12.

Meyer, R. (1927) : *Ibid.*, A., 10, p. 329.

Mothes, K. (1932) : Biol. Centr., 52, p. 193.

Müller, D. (1932) : Planta, 16, p. 1.

Müller, D. and Larsen, P. (1935) : *Ibid.*, 23, p. 501.

Neidig, R. E. and Snyder, R. S. (1922) : Idaho Agric. Res. Sta. Bull. No. 1.

Rackmann, K. (1935) : Z. Pflanzenernähr., 40, p. 148.

Rippel, A. and Ludwig, O. (1926) : Biochem. Z., 177, p. 318.

Rippel, A. and Meyer, R. (1933) : Z. Pflanzenernähr., A, 27, p. 257.

Singh, B. N. and Lal, K. N. (1935) : Ann. Bot.. 49, p. 291.

Smirnow, A. I. (1928) : Planta, 6, p. 687.

Stone, W. E. (1933) : J. Agric. Res., 46, p. 565.

Turner, T. W. (1922) : Amer. J. Bot., 9, p. 415.

Wagner, H. (1932) : Z. Pflanzenernähr., A, 25, p. 48.

Yates, F. (1933) : J. Agric. Sci., 23, p. 108.

[*From the* QUARTERLY JOURNAL OF THE ROYAL METEOROLOGICAL SOCIETY.
Vol. LXII. No. 267. October, 1936.]

551·577·34

ON THE SECULAR VARIATION OF THE RAINFALL AT ADELAIDE, SOUTH AUSTRALIA

By E. A. CORNISH, B.Agr.Sc.

(Communicated by J. GLASSPOOLE, M.Sc., Ph.D.)

[Manuscript received May 28, 1935—read May 20, 1936]

SUMMARY

A detailed analysis of the rainfall of Adelaide, South Australia, has shown that throughout the 95 years 1839-1933 there has been a definite oscillation, with a period and amplitude of approximately 23 years and 30 days respectively in the incidence and duration of the winter rains. The amplitude, though small, is nevertheless about 20 per cent of the length of the rainfall season. The total quantity precipitated has shown no statistically significant changes.

INTRODUCTORY

The general characteristics of the Australian rainfall with respect to geographical and seasonal distribution and reliability have been the subject of numerous investigations by different workers. In dealing with the broad mass of data, only the simpler statistics are normally made use of, but it was considered desirable to ascertain how far recent methods developed by Professor R. A. Fisher (1) could be applied to the specific case of the rainfall record of Adelaide, particularly as a clearer knowledge of its characteristics is of some considerable value from the point of view of agricultural production.

DATA

On January 1st, 1839, Sir George Kingston established a daily rainfall record in Adelaide on a site approximately 500 yards from the present position of the Observatory. This record was continued until November, 1879. From May, 1860, to the present day, the readings have been taken at the Observatory so that over 19 years the two sets of observations were concurrent. During this interval the average annual difference between the gauges was 0·26 inch (Kingston's being the higher), and considering the proximity of the sites it may be taken that the two series in combination give a continuous and practically uniform record of the Adelaide rainfall.

No definite statement could be found as to the diameter of the gauge employed by Kingston, nor, in fact, the size of the gauge used in the early days of the existence of the Observatory. It is fairly certain, however, that no radical departure could have been made from the standard 8 in. gauge which has been in use since 1870.

It is noteworthy that this record is probably unique, both with respect to its continuity and length, and constitutes valuable material for statistical analysis.

At this juncture a word might be added on the character of the climate. The most prominent feature is the marked winter incidence of the rains. Approximately 70-80 per cent of the annual precipitation, the average of which is in the neighbourhood of 21 in., occurs within the period April to October, the summer months being characterised by hot dry atmospheric conditions, low rainfall and high evaporation.

Analysis

The method employed in this analysis was devised by R. A. Fisher (*loc. cit.*) and in brief consists of the following steps.

The rainfall of each year was divided into 61 six-day totals and to these was fitted a series of orthogonal polynomial functions of the fifth degree in time, thus furnishing six constants with which to express the quantity and distribution of rainfall in each year. These distribution constants, designated a', b', . . . f', are given in the appendix where, for convenience of presentation in tabular form, they have been multiplied by 10^3. To obtain the coefficients of the polynomial terms they must be divided by factors of the form :—

$$\frac{(r!)^2\, 60 \,.\, 59 \,.\,.\,.\, (61-r)}{(2r+1)!}$$

where r is the degree of the term fitted.

The first, a', represents the average rainfall over the year in 10^{-3} inches per six-day period, while b' . . . f' are constants specifying the distribution of the rainfall throughout the year.

The method was originally developed to represent an almost continuous distribution of rain. In consequence, it might be anticipated that in these data where sporadic rains, not infrequently separated by lengthy rainless spells, occur during the summer months the unit of subdivision is unsuitable. However, the only consequence of the use of this type of division is that negative values for the polynomial curves may occur during the long dry periods, but this should not be taken to indicate that the process of analysis is unsuitable.

To examine the behaviour of the six constants a' . . . f' over the period under review, the series of 95 values of each was analysed in exactly the same manner as the rainfall record of each year had been treated.

The analysis of the sequence of values of b' is presented as an illustration of the procedure next adopted. The sum of the squares of the deviations of the b' values from their mean may be analysed into two portions :—

(*a*) A sum of squares associated with regression on time and due to a comparatively simple trend predominating over the random fluctuations, and

(*b*) The remaining sum of squares which may be attributed to random annual variation.

The sum of squares of the regression formula may be analysed further to show the contribution of each term to the total for regression.

TABLE I.—ANALYSIS OF THE SEQUENCE OF VALUES OF b'

Source of variation	Degrees of freedom	Sum of squares	Mean squares
1st order regression	1	454·12	454·12
2nd " "	1	4427·57	4427·57
3rd " "	1	5050·95	5050·95
4th " "	1	156·25	156·25
5th " "	1	1953·64	1953·64
Total for regression	5	12042·53	2408·51
Deviations from regression	89	72311·47	812·49
Total	94	84354·00	

The mean square of the random fluctuations provides a basis with which to test the significance of the slow changes represented by the constants of the regression formula. The values of z (2) for the quadratic and cubic terms are individually significant at the 5 per cent point while the regression as a whole has the same significance.

The constant b' is one measure of the distribution of the yearly rainfall being, in fact, proportional to the slope of a straight line fitted to the rainfall figures of each year. It may be concluded that the variations from year to year in the incidence of the rains has not been wholly due to chance, but that some slowly changing cause affects in the same manner the rainfall distribution of a number of consecutive years.

The components of change specifying the secular trend of the remaining constants are given in Table II.

TABLE II.—SUMMARY OF ANALYSIS OF THE RAINFALL DISTRIBUTION CONSTANTS $a' \ldots f'$

	a'	b'	c'	d'	e'	f'
Mean	347·50	+14·21	−59·73	− 2·13	+17·64	− 3·53
x'_2	− 10·93	−21·31	+14·61	−10·83	+ 1·40	+15·37
x'_3	+ 80·81	+66·54	− 1·14	−17·40	+14·11	+39·54
x'_4	+ 19·73	−71·07	−32·63	+ 5·20	+ 1·62	− 2·74
x'_5	−103·58	−12·50	+21·99	− 2·25	− 3·34	+ 7·81
x'_6	+ 46·94	−44·20	−46·75	+24·40	+12·87	−26·50
Standard Residue	72·12	28·50	25·87	18·34	18·12	14·01

It is apparent that no changes have taken place in a', c', d' and e'; with f', however, the mean appears to show gradual change, the x'_3 term being significant. It will be observed that in all rainfall features the annual variability is very great.

Owing to the seasonal variations of rainfall, the mean values of b', c', e' and f' differ significantly from zero. The mean values given in Table II specify the average seasonal distribution of rain for the 95 years 1839-1933 and the curve which they represent is shown in Fig. 1, where the monthly means, calculated on a per day

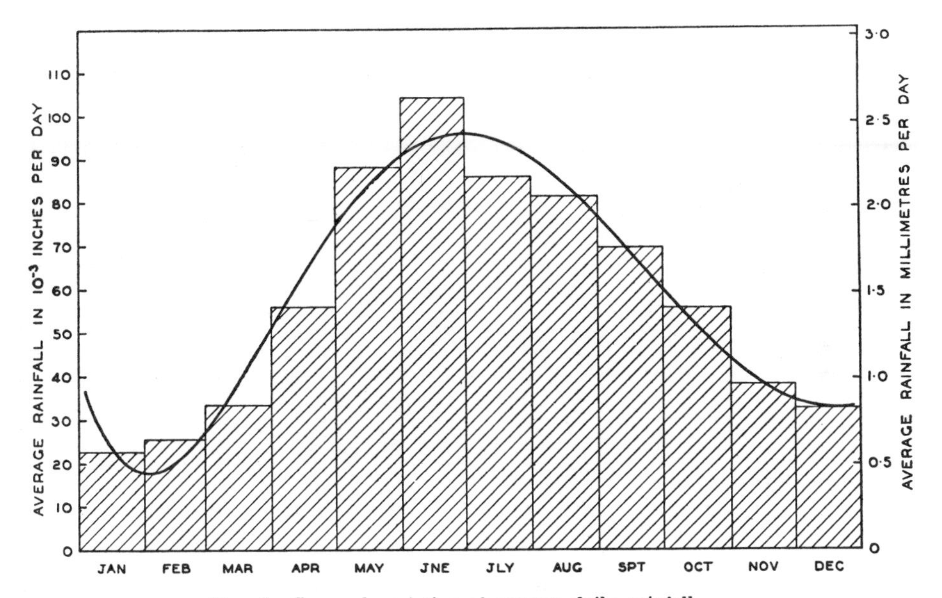

FIG. 1.—Seasonal variation of average daily rainfall.

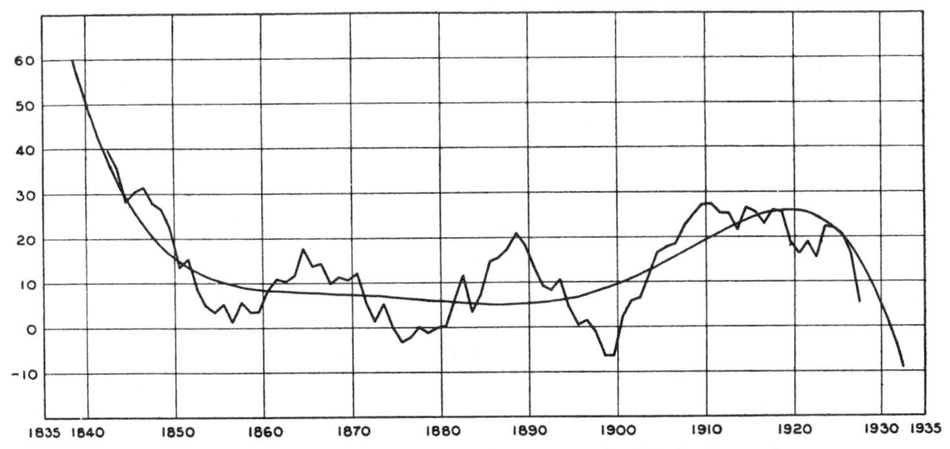

FIG. 2.—Ten-year means and polynomial curve of the distribution value b'.

basis, have been included for comparison. The rainfall sequence is represented more accurately by the smooth curve since the monthly means are subject to rather high standard errors.

In Fig. 2 the jagged line is a graph of successive 10-year means of b', while the smooth curve shows the course of the mean as represented by the first five terms of the polynomial expansion.* The polynomial curve does not show the well defined oscillations in the middle of the series.

Although the changes in the total precipitation have been found statistically insignificant, reference to Fig. 3 in which 10-year means of a' are plotted, demonstrates a change over the period under review, which is nevertheless remarkable.

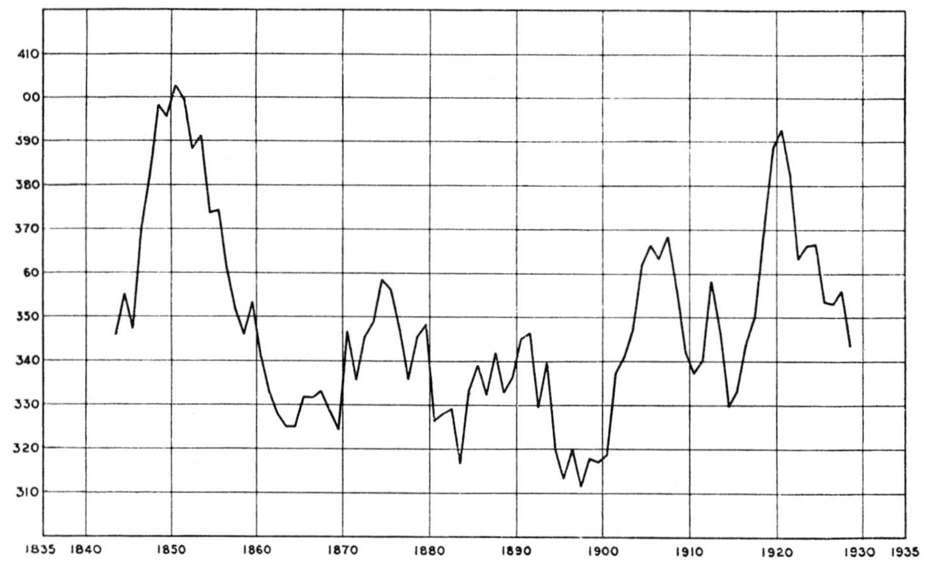

FIG. 3.—Ten-year means of the distribution value a^1.

The tests of significance employed above are based upon the assumption of the normality of the distribution of the quantities $a' \ldots f'$. Table III, which summarises a test of normality, shows the measures of departure with standard errors appropriate to a normal distribution.

TABLE III.—TEST OF THE NORMALITY OF $a' \ldots f'$

	a'	b'	c'	d'	e'	f'	Standard error
g_1	+0·224	+0·190	+0·312	−0·245	+0·385	+0·396	±0·247
g_2	−0·468	−0·372	+0·187	+0·669	+0 592	+3·130	±0·490

It will be observed that f' is the only quantity showing a significant departure from normality; the positive value of g_2 indicating the departure to be of a symmetrical kind, such that the apex and

* The analysis has been carried as far as the time of the 7th degree, yielding for the coefficients x'_7 and x'_8 the values −10·55 and −2·44 respectively. As these are insignificant in comparison with 28·50, the standard residue of b′, the polynomial represented by the 7 times of the expansion has not been computed.

two tails of the distribution curve are increased at the expense of the shoulders. The disturbance which this departure will cause in the distribution of z being unknown, it is safer to attach little reliability to the value of z calculated from the quadratic term of the series for f'.

As a matter of interest it seemed worth while to determine whether rainfall at any time of the year was correlated with rainfall at any other time. If such a relationship existed it would appear on correlating the rainfall distribution constants, since these have been computed from uncorrelated functions of time. After making allowance for secular trend, the correlations were evaluated, and their transformed values, $z = \tanh^{-1} r$, are presented in Table IV where each figure has a standard error of $\pm 0{\cdot}1072$.

TABLE IV.—VALUES OF $z=\tanh^{-1} r$

	a'	b'	c'	d'	e'
b'	+0·0563				
c'	−0·5695	−0·0438			
d'	+0·0110	−0·1951	+0·1107		
e'	+0·2524	+0·2534	−0·2432	−0·1210	
f'	−0·3236	+0·1547	+0·3083	+0·0292	−0·2110

With the exception of that between a' and c' all relationships are weak, and since this correlation is the largest of 15 it cannot be regarded as proof of association.

CHANGES IN THE RAINFALL SEQUENCE

In order to express in a simpler form the changes in the rainfall sequence which the variations in the coefficient b' represent, and which the analysis has shown to exist, the problem was approached from another angle. An examination of the monthly records revealed the fact that the main fluctuation seemed to concern the date of incidence of the winter rains. It therefore appeared desirable to ascertain to what extent the oscillation in the mean date of attainment of the yearly maximum of rainfall would account for the disturbance observed.

Consider the total rainfall of any two successive periods of 183 days; if the date dividing them falls in the spring, the second will generally contain the smaller quantity of rain since the climate has a pronounced winter rainfall, and vice-versa if the day of division is situated in the autumn. If daily differences of such totals are taken, a sequence of values should be obtained which changes sign regularly twice per rainfall year, once in the winter from negative to positive and again in the summer from positive to negative.

The 365 (or 366) differences between the rainfall of the 183 days preceding and following each day of the year for the 95 years were evaluated. With the exception of several years, the change of sign in the winter and summer was defined very clearly. In the exceptional years, the differences alternated in sign for short periods of varying lengths before definitely adopting the opposite sign. In order to find the day of zero difference in these cases the daily differences were smoothed by taking successive 10-day means.

The date in the winter at which the preceding six months had received as much rain as the six months following may be regarded as an empirical median of the rainfall sequence; two quartiles were then located, one on either side of the median, between each of which and the median one quarter of the rainfall of the year surrounding the median date, had fallen. Fig. 4 is a

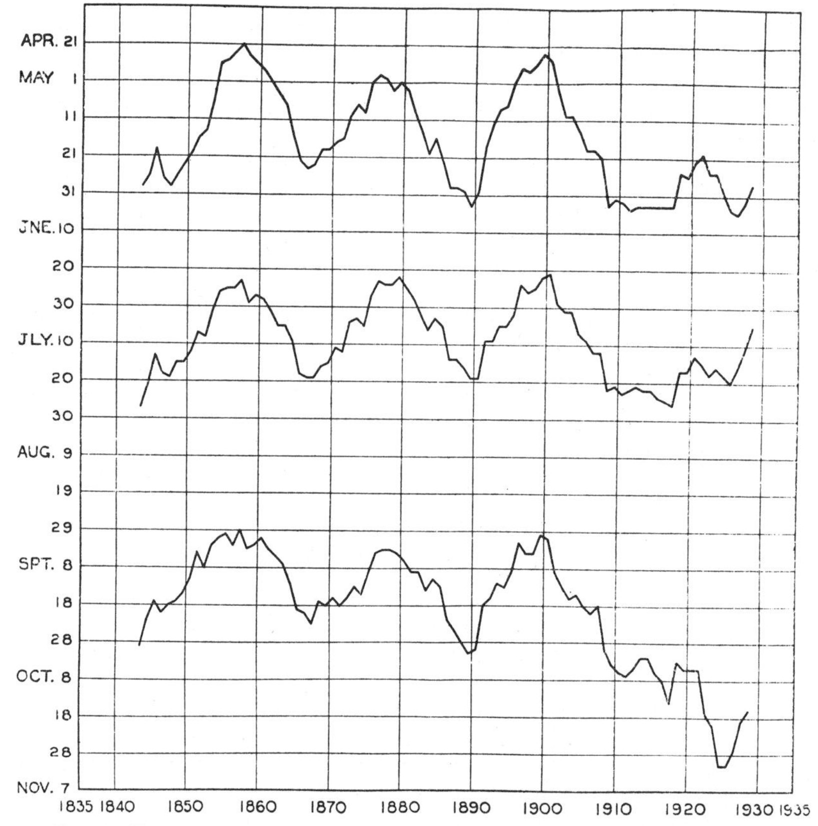

FIG. 4.—Ten-year means of the median and quartiles of the rainfall sequence.

diagram of running 10-year averages of these three dates, the centre curve showing the course of the median, while the upper and lower curves illustrate the variations in the quartile dates.

On comparing the diagrams it will be observed how excellent is the agreement of the three curves of Fig. 4 with the graph of 10-year means of b' of Fig. 2. The periods closely approximate each other and the movements are in phase, but the displacements are in opposite directions, which must necessarily be the case owing to the physical nature of the coefficient b'. The abnormal behaviour of b' subsequent to 1912 is reflected in the departure of the curves of Fig. 4 from their law of change, the extreme lateness of the date of attainment of the third quarter of the rainfall total in the vicinity of 1925 being due to the dominance of several very late years.

The diagrams of Fig. 4 are really self-explanatory for the significance of the slow changes in the coefficient b' is at once apparent. There has been a very regular oscillation, of which the period and amplitude are approximately 23 years and 30 days respectively in the duration and incidence of the winter rains, but in recent years there has been a definite tendency to depart from this course.

The oscillation has been confined mainly to a portion of the rainfall curve rather than to the curve as a whole, for the date of minimum has remained practically constant over the 95 years, the section showing the greatest movement being that extending from April to November.

The mutual agreement of the curves of Fig. 4 brings into prominence one other point and supports the contention put forward at the beginning of this section of the paper. In the first place, the relative constancy of the length of the rainy season is shown clearly, and secondly the agreement of the curves substantiates the view that a shift of date, regardless of quantity of seasonal precipitation, provides an adequate description of the changes in progress, for whether the rainfall be 16 in. or 25 in. the quartiles and median tend to follow an orderly sequence.

It is a pleasure to gratefully acknowledge my indebtedness to Professor R. A. Fisher, F.R.S., for invaluable advice during the preparation of the paper and for his subsequent active interest. I also wish to record my thanks to Professor J. A. Prescott and Miss F. E. Allan for opportunities of discussing the work during its progress, and to Mr. E. Bromley, Divisional Meteorologist for South Australia, who kindly placed the record at my disposal.

References

1. Fisher, R. A. *Phil. Trans. R. Soc.*, (B), **213,** 1924, p. 89.
2. Fisher, R. A. Statistical Methods for Research Workers, 5th edition, 1934.

Appendix

Rainfall Distribution Constants, 1839-1933

Year	a'	b'	c'	d'	e'	f'
1839	325	+69	− 49	−35	+ 1	− 2
1840	397	+70	− 23	+11	+48	+53
1	294	+ 9	− 44	+11	+ 6	− 2
2	333	+37	− 49	−10	− 9	− 4
3	280	+27	− 61	− 1	+21	− 7
4	279	+24	− 56	−11	+15	− 2
5	308	+49	− 67	− 8	+28	0
6	441	+44	− 46	+ 9	+17	+ 3
7	477	− 9	− 89	+13	+65	−36
8	324	+76	− 51	−25	− 8	− 2
9	417	+32	−120	− 2	+41	−12

Rainfall Distribution Constants, 1839-1933—*Continued*

Year	a'	b'	c'	d'	e'	f'
1850	319	− 6	− 8	− 3	+14	+12
1	522	+31	−114	−22	+40	−27
2	450	+46	− 82	−24	+45	+ 5
3	445	− 8	−106	+22	+18	−16
4	252	+12	− 61	+ 3	+ 8	+ 2
5	380	+ 5	− 76	+11	0	+ 3
6	409	−43	− 73	+ 9	+15	−13
7	364	+ 7	− 55	+ 9	+ 5	+ 6
8	353	+11	− 19	+20	+19	− 2
9	244	− 5	− 51	+ 9	+ 9	−15
1860	323	−21	− 65	+34	− 9	−19
1	387	+48	− 42	+32	+20	−17
2	358	+ 7	−106	− 8	+35	+ 1
3	390	+36	− 83	−13	+20	− 8
4	325	−10	− 70	−34	+44	− 4
5	254	+ 6	− 78	− 3	+28	+ 8
6	330	+ 4	− 77	−10	+16	− 6
7	313	+33	− 58	−20	− 8	+ 4
8	327	+ 5	− 58	−17	+20	−11
9	243	+ 9	− 37	+ 6	− 4	−10
1870	391	+39	− 43	−54	+24	+17
1	386	+ 8	− 10	−27	+29	−18
2	372	+13	− 70	− 8	+17	− 5
3	345	− 9	− 76	+ 8	+ 9	− 2
4	282	+ 4	− 73	+ 2	+18	− 6
5	479	0	− 60	+53	+29	−15
6	220	+10	− 35	+ 1	+ 5	+ 3
7	409	−32	− 59	+17	−31	0
8	362	−39	− 66	+27	−15	− 2
9	339	+49	− 47	+19	+27	+ 1
1880	368	−12	− 70	− 1	+ 6	−10
1	296	−25	− 39	−26	+44	−21
2	259	+23	− 64	− 4	+19	− 1
3	442	+15	−101	+19	+24	−21
4	308	−10	− 47	+ 6	+12	−11
5	260	+ 3	− 67	0	+23	+ 2
6	236	+32	− 34	−23	+11	− 2
7	421	+26	− 79	+10	+25	−21
8	239	+14	− 75	−10	+36	+ 5
9	506	−32	− 58	−21	+24	−59
1890	423	+27	− 77	−31	+21	+ 6
1	230	+49	− 20	−16	+11	− 3
2	353	+30	− 50	−21	+10	− 4
3	353	+37	− 89	− 2	+20	− 6
4	341	+23	− 60	+ 4	+19	+ 1
5	349	−20	− 65	+ 5	+36	−12
6	249	−21	− 15	+33	+31	+ 2
7	253	−15	− 51	−18	+11	+12
8	340	+ 6	− 76	+23	+11	−26
9	309	−11	− 34	+15	− 6	−11
1900	357	−32	− 81	+20	+ 7	+ 3
1	295	+ 8	− 45	+ 7	+27	− 9
2	269	+40	− 20	+ 4	+21	−10
3	417	+10	− 51	+ 9	+ 8	− 1
4	333	−29	− 57	−25	+39	−24
5	365	−19	− 80	− 4	+27	−29
6	435	+65	− 84	− 5	+16	0
7	291	+20	− 56	+14	+13	− 7
8	403	+13	− 80	+ 2	+ 1	− 1
9	454	+40	− 98	−25	+21	−10

RAINFALL DISTRIBUTION CONSTANTS, 1839-1933—*Continued*

Year	a^I	b^I	c^I	d^I	e^I	f^I
1910	404	+18	− 76	+ 2	+ 9	+11
1	262	+21	− 37	− 7	+ 4	+ 2
2	321	+47	− 55	−12	+ 9	− 5
3	298	+50	+ 16	+ 1	−10	+23
4	188	− 6	− 1	+18	− 6	− 8
5	318	+ 4	− 95	− 9	+25	+ 2
6	462	+67	− 92	−18	+47	−17
7	474	0	− 82	+ 5	+ 6	+11
8	289	+12	− 70	− 8	+27	− 6
9	282	+ 3	− 35	−11	+19	+36
1920	438	+69	− 67	+19	+41	− 1
1	371	+10	− 44	−32	+17	−23
2	380	+20	− 31	− 4	+67	−16
3	488	+81	−101	−21	+48	−10
4	384	− 9	− 44	− 7	−18	+ 4
5	359	−63	− 28	−13	−11	+30
6	364	+40	− 75	− 7	+ 8	+ 3
7	277	+26	− 50	−13	+17	− 1
8	319	−24	− 44	−15	+ 3	+ 7
9	287	+73	− 11	+32	+53	+27
1930	306	+65	− 66	−32	+13	+13
1	366	− 4	−100	− 9	+43	− 4
2	410	−23	− 92	+12	+ 4	− 4
3	363	−29	− 58	− 3	+21	+ 4
Average 1839–1933	347·50	+ 14·21	−59·73	−2·13	+17·64	−3·53

DISCUSSION

The President, Dr. F. J. W. WHIPPLE, opened the discussion by giving a summary of the note printed on p. 492 with the heading "Note on the analysis of variation by the use of orthogonal polynomials."

Sir GILBERT WALKER said he would like to ask Prof. Fisher to explain a little more fully the justification for regarding this method of treating data as satisfactory. If the representation of a single year's rainfall was good, we should expect the mean of 95 such representations to be decidedly closer to the mean of the actual rainfall of 95 years, which would to a high degree of approximation be continuous at the beginning and end of the year. But Fig. 1 showed that this result is not attained.

He had found it difficult to accept the statement that "this gives a 23-year period." He did not think in any case that the author meant to imply that there was an established period of 23 years; but he would like to point out that in the data in Fig. 4 there were well-marked maxima in 1857, 1878 and 1889 which gave two whole fluctuations in 42 years.

Sir Gilbert said he would like to congratulate the author on the amount of work he had put into this useful paper.

Lieut.-Colonel E. GOLD asked Prof. Fisher whether similar results to those illustrated in Fig. 4 would have been obtained if the author had taken the phase of the sine curve for the annual variation.

He had been interested to find that the author had arrived at his main point by a simple method which any meteorologist could apply without being a statistician. He wondered whether an equally simple method could have been devised if it had been c' which had been of real significance. Adelaide happened to possess this particular peculiarity, but other places might have other peculiarities and these might be brought to light by similar methods.

Dr. A. T. DOODSON said that he hoped that an explanation of the variables might be included in the paper as he had found great difficulty in reading the paper and had had to deduce certain things which ought to have been explained in the paper. What interested him most was the effect of any analysis on the extremes of the data. He had done a good deal of harmonic analysis, and in theoretical problems one found that the greatest difficulties occurred at the boundaries. This had been the case in this particular analysis, as is evident from the first two figures. The formula could not be used for extrapolation beyond 1930. There was a point to be considered in relation to apparent periodicities in the residues, for a polynomial to the fifth degree applied to a harmonic term would leave a residue having an apparent but spurious period of about one-sixth of the original period. The author, however, had avoided this difficulty in his analysis of periodicities, but the three curves of Fig. 4 were not independent proofs of the periodicity since there is a large correlation between the medians and quartiles arising from the fact that the bulk of the annual precipitation occurred in the winter months.

Mr. E. G. BILHAM said it appeared to him that as the " median date " as defined in the paper was not necessarily a unique date there might be advantages in working with the date up to which half the year's rainfall had occurred, from January 1.

Mr. L. H. G. DINES pointed out that it is possible to find any number of periods in a series of perfectly chance figures. It would have been far more interesting in this case if, instead of looking for periodicities without regard to any likely cause, the author had correlated the rainfall with sunspots, for example.

Professor R. A. FISHER, in reply, said that it was clear from the discussion that the opinion of the speakers was divided as to how far it was necessary to bow down to the sanctity of harmonic analysis. The President, Dr. Whipple, had been at great pains to show that if polynomial forms were used to give approximations to harmonic curves, the fit might be very inferior. His diagram demonstrated with equal force how very badly harmonic forms would work if used to give approximations to polynomials. It should be remembered that both sets of orthogonal functions were artifacts imposed by the mathematician on the data, for his own convenience. Mr. Cornish had fitted polynomials, not to harmonic curves, but to sequences of rainfall.

He regretted that it should be necessary again to refute the misleading suggestion that the purpose of the polynomials was to represent the average annual sequence of rainfall; by this purpose harmonic curves would be appropriate since they were periodic, as the average annual sequence was bound to be. This was not true of the individual annual sequences, since it was not in fact true that the weather repeated itself at intervals of 12 months. Consequently, in any study of variation from year to year harmonic curves would be bound to misrepresent the data by imposing a condition which was in fact false. It was one great advantage of polynomials that they did not impose this artificial restriction on the data.

To write down, as Dr. Whipple had done, " the approximation to sines and cosines " is therefore a very unfortunate method of judging of the competence of the method. It is also inaccurate to speak of the coefficients as approximations, since they are directly calculated from the observational record. Dr. Whipple's statement that the 12 monthly totals of rainfall would give twice as much information about the distribution of rain is unsupported by any evidence. Dr. Whipple found it curious that after using harmonic analysis for a purpose for which it seemed appropriate, he (Prof. Fisher) had adopted other methods when faced with a problem of a different type. He himself thought that mathematicians would be of more use in scientific research, if they did not so frequently attempt to impose their own favourite fads, but sought out methods appropriate to their material.

515.5 : 517

A NOTE ON THE ANALYSIS OF VARIATION BY THE USE OF ORTHOGONAL POLYNOMIALS

By F. J. W. WHIPPLE, Sc.D., F.Inst.P.

The publication in the *Quarterly Journal* of a paper (p. 481) in which Prof. Fisher's scheme of analysis of meteorological data is applied necessitates an examination of the principles on which the scheme is based. Details of his method are to be found in a paper* by Prof. Fisher " On the influence of rainfall on the yield of wheat at Rothamsted." The essential idea is that the characteristics of the rainfall of any year can be represented by half a dozen coefficients, that these can be correlated in turn with the magnitude of the harvest and that a function can be constructed to represent the influence on the crop of rainfall at any time of year.

In Mr. Cornish's paper on Adelaide rainfall there is no application of the statistics to the study of crops and the method of analysis must be judged by the extent to which comprehension of changes in the distribution through the year is facilitated.

In harmonic analysis we express a function of the time in terms of multiples of sines and cosines. The process can be generalised by the use of other orthogonal functions. The simplest functions which can be used are the Legendre Polynomials, which are familiar as Zonal Harmonics. These functions may be defined by the equations

$$P_0(x)=1,\ P_1(x)=x,\ P_2(x)=\tfrac{3}{2}\left(x^2-\tfrac{1}{3}\right),$$

$$P_n(x)=\frac{1,\,3,\,5,\,\ldots\,(2n-1)}{1,\,2,\,3,\,\ldots\,n}\left[x^n-\frac{n(n-1)}{2(2n-1)}x^{n-2}\right.$$

$$\left.+\frac{n(n-1)(n-2)(n-3)}{2,\,4,\,(2n-1)(2n-3)}x^{n-4}-\ldots\right]$$

The functions are significant over the range from $x=-1$ to $x=+1$ and satisfy the orthogonal condition

$$\int_{-1}^{1} P_m(x)\,P_n(x)\,dx=0.$$

* *London, Phil. Trans. R. Soc.*, (B), **213,** 1924, p. 89.

Further

$$\int_{-1}^{1} [P_n(x)]^2 \, dx = \frac{2}{2n+1}$$

The Legendre functions are graphed in Fig. 1. It will be noticed that in general they are oscillatory functions and that the oscillations are more violent near the ends of the range. Any function which exists over the range from $x = -1$ to $x = +1$ can be expressed as the sum of a series involving Legendre polynomials. If the series representing $f(x)$ is $a_0P_0 + a_1P_1 + \ldots a_nP_n + \ldots$ then a_n is determined by the equation

$$a_n = \frac{2n+1}{2} \int_{-1}^{1} f(x) \, P_n(x) \, dx.$$

Further, the first few terms of the series will serve as an approximation to the function and it can be shown that $n+1$ terms give the best approximation expressible as a polynomial of degree n, best in the sense that the average value of the square of the difference between function and approximation is as small as possible.

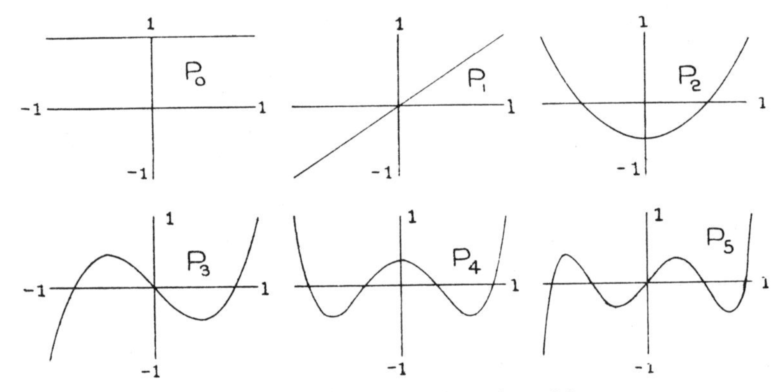

FIG. 1.—Graphs of Legendre polynomials.

To judge the competence of such approximations we may write down the approximations to sines and cosines, allowing the range $-\pi$ to $+\pi$ of the argument θ of the trigonometrical functions to correspond with the range -1 to $+1$ of the argument x of the Legendre polynomials, so that $x = \theta/\pi$. In such cases the coefficients a_n can be found by integration by parts, the successive differential coefficients of P_n at $x = 1$ being known, viz.,

$$\frac{n(n+1)}{2}, \; \frac{(n-1)\,n\,(n+1)(n+2)}{2,\,4}, \; \frac{(n-2)(n-1)\,n\,(n+1)(n+2)(n+3)}{2,\,4,\,6}$$

etc.

Adopting a more convenient notation, we write these coefficients in the form

$$\frac{n_2}{2}, \; \frac{(n-1)_4}{2,\,4}, \; \frac{(n-2)_6}{2,\,4,\,6}, \text{ etc.}$$

Thus it may be proved that

$$(-1)^{r-1} \sin r\theta = \Sigma_{n \text{ odd}} (2n+1)\left[\frac{1}{r\pi} - \frac{(n-1)_4}{2^2 2!} \cdot \frac{1}{(r\pi)^3} + \frac{(n-3)_8}{2^4 4!} \frac{1}{(r\pi)^5} \cdots\right] P_n(x)$$

and

$$(-1)^r \cos r\theta = \Sigma_{n \text{ even}} (2n+1)\left[\frac{(n)_2}{2(r\pi)^2} - \frac{(n-2)_6}{2^3 3!} \frac{1}{(r\pi)^4} + \frac{(n-4)_{10}}{2^5 5!} \frac{1}{(r\pi)^6} \cdots\right] P_n(x)$$

In particular

$$\sin \theta = \frac{3}{\pi} P_1 + \frac{7}{\pi}\left(1 - \frac{15}{\pi^2}\right) P_3 + \frac{11}{\pi}\left(1 - \frac{105}{\pi^2} + \frac{945}{\pi^4}\right) P_5 \cdots$$

and

$$\cos \theta = -\frac{15}{\pi^2} P_2 - \frac{9}{\pi^2}\left(10 - \frac{105}{\pi^2}\right) P_4 + \cdots$$

and by evaluating the coefficients we find

$$\sin \theta = 0{\cdot}955\, P_1 - 1{\cdot}158\, P_3 + 0{\cdot}217\, P_5 \cdots$$
$$\sin 2\theta = -0{\cdot}4775\, P_1 - 0{\cdot}691\, P_3 + 1{\cdot}85\, P_5 - 0{\cdot}825\, P_7 \cdots$$
$$\cos \theta = -1{\cdot}520\, P_2 + 0{\cdot}583\, P_4 \cdots$$
$$\cos 2\theta = 0{\cdot}380\, P_2 + 1{\cdot}673\, P_4 - 1{\cdot}399\, P_6 \cdots$$

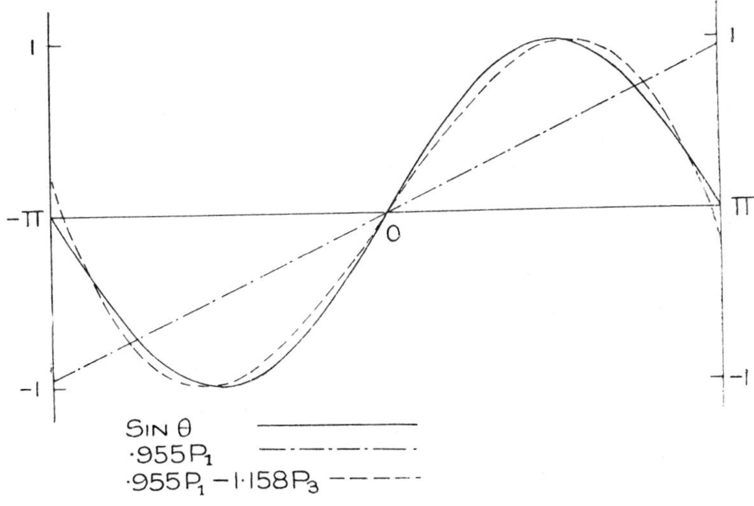

FIG. 2.—Graphs of sin θ and of approximations by Legendre functions.

In Figs. 2 to 5 the trigonometrical functions are graphed as well as successive approximations calculated by using values of the P's taken from Dale's five-figure tables. For $\sin \theta$ the second approximation is quite close. The third approximation, $0{\cdot}955\, P_1 - 1{\cdot}158\, P_3 + 0{\cdot}217\, P_5$, would be indistinguishable from $\sin \theta$. The polynomial of the fourth degree is an excellent representation of $\cos \theta$. The approximations to $\sin 2\theta$ and $\cos 2\theta$ by polynomials of the fifth and fourth degrees respectively are not nearly so good. In the case of cos 2θ there is an error of 50 per cent in the approximation in the middle of the range; such an error was to be expected since the first of the neglected terms in the expansion for cos 2θ has a large coefficient, nearly 1·4.

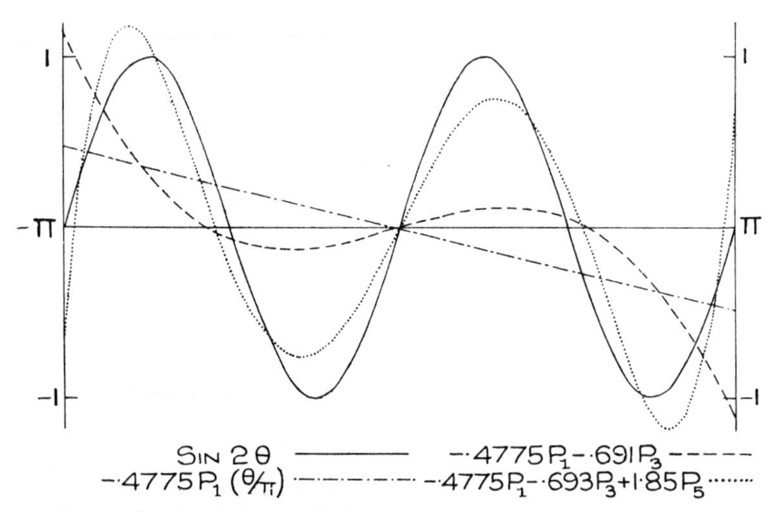

FIG. 3.—Graphs of sin 2θ and of approximations by Legendre functions.

In Fisher's paper (and the precedent is followed by Cornish) rainfall distribution throughout the year is expressed in terms of polynomials up to the fifth degree and the goodness of fit may be expected to be comparable with that shown in our examples. The approximation is likely to be about as good as that obtained by

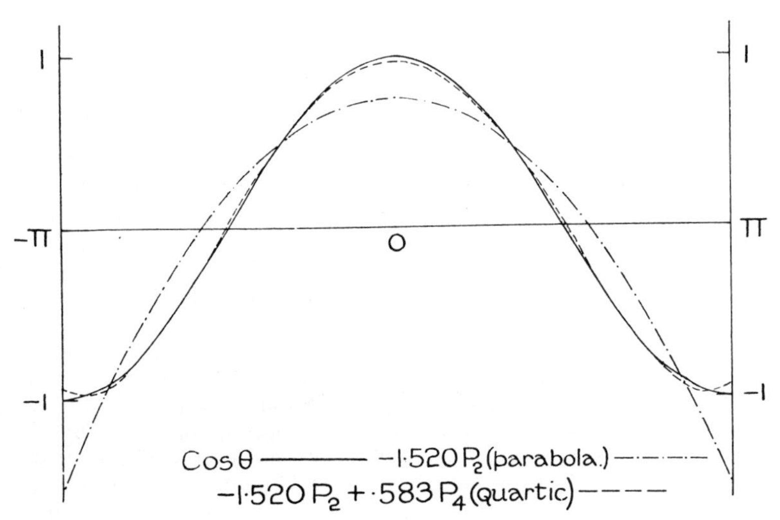

FIG 4.—Graphs of cos θ and of approximations by Legendre functions.

adopting harmonic analysis and including terms up to those in sin 2θ and cos 2θ. The 12 monthly totals of rainfall* would give twice as much information about the distribution of rain through the year as six terms of the series of polynomials.

In the foregoing discussion we have used integrals. In Fisher's work sums take the place of integrals and instead of Legendre functions he has the polynomials which were defined by Tchebycheff† in 1854. When the number of values of the independent variable is large the Tchebycheff functions tend to take the form of Legendre functions. Accordingly we can interpret Fisher's results in terms of the latter functions.

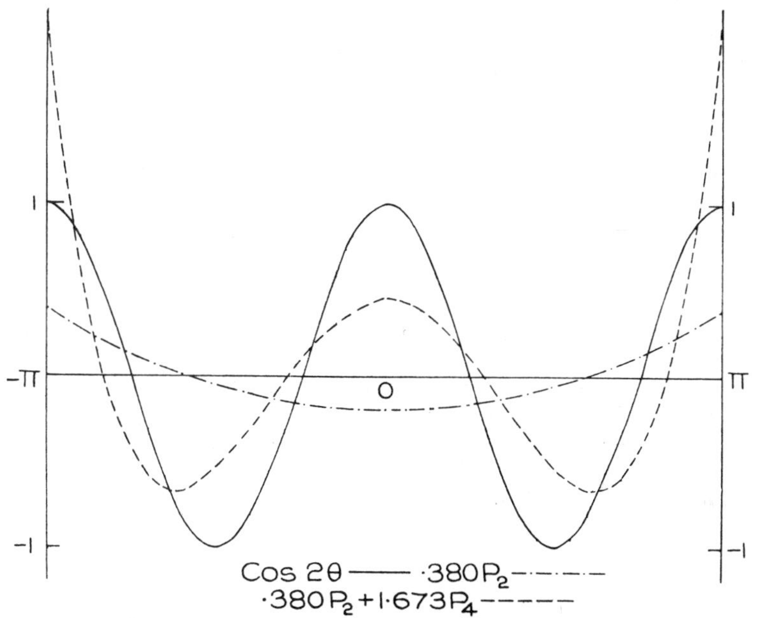

FIG. 5.—Graphs of cos 2θ and of approximations by Legendre functions.

Thus in the Appendix to his paper Cornish tabulates for the rainfall measured in thousandths of an inch per 6 days the values of certain coefficients a', b', . . . f'. It is more convenient to think of the rainfall as so many, say R, thousandths of an inch per day. R is given by

$$6R = a'P_0 + 3\,b'P_1 + 5\,c'P_2 + 7\,d'P_3 + 9\,e'P_4 + 11\,f'P_5 \ldots$$

the argument x of the polynomials ranging from -1 at the

* Totals for Adelaide are in "World Weather Records" (*Smithsonian Misc. Coll.* **79**, 1927, p. 417, and **90**, 1934, p. 195).

† *Bull. Phys. Math. Acad. Imp. Sci., St. Petersbourg,* **13,** 1854, p. 210. Cf. Jordan. *Proc. London Math. Soc.*, (2), **20,** 1921, p. 297. F. E. Allan. *Proc. R. Soc., Edin.*, **50,** 1930, p. 310. In Fisher's notation the normal orthogonal function is T_r and $\underset{n\to\infty}{L}[n^{\frac{1}{2}}(2r+1)^{-\frac{1}{2}}\,T_r(t)] = P_r(x)$

when $$\underset{n\to\infty}{L}\frac{t}{n} = \frac{x}{2}$$

beginning of the calendar year to $+1$ at the end. Since the standard deviation of P_n from zero is $(2n+1)^{-\frac{1}{2}}$ the standard deviations of the successive terms in the series for $6R$ are given in Fisher's notation by equations,

$$x'_2 = \sqrt{3}\, b', \quad x'_3 = \sqrt{5}\, c', \quad x'_4 = \sqrt{7}\, d', \quad x'_5 = \sqrt{9}\, e', \quad x'_7 = \sqrt{11}\, f'$$

This notation is used in Table II of Cornish's paper in which variations of the original a', b', . . . f' over a period of years are summarised. If Q is the coefficient designated at the head of a column, then the departure from the mean of the value of Q at any date is equal to

$$\sqrt{3}\, x'_2 P_1 + \sqrt{5}\, x'_3 P_2 + \sqrt{7}\, x'_4 P_3 + \sqrt{9}\, x'_5 P_4 + \sqrt{11}\, x'_6 P_5,$$

the argument of the polynomials running from -1 in 1839 to $+1$ in 1933.

The disadvantage of approximation by polynomials is that there is likely to be much distortion at the ends of the range. This is illustrated by our investigation of the sine and cosine curves and by the graphs in Cornish's paper. There are some striking examples in Fisher's paper. For example, according to his Fig. 6 it is alleged that the crop on plot 7 at Rothamsted is reduced by 0·7 bushels per acre by an inch of rain at the beginning of the previous September, whilst rain at the end of that month has no effect at all. When it is realised that the regression equation represented by the figure in question had to involve an expression of the fifth degree, it is seen that no attention need be paid to the vagaries of the ends of the curve.

Harmonic analysis would seem to be a preferable method in all such problems. The arithmetic of Fisher's method may possibly be less laborious, but it is doubtful whether that would be the case if discretion were used in deciding how far the year was to be subdivided. One disadvantage in the use of polynomials is that the representation fails to reproduce a periodic curve. The mean value for a long run of years of daily rainfall should be represented (after smoothing) by a curve the two ends of which fit together without any discontinuity of slope or curvature. This criterion is not satisfied by the curve in Cornish's Fig. 1. The approximation represented by his curve may perhaps have some scientific interest, but cannot be regarded as so generally serviceable as the approximation which would have been obtained by harmonic analysis. It is curious that after using harmonic analysis in a paper which he wrote with Miss Mackenzie in 1922, Fisher should have turned in 1923 to a method which seems so much less fitted for its purpose.

Finally attention must be called to the remarkable oscillation discovered by Cornish in the dates of occurrence of the winter rains at Adelaide. As he says, the period of this oscillation was about 23 years.

It may be more than a coincidence that the oscillation kept step with the variations in the number of sunspots. During the period covered by the Adelaide records the minima in the sunspot numbers occurred in the following years, 1843, 1856, 1867, 1878,

1889, 1901, 1913 and 1923. From 1856 to 1913 these dates mark as nearly as may be the turning points of the curves in Cornish's Figs. 2 and 4.

Now it was discovered by Hale that there is a curious alternation from cycle to cycle in the magnetic polarity of sunspots. Sunspots usually occur in pairs with opposite polarity. When this was first investigated the preceding spot of a pair in the northern hemisphere of the sun had round it a magnetic field like that round a south-seeking magnetic pole, at the same time the preceding spot of a pair in the southern hemisphere was like a north-seeking pole. In the new cycle which began in 1912 the preceding spots were N poles in the northern hemisphere and S poles in the southern hemisphere. The reversal was observed again in 1923 and 1934; in all probability the alternation has persisted for centuries.

To distinguish between the two types of cycle we may refer to them as P cycles and E cycles. During a P cycle the magnetic field round the leading sunspot of a pair is such as would be produced by the cyclonic movement of positive electricity round the spot. During an E cycle the field round the leading spot could be produced by the cyclonic movement of electrons.

The sun is in a P cycle now as it was during the years 1843 to 1856, 1867 to 1878, 1889 to 1901 and 1913 to 1923.

Cornish's work shows that during the period of the Adelaide rainfall record the rainy season tended to get earlier year by year during a P cycle, and later year by year during an E cycle.

That Mr. Cornish was led to this very remarkable discovery appears, however, to have been a happy accident. He noticed that the coefficients x'_3 and x'_4 under b' in Table II were significant, so that a polynomial of degree 3 and, a fortiori, a polynomial of degree 5 would serve for representing the secular variation of b'. This was confirmed by graphing b' and at the same time the regular oscillations which were *not* associated with the polynomial became conspicuous.

It may be remarked, however, that the 23-year period would have been discovered if Mr. Cornish had used harmonic analysis, for the phase angle of the first term of the harmonic series would have been subject to just such changes as were found for b'. If periodic shifts of the seasons are to be sought in statistics for places other than Adelaide it seems likely that Cornish's analysis by medians and quartiles will lead to significant results with least labour.

Reprinted from La Revue de l'Institut International de Statistique, Vol. 5, pp. 307-22, 1937.

MOMENTS AND CUMULANTS IN THE SPECIFICATION OF DISTRIBUTIONS.

By E. A. Cornish and R. A. Fisher F. R. S.

1.

The very considerable statistical literature which has grown up on the use of the moments of populations and samples, and on other quantities allied to these, is rendered confusing by variations in notation and terminology, and by the different aims which authors have had in view in using these quantities. The following notes aim at clarifying the subject by suggesting a uniform and consistent notation, specifying briefly the relations between the different quantities ordinarily used, and summarising the results which have been obtained.

The distribution of a variable quantity x can be specified by means of a frequency function f, often termed the probability integral, specifying the total frequency in the population for which the variate is less than an assigned value x. For discontinuous distributions f will be a step function, increasing discontinuously at the values of x at which finite fractions of the total frequency are concentrated, and remaining constant between these values. For certain other distributions f is continuous and differentiable so that $\frac{df}{dx}$ represents the frequency density in the element of range dx, or, the ordinate of a frequency curve at this point. These are the two common cases, but it is also possible for f to be continuous, but not differentiable, and so incapable of representation by a frequency curve.

2. The Characteristic Function.

In all cases we may define a function of a real variable t in the form

$$M(t) = \int_{-\infty}^{\infty} e^{itx}\, df,$$

which is known as the characteristic function of the distribution. The absolute value of M never exceeds unity for any real value of t. $M(t)$ and $M(-t)$ are equal, if real, and conjugate quantities, if complex. If, in the neigbourhood of $t = 0$, M can be expanded in a series of powers of t, this series will be

$$\sum_{r=0}^{\infty} \frac{(it)^r}{r!} \int_{-\infty}^{\infty} x^r\, df,$$

or

$$\sum_{r=0}^{\infty} \frac{(it)^r}{r!} \mu'_r,$$

where μ'_r is the rth moment of the distribution of x about the origin. The characteristic function may, therefore, be spoken of as the moment generating function. μ'_r is, of course, the average value of x^r and when this is finite, the characteristic function is differentiable r times at the origin.

If μ'_1 is the mean, the factor e^{itx} may be resolved into the product

$$e^{it\mu'_1} \cdot e^{it(x-\mu'_1)}$$

of which the first factor is constant, while the average value of the second factor gives a characteristic function referred to the mean of the distribution, and therefore formally expansible as a generating function of the moments about the mean. The relation between the moments about zero and the moments about the mean of the distribution may therefore be obtained by equating coefficients of powers of t in the identity

$$1 + \mu_2 \frac{t^2}{2!} + \mu_3 \frac{t^3}{3!} + \mu_4 \frac{t^4}{4!} + \ldots\ldots$$

$$= e^{-\mu'_1 t}\left(1 + \mu'_1 t + \mu'_2 \frac{t^2}{2!} + \mu'_3 \frac{t^3}{3!} + \mu'_4 \frac{t^4}{4!} + \ldots\ldots\right)$$

giving the series of relations

$$\begin{aligned}\mu_2 &= \mu'_2 - \mu'^2_1, \\ \mu_3 &= \mu'_3 - 3\mu'_2\mu'_1 + 2\mu'^3_1, \\ \mu_4 &= \mu'_4 - 4\mu'_3\mu'_1 + 6\mu'_2\mu'^2_1 - 3\mu'^4_1, \\ &\ldots\ldots\ldots\ldots\ldots\ldots\end{aligned}$$

by which the moments about the mean may be obtained from those about any other origin.

3. **The Cumulative Function.**

In studying the distributions of quantities compounded of ingredients, each distributed independently in a known distribution, Laplace was led to introduce a function known as the cumulative function, which is simply the logarithm of the characteristic function.

If x is distributed in a distribution specified by the frequency element df_1, and y is independently distributed in a distribution specifed by the element df_2, the frequency of the simultaneous occurrence of any particular pair of values x and y will be $df_1\, df_2$, and the characteristic function of the sum, $x + y$, will be

$$\int_{-\infty}^{\infty}\int_{-\infty}^{\infty} e^{it(x+y)}\, df_1\, df_2,$$

which is clearly the product of the characteristic functions of x and y separately. Consequently, if $K = \log M$, be written for the cumulative function, the cumulative function of $x + y$ is simply the sum of the cumulative functions of x and y separately. Evidently this relationship holds for any number of ingredients and is fundamental in the study of distributions of compound quantities.

The identity of these functions for all values of t carries with it the identity of their coefficients when M and K are expressible by power series. We are therefore led to recognise the coefficients of the expansion of K in powers of t as quantities of peculiar significance in the specification of the distribution. We shall call these quantities cumulants, denoted by $\kappa_1, \kappa_2, \kappa_3 \ldots\ldots$, and defined by the identity

$$\kappa_1 t + \kappa_2 \frac{t^2}{2!} + \kappa_3 \frac{t^3}{3!} + \kappa_4 \frac{t^4}{4!} + \ldots\ldots$$
$$= \log \left\{ 1 + \mu'_1 t + \mu'_2 \frac{t^2}{2!} + \mu'_3 \frac{t^3}{3!} + \mu'_4 \frac{t^4}{4!} + \ldots\ldots \right\},$$

by which the moments about zero may be expressed in terms of cumulants, or *vice versa*, or by

$$\kappa_2 \frac{t^2}{2!} + \kappa_3 \frac{t^3}{3!} + \kappa_4 \frac{t^4}{4!} + \ldots\ldots$$
$$= \log \left\{ 1 + \mu_2 \frac{t^2}{2!} + \mu_3 \frac{t^3}{3!} + \mu_4 \frac{t^4}{4!} + \ldots\ldots \right\}$$

giving the corresponding relations with the moments about the mean. The latter are of great simplicity for

$$\mu_2 = \kappa_2 ,$$
$$\mu_3 = \kappa_3 ,$$
$$\mu_4 = \kappa_4 + 3\,\kappa_2^2 ,$$
$$\mu_5 = \kappa_5 + 10\,\kappa_3\,\kappa_2$$

and so on. The numerical coefficients may be written down at sight for the coefficient of κ_2^2 is the number of ways in which 4 objects may be divided into two groups of 2 each, and that of $\kappa_3\,\kappa_2$ is the number of ways in which 5 objects may be divided into a group of 3 and a group of 2. The same rule holds generally.

For, if

$$(p_1^{\pi_1}\, p_2^{\pi_2} \ldots\ldots)$$

stand for any partition of a number r, the coefficient of $\kappa_{p_1}^{\pi_1}\, \kappa_{p_2}^{\pi_2} \ldots\ldots$ in the expansion of

$$e^{\kappa_2 \frac{t^2}{2!} + \kappa_3 \frac{t^3}{3!} + \ldots\ldots}$$

is seen to be

$$\frac{t^r}{(p_1!)^{\pi_1}\, \pi_1!\, (p_2!)^{\pi_2}\, \pi_2! \ldots\ldots};$$

so that the coefficient of $\kappa_{p_1}^{\pi_1}\, \kappa_{p_2}^{\pi_2} \ldots\ldots$ in μ_r is

$$\frac{r!}{(p_1!)^{\pi_1}\, \pi_1!\, (p_2!)^{\pi_2}\, \pi_2! \ldots\ldots}$$

or the number of ways of distributing r objects into undifferentiated receptacles π_1 containing p_1 each, π_2 containing p_2 each, and so on.

3. 1. **Average effects of grouping.**

When, by reason of the limited accuracy of instrumental measurements, for convenience of record, or to simplify the calculations, variates are grouped so that to all values lying in the range $x - \frac{1}{2}h$ to $x + \frac{1}{2}h$ is assigned the conventional value x, the cumulants of the distribution will be somewhat affected.

To any true value ξ, the process of grouping adds a grouping error $x-\xi$, where x is the centre of the group in which ξ falls. Knowing the group limits, we know also the actual error introduced for each possible value of ξ. With equal intervals this error will be a periodic function of ξ, and an exact study of the effects of grouping must involve the phase relationship between the group limits and any such characteristic of the population as its mean. With moderately fine grouping, the periodic corrections are, however, small; and it is often sufficiently accurate to consider only the average effects of grouping, when for a given grouping interval, h, the group limits are supposed to fall with equal frequency in any equal lengths in which the interval may be divided.

In this case the error of grouping is distributed with uniform frequency over the range from $-\frac{1}{2}h$ to $+\frac{1}{2}h$ so that its frequency distribution is

$$df = \frac{1}{h}\,dx, \qquad -\tfrac{1}{2}h < x < \tfrac{1}{2}h$$

for all values of ξ independently.

The average cumulants of the grouped distribution will therefore differ from those of the original ungrouped distribution by the cumulants of the grouping error.

The characteristic function is

$$M = \frac{1}{h}\int_{-\frac{1}{2}h}^{\frac{1}{2}h} e^{itx}\,dx$$

$$= \frac{2}{ht}\sin\tfrac{1}{2}ht.$$

$$= 1 + \frac{h^2}{12}\frac{(it)^2}{2!} + \ldots\ldots + \frac{h^{2r}}{2^{2r}(2r+1)}\frac{(it)^{2r}}{(2r)!} + \ldots\ldots$$

Hence

$$K = \frac{h^2}{12}\frac{(it)^2}{2!} - \frac{h^4}{120}\frac{(it)^4}{4!} + \frac{h^6}{252}\frac{(it)^6}{6!} - \ldots\ldots$$

Since, with group interval smaller than the standard deviation, the higher cumulants expressed in group units increase rapidly, the effects of grouping on them are extremely small, even when the second cumulant is materially affected.

The deduction of the coefficients $\frac{h^2}{12}, -\frac{h^4}{120}, \frac{h^6}{252}, \ldots\ldots$ from the cumulants estimated from grouped data is equivalent to Sheppard's adjustments of the moments.

4. The symmetric function of a finite sample of observations of which the mean value is κ_w.

It is easy to see that the condition that the mean value of a symmetric function of the observations shall be equal to one of the cumulants, or some function of the cumulants, for samples of all sizes, is sufficient to determine the symmetric function completely. This property was, however, long overlooked, and the series of statistics which afford unbiassed estimates of the cumulants was, in fact, only introduced in connection with a study of the sampling distributions of such estimates, which are found to be greatly simplified, both in their form and their derivation, by using the appropriate series of statistics. Corresponding with any partition

$$P = (p_1^{\pi_1} p_2^{\pi_2} \ldots\ldots), \qquad S(\pi) = \rho, \qquad S(p\pi) = w,$$

of the partible number w there exists a monomial symmetric function of a sample of n observations. If of the n observations π_1 are chosen to be raised to the power of p_1; of the remainder we choose π_2 to be raised to the power of p_2, and so on, leaving $n-\rho$ observations not involved, the product of the powers of the chosen observations constitutes a typical term of the symmetric function. The number of similar terms that can be formed is

$$\frac{n!}{\pi_1!\, \pi_2! \ldots\ldots (n-\rho)!},$$

and the sum of these will be designated by the symbol $G(P)$.

Thus $G(2^2 1)$ stands for

$$\tfrac{1}{2} \overset{n}{\underset{r=1}{S}} \overset{n}{\underset{s=1}{S}} \overset{n}{\underset{t=1}{S}} (x^2_r\, x^2_s\, x_t)$$

in which r, s and t may take any three different values from 1 to n. The factor $\frac{1}{2}$ is required since interchange of the values r and s leaves the monomial function unaltered.

Since the observations are independent, the mean value of any term is

$$\mu'^{\pi_1}_{p_1} \mu'^{\pi_2}_{p_2} \ldots\ldots\ldots$$

so that

$$\overline{G(P)} = \frac{n!}{\pi_1!\, \pi_2! \ldots\ldots (n-\rho)!} \mu'^{\pi_1}_{p_1} \mu'^{\pi_2}_{p_2} \ldots\ldots$$

But we know that

$$\kappa_w = \sum_P \frac{(-)^{\rho-1} (\rho-1)!\, w!}{(p_1!)^{\pi_1} (p_2!)^{\pi_2} \ldots\ldots \pi_1!\, \pi_2! \ldots\ldots} \mu'^{\pi_1}_{p_1} \mu'^{\pi_2}_{p_2} \ldots\ldots,$$

the summation extending over all partitions of w.

Hence to obtain a statistic k_w such that $\overline{k_w} = \kappa_w$, it is sufficient to put

$$k_w = \sum_P \frac{(-)^{\rho-1} (\rho-1)!}{n(n-1) \ldots\ldots (n-\rho+1)} \frac{w!}{(p_1!)^{\pi_1} (p_2!)^{\pi_2} \ldots\ldots} G(P).$$

The set of symmetric functions most easily calculated from the observations, at least if these are grouped, or have many repetitions of the same value, are the sums of powers $s_q = S(x^q)$.

From these, corresponding with any partition Q of the partible number w, where

$$Q = (q_1^{\chi_1} q_2^{\chi_2} \ldots\ldots)$$

it is easy to construct the symmetric function

$$S(Q) = s^{\chi_1}_{q_1} s^{\chi_2}_{q_2} \ldots\ldots$$

To express $G(P)$ in terms of $S(Q)$ we require the bipartitional function $Gs(P, Q)$ defined by the identity

$$G(P) = \sum_Q Gs(P, Q)\; S(Q).$$

Gs is found to be an integer divided by $\pi_1!\, \pi_2! \ldots\ldots$; for values of w from 2 to 6 the values of $\pi_1!\, \pi_2! \ldots\ldots Gs(P, Q)$ are tabulated below.

$w = 2$.

P	$S(Q)$ (2)	(1^2)
$G(2)$	1	.
$2G(1^2)$	—1	1

$w = 3$.

P	$S(Q)$ (3)	(21)	(1^3)
$G(3)$	1	.	.
$G(21)$	—1	1	.
$6G(1^3)$	2	—3	1

$w = 4$

P	$S(Q)$ (4)	(31)	(2^2)	(21^2)	(1^4)
$G(4)$	1	.	.	.	.
$G(31)$	—1	1	.	.	.
$2G(2^2)$	—1	.	1	.	.
$2G(21^2)$	2	—2	—1	1	.
$24G(1^4)$	—6	8	3	—6	1

$w = 5$

P	$S(Q)$ (5)	(41)	(32)	(31^2)	(2^21)	(21^3)	(1^5)
$G(5)$	1	.	.	.	.	.	.
$G(41)$	—1	1	.	.	.	.	.
$G(32)$	—1	.	1	.	.	.	.
$2G(31^2)$	2	—2	—1	1	.	.	.
$2G(2^21)$	2	—1	—2	.	1	.	.
$6G(21^3)$	—6	6	5	—3	—3	1	.
$120G(1^5)$	24	—30	—20	20	15	—10	1

$w = 6$

P	$S(Q)$ (6)	(51)	(42)	(3^2)	(41^2)	(321)	(2^3)	(31^3)	(2^21^2)	(21^4)	(1^6)
$G(6)$	1	.	.	.	.	.	.	.	.	.	.
$G(51)$	—1	1	.	.	.	.	.	.	.	.	.
$G(42)$	—1	.	1	.	.	.	.	.	.	.	.
$2G(3^2)$	—1	.	.	1	.	.	.	.	.	.	.
$2G(41^2)$	2	—2	—1	.	1	.	.	.	.	.	.
$G(321)$	2	—1	—1	—1	.	1	.	.	.	.	.
$6G(2^3)$	2	.	—3	.	.	.	1	.	.	.	.
$6G(31^3)$	—6	6	3	2	—3	—3	.	1	.	.	.
$4G(2^21^2)$	—6	4	5	2	—1	—4	—1	.	1	.	.
$24G(21^4)$	24	—24	—18	—8	12	20	3	—4	—6	1	.
$720G(1^6)$	—120	144	90	40	—90	—120	—15	40	45	—15	1

In the expression for k_w the coefficient of $G(P)$ for partitions of a fixed number of parts ρ, is proportional to

$$\frac{w!}{(p_1!)^{\pi_1}(p_2!)^{\pi_2}\ldots\ldots} = \pi_1!\,\pi_2!\ldots\ldots a(P)$$

where $a(P)$ is the elementary partitional function

$$\frac{w!}{(p_1!)^{\pi_1}\pi_1!\,(p_2!)^{\pi_2}\pi_2!\ldots\ldots}.$$

We may therefore use the property of the function $Gs(P, Q)$, namely that

$$\sum_{P/\rho} \pi_1!\,\pi_2!\ \ldots\ldots\ a(P)\ Gs(P, Q)$$

is $a(Q)$ times the coefficient of x^ρ in the product, for all parts q of Q, of the polynomials

$$F(q) = x \sum_{r=0} (-x\,\triangle)^r\,(1^{q-1}).$$

For the smaller values of q we have

q	F
1	x.
2	$x - x^2$
3	$x - 3x^2 + 2x^3$
4	$x - 7x^2 + 12x^3 - 6x^4$
5	$x - 15x^2 + 50x^3 - 60x^4 + 24x^5$
6	$x - 31x^2 + 180x^3 - 390x^4 + 360x^5 - 120x^6$
7	$x - 63x^2 + 602x^3 - 2100x^4 + 3360x^5 - 2520x^6 + 720x^7$.

The expression for k_w thus reduces to

$$k_w = \sum_Q \left\{ \sum_\rho \frac{(-)^{\rho-1}(\rho-1)!}{n(n-1)\ldots\ldots(n-\rho+1)} u_\rho\, a(Q)\, S(Q) \right\}$$

where u_ρ is given by

$$F^{\chi_1}(q_1)\, F^{\chi_2}(q_2) \ldots\ldots = \sum u_\rho x^\rho.$$

The process of simplification may be illustrated by determining the coefficient of $s_3 s_2^2$ in k_7. We have
$Q = (32^2)$, $a(Q) = 105$, $F(3)\,F^2(2) = x^3 - 5x^4 + 9x^5 - 7x^6 + 2x^7 = \sum u_\rho x^\rho$,
so that the coefficient of $s_3 s_2^2$ is

$$a(Q) \sum_\rho \frac{(\rho-1)!\,(-)^{\rho-1}}{n(n-1)\ldots\ldots(n-\rho+1)} u_\rho$$

$$= 105 \left(\frac{2!}{n(n-1)(n-2)} 1 + \frac{3!}{n(n-1)(n-2)(n-3)} 5 \right.$$

$$+ \frac{4!}{n(n-1)(n-2)(n-3)(n-4)} 9 + \frac{5!}{n(n-1)(n-2)(n-3)(n-4)(n-5)} 7$$

$$\left. + \frac{6!}{n(n-1)(n-2)(n-3)(n-4)(n-5)(n-6)} 2 \right) = \frac{210\,n}{(n-3)(n-4)(n-5)(n-6)}.$$

5. Transformation of the characteristic function.

If ξ is any function of x capable of expansion in a power series

$$\xi(x) = a_0 + a_1 x + a_2 x^2 + \ldots\ldots$$

then the characteristic function of ξ, is the average value of $e^{i\tau\xi}$.

The coefficient of $(i\tau)^r/r!$ is the average value of

$$(a_0 + a_1 x + a_2 x^2 + \ldots\ldots)^r$$

or, if the characteristic function $M(t)$ is differentiable

$$\left\{ a_0 + a_1 \frac{d}{d(it)} + a_2 \frac{d^2}{d(it)^2} + \ldots\ldots \right\}^r M(t).$$

Hence, the characteristic function of any function $\xi(x)$ may be expressed in terms of that of x in the form

$$e^{\tau\xi\{d/d(it)\}}\, M_x(t).$$

6. The operational properties of the cumulants.

If the element of frequency is $y\,dx$, and y and its differential coefficients vanish at the limits of the range

$$\int e^{itx}\left(\frac{d^r}{dx^r}y\right)dx=\left[e^{itx}\frac{d^{r-1}}{dx^{r-1}}y\right]-it\int e^{itx}\left(\frac{d^{r-1}}{dx^{r-1}}y\right)dx$$

$$=-it\int e^{itx}\left(\frac{d^{r-1}}{dx^{r-1}}y\right)dx$$

....

$$=(-it)^r M.$$

Hence

$$\int e^{itx}\, e^{a_q\left(-\frac{d}{dx}\right)^q}\, ydx=Me^{a_q(it)^q}$$

It thus appears that the cumulative function of the distribution

$$df=e^{a_q\left(-\frac{d}{dx}\right)^q}\,ydx$$

differs from that of

$$df=ydx$$

by

$$a_q\,(it)^q\,;$$

or that the operator

$$e^{\frac{a_q}{q!}\left(-\frac{d}{dx}\right)^q}$$

merely increases the qth cumulant by a_q.

The action of the operator $e^{-a_1\frac{d}{dx}}$ merely transforms a function $f(x)$ into $f(x-a_1)$; when acting on a frequency function, it thus simply increases the mean by a_1, leaving the distribution otherwise unchanged. Similarly it appears that

$$e^{\frac{a_2}{2!}\frac{d^2}{dx^2}}$$

simply increases the variance by a_2, leaving the mean and other cumulants unchanged, as would be done by scattering each element of frequency in a normal distribution with variance a_2. Similarly, the other operators of the form

$$e^{\frac{a_q}{q!}\left(-\frac{d}{dx}\right)^q}$$

may be used to adjust any of the other cumulants to desired values.

7. The probability integral of a distribution having given cumulants.

Since the frequency element of the distribution of a variable ξ having given cumulants κ_1, κ_2, κ_3, ... can be represented formally by

$$dp=exp\left\{-(\kappa_1-m)\frac{d}{d\xi}+\frac{1}{2}(\kappa_2-v)\frac{d^2}{d\xi^2}-\frac{1}{6}\kappa_3\frac{d^3}{d\xi^3}+\ldots\ldots\right\}\frac{1}{\sqrt{2\pi v}}e^{\frac{-(\xi-m)^2}{2v}}d\xi$$

where m and v are the mean and variance of any normal distribution chosen for convenience, we may use this choice to simplify the determination of the probability integral.

Sometimes the exact values of κ_1 and κ_2 may be used; in this case the expression will only involve the higher cumulants κ_3, κ_4 ...... More frequently the successive cumulants are expressed in power series of the reciprocal of some number n, so that the order of magnitude of κ_r is that of $n^{-(r-1)}$.

If m and v be chosen to agree with the leading terms of the series for κ_1 and κ_2, the ratio of κ_r to $v^{\frac{1}{2}r}$ will be of the order of $n^{-(\frac{1}{2}r-1)}$, when r exceeds 2, and the expansion takes the form

$$dp = exp\left\{-av^{\frac{1}{2}}\frac{d}{d\xi}+\frac{1}{2}bv\frac{d^2}{d\xi^2}-\frac{1}{6}cv^{\frac{3}{2}}\frac{d^3}{d\xi^3}+\ldots\ldots\right\}\frac{1}{\sqrt{2\pi v}}e^{\frac{-(\xi-m)^2}{2v}}d\xi$$

where a and c are of order $n^{-\frac{1}{2}}$, b and d of order n^{-1}, e of order $n^{-\frac{3}{2}}$, f of order n^{-2}, and so on.

Expanding the operator and integrating, we have for the frequency less than $m+\xi v^{\frac{1}{2}}$, the expansion, of which the first four adjustment terms have been retained,

$$
\begin{aligned}
& p \\
& -z\left(a+\frac{1}{6}c\xi_2\right) \\
& +z\left(\frac{1}{2}a^2\xi_1+\frac{1}{2}b\xi_1+\frac{1}{6}ac\xi_3+\frac{1}{24}d\xi_3+\frac{1}{72}c^2\xi_5\right) \\
& -z\left(\frac{1}{6}a^3\xi_2+\frac{1}{2}ab\xi_2+\frac{1}{12}a^2c\xi_4+\frac{1}{12}bc\xi_4+\frac{1}{24}ad\xi_4+\frac{1}{120}e\xi_4+\frac{1}{72}ac^2\xi_6\right. \\
& \qquad \left.+\frac{1}{144}cd\xi_6+\frac{1}{1296}c^3\xi_8\right) \\
& +z\left(\frac{1}{24}a^4\xi_3+\frac{1}{8}b^2\xi_3+\frac{1}{4}a^2b\xi_3+\frac{1}{36}a^3c\xi_5+\frac{1}{12}abc\xi_5+\frac{1}{48}a^2d\xi_5+\frac{1}{48}bd\xi_5\right. \\
& \qquad +\frac{1}{120}ae\xi_5+\frac{1}{720}f\xi_5+\frac{1}{144}a^2c^2\xi_7+\frac{1}{144}bc^2\xi_7+\frac{1}{1152}d^2\xi_7+\frac{1}{144}acd\xi_7 \\
& \qquad \left.+\frac{1}{720}ce\xi_7+\frac{1}{1296}ac^3\xi_9+\frac{1}{1728}c^2d\xi_9+\frac{1}{31104}c^4\xi_{11}\right)
\end{aligned}
$$

where

$$z=\frac{1}{\sqrt{2\pi}}e^{-\frac{1}{2}\xi^2},$$

$$p=\int_{-\infty}^{\xi} z\,d\xi$$

and ξ_r is the Hermite polynomial given by

$$\frac{d^r}{d\xi^r}z=\xi_r z,$$

or in full,

$\xi_1=-\xi$ $\qquad\qquad$ $\xi_2=\xi^2-1$

$\xi_3=-\xi^3+3\xi$ $\qquad\qquad$ $\xi_4=\xi^4-6\xi^2+3$

$\xi_5=-\xi^5+10\xi^3-15\xi$ $\qquad\qquad$ $\xi_6=\xi^6-15\xi^4+45\xi^2-15$

$\xi_7=-\xi^7+21\xi^5-105\xi^3+105\xi$ $\qquad\qquad$ $\xi_8=\xi^8-28\xi^6+210\xi^4-420\xi^2+105$

$\xi_9=-\xi^9+36\xi^7-378\xi^5+1260\xi^3-945\xi$ $\qquad$ $\xi_{10}=\xi^{10}-45\xi^8+630\xi^6-3150\xi^4+4725\xi^2-$

$$\xi_{11}=-\xi^{11}+55\xi^9-990\xi^7+6930\xi^5-17325\xi^3+10395\xi.$$

8. The expansion for the abscissa corresponding to any given level of probability.

Although it is sometimes of interest to work out the actual value of the probability corresponding with a given deviation, it is of much more general utility to know the values of the deviates corresponding with the assigned levels of probability. If now we write x for the normal deviate having the same probability integral, the difference $\xi - x$ may be found by equating the expression above for the probability to

$$p - (\xi - x)\, z + \frac{1}{2}\,(\xi - x)^2\, \xi_1 z - \frac{1}{6}\,(\xi - x)^3\, \xi_2 z + \frac{1}{24}\,(\xi - x)^4\, \xi_3 z.$$

By equating terms of each order of magnitude in succession, as in the inversion of power series, we find the polynomials are much simplified, giving

$$\xi - x = a + \frac{1}{6}\, c\, (\xi^2 - 1)$$

$$+ \frac{1}{2}\, b\, \xi - \frac{1}{3}\, ac\, \xi + \frac{1}{24}\, d\, (\xi^3 - 3\,\xi) - \frac{1}{36}\, c^2\, (4\,\xi^3 - 7\,\xi)$$

$$- \frac{1}{2}\, ab + \frac{1}{6}\, a^2 c - \frac{1}{12}\, bc\, (5\,\xi^2 - 3) - \frac{1}{8}\, ab\, (\xi^2 - 1)$$

$$+ \frac{1}{120}\, e\, (\xi^4 - 6\,\xi^2 + 3) + \frac{1}{36}\, ac^2\, (12\,\xi^2 - 7)$$

$$- \frac{1}{144}\, cd\, (11\,\xi^4 - 42\,\xi^2 + 15) + \frac{1}{648}\, c^3\, (69\,\xi^4 - 187\,\xi^2 + 52)$$

$$- \frac{3}{8}\, b^2 \xi + \frac{5}{6}\, abc\, \xi + \frac{1}{8}\, a^2 d\, \xi - \frac{1}{48}\, bd\, (7\,\xi^3 - 15\,\xi) - \frac{1}{30}\, ae\, (\xi^3 - \xi)$$

$$+ \frac{1}{720}\, f\, (\xi^5 - 10\,\xi^3 + 15\,\xi) - \frac{1}{3}\, a^2 c^2 \xi + \frac{1}{72}\, bc^2\, (36\,\xi^3 - 49\,\xi)$$

$$- \frac{1}{384}\, d^2\, (5\,\xi^5 - 32\,\xi^3 + 35\,\xi) + \frac{1}{36}\, acd\, (11\,\xi^3 - 21\,\xi)$$

$$- \frac{1}{360}\, ce\, (7\,\xi^5 - 48\,\xi^3 + 51\,\xi) - \frac{1}{324}\, ac^3\, (138\,\xi^3 - 187\,\xi)$$

$$+ \frac{1}{864}\, c^2 d\, (111\,\xi^5 - 547\,\xi^3 + 456\,\xi) - \frac{1}{7776}\, c^4\, (948\,\xi^5 - 3628\,\xi^3 + 2473\,\xi).$$

In these expressions it should be noticed that the polynomials involved are in the deviate ξ. It is in many ways more convenient to use instead polynomials in the normal deviate x, corresponding to the probability required. This involves an awkward substitution, which may, perhaps, best be carried out by observing that if

$$\xi - x = f(\xi) = f\{x + (\xi - x)\},$$

then it may be written to the required degree of approximation as

$$f(x) + f(x)\, f'(x) + f(x)\, f'^2(x) + \frac{1}{2}\, f^2(x)\, f''(x)$$

$$+ f(x)\, f'^3(x) + \frac{3}{2}\, f^2(x)\, f'(x)\, f''(x) + \frac{1}{6}\, f^3(x)\, f'''(x).$$

In the result of this substitution all but one of the terms involving a must be eliminated, since a change of the mean leaving all other cumulants unchanged, changes all points of fixed probability by the same amount. The adjustment to the normal deviate x having the required probability integral is:

$$a + \frac{1}{6} c (x^2 - 1)$$

$$+ \frac{1}{2} bx + \frac{1}{24} d (x^3 - 3x) - \frac{1}{36} c^2 (2x^3 - 5x)$$

$$- \frac{1}{6} bc (x^2 - 1) + \frac{1}{120} e (x^4 - 6x^2 + 3) - \frac{1}{24} cd (x^4 - 5x^2 + 2)$$

$$+ \frac{1}{324} c^3 (12x^4 - 53x^2 + 17)$$

$$\left\{ \begin{array}{c} - \frac{1}{8} b^2 x - \frac{1}{16} bd (x^3 - 3x) + \frac{1}{720} f (x^5 - 10x^3 + 15x) \\ + \frac{1}{72} bc^2 (10x^3 - 25x) - \frac{1}{384} d^2 (3x^5 - 24x^3 + 29x) \\ - \frac{1}{180} ce (2x^5 - 17x^3 + 21x) + \frac{1}{288} c^2 d (14x^5 - 103x^3 + 107x) \\ - \frac{1}{7776} c^4 (252x^5 - 1688x^3 + 1511x). \end{array} \right.$$

For numerical work the polynomials in x required may very easily be tabulated for chosen levels of probability (Table I). In Table II are given the numerical values of the first five Hermite Polynomials over the range of levels of probability chosen for Table I.

8. 1. The cumulants of the distribution of the test of significance, z, and the approximate values for different levels of significance.

For the purpose of obtaining its cumulants, we may write z, which is half the logarithm of the ratio of two estimated variances, in the form

$$z = \tfrac{1}{2} \log \chi_1^2/n_1 - \tfrac{1}{2} \log \chi_2^2/n_2 .$$

Now the distribution of χ_1 is given by

$$df = \frac{1}{\frac{n_1 - 2}{2}!} e^{-\frac{1}{2}\chi_1^2} \left(\frac{\chi_1^2}{2}\right)^{\frac{n_1-2}{2}} d\left(\frac{\chi_1^2}{2}\right)$$

so that the mean value of

$$exp \left\{ \tfrac{1}{2} it \log \chi_1^2/n_1 \right\}$$
$$= exp \left\{ \tfrac{1}{2} it \log (\tfrac{1}{2}\chi_1^2) - \tfrac{1}{2} it \log \tfrac{1}{2} n_1 \right\}$$

is

$$\frac{\frac{n_1 + it - 2}{2}!}{\frac{n_1 - 2}{2}!} exp (-\tfrac{1}{2} it \log \tfrac{1}{2} n_1),$$

whence the cumulative function of z is seen to be

$$K = \log \frac{n_1 + it - 2}{2}! - \log \frac{n_1 - 2}{2}! - \tfrac{1}{2} it \log \tfrac{1}{2} n_1$$
$$- \left(\log \frac{n_2 + it - 2}{2}! - \log \frac{n_2 - 2}{2}! - \tfrac{1}{2} it \log \tfrac{1}{2} n_2 \right).$$

TABLE I.

	P								
	,25	,10	,05	,025	,01	,005	,0025	,001	,0005
c	— ,09084	,10706	,28426	,47358	,73532	,93915	1,14657	1,42491	1,63793
b	,33724	,64078	,82243	,97998	1,16317	1,28791	1,40352	1,54511	1,64526
d	— ,07153	— ,07249	— ,02018	,06872	,23379	,39012	,57070	,84331	1,07320
c^2	,07663	,06106	— ,01878	— ,14607	— ,37634	— ,59171	— ,83890	— 1,21025	— 1,52234
bc	,09084	— ,10706	— ,28426	— ,47358	— ,73532	— ,93915	— 1,14657	— 1,42491	— 1,63793
e	,00398	— ,03464	— ,04928	— ,04410	— ,00152	,06010	,14841	,30746	,46059
cd	,00282	,14644	,17532	,10210	— ,17621	— ,53531	— 1,02868	— 1,89355	— 2,71243
c^3	— ,01428	— ,11629	— ,11900	— ,02937	,25195	,59757	1,06301	1,86787	2,62337
b^2	— ,08431	— ,16019	— ,20561	— ,24500	— ,29079	— ,32198	— ,35088	— ,38628	— ,41132
bd	,10729	,10874	,03027	— ,10308	— ,35068	— ,58518	— ,85605	— 1,26496	— 1,60980
f	,00998	,00227	— ,01082	— ,02357	— ,03176	— ,02621	— ,00666	,04591	,10950
bc^2	— ,19158	— ,15265	,04696	,36517	,94084	1,47928	2,09726	3,02562	3,80584
d^2	— ,03285	,00776	,05985	,09659	,07888	— ,01226	— ,19116	— ,59060	— 1,03555
ce	— ,05126	,01086	,09462	,16106	,16058	,05366	— ,17498	— ,70464	— 1,30531
c^2d	,14764	— ,10858	— ,39517	— ,55856	— ,32621	,35696	1,60445	4,29304	7,23307
c^4	— ,06898	,09585	,25623	,31624	,07286	— ,46534	— 1,39199	— 3,32708	— 5,40702

TABLE II.

Hermite Polynomials.

P	x	$x^2—1$	$x^3—3x$	$x^4—6x^2+3$	$x^5—10x^3+15x$
,25	,6744897943	— ,5450635174	— 1,716619368	,4773483076	7,188444035
,10	1,281551565	,6423744138	— 1,739867195	— 4,156852768	1,632247612
,05	1,644853627	1,705543454	— ,484337918	— 5,913295343	— 7,789153620
,025	1,959963985	2,841458821	1,649228983	— 5,291947053	— 16,96894156
,01	2,326347874	4,411894431	5,610905483	— ,182765252	— 22,86879749
,005	2.575829303	5,634896598	9,362873171	7,212473280	— 18,87339266
,0025	2,807033768	6,879438576	13,69674885	17,80892082	— 4,79675330
,001	3,090223231	8,549479614	20,23935405	36,89568324	33,05848120
,0005	3,290526731	9,827566171	25,75681573	55,27079216	78,84275615

The expansion of K in powers of it, may therefore be found from the differential coefficients with respect to n of

$$\log \frac{n-2}{2}\,!\,.$$

Now with sufficient approximation

$$\log \frac{n-2}{2}\,! = \tfrac{1}{2}(n-1)\log \tfrac{1}{2}n - \tfrac{1}{2}n + \tfrac{1}{2}\log(2\pi) + \frac{1}{6n},$$

and its successive differential coefficients are:

$$\frac{1}{2}\log \tfrac{1}{2}n - \frac{1}{2n} - \frac{1}{6n^2}$$

$$\frac{1}{2n} + \frac{1}{2n^2} + \frac{1}{3n^3}$$

$$-\frac{1}{2n^2} - \frac{1}{n^3}$$

$$\frac{1}{n^3} + \frac{3}{n^4}$$

$$-\frac{3}{n^4}$$

$$\frac{12}{n^5}$$

From these, writing r_1 and r_2 for the reciprocals of n_1 and n_2 respectively, we may obtain the cumulants

$$\begin{aligned}
\kappa_1 &= -\tfrac{1}{2}(r_1 - r_2) - \tfrac{1}{6}(r_1^2 - r_2^2),\\
\kappa_2 &= \tfrac{1}{2}(r_1 + r_2) + \tfrac{1}{2}(r_1^2 + r_2^2) + \tfrac{1}{3}(r_1^3 + r_2^3),\\
\kappa_3 &= -\tfrac{1}{2}(r_1^2 - r_2^2) - (r_1^3 - r_2^3),\\
\kappa_4 &= (r_1^3 + r_2^3) + 3(r_1^4 + r_2^4),\\
\kappa_5 &= -3(r_1^4 - r_2^4),\\
\kappa_6 &= 12(r_1^5 + r_2^5).
\end{aligned}$$

If we write σ for the sum $(r_1 + r_2)$ and δ for the difference $(r_1 - r_2)$ of these reciprocals, and if we choose

$$m = 0, \quad v = \tfrac{1}{2}\sigma$$

then

$$\begin{aligned}
a &= -\sqrt{\frac{2}{\sigma}}\left(\tfrac{1}{2}\delta + \frac{1}{6}\delta\sigma\right),\\
b &= \tfrac{1}{2}\left(\sigma + \frac{\delta^2}{\sigma}\right) + \frac{1}{6}\left(\sigma^2 + 3\delta^2\right),\\
c &= -\sqrt{\frac{2}{\sigma}}\left\{\delta + \tfrac{1}{2}\left(3\delta\sigma + \frac{\delta^3}{\sigma}\right)\right\},\\
d &= \left(\sigma + \frac{3\delta^2}{\sigma}\right) + \frac{3}{2}\left(\sigma^2 + 6\delta^2 + \frac{\delta^4}{\sigma^2}\right),\\
e &= -\sqrt{\frac{2}{\sigma}}\left\{6\delta\left(\sigma + \frac{\delta^2}{\sigma}\right)\right\},\\
f &= 6\left(\sigma^2 + 10\delta^2 + \frac{5\delta^4}{\sigma^2}\right).
\end{aligned}$$

For any special values of n_1 and n_2 the six cumulant adjustments can be evaluated numerically, and the values substituted in the general formula given above. It is, however, also of interest to make the substitution algebraically and obtain the general form of the z values at different levels of significance in terms of σ and δ. We then have

$$
\begin{aligned}
& x\sqrt{\tfrac{1}{2}\sigma} \\
& -\frac{1}{6}\,\delta\,(x^2+2) \\
& +\sqrt{\tfrac{1}{2}\sigma}\left\{\frac{\sigma}{24}(x^3+3x)+\frac{1}{72}\frac{\delta^2}{\sigma}(x^3+11x)\right\} \\
& -\frac{\delta\sigma}{120}(x^4+9x^2+8)+\frac{\delta^3}{3240\,\sigma}(3x^4+7x^2-16) \\
& +\sqrt{\tfrac{1}{2}\sigma}\left\{\frac{\sigma^2}{1920}(x^5+20x^3+15x)+\frac{\delta^2}{2880}(x^5+44x^3+183x)\right. \\
& \qquad \left. +\frac{\delta^4}{155520\,\sigma^2}(9x^5-284x^3-1513x)\right\}.
\end{aligned}
$$

Example.

When $n_1=24$ and $n_2=60$ the 5 % value of z is .26534844. The value obtained from the first approximation and the four correction terms above is

$$
\begin{aligned}
&.2809\,1224 - .0196\,0643 + (.0038\,9559 + .0005\,7292) - (.0004\,8210 - .0000\,0206) \\
&\qquad + (.0000\,3805 + .0000\,1886 - .0000\,0046) \\
&= .2653\,5073.
\end{aligned}
$$

The successive errors in this case diminish progressively, with alternating sign.

Résumé: Les cumulants d'une distribution sont définis et leurs relations avec les moments ordinaires sont indiquées. On insiste sur l'importance de ces fonctions dans la description des distributions.

La fonction symétrique d'un échantillon fini d'observations dont la valeur moyenne est égale au cumulant correspondant de la population est défini et une expression convenable pour son évaluation est développée en termes de sommes de puissances des observations de l'échantillon.

Les propriétés opératives des cumulants sont discutées et le développement de l'intégrale des probabilités d'une distribution ayant des cumulants donnés en est dérivé. Ce développement est transformé en une expression plus commode, qui donne la valeur des déviations correspondant à un niveau donné de probabilité, en fonction de la déviation normale ayant la même intégrale des probabilités. Les valeurs numériques des polynômes dans la déviation normale considérés dans le dernier développement sont calculés pour certains niveaux choisis de probabilité. De plus, les valeurs numériques des 5 premiers polynômes d'Hermite sont donnés pour les mêmes niveaux de probabilité. Un exemple de l'emploi des formules développées est donné par le calcul de la valeur approximative de z pour le niveau de probabilité de 5 p.cent.

Reprinted from *The Journal of the Australian Institute of Agricultural Science*, Vol. 4, No. 4, pp.199-203, December, 1938.

Factorial Treatments in Incomplete Randomized Blocks

By

E. A. CORNISH,
Waite Agricultural Research Institute, S.A.

The method of balanced incomplete blocks described in [1], [2] and referred to in [3], section 17e was designed by Yates to meet the special requirements of observational data in which the nature of the material imposes a definite restriction on the number of units in naturally occurring groups, with a consequent restriction on the number of treatments which may be tested. Examples of situations in which this arrangement is of particular use are: litters of animals, paired material and in the local lesion method of testing the infectivity of virus preparations where there are usually not more than six suitable leaves on a plant. It is also of value in trials involving large numbers of varieties.[4]

This novel experimental arrangement is a modification of the familiar randomized block, analogous to confounding, which permits dispensation of the restriction that the number of units per block shall equal the number of treatments.

If t treatments are to be compared in b randomized blocks, each of k experimental units, $k<t$, block differences can be eliminated if it is arranged that every two treatments occur together in the same number, λ, of blocks. If there are r replicates of each treatment, the total number of experimental units is $tr=bk=N$, and $\lambda=r(k-1)/(t-1)$. With seven treatments a, b, c, d, e, f and g, in blocks of three, the following grouping of the treatments in seven blocks satisfies the conditions. In this case $\lambda=1$.

Block				Block			
1	*a*	*b*	*c*	5	*b*	*e*	*g*
2	*a*	*d*	*e*	6	*c*	*d*	*g*
3	*a*	*f*	*g*	7	*c*	*e*	*f*
4	*b*	*d*	*f*				

The treatment effects, free from block differences, are calculated from the t quantities

$$Q_i=kT_i-S, \qquad i=1, 2, \ldots, t$$

by dividing by krE. T_i is the total of all replicates of the ith treatment, S is the total of blocks containing the ith treatment and E is the efficiency factor of the incomplete block arrangement. It is equal to $(1-1/k)/(1-1/t)$ and measures the fractional loss of information when there is no reduction in the error variance per experimental unit when blocks of k, instead of blocks of t units, are used. Thus the effect of treatment i is

$$t_i = \frac{1}{krE} Q_i.$$

It should be noted that the quantity Q defined above is k times the analogous quantity defined in [2]. The adjusted mean of the ith treatment is

$$\frac{1}{krE} Q_i + \text{general mean.}$$

The sums of squares of treatments and blocks are as follow:

Treatments $\frac{1}{k^2rE} \overset{t}{\underset{i=1}{S}} Q^2_i$

Blocks $\frac{1}{k} \overset{b}{\underset{j=1}{S}} B_j^2, - GT^2/N.$

B_j is the total of experimental units in the jth block. The error sum of squares is obtained in the usual manner by subtraction from the total sum of squares. The estimated standard error of the difference between any two adjusted treatment means is the square root of $2s^2/rE$, where s^2 is the estimate of the error variance per experimental unit. Note that these expressions only differ from the standard formulæ used in ordinary randomized block analyses by the substitution of the Q's for the treatment totals and the introduction of the factors kE and k^2E.

In the literature cited above, no reference is made to the use of factorial systems of treatments in incomplete blocks. The purpose of this note is to point out that they may be employed, but at the same time it must be strongly emphasized that their indiscriminate use in this design is not being advocated. There are many eminently suitable designs for use in situations where it is desirable to resort to confounding, and the reader is referred particularly to [3] in which Yates has given a complete account of them. It is incumbent upon the experimenter first, to decide on the information he is desirous of obtaining, secondly, to devise a system of treatments appropriate to the questions being asked, and thirdly, to determine what experimental arrangement is best adapted to provide answers to his questions with the greatest possible precision, due consideration being given to the nature of the material with which he is dealing and the proposed system of treatments.

If the group of t treatments constitute a factorial system arranged in incomplete blocks, it may be readily shown that the only manner in which the functions representing main effects and interactions differ from the standard expressions of orthogonal data, lies in the substitution of the Q's for the treatment totals and the introduction of the factor kE. Similarly, the functions giving the contributions of the main effects and interactions to the treatment sum of squares differ from the standard expressions only in the use of the Q's and the addition of the factor k^2E.

As an example, consider the simple case in which four treatments, 2 factors a and b at 2 levels each, are arranged in four incomplete blocks of 3 units each;

$t=4$, $b=4$, $k=3$, $r=3$ and $E=\frac{8}{9}$. The arrangement of the treatments satisfying the incomplete block conditions is

$$\begin{array}{cccc} a_0b_0 & a_1b_0 & a_0b_1 & a_1b_1 \\ a_1b_0 & a_0b_1 & a_1b_1 & a_0b_0 \\ a_0b_1 & a_1b_1 & a_0b_0 & a_1b_0, \end{array}$$

where the columns constitute the blocks. The table of Q's is

$$\begin{array}{cccc} & b_0 & b_1 & \\ a_0 & Q_{00} & Q_{01} & Q_{00}+Q_{01} \\ a_1 & Q_{10} & Q_{11} & Q_{10}+Q_{11} \\ & Q_{00}+Q_{10} & Q_{01}+Q_{11} & 0 \end{array}$$

The estimates of the main effects of the factors a and b are respectively,

$$\tfrac{1}{32}(Q_{11}+Q_{10}-Q_{01}-Q_{00}) \text{ and } \tfrac{1}{32}(Q_{11}+Q_{01}-Q_{10}-Q_{00})$$

and the estimate of their interaction is

$$\tfrac{1}{32}(Q_{11}+Q_{00}-Q_{01}-Q_{10}).$$

The divisor 32 is the product of two factors,

(i) $4r=12$, the factor which would ordinarily be used,

(ii) $kE=\frac{24}{9}$. depending on the incomplete block design.

The estimated standard error of each of the above quantities is the root of $\frac{1}{4rE}$ times the error variance per experimental unit.

The contributions of the three items to the treatment sum of squares are

$$\frac{1}{96}\{(Q_{11}+Q_{10}-Q_{01}-Q_{00})^2+(Q_{11}+Q_{01}-Q_{10}-Q_{00})^2+(Q_{11}+Q_{00}-Q_{01}-Q_{10})^2\}$$

$$=\frac{1}{24}(Q_{00}{}^2+Q_{10}{}^2+Q_{01}{}^2+Q_{11}{}^2) \text{ since } S(Q)=0.$$

As before, the divisor is the product of two factors, (i) $4r=12$, and (ii) $k^2E=\frac{72}{9}$.

TABLE I.

Values of Q.

Concentration of virus.

Medium.	1	1/2	1/4	1/8	1/16	1/32	1/64	*Total.*
Water ..	711	170	−335	−246	−247	−357	−883	−1187
0·05M. PO_4	963	385	731	315	−426	−417	−452	1099
0·10M. PO_4	591	599	−110	385	−426	−246	−705	88
Total ..	2265	1154	286	454	−1099	−1020	−2040	0

The following example, based on data given by Youden,[5] shows the numerical details involved in evaluating the contributions to the treatment sum of squares in the case of a 2-way classification of 21 treatments.

Table I summarizes the relevant information. In this example $b=21$, $r=5$, $k=5$ and $E=\frac{21}{25}$.

Treatment sum of squares:

$$\frac{1}{k^2 rE} \mathop{S}_{i=1, j=1}^{3, 7} Q^2_{ij} = \frac{1}{105}[711^2+963^2+ \ldots +(-705)^2] = 52872 \cdot 2$$

Concentrations:

$$\frac{1}{3k^2 rE} \mathop{S}_{j=1}^{7} \left(\mathop{S}_{i=1}^{3} Q_{ij} \right)^2 = \frac{1}{3 \times 105}[2265^2+1154^2+ \ldots +(-2040)^2] = 41776 \cdot 7$$

Media:

$$\frac{1}{7k^2 rE} \mathop{S}_{i=1}^{3} \left(\mathop{S}_{j=1}^{7} Q_{ij} \right)^2 = \frac{1}{7 \times 105}[(-1187)^2+1099^2+88^2] = 3570 \cdot 8$$

Interaction:

$$52872 \cdot 2 - (41776 \cdot 7 + 3570 \cdot 8) = 7524 \cdot 7.$$

The complete analysis is:

Variation due to	Deg. freed.	Sum Sq.	Mean Sq.	S.D.
Plants (blocks)	20	34153·3	1707·7	
Leaf positions	4	6232·6	1558·2	
Concentrations	6	41776·7	6962·8	
Media	2	3570·8	1785·4	
Concen. × media	12	7524·7	627·1	
Error	60	23014·8	353·6	18·8
Total	104	116272·9		

TABLE II.
Adjusted lesion counts.
Concentration of virus.

Medium.	1	1/2	1/4	1/8	1/16	1/32	1/64	*Mean.*
Water ..	80·8	55·1	31·0	35·3	35·2	30·0	4·9	38·9
0·05M. PO_4	92·8	65·3	81·8	62·0	26·7	27·1	25·5	54·5
0·10M. PO_4	75·1	75·5	41·7	65·3	26·7	35·3	13·4	47·6
Mean ..	82·9	65·3	51·5	54·2	29·5	30·8	14·6	46·97

The reader's attention is drawn to the ingenious device, due to Youden (*loc. cit.*) for eliminating from the analysis, the variation caused by the ordinal position of the leaves on the plants. It is worth noting that this modification of the incomplete block design is of general application. A more recent discussion of the Youden square has been given by Fisher.[6]

The adjusted lesion counts are given in Table II.

The estimated standard error of the difference between

(i) any two adjusted concentration means is:

$$\sqrt{\frac{2}{3}\times\frac{5}{21}\times 353\cdot 6}=7\cdot 49 \text{ lesions,}$$

(ii) any two adjusted media means is:

$$\sqrt{\frac{2}{7}\times\frac{5}{21}\times 353\cdot 6}=4\cdot 90 \text{ lesions,}$$

(iii) any two adjusted means within the table is:

$$\sqrt{2\times\frac{5}{21}\times 353\cdot 6}=12\cdot 98 \text{ lesions.}$$

The procedure to be adopted with other factorial systems is obtainable as a mere extension of the above reasoning. With the aid of the following formulæ the variance of any comparison is easily evaluated. The variance of $Q_i = rk(k-1)\sigma^2$ and the covariance of Q_i and $Q_j = -\lambda k\sigma^2$, where σ^2 is the error variance per experimental unit of the hypothetical infinite population from which the sample was drawn.

References.

[1] Yates, F.: Incomplete randomized blocks. Annals of Eugenics VII, 121-140.

[2] Fisher, R. A., and Yates, F.: Statistical Tables for Biological, Agricultural and Medical Research, pp. 90. Oliver and Boyd, 1938.

[3] Yates, F.: The design and analysis of factorial experiments. Imp. Bur. Soil Sc., Harpenden.

[4] Yates, F.: A new method of arranging variety trials involving a large number of varieties. Journ. Agric. Sc. XXVI, 424-455.

[5] Youden, W. J.: Use of incomplete block replications in estimating tobacco mosaic virus. Contrib. Boyce Thompson Inst. IX, 41-48.

[6] Fisher, R. A.: The mathematics of experimentation. Nature CXLII, 442-443.

AUSTRALASIAN MEDICAL PUBLISHING COMPANY LIMITED

Reprinted from *The Journal of the Australian Institute of Agricultural Science,* Vol. 6, No. 1, pp. 31-39, March, 1940.

The Analysis of Quasi-Factorial Designs with Incomplete Data

By

E. A. Cornish, B.Ag.Sc.,

Waite Agricultural Research Institute, Adelaide, S.A.

In recent years Yates[5, 6, 7] has developed the quasi-factorial designs for the special purpose of testing large numbers of treatments, for example, varieties of a crop in agricultural trials. These designs are of three types, namely, quasi-factorial designs in randomized blocks, balanced incomplete blocks and quasi-Latin squares, each of which is characterized by a special arrangement of the treatments within the groups of units (blocks) into which the experimental material is divided. The several types of symmetrical arrangement are such that they facilitate the elimination of group differences from the treatment comparisons and, at the same time, easily lend themselves to statistical analysis.

In cases where some of the observations are non-existent the appropriate analytical procedure can no longer be employed. The formulæ which are required to estimate missing values in the particular case where several isolated observations are missing have been given in papers now in press.[1, 2] The present paper is concerned with methods of analysing data derived from balanced incomplete block designs in which complete sets of experimental values are missing, i.e. a block or a treatment.

The notation introduced by Yates[6] will be followed as closely as possible. This notation, with slight modifications, has also been used by Fisher and Yates.[3]

The formulæ given below may all be derived by a straightforward application of the method of least squares.[4]

1. Missing Block.

(*a*) $k<\frac{1}{2}t$.

Without loss of generality it may be assumed that block 1, containing treatment 1, 2, ..., k, is missing. The set of constants which has to be determined is

mean: m.
blocks: $b_2, b_3, \ldots, b_b$; $S(b)=0$.
treatments: $t_1, t_2, \ldots, t_t$; $S(t)=0$.

If G is the grand total of existing observations and B and T stand for block and treatment totals, respectively, the normal equations are

$$(N-k)m-S(t_v)=G, \quad v=1, 2, \ldots, k.$$
$$k(b_u+m)+S(t)=B_u, \quad u=2, 3, \ldots, b.$$
$$(r-1)(t_v+m)+S(b)=T_v,$$
$$r(t_w+m)+S(b)=T_w, \quad w=k+1, \ldots, t.$$

In any b equation the summation is taken over the treatments contained in the block concerned and, in any t equation, it is taken over the existing blocks in which replicates of the treatments occur.

The equations are not independent but the solution is easily derived. Since treatments $k+1, \ldots, t$ are unaffected by the missing block the quantities $Q_{k+1}, Q_{k+2}, \ldots, Q_t$ retain their normal values, so that the corresponding t may be obtained directly from the relations

$$t_w=Q_w/krE.$$

The equations in $t_1, t_2, \ldots, t_k$ reduce to

$$Pt_1+t_2+\ldots+t_k=Q'_1,$$
$$t_1+Pt_2+\ldots+t_k=Q'_2,$$
$$\ldots\ldots\ldots\ldots$$
$$t_1+t_2+\ldots+Pt_k=Q'_k,$$

where $P=\{(k-1)(r-1)+\lambda\}$. The Q' are calculated in the ordinary manner except, of course, for the omission of the observations of the missing block. The solution of this set of equations is

$$t_v=AQ'_v+B(Q'_1+Q'_2+\ldots+Q'_k),$$

where

$$A=(krE-1)/k^2rE(rE-1) \text{ and } B=-1/k^2rE(rE-1)$$

and Q'_v is omitted from the bracket on the right hand side.

This gives the complete set of values of the t. Direct substitution in the b equations yields the values of $b+m$ and the simplest method of determining m is by substitution of the t in the m equation.

The treatment comparisons, in terms of a single experimental unit, are given by the differences of the quantities t and the adjusted treatment means by $t+m$.

The reduction in the sum of squares due to fitting the constants b and t is

$$SB(b+m)+StT-G^2/(N-k),$$

from which the sum of squares for testing the significance of treatments is obtained by deducting the sum of squares ascribable to blocks when treatments are disregarded, namely,

$$\frac{1}{k}S(B^2)-G^2/(N-k).$$

(*b*) $k>\frac{1}{2}t$.

If $k>\frac{1}{2}t$ the following procedure is, arithmetically, the more convenient to use in practice. It is first necessary to calculate for each treatment a quantity, Q', defined as k times the treatment total plus the totals of blocks in which that treatment does *not* occur. The t of treatments not contained in the missing block are then given by

$$t_w=(Q'_w-G)/krE,$$

m by

$$m=\{G-S(t_w)\}/k(b-1)$$

and the t of treatments belonging to the missing block by the relations

$$t_v=\{Q'_v-k(b-1)m\}/k(rE-1).$$

Direct substitution in the b equations, which are now written in terms of the t of treatments not in the blocks, gives the values of $b+m$. In other respects the analysis follows the steps given in section 1 (*a*).

When a complete block of units is missing all treatment comparisons are not of the same accuracy. There are three types of comparison:

(i) between two treatments neither of which has a replicate in the missing block: variance $2\sigma^2/rE$,

(ii) between two treatments both of which have a replicate in the missing block: variance $\frac{2\sigma^2}{rE}\left(1+\frac{1}{rE-1}\right)$,

(iii) between two treatments one of which has a replicate in the missing block: variance $\frac{2\sigma^2}{rE}\left\{1+\frac{k-1}{2k(rE-1)}\right\}$.

Example.

Fisher and Yates[3] have given an example showing the analysis of an incomplete block experiment involving 9 treatments in 18 blocks of 4 units each. In this design $t=9$, $b=18$, $k=4$, $r=8$, $\lambda=3$ and $E=27/32$. The reader is referred to the book quoted for the original data.

The values of B and $b+m$ are given in Table 1, and in Table 2 are given the values of T, Q or Q' and the t.

TABLE 1.

B		20·7	19·8	11·8	32·1	33·0	20·7	17·8	11·5
$b+m$ ..		5·4992	4·8540	3·1843	8·0490	8·0167	5·0793	4·7623	2·9008
B	26·5	22·1	11·4	16·7	13·7	12·7	18·4	18·1	22·8
$b+m$ ..	6·6965	5·5417	2·5972	4·1499	3·1362	3·1738	4·7492	4·6210	5·4971

TABLE 2.

	Treatment.									
	1	2	3	4	5	6	7	8	9	
T	43·9	39·1	35·9	33·9	34·8	33·0	28·6	42·8	37·8	329·8
Q or Q' ..	23·4	0	−1·4	8·5	4·6	−5·4	−23·9	−1·3	−4·5	
t	0·8667	0	−0·0710	0·3594	0·1899	−0·2449	−0·8852	−0·0481	−0·1667	
$t+m$	5·72	4·85	4·78	5·21	5·04	4·61	3·97	4·81	4·69	

The normal equations are

$$68m-(t_3+t_4+t_5+t_6)=329\cdot8,$$
$$4(b_2+m)+t_2+t_6+t_7+t_9=20\cdot7,$$
$$\cdots\cdots\cdots\cdots\cdots$$
$$7(t_3+m)+b_3+b_5+b_8+b_{10}+b_{13}+b_{14}+b_{16}=35\cdot9,$$
$$\cdots\cdots\cdots\cdots\cdots$$
$$8(t_1+m)+b_3+b_6+b_{11}+b_{12}+b_{13}+b_{14}+b_{15}+b_{18}=43\cdot9,$$
$$\cdots\cdots\cdots\cdots\cdots$$

The quantities krE, A and B have, respectively, the values 27, 26/621 and −1/621, so that, for example,

$$t_1=23\cdot4/27=0\cdot8667$$

and

$$t_3=\frac{1}{621}\{26\times(-1\cdot4)-8\cdot5-4\cdot6+5\cdot4\}=-0\cdot0710.$$

Also

$$(b_2+m)=\frac{1}{4}(20\cdot7+0\cdot2449+0\cdot8852+0\cdot1667)=5\cdot4992, \text{ etc.,}$$

and, finally,

$$m=\frac{1}{68}(329\cdot8-0\cdot0710+0\cdot3594+0\cdot1899-0\cdot2449)=4\cdot8534.$$

The sum of squares attributable to the constants b and t is

$$5\cdot4992\times20\cdot7+\ldots+0\cdot8667\times43\cdot9-0\cdot0710\times35\cdot9+\ldots-329\cdot8^2/68$$
$$=1784\cdot7148,$$

while the sum of squares taken out by blocks alone is 173·2850. Table 3 shows the analysis of variance.

Examples of the three types of treatment comparison with their estimated standard deviations are

(i) t_1-t_2 $0{\cdot}87\pm0{\cdot}83$,
(ii) t_4-t_3 $0{\cdot}43\pm0{\cdot}90$,
(iii) t_1-t_3 $0{\cdot}94\pm0{\cdot}86$.

TABLE 3.
Analysis of Variance.

	Deg. Freed.	Sum. Sq.	Mean Sq.
Total	67	286·0100	
Residual	43	100·8252	2·3448
Constants b and t	24	185·1848	
Blocks	16	173·2850	
Remainder (treatment)	8	11·8998	1·4875

2. Missing Treatment.

In dealing with the case of a missing treatment it is convenient to consider the designs in which $\lambda=1$ separately from those in which $\lambda>1$.

(A) Designs in which $\lambda=1$.

Without loss of generality it may be assumed that treatment 1 is missing and that its first replicate occurs in block 1 together with treatments $2, 3, \ldots, k$ while its remaining replicates occur in blocks $2, 3, \ldots, r$.

The normal equations are

$$\begin{aligned}(N-r)m-\mathrm{S}(b_u) &= G, \qquad u=1, 2, \ldots, r.\\ (k-1)(b_u+m)+\mathrm{S}(t) &= B_u,\\ k(b_v+m)+\mathrm{S}(t) &= B_v, \qquad v=r+1, \ldots, b.\\ r(t_w+m)+\mathrm{S}(b) &= T_w, \qquad w=2, 3, \ldots, t.\end{aligned}$$

The equations in the t of the first block containing a replicate of the missing treatment reduce to

$$\begin{aligned}tt_2+b_1+m&=Q'_2,\\ tt_3+b_1+m&=Q'_3,\\ &\ldots\ldots\\ tt_k+b_1+m&=Q'_k.\end{aligned}$$

There are, in all, r such sets of equations involving r sets of the t, no two of which have a common t and, further, each set involves one of the first r values of the b.

The b_u+m are given by

$$b_u+m=\{tB_u-(Q'_i+Q'_j+\ldots)\}/(k-1)(t-1)$$

where Q'_i, Q'_j, ... are the Q' of treatments in block u, so that the values of the t are given by

$$t_w=\frac{1}{t}Q'_w-\left\{B_u-\frac{1}{t}(Q'_i+Q'_j+\ \ldots\ +Q'_w)\right\}/(k-1)(t-1)$$

assuming that treatment w also occurs in block u. The remaining $b+m$ can now be obtained by direct substitution of the t in the original equations of the blocks concerned and m is obtained by substitution in the m equation. The remaining steps proceed as in section 1.

There are two types of comparison:

(i) between two treatments which occur together in one of the r blocks containing the missing treatment: variance $\frac{2k}{t}\sigma^2$,

(ii) between two treatments which occur in different members of the set of r blocks containing the missing treatment: variance

$$\frac{2k}{t}\sigma^2\left\{1+\frac{1}{(k-1)(t-1)}\right\}.$$

It should be noted that the variance of the first type retains its normal value.

Any design in which $\lambda=1$, $b=t$ and $k=r$ may also be analysed by the method given below in section 2 (B).

(B) *Designs in which* $\lambda>1$.

(a) Designs in which $b=t$ and $k=r$.

These designs are symmetrical with respect to both the blocks and the treatments. Consider first the case $r<\frac{1}{2}t$.

The normal equations are

$$\begin{aligned}
(N-r)m-S(b_u) &= G, & u&=1, 2, \ldots, r. \\
(r-1)(b_u+m)+S(t) &= B_u, & & \\
r(b_v+m)+S(t) &= B_v, & v&=r+1, \ldots, t. \\
r(t_w+m)+S(b) &= T_w, & w&=2, 3, \ldots, t.
\end{aligned}$$

The quantities Q or Q' are now calculated for the blocks instead of the treatments, the Q of any block being obtained by deducting from r times the block total, the totals of treatments occurring in the block. The b of blocks not containing the missing treatment are then given by

$$b_v=Q_v/r^2E$$

and the b of blocks containing the missing treatment by

$$b_u=AQ'_u+B(Q'_1+Q'_2+\ldots+Q'_r),$$

where A and B have their former values and Q'_u is omitted from the bracket on the right hand side. The values of the $(t+m)$ and m are then obtained by substitution.

When $\lambda=2$, there are, in general, two types of treatment comparison similar to those given for the case $\lambda=1$, with variances

$$\frac{r\sigma^2}{t}\left\{1+\frac{1}{r(2t-r)}\right\} \text{ and } \frac{r\sigma^2}{t}\left\{1+\frac{2}{r(2t-r)}\right\},$$

respectively. The first type only is possible when $t=4$ and $r=3$.

If $r>\frac{1}{2}t$, the values of the constants are best obtained by the method of section 1 (b), solving first for the block constants and subsequently determining the treatment constants by substitution. The following example illustrates the method.

TABLE 4.—*Square Roots of Scores.*

Block 1	2	3	4	5	6	7
(2) 0·0	(7) 2·4	(6) 2·4	(2) 3·6	(7) 4·7	(2) 5·1	(5) 7·5
(7) 2·2	(4) 2·6	(3) 2·4	(1) —	(6) 5·3	(4) 4·7	(1) —
(6) 3·3	(1) —	(4) 3·0	(6) 5·1	(5) 5·5	(5) 7·6	(7) 2·6
(3) 5·0	(2) 1·4	(1) —	(5) 6·5	(4) 6·6	(3) 9·6	(3) 4·4

Treatments are indicated by the numbers in parentheses; treatment 1 is missing.

TABLE 5.

Block	1	2	3	4	5	6	7	
B	10·5	6·4	7·8	15·2	22·1	27·0	14·5	103·5
rB	42·0	25·6	31·2	60·8	88·4	108·0	58·0	
Q'	86·0	90·2	80·3	111·0	119·9	136·0	101·1	724·5
b	−1·2500	−1·1057	−2·0957	·9743	1·1714	2·3214	−0·0157	

Treatment	2	3	4	5	6	7
T	10·1	21·4	16·9	27·1	16·1	11·9
$t+m$	2·2900	5·6100	4·1521	5·6621	4·3250	3·2750

In Table 4 are given data which have been taken from [6]. In this design $b=t=7$, $r=k=4$, $\lambda=2$ and $E=7/8$.

Table 5 summarizes the data required to calculate the constants and complete the analysis. The first step is to calculate the quantities Q' corresponding to the blocks. For example,

$$Q'_1=4B_1+T_4+T_5=42\cdot0+16\cdot9+27\cdot1=86\cdot0.$$

Next, the b of blocks not containing the missing treatment are evaluated; thus

$$b_1=(Q'_1-G)/r^2E=(86\cdot0-103\cdot5)/14=-1\cdot2500$$

and, using these, the value of $24m$ is

$$103\cdot5+1\cdot2500-1\cdot1714-2\cdot3214=101\cdot2572.$$

The remaining b may now be determined. For example,

$$b_2=\{Q'_2-r(t-1)m\}/r(rE-1)=(90\cdot2-101\cdot2572)/10=-1\cdot1057.$$

Finally, the $t+m$ are obtained by substituting the numerical values of the b in the t equations, thus

$$t_2+m=\frac{1}{4}(T_2+b_3+b_5+b_7)$$

$$=\frac{1}{4}(10\cdot1-2\cdot0957+1\cdot1714-0\cdot0157)$$

$$=2\cdot2900.$$

The residual sum of squares is 19·3264 with 12 degrees of freedom. Examples of the two types of treatment comparison with their estimated standard deviations are

(i) t_4-t_2 $\quad 1\cdot86\pm0\cdot97$,

(ii) t_3-t_2 $\quad 3\cdot32\pm0\cdot98$.

The generalized form of the variance of any treatment comparison is

$$\left\{\frac{2}{r}+\frac{2\alpha}{r^3E}\left(1+\frac{1}{rE-1}\right)+\frac{2\beta}{r^3E}\right\}\sigma^2$$

where $\alpha=\lambda-x$, $\beta=r-2\lambda+x$ and x is the number of times that pair of treatments occur together in the r blocks containing the missing treatment. For example, in the designs ($\lambda=3$) given by Fisher and Yates,[3] the numerical values of the factor within the brackets are

x	Design (4.5.5.4)	(6.11.11.6)	(7.15.15.7)
1	—	328/891	268/855
2	6/11	326/891	—
3	—	—	14/45

(*b*) Designs in which $b\neq t$, $k\neq r$.

Owing to the asymmetrical relations of the blocks within the treatments, the solution of the normal equations is not as straightforward as in the above cases. The equations in the t are, however, easily reduced to the form

$$krEt_i+S(b+m)=Q'_i,$$

where the summation is taken over the blocks in which treatment i occurs with the missing treatment. If the members of this set of equations are multiplied by $(k-1)$ they can easily be expressed in terms of the t. After reduction to this form, the solution is best obtained in the manner given in [8]. The case, $k>\frac{1}{2}t$, is probably best dealt with in the same manner.

If exact values of the variances of treatment comparisons are required for special purposes they can be obtained by the method given in [8]. However, this procedure involves lengthy computations which will, in many cases, not be worth while. The exact variances given above may be used as a guide to the magnitude of the increases to be expected.

3. Youden Square.

This modification of the incomplete block design has been described in [9].

(*a*) *Missing block.*

In addition to the conditions stated in section 1 (*a*) it may be assumed that row 1 contains treatment 1, etc., so that, when $r<\frac{1}{2}t$, the normal equations are

$$\begin{aligned}
&r(t-1)m-S(t_v) &&=G, && v=1, 2, \ldots, r,\\
&r(b_u+m)+S(t) &&=B_u, && u=2, 3, \ldots, t,\\
&(r-1)(t_v+m)+S(b)-r_v &&=T_v, &&\\
&r(t_w+m)+S(b) &&=T_w, && w=r+1, \ldots, t,\\
&(t-1)(r_v+m)-t_v &&=R_v, &&
\end{aligned}$$

where the r_v, $S(r)=0$, are the constants representing the rows and R_v stands for the total of existing values in the vth row. The solution of the equations is obtained by the method of section 1 (a) except for the determination of the r_v which are given by

$$r_v = \frac{rE-1}{rt(r-2)}\left[rR_v - G + \frac{1}{r(rE-1)}\{(r-1)Q'_v - (Q'_1+Q'_2+\ldots+Q'_r)\}\right],$$

omitting Q'_v from the second term in { }, and the t_v which are given by

$$t_v = AQ'_v + B(Q'_1+Q'_2+\ldots+Q'_r) + r(A-B)r_v,$$

where A and B have their former values.

If $r > \frac{1}{2}t$, the t_v are given by

$$t_v = \{(t-1)Q'_v + rR_v - rt(t-1)m\}/rt(r-2)$$

and the $r_v + m$ by

$$r_v + m = (R_v + t_v)/t - 1,$$

otherwise, the analysis proceeds in the manner given in section 1 (b).

(b) *Missing treatment.*

In these cases the solution of the normal equations is obtained by substituting the blocks for the treatments and following the method of section 3 (a).

In the Youden square the variance of any treatment comparison will be slightly greater than the variance of the corresponding comparison in incomplete blocks. There are, however, two exceptional cases. In both of these the variances have the values given above.

(i) Missing block; comparisons of type (i).

(ii) Missing treatment; comparisons between treatments which occur together λ times in the r blocks affected by the missing treatment. In this case $\alpha = 0$.

References.

[1] Cornish, E. A. Ann. Eugen. (In press.)
[2] ——— Ann. Eugen. (In press.)
[3] Fisher, R. A., and Yates, F. Statistical Tables for Biological, Agricultural and Medical Research. Oliver and Boyd, Edinburgh, 1938.
[4] Yates, F. J. Agric. Sc., 23: 108-145, 1933.
[5] ——— J. Agric. Sc., 26: 424-455, 1936.
[6] ——— Ann. Eugen, 7: 121-140, 1936.
[7] ——— Ann. Eugen., 7: 319-332, 1937.
[8] Yates, F., and Hale, R. W. Supp. J. Roy. Stat. Soc., 6: 67-79, 1939.
[9] Youden, W. J. Contrib. Boyce Thompson Inst., 9: 41-48, 1937.

AUSTRALASIAN MEDICAL PUBLISHING COMPANY LIMITED

Reprinted from Annals of Eugenics, Vol. 10, Part 1, pp. 112-118, 1940.

THE ESTIMATION OF MISSING VALUES IN INCOMPLETE RANDOMIZED BLOCK EXPERIMENTS

By E. A. CORNISH
Waite Agricultural Research Institute, South Australia

DURING recent years Yates has developed the series of experimental arrangements known as quasi-factorial designs. These designs are likely to be of great utility in all experimental work in which large numbers of treatments have to be compared and in which the experimental material contains only limited numbers of units in naturally occurring, homogeneous groups. A prominent member of this series is the design called incomplete randomized blocks (Yates, 1936).

The balanced structure of this experimental design confers on it the great practical advantage of a simple analysis, but if, through some unforeseen circumstance, one or more of the observations is rendered unreliable or non-existent, the symmetry is destroyed. The introduction of asymmetry into a design in which confounding is already extensively employed will tend, in some circumstances, to increase the complexity of a general analysis for non-orthogonal data made by the methods given in Yates (1933*a*). In the case of the ordinary randomized block or Latin square, it was found (Allan & Wishart, 1930; Yates, 1933*b*) that the most satisfactory method of dealing with the problem presented by the occurrence of missing values was to estimate these values and analyse the completed set of observations. The equations of estimation were obtained by minimizing the residual variance after substituting unknowns for the missing values (Yates, 1933*b*). The same method is applicable in the case of balanced incomplete blocks. It will be found that the formulae are very little more complicated than those appropriate to the ordinary randomized block or Latin square.

Throughout this paper the notation first employed in Yates (1936) will be used. It is convenient, however, to introduce slight modifications in some of the symbols:

Value of a missing observation	x
Total of existing values in the block containing x	B_x
Total of existing replicates of the treatment containing x	T_x
Total of blocks containing this treatment other than the block containing x	S_x
Total of existing values	G_x

Then $$Q_x = k(T_x + x) - (B_x + x) - S_x.$$

If a replicate of treatment i occurs in the same block as x, then

$$Q_i = kT_i - (B_x + x) - S_i.$$

Single missing value

(a) Incomplete block

When only one value is missing the function to be minimized is

$$x^2-\frac{1}{k}(B_x+x)^2-\frac{1}{k^2rE}(Q_x^2+Q_i^2+Q_j^2+\dots),$$

where Q_i, Q_j, ... are the Q's of the remaining $(k-1)$ treatments in the same block as the missing value. This expression is a minimum for variations in x, when

$$x=\frac{1}{N-b-t+1}\left[bB_x+\frac{t-1}{k(k-1)}\{k(k-1)T_x-(k-1)S_x-k(T_i+T_j+\dots)+S_i+S_j+\dots\}\right].$$

(b) Youden square

In the Youden (1937) square, $b=t$ and $k=r$. If R_x stands for the total of existing values in the row containing x, the formula for a single missing value is

$$x=\frac{1}{(b-1)(r-2)}$$
$$\times\left[bB_x+rR_x-G_x+\frac{(b-1)}{r(r-1)}\{r(r-1)T_x-(r-1)S_x-r(T_i+T_j+\dots)+S_i+S_j+\dots\}\right].$$

Several missing values

When more than one value is missing there will be two or more unknowns and the process of minimizing the residual sum of squares will yield as many equations as there are unknowns, analogous to the simultaneous equations of partial regression. The form assumed by these equations, however, depends on the structural relationships between the treatments and blocks from which the values are missing.

Consider the case of the ordinary incomplete block and denote the missing observations by x, y, ... If the first two values, x and y, belong to different blocks but are both replicates of the same treatment, and assuming these classes involve no other missing value, the quadratic function, in so far as it involves x and y, is of the form

$$x^2+y^2+\dots-\frac{1}{k}\{(B_x+x)^2+(B_y+y)^2+\dots\}-\frac{1}{k^2rE}\{Q_{xy}^2+Q_i^2+Q_j^2+\dots+Q_l^2+Q_m^2+\dots\},$$

where Q_i, Q_j, ... are the Q functions of the $(k-1)$ remaining treatments in the block containing x and Q_l, Q_m, ... are similar functions for the treatments in the block containing y. The first two equations in this case are

$$Ax-By+\dots=bB_x+\frac{(t-1)}{k(k-1)}$$
$$\times\{k(k-1)T_{xy}-(k-1)S_{xy}-(k-1)B_y-k(T_i+T_j+\dots)+S_i+S_j+\dots\},$$

$$-Bx+Ay+\dots=bB_y+\frac{(t-1)}{k(k-1)}$$
$$\times\{k(k-1)T_{xy}-(k-1)S_{xy}-(k-1)B_x-k(T_l+T_m+\dots)+S_l+S_m+\dots\},$$

The formulae of the text have been generalised to cover the case where the whole experiment has been replicated **n** times. General formulae, more convenient for use when the number of units per block exceeds half the number of treatments, have also been added.

Incomplete block

$k < \frac{1}{2}t$

$$x = \frac{1}{n(N-1)-t+1}\left[n b B_x - \frac{t-1}{k(k-1)}\left\{(k-1)Q'_x - Q'_i - Q'_j - \cdots\right\}\right]$$

$$Q'_x = kT_x - S_x$$

$$Q'_i = kT_i - S_i$$

$$Q'_j = kT_j - S_j \text{ and so on.}$$

$k > \frac{1}{2}t$

$$x = \frac{1}{n(N-1)-t+1}\left[n b B_x + \frac{t-1}{k(k-1)}\left\{(k-1)Q'_x - S(Q)\right\}\right]$$

$$Q'_x = kT_x - S_x$$

and S(Q) is the sum of the Q functions of the remaining treatments in the block containing the missing value.

Youden Square

$r < \frac{1}{2}t$

$$x = \frac{1}{(t-1)\{n(r-1)-1\}}\left[n(tB_x + rR_x - G_x) + \frac{t-1}{r(r-1)}\left\{(r-1)Q'_x - Q'_i - Q'_j - \cdots\right\}\right]$$

$r > \frac{1}{2}t$

$$x = \frac{1}{(t-1)\{n(r-1)-1\}}\left[n(tB_x + rR_x - G_x) + \frac{t-1}{r(r-1)}\left\{(r-1)Q'_x - S(Q)\right\}\right]$$

In the general formulae for the Youden Square G_x stands for the total of existing values of the particular square containing x.

where $$A = (N-b-t+1)$$

and $$B = \frac{1}{k}(k-1)(t-1).$$

The labour of deriving these equations and solving them can be avoided by repeated application of the formula for a single missing value, substituting approximations for the remaining missing values (Yates, 1933*b*). If care is taken in selecting the approximations the solutions converge rapidly.

A special case is worth noting. If the second missing value is neither a member of the same block, nor occurs in a block containing a replicate of any of the k treatments in the block containing the first missing value, then each equation involves only its leading unknown. Each equation thus degenerates to the formula for a single missing value, which can be solved directly. If a third value is missing and account is taken of the first two values, each equation will involve only its appropriate unknown; similarly for the fourth and so on. As an example, consider the arrangement given in Table 1.

Table 1

($t=9$, $b=12$, $k=3$, $r=4$ and $\lambda=1$)

Block											
1	2	3	4	5	6	7	8	9	10	11	12
$a(x)$	a	a	a	b	b	b	c	c	c	$d(y)$	$g(z)$
b	f	d	e	d	e	f	d	e	f	e	h
c	h	g	i	i	h	g	h	g	i	f	i

Suppose the missing values are those corresponding to treatments a, d and g in blocks 1, 11 and 12 respectively (they might equally have belonged to the remaining treatments of these blocks). Clearly, Q_a does not involve the totals of blocks 11 and 12, and therefore the equation with x as leading term involves neither y nor z; similarly for Q_d and Q_g. Hence, in this particular case, as many as three missing values can be estimated directly.

Example 1. Fisher & Yates (1938) have given an example showing the analysis of an incomplete block experiment involving nine treatments in eighteen blocks of four units each. Table 2, quoted from their book, gives the scores; x, y and z stand for the observations assumed missing and $a, b, \ldots, i$ for the treatments.

Table 3 gives the additional data required for estimating the missing values.

Using the direct method of solution, the quadratic function to be minimized is

$$x^2+y^2+z^2-\tfrac{1}{4}\{(12{\cdot}8+x)^2+(27{\cdot}3+y)^2+(14{\cdot}8+z)^2\}$$
$$-\frac{1}{108}[\{4(36{\cdot}9+x)-(12{\cdot}8+x)-(27{\cdot}3+y)-99{\cdot}4\}^2+\ldots$$
$$+\{4(31{\cdot}9+z)-(14{\cdot}8+z)-(12{\cdot}8+x)-115{\cdot}2\}^2].$$

Table 2

($t=9$, $b=18$, $k=4$, $r=8$, $\lambda=3$ and $E=27/32$)

f 2·6	f 5·9	a x	i 2·4	i 5·0	d 10·1	b 3·9	h 4·0	b 2·8
d 9·7	g 2·6	f 4·6	d 4·0	h 7·4	a 9·7	d 4·1	f 6·1	f 2·6
c 5·4	i z	i 4·9	g 3·0	e 10·3	f y	e 6·4	g 4·4	e 2·8
e 6·9	b 6·3	c 3·3	f 2·4	c 9·4	h 7·5	i 6·3	c 3·3	h 3·3
24·6	$14·8+z$	$12·8+x$	11·8	32·1	$27·3+y$	20·7	17·8	11·5
b 5·7	b 4·7	a 3·0	c 7·5	c 3·7	i 3·0	b 4·5	g 2·6	b 7·3
h 9·3	g 6·6	h 1·4	g 2·2	a 5·2	g 2·6	d 6·0	e 4·9	e 5·4
c 5·4	a 5·5	i 4·2	e 2·6	d 2·4	e 4·7	g 4·6	d 6·0	f 5·7
i 6·1	h 5·3	d 2·8	a 4·4	b 2·4	a 2·4	c 3·3	h 4·6	a 4·4
26·5	22·1	11·4	16·7	13·7	12·7	18·4	18·1	22·8

Table 3. *Treatment*

	a	b	c	d	f	g	h	i
T	$36·9+x$	39·1	41·3	43·6	$29·9+y$	28·6	42·8	$31·9+z$
S	99·4	135·7	149·8	118·7	88·5	117·6	139·5	115·2

Differentiating with respect to x, y and z in turn and simplifying, the following set of simultaneous equations is obtained:

$$69x+6y+2z = 406·3,$$

$$6x+69y+3z = 624·8,$$

$$2x+3y+69z = 389·9,$$

of which the solution is $\quad x = 5·0, \quad y = 8·4, \quad z = 5·1.$

Now consider the estimation of these values by successive approximation. In this example there are no differences due to treatment (the "treatments" were a dummy set superimposed on the data), but there are marked differences between the blocks. If the means of block 3, (4·3), and block 2, (4·9), are taken as the values of x and z respectively, the first approximation for y is

$$\frac{18\times 27·3}{46}+\frac{8}{4\times 3\times 46}[12\times 29·9-3\times 125·3-4(43·6+41·2+42·8)$$
$$+118·7+116·5+139·5] = 8·5.$$

Substituting 8·5 for y and retaining z as 4·9, x is found to be 5·0, and finally using $y = 8·5$, $x = 5·0$, z is found to be 5·1. The second approximations are $x = 5·0$, $y = 8·4$ and $z = 5·1$.

The new set of Q's is

	a	b	c	d	e	f	g	h	i	Total
Q	14·7	0·8	−2·4	20·0	7·6	−8·7	−23·1	−4·0	−4·9	0

from which a new series of adjusted intra-block estimates of the treatment means may be derived.

Example 2. Table 4 gives a set of estimates of the percentage of digestible fibre in a series of dietary treatments fed to sheep (A. E. Scott, unpublished data). The design consists of nine treatments denoted by $a, b, \ldots, i$, in twelve blocks of three units each; x, y and z stand for the values assumed missing.

Table 4

($r=4$, $b=12$, $t=9$, $k=3$, $\lambda=1$ and $E=3/4$)

g z h 63·1 i 65·3	d y e 63·9 f 67·8	a x b 66·2 c 71·1	a 61·0 f 67·0 h 62·9	b 66·0 d 61·1 i 64·9	c 69·9 d 62·0 h 62·6
128·4 + z	131·7 + y	137·3 + x	190·9	192·0	194·5
b 65·4 e 64·8 h 63·9	c 69·2 e 64·3 g 60·2	b 66·6 f 67·2 g 61·0	a 56·9 d 61·9 g 61·9	c 69·1 f 67·3 i 65·8	a 58·8 e 63·4 i 65·1
194·1	193·7	194·8	180·7	202·2	187·3

Table 5 gives the additional data required for the estimation of the missing values.

Table 5. *Treatment*

	a	b	c	d	e	f	g	h	i
T	176·7 + x	264·2	279·3	185·0 + y	256·4	269·3	183·1 + z	252·5	261·1
S	558·9	580·9	590·4	567·2	575·1	587·9	569·2	579·5	581·5

Applying the formula for a single missing value,

$$x = \frac{12 \times 137{\cdot}3}{16} + \frac{8}{3 \times 2 \times 16}[6 \times 176{\cdot}7 - 2 \times 558{\cdot}9 - 3(264{\cdot}2 + 279{\cdot}3) + 580{\cdot}9 + 590{\cdot}4] = 59{\cdot}9;$$

similarly $y = 62{\cdot}2$ and $z = 61{\cdot}3$.

Tests of significance

The steps in the completion of the analysis and the test of significance of the treatment mean square proceed in exactly the same manner as described by Yates (1933*b*). The calculations needed to complete the analysis of the data of example 1, above, are given in Tables 6 and 7.

Table 6. *Analysis of completed values* (*test for treatment*)

	Degrees of freedom	Sum squares	Mean squares
Total	68	324·9787	
Residual	43	112·5618	2·6177
Block + treatment constants	25	212·4169	
Blocks	17	200·1062	11·7710
Treatment	8	12·3107	1·5388

Table 7. *Analysis of existing values (test for treatment)*

	Degrees of freedom	Sum squares	Mean squares
Total	68	312·6333	
Residual	43	112·5618	2·6177
Block + treatment constants	25	200·0715	
Blocks	17	188·5525	
Remainder	8	11·5190	1·4399

Note the reduction (0·7917) in the sum of squares for testing treatments. It can be shown that the correct mean square for testing treatments is always less than the treatment mean square in the analysis of the completed values. Yates (1933*b*) has given a formula for evaluating the corresponding reduction in the case of the ordinary randomized block. A similar formula is applicable in the present case. Denote the estimated yield of a missing value in a block containing only one such value by a and the block total including this value by V_a. If the corresponding quantities in a block containing two missing values are b_1, b_2 and V_b, and so on, the formula is

$$\frac{1}{k(k-1)}S_1(V_a-ka)^2+\frac{1}{2k(k-2)}S_2\{2V_b-k(b_1+b_2)\}^2+\tfrac{1}{2}S_2(b_1-b_2)^2+\dots.$$

S_1, S_2, ... denote summation over all blocks containing one, two or more missing values.

In the majority of cases a test of significance for blocks is not necessary. If required, it is made in a similar manner to the test for treatment. In this case, also, the correct mean square is always less than the block mean square in the analysis of the completed set of observations. The necessary reduction is calculable by a formula similar to that above. V will now stand for a treatment total and r is substituted for k. In the example given, this reduction is 15·2368, giving a corrected mean square of 10·3632.

In the Youden square with a single missing value the reduction in the sum of squares of treatments is given by the expression

$$\frac{1}{br(b-1)(r-1)}(bB_a+rR_a-G_a-bra)^2,$$

where a is the estimate of the missing value, B_a and R_a are the totals of the corresponding block and row including the estimate, and G_a is the grand total including the estimate.

The exact analysis of the Youden square with more than one missing value is made in a similar manner to the analysis given by Yates (1933*b*) for the ordinary Latin square.

Errors of treatment differences

The variance of the difference between treatments whose Q's do not involve estimates of the missing values is $2s^2/rE$. If exact values of the variances of other treatment comparisons are required they can be obtained by the method given in Fisher (1936, § ~~19.1~~ 29).

Slight modifications have to be introduced owing to the presence of redundant constants (Yates & Hale, 1939). This procedure involves lengthy computations which will, in general, not be worth while. If only a few values are missing and the exact test of significance is made in the analysis of variance, the use of the formula $2s^2/rE$ will not result in serious error.

REFERENCES

F. E. Allan & J. Wishart (1930). "A method of estimating the yield of a missing plot in field experimental work." *J. Agric. Sci.* **20**, 399–406.

R. A. Fisher (1936). *Statistical Methods for Research Workers*, 6th ed. Edinburgh: Oliver and Boyd.

R. A. Fisher & F. Yates (1938). *Statistical Tables for Biological, Agricultural and Medical Research.* Edinburgh: Oliver and Boyd.

F. Yates (1933*a*). "The principles of orthogonality and confounding in replicated experiments." *J. Agric. Sci.* **23**, 108–45.

—— (1933*b*). "The analysis of replicated experiments when the field results are incomplete." *Emp. J. Exp. Agric.* **1**, 129–42.

—— (1936). "Incomplete randomized blocks." *Ann. Eugen., Lond.*, **7**, 121–40.

F. Yates & R. W. Hale (1939). "The analysis of Latin squares when two or more rows, columns or treatments are missing." *Supp. J.R. Stat. Soc.* **6**, 67–79.

W. J. Youden (1937). "Use of incomplete block replications in estimating tobacco mosaic virus." *Contr. Boyce Thompson Inst.* **9**, 41–8.

REPRINTED FROM
ANNALS OF EUGENICS, VOL. 10, PART 2, pp. 137–143, 1940

THE ESTIMATION OF MISSING VALUES IN QUASI-FACTORIAL DESIGNS

BY E. A. CORNISH
Waite Agricultural Research Institute

IN recent years Yates (1937 *a*, *b*, 1939) has developed the quasi-factorial designs for the special purpose of testing large numbers of treatments, e.g. varieties of a crop in agricultural trials. These designs are of three types, namely, quasi-factorial designs in randomized blocks, balanced incomplete blocks and quasi-Latin squares, each of which is characterized by a special arrangement of the treatments within the groups of units (blocks) into which the experimental material is divided. The several types of symmetrical arrangement are such that they facilitate the elimination of group differences from the treatment comparisons and, at the same time, easily lend themselves to statistical analysis.

In cases where some of the observations are non-existent the appropriate analytical procedure can no longer be employed. Of the various methods available for dealing with such data the most convenient is to determine estimates of the missing observations which minimize the intra-block variance (Yates, 1933). These estimates are used to form a complete set of values which may be analysed in the manner appropriate to the particular design, and from which efficient and unbiased intra-block estimates of the treatment effects may be derived.

In the original descriptions of these designs it was pointed out that loss of efficiency, as compared with designs in ordinary randomized blocks, would result if the subdivision into small groups did not effect a reduction in the intra-block error. Recently, it has been shown (Yates, 1939) that information on the treatments, contained in the inter-block comparisons, can be recovered if the relative accuracy of the inter- and intra-block comparisons is well determined. In certain cases where observations are missing, it might be considered desirable to recover the inter-block information. The formulae given below provide only intra-block estimates of the missing values and, as such, make no use of the inter-block information. The determination of estimates which will enable inter-block information to be recovered presents a different problem which is now under consideration.

TWO- AND THREE-DIMENSIONAL QUASI-FACTORIAL DESIGNS IN RANDOMIZED BLOCKS

A. *Two-dimensional designs*

(i) *Two equal groups of sets. Square lattice.* The groups of sets are denoted by X and Y. Each treatment is designated by a pair of numbers uv such that u indicates the set to which it belongs in group X and v the set to which it belongs in group Y, where $u, v = 1, 2, \ldots, p$ and p^2 is the number of treatments. Table 1 explains the notation adopted for the group and marginal totals.

Table 1

X group

		v			Total $X_{u.}$
u	x_{11}	x_{12}	$\cdots$	x_{1p}	$X_{1.}$
	x_{21}	x_{22}	$\cdots$	x_{2p}	$X_{2.}$
	$\cdots$	$\cdots$	$\cdots$	$\cdots$	$\cdots$
	x_{p1}	x_{p2}	$\cdots$	x_{pp}	$X_{p.}$
Total $X_{.v}$	$X_{.1}$	$X_{.2}$	$\cdots$	$X_{.p}$	$X_{..}$

Y group

		v			Total $Y_{u.}$
u	y_{11}	y_{12}	$\cdots$	y_{1p}	$Y_{1.}$
	y_{21}	y_{22}	$\cdots$	y_{2p}	$Y_{2.}$
	$\cdots$	$\cdots$	$\cdots$	$\cdots$	$\cdots$
	y_{p1}	y_{p2}	$\cdots$	y_{pp}	$Y_{p.}$
Total $Y_{.v}$	$Y_{.1}$	$Y_{.2}$	$\cdots$	$Y_{.p}$	$Y_{..}$

In this type of arrangement the X and Y groups of sets are replicated an equal number of times each, say r. The total of the r replicates of treatment uv in group X is x_{uv}. The total of the $2r$ replicates is

$$x_{uv}+y_{uv} = T_{uv};$$

also

$$X_{u.}+Y_{u.} = T_{u.}, \text{ etc.}$$

Let a stand for the missing observation and B for the total of existing values in the block containing a. Let x'_{uv}, $X'_{u.}$, etc. stand for the several treatment totals involving the missing value. The estimate of a missing observation belonging to treatment uv in group X is then

$$a = \{2rpB+p^2T'_{uv}-p(X'_{u.}+Y_{u.})+p(X'_{.v}-Y_{.v})-(X'_{..}-Y_{..})\}/(p-1)(2rp-p-1).$$

The corresponding formula for a missing value in the Y group is obtained by a cyclic substitution of the symbols X and Y and a simultaneous interchange of the subscripts according to the scheme $u. \rightarrow .v \rightarrow u..$

(ii) *Three equal groups of sets. Triple lattice.* In this case the treatments are divided into three equal groups of sets. With an obvious extension of the previous notation, the formula for a single missing value belonging to treatment uvw in group X is

$$a = \{3rpB+p^2T'_{uvw}-p(X'_{u..}+Y_{u..}+Z_{u..})+\tfrac{1}{2}p(X'_{.v.}+Z_{.v.}-2Y_{.v.})$$
$$+\tfrac{1}{2}p(X'_{..w}+Y_{..w}-2Z_{..w})-\tfrac{1}{2}(2X'_{...}-Y_{...}-Z_{...})\}/(p-1)(3rp-p-1).$$

Formulae for the groups Y and Z are obtained by a cyclic substitution of the symbols X, Y and Z and an interchange of the subscripts according to the scheme $u.. \rightarrow .v. \rightarrow ..w \rightarrow u...$

(iii) $(p+1)$ *equal groups of sets. Balanced lattice.* This is a special case of the balanced incomplete block designs. Formulae for missing values have been given in my earlier paper (1940).

B. *Three-dimensional designs*

Three equal groups of sets. Cubic lattice. With the same notation* as used for the triple lattice, the formula for a single missing value belonging to treatment uvw in group X is

$$a = \{3rp^2B+p^3T'_{uvw}-\tfrac{1}{2}p(Y_{u..}+Z_{u..}-2X'_{u..})-\tfrac{1}{2}p(X'_{.v.}+Y_{.v.}-2Z_{.v.})$$
$$-\tfrac{1}{2}p(X'_{..w}+Z_{..w}-2Y_{..w})+\tfrac{1}{2}p^2(X'_{uv.}+Y_{uv.}-2Z_{uv.})+\tfrac{1}{2}p^2(X'_{u.w}+Z_{u.w}-2Y_{u.w})$$
$$-p^2(X'_{.vw}+Y_{.vw}+Z_{.vw})-\tfrac{1}{2}(2X'_{...}-Y_{...}-Z_{...})\}/(p-1)(3rp^2-p^2-p-1).$$

Formulae for the groups Y and Z are obtained by cyclic substitution of the symbols and interchange of the subscripts in the 3rd, 4th and 5th terms. The 6th term is constant (except

*This is not strictly true. If the designs are regarded as factorial and the factors are denoted by x, y, and z, then, according to the notation actually adopted in the cubic lattice, the factor x, for example, is unconfounded in group X but confounded in groups Y and Z, whereas, in the triple lattice the reverse is the case.

for a shift of the prime) in groups X and Y, and similarly, the 7th term is constant in groups X and Z. The term in $(Y_{.vw}+Z_{.vw}-2X_{.vw})$ is constant in groups Y and Z; it is then clear how the remaining terms change.

Example 1. Dawson (1939, Table 2) has given data obtained from a cubic lattice experiment ($p=3$, $r=1$) on corn varieties. Assume that the observation (582) corresponding to treatment 111 in group X is missing. The data required to estimate the missing value are given in Table 2.

Table 2

Varietal totals	Group		
	X	Y	Z
111	—	600	608
1..	3868	4530	4449
.1.	4218	4882	4738
..1	4104	4758	4669
11.	1082	1714	1660
1.1	1072	1690	1685
.11	1166	1788	1732
...	12920	13548	13330

$T'_{111}=1208$; $B=1166$.

Hence

$$a=\frac{1}{2\times14}\{27\times1166+27\times1208-\tfrac{3}{2}(4530+4449-2\times3868)-\tfrac{3}{2}(4218+4882-2\times4738)$$
$$-\tfrac{3}{2}(4104+4669-2\times4758)+\tfrac{9}{2}(1082+1714-2\times1660)$$
$$+\tfrac{9}{2}(1072+1685-2\cdot1690)-9(1166+1788+1732)$$
$$-\tfrac{1}{2}(2\times12920-13548-13330)\}=611.$$

Balanced set of quasi-Latin squares. Lattice squares

The number of treatments must be a perfect square, say p^2. The treatments are divided into $(p+1)$ orthogonal groups of p sets, each set containing p treatments. If p is odd, balance can be attained with a minimum of $\frac{1}{2}(p+1)$ replications, but if p is even, every group of sets must be included twice, giving $(p+1)$ squares. The restrictions on the values which p may take are given in the original description of this design.

Let a stand for the missing value. Let R_a, C_a and T_a be the totals of existing values in the row, column and square respectively in which a occurs. Let V_a be the total of existing replicates of the treatment containing a and S_a the total of rows and columns in which this treatment occurs other than the row and column involving a. Then

$$pQ_a=p(V_a+a)-(R_a+a)-(C_a+a)-S_a=Q'_a-(p-2)a-R_a-C_a,\text{ say,}$$

where

$$Q'_a=pV_a-S_a.$$

Similarly, if a replicate of treatment i occurs in the same row as a,

$$pQ_i=pV_i-(R_a+a)-S_i=Q'_i-(R_a+a),$$

and in the same column,

$$pQ_j=pV_j-(C_a+a)-S_j=Q'_j-(C_a+a).$$

If p is odd and the set of $\frac{1}{2}(p+1)$ squares is replicated r times, the estimate of the missing value is given by

$$a = \frac{r\{p(R_a+C_a)-T_a\}}{(p-1)\{r(p-1)-2\}} + \frac{2}{p(p-1)^2\{r(p-1)-2\}} \times [(p-1)^2 Q'_a - (p-1)\{S(Q'_i)+S(Q'_j)\} + S(pQ)],$$

where the last summation within the square bracket is taken over the remaining treatments. If p is even, replace $\{r(p-1)-2\}$ by $\{r(p-1)-1\}$ and omit the factor 2 from the multiplier of the second term.

Example 2. Yates (1937*a*) has given an example showing the analysis of a balanced set of lattice squares involving 25 varieties ($r=1$). Table 3 shows the yields.

Table 3. *Yields (less* 100)

Square I

a	15	−21	29	−1	22+a
6	4	−5	20	19	44
13	10	−2	22	16	59
10	−1	11	40	9	69
3	3	−13	13	−4	2
32+a	31	−30	124	39	196+a

Square II

−2	−7	2	−7	0	−14
27	12	22	37	29	127
25	20	13	8	13	79
35	19	28	28	19	129
16	15	9	26	16	82
101	59	74	92	77	403

Square III

19	19	29	29	32	128
−3	21	4	17	10	49
−2	38	32	38	37	143
3	18	21	15	19	76
8	31	48	52	30	169
25	127	134	151	128	565

The yield assumed missing is 112.

The squares have been arranged systematically to correspond with the sets of varieties given in (Yates, 1937*b*, p. 88). The value of $S(pQ)$ is -3750. Table 4 gives the remaining data needed to calculate the missing value.

Table 4. *Variety*

	1	2	3	4	5	6	11	16	21
V	17	48	24	72	45	33	36	82	77
S	240	393	400	572	495	403	448	527	492
$-Q'$	155	153	280	212	270	238	268	117	107

$S(Q'_i) = -915$; $S(Q'_j) = -730$.

Hence, $$a = \frac{1}{4\times 2}\{5(32+22)-196\} + \frac{2}{5\times 16\times 2}(-16\times 155 + 4\times 1645 - 3750) = 13\cdot 6.$$

SEVERAL MISSING VALUES

When several values are missing, the process of minimizing the residual sum of squares will yield a set of simultaneous equations containing as many equations as there are unknowns. The form assumed by these equations depends on the structural relationships between the treatments and blocks (or rows and columns) from which the values are missing. Estimates of the unknowns are given by the solution of these equations. In practice, it is more convenient to utilize a process of successive approximation to determine the values of the unknowns. This is accomplished by repeated application of the formula for a single missing value, substituting approximations for the remaining unknowns. Careful selection of the first approximation ensures rapid convergence of the solution.

TESTS OF SIGNIFICANCE

In all these quasi-factorial designs the steps in the completion of the analysis and the test of significance of the treatment mean square proceed in a similar manner to those described in Yates, 1933. In the analysis of the completed set of observations the treatment mean square is always exaggerated. The deduction required to determine the correct treatment sum of squares for quasi-factorial designs in randomized blocks may be calculated by the formula given by Yates (1933); p will now stand for the number of treatments per block and not the total number of treatments.

The reduction in the sum of squares of treatments in a set of lattice squares having a single missing value is given by the expression

$$\{p(R+C)-T-p^2a\}/p^2(p-1)^2,$$

where a is the estimate of the missing value and R, C and T are the totals of the corresponding row, column and square including the estimate. The formula applies for both odd and even integer values of p.

The calculations needed to complete the analysis of example 2, above, are given in Tables 5 and 6.

Table 5. *Analysis of completed set*

	Degrees of freedom	Sum of squares
Squares	2	2532·76
Rows	12	2691·62
Columns	12	7110·02
Varieties	24	1584·23
Residual	23	1380·44
Total	73	15299·07

Table 6. *Analysis of existing set*

	Degrees of freedom	Sum of squares
Total	73	15294·59
Residual (from Table 5)	23	1380·44
Square, row, column and variety constants	50	13914·15
Square, row and column constants	26	12381·48
Remainder (varieties)	24	1532·67

The reduction in the sum of squares due to fitting constants representing the mean, rows, columns and squares is 30690·89, while the ordinary correction for the mean is 18309·41. The difference, 12381·48, is the reduction in the sum of squares due to fitting constants for rows, columns and squares only. The correct sum of squares for varieties is 1532·67, which agrees with the value obtained by deducting the correction, calculated from the formula given above, from 1584·23. The same result could have been obtained by the procedure given below.

The exact analysis of a set of lattice squares with more than one missing value is more complicated because the reduction in the treatment sum of squares cannot be expressed as a simple formula. However, it is not necessary to determine, directly, the constants representing rows, columns and squares. An alternative and simpler procedure is to evaluate an auxiliary set of estimates of the missing observations by successive approximation, using the formula

$$a = \frac{p(R_a + C_a) - T_a}{(p-1)^2},$$

and disregarding, entirely, the treatment classification. The total variation of this new set is analysed into components for rows, columns, squares and residual. The deduction of this residual from the total sum of squares of existing values gives the sum of squares due to fitting constants for rows, columns and squares only.

Errors of treatment differences

In the lattice designs, the variances of treatment comparisons which do not involve estimates of missing observations retain the general values given in the original description. The general formulae for the corresponding variances in a set of lattice squares, replicated r times, are $\frac{4}{r(p-1)}$ (p odd) and $\frac{2}{r(p-1)}$ (p even) multiplied by the error variance of a single plot.

When the treatment comparisons involve estimates of missing values, considerable labour is required to determine the exact values of the variances. In the majority of cases the use of the normal values will suffice. Greater accuracy may be obtained by using the approximate method given by Yates (1936*a*).

As a rough guide to the magnitude of the increases to be expected, the exact values of the variances in the case of a square lattice may be cited. The variance of the difference between two adjusted treatment means, one of which contains a missing value is

for two treatments in the same set:

$$\frac{s^2}{r}\left(\frac{p+1}{p}\right)\left\{1+\frac{p(p+1)}{2(p-1)(2pr-p-1)}\right\},$$

and for two treatments not in the same set:

$$\frac{s^2}{r}\left(\frac{p+2}{p}\right)\left\{1+\frac{p^3}{2(p+2)(p-1)(2pr-p-1)}\right\},$$

where s^2 is the estimate of the intra-block variance.

REFERENCES

E. A. Cornish (1940). "The estimation of missing values in incomplete randomized block experiments." *Ann. Eugen., Lond.*, **10**, 112–118.

C. D. R. Dawson (1939). "An example of the quasi-factorial design applied to a corn breeding experiment." *Ann. Eugen., Lond.*, **9**, 157–73.

F. Yates (1933). "The analysis of replicated experiments when the field results are incomplete." *Emp. J. Exp. Agric.* **1**, 129–42.

—— (1936*a*). "A new method of arranging variety trials involving a large number of varieties." *J. Agric. Sci.* **26**, 424–55.

—— (1936*b*). "Incomplete randomized blocks." *Ann. Eugen., Lond.*, **7**, 121–40.

—— (1937*a*). "A further note on the arrangement of variety trials: Quasi-Latin squares." *Ann. Eugen., Lond.*, **7**, 319–32.

—— (1937*b*). *Design and analysis of factorial experiments.* Imp. Bur. Soil Sci. Harpenden.

—— (1939). "The recovery of inter-block information in variety trials arranged in three dimensional lattices." *Ann. Eugen., Lond.*, **9**, 136–56.

REPRINTED FROM
ANNALS OF EUGENICS, VOL. 10, PART 3, pp. 269–279, 1940

THE ANALYSIS OF COVARIANCE IN QUASI-FACTORIAL DESIGNS

By E. A. CORNISH
Waite Agricultural Research Institute

THE value and application of the analysis of covariance as a means of increasing the precision of well-designed experiments have been discussed at length by Fisher (1937, 1938) and others. In appropriate cases the same advantages may be gained with experiments arranged in the quasi-factorial designs recently developed by Yates (1936*a*, 1936*b*, 1937). It is the object of this paper to discuss the application of the analysis of covariance in the three types of these designs, namely, balanced incomplete blocks, quasi-Latin squares and quasi-factorial designs in randomized blocks. The cubic lattice has been chosen as a typical example of the unbalanced lattice designs in randomized blocks.

The results of a uniformity trial on sugar beet, given by Immer (1932), will serve to illustrate the methods of calculation. The apparent purity of the sugar extracted from the beet has been taken as the dependent variable, (y), and the sugar percentage as the independent variable, (x).

The data used in the three examples given below comprise the following blocks and rows from Immer's Tables VIII and X:

Balanced incomplete blocks		Lattice squares	Cubic lattice
Blocks		Blocks	Blocks
1–6	7	1–5	1–9
Rows 1–8	1–4	Rows 1–15	Rows 1–9

The notation introduced by Yates will be followed as closely as possible.

In all cases the analyses of variance for the dependent and independent variables are carried out in the manner given in the original descriptions.

BALANCED INCOMPLETE BLOCKS

With an obvious extension of Yates's notation the sums of products for blocks and treatments in the analysis of covariance are

$$\frac{1}{k} SB(y)\, B(x) - G(y)\, G(x)/N \quad \text{and} \quad \frac{1}{k^2 r E} SQ(y)\, Q(x),$$

respectively. The residual sum of products is obtained by subtraction from the total sum of products.

The test of significance of treatments in the analysis of the adjusted values of the dependent variable is made in the manner given by Fisher (1938, § 49·1).

If b stands for the regression of y on x, as determined from the entries in the line corresponding to the residual in the table of the analysis of variance and covariance, the treatment effects, adjusted to a constant value of the independent variable on the basis of the regression, are given by the differences of the quantities

$$t_i = \frac{1}{krE}\{Q_i(y) - bQ_i(x)\},$$

and the adjusted treatment means by

$$t_i + \bar{y},$$

where $\bar{y} = G(y)/N$.

If $k > \frac{1}{2}t$, the sum of products due to treatments is best obtained from the expression

$$\frac{1}{k^2rE}\{SQ'(y)\,Q'(x) - SQ'(y)\,SQ'(x)/t\}.$$

The substitution of the functions Q' for the functions Q necessitates a slight alteration in the method of determining the adjusted treatment effects, which are now given by the differences of the quantities

$$t_i = \frac{1}{krE}\left[Q_i'(y) - b\left\{Q_i'(x) - \frac{t-1}{k-1}\bar{x}\right\}\right], \quad \bar{x} = G(x)/N$$

and the adjusted treatment means by

$$t_i - \frac{t-k}{k-1}\bar{y}.$$

The estimated variance of the difference between any two adjusted treatment means is

$$s^2\left(\frac{2}{rE} + \frac{d^2}{R}\right),$$

where d is the difference between the corresponding means of the independent variable, s^2 the estimated residual variance in the analysis of the adjusted values and R the residual sum of squares in the analysis of variance of the independent variable (Wishart, 1936).

Example. Table 1 shows the dummy trial of thirteen treatments in thirteen blocks of four plots which has been superimposed on the chosen observations. The letters a, b, ... indicate the hypothetical treatments.

Table 2 gives, for both variables, the treatment totals, T, the totals of blocks containing each treatment, S, and the functions Q.

The sum of products due to blocks is

$$\tfrac{1}{4}(19{\cdot}0 \times 1{\cdot}71 + \ldots - 15{\cdot}1 \times 0{\cdot}71) - 211{\cdot}6 \times 6{\cdot}21/52 = 10{\cdot}355,$$

and the sum of products due to treatments is

$$(28{\cdot}4 \times 4{\cdot}32 + \ldots + 13{\cdot}1 \times 2{\cdot}50)/52 = 9{\cdot}397.$$

Table 3 shows the analysis of variance and covariance and Table 4 gives the test of significance of the reduced treatment mean square. The regression of apparent purity on sugar percentage is $9{\cdot}438/3{\cdot}8776 = 2{\cdot}434$.

Table 1. *Apparent purity % (less 80), y; sugar % (less 14), x*

	y	x		y	x		y	x		y	x		y	x		y	x		y	x
(*k*)	5·4	0·60	(*j*)	4·8	0·41	(*i*)	1·0	−0·15	(*a*)	4·8	0·24	(*d*)	3·6	0·22	(*h*)	1·9	−0·82	(*m*)	2·3	−0·34
(*c*)	4·6	0·22	(*g*)	4·8	0·58	(*j*)	6·1	0·29	(*f*)	2·0	−0·39	(*f*)	2·8	−0·34	(*b*)	7·6	0·20	(*l*)	3·4	−0·24
(*f*)	4·7	0·31	(*m*)	3·0	0·52	(*h*)	5·7	0·99	(*g*)	1·4	0·09	(*h*)	3·9	−0·24	(*e*)	6·0	0·80	(*k*)	4·2	−0·10
(*j*)	4·3	0·58	(*b*)	3·0	0·19	(*a*)	8·4	0·85	(*e*)	2·9	−0·32	(*m*)	1·5	−0·63	(*k*)	4·8	−0·10	(*a*)	5·2	−0·03
	19·0	1·71		15·6	1·70		21·2	1·98		11·1	−0·38		11·8	−0·99		20·3	0·08		15·1	−0·71
(*l*)	6·5	0·14	(*c*)	3·8	−0·34	(*a*)	4·2	0·40	(*i*)	4·4	0·17	(*l*)	2·3	−0·17	(*d*)	3·0	−0·80			
(*b*)	6·3	0·84	(*h*)	1·1	−0·13	(*c*)	5·8	0·41	(*c*)	0·8	−0·26	(*e*)	6·3	0·41	(*k*)	6·1	−0·31			
(*f*)	6·4	1·38	(*l*)	4·2	0·14	(*d*)	0·5	−0·56	(*m*)	4·0	−0·05	(*d*)	1·4	−0·65	(*i*)	4·6	−0·29			
(*i*)	4·7	0·82	(*g*)	2·4	0·09	(*b*)	4·1	0·38	(*e*)	4·6	0·64	(*j*)	5·1	0·29	(*g*)	4·9	0·27			
	23·9	3·18		11·5	−0·24		14·6	0·63		13·8	0·50		15·1	−0·12		18·6	−1·13			

Table 2. *Treatment*

	a	*b*	*c*	*d*	*e*	*f*	*g*	*h*	*i*	*j*	*k*	*l*	*m*	Total
$T(y)$	22·6	21·0	15·0	8·5	19·8	15·9	13·5	12·6	14·7	20·3	20·5	16·4	10·8	211·6
$S(y)$	62·0	74·4	58·9	60·1	60·3	65·8	56·8	64·8	77·5	70·9	73·0	65·6	56·3	846·4
$Q(y)$	28·4	9·6	1·1	−26·1	18·9	− 2·2	− 2·8	−14·4	−18·7	10·3	9·0	0·0	−13·1	0
$T(x)$	1·46	1·61	0·03	− 1·79	1·53	0·96	1·03	− 0·20	0·55	1·57	0·09	− 0·13	− 0·50	6·21
$S(x)$	1·52	5·59	2·60	− 1·61	0·08	3·52	− 0·05	0·83	4·53	5·27	− 0·05	2·11	0·50	24·84
$Q(x)$	4·32	0·85	− 2·48	− 5·55	6·04	0·32	4·17	− 1·63	− 2·33	1·01	0·41	− 2·63	− 2·50	0

Table 3. *Analysis of variance and covariance*

	D.F.	y^2	xy	x^2
Blocks	12	48·30	10·355	5·1278
Treatments	12	55·13	9·397	2·5529
Residual	27	63·24	9·438	3·8776
Total	51	166·67	29·190	11·5583

Table 4. *Test of significance with reduced variance*

	D.F.	y^2	xy	x^2	D.F.	Reduced y^2	Mean sq.
Treatments	12	55·13	9·397	2·5529	12	22·93	1·911
Residual	27	63·24	9·438	3·8776	26	40·27	1·549
Treatment + residual	39	118·37	18·835	6·4305	38	63·20	

The treatment means, before and after adjustment to constant sugar percentage, are as follows:

	a	b	c	d	e	f	g	h	i	j	k	l	m	
Uncorrected	86·25	84·81	84·15	82·06	85·52	83·90	83·85	82·96	82·63	84·86	84·76	84·07	83·06	84·07
Corrected	85·45	84·65	84·62	83·10	84·39	83·84	83·07	83·27	83·07	84·67	84·68	84·56	83·53	84·07

Thus, for example, the adjusted mean of treatment a is

$$\frac{1}{13}\{Q_a(y) - 2{\cdot}434 \times Q_a(x)\} + \bar{y} = \frac{1}{13}\{28{\cdot}4 - 2{\cdot}434 \times 4{\cdot}32\} + 84{\cdot}07 = 85{\cdot}45\ \%.$$

It is apparent that the effect of the adjustment has been to reduce the variation between the treatment means.

The estimated variance of the difference between the corrected means of treatments a and b, for example, is

$$1{\cdot}549\left\{\frac{8}{13} + \left(\frac{3{\cdot}47}{13}\right)^2 \frac{1}{3{\cdot}8776}\right\} = 0{\cdot}982,$$

giving a standard deviation of 0·991.

Occasionally, the analysis of covariance may be of value in designs arranged in Youden squares. The analysis is similar to that given above except for the additional calculations involved in determining the contributions of "rows" in the analysis of variance and covariance.

Quasi-latin squares: lattice squares

The sums of products in the analysis of covariance due to rows, columns and squares are calculated in the usual way and, if p is odd, the sum corresponding to treatments is given by

$$\frac{2}{p^2(p-1)}\{SpQ(y)\,pQ(x) - SpQ(y)\,SpQ(x)/p^2\}.$$

The residual sum of products is obtained by subtraction from the total.

Treatment effects, adjusted to a constant value of the independent variable, are given by the differences of the quantities

$$t_i = \frac{2}{p(p-1)} pQ_i(y) - b\left\{\frac{2}{p(p-1)} pQ_i(x) + \frac{p+1}{p-1}\bar{x}\right\},$$

and the adjusted treatment means by

$$t_i + \frac{2p}{p-1}\bar{y}.$$

The estimated variance of the difference between any two adjusted treatment means is

$$s^2\left(\frac{4}{p-1} + \frac{d^2}{R}\right),$$

where d, s^2 and R represent the same quantities as before.

If p is even, replace the numerators of the factors $2/\{p^2(p-1)\}$ and $2/\{p(p-1)\}$ by unity and the factor $4/(p-1)$ by $2/(p-1)$.

Example. Table 5 shows the arrangement of twenty-five treatments superimposed on the selected data, the numbers in the brackets indicating the treatments.

The quantities $5Q$, for both variables, are shown in Table 6.

The sum of products due to treatments is

$$\frac{1}{50}\{46{\cdot}0 \times 2{\cdot}53 - 35{\cdot}2 \times 2{\cdot}15 + \ldots + 74{\cdot}5 \times 3{\cdot}50 - 1398{\cdot}5 \times 61{\cdot}45/25\} = 12{\cdot}631.$$

Table 7 shows the analysis of variance and covariance and Table 8 the test of significance of the reduced treatment mean square.

The regression of apparent purity on sugar percentage is $29{\cdot}088/6{\cdot}9405 = 4{\cdot}191$.

The adjusted means are shown in Table 9. Thus, for example, the adjusted mean of treatment 1 is

$$\frac{2}{5\times 4} 5Q_1(y) + \frac{10}{4}\bar{y} - 4{\cdot}191\left\{\frac{2}{5\times 4} 5Q_1(x) + \frac{3}{2}\bar{x}\right\} = 84{\cdot}72 + 4{\cdot}191 \times 0{\cdot}0072 = 84{\cdot}75\ \%.$$

As before, the effect of the adjustment has been to reduce the variation among the treatment means.

The estimated variance of the difference between the corrected means of treatments 1 and 2, for example, is

$$1{\cdot}342\left(1 + \frac{0{\cdot}4680^2}{6{\cdot}9405}\right) = 1{\cdot}385,$$

giving a standard deviation of $1{\cdot}177$.

Unbalanced lattices in randomized blocks

The sums of products in the analysis of covariance due to blocks and groups are calculated in the usual manner. The sum of products due to treatments can be built up from its several components by isolating each of them from the particular region of the experiment in which it is unconfounded (*vide* Dawson, 1939); but, for machine calculation, it is best determined

Table 5. *Apparent purity % (less 80), y; sugar % (less 14), x*

	y	x		y	x		y	x		y	x		y	x	Totals	
(17)	5·4	0·60	(16)	4·8	0·41	(19)	1·0	−0·15	(20)	4·8	0·24	(18)	3·6	0·22	19·6	1·32
(7)	4·6	0·22	(6)	4·8	0·58	(9)	6·1	0·29	(10)	2·0	−0·39	(8)	2·8	−0·34	20·3	0·36
(22)	4·7	0·31	(21)	3·0	0·52	(24)	5·7	0·99	(25)	1·4	0·09	(23)	3·9	−0·24	18·7	1·67
(2)	4·3	0·58	(1)	3·0	0·19	(4)	8·4	0·85	(5)	2·9	−0·32	(3)	1·5	−0·63	20·1	0·67
(12)	6·5	0·14	(11)	3·8	−0·34	(14)	4·2	0·40	(15)	4·4	0·17	(13)	2·3	−0·17	21·2	0·20
Total	25·5	1·85		19·4	1·36		25·4	2·38		15·5	−0·21		14·1	−1·16	99·9	4·22
(12)	6·3	0·84	(20)	1·1	−0·13	(1)	5·8	0·41	(9)	0·8	−0·26	(23)	6·3	0·41	20·3	1·27
(6)	6·4	1·38	(14)	4·2	0·14	(25)	0·5	−0·56	(3)	4·0	−0·05	(17)	1·4	−0·65	16·5	0·26
(18)	4·7	0·82	(21)	2·4	0·09	(7)	4·1	0·38	(15)	4·6	0·64	(4)	5·1	0·29	20·9	2·22
(5)	9·7	1·87	(8)	−0·4	−0·63	(19)	0·4	−0·34	(22)	0·8	−0·34	(11)	2·2	−0·02	12·7	0·54
(24)	0·8	−0·26	(2)	2·7	−0·10	(13)	5·8	0·24	(16)	1·9	−0·24	(10)	2·9	−0·32	14·1	−0·68
Total	27·9	4·65		10·0	−0·63		16·6	0·13		12·1	−0·25		17·9	−0·29	84·5	3·61
(21)	1·7	−0·12	(3)	−2·3	−0·71	(19)	6·0	0·87	(10)	4·7	0·48	(12)	1·1	−0·29	11·2	0·23
(17)	5·6	0·46	(24)	4·2	0·31	(15)	3·9	0·17	(1)	6·5	0·02	(8)	3·3	0·16	23·5	1·12
(13)	4·2	−0·19	(20)	6·2	0·48	(6)	−0·1	−0·49	(22)	6·2	0·82	(4)	2·8	−0·51	19·3	0·11
(5)	5·4	0·77	(7)	4·4	0·60	(23)	5·4	0·34	(14)	3·0	−0·05	(16)	3·8	0·07	22·0	1·73
(9)	2·8	0·07	(11)	2·4	0·26	(2)	8·5	0·78	(18)	2·2	−0·19	(25)	3·4	0·35	19·3	1·27
Total	19·7	0·99		14·9	0·94		23·7	1·67		22·6	1·08		14·4	−0·22	95·3	4·46

Table 6. *Values of* $-5Q$

Treatments	y	x	y	x	y	x	y	x	y	x
1–5	46·0	2·53	35·2	−2·15	72·9	7·64	36·5	1·72	27·9	−3·23
6–10	71·6	1·06	54·7	1·23	66·5	4·06	68·6	5·52	53·6	1·64
11–15	63·4	4·52	51·0	4·53	43·5	0·19	60·7	2·57	52·4	−0·15
16–20	49·1	2·06	60·7	3·20	71·9	5·13	72·2	4·37	39·1	−0·15
21–25	64·4	3·39	52·4	1·05	38·7	2·34	71·0	4·88	74·5	3·50

Table 7. *Analysis of variance and covariance*

	D.F.	y^2	xy	x^2
Squares	2	5·00	0·233	0·0154
Rows	12	29·37	3·823	1·6583
Columns	12	76·21	19·155	6·0460
Treatments	24	91·99	12·631	2·9138
Residual	24	152·79	29·088	6·9405
Total	74	355·36	64·930	17·5740

Table 8. *Test of significance with reduced variance*

	D.F.	y^2	xy	x^2	D.F.	Reduced y^2	Mean sq.
Treatments	24	91·99	12·631	2·9138	24	37·28	1·553
Residual	24	152·79	29·088	6·9405	23	30·87	1·342
Treatments + residual	48	244·78	41·719	9·8543	47	68·15	

Table 9

Treatments	Adjusted treatment means				
1–5	84·75	83·87	84·20	85·36	84·15
6–10	81·58	83·34	83·34	83·75	83·62
11–15	83·85	85·09	84·02	83·30	82·99
16–21	84·25	83·56	83·25	82·90	84·32
21–25	83·27	83·49	85·40	83·24	82·31

by a formula analogous to the formula given by Yates for calculating the sum of squares due to treatments.

Consider first the two-dimensional lattice. Let the treatment means of the dependent variable, corrected for block differences only, be t_{uv}. Let T_{uv} be the total of all replicates of treatment uv for the independent variable and, similarly, let $X_{u.}$ and $Y_{.v}$ be subtotals of groups X and Y for fixed u and v, respectively. The sum of products due to treatments is then

$$St_{uv}T_{uv} - (St_{u.}X_{u.} + St_{.v}Y_{.v}).$$

With an obvious extension of the notation, the sum of products due to treatments in a two-dimensional lattice, divided into three groups of sets, is

$$St_{uvw}T_{uvw} - (St_{u..}X_{u..} + St_{.v.}Y_{.v.} + St_{..w}Z_{..w}).$$

In a three-dimensional lattice, in three groups of sets, the corresponding formula is

$$St_{uvw}T_{uvw} - (St_{.vw}X_{.vw} + St_{u.w}Y_{u.w} + St_{uv.}Z_{uv.}).$$

Alternatively, we may use the t to refer to the independent variable and the totals T, etc., to the dependent variable. The residual sum of products is obtained by subtraction from the total.

In order to adjust the treatment means of the dependent variable to a constant value of the independent variable the two series of means, corrected for block differences, are required. Let t and t' refer to the dependent and independent variables, respectively. The adjusted treatment means are then given by

$$t - b(t' - \bar{t'}),$$

where b is the regression coefficient.

The estimated variance assignable to the difference between two adjusted treatment means is

$$s^2\left(f + \frac{d^2}{R}\right),$$

where s^2, d and R have the same meanings as before and f is a function determined by the design and the particular comparison under examination (Yates, 1936*a*).

Example. The treatments will be referred to in the notation 111, 112, ..., 333, each set of three numbers indicating the position of the treatment in the basic cube. Table 10 shows the random arrangement of the treatments superimposed on the data.

Table 10. *Arrangement of treatments*

Group X	221	313	122	311	232	131	123	112	333
	121	213	322	111	332	231	223	312	233
	321	113	222	211	132	331	323	212	133
Group Y	321	133	221	232	223	132	332	121	323
	331	113	231	222	213	122	322	111	333
	311	123	211	212	233	112	312	131	313
Group Z	232	321	133	113	311	221	332	211	121
	233	322	131	111	313	222	331	212	122
	231	323	132	112	312	223	333	213	123

In Table 11 the observations have been arranged systematically to show the values from each of the groups X, Y and Z and also the totals of the three replicates, T.

Tables 12 and 13 show the remaining data needed to complete the analysis. The totals $X_{.vw}$, $Y_{u.w}$ and $Z_{uv.}$ of Table 13 are the block totals.

The antepenultimate column of Table 11 gives the treatment means of the dependent variable corrected for block differences only, and the penultimate column the corresponding means of the independent variable. The averages of the t of the dependent variable for fixed u, v and w are given in Table 13.

The sum of products due to treatments is obtained from the appropriate values of Tables 11 and 13. Thus, from columns 9 and 10 (Table 11), 3 and 4, 9 and 10 and 15 and 16 (Table 13), this sum is

$$5{\cdot}79 \times 0{\cdot}59 - 3{\cdot}47 \times 0{\cdot}63 + \ldots - \{5{\cdot}63 \times (-0{\cdot}06) + \ldots + 4{\cdot}96 \times 0{\cdot}11 - \ldots + 4{\cdot}46 \times 0{\cdot}25 + \ldots\}$$

$$= 76{\cdot}577 - 58{\cdot}471 = 18{\cdot}106.$$

Table 14 shows the analysis of variance and covariance and Table 15 the test of significance of the reduced treatment mean square.

Table 11. *Apparent purity % (less 80), y; sugar % (less 14), x; adjusted treatment means*

uvw	*X*		*Y*		*Z*		*T*		*t*		
	y	*x*	*y*	*x*	*y*	*x*	*y*	*x*	*y*	*x*	*y*
111	2.0	−0.39	4.8	0.34	4.6	0.64	11.4	0.59	5.79	0.423	5.03
112	5.5	0.02	6.1	−0.31	0.8	−0.34	12.4	−0.63	3.47	−0.162	4.34
113	3.0	0.52	3.8	−0.34	4.0	−0.05	10.8	0.13	4.11	−0.140	4.93
121	4.6	0.22	0.4	−0.49	9.2	1.87	14.2	1.60	3.99	0.228	3.77
122	1.0	−0.15	3.0	−0.80	3.3	0.24	7.3	−0.71	−0.10	−0.702	2.29
123	2.3	−0.34	1.1	−0.13	5.7	0.48	9.1	0.01	1.40	−0.392	2.92
131	1.9	−0.82	4.4	0.26	4.1	0.38	10.4	−0.18	5.10	0.252	4.82
132	3.9	−0.24	4.8	−0.10	0.4	−0.34	9.1	−0.68	2.94	−0.057	3.52
133	5.4	0.51	3.0	0.19	0.5	−0.56	8.9	0.14	3.15	0.046	3.45
211	1.4	0.09	5.8	0.41	5.9	0.43	13.1	0.93	5.04	0.441	4.23
212	2.2	−0.19	0.8	−0.26	5.8	0.58	8.8	0.13	2.88	0.249	2.60
213	4.8	0.58	2.3	−0.17	0.9	−0.49	8.0	−0.08	2.36	−0.135	3.16
221	5.4	0.60	8.4	0.85	4.6	−0.29	18.4	1.16	6.15	0.398	5.46
222	5.7	0.99	4.4	0.17	4.9	0.27	15.0	1.43	5.16	0.581	3.96
223	3.4	−0.24	1.5	−0.63	5.3	0.58	10.2	−0.29	3.03	−0.004	3.46
231	7.6	0.20	4.2	0.40	9.7	1.87	21.5	2.47	6.91	0.591	5.68
232	3.6	0.22	2.9	−0.32	6.4	1.38	12.9	1.28	4.24	0.306	3.81
233	5.8	0.07	6.3	0.41	4.7	0.82	16.8	1.30	4.39	0.059	4.65
311	4.8	0.24	6.3	0.84	1.4	−0.65	12.5	0.43	6.05	0.450	5.22
312	4.8	0.17	3.9	−0.32	2.2	−0.02	10.9	−0.17	3.79	0.184	3.70
313	4.8	0.41	8.3	1.19	5.1	0.29	18.2	1.89	6.00	0.469	5.11
321	4.7	0.31	4.3	0.58	4.2	0.14	13.2	1.03	5.48	0.420	4.73
322	6.1	0.29	7.2	0.22	2.4	0.09	15.7	0.60	5.46	0.229	5.24
323	4.2	−0.10	5.8	0.41	−0.4	−0.63	9.6	−0.32	2.92	−0.178	3.84
331	6.0	0.80	6.5	0.14	5.2	−0.03	17.7	0.91	6.71	0.592	5.48
332	2.8	−0.34	5.2	−0.03	3.0	−0.80	11.0	−1.17	3.69	−0.131	4.47
333	4.2	0.14	5.0	0.36	1.8	−0.10	11.0	0.40	2.56	0.051	2.84

Table 12

u..	$X_{u..}$		$T_{u..}$	
	y	*x*	*y*	*x*
1..	29.6	−0.67	93.6	0.27
2..	39.9	2.32	124.7	8.33
3..	42.4	1.92	119.8	3.60
...	111.9	3.57	338.1	12.20

.v.	$Y_{.v.}$		$T_{.v.}$	
	y	*x*	*y*	*x*
.1.	42.1	1.38	106.1	3.22
.2.	36.1	0.18	112.7	4.51
.3.	42.3	1.31	119.3	4.47
...	120.5	2.87		

..w	$Z_{..w}$		$T_{..w}$	
	y	*x*	*y*	*x*
..1	48.9	4.36	132.4	8.94
..2	29.2	1.06	103.1	0.08
..3	27 6	0.34	102.6	3.18
...	105.7	5.76		

Table 13

.vw	$X_{.vw}$		$t_{.vw}$	$T_{.vw}$	
	y	x	y	y	x
.11	8·2	−0·06	5·63	37·0	1·95
.12	12·5	0·00	3·38	32·1	−0·67
.13	12·6	1·51	4·16	37·0	1·94
.21	14·7	1·13	5·21	45·8	3·79
.22	12·8	1·13	3·51	38·0	1·32
.23	9·9	−0·68	2·45	28·9	−0·60
.31	15·5	0·18	6·24	49·6	3·20
.32	10·3	−0·36	3·62	33·0	−0·57
.33	15·4	0·72	3·37	36·7	1·84

u.w	$Y_{u.w}$		$t_{u.w}$	$T_{u.w}$	
	y	x	y	y	x
1.1	9·6	0·11	4·96	36·0	2·01
1.2	13·9	−1·21	2·11	28·8	−2·02
1.3	7·9	−0·28	2·89	28·8	0·28
2.1	18·4	1·66	6·03	53·0	4·56
2.2	8·1	−0·41	4·09	36·7	2·84
2.3	10·1	−0·39	3·26	35·0	0·93
3.1	17·1	1·56	6·08	43·4	2·37
3.2	16·3	−0·13	4·31	37·6	−0·74
3.3	19·1	1·96	3·83	38·8	1·97

uv.	$Z_{uv.}$		$t_{uv.}$	$T_{uv.}$	
	y	x	y	y	x
11.	9·4	0·25	4·46	34·6	0·09
12.	18·2	2·59	1·76	30·6	0·90
13.	5·0	−0·52	3·73	28·4	−0·72
21.	12·6	0·52	3·43	29·9	0·98
22.	14·8	0·56	4·78	43·6	2·30
23.	20·8	4·07	5·18	51·2	5·05
31.	8·7	−0·38	5·28	41·6	2·15
32.	6·2	−0·40	4·62	38·5	1·31
33.	10·0	−0·93	4·32	39·7	0·14

Table 14. *Analysis of variance and covariance*

	D.F.	y^2	xy	x^2
Blocks	24	147·90	33·728	12·0814
Groups	2	4·09	−0·770	0·1684
Treatments	26	109·86	18·106	4·0074
Residual	28	94·87	18·168	6·5021
Total	80	356·72	69·232	22·7593

The regression of apparent purity on sugar percentage is

$$18{\cdot}168/6{\cdot}5021 = 2{\cdot}794.$$

Table 15. *Test of significance with reduced variance*

	D.F.	y^2	xy	x^2	D.F.	Reduced y^2	Mean square
Treatments	26	109·86	18·106	4·0074	26	35·42	1·362
Residual	28	94·87	18·168	6·5021	27	44·11	1·634
Treatments + residual	54	204·73	36·274	10·5095	53	79·53	

The treatment means, adjusted to a constant value of sugar percentage, are given in the last column of Table 11. Thus, for example, the adjusted mean of treatment 111 is

$$85{\cdot}79 - 2{\cdot}794(0{\cdot}423 - 0{\cdot}151) = 85{\cdot}03\ \%.$$

Examples of the variances of the three types of comparison between treatments, corrected on the basis of the regression, are:

$$V(t_{211} - t_{111}) = 1{\cdot}634\left\{\frac{26}{27} + \frac{0{\cdot}018^2}{6{\cdot}5021}\right\} = 1{\cdot}573, \quad \text{S.D.} = 1{\cdot}254,$$

$$V(t_{122} - t_{111}) = 1{\cdot}634\left\{\frac{31}{27} + \frac{1{\cdot}125^2}{6{\cdot}5021}\right\} = 2{\cdot}193, \quad \text{S.D.} = 1{\cdot}481,$$

$$V(t_{222} - t_{111}) = 1{\cdot}634\left\{\frac{33}{27} + \frac{0{\cdot}158^2}{6{\cdot}5021}\right\} = 2{\cdot}003, \quad \text{S.D.} = 1{\cdot}415.$$

It is obvious that the technique can be extended to cover the case where it is desired to make allowance, simultaneously, for two or more concomitant measurements.

REFERENCES

C. D. R. Dawson (1939). "An example of the quasi-factorial design applied to a corn breeding experiment." *Ann. Eugen., Lond.*, **9**, 157–73.

R. A. Fisher (1937). *The Design of Experiments.* Edinburgh: Oliver and Boyd.

—— (1938). *Statistical Methods for Research Workers.* Edinburgh: Oliver and Boyd.

F. R. Immer (1932). "Size and shape of plot in relation to field experiments with sugar beets." *J. Agric. Res.* **44**, 649–68.

J. Wishart (1936). "Tests of significance in analysis of covariance." *Supp. J. Roy. Statist. Soc.* **3**, 79–82.

F. Yates (1936*a*). "A new method of arranging variety trials involving a large number of varieties." *J. Agric. Sci.* **26**, 424–55.

—— (1936*b*). "Incomplete randomized blocks." *Ann. Eugen., Lond.*, **7**, 121–40.

—— (1937). "A further note on the arrangement of variety trials: quasi-Latin squares." *Ann. Eugen., Lond.*, **7**, 319–32.

Reprinted from *The Journal of the Australian Institute of Agricultural Science*, Vol. 7, No. 1, pp. 19-26, March, 1941.

The Analysis of Quasi-Factorial Designs with Incomplete Data

2. LATTICE SQUARES

By

E. A. CORNISH, B.Agr.Sc.,

Waite Agricultural Research Institute, S.A.

The analysis of experimental data derived from balanced incomplete block designs, the internal structure of which has been impaired by the loss, or unreliability, of the observations contained in a block or treatment, has been described in a previous paper in this Journal.[1] The present paper is concerned with the analysis of quasi-Latin square designs (lattice squares),[3] [4] which have also been impaired by the loss of complete sets of experimental values, e.g. a row (or column) or a treatment.

Structure of Lattice Squares.

In the lattice square designs, the quasi-factorial principle has been extended in order to allow for the simultaneous elimination from the treatment comparisons, of differences associated with two different groupings of the experimental material, as in the ordinary Latin square.

The number of treatments or varieties must be a perfect square. If the number of varieties is, say p^2, it is possible, for certain values of p[2], to divide the varieties into $(p+1)$ orthogonal groups of p sets, each set containing p varieties, i.e. in such a way that each set of any given group of sets contains only one variety from each set of any other group of sets. Each complete replication of the varieties is arranged in the field in the form of a square, of which the rows correspond to one group of sets and the columns to a second, thus confounding the varietal comparisons corresponding to two groups of sets with row and column differences in each replication.

If p is odd, balance can be attained with a minimum of $\frac{1}{2}(p+1)$ replications but, if p is even, each group of sets must be included twice, giving $(p+1)$ replications and squares.

In the analysis of these designs it is first necessary to calculate for each variety a quantity, (pQ), defined as p times the variety total less the totals of each row and column in which that variety occurs. The computation of the sum of squares ascribable to varieties is the only new feature of the analysis of variance, since items for lattice squares, rows and columns are determined in the ordinary manner. When p is odd, the sum of squares for varieties is

$$\frac{2}{p^2(p-1)} S\{(pQ)-(\overline{pQ})\}^2$$

and the adjusted varietal means are given by

$$\frac{2}{p(p-1)}(pQ)+\frac{2p}{p-1}\times\text{general mean.}$$

The estimated variance of the difference between any two adjusted means is

$$\frac{4}{p-1}\times\text{estimated variance of a single plot.}$$

If p is even, all constant multipliers in the above expressions, except the factor $2p/(p-1)$, must be divided by 2.

In the analysis given below, only the case when p is odd has been considered in detail. When p is even, the analysis, though somewhat more complex, follows similar lines.

Missing Treatment.

Without loss of generality it may be assumed that the replicates of variety 1 are missing from the first row and first column of each square, and further, that in the first square variety 1 is associated with varieties 2, 3, ..., p in row 1, and with varieties $p+1$, $2p+1$, ..., p^2-p+1 in column 1. The set of constants which has to be determined is as follows:

general mean	: m		
lattice squares	: l_i	$S(l_i)=0$	$i=1, 2, \ldots, \frac{1}{2}(p+1)$
rows	: r_{ij}, for a given value of i	$S(r_{ij})=0$	$j, k=1, 2, \ldots, p$
columns	: c_{ik}, for a given value of i	$S(c_{ik})=0$	
varieties	: v_t	$S(v_t)=0$	$t=2, 3, \ldots, p^2$

If G is the grand total of existing observations and L, R, C and V stand for the totals of lattice squares, rows, columns and varieties, respectively, the normal equations are

$$\begin{aligned}
\tfrac{1}{2}(p+1)(p^2-1)m-S(r_{i1}+c_{i1})&=G,\\
(p^2-1)(m+l_i)-(r_{i1}+c_{i1})&=L_i,\\
(p-1)(m+r_{i1})+(p-1)l_i-c_{i1}+S(v)&=R_{i1},\\
p(m+r_{i2})+pl_i+S(v)&=R_{i2},\\
\cdot\quad\cdot\quad\cdot\quad\cdot\quad\cdot&\\
(p-1)(m+c_{i1})+(p-1)l_i-r_{i1}+S(v)&=C_{i1},\\
p(m+c_{i2})+pl_i+S(v)&=C_{i2},\\
\cdot\quad\cdot\quad\cdot\quad\cdot\quad\cdot&\\
\tfrac{1}{2}(p+1)(m+v_t)+S(r+c)&=V_t.
\end{aligned}$$

In any r or c equation the summation is taken over the varieties contained in the row or column concerned, while in any v equation, it is taken over the rows and columns in which replicates of the particular variety occur.

The value of m is calculated from the relation

$$\tfrac{1}{2}p(p-1)(p^2-1)m=(p-2)G+S(R_{i1}+C_{i1}).$$

The equations in the v of varieties associated with the missing variety in row 1 and column 1 of the first square reduce to

$$\left.\begin{array}{l} Av_2+B(v_3+\ldots\ldots+v_p+v_{p+1}+\ldots\ldots+v_{p^2-p+1})=Q'_2, \\ Bv_2+Av_3+B(v_4+\ldots+v_p+v_{p+1}+\ldots\ldots+v_{p^2-p+1})=Q'_3, \\ \quad\cdot\quad\cdot\quad\cdot\quad\cdot\quad\cdot\quad\cdot\quad\cdot\quad\cdot \\ B(v_2+v_3+\ldots\ldots+v_p+v_{p+1}+\ldots)+Av_{p^2-p+1}=Q'_{p^2-p+1}, \end{array}\right\} \begin{array}{c}2(p-1)\\ \text{equations}\end{array}$$

where $A=\tfrac{1}{2}p^2(p+1)(p-2)$, $B=-\tfrac{1}{2}p(p+1)$ and

$$Q'_p=(p^2-1)(pQ_p)-p(V_2+\ldots+V_p+V_{p+1}+\ldots+V_{p^2-p+1})+2pG-\tfrac{1}{2}p(p-1)(p^2-1)m-L_1.$$

The values of (pQ) are obtained in the ordinary manner, as indicated above, except, of course, for the omission of the observations of the missing variety. The solutions of this set of equations are

$$v_p=aQ'_p+b(Q'_2+Q'_3+\ldots+Q'_{p^2-p+1})$$

where

$$a=2(p-2)^2/p(p+1)(p-1)^2(p^2-4p+3),$$
$$b=a/(p-2)^2,$$

and Q'_p is omitted from the bracket on the right-hand side.

There are, in all, $\tfrac{1}{2}(p+1)$ such sets of equations, each of which involves a different selection of the v. Thus the second set involves the v of varieties associated with the missing variety in the second square and so on.

Although the Q' are expressible in other forms, it will be found that the above form is very convenient to use in practice.

When all the v have been obtained the r_{i1} and c_{i1} may be evaluated; thus

$$pr_{11}=V_2+\ldots+V_p-G+\tfrac{1}{2}p(p^2-1)m-\tfrac{1}{2}(p+1)(v_2+\ldots+v_p)$$

and

$$pc_{11}=V_{p+1}+\ldots+V_{p^2-p+1}-G+\tfrac{1}{2}p(p^2-1)m-\tfrac{1}{2}(p+1)(v_{p+1}+\ldots+v_{p^2-p+1}).$$

The remaining r_{i1} and c_{i1} are obtained from similar expressions involving the appropriate variety totals and variety constants. The $(m+l_i)$ may now be found by substitution in the normal equations of the l_i and, finally, the remaining r and c are obtained by substitution in the original r and c equations.

The varietal comparisons, in terms of a single plot, are given by the differences of the quantities v, and the adjusted varietal means by $v+m$.

The reduction in the sum of squares due to fitting the constants l, r, c and v is

$$S(m+l)L+SrR+ScC+SvV-2G^2/(p+1)(p^2-1),$$

from which the sum of squares for testing the significance of varieties is obtained by deducting the sum of squares ascribable to lattice squares, rows and columns, when varieties are disregarded. To determine the latter sum of squares, a new set of constants representing lattice squares, rows and columns must be evaluated. The normal equations take the form given above omitting the v. The general mean is obtained from the expression already given and the $(m+l_i)$ from the relations

$$p(p-1)^2(m+l_i)=(p-2)L_i+R_{i1}+C_{i1}.$$

The r_{i1} and c_{i1} can now be determined from

$$pr_{i1}=R_{i1}-L_i+p(p-1)(m+l_i),$$
$$pc_{i1}=C_{i1}-L_i+p(p-1)(m+l_i),$$

and finally the remaining r and c are obtained by substitution.

Varietal comparisons are not all of the same accuracy ; when all replicates of one variety are missing, the types of comparison with their variances are :

(i) between varieties associated with the missing variety in any given square : variance $4\sigma^2/(p-1)$,

(ii) between varieties associated with the missing variety in different squares : variance $\frac{4\sigma^2}{p-1}\left\{1+\frac{2}{(p+1)(p^2-4p+3)}\right\}$.

Example 1.

Yates[2] has given an example showing the analysis of a balanced set of quasi-Latin squares involving 25 varieties ($p=5$). Table I shows the yields and the totals of rows, columns and lattice squares. Assume that variety 1 is missing. The normal equations are then

$$72m-r_{14}-c_{12}-r_{25}-c_{21}-r_{34}-c_{31}=1147,$$

$$24(m+l_1)-r_{14}-c_{12}=196,$$

. . . . .

$$5(m+r_{11})+5l_1+v_{21}+v_{22}+v_{23}+v_{24}+v_{25}=2,$$

. . . . .

$$5(m+c_{11})+5l_1+v_3+v_8+v_{13}+v_{18}+v_{23}=-30,$$

. . . . .

$$3(m+v_2)+r_{14}+c_{14}+r_{24}+c_{23}+r_{33}+c_{32}=48,$$

. . . . .

In Table II, the variety totals, V, the values of $-24(5Q)$ and the v are given in sets as they occur together in the rows and columns containing the missing variety.

TABLE I.

Yields (less 100) and Arrangement of Varieties.

							r
Square 1	(23) −13	(21) 3	(25) −4	(22) 3	(24) 13	2	−10·5250
	(8) −5	(6) 6	(10) 19	(7) 4	(9) 20	44	1·9750
	(18) 11	(16) 10	(20) 9	(17) −1	(19) 40	69	0·6750
	(3) −21	(1) 12	(5) −1	(2) 15	(4) 29	34	−0·3000
	(13) −2	(11) 13	(15) 16	(12) 10	(14) 22	59	8·1750
	−30	44	39	31	124	208	
c	−15·4250	−1·2000	3·0750	−3·2250	16·7750		
Square 2	(25) 2	(6) 9	(14) 22	(17) 28	(3) 13	74	2·5375
	(7) −7	(18) 26	(21) 37	(4) 28	(15) 8	92	−0·8625
	(19) 0	(5) 16	(8) 29	(11) 19	(22) 13	77	−1·3625
	(13) −7	(24) 15	(2) 12	(10) 19	(16) 20	59	−6·9625
	(1) −2	(12) 16	(20) 27	(23) 35	(9) 25	101	6·6500
	−14	82	127	129	79	403	
c	−19·7500	−0·0625	8·0375	11·7375	0·0375		
Square 3	(10) −2	(19) 38	(3) 32	(21) 37	(12) 38	143	5·3875
	(22) 3	(6) 18	(20) 21	(13) 19	(4) 15	76	−5·9125
	(18) −3	(2) 21	(11) 4	(9) 10	(25) 17	49	−13·7125
	(1) 19	(15) 19	(24) 29	(17) 32	(8) 29	128	0·8500
	(14) 8	(23) 31	(7) 48	(5) 30	(16) 52	169	13·3875
	25	127	134	128	151	565	
c	−22·9500	2·9875	8·6875	4·1875	7·0875		

Varieties are indicated by the numbers in parentheses.

TABLE II.

Variety.	V	$-24(5Q)$	v	Variety	V	$-24(5Q)$	v
2	48	4200	4·9708	6	33	6480	−4·5292
3	24	7248	−7·7292	11	36	7200	−7·5292
4	72	5616	−0·9292	16	82	3576	7·5708
5	45	7008	−6·7292	21	77	3336	8·5708
	189		−10·4168		228		4·0832
9	55	6048	−2·3917	7	45	5592	−0·4917
12	64	5976	−2·0917	13	10	5520	−0·1917
20	57	6312	−3·4917	19	78	3312	9·0083
23	53	5640	−0·6917	25	15	5472	0·0083
	229		−8·6668		148		8·3332
8	53	5112	1·8583	10	36	5760	−0·8417
15	43	6960	−5·8417	14	52	7176	−6·7417
17	59	5880	−1·3417	18	34	2352	13·3583
24	57	5400	0·6583	22	19	4224	5·5583
	212		−4·6668		141		11·3332

Now $G=1147$ and the total of rows and columns containing the missing variety is

$$22+32+103-12+109+6=260,$$

so that,

$$m=\frac{1}{240}(3\times 1147+260)=15{\cdot}4208.$$

The next step is to determine the quantities Q'. In the expression for Q', the first term is the only one which varies over any set of $2(p-1)$ varieties. Consequently, the algebraic sum of the remaining terms need only be determined once. For the varieties associated with variety 1 in the first square this quantity is

$$-5(189+228)+11470-3701-196=5488,$$

the values of the corresponding quantities in the second and third squares being 5479 and 5458, respectively, so that, for example,

$$Q'_2=-4200+5488=1288,$$
$$Q'_7=-5592+5479=-113,$$
and $$Q'_8=-5112+5458=346.$$

The quantities a and b have the values $a=9/1920$ and $b=1/1920$ and so

$$v_2=\frac{1}{1920}(9\times 1288-1760-\ldots.+2152)=4{\cdot}9708, \text{ etc.}$$

The constants of the rows and columns containing the missing variety may now be determined. For example,

$$r_{14}=\frac{1}{5}(189-1147+925{\cdot}25+31{\cdot}25)=-0{\cdot}3000$$

and $$c_{12}=\frac{1}{5}(228-1147+925{\cdot}25-12{\cdot}25)=-1{\cdot}2000.$$

After substituting these quantities in the l equations, the values $l_1=-7\cdot3166$, $l_2=0\cdot9084$ and $l_3=6\cdot4084$ are obtained. The values of the r and c are given in Table I.

The sum of squares attributable to the constants l, r, c and v is

$$8\cdot1042\times196+\ldots.-10\cdot5250\times2+\ldots.-15\cdot4250\times(-30)-\ldots. +4\cdot9708\times48-\ldots.-1147^2/72=13730\cdot28,$$

while the sum of squares due to lattice squares, rows and columns when varieties are ignored is

$$8\cdot0250\times196+\ldots.-7\cdot625\times2+\ldots.-14\cdot025\times(-30)-\ldots.-1147^2/72 =12259\cdot42.$$

The analysis of variance is given in Table III.

TABLE III.
Analysis of Variance.

	Deg. Freed.	Sum Sq.	Mean Sq.
Total	71	14966·65	
Residual	22	1236·37	56·20
Constants l, r, c and v ..	49	13730·28	
Constants l, r and c (ignoring varieties)	26	12259·42	
Remainder (varieties) ..	23	1470·86	63·95

Examples of the two types of varietal comparison with their estimated standard deviations are

(i) v_2-v_3 12·70, v_2-v_6 9·50, v_6-v_{11} 3·00 } $\pm7\cdot50$ (ii) v_2-v_7 $5\cdot46\pm7\cdot65$.

Missing Row.

Without loss of generality it may be assumed that the first row, containing varieties 1, 2, ..., p, of the first square, is missing and that the first column contains varieties 1, $p+1$, $2p+1$, ..., p^2-p+1. The normal equations are now

$$\tfrac{1}{2}p(p-1)(p+2)m-pl_1-S(v_t)=G, \quad t=1,2,\ldots.,p$$
$$p(p-1)(m+l_1)-S(v_t)=L_1,$$
$$p^2(m+l_2) \quad =L_2,$$
$$\cdot \quad \cdot \quad \cdot \quad \cdot$$
$$p(m+r_{1i})+pl_1+S(v)=R_{1i}, \quad i=2,3,\ldots.,p$$
$$p(m+r_{2j})+pl_2+S(v)=R_{2j}, \quad j=1,2,\ldots.,p$$
$$\cdot \quad \cdot \quad \cdot \quad \cdot$$
$$(p-1)(m+c_{1k})+(p-1)l_1+S(v)=C_{1k}, \quad k=1,2,\ldots.,p$$
$$p(m+c_{2k})+pl_2+S(v)=C_{2k},$$
$$\cdot \quad \cdot \quad \cdot \quad \cdot$$
$$\tfrac{1}{2}(p-1)(m+v_1)-l_1+S(r+c)=V_1,$$
$$\cdot \quad \cdot \quad \cdot \quad \cdot$$
$$\tfrac{1}{2}(p+1)(m+v_{p+1})+S(r+c)=V_{p+1},$$
$$\cdot \quad \cdot \quad \cdot \quad \cdot$$

The general mean is obtained directly from the relation

$$\tfrac{1}{4}p^2(p-1)(p^2-1)m=p\{\tfrac{1}{2}(p-1)G+V_1+V_2+\ldots+V_p\}$$
$$-\tfrac{1}{2}(p+1)\{T_2+T_3+\ldots+T_{\frac{1}{2}(p+1)}\}$$

and v_1 from

$$\tfrac{1}{2}p(p-3)v_1=G+(pQ_1)-\tfrac{1}{2}p(p^2-1)m-S(Q'),$$

where the summation is taken over the varieties associated with variety 1 in the first column of the first square, and the Q' are given by

$$Q'=\frac{2}{p(p-1)^2}\{G+(pQ_1)+(p-1)(pQ)-C_{11}\}.$$

v_2, v_3, ... are obtained from expressions similar to that of v_1, replacing (pQ_1) and C_{11} by the corresponding (pQ) and C and using the Q' functions of treatments from the appropriate column.

When $v_1, \ldots, v_p$ have been found, the remaining v may be determined; thus

$$v_{p+1}=Q'_{p+1}-\frac{1}{p-1}v_1,$$
$$v_{2p+1}=Q'_{2p+1}-\frac{1}{p-1}v_1$$

and so on.

The l, r and c may now be obtained by substitution in the normal equations.

There are five types of varietal comparison:

(i) between varieties of the missing row: variance $\frac{4\sigma^2}{p-1}\left\{1+\frac{2(p-1)}{p(p-3)}\right\}$

(ii) between existing varieties which occur together in the same column of the square with the missing row: variance $4\sigma^2/(p-1)$

(iii) between existing varieties which occur in different columns of the square with the missing row: variance $\frac{4\sigma^2}{p-1}\left\{1+\frac{2}{p(p-1)(p-3)}\right\}$

(iv) between a variety of the missing row and existing varieties in the same column of the square with the missing row

(v) between a variety of the missing row and existing varieties not in the same column in the square with the missing row.

The variances of the last two types of comparison probably lie within the limits given by the variances of comparisons of types (i) and (iii).

Example 2.

Assume that the fourth row of the first square in Table I is missing. The new grand total is 1142 and the new totals for varieties 1 to 5, in order, are 17, 33, 45, 43 and 46, so that

$$m=\frac{1}{600}\{5(2\times 1142+184)-3(403+565)\}=15\cdot 7267.$$

Table IV shows the values of $-(5Q)$ for varieties 1 to 5 and the values of $-4(5Q)$ for the remaining varieties in sets as they occur together in the columns of the first square.

TABLE IV.

Variety.	$-(5Q)$	$-4(5Q)$			
1, 6, 11, 16, 21	155	1080	1200	596	556
2, 7, 12, 17, 22	197	864	928	996	720
3, 8, 13, 18, 23	205	1012	996	552	1016
4, 9, 14, 19, 24	233	884	1156	428	860
5, 10, 15, 20, 25	226	1040	1240	1048	908

Now

$$G+(5Q_1)-C_{12}=1142-155-32=955,$$

so that

$$Q'_6=\frac{1}{40}(955-1080)=-3\cdot125,$$

$$Q'_{11}=\frac{1}{40}(955-1200)=-6\cdot125,$$

$$Q'_{16}=\frac{1}{40}(955-\ 596)=\ 8\cdot975,$$

$$Q'_{21}=\frac{1}{40}(955-\ 556)=\ 9\cdot975,$$

whence

$$v_1=\frac{1}{5}\{G+(5Q_1)-60m-(Q'_6+Q'_{11}+Q'_{16}+Q'_{21})\}$$

$$=\frac{1}{5}(1142-155-943\cdot6+3\cdot125+6\cdot125-8\cdot975-9\cdot975)=6\cdot740$$

and, since,

$$\tfrac{1}{4}v_1=1\cdot685,$$

$$v_6=-3\cdot125-1\cdot685=-4\cdot810.$$

Similarly $v_{11}=-7\cdot810$, $v_{16}=7\cdot290$ and $v_{21}=8\cdot290$.

(1) Cornish, E. A. J. Aust. Ins. Agric. Sc., 6: 31-39, 1940.
(2) Fisher, R. A., and Yates, F. Statistical Tables for Biological, Agricultural and Medical Research. Oliver & Boyd, Edinburgh, 1938.
(3) Yates, F. Ann. Eugen., 7: 319-332, 1937.
(4) ——— The Design and Analysis of Factorial Experiments. Imp. Bur. Soil Sc., Harpenden. 1937.

(Received for publication, 1st November, 1940.)

AUSTRALASIAN MEDICAL PUBLISHING COMPANY LIMITED

The Recovery of Inter-block Information in Quasi-Factorial Designs with Incomplete Data

S U M M A R Y.

This is the first of a series of papers concerned with the recovery of information in quasi-factorial designs that have incomplete data. The paper describes an approximate method for dealing with square, triple, and cubic lattices and discusses the effect of the approximations on the estimation of the adjusted treatment effects and their errors and the analysis of variance.

Commonwealth of Australia, Council for Scientific and Industrial Research, Bulletin No. 158. Melbourne, 1943.

The Recovery of Inter-block Information in Quasi-Factorial Designs with Incomplete Data

1.—Square, Triple, and Cubic Lattices*

[By E.A. Cornish, M.Sc., B.Agr.Sc.]

I. INTRODUCTION.

The quasi-factorial designs, introduced by Yates (1936a,b; 1937), were designed to overcome the serious difficulties that are encountered in various types of biological experimentation when large numbers of treatments have to be tested simultaneously. These designs, which are of three types

(a) unbalanced lattices in randomised blocks,

(b) incomplete randomised blocks,

(c) quasi-Latin squares (lattice squares),

have, as their principal object, the elimination of heterogeneity in experimental material to a greater degree than is possible by the use of ordinary randomised blocks.

When these designs were first proposed, attention was directed to methods of analysis which made full adjustment for group (block, row, or column) differences, but, at the same time, it was pointed out that loss of efficiency, as compared with designs in ordinary randomised blocks, would result if these differences were negligibly small or non-existent. This loss of efficiency, or alternatively information, is due to the fact that the differences between the group means comprise either partly, or entirely, differences between treatments. In the limiting case, where there are no block differences the loss amounts to $1-E$ where E is the efficiency factor of the particular design. Recently it has been shown (Fisher and Yates 1938; Yates 1939, 1940b,c; Cox *et al.* 1940) that information on the treatments, contained in the inter-block comparisons, can be recovered if the relative accuracy of the inter- and intra-block comparisons is well determined.

When there is a full complement of observations, the several types of symmetrical arrangement easily lend themselves to either method of statistical analysis, but in cases where the data are incomplete, or unreliable, the appropriate analytical procedure can no longer be employed. The procedure to be followed when the inter-block formation is rejected has already been given (Cornish 1940a,b,c; 1941a,b), and in the present series, the first of which deals with square, triple, and cubic lattices, consideration is given to methods of analysis appropriate to the recovery of inter-block information.

C.13851-Z.   * Typescript received September 2, 1942.

II. THE LATTICE DESIGNS.

In order to illustrate the construction of these designs, for example in variety trials, consider the simplest type, namely, the square lattice. This design, with equal groups of sets, is applicable where the number of varieties, p^2 say, is a perfect square. To construct this design the varieties are first assigned at random to the p^2 intersections of a $p \times p$ lattice. In both directions of the lattice there are p lines which divide the varieties into p sets each containing p varieties. After allocating the p sets of each group to p blocks at random, the groups are set out in the field in the form of p randomised blocks of p plots, in such a way that the blocks constituting complete replications are arranged in compact units. The two groups must be equally replicated, say r times.

Fig. 1 illustrates a 5 X 5 lattice for 25 varieties, the position of each variety within the lattice being determined by two numbers u and v, both of which may take any integer value from 1 to 5.

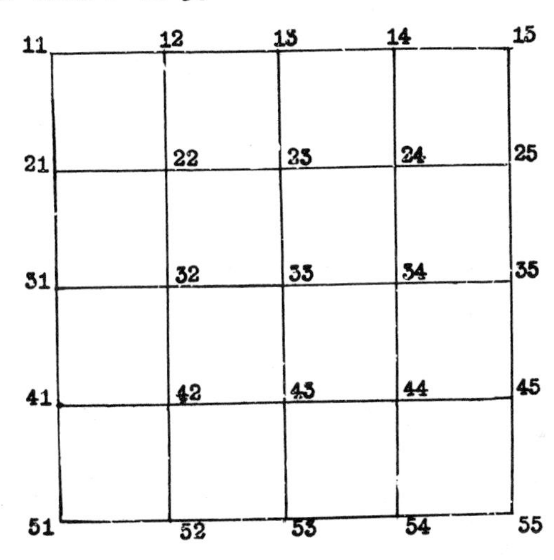

Fig. 1 - 5 x 5 Lattice

The analogy with an ordinary factorial design involving two factors x and y each at 5 levels is immediately apparent, for the intersections of the lattice may be taken to represent the 25 combinations of the factors, and the levels associated with any intersection as the co-ordinates of that intersection referred to the axes of the lattice.

The degrees of freedom for the varieties, p^2-1, may, therefore, be set out as if arising from the main effects and interaction of two hypothetical factors x and y; thus

		Deg. freed.
Main effects	X	$p-1$
	Y	$p-1$
Interaction	XY	$(p-1)^2$
	Total	p^2-1

and the varietal means may be expressed in terms of the main effects and interactions in the following way. Let V_{uv} stand for the mean yield of variety uv, X_u for the mean yield of factor x at the uth level, Y_v for the mean yield of factor y at the vth level, and M for the general mean, then

$$V_{uv} = M + (X_u - M) + (Y_v - M) + (V_{uv} - X_u - Y_v + M).$$

III. NOTATION.

The group in which the vertical lines (of Fig. 1) show the constitution of the blocks will be designated group X, and the group of horizontal lines showing the other types of block will be designated group Y.

The total of the r replicates of variety uv in group X will be denoted by the x_{uv} and in group Y by y_{uv}. Totals and subtotals will be denoted thus

$$x_{11} + x_{12} + \ldots\ldots + x_{1p} = X_{1.}$$
$$X_{1.} + X_{2.} + \ldots\ldots + X_{p.} = X_{..}$$

with similar relations for group Y; and totals over the two groups thus

$$x_{uv} + y_{uv} = T_{uv},$$
$$X_{p.} + Y_{p.} = T_{p.}\ ,$$
$$X_{..} + Y_{..} = T_{..}\ .$$

These relations are set out in tabular form below.

TABLE 1.

Group X						Group Y						Groups (X + Y)					
$u \backslash v$	1	2	....	p	Total	$u \backslash v$	1	2	....	p	Total	$u \backslash v$	1	2	....	p	Total
1	x_{11}	x_{12}	....	x_{1p}	$X_{1.}$	1	y_{11}	y_{12}	....	y_{1p}	$Y_{1.}$	1	T_{11}	T_{12}	....	T_{1p}	$T_{1.}$
2	x_{21}	x_{22}	....	x_{2p}	$X_{2.}$	2	y_{21}	y_{22}	....	y_{2p}	$Y_{2.}$	2	T_{21}	T_{22}	....	T_{2p}	$T_{2.}$
.	.	.	. . .		.	.	.	.	. . .		.	.	.	.	. . .		.
.	.	.	. . .		.	.	.	.	. . .		.	.	.	.	. . .		.
p	x_{p1}	x_{p2}	....	x_{pp}	$X_{p.}$	p	y_{p1}	y_{p2}	....	y_{pp}	$Y_{p.}$	p	T_{p1}	T_{p2}	....	T_{pp}	$T_{p.}$
Total	$X_{.1}$	$X_{.2}$	....	$X_{.p}$	$X_{..}$	Total	$Y_{.1}$	$Y_{.2}$	....	$Y_{.p}$	$Y_{..}$	Total	$T_{.1}$	$T_{.2}$	....	$T_{.p}$	$T_{..}$

IV. ESTIMATES OF VARIETAL EFFECTS.

It is apparent from Fig. 1 that the varietal contents of the blocks are arranged so that the main effect X is unconfounded in group X but the main effect Y is confounded, whereas, in group Y the position is reversed, and that the interaction XY is not confounded in either group. Three estimates of the confounded effects may thus be determined:-

(i) Intra-block estimates derived from regions where the effects are unconfounded with block differences,

(ii) Inter-block estimates derived from regions where the effects are confounded with block differences,

(iii) Unadjusted estimates derived from the whole experiment.

For example, if X_i, X_b, and X_o stand for the intra-block, inter-block, and unadjusted estimates respectively of the main effects of factor x at the uth level, then X_i will be derived from the total $X_{u\cdot}$, X_b from the total $Y_{u\cdot}$, and X_o from the total $T_{u\cdot}$.

V. METHODS OF ANALYSIS.

It will be noted that, apart from the restrictions imposed by the lattice design, the arrangement is similar to an ordinary randomised block design of p^2 varieties in blocks of p^2 plots. If the confounding is ignored, and the experiment is treated as if it were an ordinary randomised block design,* the analysis will be that of the unadjusted yields, and will give equal weight to all intervarietal comparisons. It is clear that if the small blocks are really differentiated, the block differences will affect the inter-block comparisons and increase their error relative to that of the intra-block comparisons, thus introducing an inequality in the weights. However, the special features of the arrangement are such that, if this is the case, the block differences may be entirely eliminated from the varietal comparisons. This attribute of the quasi-factorial designs was emphasised when they were first described, and in the analysis originally proposed (Yates 1936a), only the unconfounded or intra-block estimates of the main effects and interactions were used in determining the varietal means. The analysis of variance took the form shown in Table 2.

TABLE 2.

Variation due to -	Deg. freed.
Groups	$2r-1$
Blocks (ignoring varieties)	$2r(p-1)$
Varieties (eliminating blocks)	p^2-1
Residual (intra-block error)	$(p-1)(2rp-p-1)$
Total	$2rp^2-1$

The sums of squares for groups, and blocks within groups, were obtained in the ordinary way, but in assessing the block sum of squares no account was taken of the fact that the varietal contents of the blocks varied. The sum of squares for varieties was determined by evaluating its components from unconfounded regions of the experiment, using the methods applicable to ordinary factorial designs, or by a more direct process equivalent to this method. Since all variation associated with the small blocks has been entirely eliminated, all information on the varietal comparisons, contained in the inter-block comparisons, has been rejected. Yates (1940a) refers to the resultant varietal means as fully adjusted means.

In general, the most accurate estimates will lie somewhere between the unadjusted and fully adjusted estimates, because the inter-block comparisons will still provide some information even though they are less accurate than the intra-block comparisons. A weighted mean of the inter- and intra-block estimates will therefore provide the best estimate of the varietal effects. Yates (1940a) has termed these the partially adjusted estimates, and has shown (1939), that the process of weighting can only be carried out efficiently if

* This point is discussed at a later stage in this section.

the relative accuracy of the inter- and intra-block comparisons is well determined.

Let w' and w stand for the reciprocals of the true error variances per plot of the inter- and intra-block estimates, respectively. If the yields, apart from the varietal effects, are regarded as the sum of two independent and normally distributed quantities, the first constant for all plots of a block, but varying from block to block with variance A, and the second varying independently from plot to plot with variance B, then

$$w = 1/B \quad \text{and} \quad w' = 1/(pA + B).$$

Let X stand for the partially adjusted estimate of the main effect of factor x at the uth level, then, since X_t and X_b are derived from equal numbers of replicates

$$X_0 = \tfrac{1}{2}\,(X_t + X_b) \qquad (1)$$

and so

$$X = (wX_t + w'X_b)/(w + w'),$$

and, on substituting for X_t from equation (1)

$$X = X_0 + \lambda(X_0 - X_b), \qquad (2)$$

where $\lambda = (w - w')/(w + w')$. There will be a similar expression for the main effect Y. Since the interaction is unconfounded

$$XY = XY_0.$$

In order to estimate w' it is necessary to have an estimate of the inter-block variance. This variance cannot be derived from the sum of squares for blocks calculated directly from the block totals since the latter are, to some extent, affected by varietal differences. The varietal effects must be eliminated from the components of the block sum of squares, and this slightly alters the analysis of variance, which now takes the form shown in Table 3.

TABLE 3.

Variation due to -	Deg. freed.
Groups	$2r-1$
Blocks (eliminating varieties)	$2r(p-1)$
Varieties (ignoring blocks)	p^2-1
Residual (intra-block error)	$(p-1)(2rp-p-1)$
Total	$2rp^2-1$

The sums of squares for groups, residual, and total, stand as before so that the only alteration involves a different partitioning of the sum of squares for blocks plus varieties. The varietal sum of squares is obtained in the usual way, while the block sum of squares is derived from block totals adjusted to allow for change in varietal content. The mean squares for blocks and residual in this analysis provide the basis for determining estimates of the weighting coefficients w and w'.

As mentioned at the beginning of this section, there is a third way in which the analysis can be made. Yates (1939) has demonstrated that the quasi-factorial designs possess a remarkable property, namely, that the usual randomised block analysis always provides an unbiased estimate of the average error of the unadjusted yields, despite the restrictions on the arrangement. This analysis of variance is derived from that in Table 3 by pooling the sums of squares for blocks and residual and the corresponding degrees of freedom. The analysis of the unadjusted yields then takes the form shown in Table 4.

TABLE 4.

Variation due to -	Deg. freed.
Groups	$2r-1$
Varieties (ignoring blocks)	p^2-1
Residual	$(2r-1)(p^2-1)$
Total	$2rp^2-1$

If the experiment were arranged in ordinary randomised blocks of p^2 plots, the average error variance of all varietal comparisons would be

$$\frac{1}{r(p+1)}\left\{\frac{p}{w}+\frac{1}{w'}\right\},$$

whereas, the average error variance of the lattice design, on the basis of an analysis of the type in Table 2, is

$$\frac{1}{rw}\;\frac{p+3}{p+1}\,.$$

When the inter-block information is rejected the efficiency of the square lattice is, therefore,

$$\frac{1}{r(p+1)}\left\{\frac{p}{w}+\frac{1}{w'}\right\}\bigg/\frac{1}{rw}\;\frac{p+3}{p+1}$$

$$= (p+w/w')/(p+3).$$

On the other hand, the average error variance of all varietal comparisons, on the basis of an analysis of the type shown in Table 3, is

$$\frac{1}{rw(p+1)}\left\{(p-1)+4w/(w+w')\right\},$$

so that, if the inter-block information is recovered, the efficiency of the square lattice design is

$$\frac{p+w/w'}{p-1+4w/(w+w')}\,.$$

In assessing these efficiencies the slight loss of information due to inaccuracies of weighting has been ignored. Table 5 which has been extracted from data given by Cox et *al.* (1940) and Yates (1939), shows the percentage efficiency of the square and cubic lattices for various values of the ratio w/w', with and without the recovery of inter-block information. The table shows clearly the extent of the gain in efficiency when the inter-block information is recovered.

TABLE 5.

Percentage efficiencies of lattice designs relative to randomised blocks.

	p \ w/w'	1	2	3	4	6	8	10
Square lattice	5	100 75	105.0 87.5	114.3 100	125.0 112.5	148.1 137.5	172.0 162.5	196.4 187.5
	6	100 77.8	104.3 88.9	112.5 100	122.0 111.1	142.4 133.3	163.6 155.6	185.3 177.8
	7	100 80	103.8 90.0	111.1 100	119.6 110.0	137.9 130.0	157.0 150.0	176.4 170.0
Cubic lattice	3	100 59.1	108.3 77.3	122.7 95.5	138.9 113.6	173.2 150.0	208.4 186.4	244.1 222.7
	4	100 66.7	107.0 82.5	119.4 98.4	133.3 114.3	163.1 146.0	193.7 177.8	224.8 209.5

The upper figures represent the efficiencies when the inter-block information is recovered, the lower figures those when it is ignored.

The circumstances under which each of the three methods of analysis mentioned above is appropriate, may be derived from a consideration of the functional relationship between the additional gain in efficiency, due to the recovery of the inter-block information, and the weighting coefficients w and w'.

(i) If $w' = w$ there are no real differences between blocks, consequently no adjustment is applied to the unadjus ed yields (*vide* equation (2)) and the recoverable inter-block information is a maximum. In this case the ordinary randomised block analysis is the proper method.

(ii) When the differences between the blocks are very large, w' approaches zero and the adjustment to be applied to the unadjusted yields approaches its maximum value. If w' is put equal to 0, full adjustment is made and the recoverable inter-block information is zero. In this case the analysis designed to make full adjustment for the block effects is appropriate.

(iii) For values of w' in the range between these extremes, the adjustment to be applied to the unadjusted yields, and the amount of recoverable inter-block information vary steadily as $w' \longrightarrow 0$. In these cases the partially adjusted yields are more accurate than either the fully adjusted, or the unadjusted yields, and the appropriate form of the an lysis is that given in Table 3.

Three situations may therefore arise in problems of the type presented by these designs with incomplete data, each of which requires specific treatment. In two of these cases the exact solution is already known. In case (i) above, where the ordinary randomised block analysis is appropriate, there is no difficulty in making an analysis when the data are incomplete. Estimates of the missing values may be calculated by well established formulae and the exact analysis completed in the manner described by Yates (1933). Again, in case (ii), the exact analysis is known. The formulae for estimating the missing values and the methods of analysis have been given by the author (Cornish 1940b). For the third situation a new method of analysis is required, and this is the case with which the present paper is concerned.

The problem of analysis may be divided into two parts:-

(i) the estimation of the magnitude of the effects the experiment was designed to test;

(ii) the provision of appropriate tests of significance and estimates of error.

The first requirement necessitates the calculation of efficient and unbiased estimates of the partially adjusted varietal means, and this, in turn, involves the determination of efficient and unbiased inter- and intra-block estimates of the varietal effects. Estimates of the weighting coefficients w and w' will also be needed in determining the adjusted means, in making tests of significance, and in evaluating the standard deviations of the intervarietal comparisons.

Inasmuch as a strictly exact analysis is impossible, even with a complete set of observations, an exact solution is impossible when the data are incomplete, but, apart from the inaccuracies arising from the substitution of estimates of w and w' for the parameters themselves, I can put forward only an approximate solution. This solution should, however, be sufficiently accurate for all practical purposes. If an exact analysis is required, the two methods mentioned above may be used, but it is obvious that either course may result in an appreciable loss of information. As $w' \longrightarrow w$ the loss of information resulting from the analysis of the unadjusted yields tends to zero, and, at the other extreme as $w' \longrightarrow 0$, the loss resulting from the analysis of the fully adjusted yields also tends to zero, so that near the ends of the range the loss will be sufficiently small to make the exact analysis worth while.

VI. ESTIMATION OF MISSING VALUES.

As in the cases previously discussed, the problem can best be approached through the estimation of values for the missing observations. For simplicity the square lattice is used for purposes of illustration. Assume that one of the observations belonging to variety uv is missing from group X.

In Section V it was pointed out that the ordinary randomised block analysis always provides an unbiased estimate of the average error of the unadjusted varietal yields. An estimate of the missing value may therefore be determined from the formula given by Yates (1933). This estimate, which will be designated an unadjusted estimate, may be included with the existing values of the totals $T_{u\cdot}$ and $T_{\cdot v}$ to determine efficient and unbiased estimates of the main effects, X_0 at the uth level, and Y_0 at the vth level.

If only the intra-block information is used, another estimate of the missing observation may be obtained from the formula appropriate to the analysis of the fully adjusted yields (Cornish 1940b). This estimate, designated an intra-block estimate, may be combined with the existing values of the total $X_{u\cdot}$ to provide an efficient and unbiased estimate of X_i at the uth level.

Neither of these estimates is, however, suitable for inclusion with the existing values of the total $X_{\cdot v}$ to determine Y_b at the vth level, but since efficient and unbiased estimates of Y_o and Y_t (taken individually) are known, Y_b can be determined with sufficient accuracy from the relation

$$Y_b = 2Y_o - Y_t.$$

X_b and X_t are unaffected since the observations of group Y have been assumed to be intact.

If more than one value is missing this procedure need only be slightly modified to determine the inter- and intra-block estimates of the main effects and interactions (in the case of the cubic lattice).

The formulae for determining the intra-block estimates of the missing values are given below.

Square Lattice.

Let a stand for the missing observation and B for the total of existing values in the block containing a. Let x'_{uv}, $X'_{u\cdot}$, etc. stand for the several totals involving the missing value.

The estimate of a missing observation belonging to variety uv in group X is then

$$a = \left\{2rpB + p^2T'_{uv} + p(X'_{u\cdot} - Y_{u\cdot}) - p(X'_{\cdot v} + Y_{\cdot v}) - (X'_{\cdot\cdot} - Y_{\cdot\cdot})\right\} \Big/ (p-1)(2rp-p-1) \tag{3}$$

The corresponding formula for a missing value in group Y is obtained by a cyclic substitution of the symbols X and Y, and a simultaneous interchange of the subscripts according to the scheme $u\cdot \longrightarrow \cdot v \longrightarrow u\cdot$.

Triple Lattice.

With an extension and slight modification of the notation* used above, the formula for a single missing value belonging to variety uvw in group X is

$$a = \left\{3rpB + p^2T'_{uvw} - p(X'_{u\cdot\cdot} + Y_{u\cdot\cdot} + Z_{u\cdot\cdot}) + \tfrac{1}{2}p(X'_{\cdot v\cdot} + Z_{\cdot v\cdot} - 2Y_{\cdot v\cdot}) + \tfrac{1}{2}p(X'_{\cdot\cdot w} + Y_{\cdot\cdot w} - 2Z_{\cdot\cdot w}) - \tfrac{1}{2}(2X'_{\cdot\cdot\cdot} - Y_{\cdot\cdot\cdot} - Z_{\cdot\cdot\cdot})\right\} \Big/ (p-1)(3rp-p-1). \tag{4}$$

Formulae for groups Y and Z are obtained by a cyclic substitution of the symbols X, Y, and Z, and an interchange of the subscripts according to the scheme $u\cdot\cdot \longrightarrow \cdot v\cdot \longrightarrow \cdot\cdot w \longrightarrow u\cdot\cdot$.

Cubic Lattice.

The formula for a single missing value belonging to variety uvw in group X is

$$a = \Big\{3rp^2B + p^3T'_{uvw} - \tfrac{1}{2}p(Y_{u\cdot\cdot} + Z_{u\cdot\cdot} - 2X'_{u\cdot\cdot}) - \tfrac{1}{2}p(X'_{\cdot v\cdot} + Y_{\cdot v\cdot} - 2Z_{\cdot v\cdot}) - \tfrac{1}{2}p(X'_{\cdot\cdot w} + Z_{\cdot\cdot w} - 2Y_{\cdot\cdot w}) + \tfrac{1}{2}p^2(X'_{uv\cdot} + Y_{uv\cdot} - 2Z_{uv\cdot}) + \tfrac{1}{2}p^2(X'_{u\cdot w} + Z_{u\cdot w} - 2Y_{u\cdot w}) - p^2(X'_{\cdot vw} + Y_{\cdot vw} + Z_{\cdot vw}) - \tfrac{1}{2}(2X'_{\cdot\cdot\cdot} - Y_{\cdot\cdot\cdot} - Z_{\cdot\cdot\cdot})\Big\} \Big/ (p-1)(3rp^2 - p^2 - p - 1). \tag{5}$$

* See footnote on next page.

Formulae for groups Y and Z are obtained by a cyclic substitution of the symbols and interchange of the subscripts in the 3rd, 4th, and 5th terms. The 6th term is constant (except for the shift of the prime) in groups X and Y, and the 7th term is constant in groups X and Z. The term in ($Y._{vw}$ + $Z._{vw}$ − $2X._{vw}$) is constant in groups X and Z; it is then clear how the remaining terms change.

If several values are missing, the estimates are most conveniently determined by a process of successive approximation. This is accomplished by repeated application of the formula for a single missing value, substituting approximations for the unknowns. Careful selection of the first approximations ensures rapid convergence of the solution.

Example.

Cox *et al.* (1940) have given an example showing the analysis of a square lattice involving 81 varieties in which the X and Y groups were duplicated ($p = 9$ and $r = 2$). Since the original observations are too extensive for reproduction, only the quantities needed to illustrate the analysis are given here. Table 6 shows the block totals.

TABLE 6.

Block Totals.

Group X			Group Y		
$.v$	Replication 1	Replication 2	$u.$	Replication 1	Replication 2
.1	277.2 + a_1	233.3	1.	272.4	215.2
.2	284.1	256.7	2.	233.9 + a_2	243.7
.3	315.5	266.8	3.	281.6	222.5
.4	255.4	260.9	4.	309.8	237.4
.5	310.0	247.2	5.	293.5	223.5
.6	316.0	235.6	6.	229.4	257.2
.7	312.0	305.9	7.	324.4	265.2
.8	309.0	288.8	8.	321.5	240.9
.9	311.4	235.5	9.	327.3	269.9
..	2690.6 + a_1	2330.7	..	2593.8 + a_2	2173.5

Two observations have been assumed missing; that of variety 11 in the first replicate of group X (actual value 28.3) and that of variety 21 in the first replicate of group Y (actual value 29.3). Table 7 includes the remaining totals needed in the analysis.

TABLE 7.

$u.$	$X_{u.}$	$.v$	$X._{v}$	$u.$	$Y_{u.}$	$.v$	$Y._{v}$	$u.$	$T_{u.}$	$.v$	$T._{v}$
1.	492.8+a_1	.1	510.5+a_1	1.	485.6	.1	499.2+a_2	1.	978.4+a_1	.1	1009.7 +a_1+a_2
2.	551.6	.2	540.8	2.	477.6+a_2	.2	532.2	2.	1029.2+a_2	.2	1073.0
3.	562.7	.3	582.3	3.	504.1	.3	529.9	3.	1066.8	.3	1112.2
4.	572.6	.4	516.3	4.	547.2	.4	527.8	4.	1119.8	.4	1044.1
5.	529.6	.5	557.2	5.	517.0	.5	536.0	5.	1046.6	.5	1093.2
6.	550.5	.6	551.6	6.	486.6	.6	529.1	6.	1037.1	.6	1080.7
7.	595.2	.7	617.9	7.	589.6	.7	545.0	7.	1184.8	.7	1162.9
8.	591.9	.8	597.8	8.	562.4	.8	534.0	8.	1154.3	.8	1131.8
9.	574.4	.9	546.9	9.	597.2	.9	534.1	9.	1171.6	.9	1081.0
..	5021.3+a_1	..	5021.3+a_1	..	4767.3+a_2	..	4767.3+a_2	..	9788.6+a_1 +a_2	..	9788.6 +a_1+a_2

$T_{11} = 65.9 + a_1$, $T_{21} = 87.7 + a_2$

* In the case of the triple lattice it is convenient to modify the notation so that the factor x is confounded in group X, but unconfounded in groups Y and Z. This is, of course, the notation originally adopted by Yates for this design. On the other hand, in the square lattice x is unconfounded in group X but confounded in group Y. In the cubic lattice a similar notation is adopted. The factor x is unconfounded in group X but confounded in groups Y and Z.

Unadjusted estimates.

The value 26.18 has been taken as the first approximation to a_1. Applying the ordinary randomised block formula, the first approximation to a_2 is

$$a_2 = (81 \times 87.7 + 4 \times 2593.8 - 9814.78)/240 = 31.93.$$

This is substituted for a_2, and the second approximation to a_1 is determined and so on. The final values are

$$a_1 = 26.17, \quad a_2 = 31.93.$$

Intra-block estimates.

The first approximation to a_1 has been taken as 27.91. Using this value,

$$T'_{21} = 87.7, \quad Y'_{2.} = 477.6, \quad Y'_{.1} = 499.2, \quad Y'_{..} = 4767.3,$$

$$B' = 233.9, \quad X_{2.} = 551.6, \quad X_{.1} = 538.41, \quad X_{..} = 5049.21.$$

and $p = 9$, $r = 2$, so that, from equation (3), the first approximation to a_2 is

$$(36 \times 233.9 + 81 \times 87.7 - 9 \times 1029.2 - 9 \times 39.21 + 281.91)/208 = 29.76$$

The final values are

$$a_1 = 27.90, \quad a_2 = 29.76.$$

VII. ESTIMATION OF w AND w'

If the intra-block estimates of the missing values, calculated from the formulae of Section V, are included with the existing observations, an analysis of the type shown in Table 3 may be completed. The mean squares for blocks (eliminating varieties) and residual in this analysis provide the basis for estimating w and w. From this analysis, the intra-block error variance is correctly estimated by dividing the residual sum of squares by the corresponding number of degrees of freedom, reduced by the number of missing observations, but the inter-block variance requires further consideration before it can be used. It can be shown that the inter-block variance in the analysis of the completed set of observations is always too large. When only one observation is missing, the reduction in the sum of squares for blocks (eliminating varieties) is directly calculable from the expression

$$(mV + nR - G - mna)^2/mn(m - 1)(n - 1), \qquad (6)$$

where a is the estimate of the missing value and V, R, and G are the variety, group and grand totals, respectively, including the estimate; m is the number of varieties (and equals p^2 or p^3) and n the number of groups.

In order to determine the correct sum of squares for blocks when more than one value is missing, constants representing groups and varieties only must be fitted, and the consequent reduction in the sum of squares determined, but it is not necessary to calculate these constants directly. An alternative and simpler procedure consists of evaluating an auxiliary set of estimates of the missing observations by successive approximation using the ordinary randomised block formula. These estimates are, in fact, the unadjusted estimates of the missing values and have already been determined for the purpose of estimating the unadjusted varietal effects. The auxiliary estimates are substituted for the missing values and the completed set is analysed by the ordinary randomised block procedure. The residual sum of squares in this analysis is deducted from the total sum of squares of existing values to give the sum of squares due to fitting constants for groups and varieties only, and the deduction of the latter sum, from the sum of squares due to fitting constants for groups, varieties, and blocks, gives the correct sum of squares for blocks (eliminating varieties). The reduction in the block sum of squares is never likely to be large, and if only a few observations are missing, it will be negligible.

Following Yates, the estimate of w is given by the reciprocal of the intra-block error variance, but before the corrected inter-block variance can be used in the estimation of w' it requires still further consideration. When the observations are complete the mean square for blocks (eliminating varieties) has the expectation

$$F(A) + B,$$

where the function F depends on the type of lattice and the number of times the complete lattice is replicated. For example, with a single replication of a complete square lattice, this expectation is

$$\tfrac{1}{2}pA + B.$$

If certain observations are missing, the reduced inter-block variance has a different expectation, say,

$$f(A) + B$$

and $F(A) > f(A)$. Considerable labour is required to determine the exact value of the function $f(A)$, and, unless it is essential to have its true value, the work involved will not be worthwhile. For all practical purposes the function $F(A)$ can be used and this approximation is never likely to introduce serious errors. As a rough guide to the magnitude of the difference to be expected, the exact value of the expectation, $f(A) + B$, has been determined for the smallest square lattice that can be analysed when the data are incomplete. This is a square lattice with 9 varieties ($p = 3$, $r = 1$). With a complete set of observations

$$F(A) + B = 1.5\,A + B.$$

Two values were assumed to be missing, one in each group, and for this particular case

$$f(A) + B = 1.1429\,A + B.$$

The error introduced by counting the inter-block variance as having the expectation $1.5\,A + B$ instead of $1.1429\,A + B$ is relatively high in this instance, but it must be remembered that 11 per cent. of the observations have been assumed to be missing. It seems likely that an experiment would only be worth analysing in very special circumstances when as much as, or more than, 10 per cent. of the observations are missing. For larger square lattices, and for triple and cubic lattices in which only a few observations are missing, the error involved in taking $F(A) \div B$ instead of the true expectation is never likely to be appreciable.

The estimation of w and w' may be illustrated by further consideration of the example given in the previous section. The crude sum of squares of the original 324 observations is 305930.06 so that, after deducting $28.3^2 + 29.3^2$, the sum of squares of the supposedly existing set is 304270.68.

The unadjusted estimates, 26.17 and 31.93 are substituted for a_1 and a_2 respectively, and the experiment is analysed as an ordinary randomised block design of 81 varieties in quadruplicate. This gives the analysis of variance of Table 8.

TABLE 8.

Analysis of completed set (unadjusted values).

Variation due to -	Deg. freed.	Sum Sq.
Groups	3	2372.67
Varieties	80	2180.68
Residual	238	2170.17
Total	321	6723.52

The intra-block estimates, 27.90 and 29.76, are now substituted for a_1 and a_2 and the analysis of variance is completed according to the scheme of Table 3. This analysis is given in Table 9.

TABLE 9.

Analysis of completed set (intra-block estimates).

Variation due to -	Deg. freed.		Sum Sq.	
Groups		3		2374.88
Component (a)	16		961.80	
Component (b)	16		343.03	
Blocks (eliminating varieties)		32		1304.83
Varieties (ignoring blocks)		80		2159.19
Residual (intra-block error)		206		871.05
Total		321		6709.95

The total sum of squares of deviations of the existing set from their mean is

$$304270.68 - 9788.6^2/322 = 6703.32,$$

and from this sum, the residual sum of squares of Table 8, 2170.17, is deducted to obtain the reduction in the sum of squares due to fitting constants for groups and varieties only. This gives

$$6703.32 - 2170.17 = 4533.15.$$

Finally this quantity is deducted from the sum of squares due to group, variety, and block constants to give the reduced sum of squares due to blocks (eliminating varieties). The steps in these calculations are set out in Table 10.

TABLE 10.

Analysis of existing set (intra-block estimates).

Variation due to	Deg. freed.	Sum sq.	Mean sq.
Total	321	6703.32	
Residual (from Table 9)	206	871.05	4.2284
Group, variety and block constants	115	5832.27	
Group and variety constants	83	4533.15	
Remainder (blocks elim. var.)	32	1299.12	40.5975

The expectation of the block mean square with a complete set of data is $\frac{3}{4}9A + B$. Since only two observations are missing, the mean square 40.5975 will have an expectation nearly equal to $\frac{3}{4}9A+B$ and using this approximation

$$w = 1/(4.2284) = .23650,$$

$$w' = 3/(4 \times 40.5975 - 4.2284) = .01897,$$

and

$$\frac{w - w'}{w + w'} = \lambda = .8515.$$

VIII. APPROXIMATE TEST OF SIGNIFICANCE.

An analysis of variance of the form given in Table 3 does not directly provide the appropriate varietal sum of squares for use in a test of significance of the partially adjusted yields. The process of weighting described in Section V alters the sums of squares for the main effects (and two-factor interactions in the case of the cubic lattice) but does not influence interactions that are unconfounded. For example, in the square lattice, the sums of squares for the main effects X and Y are derived from the sums of squares of deviations of the weighted totals,

$$wX_{u\cdot} + w'Y_{u\cdot}\,,$$
$$wY_{\cdot v} + w'X_{\cdot v}\,,$$

respectively, and the sum of squares for the two-factor interaction XY from the unadjusted totals T_{uv}, $T_{u\cdot}$, and $T_{\cdot v}$. The test of significance is not exact owing to the inaccuracy of weighting, and when the data are incomplete further errors are introduced by the method of determining the inter- and intra-block estimates described in Section VI.

Considering only the unadjusted values, an exact test of significance of the varietal mean square can be made, but, if both the varieties and the small blocks are really differentiated, this test will, in general, underestimate the significance of the partially adjusted yields. The test has, however, the advantages that it is easily made and that it will suffice in the majority of cases. In the analysis of the completed set of observations, including the auxiliary or unadjusted estimates, the sum of squares for varieties is exaggerated. The amount by which this sum must be reduced to make the test exact is directly expressible in terms of a simple formula. Denote the estimated yield of a missing value in a group containing only one such value by a, and the group total including this value by R_a. If the corresponding quantities in a group containing two missing values are b_1, b_2, and R_b and so on, the reduction is

$$\frac{1}{m(m-1)} S_1(R_a - ma)^2 + \frac{1}{2m(m-2)} S_2 [2R_b - m(b_1 + b_2)]^2 + \frac{1}{2} S_2 (b_1 - b_2)^2 + \ldots \qquad (7)$$

where S_1, S_2, ... denote summation overall groups containing one, two... missing values. It is obvious that if the unadjusted means are significantly different as judged by this test, no further test is necessary.

If the inter-block information is rejected and only the intra-block information considered, an exact test of the significance of the intra-block varietal effects can be made. In order to make this test, the sum of squares for varieties (eliminating blocks) must be evaluated. This is most conveniently obtained in the following way. Using the intra-block estimates of the missing values, the sum of squares for blocks (ignoring varieties) is determined and from it is deducted the sum of squares for blocks (eliminating varieties) from Table 3. The difference between these two quantities is now deducted from the sum of squares for varieties (ignoring blocks) to give the sum of squares for varieties (eliminating blocks). As it stands, this quantity is not suitable for use in an exact test of significance since it also would yield an exaggerated estimate of the varietal mean square. The necessary reduction is, however, easily calculated from a formula similar to that given above. R will now stand for a block total and p is substituted for m. This test also underestimates the true significance of the partially adjusted yields, but in cases where the blocks are differentiated the degree of underestimation is much smaller than with the first test.

These tests may be illustrated with the data of the example discussed above. The sum of squares for varieties, 2180.68 of Table 8, has first to be reduced. Substitute the unadjusted estimates 26.17 and 31.93 for a_1 and a_2 in the appropriate group totals of Table 6 and apply the formula (7). The reduction is

$$\frac{1}{m(m-1)} S(R_a - ma)^2$$

$$= \frac{1}{81 \times 80} [(2716.77 - 81 \times 26.17)^2 + (2625.73 - 81 \times 31.93)^2]$$

$$= 55.24,$$

and the first approximate test of significance is as follows:-

	Deg. freed.	Sum sq.	Mean sq.	Ratio
Varieties (ignoring blocks)	80	2125.44	26.568	2.91
Residual	238	2170.17	9.118	

The varieties are significantly different so that actually no further test is necessary.

For the second test the sum of squares for varieties (eliminating blocks) is required. Substitute the intra-block estimates 27.90 and 29.76 for a_1 and a_2 in the appropriate block totals of Table 6, and determine the sum of squares for blocks (ignoring varieties). This sum is

$$\frac{1}{9}(305.1^2 + \ldots + 233.3^2 + \ldots + 272.4^2 + \ldots + 213.2^2 + \ldots)$$
$$- \frac{1}{81}(2718.50^2 + 2330.70^2 + 2623.56^2 + 2173.50^2)$$
$$= 2228.66$$

From Table 9, the sum of squares for blocks (eliminating varieties) is 1304.83 and so the difference between these two sums for blocks is 923.83. This quantity is now deducted from the sum of squares for varieties (ignoring blocks), 2159.19, also from Table 9, to give 1235.36 as the sum of squares for varieties (eliminating blocks). Substituting for a_1 and a_2 and applying formula (7) the necessary reduction in this sum of squares is

$$\frac{1}{9 \times 8}\left\{(305.10 - 9 \times 27.90)^2 + (263.66 - 9 \times 29.76)^2\right\} = 40.74$$

The second approximate test is then

	Deg.freed.	Sum sq.	Mean sq.	Ratio
Varieties (eliminating blocks)	80	1194.62	14.933	3.53
Residual	206	871.05	4.228	

IX. CALCULATION OF ADJUSTED VARIETAL MEANS.

In order to adjust the varietal means, the standard method of adjustment has to be modified only for the first two levels of factor x, and the first level of factor y, since these are the only terms affected by the occurrence of the missing values. Consider the hypothetical factor x first, and let X_o, X_i, and X_b stand respectively for the unadjusted, intra-block, and inter-block estimates of the mean yields of factor x at the uth level. X_o at the first level is obtained from $T_{1.}$ after including the unadjusted estimate of a_1, i.e., 26.17, so that

$$X_o = 1004.57/36 = 27.9047.$$

Since the total $Y_{1.}$ contains no missing value

$$X_b = 485.6/18 = 26.9778.$$

Thus the adjustment term for the first level of factor x is

$$.8515 \times .9269 = .7893$$

For the second level of factor x, the unadjusted estimate of a_2, viz. 31.93, is substituted for the missing value in $T_{2.}$ and so, at the second level,

$$X_o = 1061.13/36 = 29.4758.$$

Since $Y_{2.}$ contains a missing value, X_b at the second level must be obtained from the relation

$$X_b = 2X_o - X_i$$

in which X_o and X_t are known. $X_t = 551.6/18 = 30.6444$, hence $X_b = 28.3072$, and the adjustment is

$$.8515 \times 1.1686 = .9951.$$

Y_o at the first level is obtained from $T_{.1}$, including both unadjusted estimates, and is

$$1067.80/36 = 29.6611.$$

$Y_t = 29.3867$, and is obtained from $Y_{.1}$, including the intra-block estimate of a_2. From the values of Y_o and Y_t, the value of Y_b is 29.9355, and the adjustment is

$$.8515 \times (-.2744) = -.2337.$$

Using these quantities, the partially adjusted varietal means may now be determined by the standard procedure.

X. ERRORS OF VARIETAL DIFFERENCES.

If the main tests of significance are made by means of the analysis of variance, precise estimates of the variances of the varietal comparisons are not necessary, and provided care is taken in claiming significance for differences near the border line (for a given level of significance), the normal values should suffice. A guide to the magnitude of the increases to be expected in the case of the fully adjusted yields has been given (Cornish 1940b) and from the expressions quoted, it is obvious that for all lattices likely to be used in practice the increases are negligibly small.

XI. REFERENCES.

Cornish, E.A. (1940a). - The estimation of missing values in incomplete randomised blocks experiments. *Ann. Eugen.* 10: 112-118.

______(1940b). - The estimation of missing values in quasi-factorial designs. Ibid. 10: 137-143.

______(1940c). - The analysis of quasi-factorial designs with incomplete data. 1. Incomplete Randomised Blocks. *J. Aust. Inst. Agric. Sci.* 6: 31-39.

______(1941a). -The analysis of quasi-factorial designs with incomplete data. 2. Lattice Squares. Ibid. 7: 19-26.

______(1941b). - The analysis of quasi-factorial designs with incomplete data. 3. Square, triple and cubic lattices. (unpublished).

Cox, G.M., Eckhardt, R.C., and Cochran, W.G. (1940). - The analysis of lattice and triple lattice experiments in corn varietal tests. Iowa Agr. Expt. Sta. Res. Bull. 281

Fisher, R.A., and Yates F. (1938). - "Statistical Tables for Biological, Agricultural, and Medical Research." (Oliver and Boyd : Edinburgh).

Yates, F. (1933). - The analysis of replicated experiments when the field results are incomplete. *Emp. J. Expt. Agric.* 1: 129-142.

Yates, F. (1936a). - A new method of arranging variety trials involving a large number of varieties. *J. Agric. Sci.* 26: 424-455.

______(1936b). - Incomplete Randomised Blocks. *Ann. Eugen.* 7: 121-140.

______(1937). - A further note on the arrangement of varietal trials: Quasi-Latin squares. Ibid. 7: 319-332.

______(1939). - The recovery of inter-block information in variety trials arranged in three dimensional lattices. Ibid. 9: 136-156.

______(1940a). - Modern experimental design and its function in plant selection. *Emp. J. Expt. Agric.* 8: 223-230.

______(1940b). - Lattice Squares. *J. Agric. Sci.* 30: 672-687.

______(1940c). - The recovery of inter-block information in balanced incomplete block designs. *Ann. Eugen.* 10: 317-325.

Transactions of Institution of Engineers, Aust., Vol. 24, Mar. 1943.

TRANSACTIONS OF THE INSTITUTION

The Application of Statistical Methods to the Quality Control of Materials and Manufactured Products.

A Symposium of Papers* by—
ASSOC. PROFESSOR M. H. BELZ, M.A., M.SC.,
E. A. CORNISH, M.SC., B.AGR.SC.,
A. L. STEWART, B.SC., B.E., A.M.I.E.AUST.

I—ASSOC. PROF. M. H. BELZ.

The subject of the symposium of papers is an ambitious sounding title, but I do not wish it to convey the impression that it will solve all troubles. We believe that it will solve some in a way that is thoroughly scientific and so help the war effort, and on this account we believe it to be desirable to bring the subject before your notice.

I should like at the outset to enlarge on the adjective " statistical " in the title.

I am well aware that all of you who are interested in manufacturing piece parts are also vitally interested in maintaining the quality of your product, whether this is decided by a direct measurement, or by observing the fraction defective. For this purpose you have introduced elaborate systems of inspection and have in some cases imposed restrictions on the quality more severe than those required by the specification limits. But control of quality in this way has been, I suggest, an empirical process, a guess-work sort of affair. You may argue that you have collected statistics at all stages of your manufacturing process, and that you have used these in determining the details of your production. You may therefore claim that, after you have decided those details, the rest is purely routine. The way in which I use the word " statistical " is however much more significant. If you have read the notes which were printed with the notice of this meeting, you will have perceived that they are wound round the underlying mathematical theory of the subject, in which probability considerations play the dominant role. The application of these statistical principles to quality control represents an entirely new departure in manufacturing technique, and any method for controlling quality which does not make use of them, is, I repeat, a guess-work affair.

This meeting is the Melbourne counterpart of a highly successful joint meeting of The Institutions of Civil, Mechanical, and Electrical Engineers, held in London in April, 1942. For a long time in Australia it has been felt that a wider appreciation of the results of statistical theory, in particular, those relating to quality control, could not fail to be beneficial in engineering practice. When Mr. McCleery, a statistician of the N.S.W. Department of Agriculture, was attached to the Small Arms Factory at Footscray to look after quality control there, we thought that the habit would spread and that both manufacturers and purchasers would be quick to see its advantages. But things did not move at all fast, and when we read about the London meeting, which gave to Quality Control a garb of the utmost respectability, we at once realised that a similar meeting here would be the best means for achieving the same objective in the shortest possible time. Our sincere thanks are due to the various Institutions, for falling in so readily with our suggestions.

It is a commonplace that all parts made by a repetitive process are subject to variability. In all specifications there are set out tolerance ranges, prescribing the permissible amount of variability. If this is exceeded, the product is usually scrapped—in some rare cases, regraded. What is not set out, however, is how the tolerance limits are arrived at. To quote an actual case, can anyone tell me how the standardizer who drew up the specification for lavatory seats for a certain branch of the Services arrived at the tolerances on the dimensions of the hole? Except in a few rare instances, as for example, in the construction of lenses where images from two faces cease to coincide if the tolerance is beyond a certain calculable amount, it is pretty safe to assume that tolerances are arrived at empirically. If a rod is to be 0.125 in. in diameter, the standardizer who specifies a tolerance of 0.003 in. is undoubtedly putting down a figure which merely " looks right " to him, and the manufacturer must, in general, accept this figure and keep within the range. He sets up his machinery for producing the part and ultimately reaches one of two stages : either it is uneconomic to work to the specifications imposed because the number of rejections is too large, or a state is reached where there appears to be control over the quality.

The ideal situation here is that in which there are no rejections and if this is realized, even for a short time, the variations which certainly occur but which do not disturb the manufacturer, are ascribed to that nebulous thing called " chance." It is only when the variation is so pronounced as to fall foul of the specifications that an extraneous cause for it is sought. Now this sort of routine is unsound and has no scientific basis whatever. It merely shows the red light indicating that things are wrong. What is wanted, I suggest, is the warning amber light, and the only satisfactory means of supplying this is by statistical methodology. This conclusion applies also to those cases where the red light of intolerance is anticipated by more severe empirical gauging than the specifications require. This procedure will certainly assist in keeping down rejection, but it may be altogether too severe, or not severe enough, and accordingly

*This Symposium of papers, No. 802, was presented before a Joint Meeting of the Melbourne Division of The Institution, the Victorian Branch of The Australian Chemical Institute, The Institute of Industrial Management (Aust.), and the Victorian Division of The Institution of Automotive Engineers, with the cooperation of the Standards Association of Australia. The Meeting was held at the University of Melbourne on 9th December, 1942.

The firstnamed author is Associate Professor of Mathematics, University of Melbourne; Mr. Cornish is Officer-in-Charge, Biometrics Section, Council for Scientific and Industrial Research; and Mr. Stewart is a Technical Officer of the Standards Association of Australia.

uneconomic. How to determine the economic degree of severity, we shall proceed to demonstrate.

We all have a general idea of what we mean by chance variations. Take the familiar case of a hair cut. Sometimes I go for 28 days, others 36 and so on, and the multitudinous contributing factors to this variation provide the chance elements which over a very long period have kept me a regular patron of the barber, with a mean of about 30 days and a variability up to 5 days. There is a sort of comfortable stability about this procedure and if it is seriously interfered with, I can certainly nail it down to a major circumstance. In manufacturing processes, the chance element is always present, but here too there may be a major disturbing factor imposed on the chance factors. To test this we must examine the variability of the product. If the results show what I have called a comfortable stability, nicely grouped about the mean with no pronounced outlying values at either end, this is taken to indicate that chance only is operating.

This regular grouping about the mean can be predicted mathematically under ideal conditions, such as the ideal tossing of ideal coins ; and people have been amused from time to time to verify the theory by showing that the predicted form is realized very closely when actual coins are tossed some thousands of times. I have here an experiment to demonstrate this point, which is much quicker in action than coin tossing.

It consists of a funnel, with 120° angle and ½ in. opening filled with steel balls of ¼ in. diameter. There is a set of channels at the bottom, each ½ in. wide, and also several rows of pins, all at ½ in. spacing. If the hole is above an open channel of pins, all the balls fall into this channel. Suppose now we move the top and every second row of pins ¼ in. to the right. A ball coming to the first pin now has an equal chance of going left as right and so we expect

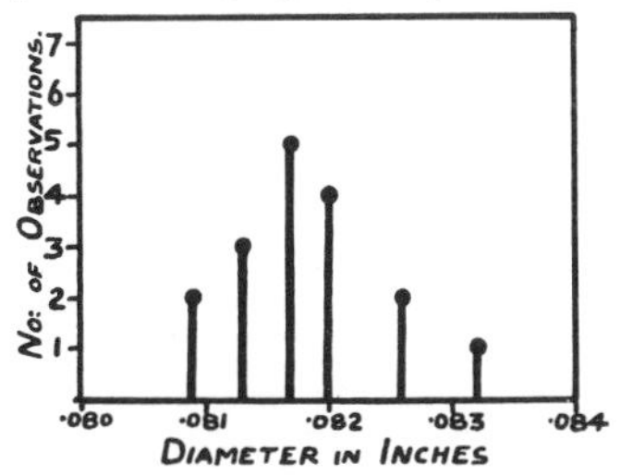

Fig. 1 (*I*).—*Frequency Diagram.*

that half the balls will go left and half right at this pin. Those that go left are similarly halved at the next row ; and so on. The result is a distribution of balls over a certain range, with varying numbers of balls in the compartments. If the chance of going left or right at each pin were exactly ½, we should theoretically arrive at a symmetical distribution, which is supported by the results here obtained.

If we had arranged the pins so that each trisected the space between the pins above it, one-third of the balls would pass in one direction and two-thirds in the other at each pin, and we should now get an unsymmetrical distribution. Once again, however, we have that comfortable stability about the distribution.

When we come to manufactured products, a distribution of the observed quality, e.g., a length, which conforms in general character to the above shapes, is taken to indicate that chance only is operating and the process is said to be under statistical control.

Let us illustrate by a cylindrical piece part, the specifications for which prescribe that the diameter shall be between 0.080 in. and 0.084 in. A batch of 17 of these is measured for diameter with the results shown in Fig. 1. The horizontal scale is inches. The vertical scale gives the number of parts with corresponding diameter.

	2	with diameter	0.0809 in.
	3	,, ,,	0.0813 in.
	5	,, ,,	0.0817 in.
	4	,, ,,	0.0820 in.
	2	,, ,,	0.0826 in.
	1	,, ,,	0.0832 in.
Total :	17	*Mean diameter*:	0.0818 in.

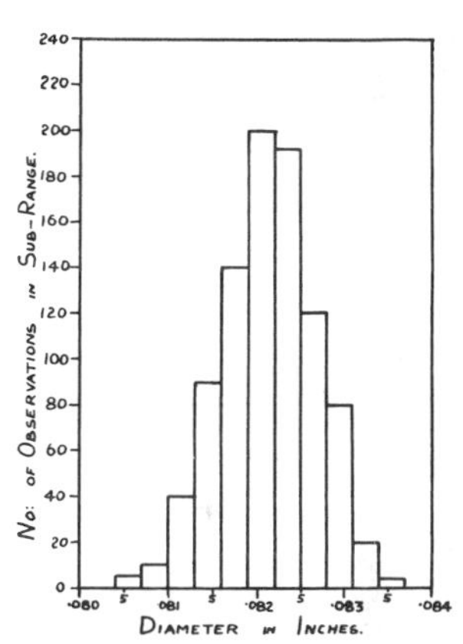

Fig. 2 (*I*).—*Histogram.*

Suppose now that we have measured 1,000 of these parts. Instead of plotting ordinates for each of the different diameters, which might be very numerous, a more informative procedure would be to divide the range of variation up into sub-ranges, say 10 to 15, and group together all the observations falling in a sub-range. The corresponding diagram is now composed of a number of contiguous rectangles, as in Fig. 2, in which there are actually 903 observations and 13 sub-ranges. The important feature of this diagram is that the area of each rectangle is proportional to the number of observations falling within the corresponding sub-range. If, by a simple change of scale, we say that the total area of all the rectangles is 1, the area of any one rectangle becomes the *proportion* of observations falling within the corresponding sub-range. The area to the right of any lower corner is then the proportion of observations with diameters exceeding this value, and so on.

On increasing the size of the sample, and at the same time increasing the number of sub-ranges, the rectangles become finer and finer and ultimately merge into a smooth curve. We shall suppose that the device mentioned above has been followed so that at each stage we have a set of rectangles of total area 1. The limiting smooth curve, which refers to the infinite bulk from which the samples are drawn, will likewise have unit area. We may get the curve shown in Fig. 3, and it is called a *probability curve.* Areas beneath this curve play a similar role to the areas of the above rectangles. For example, the area shaded becomes the probability that a test piece chosen haphazardly from the bulk will have a diameter exceeding 0.0825 in. This area is about 0.246. This means that if we were to take 1000 test pieces at random, we would expect to find 246 of them with diameters greater than 0.0825 in. and 754 of them with diameters less than 0.0825 in. Some divergence from these expectations would, of course, show up in an actual experiment due to what are called "fluctuations of sampling," but we should be rather surprised if the divergence were at all large.

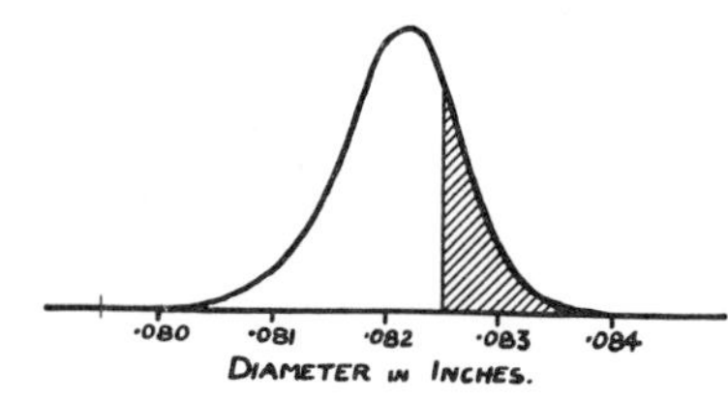

Fig. 3 (*I*).—*Probability Curve.*

Mathematicians have devoted a great deal of time to the study of probability curves of the above and similar forms. By calculating areas for special forms and shapes and tabulating them, they have made available all that is required for action in practical applications. The most useful of all the curves that are employed is the so-called normal curve.

This is a symmetrical curve, as shown in Fig. 4, having a compact mathematical equation, and if we denote by σ the distance from the centre to the point where the curve crosses its tangent, then the following results are known :

The area lying outside the limits centre $\pm$ 1.96σ, is 0.05.
" " " " " " centre $\pm$ 2.58σ, is 0.01.
" " " " " " centre $\pm$ 3σ , is 0.0027.
" " " " " " centre $\pm$ 3.09σ, is 0.002.

Thus, in the long run, we should expect only 1 in 20 of the observations to fall outside the first range, 1 in 100 outside the second range, 1 in 370 outside the third range, and only 1 in 500 outside the fourth range.

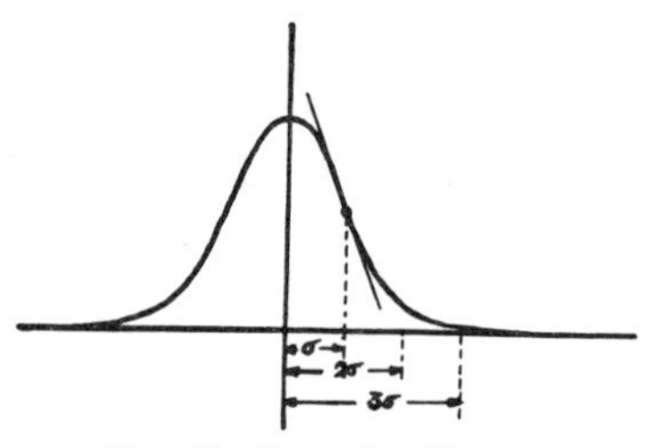

Fig. 4 (*I*).—*Normal Probability Curve.*

In practice we do not know the centre point of the smooth curve or the yardstick σ (called the standard deviation), because we have not the time to measure the whole bulk, so we estimate these values from the samples measured, which are often not large. The centre is estimated by the arithmetic mean (just add all the observations together and divide by their sum) and σ, either by multiplying the range of the observations (greatest value — least value) by an appropriate factor, depending on the size of the sample (which is tabulated), or by taking the root-mean square of the deviations from the mean. If we now calculate the numbers

$$\text{mean} + 3 \times \text{estimate of } \sigma,$$
$$\text{and mean} - 3 \times \text{estimate of } \sigma,$$

and if we find that an actual observation falls outside this range, we have an event which would occur by chance alone only once in 370 trials. That it has happened is, accordingly, felt to be due to something other than chance ; it is the case of a major circumstance, and without any hesitation at all we would say that there is some assignable cause at the bottom of it. Of course, if we wished, we could be even more conservative and draw the lines of reasonableness at the 2.58 figure above and below the mean ; or we could be more lenient and choose the 3.09 figure. These considerations form the statistical basis of quality control. The coefficient which is recommended by the American Standards Association is 3, a figure chosen on economic grounds as a result of many years of study and use of Quality Control in the Bell Telephone Co. and elsewhere. In England the coefficient 3.09 is favoured, chiefly because of its simple interpretation, namely, 1 in 1000 above the upper limit and 1 in 1000 below the lower limit ; for convenience we shall adopt the American standard practice, although there is no significant difference between the two figures.

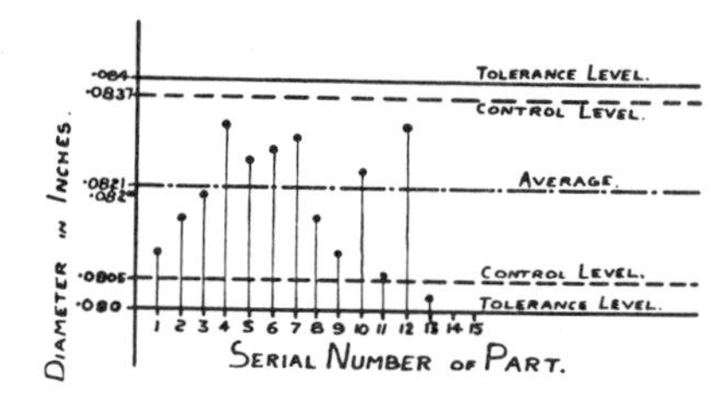

Fig. 5 (*I*).—*Typical Control Chart.*

As soon as we have our limits mean $\pm$ 3 $\times$ estimate of σ, we draw them on a chart and plot subsequent observations on the chart by means of a dot, as in Fig. 5, and we can see at a glance whether the process is under control by simply looking at the successive dots. So long as they continue to fall between the control lines shown, nothing needs attention (items 1 to 12). It does not matter how they lie inside—up or down or anyway—so long as they do not get outside. If they do, as item 13, then the order of the day is—"Look for the cause."

Now I know what you gentlemen will say to this—that what you want is something to tell you the cause of the trouble. That, however, is another problem which does not fall within our more or less academic field. But we are able in the above way to tell you that you must look, and if you will train your technical staff in the statistical technique, or will train statisticians in the technical processes, I feel sure that the combination will lead quickly to the cause, at any rate in many cases. In the past you have taken action only when you had to stop, because the specification limits were exceeded. Now it is a universal experience that when economic statistical control exists the control limits should always fall within the tolerance limits of the specifications. Since action would be taken as soon as a dot falls outside the control lines, we have here a reasoned method which will keep the quality always within safe limits, with practically no rejections.

I have based the above discussion on the normal curve. You may object that this pre-supposes that the observations in all possible samples will possess a symmetrical distribution, which is certainly not always the case. Skew distributions do often occur and the results of the normal curve are not then applicable. This difficulty, however, is overcome by an extremely simple device. Instead of working with the individual observations themselves, we group them in small batches, of four or five at a time, and record their mean values on the chart. An elegant mathematical proof shows that these means themselves are distributed, for all practical purposes, in the normal form, and so all we have to do is to work with these sample means. They will vary about the same general mean as the individual observations, but, of course, the yardstick that measures the variability is now greatly reduced. Its value is immediately deductible from that for the individual observations. Apart from this the procedure is exactly as before.

In addition to control charts for the variability of means of small samples, we can also construct control charts for the range of observation. Suppose we take samples of five at a time; the range of the observations will vary from one sample to the next, but we should not expect more variability in these ranges than can reasonably be ascribed to chance. The theory again shows how to assign limits to the variability that may be permitted statistically, and in practice all one has to do is, as before, to look up the proper entry in a set of tables and construct the control chart. This chart, in conjunction with the chart of means, will usually give all the warning that is required to indicate the need for action.

Fig. 6 shows how the two charts operate together. It is clear from the sub-division of the diagram :

A—Samples 1—10, complete control.
B—Samples 11—20, out of control, because all measurements consistently high; variability unaffected.
C—Samples 21—30, out of control—measurements brought back to control, but range excessive.
D—Samples 31—40, both average and range out of control.

I shall conclude my expository remarks by referring to another type of observation showing how the control chart is applicable to percentage defective.

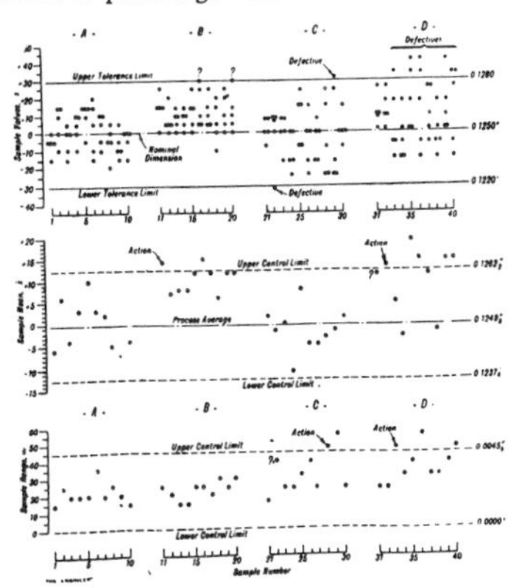

Fig. 6 (I).—Control Charts of Mean and Range.

Suppose that over a long period of observation it is found that the product from one machine is 10 per cent. defective, a figure which the management regards as not uneconomic. A certain batch of parts is measured and the percentage defective is 12. Does this difference indicate that a major circumstance is operating, or is it due solely to chance? The yardstick required to settle this question will, of course, depend on the size of the batch. If there are 10 in it, we expect one defective, but we would not be greatly surprised if we found three, i.e., 30 per cent. defective; but if there are 1000 in it, we should definitely object to 300 rejects when we only expect 100; 12 per cent. defective in this case might be considered satisfactory but 30 per cent. would not. The larger the batch, therefore, the smaller the yardstick.

The yardstick also depends on the value of the percentage defective. This, we may hope, is the average of a large number of percentages calculated on an hourly, daily, or some planned basis, and if it is quite small, say 1 per cent., it means that there cannot be much variability from this figure; any great variability would quickly send up the average. If the proportion is 10 per cent., it means that the individual proportions can range from 0 to 10 on one side, and from 10 upwards on the other. The latter may be up to 30, say, the mean being always at 10, but you see that there is now a much greater variability in the proportion. The larger the proportion, therefore, the larger the yardstick. This can be expressed mathematically.—

Let p denote the fraction defective (so that $100p$ is the percentage defective); thus, if $p = 0.01$, then the percentage defective is 1; if $p = 0.1$, the percentage defective is 10. Then $1 - p$ denotes the fraction effective. Also, let n be the size of the batch. Then the theory shows that the yardstick applicable to p is estimated by

$$\sqrt{\frac{p(1-p)}{n}}$$

Using the American convention, the control limits for p, based on this yardstick, are thus—

$$p + 3\sqrt{\frac{p(1-p)}{n}}$$

and

$$p - 3\sqrt{\frac{p(1-p)}{n}}$$

Taking the above case, suppose that 1000 is the size of the batch per shift coming up for inspection, so that $n = 1000$. With $p = 0.1$, the formula gives—

$$\sqrt{\frac{p(1-p)}{n}} = \sqrt{\frac{0.1 \times 0.9}{1000}} = 0.0095.$$

The control lines are thus set at $0.1 + 3 \times 0.0095 = 0.128$, and $0.1 - 3 \times 0.0095 = 0.072$.

When, now, subsequent batches of 1000 are measured, the proportion defective must lie between these two figures, otherwise the process is judged to be out of control. A chart to illustrate all of this is constructed exactly as before, and if the point is plotted on the chart as soon as the operation is completed, you have the green or amber light immediately in service.

This brings me to the end of my introductory account of the subject, and it is clear that I have only touched the central theme of the method.

II—E. A. CORNISH.

Professor Belz has introduced you to the elements of statistical theory underlying this so-called method of quality control, and in this brief talk I propose to continue the discussion by showing how statistical methods can be used in framing a certain type of acceptance specification.

In judging the quality or, in other words, in testing the conformity of a particular product with specification, two types of test are encountered. In the first place, the nature of the test may be such that all articles produced can be tested, e.g., in measuring the diameter of a shaft; and in the second place, the nature of the test may be such that the product has to be destroyed, e.g., in the final testing of ammunition, fuses, etc.

I propose to consider an example of the second type. Since this type of testing is both costly and destructive it must necessarily be done by sampling. It is thus highly desirable that all possible information should be extracted from the test data. In order to show how statistical theory can be used in the extraction of this information, it will be best if we begin by considering what happens when statistical methods are not applied.

If the job of framing this specification is to be tackled properly three steps are involved :

(i) The determination of a standard of quality.
(ii) The statement of a design specification.
(iii) The statement of an appropriate acceptance specification.

Frequently these are avoided by what appears on the surface to be a very clever and practical stratagem—it is very simple. Just write out a good stiff acceptance specification with the object of getting the very best obtainable. Well, let us see where this sort of approach leads us.

The acceptance specification calls for inspection and, since we are considering the case where the test is destructive, sampling must suffice. The specification, therefore, reduces to one or more requirements such as the following :—

"a test sample of n shall be selected at random from each batch or consignment and not more than C shall fail to pass the test."

Many of you have, or will, come in contact with this type of specification, and those who have will know that the value for n is often small—of the order of 10 to 20. I want to show that this popular and well-known type of specification offers practically no guarantee of the quality of the product which will be accepted under it, and that the quality of product so accepted is practically that which happens to be offered for inspection.

Suppose that a manufacturer submits for acceptance a large consignment containing 10 per cent. defectives and 90 per cent. effectives. If a random sample of 10 articles is selected from this consignment, the probability that the sample will contain :

no defective is	0.35
one defective is	0.39
two or more defectives is ...	0.26

We can state this another way. Suppose repeated samples of 10 are selected from this consignment. In the long run 35 out of 100 samples will be better than the consignment, 39 will be of the same quality as the consignment, and 26 will be lower than the consignment. Notice that those samples better than the batch occur with greater frequency than those poorer than the batch *and consequently there is a marked tendency for the small single sample to misrepresent the batch.* That is the first point I want to establish.

These points, together with some others, are illustrated in Fig. 1. This diagram shows the effects of random sampling in the case of samples of 10. Along the abscissa is plotted the fraction defective (or effective) observed in the sample. The ordinates show the relative frequency with which each type of sample occurs. The asymmetrical curve on the right refers to the hypothetical case of which we have been speaking. The curve centred at 50 per cent. defective shows the distribution of samples from a lot which is 50 per cent. defective. Note that this curve is symmetrical; there is no tendency for the sample to misrepresent the lot, since samples better than the lot occur just as frequently as samples which are poorer than the lot. The dotted curve shows the distribution for samples of 50 from the lot containing 10 per cent. defectives. It shows that, with larger samples, the curve is much more symmetrical and that chance does not play such havoc in the selection of samples.

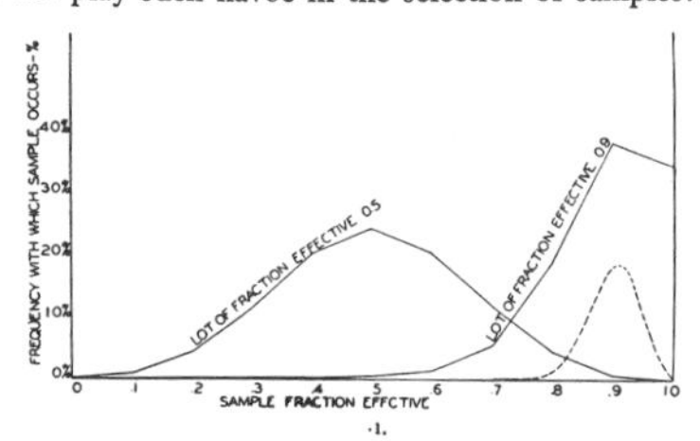

Fig. 1 *(II).—Chart Showing the Effect of Chance in the Selection of Samples of* 10.

It is thus obvious that, if the sample is large enough, or if the fraction defective is near to 0.5, the difficulty of misrepresentation may be overcome, but these conditions are seldom encountered in practice.

Hence with the specification we are considering, small single samples offer no reasonable guarantee of the quality of the product accepted.

Now let us return to the specification and restate it, giving n and c particular values; thus, a sample of 10 shall be selected at random and not more than one shall fail to pass the test.

Suppose two manufacturers submit their products for acceptance. The product of the first, we will suppose, has 10 per cent. defectives. If a sample of 10 is subjected to test, then the chance that his product will be accepted is :

$$0.35 + 0.39 = 0.74$$

i.e., it is the sum of the chances that there will be no defective and one defective in the sample.

If the product of the second manufacturer contains 20 per cent. defectives, then the chance that a sample of 10 drawn from his consignment will contain :

no defective is	0.11,
one defective is	0.27,
two or more defectives is ...	0.62;

so that the chance that his product will pass specification is 0.38.

For the purpose of illustration, we can consider 0.74 as near enough to $\frac{3}{4}$, so that on the average, 3 out of 4 consignments presented by the first manufacturer will pass the acceptance specification, while 1 will be rejected. Notice that the batch rejected is actually acceptable because it really contains only 10 per cent. defectives.

Further, if we take 0.38 as near enough to $\frac{2}{5}$, on the average 2 out of 5 consignments presented by the second manufacturer will be accepted, and 3 rejected.

Thus, sometimes good batches will be rejected and sometimes defective batches accepted, but the really important point to note is that the *batches accepted in the latter case are really no better than those rejected.*

Fig. 2 shows the relationship between probability of acceptance and rejection, and the proportion of defectives in the batch under test. The solid curve refers to the specification about which we have just been speaking. In the shaded area the average quality level of product presented —assuming, of course, that all qualities between 0.80 and 0.90 are presented equally frequently—is ·0.85, and the average quality accepted under this type of specification is 0.86.

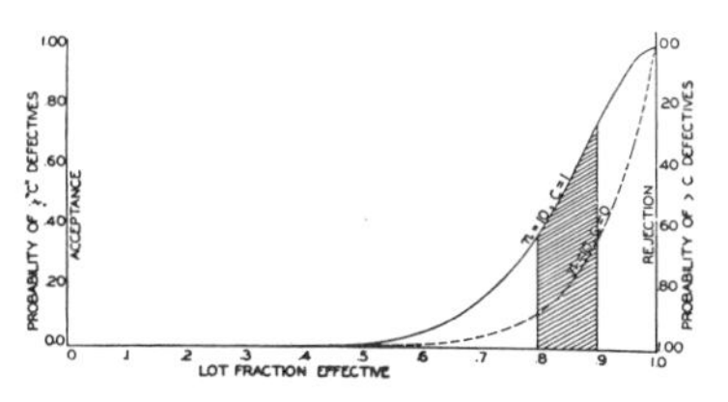

Fig. 2 (*II*).—*Curves Showing the Probability of C or less Defectives in a Sample of n, Drawn from Lots of Various Fractions Effective.*

Hence quality of product accepted is very nearly equal to that of the product presented for inspection.

Note also that even when only 5 per cent. of the product is defective—and this is not by any means generally regarded as good—the product will pass specification 91 per cent. of the time.

The dotted curve shows these relations when c is reduced to 0, in order to make the specification more stringent. Apart from reducing the chance of acceptance of poor quality, the same criticisms can be levelled at it.

You can thus see that, under this type of specification, the only way to prevent the acceptance of poor quality consists in not allowing it to be presented for inspection.

So far the analysis has been wholly destructive. It neither points out the basic faults nor suggests their remedies, so let us now see how a little engineering sense and a few elementary statistical methods can accomplish these constructive ends.

Suppose we return to the hypothetical case of the two manufacturers, to which we have already referred. All consignments of the first manufacturer should have been accepted, but you will remember that some were rejected, whereas all those of the second manufacturer should have been rejected but some were accepted.

The first basic fault of the specification then lies in judging each batch on its own merits on a small sample basis, which even in the case of homogenous batches is obviously not an efficient basis for discrimination between good and bad lots.

The second basic fault (this is a corollary) of the specification is its failure to make use of available knowledge other than that supplied by the sample, but the cogency of this criticism may not be apparent until we describe how statistical methods make efficient use of this knowledge.

Earlier in this talk, I pointed out that the logical approach to a good specification involved three steps :

(i) A standard of quality.
(ii) A design specification.
(iii) An appropriate acceptance specification.

Now let us consider these points in order.

The standard of quality may be regarded as providing a datum point from which quality can be reckoned. We must know the manufacturer's quality level, because in the first place we want to stop the influx of the product of faulty manufacture, and in the second place it is impossible to predict, except approximately, the quality of a batch from a small sample, even though the sample may be perfect.

If the quality cannot be obtained from a large sample we must turn to some other source of information. This is, in general, abundantly available and exists in the past performance of the manufacturer. It is part and parcel of a very valuable asset which is generally called experience and expert engineering judgment. When data are costly and scant, no scrap of pertinent information should be allowed to escape, and statistics supply the competent means of :

(i) determining the validity of the prior knowledge,
(ii) combining mathematically the accumulated knowledge with that supplied by the sample itself.

To develop a technique for accomplishing these objectives, we need to make use of the idea put forward by Professor Belz, viz., the concept of statistical quality control.

As soon as we have established that a state of control exists and have determined the general quality level and its variation from sample to sample, we can predict the limits within which practically all samples of any given size will lie. The small sample then becomes a significant index of batch quality.

To sum up, we may say that, although we cannot judge, except most approximately, the quality of a batch from a small sample, we can judge by statistical methods whether or not there is any reason for believing a batch is different from its predecessors, and knowing the quality level from the accumulated data we can infer the quality of the lot with considerable assurance.

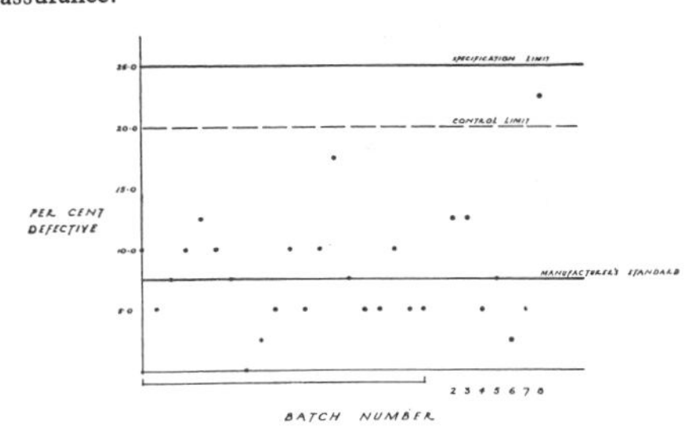

Fig. 3 (*II*).—*Control Chart.*

The design specification is merely a statement of what is wanted. The acceptance specification states the quantity and kind of evidence which will be accepted as satisfactory proof that the product will meet the design specification. If we frame this specification on a statistical base we can consciously minimise the chance rejection of good lots and the chance acceptance of defective lots.

Fig. 3 illustrates the operation of the three sections of the specification and shows the percentage of defectives in successive samples of 40. The large fraction defective does not necessarily indicate poor quality, for the test conditions were made more severe than the working conditions ; otherwise only slightly defective batches would be likely to show no failures at all.

The steps are :—

(i) The state of statistical uniformity, level of quality, and measure of variation are inferred from an accumulation of samples. The general level of quality must show that the product conforms with the design specification, and the statistical uniformity must exist, otherwise it is impossible to discriminate efficiently between batches of good and bad quality by means of small samples. The first 20 points represent this step. From the first batch of articles, 20 samples, each of 40, are selected as nearly as possible in order of manufacture. The fraction defective in each sample is computed and its representative point placed on the chart. The average percentage defective from all 20 samples is computed and shows that the manufacturer's standard meets the design specification. The control limit is calculated and placed on the chart, and we note that no point falls outside this limit, showing that statistcial uniformity exists. At this stage the product meets with the terms of the first part of the acceptance specification.

(ii) Assuming that uniformity exists and the quality level is satisfactory, there must be no change in quality as shown by statistical tests when successive samples are taken from successive batches.

This step is represented by the next six points on the chart. Note that they all fall within the control limit and do not cause the average percentage defective to exceed the design specification. Batches 1 to 7 would, therefore, be acceptable under the conditions of the specification.

(iii) In the event of a rejection, the whole process is started again with the next batch.

This is represented by the last point on the diagram. This batch is rejected because its percentage defective exceeds the control limit showing that statistical uniformity has been lost, even though the percentage defective does not exceed the design specification. The whole process is started again with the next batch.

III.—A. L. STEWART.

The previous speakers have demonstrated the statistical line of thought adopted in methods of quality control as well as some aspects of sampling. I propose to deal with the practical application by means of one or two specific illustrative examples. For the details of these examples I am indebted to overseas sources of information, for, to my knowledge, the methods have not as yet reached a stage of development in Australia to make available similar data from our own manufacture. Investigations which have been, and are being made, may be referred to by some speakers during the discussion to follow.

The work involved in applying the methods is largely engineering based on judgment, knowledge of the processes, and technical skill in tracking down unwanted causes of variation to their source. Problems may possibly arise, however, which can best be solved through co-operation between a technical man conversant with statistical methods, and a statistician having a good knowledge of the production processes. Thus, it is most important that the statistician and the engineer should learn to speak the same language insofar as they are dealing with a common problem.

I have chosen an example to illustrate the practical application of quality control in the manufacture, on a quantity basis, of electro-mechanical apparatus and the component piece parts. The example is taken from the recent experience of one of the well-known British workers in this field, viz., Mr. H. Rissik of Standard Telephones and Cables Ltd. Mr. Rissik collaborated with the chief inspector of Creed & Co. Ltd., of Croydon, in a thorough analysis of their inspection set-up with a view to increasing its effectiveness as an aid to the production organisation. As a result, this long-established and progressive manufacturing firm has been the first production engineering organisation in England actually to apply the technique of quality control to its machine shop processes.

The machine shop at the Creed works deals in the aggregate with some 12,000 separate components, having an average of five distinct machining operations on each. Many of these components are of small size and intricate shape, and the machining limits are of the order of plus and minus 0.001 in.

Due to general production difficulties arising out of the war situation, such as the gradual dilution of labour and the lack of trained inspectors, the firm's accelerated output was being impaired by the increasing proportion of defective piece parts which came to the assembly lines. The actual proportion was not high, but being concerned with the manufacture of precision mechanisms, the firm tried quality control methods as a means of increasing the rate of effective production in the machine shop.

The inspection set-up in the shop provided for the usual "first-off" inspection, followed by a "floating" inspection of the piece parts during actual manufacture, and ending with an independent final inspection of the completed batch at the conclusion of each process operation. In accordance with practice which is by no means uncommon in production engineering work, no records of any consequence were kept of the results of piece part inspection, nor was the inspection of the process operations in the machine shop carried out according to any particular plan. The floor inspector simply went round his group of machines and from time to time measured *some* of the piece parts produced. If they were satisfactory the job went on; if they were wrong, the tool-setter was called upon to give attention to the machine.

To obtain a better inspection control as a preliminary to the statistical control of piece part quality, two changes were made in the inspection organisation and routine:

(1) The system of "floating" inspection was organised so as to ensure regular visits to the machines, a fixed number of four piece parts being inspected at each visit.

(2) A system of Process Inspection Record cards was introduced, each card becoming, in effect, a complete case history of the relevant process.

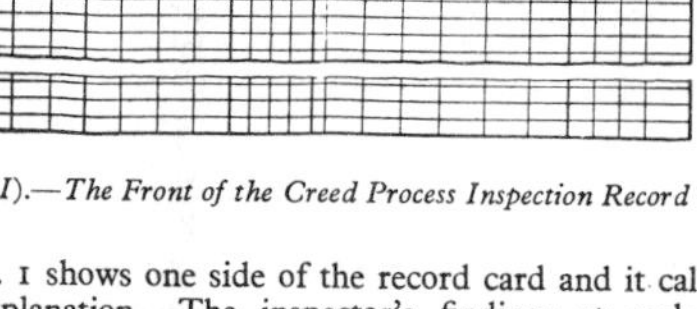

Fig. 1 (III).—The Front of the Creed Process Inspection Record Card.

Fig. 1 shows one side of the record card and it calls for little explanation. The inspector's findings at each visit are recorded in columns, *A*, *B*, *C*, and *D*. The next two columns, headed "mean" and "range," are for use in plotting quality control charts in accordance with the standard methods now laid down. The inspector's findings for any particular points to be watched and which are summarised on the back of the card, are recorded in the four precautionary inspection columns marked 1 to 4. The inspection interval is decided by the production rate and the nature of the process operation, the amount of process inspection varying from about 5 to 10 per cent.

The next step taken was to subject the information given on the record cards to routine statistical analysis by the control chart method, so as to gain some idea as to the stability or otherwise of the various processes. As the result of this routine analysis, confirmation was soon obtained that certain processes known to be troublesome, in that thay had consistently given unusually large numbers of rejects in the batches coming to the assembly lines, were, in fact, difficult to keep within the limits of tolerance specified in the piece part drawing. Quality control charts were therefore introduced for these processes with the object of stabilizing the product by the elimination, where possible, of causes of variation other than those due to chance. It is emphasised here that it is the logical thing to take first a few particular jobs which experience or analysis has shown are difficult

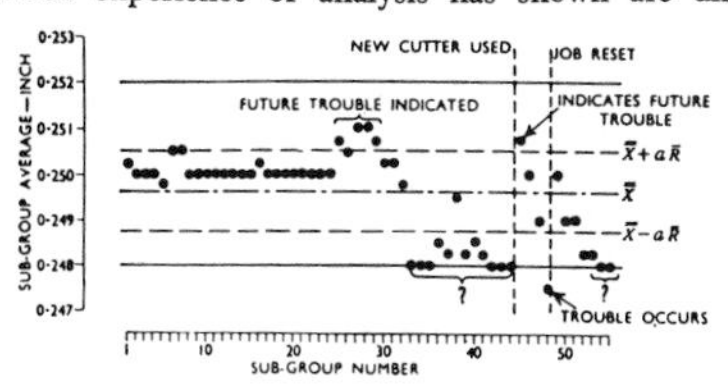

Fig. 2 (III).—Analysis of Milling Operation.

to control. Then, if benefits are obtained from maintaining control charts for these few cases, the system can be extended gradually. No practical engineer would litter a machine shop with paper by introducing two control charts for every dimension of every component for each process.

The two main sections of the shop which had been known to be troublesome were the semi-skilled milling section and the capstan section, and the Inspection Record cards were applied for a start to one or two components from these processes only.

Fig. 2 shows an actual analysis of a milling operation. In the early stages, before a quality control chart was actually used during production, the records from the process inspection were analysed, and interesting information was obtained in this way. It will be noticed that for the first twenty or more points taken over two day shifts and one night shift, the job ran stably, and then suddenly a number of points were obtained above the upper control limit which indicated future trouble. Then, during the next day and night there was a steady drift toward the bottom limit. The inspector knew nothing about control charts at that time, but he realised that something was going wrong and when he

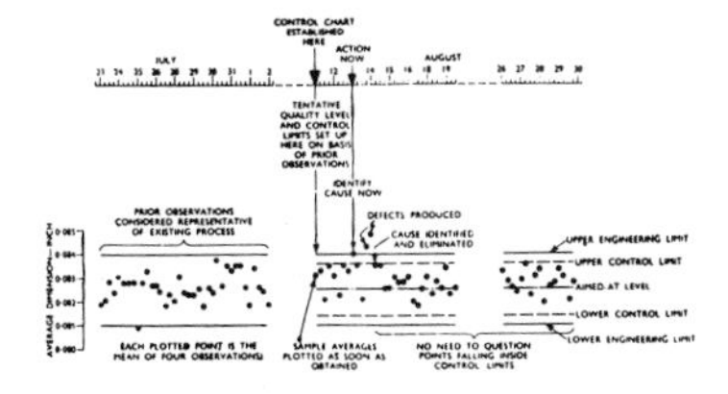

Fig. 3 (III).—The Control Chart as a Guide to Action.

consistently got readings at the lower tolerance limit, he stopped the job and, suspecting tool wear, arranged for a new cutter to be used. In spite of that, the downward trend continued and rejects occurred. The job was then reset and it ran satisfactorily for the rest of the batch except that the last couple of readings were again on the bottom limit. In this case the analysis shows that there was clear indication some 24 hours in advance that trouble was impending and that it was the machine itself that was at fault because the trouble was attributable neither to tool wear nor to faulty setting.

Figs. 3 and 4 illustrate the operation of quality control as an active process, by which inspection becomes effectively productive, as it should be. Only too often in manufacturing organisations the idea prevails that inspection is non-productive, and that is because the function of inspection has degenerated into the mere sorting out of bad work from good, instead of assisting production by the interception of tendencies towards the creation of defective work.

Fig. 3 is an example of a control chart for a capstan operation. On the left side of the chart is shown the plotting of points which are representative of the process, and by using this information control limits were set for the future run. It will be seen that the first few points in the control chart were satisfactory, but there follows a distinct upward drift in average dimensions. When the 9th point fell outside the control limit, an effort was made to find the cause before defectives were produced. The attempt was not at first successful and a certain number of defectives were produced, but later the source of trouble was located and eliminated by the tool-setter. Subsequently, the process continued under stable conditions until the end of the batch.

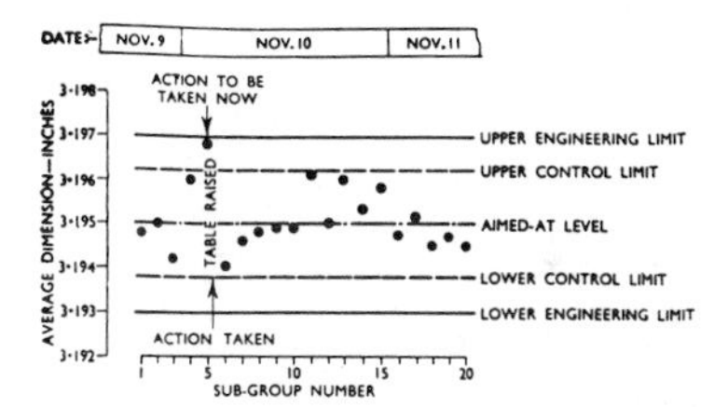

Fig. 4 (III).—Occurrence of Trouble (afterwards put right) in Milling a Difficult Casting.

Fig. 4 shows, perhaps more clearly, a similar application to a milling operation on a difficult casting. The control limits were, of course, set from past records for the process. The points soon showed that something was going wrong and, in the case of a milling operation, it was not as a rule difficult to locate the trouble. It was suspected that either the set-up or the table had shifted. Before rejects were produced the table was raised slightly, and the remainder of the run continued within the control limits.

Startling results were, naturally, not immediate, but as the personnel became accustomed to the use of control charts, it soon became so evident that the amount of defective work was diminishing, that after a trial period of several months, it was decided to extend the system throughout the machine shop as a whole. The inspectors required for this work were those released from the system of independent final inspection which, it had been found, could be abolished.

Fig. 5 gives a general summary of the results recently achieved. The graph shows the labour time spent on producing scrap and on the rectification of defective parts as a percentage of the machine shop or section output in man-hours. The milling section had shown a very satisfactory improvement. The capstan section was semi-automatic, but even so, there had been a certain reduction in defects. The machine shop aggregate had shown in six months a decline of 50 per cent. in defective work. (It should be mentioned here that during the period in question the machine shop production had increased by over 25 per cent.).

In this particular application quality control has undoubtedly assisted production. Apart from the question of minimizing defective work, the following important benefits

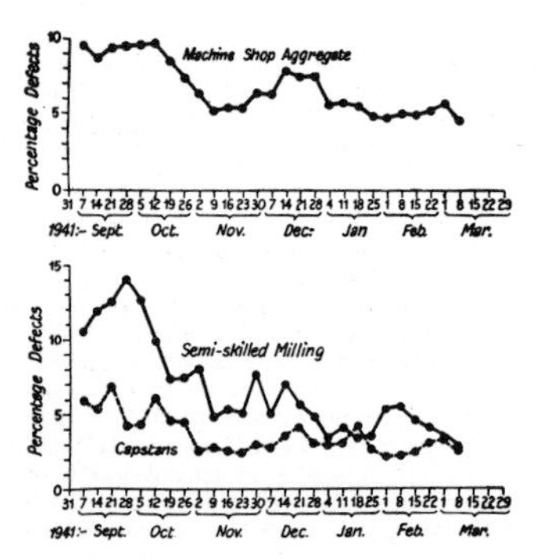

Fig. 5 *(III).—Summary of Six Months Operating Results with Quality Control at Creed and Co. during* 1941-1942.

have been obtained which, in themselves, have more than justified the introduction of the system described.—

(1) The control charts provide factual evidence of the running of the jobs to which they relate. This evidence has proved invaluable in determining the root causes of trouble, e.g., faulty tooling, machine wear, difficulties imposed by design, and so on.

(2) Such evidence has enabled the chief inspector to issue frequent reports on the difficulties encountered with specific jobs and on the condition of machines. This has enabled the planning department to take immediate action before production had to be held up.

(3) On the evidence of control charts it is possible to advise the design department whether or not, in any particular case, it is possible to work to certain narrower limits if that should become necessary at any time.

(4) The use of control charts in the machine shop where they can be seen and studied by inspectors, tool-setters, and operatives alike, has made for closer co-operation between the inspection and production organisations.

This last benefit is, in the long run, the most important of all. The inspector will come to be judged, not by the amount of material he scraps, but by the degree of help he gives to the production engineer in keeping rejections to a low level.

It is of the highest importance, as many of you will agree, that the psychological aspect of introducing a new system into an existing organisation should be carefully handled. It is also desirable that it should be introduced very gradually. In the application just described, before the introduction of quality control as a definite instruction from the management, the firm had held meetings of the inspectors and tool-setters and told them what was being aimed at and how it worked, asking them for their co-operation and also for constructive suggestions. This had proved most successful. The inspectors and setters and the production people on the job who were watching quality and had it in their care, became definitely interested in the new system. The man responsible for a particular job, having had the chart referring to that job interpreted for him, was obviously interested in the graphical display of his past experience and gained great satisfaction as he watched the chart settle down again after corrective action had been taken. Also the maintaining of the charts had a stimulating effect on members of the inspection staff, giving them a greater interest in their work and a due sense of its importance.

In regard to American war production, the control chart method of quality control recommended and described in the three American War Standards now published, is being put to a considerable and growing use in the Ordnance Department of the U.S. Army.

Experience of the Ordnance Department in applying quality control techniques to wartime production of munitions, has conclusively demonstrated the value of the method. The beneficial results are now becoming obvious not only by way of more uniform products, but also in economy and speed in production.

I would like briefly to illustrate an example that has arisen in America in the manufacture of time fuses. Quality control methods were not at first applied, and practically all the makers failed to produce acceptable fuses.

The characteristics demanded in this time fuse are as shown in Fig. 6 by the diverging tolerance limits. Colonel Simon, of the U.S. Army, one of the foremost workers in statistical methods, was consulted and, by applying these methods to the sampling and testing of the fuses, he obtained the full curved line as the characteristic of one make of fuses by joining up points determined from the control charts for the various time settings. The dotted curves were obtained from the control limits of the same charts. The curves for some makes turned upwards. As all the timings would fall between the curved dotted lines, it was evident that the tolerances achieved were at all settings a good deal finer than those specified, but many of the fuses would fail to pass the specified requirements at the longer time settings as had originally been found to be the case.

It was a simple matter to take advantage of this by altering the time setting graduations, e.g., 30 sec. less the error 4/10 sec., i.e. 29.6 on the re-graduated scale to replace the 30 sec. marking on the existing fuses, etc. ; in this way

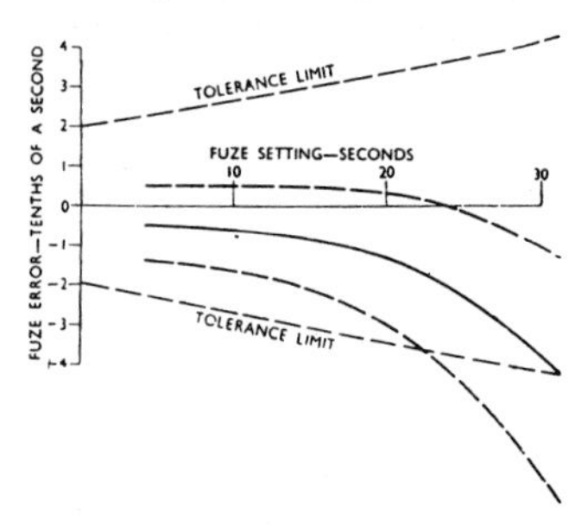

Fig. 6 *(III).—Limits Demanded and Results Obtained with Time Fuses.*

it was made possible to use the fuses which had repeatedly failed to pass the prescribed test, and which actually were more perfect than had been demanded.

I shall outline very briefly a typical control procedure. This example is dealt with very fully in the latest American Standard Z1.3. Typical details are shown in Figs. 7 and 8.

Fig. 7 shows the control chart data sheet which is self-explanatory. The quality characteristic illustrated is the measurable " time of blow " of an electrical fuse whose specification requires that the fuse, when tested under specified circuit conditions, shall blow within a maximum time of 150 seconds. The test is thus destructive, and small equal periodic samples of five fuses per hour are tested in carrying out the control procedure.

Fig. 8 shows the control chart record for the average and range of observations in the sample. It will be seen from the analysis of the preliminary data, comprising samples No. 1-25, that lack of control existed and the results from samples Nos. 10-14, being attributable to a faulty lot of raw material, were disregarded in computing the control limits for actual control operation. The cause for indications of lack of control at sample Nos. 31 and 40 was investigated and discovered at sample No. 45, and the process changed accordingly. The subsequent 25 samples indicate relatively good control, and at sample No. 71 the level of control was reviewed and, using data from samples No. 46-70 (i.e., subsequent to the elimination of a major cause), new control limits were established and extended into the future. The manner in which the plotted values of the range from samples No. 80 onwards have clustered about the standard value, indicates, in conjunction with the control shown in the average chart, how the uniformity in quality has improved.

It is to be noted that if the output of fuses is considered as a succession of hourly lots, the inspection required for the disposition of the product, lot by lot, may amount to as much as 8 samples of 5 fuses during the hour. However, once control has been established, one hourly sample of 5 suffices for giving the needed quality assurance unless, for the current hourly sample, lack of control is indicated, or one or more defects are found. Then an additional 35 fuses from the same lot are tested and if not more than one defect is found in the total of 8 samples of 5 tested, the lot is released for shipment. This example gives an idea of the amount by which inspection can be reduced for deciding the disposition of each lot when some acceptable evidence of control, as given by the control chart procedure, is available. This reduction in inspection means savings in time, labour, and materials.

In conclusion, it has been pointed out that the statistical approach to quality control is quite a recent development. In the opinion of Dr. Shewhart, of the Bell Telephone Laboratories, who originated this technique, its future importance might well be visualised by comparing its present stage with the first introduction into practice of limit gauges some 70 years ago. The adoption of manufacturing limits embodied in *Go* and *Not-Go* gauges, was a major development which helped to make mass production what it is today. Similarly, the use of action or control limits in addition to manufacturing or tolerance limits is of great significance because, by giving a firmer grasp on the problem of quality control, they enable us to go to the extreme in efficient use of materials and component parts. The more confidence the engineer has in the control of quality of a material, the lower will be the safety factor he should adopt in his designs, for material is wasted, obviously, in all cases where the safety factor could have been lower than the one adopted.

From this, and in the light of the considerable interest which statistical quality control has recently created, perhaps it is destined to fill an important role in the make-up of the engineer and industrial executive of tomorrow.

Acknowledgment is made of the charts used and the assistance received from the American War Standard Z1.3, L. E. Simon's *An Engineers' Manual of Statistical Methods*, The Journal of The Institution of Mechanical Engineers, a report by H. Rissik, Esq., on the application of quality control in the Creed organisation, and an article under his name in *The Engineer*, 31st July, 1942.

CONTROL CHART DATA SHEET

PRODUCT Type XX fuse
CHARACTERISTIC Time of blow
UNIT OF MEASUREMENT seconds
SPECIFIED LIMITS - MIN.
150 sec. MAX.
SPECIFICATION NO. XXXX

PRODUCTION ORDER NO. 00-000
PRODUCTION DEPT. NO. 000
NORMAL DAILY OUTPUT 0000
SAMPLE 5 PER hour
TEST SET NO. 00

DATE Nov. 3-4, 1941
INSPECTOR'S CLOCK NO. 000

SIGNED: Jack Doe
INSPECTOR

IDENTIFICATION	OBS. VALUE	IDENT.	OBS. VALUE	IDENT.	OBS. VALUE	IDENT.	OBS. VALUE	IDENT.	OBS. VALUE	IDENT.	OBS. VALUE
SAMPLE NO. 1	42	2	42	3	9	4	36	5	42	6	51
TIME 8:35 A.M.	65	9:25	45	10:30	24	11:00	54	1:00 P.M.	51	1:50	74
OPERATOR NO. 00	75	00	68	00	80	00	69	00	57	00	75
MACHINE NO. 000	78	000	72	000	81	000	77	000	59	000	78
---- NO.	87		90		81		84		78		132
TOTAL	5) 347		5) 317		5) 275		5) 320		5) 287		5) 410
AVERAGE, $\bar{X}$	69.4		63.4		55.0		64.0		57.4		82.0
LARGEST VALUE	87		90		81		84		78		132
SMALLEST VALUE	42		42		9		36		42		51
DIFFERENCE: RANGE, R	45		48		72		48		36		81
SAMPLE NO. 7	60	8	18	9	15	10	69	11	64	12	61
TIME 3:05	60	4:00	20	8:30 A.M.	30	9:25	109	10:30	91	11:35	78
OPERATOR NO. 00	72	00	27	00	39	00	113	00	93	00	94
MACHINE NO. 000	95	000	42	000	62	000	118	000	109	000	109
---- NO.	138		60		84		(153)		112		136
TOTAL	5) 425		5) 167		5) 230		5) 562		5) 469		5) 478
AVERAGE, $\bar{X}$	85.0		33.4		46.0		112.4		93.8		95.6
LARGEST VALUE	138		60		84		153		112		136
SMALLEST VALUE	60		18		15		69		64		61
DIFFERENCE: RANGE, R	78		42		69		84		48		75

NOTES: Sample No. 10. 1 fuse exceeds max. limit. 35 add'l fuses tested. Data recorded on supplementary data sheet.

Fig. 7 (III).—Control Chart Data Sheet.

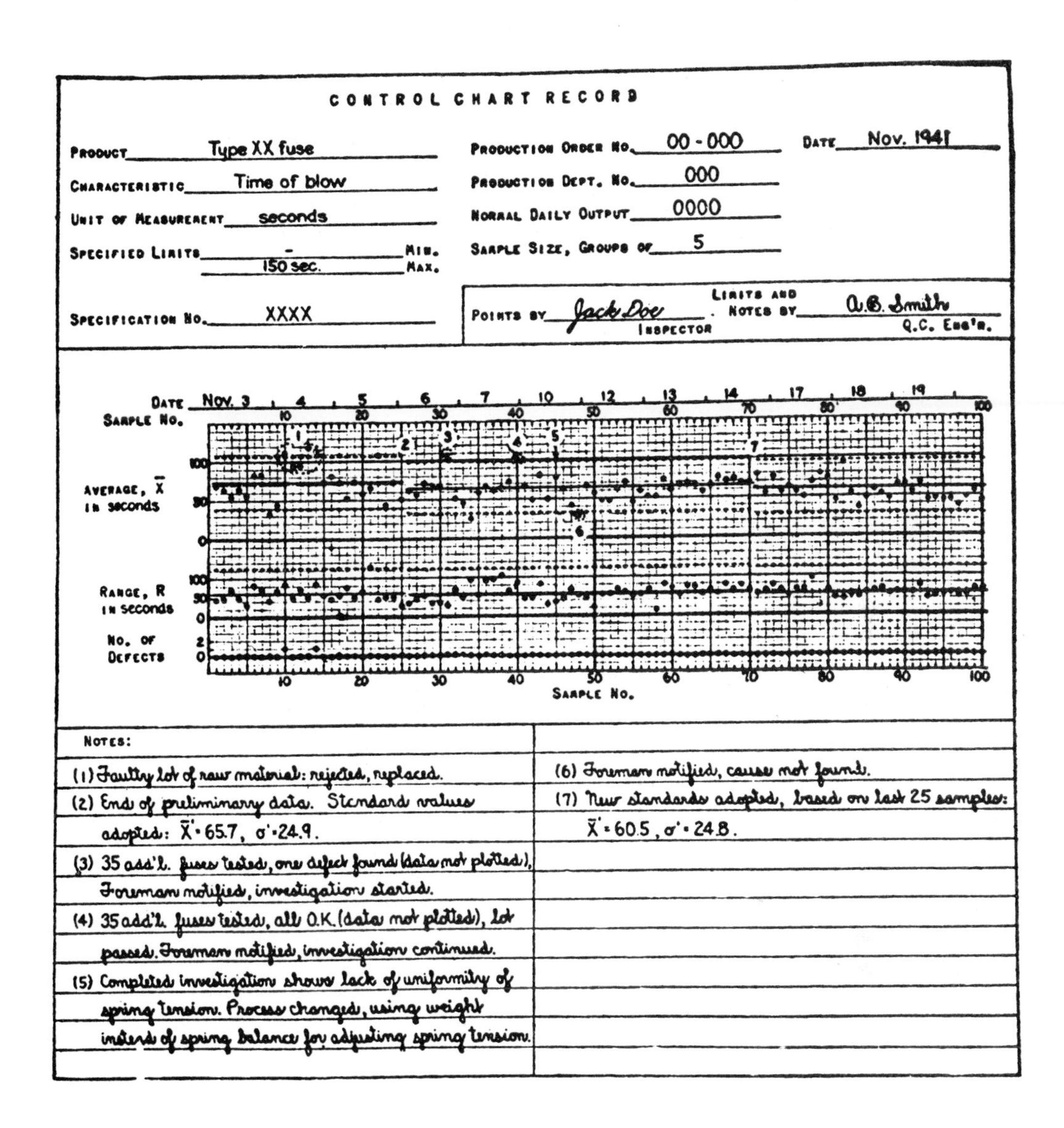

CONTROL CHART RECORD

PRODUCT: Type XX fuse
PRODUCTION ORDER NO.: 00 - 000
DATE: Nov. 1941
CHARACTERISTIC: Time of blow
PRODUCTION DEPT. NO.: 000
UNIT OF MEASUREMENT: seconds
NORMAL DAILY OUTPUT: 0000
SPECIFIED LIMITS: - MIN.; 150 sec. MAX.
SAMPLE SIZE, GROUPS OF: 5
SPECIFICATION NO.: XXXX
POINTS BY: Jack Doe, INSPECTOR
LIMITS AND NOTES BY: A.B. Smith, Q.C. ENG'R.

DATE: Nov. 3, 4, 5, 6, 7, 10, 12, 13, 14, 17, 18, 19
SAMPLE NO.: 10, 20, 30, 40, 50, 60, 70, 80, 90, 100
AVERAGE, $\bar{X}$ IN seconds: 100, 50, 0
RANGE, R IN seconds: 100, 50, 0
NO. OF DEFECTS: 2, 0
SAMPLE NO.

NOTES:

(1) Faulty lot of raw material: rejected, replaced.

(2) End of preliminary data. Standard values adopted: $\bar{X}' = 65.7$, $\sigma' = 24.9$.

(3) 35 add'l. fuses tested, one defect found (data not plotted), Foreman notified, investigation started.

(4) 35 add'l. fuses tested, all O.K. (data not plotted), lot passed. Foreman notified, investigation continued.

(5) Completed investigation shows lack of uniformity of spring tension. Process changed, using weight instead of spring balance for adjusting spring tension.

(6) Foreman notified, cause not found.

(7) New standards adopted, based on last 25 samples: $\bar{X}' = 60.5$, $\sigma' = 24.8$.

Fig. 8 (III).—Suggested Form for a Control Chart Record

Commonwealth of Australia, Council for Scientific and Industrial Research, Bulletin No. 176, Melbourne, 1944. By:–
I.F. Phipps, B.Agr.Sc., M.Sc., Ph.D. (Waite Agricultural Research Institute)
A.T. Pugsley, B.Agr. Sc. (Waite Agricultural Research Institute)
S.R. Hockley (Waite Agricultural Research Institute) and
E.A. Cornish, M.Sc., B.Agr.Sc. (Section of Biometrics)

SUMMARY.

The paper describes in detail the procedure to be adopted for recovering inter-block information in cubic lattice designs, and the computations which are required for this full analysis are illustrated by a numerical example.

The Analysis of Cubic Lattice Designs in Varietal Trials*

1. INTRODUCTION.

Among the chief problems presented to the cereal breeder are the evaluation and comparison of the yielding abilities of the numerous selected strains derived from cross-breds. As soon as reasonable purity has been attained, it is desirable that these selected strains should be tested in drill sown plots, under field conditions, but the fact that such large numbers of varieties must be tested simultaneously introduces many problems in experimental design. Ordinary designs, such as the Latin square and randomized block, can no longer be used; they become too unwieldy, require too many replications, and do not permit the efficient elimination of soil fertility differences. Similar difficulties are encountered in other types of biological experimentation where large numbers of treatments have to be tested. To overcome these difficulties, Yates (1936a, 1936b, 1937a) introduced a series of experimental designs of three types:-

(i) unbalanced lattices in randomized blocks,

(ii) balanced incomplete randomized blocks,

(iii) quasi-Latin squares (lattice squares).

The value and application of these designs as the means of increasing the precision of various types of experimental work have been discussed at length by several authors (Yates 1936a, 1936b, 1937a, 1940a; Goulden 1937a, 1937b, 1939; Cox *et al.* 1940), so that it is unnecessary to elaborate on these points in the present paper.

Although these designs were originally proposed for the purpose of completely eliminating heterogeneity in experimental material, it was pointed out that loss of efficiency, as compared with designs in ordinary randomized blocks, would result if the subdivision into small groups did not affect a reduction in the intra-block error. This loss of efficiency is due to the fact that the differences between the group means comprise either partly, or entirely, differences between varieties, and in the limiting case, where there are no group differences, the loss amounts to $1-E$ where E is the efficiency factor of the particular design. Recently it has been shown (Fisher and Yates 1938; Yates 1939, 1940b, 1940c; Cox et *al.* 1940) that information on the treatments, contained in the inter-block comparisons, can be recovered if the relative accuracy of the inter- and intra-block comparisons is well determined.

In 1940 it was decided to employ one of the unbalanced cubic lattice designs for testing cross-breds under investigation at the Waite Institute. The trial re-affirmed all claims made for this type of design, and the method was subsequently adopted for routine testing. As Yates' original description,

* Typescript received August 11, 1943.

P.11230-2.

illustrating the recovery of inter-block information in cubic lattice designs, is not readily available in this country, the opportunity has been taken to give the analysis of the 1940 yield data in detail.

2. THE CUBIC LATTICE.

The cubic lattice design, with equal groups of sets, is applicable where the number of varieties, p^3 say, is a perfect cube. In order to construct such a design, the varieties are first assigned at random to the p^3 intersections of a $p \times p \times p$ lattice. In any one of the three mutually orthogonal directions of the lattice, there are p^2 lines which divide the varieties into p^2 sets, each containing p varieties. After the random allocation of the p^2 sets of each group to p^2 blocks, the groups are set out in the field in the form of p^2 randomized blocks of p plots, in such a way that the blocks constituting complete replications are arranged in compact units on the ground. All groups must be equally replicated, say r times.

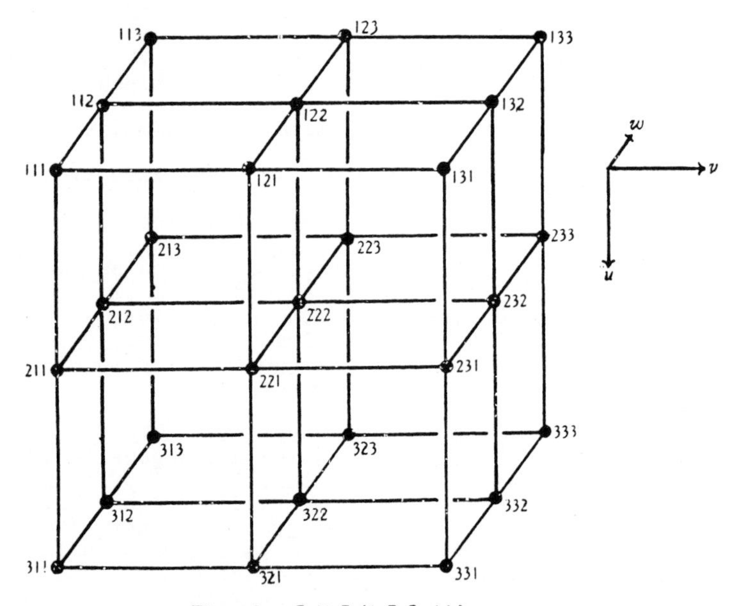

Fig. 1 - 3 X 3 X 3 lattice.

Fig. 1 illustrates a 3 X 3 X 3 lattice for 27 varieties, the position of each variety within the lattice being determined by 3 numbers, u, v, w, each of which may take any integer value from 1 to 3. The analogy with an ordinary factorial design involving three factors x, y, and z, each at 3 levels, is immediately apparent, for the intersections of the lattice may be taken to represent the 27 combinations of the factors, and the levels associated with any intersection as the co-ordinates of that intersection referred to the axes of the lattice.

The degrees of freedom, $p^3 - 1$, for the p^3 varieties, may therefore be set out as if arising from the main effects and interactions of three hypothetical factors x, y, and z; thus

		Deg. freed.
Main effects	X	$p - 1$
	Y	$p - 1$
	Z	$p - 1$
Two-factor interactions	XY	$(p - 1)^2$
	XZ	$(p - 1)^2$
	YZ	$(p - 1)^2$
Three-factor interaction	XYZ	$(p - 1)^3$
Total		$p^3 - 1$

In the 1940 trial, 125 varieties were tested in a single complete lattice so that $p = 5$ and $r = 1$. The experiment may thus be regarded as a test of 3 factors each at 5 levels.

3. FIELD PLAN AND OPERATIONS.

Fig. 2 shows the random arrangement of the trial.

The 125 strains tested comprised 100 cross-bred strains and 25 varieties, the former in the 6th and later generations from crossing. Seed of many of these cross-bred strains was harvested from pedigree rows yielding approximately 2 lb. of grain.

Size of plot is restricted by the amount of seed and area of land available. In this case the plots were four drill hoes wide and one chain long, giving an area of 1/283 acre. If land is available, a greater width of plot is desirable because of the border effect of pathways. The plots were sown with a 9-hoe drill, having the middle seed and superphosphate feeds blocked, thus allowing two plots to be sown at the one time. Sowing operations are greatly simplified if the drill follows through from group to group, sowing two plots of each group and then returning wheel on wheel mark. It is advisable to sight every 10th drill width in order to maintain the space between plots at a constant distance. Under this arrangement, it is necessary that the seed for each plot be bagged and set out in place before drilling is commenced. Using two 9-hoe drills the work can be done comfortably in one day.

4. CROSS-BRED MATERIAL.

Much of the cross-bred material in the 1940 trial, and more of it in the 1941 and 1942 trials, was of Hope parentage. This variety was produced by McFadden from a cross between Marquis and Yaroslav Emmer. Hope was introduced by the Waite Institute in 1928, and the first crosses with Australian varieties were made in the following year.

The botanical and agronomic characters of Hope are so widely different from those of Australian varieties that very

Fig. 2 - Random Arrangement of the Trial.

Group																									
Group Z	212	543	111	151	243	451	415	551	121	515	454	251	445	142	315	345	354	534	321	224	135	332	425	231	521
	213	545	113	155	242	454	414	555	122	514	455	254	444	144	313	341	351	533	325	223	133	331	424	233	523
	211	544	114	154	245	453	413	552	124	511	453	252	443	141	312	342	355	535	322	222	132	333	423	232	522
	215	542	112	152	241	455	411	553	125	512	452	255	442	143	314	344	353	532	323	221	131	334	421	235	524
	214	541	115	153	244	452	412	554	123	513	451	253	441	145	311	343	352	531	324	225	134	335	422	234	525

Group																									
Group X	511	524	312	525	545	544	521	323	141	243	153	431	134	154	135	355	342	251	313	552	314	115	133	532	322
	411	224	112	125	345	344	221	223	441	343	353	331	334	454	235	555	542	551	513	352	214	315	333	332	222
	311	324	412	425	245	444	321	123	341	543	553	531	434	354	335	455	242	151	213	452	114	515	433	132	422
	211	124	512	325	445	144	421	423	541	443	253	131	534	254	535	155	142	351	413	252	514	215	233	232	522
	111	424	212	225	145	244	121	523	241	143	453	231	234	554	435	255	442	451	113	152	414	415	533	432	122

Group																									
Group Y	532	222	341	355	122	233	534	535	452	143	125	445	111	245	154	244	324	513	413	441	323	511	454	211	322
	522	212	351	335	152	243	524	515	442	133	145	435	141	215	124	214	334	553	423	411	313	531	434	231	352
	542	242	321	345	142	223	554	545	432	153	115	425	121	225	114	254	354	533	433	421	333	541	414	241	342
	552	232	311	325	112	253	544	525	412	123	155	415	131	255	134	224	314	543	453	431	343	551	424	251	312
	512	252	331	315	132	213	514	555	422	113	135	455	151	235	144	234	344	523	443	451	353	521	444	221	332

The order of the varietal identification numbers in each block reading from top to bottom gives the order of the plots in the field layout from L. to R.

few of the progeny of crosses possess the desired combination of the agronomic characters of local wheats and the resistance to disease of the Hope parent. For this reason, large numbers of plants and progenies have to be grown and examined. By 1938 much valuable material, possessing rust and flag smut resistance, had been secured. Severe epidemics of mildew and leaf and stem rust in 1939 made possible the selection of fixed strains resistant to these diseases. Of the numerc strains of Hope parentage tested in 1940, 1941, and 1942, one strain, derived from a Nabawa x Hope cross, was outstanding. It was distributed to farmers under the name "Warigo" in 1942. This variety is resistant to stem rust, leaf rust, flag-smut, loose smut, and mildew, and is partially resistant to bunt. Accounts of the breeding work and the disease resistance trials have been given in the Annual Reports of the Waite Institute.

5. NOTATION.

The group in which the vertical lines (as in Fig. 1) show the constitution of the blocks will be designated group X and the other two groups Y and Z. The total of the r replicates of variety uvw in group X will be denoted by x_{uvw}, in group Y by y_{uvw} and in group Z by z_{uvw}. Totals and subtotals will be denoted thus:-

$$x_{111} + x_{112} + x_{113} + x_{114} + x_{115} = X_{11.} ,$$

$$X_{11.} + X_{12.} + X_{13.} + X_{14.} + X_{15.} = X_{1..} ,$$

$$X_{1..} + X_{2..} + X_{3..} + X_{4..} + X_{5..} = X_{...} ,$$

with similar relationships for groups Y and Z, and totals over the three groups thus:-

$$x_{111} + y_{111} + z_{111} = T_{111} ,$$

$$X_{11.} + Y_{11.} + Z_{11.} = T_{11.} ,$$

$$X_{1..} + Y_{1..} + Z_{1..} = T_{1..} ,$$

$$X_{...} + Y_{...} + Z_{...} = T_{...} .$$

Table 1 shows the yield of grain from each variety in lb. per plot from each group and all the subtotals and totals required in the analysis. Throughout the formation of this table numerous cross checks are available. The totals underlined are the block totals in their respective groups.

*TABLE 1 - Yield of grain from each variety in lb. per plot.

Varietal identification number	Group			Total
	X	Y	Z	
111	9.5	8.0	7.7	25.2
112	11.2	7.7	10.2	29.1
113	10.0	10.2	10.2	30.4
114	11.5	10.0	11.5	33.0
115	9.7	9.7	8.7	28.1
11.	51.9	45.6	48.3	145.8
121	9.5	10.0	11.5	31.0
122	6.7	9.5	9.7	25.9
123	10.2	11.0	10.7	31.9
124	11.7	11.2	11.2	34.1
125	9.7	8.0	9.2	26.9
12.	47.8	49.7	52.3	149.8
131	11.2	10.5	10.7	32.4
132	8.5	7.7	8.7	24.9
133	9.5	9.0	9.5	28.0
134	11.2	9.7	10.2	31.1
135	11.2	12.5	10.2	33.9
13.	51.6	49.4	49.3	150.3
141	8.7	8.2	9.2	26.1
142	10.5	9.5	10.5	30.5
143	6.5	6.0	6.7	19.2
144	11.5	10.5	10.5	32.5
145	9.2	9.7	8.7	27.6
14.	46.4	43.9	45.6	135.9
151	10.2	10.0	10.5	30.7
152	9.5	7.5	9.5	26.5
153	9.2	11.2	9.2	29.6
154	11.2	11.0	11.5	33.7
155	8.7	9.7	6.2	24.6
15.	48.8	49.4	46.9	145.1
1..	246.5	238.0	242.4	726.9

TABLE 1 (Contd.)

Varietal identification number	Group			Total
	X	Y	Z	
211	10.0	8.5	10.0	28.5
212	9.0	9.0	8.5	26.5
213	12.2	10.0	11.7	33.9
214	10.7	10.0	12.0	32.7
215	10.7	9.5	11.5	31.7
21•	52.6	47.0	53.7	153.3
221	8.0	8.5	8.5	25.0
222	8.0	8.0	8.5	24.5
223	9.5	10.0	9.0	28.5
224	6.7	6.5	6.5	19.7
225	11.2	10.2	10.7	32.1
22•	43.4	43.2	43.2	129.8
231	9.2	8.2	9.2	26.6
232	9.2	8.5	9.2	26.9
233	9.0	7.7	8.0	24.7
234	11.2	10.5	11.0	32.7
235	10.0	9.5	10.5	30.0
23•	48.6	44.4	47.9	140.9
241	9.5	10.0	9.0	28.5
242	9.0	8.0	8.0	25.0
243	9.2	9.0	10.5	28.7
244	10.2	9.7	9.7	29.6
245	10.5	8.7	10.5	29.7
24•	48.4	45.4	47.7	141.5
251	8.7	9.2	9.5	27.4
252	9.0	9.5	8.5	27.0
253	12.5	10.5	11.5	34.5
254	10.7	9.5	9.0	29.2
255	5.7	5.0	6.0	16.7
25•	46.6	43.7	44.5	134.8
2••	239.6	223.7	237.0	700.3

TABLE 1 (Contd.)

Varietal identification number	Group			Total
	X	Y	Z	
311	10.5	8.5	8.7	27.7
312	9.2	8.7	9.2	27.1
313	9.5	8.7	9.7	27.9
314	9.7	9.7	9.0	28.4
315	9.0	8.2	9.0	26.2
31•	47.9	43.8	45.6	137.3
321	8.2	8.2	9.5	25.9
322	8.5	8.2	8.5	25.2
323	6.2	6.0	6.2	18.4
324	10.5	11.2	11.2	32.9
325	9.2	9.7	9.2	28.1
32•	42.6	43.3	44.6	130.5
331	10.0	9.2	7.7	26.9
332	9.5	8.5	8.7	26.7
333	10.5	9.7	8.5	28.7
334	11.0	10.7	8.7	30.4
335	8.2	7.5	7.2	22.9
33•	49.2	45.6	40.8	135.6
341	9.0	8.7	7.7	25.4
342	9.5	9.0	8.5	27.0
343	9.0	9.0	8.2	26.2
344	8.0	9.2	7.7	24.9
345	8.5	8.7	8.2	25.4
34•	44.0	44.6	40.3	128.9
351	9.0	8.7	8.5	26.2
352	9.0	8.0	7.0	24.0
353	8.0	8.2	7.0	23.2
354	10.5	9.5	9.2	29.2
355	8.5	8.2	8.5	25.2
35•	45.0	42.6	40.2	127.8
3••	228.7	219.9	211.5	660.1

TABLE 1 (Contd.)

Varietal identification number	Group			Total
	X	Y	Z	
411	10.5	8.2	9.2	27.9
412	12.5	11.7	10.2	34.4
413	9.0	7.0	8.0	24.0
414	9.5	9.2	9.0	27.7
415	11.7	11.5	11.0	34.2
41•	53.2	47.6	47.4	148.2
421	6.5	8.5	6.7	21.7
422	11.2	10.7	10.5	32.4
423	10.5	9.7	8.7	28.9
424	8.5	9.2	9.2	26.9
425	10.0	8.7	8.7	27.4
42•	46.7	46.8	43.8	137.3
431	9.0	9.7	10.2	28.9
432	11.7	10.7	10.2	32.6
433	8.2	8.2	8.0	24.4
434	10.0	9.2	8.7	27.9
435	11.5	11.7	11.5	34.7
43•	50.4	49.5	48.6	148.5
441	9.0	9.2	9.0	27.2
442	10.7	10.2	10.7	31.6
443	9.0	9.0	9.2	27.2
444	8.7	8.7	9.0	26.4
445	11.5	11.5	11.7	34.7
44•	48.9	48.6	49.6	147.1
451	7.2	9.7	10.0	26.9
452	12.0	12.0	9.7	33.7
453	8.7	8.2	7.7	24.6
454	5.5	5.2	4.7	15.4
455	6.7	6.7	6.0	19.4
45•	40.1	41.8	38.1	120.0
4••	239.3	234.3	227.5	701.1

TABLE 1 (Contd.)

Varietal identification number	Group			Total
	X	Y	Z	
511	7.0	8.5	7.0	22.5
512	8.7	9.0	7.0	24.7
513	8.2	7.2	7.7	23.1
514	9.5	9.0	8.7	27.2
515	10.5	9.2	9.7	29.4
51•	43.9	42.9	40.1	126.9
521	8.2	8.5	7.5	24.2
522	8.5	9.0	8.2	25.7
523	9.0	9.0	9.0	27.0
524	8.2	9.5	9.2	26.9
525	8.0	8.5	7.2	23.7
52•	41.9	44.5	41.1	127.5
531	9.5	9.5	9.7	28.7
532	9.5	7.0	8.7	25.2
533	8.7	9.7	10.7	29.1
534	10.0	9.2	10.5	29.7
535	9.0	8.7	9.7	27.4
53•	46.7	44.1	49.3	140.1
541	8.7	9.5	8.5	26.7
542	11.7	11.7	10.0	33.4
543	8.2	9.2	7.0	24.4
544	7.7	7.2	7.5	22.4
545	9.5	7.5	6.5	23.5
54•	45.8	45.1	39.5	130.4
551	9.5	10.2	8.2	27.9
552	9.5	9.0	8.5	27.0
553	9.0	9.2	9.0	27.2
554	10.7	10.7	10.7	32.1
555	10.5	10.0	9.7	30.2
55•	49.2	49.1	46.1	144.4
5••	227.5	225.7	216.1	669.3

TABLE 1 (Contd.)

Varietal identification number	Group			Total
	X	Y	Z	
•11	47.5	41.7	42.6	131.8
•12	50.6	46.1	45.1	141.8
•13	48.9	43.1	47.3	139.3
•14	50.9	47.9	50.2	149.0
•15	51.6	48.1	49.9	149.6
•1•	249.5	226.9	235.1	711.5
•21	40.4	43.7	43.7	127.8
•22	42.9	45.4	45.4	133.7
•23	45.4	45.7	43.6	134.7
•24	45.6	47.6	47.3	140.5
•25	48.1	45.1	45.0	138.2
•2•	222.4	227.5	225.0	674.9
•31	48.9	47.1	47.5	143.5
•32	48.4	42.4	45.5	136.3
•33	45.9	44.3	44.7	134.9
•34	53.4	49.3	49.1	151.8
•35	49.9	49.9	49.1	148.9
•3•	246.5	233.0	235.9	715.4
•41	44.9	45.6	43.4	133.9
•42	51.4	48.4	47.7	147.5
•43	41.9	42.2	41.6	125.7
•44	46.1	45.3	44.4	135.8
•45	49.2	46.1	45.6	140.9
•4•	233.5	227.6	222.7	683.8
•51	44.6	47.8	46.7	139.1
•52	49.0	46.0	43.2	138.2
•53	47.4	47.3	44.4	139.1
•54	48.6	45.9	45.1	139.6
•55	40.1	39.6	36.4	116.1
•5•	223.7	226.6	215.8	672.1

TABLE 1 (Contd).

Varietal identification number	Group			Total
	X	Y	Z	
1·1	49.1	46.7	49.6	145.4
1·2	46.4	41.9	48.6	136.9
1·3	45.4	47.4	46.3	139.1
1·4	57.1	52.4	54.9	164.4
1·5	48.5	49.6	43.0	141.1
2·1	45.4	44.4	46.2	136.0
2·2	44.2	43.0	42.7	129.9
2·3	52.4	47.2	50.7	150.3
2·4	49.5	46.2	48.2	143.9
2·5	48.1	42.9	49.2	140.2
3·1	46.7	43.3	42.1	132.1
3·2	45.7	42.4	41.9	130.0
3·3	43.2	41.6	39.6	124.4
3·4	49.7	50.3	45.8	145.8
3·5	43.4	42.3	42.1	127.8
4·1	42.2	45.3	45.1	132.6
4·2	58.1	55.3	51.3	164.7
4·3	45.4	42.1	41.6	129.1
4·4	42.2	41.5	40.6	124.3
4·5	51.4	50.1	48.9	150.4
5·1	42.9	46.2	40.9	130.0
5·2	47.9	45.7	42.4	136.0
5·3	43.1	44.3	43.4	130.8
5·4	46.1	45.6	46.6	138.3
5·5	47.5	43.9	42.8	134.2
··1	226.3	225.9	223.9	676.1
··2	242.3	228.3	226.9	697.5
··3	229.5	222.6	221.6	673.7
··4	244.6	236.0	236.1	716.7
··5	238.9	228.8	226.0	693.7
···	1181.6	1141.6	1134.5	3457.7

6. ESTIMATION AND WEIGHTING OF VARIETAL EFFECTS.

It is apparent from Fig. 1 that the varietal contents of the blocks in group X are arranged so that the main effect X and the interactions XY, XZ, and XYZ are unconfounded, whereas the main effects Y and Z and the interaction YZ are confounded with block differences. In the Y and Z groups X, Z, and XZ, and X, Y, and XY are confounded respectively.

Three estimates of the confounded effec' ı may thus be determined:-

(i) Intra-block estimates derived from regions where the effects are unconfounded with block differences.

(ii) Inter-block estimates derived from regions where the particular effects are confounded with block differences.

(iii) Unadjusted estimates derived from the whole experiment.

Let X_i, X_b, and X_o stand for the intra-block, inter-block, and unadjusted estimates of the main effect* of factor x at the uth level, respectively. Then, for example, from Table 1

$$\begin{aligned} X_i &= X_{1..}/p^2 - X_{...}/p^3 \\ &= 246.5/25 - 1181.6/125 \\ &= .41, \end{aligned}$$

$$\begin{aligned} X_b &= (Y_{1..} + Z_{1..})/2p^2 - (Y_{...} + Z_{...})/2p^3 \\ &= (238.0 + 242.4)/50 - (1141.6 + 1134.5)/250 \\ &= .51, \end{aligned}$$

$$\begin{aligned} X_o &= T_{1..}/3p^2 - T_{...}/3p^3 \\ &= 726.9/75 - 3457.7/375 \\ &= .47. \end{aligned}$$

The corresponding estimates* of the contribution of the first level of factors x and y to the interaction XY are:-

$$\begin{aligned} XY_i &= (X_{11.} + Y_{11.})/2p - (X_{1..} + Y_{1..})/2p^2 - (X_{.1.} + Y_{.1.})/2p^2 \\ &\qquad + (X_{...} + Y_{...})/2p^3 \\ &= 97.5/10 - (246.5 + 238.0)/50 - (249.5 + 226.9)/50 \\ &\qquad + (1181.6 + 1141.6)/250 \\ &= -.18, \end{aligned}$$

$$\begin{aligned} XY_b &= Z_{11.}/p - Z_{1..}/p^2 - Z_{.1.}/p^2 + Z_{...}/p^3 \\ &= 48.3/5 - 242.4/25 - 235.1/25 + 1134.5/125 \\ &= -.36, \end{aligned}$$

$$\begin{aligned} XY_o &= T_{11.}/3p - T_{1..}/3p^2 - T_{.1.}/3p^2 + T_{...}/3p^3 \\ &= 145.8/15 - 726.9/75 - 711.5/75 + 3457.7/375 \\ &= -.24. \end{aligned}$$

---

* Main effects and interactions defined in the manner given by Yates (1937b).

It will be noted that apart from the restrictions imposed by the lattice design, the arrangement is similar to an ordinary randomized block design of p^3 varieties in blocks of p^3 plots. If the confounding is ignored, and the experiment is treated as if it were an ordinary randomized block design*, the analysis will be that of the unadjusted yields, and will give equal weight to all intervarietal comparisons. It is clear that if the small blocks are really differentiated, the block differences will affect the inter-block comparisons and increase their error relative to that of the intra-block comparisons, thus introducing an inequality in the weights. However, the special features of the arrangement are such that, if this is the case, the block differences may be eliminated entirely from the varietal comparisons. This attribute of the quasi-factorial designs was emphasised when they were first described (Yates 1936a). In the analysis originally proposed, only the unconfounded or intra-block estimates of the main effects and interactions, were used in determining the varietal means. Since all variations associated with the small blocks have been eliminated, all information on the varietal comparisons, contained in the inter-block comparisons, has been rejected. Yates (1940a) refers to the resultant varietal means as fully adjusted means.

In general, the most accurate estimates will lie somewhere between the unadjusted and fully adjusted estimates, because the inter-block comparisons will still provide some information even though they are less accurate than the intra-block comparisons. A weighted mean of the inter- and intra-block estimates will therefore provide the best estimate of the varietal effects. Yates (1940a) has termed these the partially adjusted estimates, and has shown (1939), that the process of weighting can only be carried out efficiently if the relative accuracy of the inter- and intra-block comparisons is well determined.

Let w' and w stand for the reciprocals of the true error variances per plot of the inter- and intra-block estimates respectively, and X for the partially adjusted estimate of the main effect of factor x at the uth level. Since the main effect of factor x is confounded in groups Y and Z, but unconfounded in group X, X_b is derived from twice as many replicates as X_t; consequently

$$3X_o = X_t + 2X_b \qquad (1)$$

and $$X = (wX_t + 2w'X_b)/(w + 2w').$$

Substituting for X_t from (1),

$$X = X_o + 2\lambda\,(X_o - X_b) \quad \text{where } \lambda = (w - w')/(w + 2w'). \qquad (2)$$

There will be similar expressions for the partially adjusted main effects of factors y and z.

Since XY is unconfounded in groups X and Y but confounded in group Z, XY_t is derived from twice as many replicates as XY_b; hence

$$3XY_o = 2XY_t + XY_b, \qquad (3)$$

and so

$$XY = (2wXY_t + w'XY_b)/(2w + w'),$$

and, substituting as before,

$$XY = XY_o + \mu\,(XY_o - XY_b) \qquad (4)$$

* This point is discussed in Section 12.

where $\mu = (w - w')/(2w + w')$, with similar expressions for XZ and YZ.

Finally, since the three-factor interaction is unconfounded,

$$XYZ = XYZ_0 .$$

Since quantities proportional to the differences between the unadjusted and inter-block estimates are required in other parts of the analysis, considerable labour is saved by expressing the partially adjusted yields in the above form.

In actual practice, the values of w and w' are not known exactly, so estimates must be made from the data and substituted for the parameters. If the estimates are based on sufficiently large numbers of degrees of freedom, the error introduced by this substitution is negligible.

7. THE ANALYSIS OF VARIANCE.

In the case of a single replication of the complete lattice, the analysis of variance for the fully adjusted yields takes the form:-

Variation due to -	Deg. freed.
Groups	2
Blocks (ignoring varieties)	$3(p^2 - 1)$
Varieties (eliminating blocks)	$p^3 - 1$
Residual (intra-block error)	$2p^3 - 3p^2 + 1$
Total	$3p^3 - 1$.

In this analysis the sum of squares for blocks is obtained by finding the sums of squares of deviations of block totals within groups in the usual way. It is obvious, however, that since the varietal content of the blocks varies from block to block within a group, differences between block totals are compounded of both fertility and varietal differences. The block mean square, therefore, does not represent pure inter-block error and so cannot be used directly in the estimation of w'.

In order to estimate w', the components of varietal effect must be eliminated, and the method of doing this depends upon the number of times the lattice is replicated. If the three groups were replicated twice, an estimate of the block variance, free of varietal effects, would be easily obtained. Thus consider any one of the p^2 blocks of group X. This block contains p varieties which occur together again in some block of the other replicate of group X. The difference between these two block totals would thus represent pure block effect. From the replications of group X, p^2 such differences would be available with, of course, p^2 differences from each of groups Y and Z, yielding, in all, $3(p^2 - 1)$ degrees of freedom for blocks in addition to those in the above analysis. Even with p as small as 3, a sufficiently accurate estimate of block variance should be obtainable from the additional degrees of freedom alone. If only one replicate of each group exists, the degrees of freedom available for blocks are those listed in the analysis above. In making use of these, the sum of squares for blocks (eliminating varieties) is divided into three components which are calculated as indicated below.

In Tables 2 a, b, and c, various totals and subtotals have been transcribed directly from Table 1. Normally this step would not be taken, but it is included here for the sake of clarity in giving the derivation of the components of the block sum of squares. Thus, Table 2a contains the block totals from group X ($X_{\cdot vw}$) and the corresponding subtotals over the whole lattice ($T_{\cdot vw}$), together with the totals $X_{\cdot v \cdot}$, $X_{\cdot\cdot w}$, $T_{\cdot v \cdot}$, and $T_{\cdot\cdot w}$.

Block Component (c).

The total $X_{\cdot 11}$ (47.5) is the total of a block from group X containing varieties v_{u11} ($u = 1,2,\ldots\ldots 5$). The effects of these varieties are therefore confounded with block fertility, and consequently $X_{\cdot 11}$ is an estimate of the combined effect of variety and block. In the Y group the same five varieties occur in 5 different blocks. The total $Y_{\cdot 11}$ therefore provides an estimate of the sum of these five varieties unconfounded with block differences; in the same way, the total $Z_{\cdot 11}$ also provides an unconfounded estimate. The quantity

$$2X_{\cdot 11} - Y_{\cdot 11} - Z_{\cdot 11} = 2 \times 47.5 - 41.7 - 42.6 = 10.7$$

then gives a measure of block variation unaffected by varietal differences.

There will be 25 such differences derivable from the totals,

$$2X_{\cdot vw} - Y_{\cdot vw} - Z_{\cdot vw} ,$$

with two additional sets of 25 differences given by

$$2Y_{u \cdot w} - X_{u \cdot w} - Z_{u \cdot w} ,$$

$$2Z_{uv \cdot} - X_{uv \cdot} - Y_{uv \cdot} .$$

Table 2 has been set up for the calculation of these differences by an alternate and more convenient method. For example, since

$$T_{\cdot vw} = X_{\cdot vw} + Y_{\cdot vw} + Z_{\cdot vw}$$

it is more convenient to derive the numerical value of the difference from the formula

$$T_{\cdot vw} - 3X_{\cdot vw}$$

Thus, the first difference is

$$131.8 - 3 \times 47.5 = -10.7$$

The three sets of differences, designated $C(YZ_{\cdot vw})$, $C(XZ_{u \cdot w})$, $C(XY_{uv \cdot})$, and calculated in this way, are given in Tables 3a, b, and c, corresponding to the X, Y, and Z groups. The reason for reversing the sign of these differences will be given later. It is obvious that the reversal of sign does not affect the attainment of the immediate objective of deriving the block sum of squares. The values in these tables may be checked by summation across the rows and down the columns to determine the marginal totals and then by finding the marginal totals independently from relations of the form

$$T_{\cdot\cdot w} - 3X_{\cdot\cdot w} .$$

TABLE 2 - Arrangement of block totals, sub-totals, and totals of Table 1 for evaluating quantities of Table 3.

(a)

	$X_{.1w}$	$T_{.1w}$		$X_{.2w}$	$T_{.2w}$		$X_{.3w}$	$T_{.3w}$		$X_{.4w}$	$T_{.4w}$		$X_{.5w}$	$T_{.5w}$		$X_{..w}$	$T_{..w}$
·11	47.5	131.8	·21	40.4	127.8	·31	48.9	143.5	·41	44.9	133.9	·51	44.6	139.1	··1	226.3	676.1
·12	50.6	141.8	·22	42.9	133.7	·32	48.4	136.3	·42	51.4	147.5	·52	49.0	138.2	··2	242.3	697.5
·13	48.9	139.3	·23	45.4	134.7	·33	45.9	134.9	·43	41.9	125.7	·53	47.4	139.1	··3	229.5	673.7
·14	50.9	149.0	·24	45.6	140.5	·34	53.4	151.8	·44	46.1	135.8	·54	48.6	139.6	··4	244.6	716.7
·15	51.6	149.6	·25	48.1	138.2	·35	49.9	148.9	·45	49.2	140.9	·55	40.1	116.1	··5	238.9	693.7
·1·	249.5	711.5	·2·	222.4	674.9	·3·	246.5	715.4	·4·	233.5	683.8	·5·	229.7	672.1	···	1181.6	3457.7

(b)

	$Y_{u.1}$	$T_{u.1}$		$Y_{u.2}$	$T_{u.2}$		$Y_{u.3}$	$T_{u.3}$		$Y_{u.4}$	$T_{u.4}$		$Y_{u.5}$	$T_{u.5}$		$Y_{u..}$	$T_{u..}$
1·1	46.7	145.4	1·2	41.9	136.9	1·3	47.4	139.1	1·4	52.4	164.4	1·5	49.6	141.1	1··	238.0	726.9
2·1	44.4	136.0	2·2	43.0	129.9	2·3	47.2	150.3	2·4	46.2	143.9	2·5	42.9	140.2	2··	223.7	700.3
3·1	43.3	132.1	3·2	42.4	130.0	3·3	41.6	124.4	3·4	50.3	145.8	3·5	42.3	127.8	3··	219.9	660.1
4·1	45.3	132.6	4·2	55.3	164.7	4·3	42.1	129.1	4·4	41.5	124.3	4·5	50.1	150.4	4··	234.3	701.1
5·1	46.2	130.0	5·2	45.7	136.0	5·3	44.3	130.8	5·4	45.6	138.3	5·5	43.9	134.2	5··	225.7	669.3
··1	225.9	676.1	··2	228.3	697.5	··3	222.6	673.7	··4	236.0	716.7	··5	228.8	693.7	···	1141.6	3457.7

TABLE 2 (Contd)

(c)

	$Z_{u1\cdot}$	$T_{u1\cdot}$		$Z_{u2\cdot}$	$T_{u2\cdot}$		$Z_{u3\cdot}$	$T_{u3\cdot}$		$Z_{u4\cdot}$	$T_{u4\cdot}$		$Z_{u5\cdot}$	$T_{u5\cdot}$		$Z_{u\cdot\cdot}$	$T_{u\cdot\cdot}$
11·	48.3	145.8	12·	52.3	149.8	13·	49.3	150.3	14·	45.6	135.9	15·	46.9	145.1	1··	242.4	726.9
21·	53.7	153.3	22·	43.2	129.8	23·	47.9	140.9	24·	47.7	141.5	25·	44.5	134.8	2··	237.0	700.3
31·	45.6	137.3	32·	44.6	130.5	33·	40.8	135.6	34·	40.3	128.9	35·	40.2	127.8	3··	211.5	660.1
41·	47.4	148.2	42·	43.8	137.3	43·	48.6	148.5	44·	49.6	147.1	45·	38.1	120.0	4··	227.5	701.1
51·	40.1	126.9	52·	41.1	127.5	53·	49.3	140.1	54·	39.5	130.4	55·	46.1	144.4	5··	216.1	669.3
·1·	235.1	711.5	·2·	225.0	674.9	·3·	255.9	715.4	·4·	222.7	683.8	·5·	215.8	672.1	···	1134.5	3457.7

TABLE 3 - Quantities required for computing component (c) of block sum of squares.

(a)

$\alpha(YZ_{\cdot uw})$

·11	-10.7	·21	6.6	·31	-3.2	·41	-0.8	·51	5.3	··1	-2.8
·12	-10.0	·22	5.0	·32	-8.9	·42	-6.7	·52	-8.8	··2	-29.4
·13	-7.4	·23	-1.5	·33	-2.8	·43	0.0	·53	-3.1	··3	-14.8
·14	-3.7	·24	3.7	·34	-8.4	·44	-2.5	·54	-6.2	··4	-17.1
·15	-5.2	·25	-6.1	·35	-0.8	·45	-6.7	·55	-4.2	··5	-23.0
·1·	-37.0	·2·	7.7	·3·	-24.1	·4·	-16.7	·5·	-17.0	···	-87.1

TABLE 3 (Contd.)

(b)

$\sigma(XZ_{u.w})$

1.1	5.3	1.2	11.2	1.3	-3.1	1.4	7.2	1.5	-7.7	1..	12.9
2.1	2.8	2.2	0.9	2.3	8.7	2.4	5.3	2.5	11.5	2..	29.2
3.1	2.2	3.2	2.8	3.3	-0.4	3.4	-5.1	3.5	0.9	3..	0.4
4.1	-3.3	4.2	-1.2	4.3	2.8	4.4	-0.2	4.5	0.1	4..	-1.8
5.1	-8.6	5.2	-1.1	5.3	-2.1	5.4	1.5	5.5	2.5	5..	-7.8
..1	-1.6	..2	12.6	..3	5.9	..4	8.7	..5	7.3	...	32.9

(c)

$\sigma(XY_{uv.})$

11.	0.9	12.	-7.1	13.	2.4	14.	-0.9	15.	4.4	1..	-0.3
21.	-7.8	22.	0.2	23.	-2.8	24.	-1.6	25.	1.3	2..	-10.7
31.	0.5	32.	-3.3	33.	13.2	34.	8.0	35.	7.2	3..	25.6
41.	6.0	42.	5.9	43.	2.7	44.	-1.7	45.	5.7	4..	18.6
51.	6.6	52.	4.2	53.	-7.8	54.	11.9	55.	6.1	5..	21.0
.1.	6.2	.2.	-0.1	.3.	7.7	.4.	15.7	.5.	24.7	...	54.2

The three grand totals of Tables 3a, b, and c, should add to zero.

The sums of squares of deviations of the differences within sets give component (c) of the block sum of squares with 48 degrees of freedom, 16 from each set.

Thus, from the first set, the sum of squares is

$$(10.7^2 + \ldots\ldots + 4.2^2)/30 \quad - 87.1^2/750.$$

This sum of squares includes the sums of squares of deviations of the marginal totals -2.8,....., -37.0, ... which will be included later in component (b) and consequently must be deducted here in the determination of component (c). The two sums are

$$(2.8^2 + \ldots\ldots + 23.0^2)/150 - 87.1^2/750.$$

and $$(37.0^2 + \ldots\ldots + 17.0^2)/150 - 87.1^2/750.$$

Hence the contribution from Table 3a to component (c) is

$$(10.7^2 + \ldots + 4.2^2)/30 - (2.8^2 + \ldots + 23.0^2 + 37.0^2 + \ldots + 17.0^2)/150 + 87.1^2/750 = 9.13.$$

There are two similar expressions derivable from Tables 3b and 3c. The divisors are obtained thus: each difference within the tables is composed of twice the sum of 5 yields from one group, minus the sums of 5 yields from each of the other two groups, and consequently the divisor is

$$5 \times 2^2 + 5 \times 1^2 + 5 \times 1^2 = 30.$$

Each marginal total is compounded in a similar way from the totals of 25 yields, and so the divisor is

$$25 \times 2^2 + 25 \times 1^2 + 25 \times 1^2 = 150,$$

and the grand total is the sum of 25 tabular differences or 5 marginal differences, and therefore the divisor of its square is 750.

The general form for the number of degrees of freedom is $3(p-1)^2$, and for the divisors $6p$, $6p^2$, and $6p^3$.

Block Component (b)

The totals $X_{..1}$ and $Y_{..1}$ are totals of 5 blocks, containing the same varieties v_{uv1} $(u, v = 1,2, \ldots,5)$ distributed differently by blocks, and their sum

$$X_{..1} + Y_{..1}$$

represents both varietal and block effect. In the Z group, the same 25 varieties occur individually in the 25 blocks of that group, so that the total $Z_{..1}$ provides an estimate of the sum of these varieties unconfounded with block differences. The quantity

$$X_{..1} + Y_{..1} - 2Z_{..1} = 226.3 + 225.9 - 2 \times 223.9 = 4.4,$$

thus provides another measure of block variation unaffected by varietal differences. There will be 5 such differences derivable from the totals $X_{..w}$, $Y_{..w}$, and $Z_{..w}$, with two other sets of 5 differences given by

$$Y_{u..} + Z_{u..} - 2X_{u..} ,$$

$$X_{.v.} + Z_{.v.} - 2Y_{.v.} .$$

These differences may be calculated by the above formulae, but they are more conveniently obtained from the marginal totals of Tables 3 a, b, and c, by the addition of the values in corresponding pairs. For example, since

$$T_{..w} - 3X_{..w} = -2X_{..w} + Y_{..w} + Z_{..w}$$

$$T_{..w} - 3Y_{..w} = X_{..w} - 2Y_{..w} + Z_{..w}$$

their sum is

$$2Z_{..w} - X_{..w} - Y_{..w}$$

which is of the form given above with reversed sign. These quantities designated $C(X_{u..})$, $C(Y_{.v.})$, $C(Z_{..w})$ are given in Table 4. Thus

$$C(Z_{..1}) = -2.8 - 1.6 = -4.4 ,$$

....

$$C(X_{1..}) = 12.9 - 0.3 = 12.6 ,$$

....

$$C(Y_{.1.}) = -37.0 + 6.2 = -30.8 ,$$

....

The totals of these three sets of differences should check with the totals of Tables 3a, b, and c, with reversed sign; thus the total of the set $C(Z_{..w})$ derived from the marginal totals of Tables 3a and b is -54.2, and this is the total of Table 3c with reversed sign.

The sums of squares of deviations of these quantities within sets give component (b) of the block sum of squares, with 12 degrees of freedom, 4 from each set.

TABLE 4 - Quantities required for computing component (b) of block sum of squares.

$C(Z_{..w})$	$C(X_{u..})$	$C(Y_{.v.})$
-4.4	12.6	-30.8
-16.8	18.5	7.6
-8.9	26.0	-16.4
-8.4	16.8	-1.0
-15.7	13.2	7.7
-54.2	87.1	-32.9

Thus from Table 4,

$$(4.4^2 + \ldots + 15.7^2 + 12.6^2 + \ldots + 13.2^2 + 30.8^2 + \ldots + 7.7^2)/150$$
$$- (54.2^2 + 87.1^2 + 32.9^2)/750$$
$$= 8.97.$$

The divisor 150 is obtained thus: each value within the sets is the sum of 25 yields from each of two groups less twice the sum of 25 yields from the third group, and so the divisor is

$$25 \times 1^2 + 25 \times 1^2 + 25 \times 2^2 = 150,$$

and the divisor for the square of the sum of 5 such quantities is 750. The general form for the number of degrees of freedom is $3(p-1)$ and for the divisors $6p^2$ and $6p^3$.

Block Component (a).

The total $X_{..1}$ (226.3 from Table 2) is a total of 5 blocks from group X containing varieties v_{uv1} ($u, v = 1,2,\ldots\ldots 5,$); the total $Y_{..1}$ (225.9) is a total of 5 blocks from group Y containing the same varieties, but distributed differently by blocks. Since $X_{..1}$ and $Y_{..1}$ contain, in the aggregate, the same varieties, the difference $X_{..1} - Y_{..1}$ is also a measure of block variation, free of varietal effect. There are 5 such differences derivable from the totals $X_{..w}$ and $Y_{..w}$, and similarly two other sets of 5 differences obtainable from the expressions

$$Y_{u..} - Z_{u..}$$

$$X_{.v.} - Z_{.v.}$$

These three sets of differences are given in Table 5; note that the totals of the three sets add to zero. The sums of squares of deviations of the values within the sets give component (a) of the block sum of squares, with 12 degrees of freedom, 4 from each set.

TABLE 5 - Quantities required for computing component (a) of block sum of squares.

$X_{..w} - Y_{..w}$	$Y_{u..} - Z_{u..}$	$X_{.v.} - Z_{.v.}$
0.4	-4.4	-14.4
14.0	-13.3	2.6
6.9	8.4	-10.6
8.6	6.8	-10.8
10.1	9.6	-13.9
40.0	7.1	-47.1

Thus, from Table 5.

$$(0.4^2 + \ldots + 10.1^2 + 4.4^2 + \ldots + 9.6^2 + 14.4^2 + \ldots + 13.9^2)/50$$
$$- (40.0^2 + 7.1^2 + 47.1^2)/250$$
$$= 13.75.$$

Each difference is made up of 25 yields with coefficient + 1 and 25 with coefficient -1, giving for the first divisor the value

$$50 \times 1^2 = 50,$$

and consequently for the square of the sum of 5 such quantities the divisor is 250.

The general form for the number of degrees of freedom is $3(p-1)$ and for the divisors $2p^2$ and $2p^3$.

The remaining sums of squares in the analysis of variance are calculated in the usual way. The sum of the squares of the 375 individual yields is 32619.49, while the ordinary correction for the mean is

$$3457.7^2/375 = 31881.84,$$

giving 737.65 as the value of the total sum of squares.

The contribution due to groups is

$$(1181.6^2 + 1141.6^2 + 1134.5^2)/125 - 31881.84 = 10.31,$$

and that due to varieties, disregarding their classification by blocks, is, from the values of Table 1,

$$(25.2^2 + 29.1^2 + \ldots + 30.2^2)/3 \quad - 31881.84 = 600.37.$$

The residual or intra-block error sum of squares is obtained by subtraction from the total sum of squares. The results are summarized in Table 6.

TABLE 6 - Analysis of variance.

Variation due to -	Degrees of Freed.		Sum Sq.		Mean Sq.
Groups		2		10.31	
Blocks (eliminating varieties)					
Component (a)	12		13.75		
Component (b)	12		8.97		
Component (c)	48	72	39.44	62.16	.8633
Varieties (ignoring blocks)		124		600.37	
Residual (intra-block error)		176		64.81	.3682
Total		374		737.65	

8. THE ESTIMATION OF w AND w'.

Owing to the elimination of varietal effects, the mean squares corresponding to the three components of the block sum of squares have different expectations. This may be shown in the following way (Fisher, 1938, section 40; Cochran 1940). The yield of any plot, apart from varietal and group effects, may be regarded as the sum of two independent and normally distributed quantities, the first, a quantity b, constant for all plots in a given block but varying from block to block with variance A, and the second a quantity e, which varies independently from plot to plot with variance B.

A block total may then be written

$$5b + e_1 + e_2 + \ldots + e_5 .$$

The variance of this total is $25A + 5B$, the variance of a total of 5 blocks is $125A + 25B$, and the variance of the difference between two such totals is $250A + 50B$. When determining component (a) the squares of these differences are divided by 50, and consequently the expectation for the mean square of component (a) is

$$5A + B.$$

Component (b) is derived from quantities made up of the sum of two totals, each of 5 blocks, minus a correction term. These three contributors are derived from different groups and are therefore independent. The sum of the first two contains 10 different b, each with coefficient + 5, and 50 different e. The correction term is twice the sum of 25 plots, each of which is taken from one of the 25 blocks in the third group, and therefore contains 25 different b, each with coefficient − 2, and 25 different e with coefficient − 2, and so has a variance of $100A + 100B$. All correction terms from the same group contain the same 25 values of b, though they contain different values of e, but since component (b) is a sum of squares of deviations, the b values of the correction terms cancel, and it follows that the variance of the total quantity is

$$250A + 150B.$$

Since the squares of these quantities are divided by 150 in determining component (b), the mean square has the expectation

$$\tfrac{1}{3}5A + B.$$

In the same way the expectation of the mean square for component (c) may be shown to be

$$\tfrac{2}{3}5A + B.$$

In the general case, the expectations are $pA + B$, $\frac{1}{3}pA + B$, and $\frac{2}{3}pA + B$, and these values are independent of the number of times the lattice is replicated.

The expectation of the intra-block error mean square is, of course, B, and the estimate of w is $1/B$.

Now $\quad w' = 1/(pA + B),$

so that the only mean square for blocks which can be used directly in the estimation of w' is that of component (a). A more precise estimate can, however, be obtained by making use of components (b) and (c) in addition. Since an independent estimate of B is available, the mean squares of components (b) and (c) can be adjusted and combined with the mean square of component (a), but the exact method of doing this (by the method of maximal likelihood) is very complicated. A simple, yet sufficiently accurate, solution has been suggested by Yates (1939). The average mean square of components (a), (b), and (c) is

$$\frac{3(p-1)(pA+B) + 3(p-1)(\frac{1}{3}pA+B) + 3(p-1)^2(\frac{2}{3}pA+B)}{3(p-1) + 3(p-1) + 3(p-1)^2} = \tfrac{2}{3}pA + B$$

and this is the expectation of the block mean square derived from the total sum of squares for components (a), (b), and (c). By substituting for B in the latter expression, an estimate of A can be determined and hence an estimate of $pA + B$. If E_i is the intra-block error mean square and E_b is the mean square derived from the total of components (a) (b), and (c), the estimated weights may be determined directly from the relations

$$w = \frac{1}{E_i} \quad \text{and} \quad w = \frac{2}{3E_b - E_i}$$

If the whole lattice is replicated twice, there will be an additional $3(p^2 - 1)$ degrees of freedom for differences between blocks of the same varietal content with the expectation $pA + B$ per degree of freedom. This sum of squares can be combined with component (a), and the resultant mean square, based on $3(p - 1)(p + 2)$ degrees of freedom, can be used directly to estimate w'. Components (b) and (c) need not be used. If the inter-block mean square is $\leqslant$ the intra-block mean square, w' is taken equal to w and the ordinary randomized block analysis is used (*vide* Section 12).

In the present example

$$E_b = .8633 \text{ and } E_i = .3682$$

and so $$w = 2.7159, \quad w' = .9002.$$

9. THE TEST OF SIGNIFICANCE.

In testing the significance of the varietal mean square, the following approximate test, which is very simply made, should suffice in the majority of cases.

The lattice is analysed as if it were an ordinary randomized block design of 125 varieties in triplicate, and the variation between the small blocks is included in experimental error. Table 6 contains all the data needed to make this approximate test. The degrees of freedom and sum of squares for blocks (eliminating varieties) are pooled with those for intra-block error to give the analysis of variance in Table 7.

TABLE 7 - Significance of unadjusted varietal means.

Variation due to -	Deg. freed.	Sum Sq.	Mean Sq.	Ratio
Groups	2	10.31		
Varieties (ignoring blocks)	124	600.37	4.8417	9.46
Residual	248	126.97	.5120	
Total	374	737.65		

It is obvious that if the unadjusted varietal means are significantly different as judged by this test, no further test is necessary.

In other cases, for which a more precise test is needed, the procedure described below may be followed. It should be pointed out, however, that an exact test of significance is unobtainable owing to the inaccuracy of weighting.

The process of weighting described in Section 6 alters the sums of squares for the main effects and two-factor interactions but does not influence the three-factor interaction, since the latter is unconfounded. The appropriate varietal sum of squares may be obtained by recalculating the sums of squares for main effects and two-factor interactions from the corresponding weighted inter- and intra- block totals for these effects, but it can be obtained more conveniently from the unadjusted varietal sum of squares by adding the quantity,

$$-\lambda\left\{\left(1 + \frac{2w'}{w}\right) B_u - B_a\right\} - 2\mu\left\{\left(1 + \frac{w'}{2w}\right) B'_u - B'_a\right\}$$

where

B_u = component (b) of block sum of squares (ignoring varieties)

B_a = " " " " " " " (eliminating varieties)

B'_u = component (c) of block sum of squares (ignoring varieties)

B'_a = " " " " " " " (eliminating varieties)

Component (b) (eliminating varieties) was obtained from the three sets of quantities,

$$Y_{u..} + Z_{u..} - 2X_{u..},\quad X_{.v.} + Z_{.v.} - 2Y_{.v.},\quad X_{..w} + Y_{..w} - 2Z_{..w},$$

in each of which an adjustment was made for the effects of the particular varieties confounded. Component (b) (ignoring varieties) is therefore obtained by omitting the adjustment, i.e. from the quantities

$$Y_{u..} + Z_{u..},\quad X_{.v.} + Z_{.v.},\quad X_{..w} + Y_{..w}\ .$$

From Tables 1 or 2a, b, c, the values of these quantities are:-

$Y_{u..} + Z_{u..}$	$X_{.v.} + Z_{.v.}$	$X_{..w} + Y_{..w}$
480.4	484.6	452.2
460.7	447.4	470.6
431.4	482.4	452.1
461.8	456.2	480.6
441.8	445.5	467.7
2276.1	2316.1	2323.2

The sum of squares of deviations within the first set is

$$(480.4^2 + \ldots\ldots + 441.8^2)/50 - 2276.1^2/250 = 29.10$$

The divisors are 50 and 250, since each $Y_{u..} + Z_{u..}$ is a total of 50 plots.

The other two sums are 28.77 and 12.23, and the total, 70.10, of the three contributors is the unadjusted value of component (b).

Component (c) (ignoring varieties) is obtained from the block totals $X_{.vw}$, $Y_{u.w}$, and $Z_{uv.}$. Thus the contribution from Table 2a is

$$(47.5^2 + 50.6^2 + \ldots + 40.1^2)/5 - (226.3^2 + \ldots + 249.5^2 + \ldots)/25 + 1181.6^2/125$$
$$= 26.61,$$

while the sums from Tables 2 b, and c, are 49.27 and 44.30 respectively. The total of the three contributions, 120.18, is the unadjusted value of component (c)*

---

* Note that if component (a) is added to the unadjusted values of components (b) and (c), the total is 204.03, which is identical with the sum of squares for blocks within groups (ignoring the variety classification) calculated in the ordinary way.

Now

$$\lambda = 0.4020, \qquad \mu = 0.2867,$$
$$1 + 2w'/w = 1.6629, \qquad 1 + w'/2w = 1.1657,$$
$$B_u = 70.10, \quad B_a = 8.97, \qquad B'_u = 120.18, \quad B'_a = 39.44;$$

hence the unadjusted varietal sum of squares of Table 6 must be reduced by

$$43.26 + 57.71 = 100.97,$$

giving the following test of significance of the varietal mean square:-

Variation due to -	Deg. freed.	Sum Sq.	Mean Sq.	Ratio
Varieties	124	499.40	4.0274	10.94
Intra-block error	176	64.81	.3682	

The variance ratio is 10.94 as against 9.46 of the approximate test.

This test is the most precise test that can be made, but it is slightly biased. The bias is due to the inaccuracy of weighting and always exaggerates the significance of the results, but it is negligible in an experiment of this size.

10. THE ADJUSTED VARIETAL YIELDS.

As explained previously, the quantity,

$$Y_{u..} + Z_{u..} - 2X_{u..},$$

used above in the evaluation of component (b) of the block sum of squares, is a measure of block or fertility difference, after the effect of varieties has been eliminated. Now, the intra-block estimate of the mean yield of factor x at the uth level is

$$X_{u..}/p^2$$

and the inter-block estimate is

$$(Y_{u..} + Z_{u..})/2p^2 .$$

The intra-block estimate less the inter-block estimate, i.e.

$$(2X_{u..} - Y_{u..} - Z_{u..})/2p^2 ,$$

is thus a measure of the effect of blocks, free from varietal effects, with reversed sign.

It is obvious that a relation similar to equation (1) holds between the inter-block, intra-block, and unadjusted estimates of the mean yield, so that the unadjusted estimate of the mean yield of factor x at the uth level minus the inter-block estimate is equal to

$$(2X_{u..} - Y_{u..} - Z_{u..})/6p^2$$

and thus provides another measure of block effect (eliminating varieties) with reversed sign.

It is now apparent why the partially adjusted yields have been expressed in terms of the unadjusted estimates and the differences between the unadjusted and inter-block estimates, for the quantities representing the effects of blocks (eliminating varieties) are needed in computing the block sums of squares, and in virtue of the relationship established above they can readily be used to determine the corrections needed to evaluate the partially adjusted yields. The reversal of sign also helps to simplify the subsequent arithmetic, for the partially adjusted yields are determined by making an adjustment to the unadjusted means. In making these adjustments, the block effects must clearly be subtracted, since a variety which occurs in a good block should have its mean reduced to make it comparable with other varieties. However in practice it is more convenient to add the adjustments, so the latter must be in their negative form, and for this reason the reversal of sign is retained.

The quantity

$$2Z_{uv\cdot} - X_{uv\cdot} - Y_{uv\cdot}$$

used in the determination of component (c) is a measure of fertility difference, free of varietal effect. The intra-block estimate of the combined effect of factors x and y at levels u and v respectively is

$$(X_{uv\cdot} + Y_{uv\cdot})/2p\ ,$$

and the inter-block estimate is

$$Z_{uv\cdot}/p\ .$$

Their difference is

$$(X_{uv\cdot} + Y_{uv\cdot} - 2Z_{uv\cdot})/2p$$

and is therefore another measure of block effect free of varietal differences, again with reversed sign. It follows, as above, that the unadjusted estimate of the mean yield of the combination of x and y at the uth and vth levels minus the inter-block estimate is equal to

$$(X_{uv\cdot} + Y_{uv\cdot} - 2Z_{uv\cdot})/3p\ ,$$

and so is a measure of block effect (eliminating varieties) with reversed sign, and the same remarks follow as before.

The partially adjusted varietal means may now be obtained by expressing them in terms of main effects and in interactions. If v_0 is the unadjusted mean of variety uvw and m the general mean,

$$v_0 - m = X_0 + Y_0 + Z_0 + XY_0 + XZ_0 + YZ_0 + XYZ_0$$

and, if v is the partially adjusted mean,

$$v - m = X + Y + Z + XY + XZ + YZ + XYZ$$

The general mean is not influenced by the adjustments, and $XYZ_0 = XYZ$; hence

$$v = v_0 + (X - X_0) + (Y - Y_0) + (Z - Z_0) + (XY - XY_0) + (XZ - XZ_0) + (YZ - YZ_0).$$

From equation (2)

$$X - X_0 = 2\lambda(X_0 - X_b) \quad \text{etc.},$$

and from equation (4)

$$XY - XY_0 = \mu(XY_0 - XY_b) \quad \text{etc..}$$

Hence

$$v = v_o + 2\lambda\left\{(X_o - X_b) + (Y_o - Y_b) + (Z_o - Z_b)\right\}$$
$$+ \mu\left\{(XY_o - XY_b) + (XZ_o - XZ_b) + (YZ_o - YZ_b)\right\}$$

If, now, X, ..., XY, ... with appropriate subscripts stand for mean yields instead of main effects and interactions,

$$v = v_o + 2\lambda\left\{(X_o - m) - (X_b - m_{X_b}) + (Y_o - m) - (Y_b - m_{Y_b}) + (Z_o - m) - (Z_b - m_{Z_b})\right\}$$
$$+ \mu\left\{(XY_o - X_o - Y_o + m) - (XY_b - X_b - Y_b + m_{XY_b})\right.$$
$$+ (XZ_o - X_o - Z_o + m) - (XZ_b - X_b - Z_b + m_{XZ_b})$$
$$\left. + (YZ_o - Y_o - Z_o + m) - (YZ_b - Y_b - Z_b + m_{YZ_b})\right\}$$

where m_{X_b} is the mean over the region from which X_b is estimated (*vide* Section 4) etc., and m_{XY_b} is the mean over the region from which XY_b is estimated, etc.

The terms in the means vanish; hence

$$v = v_o + 2(\lambda - \mu)\left\{(X_o - X_b) + (Y_o - Y_b) + (Z_o - Z_b)\right\}$$
$$+ \mu\left\{(XY_o - XY_b) + (XZ_o - XZ_b) + (YZ_o - YZ_b)\right\}$$

and since

$$X_o - X_b = (2X_{u..} - Y_{u..} - Z_{u..})/6p^2 = C(X_{u..})/6p^2 \quad \text{etc.,}$$

$$XY_o - XY_b = (X_{uv.} + Y_{uv.} - 2Z_{uv.})/3p = C(XY_{uv.})/3p \quad \text{etc.,}$$

the partially adjusted means are obtained from the unadjusted means by addition of 6 correction terms, i.e.

$$v = v_o + \frac{\lambda - \mu}{3p^2}\left\{C(X_{u..}) + C(Y_{.v.}) + C(Z_{..w})\right\} + \frac{\mu}{3p}\left\{C(XY_{uv.}) + C(XZ_{u.w}) + C(YZ_{.vw})\right.$$

The operation of adjustment can be made arithmetically simpler by combining these 6 adjustments to form 3 correction terms.

Let

$$\alpha_{u..} = \frac{\lambda - \mu}{3p^2} C(X_{u..}), \qquad \beta_{.v.} = \frac{\lambda - \mu}{3p^2} C(Y_{.v.}), \qquad \gamma_{..w} = \frac{\lambda - \mu}{3p^2} C(Z_{..w}),$$

and

$$\delta_{.vw} = \frac{\mu}{3p} C(YZ_{.vw}) + \gamma_{..w}, \quad \varepsilon_{u.w} = \frac{\mu}{3p} C(XZ_{u.w}) + \alpha_{u..}, \quad \eta_{uv.} = \frac{\mu}{3p} C(XY_{uv.}) + \beta_{.v.}$$

Each adjusted yield is then obtained by addition of the appropriate δ, ε, and η to the corresponding unadjusted yield.

The method of adjusting the means is shown in the folder at the end of this Bulletin. In order to illustrate the method clearly, particularly in showing where each quantity is placed in relation to the others, considerably more detail than is really necessary has been included in the table.

The steps are as follow :-

(i) Determine the unadjusted varietal means and enter them opposite the identification numbers. For variety 111,

$$25.2/3 = 8.4$$

(ii) Determine the values of the α, β, and γ, and enter them in the positions shown. Thus, for example,

$$\alpha_{1..} = \frac{\lambda - \mu}{3p^2} C(X_{1..}) = .001537 \times 12.6 = .0194,$$

$$\beta_{.1.} = \frac{\lambda - \mu}{3p^2} C(Y_{.1.}) = .001537 \times (-30.8) = -.0473,$$

$$\gamma_{..1} = \frac{\lambda - \mu}{3p^2} C(Z_{..1}) = .001537 \times (-4.4) = -.0068.$$

(iii) Determine the values of δ, ε, and η, and enter them in the positions shown. For example,

$$\delta_{.11} = \frac{\mu}{3p} C(YZ_{.11}) + \gamma_{..1} = .01911 \times (-10.7) - .0068 = -.2113$$

$$\delta_{.21} = \frac{\mu}{3p} C(YZ_{.21}) + \gamma_{..1} = .01911 \times 6.6 - .0068 = .1193$$

$$\delta_{.12} = \frac{\mu}{3p} C(YZ_{.12}) + \gamma_{..2} = .01911 \times (-10.0) - .0258 = -.2169$$

$$\delta_{.22} = \frac{\mu}{3p} C(YZ_{.22}) + \gamma_{..2} = .01911 \times 5.0 - .0258 = .0698$$

At this stage the adjustments may be checked using the relation

$$S\delta + S\varepsilon + S\eta = 0.$$

(iv) Each adjusted mean is now determined by adding the appropriate δ, ε, and η. Thus

(a) each δ is added to the 5 means immediately above it in the same column,

(b) each ε is added to the 5 means opposite it in the same row,

(c) each η is added to the mean occupying the same relative tabular position.

For example,

$$v_{243} = 9.5\dot{6} - .0137 + .1947 - .0321 = 9.7156.$$

The adjusted means are shown opposite the corresponding unadjusted means. The arithmetic of this step may be checked by the relation

$$3 \text{ (sum of means)} = \text{grand total}.$$

The strains used in this trial, together with their adjusted mean yields in bushels per acre, are given in Table 8.

TABLE 8 - Adjusted mean yields in bushels per acre.

Varietal Identification Number	Strain	Yield bus./acre
143	Pusa 4 x Dundee	29.8
133	Pusa 4 x Dundee	43.6
153	Nabawa x Dundee	46.5
123	Dundee	49.2
113	Nabawa x Dundee	46.7
125	Nabawa x Dundee	40.4
145	[(Florence x Nabawa)^1 x Free Gallipoli]^1 x Dundee	42.0
115	[(Florence x Nabawa)^1 x Free Gallipoli]^1 x Dundee	42.9
155	Bena	38.0
135	[(Florence x Nabawa)^1 x Free Gallipoli]^1 x Dundee	52.6
445	[(Florence x Nabawa)^1 x Free Gallipoli]^1 x Dundee	53.8
435	[(Florence x Nabawa)^1 x Free Gallipoli]^1 x Dundee	54.6
425	[(Florence x Nabawa)^1 x Free Gallipoli]^1 x Dundee	43.1
415	[(Florence x Nabawa)^1 x Free Gallipoli]^1 x Dundee	53.6
455	Daphne	30.7
111	[Nabawa x (Nabawa x Hope)^1]^1 x Dundee	39.1
141	[Nabawa x (Nabawa x Hope)^1]^1 x Dundee	41.4
121	[Nabawa x (Nabawa x Hope)^1]^1 x Dundee	49.3
131	[Nabawa x (Nabawa x Hope)^1]^1 x Dundee	51.3
151	Sword	49.7
245	[Nabawa x (Nabawa x Hope)^1]^1 x Dundee	47.0
215	[Nabawa x (Nabawa x Hope)^1]^1 x Dundee	49.5
225	Ford (R.A.C.)	51.0
255	Ford (N.S.W.)	27.1
235	Ford x (Florence x Baldry)	47.8
154	Ghurka x Nabawa	53.6
124	Ghurka	54.0
114	Ghurka (Coleman)	52.1
134	Ghurka x Nabawa	48.9
144	Ghurka x Nabawa	51.5
244	Ghurka x Nabawa	46.7
214	Ghurka x Nabawa	50.7
254	Ghurka x Gluyas	46.1
224	Late Gluyas	31.9
234	Ghurka x Gluyas	50.8
324	Ghurka x Gluyas	51.5
334	Magnet	47.8
354	Nabawa x (Riverina x Hope)^1	45.7
314	Nabawa x (Riverina x Hope)^1	43.8
344	Nabawa x (Riverina x Hope)^1	39.3
513	Nabawa x (Riverina x Hope)^1	35.9
553	Nabawa x (Riverina x Hope)^1	42.9
533	Nabawa x (Riverina x Hope)^1	44.5
543	Aussie	39.3
523	Nabawa x (Nabawa x Hope)^1	42.6
413	Nabawa x (Nabawa x Hope)^1	37.7
423	Nabawa x (Nabawa x Hope)^1	46.2
433	(Gluyas x Hope)^2 x Nabawa	38.5
453	(Gluyas x Hope)^2 x Nabawa	39.3
443	(Gluyas x Hope)^2 x Nabawa	42.9
441	(Gluyas x Hope)^2 x Nabawa	42.3
411	(Gluyas x Hope)^2 x Nabawa	43.0
421	Waratah	35.1
431	(Gluyas x Hope)^2 x Nabawa	45.1
451	(Gluyas x Hope)^2 x Nabawa	43.1
323	Comeback	28.6
313	(Gluyas x Hope)^2 x Nabawa	43.1
333	(Gluyas x Hope)^2 x Nabawa	46.0
343	(Gluyas x Hope)^2 x Nabawa	42.0
353	Gular	37.0

TABLE 8 (Contd.)

Varietal Identification Number	Strain	Yield bus./acre
511	(Gluyas x Hope)^2 x Nabawa	34.1
531	(Gluyas x Hope)^2 x Nabawa	43.3
541	(Gluyas x Hope)^2 x Nabawa	42.3
551	(Gluyas x Hope)^2 x Nabawa	44.2
521	(Hope x Gluyas)^2 x Nabawa	38.4
454	Florence	24.3
434	Florence x Nabawa	43.3
414	Florence x Nabawa	43.6
424	Florence x Nabawa	43.3
444	Nabawa	41.2
211	Nabawa x Hope	43.3
231	Nabawa x Hope	41.5
241	Nabawa x Hope	44.9
251	Nabawa x Hope	44.1
221	Nabawa x Hope	40.3
322	Eureka II	40.1
352	Nabawa x Hope	38.0
342	Nabawa x Hope	42.9
312	Nabawa x Hope	41.8
332	Nabawa x Hope	42.6
532	Nabawa x Hope	37.9
522	Nabawa x Hope	41.2
542	Ranee	52.8
552	Nabawa x Hope	42.1
512	Nabawa x Hope	38.2
222	Nabawa x Hope	39.1
212	Nabawa x Hope	39.9
242	Bencubbin	38.6
232	Nabawa x Hope	41.2
252	Nabawa x Hope	41.9
341	Nabawa x Hope	40.9
351	Nabawa x Hope	42.7
321	Nabawa x Hope	41.4
311	Nabawa x Hope	42.8
331	Nabawa x Hope	43.4
355	Nabawa x Hope	40.1
335	Hofed I	37.2
345	Nabawa x Hope	40.2
325	Nabawa x Hope	43.5
315	Nabawa x Hope	40.7
122	Fedweb I	41.6
152	Nabawa x Hope	42.3
142	Nabawa x Hope	48.2
112	Nabawa x Hope	45.7
132	Ford x Hard Federation	39.4
233	Ford x Hard Federation	39.1
243	Ford x Hard Federation	45.8
223	Ford x Hard Federation	45.6
253	Nabawa x Hard Federation	55.0
213	Nabawa x Hard Federation	52.6
534	Nabawa x Hard Federation (Riv.)	45.3
524	Federation	43.2
554	Seewari	50.7
544	Gluwari	36.2
514	Waria	43.0
535	Bobin x Gaza x Bobin	42.4
513	Bobin x Gaza x Bobin	46.3
545	Cadia x Sword	37.6
525	Cadia x Sword	37.4
555	Onas x Nabawa	47.9
452	Dundee x Ghurka	52.7
442	Waratah x Sword	48.8
432	Riverina x Ford	50.5
412	Ford x Dundee	53.4
422	Nabawa x E4	57.9
	Average S.D. of all varietal comparisons	2.55

11. STANDARD DEVIATIONS OF DIFFERENCES BETWEEN ADJUSTED VARIETAL MEANS.

There are three types of comparison between adjusted varietal means, of which typical examples are $v_{111} - v_{211}$, $v_{111} - v_{122}$, and $v_{111} - v_{222}$ with variances

$$V(v_{111} - v_{211}) = \frac{2}{3rwp^2}\left\{\frac{3w}{w + 2w'} + \frac{6(p-1)w}{2w + w'} + (p-1)^2\right\},$$

$$V(v_{111} - v_{122}) = \frac{2}{3rwp^2}\left\{\frac{6w}{w + 2w'} + \frac{3(3p-4)w}{2w + w'} + (p-1)(p-2)\right\},$$

$$V(v_{111} - v_{222}) = \frac{2}{3rwp^2}\left\{\frac{9w}{w + 2w'} + \frac{9(p-2)w}{2w + w'} + (p^2 - 3p + 3)\right\}.$$

The average variance of all varietal comparisons is

$$V_m = \frac{2}{3wr(p^2 + p + 1)}\left\{\frac{9w}{w + 2w'} + \frac{9(p-1)w}{2w + w'} + (p-1)^2\right\}.$$

If there is no additional variation ascribable to the blocks, $w = w'$, and the above variances reduce to the form appropriate to ordinary randomized blocks, $2/3wr$. On the other hand, if $w' = 0$, the variances degenerate to the forms given by Yates (1936a).

In the present example, these variances are:-

$$V(v_{111} - v_{211}) = .009819 \times 28.0977 = .2759, \quad \text{S.D.} = .53,$$

$$V(v_{111} - v_{122}) = .009819 \times 29.7619 = .2922, \quad \text{S.D.} = .54,$$

$$V(v_{111} - v_{222}) = .009819 \times 29.9926 = .2945, \quad \text{S.D.} = .54,$$

$$V_m = .007918 \times 36.8527 = .2918, \quad \text{S.D.} = .54.$$

12. ANALYSIS AS ORDINARY RANDOMIZED BLOCK DESIGN.

Yates (1939) has demonstrated that the quasi-factorial designs possess a remarkable property, namely, that the randomized block analysis provides an unbiased estimate of the error of the unadjusted varietal yields, despite the restrictions on the arrangement.

Consider a single replication of the complete lattice. The expectation of the sum of squares for blocks (eliminating varieties) is

$$3(p^2 - 1)(\tfrac{2}{3}pA + B).$$

In an ordinary analysis this sum of squares would be combined with the intra-block error sum of squares, having an expectation

$$(p-1)^2(2p+1)B,$$

to give a mean square for error with the expectation

$$\frac{3(p^2 - 1)(\frac{2}{3}pA + B) + (p-1)^2(2p+1)B}{3(p^2 - 1) + (p-1)^2(2p+1)} = \frac{p+1}{p^2+p+1}pA + B.$$

On the null hypothesis that the unadjusted varietal yields are undifferentiated, the expectation of the mean square for varieties can be obtained by weighting the expectations of its three components - main effects, two- and three-factor interactions; thus

	Deg. freed.	Expectation of mean square
Main effects	$3(p-1)$	$\frac{2}{3}pA + B$
Two-factor interaction	$3(p-1)^2$	$\frac{1}{3}pA + B$
Three-factor interaction	$(p-1)^3$	B.

Hence the expectation of the mean square for varieties is

$$\frac{3(p-1)(\frac{2}{3}pA + B) + 3(p-1)^2(\frac{1}{3}pA + B) + (p-1)^3 B}{3(p-1) + 3(p-1)^2 + (p-1)^3}$$

$$= \frac{p+1}{p^2+p+1}pA + B$$

showing that the test of significance is unbiased.

13. EFFICIENCY OF THE CUBIC LATTICE DESIGN.

If the experiment had been arranged in ordinary randomized blocks of p^3 plots, the error variance of all unadjusted varietal comparisons would have been

$$\frac{2}{3r} \cdot \frac{p^2/w + (p+1)/w'}{p^2+p+1} \quad (5)$$

and from Table 7, the estimate of this variance is

$$\tfrac{2}{3} \times .5120 = .3413.$$

The average variance of all comparisons between the completely adjusted varietal yields is

$$\frac{1}{3wr} \cdot \frac{2p^2 + 5p + 11}{p^2 + p + 1}$$

(Yates 1936a), and from Table 6 the estimate of this variance is

$$\frac{.3682}{3} \times \frac{86}{31} = .3405 .$$

The ratio of these two variances, i.e.

$$\frac{2\{p^2 + (p+1)w/w'\}}{2p^2 + 5p + 11}, \quad (6)$$

expressed as a percentage, gives the efficiency of the lattice design relative to randomized blocks when the inter-block information is ignored. This ratio is

$$\frac{.3413}{.3405} \times 100 = 100.2\%,$$

showing that such gain as results from the reduction of the block size from 125 to 5 plots is almost balanced by the reduction in the efficiency factor, $2(p^2 + p + 1)/(2p^2 + 5p + 11)$, from 1 to .721.

The ratio of the variance (5) to V_m i.e.

$$\frac{p^2 + (p+1)\, w/w'}{\left\{\dfrac{9w}{w + 2w'} + \dfrac{9(p-1)w}{2w + w'} + (p-1)^2\right\}} \qquad (7)$$

gives the efficiency of the lattice design <u>when the inter-block information is recovered</u>. This ratio is

$$\frac{.3413}{.2918} \times 100 = 117.0\%.$$

Analysis by the new technique has therefore resulted in the recovery of 17 per cent. information if losses due to the inaccuracy of weighting are neglected.

For given values of the ratio w/w', the gains or losses to be expected in lattices of various sizes may be obtained from (6) and (7). Table 9, adapted from Yates' Table VI (1939), with the addition of the values corresponding to $p = 5$, shows these percentage efficiences for the cubic lattice.

TABLE 9 - Percentage efficiences of cubic lattices relative to ordinary randomized blocks.*

p \ w/w'	1	2	3	4	6	8	10
3	100 59.1	108.3 77.3	122.7 95.5	138.9 113.6	173.2 150.0	208.4 186.4	244.1 222.7
4	100 66.7	107.0 82.5	119.4 98.4	133.3 114.3	163.1 146.0	193.7 177.8	224.8 209.5
5	100 72.1	106.0 86.0	116.8 100.0	128.9 114.0	155.0 141.9	181.9 169.8	209.1 197.7
6	100 76.1	105.3 88.5	114.7 100.9	125.5 113.3	148.5 138.1	172.4 162.8	196.6 187.6

* Losses due to inaccuracies of weighting have not been taken into account. The upper figures represent efficiences when inter-block information is recovered, the lower figures those when inter-block information is ignored.

14. CONSEQUENCES OF INACCURATE WEIGHTING.

The substitution of the estimated weights for the unknown true weights renders the analysis slightly inexact in two ways:-

(i) The significance of the varietal mean square is exaggerated and the standard deviations of the inter-varietal comparisons are under-estimated. The degree of under-estimation is, however, of insufficient magnitude to be of any importance.

(ii) There are losses in the process of recovering the inter-block information. These losses of information have been investigated by Yates (1939) and Cochran (1940).

For the case of the cubic lattice, Yates has estimated that the average percentage information lost is never likely to exceed 3%-4%, even when p is as low as 3 and the whole lattice is not replicated. In this case the mean squares from which w' and w are calculated are based upon 24 and 28 degrees of freedom; consequently, for values of $p > 3$, or a greater number of replications, the average losses will be even smaller.

Cochran has shown that, for the square lattice (unreplicated) with $p = 5$, where the degrees of freedom are 8 and 16, the average loss of information increases to a maximum of only 5%, and for the triple lattice (unreplicated) with $p = 4$, degrees of freedom 9 and 21, the maximum loss is also of the order of 4%-5%.

The gains in efficiency resulting from the recovery of inter-block information will thus always more than offset the slight losses due to inaccuracy in weighting.

The fact that w and w' are based on variances derived from different sets of degrees of freedom, also makes exact t tests impossible, but this will never seriously disturb the tests of significance of the differences between the adjusted varietal yields.

15. REFERENCES.

Cox, G.M., Eckhardt, R.C., and Cochran, W.G. (1940). - The analysis of lattice and triple lattice experiments in corn varietal tests. Iowa Agric. Exp. Stn. Res. Bull.281.

Fisher, R.A., (1938). - "Statistical Methods for Research Workers" (Edinburgh: Oliver and Boyd.)

Fisher, R.A., and Yates, F. (1938). - "Statistical Tables for Biological, Agricultural, and Medical Research." (Edinburgh: Oliver and Boyd.)

Goulden, C.H. (1937a). - Efficiency in field trials of pseudo-factorial and incomplete randomized block methods. *Canad. J. Res.* C.15: 231-241.

______(1937b). - Modern methods for testing a large number of varieties. Dom. of Canad. Dept. Agric. Tech. Bull. 9.

______(1939). - "Methods of Statistical Analysis." (New York: John Wiley & Son.)

Yates, F. (1936a). - A new method of arranging variety trials involving a large number of varieties. *J. Agric. Sci.* 26: 424-455.

______(1936b). - Incomplete randomized blocks. *Ann. Eugen.* 7: 121-140.

______(1937a). - A further note on the arrangement of variety trials: Quasi-Latin squares. *Ann. Eugen.* 7: 319-332.

______(1937b). - Design and analysis of factorial experiments. Imp. Bur. Soil Sci., Harpenden, Tech. Comm. 35.

______(1939). - The recovery of inter-block information in variety trials arranged in three dimensional lattices. *Ann. Eugen.* 9: 136-156.

______(1940a). - Modern experimental design and its function in plant selection. *Emp. J. Exp. Agric.* 8: 223-230.

______(1940b). - Lattice squares. *J. Agr.Sci.* 30: 672-687.

______(1940c). - The recovery of inter-block information in balanced incomplete block designs. *Ann. Eugen.* 10: 317-325.

CALCULATION OF ADJUSTED VARIETAL MEANS.

		v_0	v		v_0	v		v_0	v		v_0	v		v_0	v		$C(XZ_{u\cdot1})$	$\varepsilon_{u\cdot1}$	$C(X_{u\cdot\cdot})$	$a_{u\cdot\cdot}$
	111	8.4000	8.2793	121	10.3333	10.4493	131	10.8000	10.8734	141	8.7000	8.7799	151	10.2333	10.5444		5.3	.1207	12.6	.0194
	211	9.5000	9.1742	221	8.3333	8.5500	231	8.8667	8.8019	241	9.5000	9.5277	251	9.1333	9.3463		2.8	.0819	18.5	.0284
	311	9.2333	9.0663	321	8.6333	8.7832	331	8.9667	9.2078	341	8.4667	8.6780	351	8.7333	9.0592		2.2	.0820	26.0	.0400
	411	9.3000	9.1188	421	7.2333	7.4397	431	9.6333	9.5544	441	9.0667	8.9733	451	8.9667	9.1446		-3.3	-.0373	16.8	.0258
	511	7.5000	7.2235	521	8.0667	8.1340	531	9.5667	9.1804	541	8.9000	8.9598	551	9.3000	9.3789		-8.6	-.1440	13.2	.0203
$C(YZ_{\cdot v1})$		-10.7			6.6			-3.2			-0.8			5.3		$C(Z_{\cdot\cdot1})$	-4.4			
$\delta_{\cdot v1}$		-.2113			+.1193			-.0680			-.0221			+.0945		$\gamma_{\cdot\cdot1}$	-.0068			

		v_0	v		v_0	v		v_0	v		v_0	v		v_0	v		$C(XZ_{u\cdot2})$	$\varepsilon_{u\cdot2}$
	112	9.7000	9.6864	122	8.6333	8.8125	132	8.3000	8.3582	142	10.1667	10.2276	152	8.8333	8.9686		11.2	.2334
	212	8.8333	8.4656	222	8.1667	8.2976	232	8.9667	8.7377	242	8.3333	8.1930	252	9.0000	8.8882		0.9	.0456
	312	9.0333	8.8722	322	8.4000	8.5119	332	8.9000	9.0247	342	9.0000	9.0911	352	8.0000	8.0489		2.8	.0935
	412	11.4667	11.3201	422	10.8000	10.9971	432	10.8667	10.7001	442	10.5333	10.3484	452	11.2333	11.1629		-1.2	.0029
	512	8.2333	8.0945	522	8.5667	8.7278	532	8.4000	8.0291	542	11.1333	11.2047	552	9.0000	8.9337		-1.1	-.0007
$C(YZ_{\cdot v2})$		-10.0			5.0			-8.9			-6.7			-8.8		$C(Z_{\cdot\cdot2})$	-16.8	
$\delta_{\cdot v2}$		-.2169			.0698			-.1959			-.1538			-.1940		$\gamma_{\cdot\cdot2}$	-.0258	

		v_0	v		v_0	v		v_0	v		v_0	v		v_0	v		$C(XZ_{u\cdot3})$	$\varepsilon_{u\cdot3}$
	113	10.1333	9.9083	123	10.6333	10.4271	133	9.3333	9.2470	143	6.4000	6.3278	153	9.8667	9.8499		-3.1	-.0398
	213	11.3000	11.1432	223	9.5000	9.6678	233	8.2333	8.2821	243	9.5667	9.7156	253	11.5000	11.6584		8.7	.1947
	313	9.3000	9.1396	323	6.1333	6.0719	333	9.5667	9.7590	343	8.7333	8.9034	353	7.7333	7.8422		-0.4	.0324
	413	8.0000	7.9916	423	9.6333	9.7946	433	8.1333	8.1718	443	9.0667	9.0983	453	3.2000	8.3271		2.8	.0793
	513	7.7000	7.6059	523	9.0000	9.0298	533	9.7000	9.4387	543	8.1333	8.3257	553	9.0667	9.1024		-2.1	-.0198
$C(YZ_{\cdot v3})$		-7.4			-1.5			-2.8			0.0			-3.1		$C(Z_{\cdot\cdot3})$	-8.9	
$\delta_{\cdot v3}$		-.1551			-.0424			-.0672			-.0137			-.0729		$\gamma_{\cdot\cdot3}$	-.0137	

		v_0	v		v_0	v		v_0	v		v_0	v		v_0	v		$C\ XZ_{u\cdot4}$	$\varepsilon_{u\cdot4}$
	114	11.0000	11.0433	124	11.3667	11.4575	134	10.3667	10.3710	144	10.8333	10.9109	154	11.2333	11.3548		7.2	.1570
	214	10.9000	10.7497	224	6.5667	6.7697	234	10.9000	10.7776	244	9.8667	9.9036	254	9.7333	9.7682		5.3	.1297
	314	9.4667	9.2879	324	10.9667	10.9156	334	10.1333	10.1295	344	8.3000	8.3332	354	9.7333	9.6938		-5.1	-.0875
	414	9.2333	9.2391	424	8.9667	9.1709	434	9.3000	9.1750	444	8.8000	8.7273	454	5.1333	5.1446		-0.2	.0220
	514	9.0667	9.1109	524	8.9667	9.1655	534	9.9000	9.6013	544	7.4667	7.6809	554	10.7000	10.7460		1.5	.0490
$C(YZ_{\cdot v4})$		-3.7			3.7			-8.4			-2.5			-6.2		$C(Z_{\cdot\cdot4})$	-8.4	
$\delta_{\cdot v4}$		-.0836			.0578			-.1734			-.0607			-.1314		$\gamma_{\cdot\cdot4}$	-.0129	

		v_0	v		v_0	v		v_0	v		v_0	v		v_0	v		$C(XZ_{u\cdot5})$	$\varepsilon_{u\cdot5}$
	115	9.3667	9.0854	125	8.9667	8.5743	135	11.3000	11.1536	145	9.2000	8.9015	155	8.2000	8.0638		-7.7	-.1277
	215	10.5667	10.4950	225	10.7000	10.8230	235	10.0000	10.1301	245	9.9000	9.9640	255	5.5667	5.7471		11.5	.2482
	315	8.7333	8.6293	325	9.3667	9.2318	335	7.6333	7.8782	345	8.4667	8.5232	355	8.4000	8.5022		0.9	.0572
	415	11.4000	11.3716	425	9.1333	9.1447	435	11.5667	11.5814	445	11.5667	11.4083	455	6.4667	6.5107		0.1	.0277
	515	9.8000	9.8254	525	7.9000	7.9194	535	9.1333	8.9877	545	7.8333	7.9752	555	10.0667	10.1588		2.5	.0681
$C(YZ_{\cdot v5})$		-5.2			-6.1			-0.8			-6.7			-4.2		$C(Z_{\cdot\cdot5})$	-15.7	
$\delta_{\cdot v5}$		-.1235			-.1407			-.0394			-.1521			-.1044		$\gamma_{\cdot\cdot5}$	-.0241	

	$C(XY_{u1\cdot})$	$C(XY_{u2\cdot})$	$C(XY_{u3\cdot})$	$C(XY_{u4\cdot})$	$C(XY_{u5\cdot})$
	0.9	-7.1	2.4	-0.9	4.4
	-7.8	0.2	-2.8	-1.6	1.3
	0.5	-3.3	13.2	8.0	7.2
	6.0	5.9	2.7	-1.7	5.7
	6.6	4.2	-7.8	11.9	6.1
$C(Y_{\cdot v\cdot})$	-30.8	7.6	-16.4	-1.0	7.7
$\beta_{\cdot v\cdot}$	-.0473	.0117	-.0252	-.0015	.0118
	$\eta_{u1\cdot}$	$\eta_{u2\cdot}$	$\eta_{u3\cdot}$	$\eta_{u4\cdot}$	$\eta_{u5\cdot}$
	-.0301	-.1240	.0207	-.0187	.0959
	-.1964	.0155	-.0787	-.0321	.0366
	-.0377	-.0514	.2271	.1514	.1494
	.0674	.1244	.0264	-.0340	.1207
	.0788	.0920	-.1743	.2259	.1284

SUMMARY.

This is the second of a series of papers concerned with the recovery of information in quasi-factorial designs that have incomplete data. The paper describes an approximate method for dealing with lattice squares and discusses the effect of the approximations on the estimation of the adjusted treatment effects and their errors and the analysis of variance.

Commonwealth of Australia Council for Scientific and Industrial
Research Bulletin Number 175, Melbourne, 1944

The Recovery of Inter-block Information in Quasi-Factorial Designs with Incomplete Data

2. Lattice Squares*

E.A. Cornish, M.Sc., B.Agr.Sc.

1. INTRODUCTION.

The quasi-factorial and balanced incomplete block designs were introduced by Yates (1936 a,b; 1937) and designed for the purpose of eliminating heterogeneity in experimental material to a greater degree than is possible by the use of ordinary randomized blocks. In the original papers, only the complete elimination of group (block, row, or column) differences was considered, but it was pointed out that, since the inter-group comparisons contained information on the treatments, some loss of information would inevitably result from the methods of analysis advocated. It was shown that the amount of information lost depends upon the extent to which the group means differ and that, in the limiting case, where the inter-group and intra-group comparisons are of equal accuracy, this loss amounts to $1-E$, where E is the efficiency factor of the particular design.

The means for recovering the inter-group information in the several types of design have been discussed at length by Yates (1939; 1940a,b) and Cox *et al* (1940). The methods of analysis are quite straightforward provided the data are complete, but, if the designs have been impaired by the loss, or unreliability, of one or more observations, the analyses require modification. An approximate method for recovering inter-group information has been described for square, triple, and cubic lattice designs (1943), and the present paper discusses the extension of this method to lattice square designs.

2. STRUCTURE OF LATTICE SQUARES.

In these designs the quasi-factorial principle has been extended in order to allow for the simultaneous elimination, from the treatment comparisons, of differences associated with two different groupings of the experimental material, as in the ordinary Latin square.

The number of treatments must be a perfect square. If this number is, say, p^2, it is possible for certain values of p (Fisher and Yates 1938; Stevens 1939) to divide the treatments into $(p + 1)$ orthogonal groups of p sets, each containing p treatments, i.e. in such a manner that each set of any given group of sets contains only one treatment from each set of any other group of sets. Each complete replication of the treatments is arranged in the field in the form of a square, of which the rows correspond to one group of sets, and the columns to a second, thus confounding the treatment comparisons corresponding to two groups of sets with row and column differences in each square or replication. If p is odd,

* Typescript received August 24, 1943.

balance can be attained with a minimum of $\frac{1}{2}(p + 1)$ replications but, if p is even, each group of sets must be included twice, giving $(p + 1)$ replications.

The statement made above regarding the balance attainable with odd values of p requires some slight qualification. With $\frac{1}{2}(p + 1)$ replications, the designs are completely balanced with respect to intra row and column comparisons, but the inter-row comparisons provide information on only one half of the treatment degrees of freedom, and the inter-column comparisons on the other half. Consequently, if the inter-row and inter-column information is recovered, and the row and column comparisons differ in accuracy, the comparisons of the adjusted yields will also differ in accuracy. On the other hand, when $(p + 1)$ replications are used, information on every treatment degree of freedom is obtained from both inter-row and inter-column comparisons, with the result that all adjusted treatment comparisons will be of equal accuracy, even though the row and column comparisons differ in accuracy.

3. ESTIMATION OF TREATMENT EFFECTS.

If $\frac{1}{2}(p + 1)$ replications are employed, the treatment comparisons confounded in any given square are entirely unconfounded in the remaining $\frac{1}{2}(p - 1)$ squares, whereas, with $(p + 1)$ replications, each set of comparisons is confounded in two squares but remains unconfounded in the other $(p - 1)$ squares. Three estimates of the confounded effects may thus be determined:-

(i) Intra row and column estimates derived from regions where the effects are unconfounded.

(ii) Inter-row and inter-column estimates derived from regions where the effects are confounded.

(iii) Unadjusted estimates derived from the whole experiment.

Consider first the case of $\frac{1}{2}(p + 1)$ squares. The $(p + 1)$ sets to which any particular treatment, v, belongs will be denoted by $P_{1v}, P_{1'v}, P_{2v}, P_{2'v}, \ldots$ and the yields of these sets in the several squares by ${}_{I}P_{1v}$, ${}_{II}P_{1v}$, $\ldots {}_{I}P_{1'v}$, ${}_{II}P_{1'v}$, ... If, for example, the set P_{1v} is confounded in a row of the first square, then the inter-row estimate of the mean yield of these treatments will be derived from

$${}_{I}P_{1v}$$

the intra row and column estimate from

$${}_{II}P_{1v} + {}_{III}P_{1v} + \ldots$$

and the unadjusted estimate from

$${}_{S}P_{1v} = {}_{I}P_{1v} + {}_{II}P_{1v} + {}_{III}P_{1v} + \ldots .$$

Let w_r, w_c, and w_i stand for the reciprocals of the true error variances per plot of the inter-row, inter-column, and intra row and column estimates, respectively. If the yields, apart from the treatment effects, are regarded as the sum of three normally and independently distributed quantities, the first constant for all plots of a row, but varying from row to row with variance A_r, the second constant for all plots of a column, but varying from column to column with variance

A_c, and the third varying independently from plot to plot with variance B, then

$$w_i = 1/B, \quad w_r = 1/(pA_r + B) \quad \text{and} \quad w_c = 1/(pA_c + B).$$

Let $_wP_{1v}$ stand for the partially adjusted total yield of the set P_{1v}, then since the total

$$_{II}P_{1v} + {}_{III}P_{1v} + \ldots.$$

is derived from $\frac{1}{2}(p - 1)$ as many squares as the total $_IP_{1v}$, $_wP_{1v}$,(in terms of the total yield of $\frac{1}{2}(p + 1)$ replicates) is given by

$$_wP_{1v} = \tfrac{1}{2}(p+1)\left\{w_r{}_IP_{1v} + w_i({}_{II}P_{1v} + {}_{III}P_{1v} + \ldots)\right\} / \left\{w_r + \tfrac{1}{2}(p-1)w_i\right\}$$

$$= {}_SP_{1v} + \frac{w_i - w_r}{w_r + \frac{1}{2}(p-1)w_i}\left\{{}_SP_{1v} - \tfrac{1}{2}(p+1){}_IP_{1v}\right\}.$$

If

$$L_{1v} = {}_SP_{1v} - \tfrac{1}{2}(p+1){}_IP_{1v}, \quad M_{1'v} = {}_SP_{1'v} - \tfrac{1}{2}(p+1){}_IP_{1'v},$$

$$\lambda = \frac{w_i - w_r}{w_r + \frac{1}{2}(p-1)w_i}, \qquad \mu = \frac{w_i - w_c}{w_c + \frac{1}{2}(p-1)w_i},$$

then

$$_wP_{1v} = {}_SP_{1v} + \lambda L_{1v},$$

with similar expressions for the remaining sets confounded in the rows, and

$$_wP_{1'v} = {}_SP_{1'v} + \mu M_{1'v},$$

with similar expressions for the remaining sets confounded in the columns.

The adjusted total yield of treatment v is therefore obtained from the unadjusted total by adding the quantities

$$\delta_{1v} + \delta_{2v} + \ldots. + \varepsilon_{1'v} + \varepsilon_{2'v} + \ldots.$$

where

$$\delta_{1v} = \lambda L_{1v}/p,$$

$$\varepsilon_{1'v} = \mu M_{1'v}/p, \text{ etc.}$$

4. APPROXIMATE METHOD OF ANALYSIS.

In developing the approximate method of analysis for square, triple, and cubic lattice designs with incomplete data, it was found that, as in cases previously discussed, the problem could best be approached through the calculation of estimates for the missing observations. The method described depends upon the fact that these designs may be analysed in three ways:-

(i) the analysis of the unadjusted yields,

(ii) the analysis of the partially adjusted yields,

(iii) the analysis of the fully adjusted yields.

Unadjusted and intra-block estimates are obtained from the first and third analyses, respectively, and are then used to determine approximations to the inter-block estimates of the treatment effects, while from the second analysis the weighting coefficients are estimated and then used in conjunction with the inter-block and intra-block estimates to calculate approximations to the partially adjusted yields.

5. ESTIMATION OF MISSING VALUES.

Assume that one of the observations, belonging to the set P_{1v} in a row of the first square, is missing. An estimate of this value, which will be designated an unadjusted estimate, is first determined from the formula given by Yates (1933), and then included with the existing values of the total $_SP_{1v}$ to determine an efficient and unbiased estimate of the unadjusted total yield of the particular set concerned. The quantity

$$_{II}P_{1v} + {}_{III}P_{1v} + \ldots.$$

is unaffected since the remaining squares have been assumed to be intact. The unadjusted estimate is not, however, suitable for inclusion with the existing values of the total $_IP_{1v}$ in order to provide an estimate of the confounded comparisons. An estimate of this quantity can be obtained, with sufficient accuracy, from the relation

$$_SP_{1v} = {}_IP_{1v} + ({}_{II}P_{1v} + {}_{III}P_{1v} + \ldots.)$$

where $_SP_{1v}$ includes the unadjusted estimate of the missing observation. $_IP_{1v}$ is obtained in a similar manner.

The particular treatment involved does, however, belong to other sets which are confounded in the second, third, ... squares, but which are unconfounded in the first square. If only the intra row and column information is used, another estimate of the missing observation may be obtained from the formula appropriate to the analysis of the fully adjusted yields (Cornish 1940). This estimate, which is designated an intra row and column estimate, is combined with the existing values of the unconfounded sets to which this particular treatment belongs, in order to determine efficient and unbiased intra row and column estimates of the comparisons concerned. If more than one value is missing this procedure needs only slight modification to determine the partially adjusted estimates.

The formulae for determining the intra row and column estimates are given below. Let a stand for the missing value. Let R_a, C_a, and T_a be the totals of the existing values in the row, column, and square respectively in which a occurs. Let V_a be the total of the existing replicates of the treatment containing a, and S_a the total of rows and columns in which this treatment occurs other than the row and column involving a. Then (using Yates (1937) notation)

$$pQ_a = p(V_a + a) - (R_a + a) - (C_a + a) - S_a = Q'_a - R_a - C_a - (p-2)a.$$

where

$$Q'_a = pV_a - S_a.$$

Similarly, if a replicate of treatment i occurs in the same row as a

$$pQ_i = pV_i - (R_a + a) - S_i = Q'_i - (R_a + a),$$

TABLE 1*

Square I

						Total	L	δ
	sy a_1	*ny* 67.7	*n* 68.1	*sz* 65.2	*s* 59.9	260.9+a_1	12.9	1.0
	mz 69.4	*nx* 67.4	*mx* 64.9	*nw* 77.6	*nz* 67.7	347.0	25.6	2.1
	w 46.3	*x* 59.7	*c* 76.2	*cy* 80.3	*m* 71.2	333.7	-93.7	-7.5
	— 55.2	*cx* 77.0	*my* 82.4	*cw* 80.4	*mw* 77.7	372.7	-134.0	-10.6
	sw 73.9	*sx* 79.0	*z* 51.6	*y* 44.8	*cz* 78.4	327.7	-79.9	-6.4
Total	244.8+a_1	350.8	343.2	348.3	354.9	1642.0+a_1		
M	-43.4	-66.3	-40.3	-82.3	-36.8			
ε	-1.5	-2.4	-1.4	-2.9	-1.3			

Square II

						Total	L	δ
	mz 69.1	*y* 49.6	*c* 72.6	*s* 68.7	*cx* a_2	260.0+a_2	-36.2	-2.9
	cw 62.7	*n* 74.0	*nx* 78.1	*w* 40.0	*cz* 66.3	321.1	3.8	40.3
	ny 68.9	*nz* 78.1	*sw* 69.6	*my* 68.1	*cy* 68.5	353.2	20.1	1.6
	m 64.7	— 38.6	*sz* 57.4	*sx* 65.5	*mx* 67.1	293.3	60.5	4.9
	z 48.0	*x* 39.0	*mw* 64.9	*nw* 72.2	*sy* 58.7	282.8	72.5	5.7
Total	313.4	279.3	342.6	314.5	260.6+a_2	1510.4+a_2		
M	23.9	11.0	10.2	22.7	52.9			
ε	0.9	0.4	0.4	0.8	1.9			

Square III

						Total	L	δ
	sy 57.6	*my* 69.2	*nx* 64.0	*y* 31.8	*m* 58.4	281.0	98.3	7.8
	nz 72.2	*sz* 73.3	*z* 46.2	*w* 45.9	*cx* 73.4	311.0	-29.9	-2.4
	sx 61.4	*mz* 77.9	*cy* 73.1	*mw* 70.4	*n* 70.5	353.3	9.9	0.8
	c 58.6	*cz* 68.4	— 46.7	*ny* 71.3	*nw* 69.1	314.1	45.5	3.6
	cw 56.6	*x* 52.9	*s* 60.9	*mx* 71.8	*sw* 68.7	310.9	24.1	1.9
Total	306.4	341.7	290.9	291.2	340.1	1570.3		
M	103.4	-28.4	34.5	9.5	28.9			
ε	3.7	-1.0	1.2	0.3	1.0			

* The M, L, δ, and ε calculated from estimates of the missing values are underlined.

and in the same column

$$pQ_j = pV_j - (C_a + a) - S_j = Q'_j - (C_a + a).$$

In the case of $\frac{1}{2}(p + 1)$ replications, the estimate of the missing value is given by

$$a = \frac{p(R_a+C_a) - T_a}{(p-1)(p-3)} + \frac{2}{p(p-1)^2(p-3)} \left[(p-1)^2 Q'_a - (p-1)\left\{S(Q'_i)+S(Q'_j)\right\} + S(pQ)\right] \quad \ldots\ldots.(1)$$

where the last summation within the square bracket is taken over the remaining treatments. For $(p + 1)$ replications replace $(p - 3)$ by $(p - 2)$ and omit the factor 2 from the multiplier of the second term.

If several values are missing, the estimates are most conveniently determined by a process of successive approximation. This is accomplished by repeated application of the formula for a single missing value, substituting approximations for the unknowns. Careful selection of the first approximations ensures rapid convergence of the solution.

Example.

Yates (1940a) has given an example showing the analysis of a set of lattice squares involving 25 treatments, using 3 replications. Table 1 shows the yields. Two values have been assumed missing; that of treatment *sy* in the first square (actual value 61.2) and that of treatment *cx* in the second square (actual value 74.0).

Unadjusted estimates.

The value 63.27 has been taken as the first approximation to a_1. Applying the ordinary randomized block formula, the first approximation to a_2 is

$$a_2 = (25 \times 150.40 + 3 \times 1510.40 - 4785.97)/48 = 73.03.$$

The final values are

$$a_1 = 63.29 \quad \text{and} \quad a_2 = 73.03.$$

Intra row and column estimates.

The value 67.12 has been taken as the first approximation to a_1. Table 2 summarizes the data needed to calculate the first approximation to a_2.

TABLE 2.

	Treatment								
	cx	*ms*	*y*	*c*	*s*	*cs*	*cy*	*mx*	*sy*
V	150.4	216.4	126.2	207.4	189.5	213.2	221.9	203.8	183.42
S	1374.6	1667.32	1527.5	1640.0	1599.22	1659.5	1679.4	1585.6	1504.22
$-Q'$	622.6	585.32	896.5	603.0	651.72	594.0	569.9	566.6	587.12
	$S(Q'_i) = -2736.54$,			$S(Q'_j) = -2317.62$,			$S(pQ) = -15663.42$.		

Applying the formula given above, the first approximation to a_2 is

$$a_2 = \frac{5(260.0 + 260.6) - 1510.4}{8} + \frac{1}{80}\left\{-16 \times 622.6 + 4 \times 5054.16 - 15663.42\right\} = 68.98.$$

The final values are

$$a_1 = 67.29 \quad \text{and} \quad a_2 = 68.95.$$

6. ESTIMATION OF w_r, w_c, AND w_i.

If the intra row and column estimates of the missing values, calculated from the formulae of the previous section, are included with the existing observations, the analysis of the partially adjusted yields may be completed. The mean squares for rows (eliminating treatments), columns (eliminating treatments), and residual in this analysis provide the basis for estimating w_r, w_c, and w_i, respectively. The intra row and column variance is correctly estimated by dividing the residual sum of squares by the corresponding number of degrees of freedom, reduced by the number of missing observations, but the inter-row and inter-column variances require further consideration before they can be used. It can be shown that the latter two variances are always too large, and in order to determine the correct sum of squares, e.g. for columns, the reduction in the sum of squares due to constants representing lattice squares, rows, and treatments only must be evaluated. It is not, however, necessary to determine these constants directly, because the same result can be obtained by the alternative and simpler procedure of determining an auxiliary set of estimates for the missing observations by successive approximation using the formula given below.

Assume that the missing observation belongs to one of the treatments in the ith set in a row of the first square and that this treatment occurs in the jth, k^{th} .... sets respectively in rows of the second, third, ... squares. If V, T, and G stand for the totals of the existing values of the particular treatment, the first square, and the grand total, and L'_{iv}, L'_{jv}, L'_{kv} .... stand for quantities similar to those defined in Section 3, but from which the treatment containing the missing observation is omitted, then the estimate of the missing value is given by

$$a = \frac{\left\{ \frac{1}{2}p^2(p-1)V - \frac{1}{2}(p+1)T + G - \frac{1}{2}p(p-1)L'_{iv} + p(L'_{jv} + L'_{kv} + \dots) \right\}}{\frac{1}{4}(p-1)^2(p+1)(p-2)}$$

The auxiliary estimates, so determined, are substituted for the missing values, and the completed set is subjected to the analysis for the partially adjusted yields, except that all calculations relating to columns are omitted. The residual sum of squares in this analysis is deducted from the total sum of squares of existing values to give the sum of squares due to fitting constants representing lattice squares, rows, and treatments, and the deduction of this sum from the sum of squares due to fitting constants for lattice squares, rows, columns, and treatments gives the correct sum of squares for columns (eliminating treatments). The reduction in the column sum of squares is never likely to be large, and if only a few observations are missing it will be negligible.

The correct sum of squares for rows (eliminating treatments) is found in a similar manner. The rows are omitted and an auxiliary set of estimates is obtained using a formula similar to that above, in which the quantities M are substituted for the quantities L.

The application of the formula may be illustrated by using it to determine the first approximation to a_1 (omitting the columns). Taking the first approximation to a_2 as 72.86, the required data are

$V = 116.3$,	$T = 1642.0$,	$G = 4795.56$
Square I	Square II	Square III
$L_1' = 23.2$,	$L_5' = -119.1$,	$L_1' = -93.3$,

so that

$$a_1 = \left\{50 \times 116.30 - 3 \times 1642.00 + 4795.56 - 10 \times 23.20 - 5(119.10 + 93.30)\right\}/72$$

$$= 60.98 .$$

The final values are $a_1 = 60.98$, $a_2 = 72.86$, and when the rows are omitted $a_1 = 68.13$ and. $a_2 = 70.37$.

Following Yates, the estimate of w_i is given by the reciprocal of the intra row and column variance, but before the corrected inter-row and inter-column variances can be used in the determination of w_r and w_c, they require further consideration. When the data are complete, the mean square for rows (eliminating treatments) has the expectation

$$p \frac{p-1}{p+1} A_r + B ,$$

but if certain observations are missing, the reduced inter-row variance has the expectation

$$kA_r + B$$

and $k < p(p-1)/(p+1)$. Considerable labour is required to determine the exact value of k in any particular case, and unless it is essential to have its true value the work will not be worth while. For all practical purposes $p(p-1)/(p+1)$ can be substituted for k. The nature of this type of approximation has been examined briefly in the case of the square lattice (Cornish 1943) but has not been attempted here.*

The estimation of w_r, w_c, and w_i may be illustrated by further consideration of the example given in the previous section. The crude sum of squares of the original 75 observations is 324167.55, so that, after deducting $61.2^2 + 74.0^2$, the sum of squares of the supposedly existing set is 314946.11. The intra row and column estimates, 67.29 and 68.95, are substituted for a_1 and a_2 respectively, and the experiment is analysed according to the method given by Yates (1940a). Table 3 summarizes the data, supplementing that of Table 1, which is needed to complete the analysis.

TABLE 3.

	Square I		Square II		Square III
L_1	4.92	L_1	-28.00	L_1	90.29
L_4	-125.85	L_5	64.49	L_2	-21.75
M_1	-51.38	M_5	53.09	M_1	95.39
M_2	-58.15			M_5	37.05
Totals	-268.93		120.89		148.04
	1709.29		1579.35		1570.30
Treatments:	$sy = 183.59$,		$cx = 219.35$,		grand total = 4858.94.

* This point is, however, of some interest and it is hoped that a more detailed examination will be published shortly.

The analysis of variance is given in Table 4.

TABLE 4.

Analysis of completed set (intra row and column estimates).

Variation due to -	Deg. Freed.	Sum Sq.	Mean Sq.
Squares	2	483.79	
Rows (eliminating treatments)	12	1010.22	
Columns (eliminating treatments)	12	356.36	
Treatments (ignoring rows and columns)	24	7197.97	
Residual	22	389.18	17.69
Total	72	9437.52	

The estimates 60.98 and 72.86 are now substituted for a_1 and a_2 respectively, and the analysis of variance is performed omitting the columns. The necessary supplementary data are summarized below, and the analysis of v ariance is given in Table 5.

	Square I		Square II		Square III
L_1	17.54	L_1	-35.82	L_1	83.98
L_4	-121.94	L_5	58.18	L_2	-17.84
Totals	-252.40		106.76		145.64
	1702.98		1583.26		1570.30

Treatments: $sy = 177.28$, $\quad sx = 223.26$, $\quad$ grand total = 4856.54.

TABLE 5.

Analysis of completed set (omitting columns)

Variation due to -	Deg. Freed.	Sum Sq.
Squares	2	428.07
Rows (eliminating treatments)	12	1019.99
Treatments (ignoring rows)	24	7326.71
Residual	34	718.74
Total	72	9493.51

The total sum of squares of deviations of the existing observations from their mean is

$$314946.11 - 4722.70^2 / 73 = 9413.30,$$

and from this sum, the residual sum of squares of Table 5, 718.74, is deducted to obtain the reduction in the sum of squares due to fitting constants for squares, rows, and treatments. This step gives 8694.56, and finally this quantity is deducted from the sum of squares due to lattice squares, treatments, rows, and columns to give the reduced sum of squares due to columns (eliminating treatments). The steps in these calculations are set out in Table 6.

TABLE 6.

Analysis of existing set (intra row and column estimates).

Variation due to -	Deg. Freed.	Sum Sq.	Mean Sq.
Total	72	9413.30	
Residual from Table 4	22	389.18	17.69
Square, row, column and treatment constants	50	9024.12	
Square, row and column constants	38	8694.56	
Remainder (columns eliminating treatment)	12	329.56	27.46

The reduced value for rows (eliminating treatments), 1008.75 with mean square 84.06, is found in a similar manner, using the estimates 68.13 and 70.37 for a_1 and a_2 respectively. Assuming that the corrected inter-row and inter-column variances have expectations approximating closely to those given above,

$$\lambda = \frac{2(84.06 - 17.69)}{4 \times 84.06} = .3948, \qquad \lambda/p = .0790,$$

$$\mu = \frac{2(27.46 - 17.69)}{4 \times 27.46} = .1779, \qquad \mu/p = .0356.$$

7. APPROXIMATE TEST OF SIGNIFICANCE.

The analysis of variance for the partially adjusted yields does not directly provide the appropriate treatment sum of squares for use in a test of significance of those yields. The process of weighting described in Section 3 alters the sums of squares of all components of the sum for treatments, because every treatment degree of freedom is confounded in these designs. Moreover, the test of significance is not exact owing to the inaccuracy of weighting, and when the data are incomplete further errors are introduced by the method of determining the inter-row and inter-column, and intra row and column estimates described in Section 5.

Considering only the unadjusted values, an exact test of significance of the treatment mean square can be made, but, if the treatments and the rows and/or columns are really differentiated, this test will, in general, underestimate the significance of the partially adjusted yields. The test has, however, the advantages that it is easily made and that it will suffice in the majority of cases. If the set of observations is completed by inclusion of the unadjusted estimates, the data may be analysed by the ordinary randomized block procedure. It can be shown that the sum of squares for treatments in this analysis is always exaggerated, but the amount by which this sum must be reduced to make the test of significance exact is directly expressible in terms of a simple formula (Yates 1933, Cornish 1943). If the unadjusted means are significantly different as judged by this test, it is obvious that no further test is necessary.

If the inter-row and inter-column information is rejected and only the intra row and column information considered, an exact test of the significance of the intra row and column treatment effects can be made. In order to make this test, the sum of squares for treatments (eliminating rows and columns) must be evaluated. This is most conveniently obtained in the following way. Using the intra row and column estimates of the missing observations, the sums of squares for rows (ignoring treatments) and columns (ignoring treatments) are determined, and from their sum is deducted the corresponding sum when treatments are eliminated. The difference between these two quantities is now deducted from the sum of squares for treatments (ignoring rows and columns) to give the sum for treatments (eliminating rows and columns). As it stands, this quantity is not suitable for use in an exact test of significance, since it also would yield an exaggerated estimate of the treatment mean square. For a set of lattice squares the necessary reduction is only expressible as a simple formula when one observation is missing, in which case the expression for the reduction is

$$\left\{p(R + C) - T - p^2 a\right\} / p^2(p - 1)^2$$

where a is the estimate of the missing value, and R, C, and T are the totals of the corresponding row, column, and square <u>including</u> the estimate.

If more than one observation is missing, the reduction is most conveniently determined by resorting once again to the artifice of calculating an auxiliary set of estimates. The treatment classification is disregarded, and the estimates calculated from the formula

$$a = \left\{p(R_a + C_a) - T_a\right\} / (p-1)^2$$

where R_a, C_a, and T_a stand for the quantities defined in Section 5. The total variation of this new set is analysed into components for squares, rows, columns, and residual, and the deduction of this residual from the total sum of squares of existing values gives the sum of squares for rows, columns, and squares only. This test also underestimates the true significance of the partially adjusted yields, but in cases where the rows and columns are differentiated the degree of underestimation is much smaller than with the first test.

These tests may be illustrated with the data of the example discussed above. After substituting the unadjusted estimates 63.29 and 73.03 for a_1 and a_2 and analysing the experiment as an ordinary randomized block design of 25 treatments in triplicate, the following sums of squares are obtained: squares 443.26, treatments 7305.55, and residual 1735.31.

The reduction in the treatment sum of squares is then

$$\frac{1}{25 \times 24}\left\{(1705.29 - 63.29 \times 25)^2 + (1583.43 - 73.03 \times 25)^2\right\} = 123.10,$$

and the first approximate test of significance is as follows:

	Deg. Freed.	Sum Sq.	Mean Sq.	Ratio
Treatments (ignoring rows and columns)	24	7182.45	299.27	7.93
Residual	46	1735.31	37.72	

For the second test, the sum of squares for treatments (eliminating rows and columns) is first required. Substitute the intra row and column estimates 67.29 and 68.95 for a_1

and a_2 in the appropriate row and column totals of Table 1, and determine the sums of squares for rows and columns (ignoring treatments). For squares I, II, and III the sums for rows in that order are 286.30, 639.00, and 530.42, and for columns 235.90, 449.39, and 511.94, giving a total of 2652.95. From Table 4 the sum of squares for rows (eliminating treatments) and columns (eliminating treatments) is 1366.58. The difference

$$2652.95 - 1366.58 = 1286.37$$

is now deducted from the sum of squares for treatments (ignoring rows and columns), 7197.97, also from Table 4, to give 5911.60 as the sum of squares for treatments (eliminating rows and columns). The auxiliary estimates required in the calculation of the reduction in this sum are

$$a_1 = 55.41 \quad \text{and} \quad a_2 = 68.29,$$

and the analysis of variance resulting from the substitution of these values and the omission of the treatments is given in Table 7.

TABLE 7

Analysis of completed set (omitting treatments).

Variation due to -	Deg. Freed.	Sum Sq.
Squares	2	404.29
Rows	12	1539.87
Columns	12	1357.73
Residual	46	6210.10
Total	72	9511.99

Using the residual from this analysis, the sum of squares due to rows, columns, and lattice squares only is

$$9413.30 - 6210.10 = 3203.20$$

Finally this quantity is deducted from the sum of squares due to rows, columns, lattice squares, and treatments to give the reduced sum of squares for treatments. The steps in these calculations and the second approximate test of significance are set out in Table 8.

TABLE 8.

Analysis of existing set (intra row and column estimates)

Variation due to -	Deg. Freed.	Sum Sq.	Mean Sq.	Ratio
Total	72	9413.30		
Residual (from Table 4)	22	389.18	17.69	
Squares, rows, columns and treatments	50	9024.12		
Squares, rows and columns	26	3203.20		
Remainder (treatments)	24	5820.92	242.54	13.71

8. CALCULATION OF ADJUSTED TREATMENT MEANS.

In order to adjust the treatment totals it is necessary to calculate new values for the δ and ε corresponding to the rows and columns containing the treatments *sy* and *ox*. As an example consider the calculation of the δ and ε corresponding to treatment *sy* in the second square. Designate the set of treatments in the 5th row of this square by the symbol P_{2sy} and the set of treatments in the 5th column by $P_{2'sy}$.

The total of all replicates of treatments s, x, mw, and nw is 729.30, and the total of treatment sy including the unadjusted estimate, 63.29, is 179.59, so that the total for all five treatments is

$$SP_{2sy} = 179.59 + 729.30 = 908.89$$

Since the observation for treatment sy is missing from the first square, the total of the set P_{2sy} from unconfounded regions of the experiment is obtained by substituting the intra row and column estimate, 67.29, for sy in the first square; hence

$$_{I}P_{2sy} + {}_{III}P_{2sy} = 630.09.$$

The approximation for the total $_{II}P_{2sy}$ is then found from

$$SP_{2sy} - ({}_{I}P_{2sy} + {}_{III}P_{2sy}) = 908.89 - 630.09 = 278.80$$

and the value of L corresponding to the 5th row of the second square from

$$L_5 = SP_{2sy} - \tfrac{1}{2}(p + 1)\ {}_{II}P_{2sy} = 908.89 - 3 \times 278.80 = 72.49.$$

The corresponding δ is thus

$$\lambda L_5/p = 72.49 \times .0790 = 5.7$$

The value of $SP_{2'sy}$ is

$$638.80 + 179.59 + 223.43 = 1041.82$$

in which the unadjusted estimates 63.29 and 73.03 of both treatments are included. When obtaining the total

$$_{I}P_{2'sy} + {}_{III}P_{2'sy}$$

only the intra row and column estimate of sy is used, because the missing observation corresponding to treatment cx occurs in the particular column concerned. Thus

$$_{I}P_{2'sy} + {}_{III}P_{2'sy} = 712.19$$

and the approximation for $_{II}P_{2'sy}$ is

$$1041.82 - 712.19 = 329.63$$

whence

$$M_5 = 1041.82 - 3 \times 329.63$$

$$= 52.93$$

and the corresponding ε is

$$52.93 \times .0356 = 1.9 .$$

The values of the L, M, δ, and ε are given in Table 1.

9. ERRORS OF VARIETAL DIFFERENCES.

If the main tests of significance are made by means of the analysis of variance, precise estimates of the variances of the varietal comparisons are not necessary, and provided care is taken in claiming significance for differences near the border line (for a given level of significance), the normal values should suffice. A guide to the magnitude of the increases to be expected in the case of the fully adjusted yields has been given (Cornish 1940), and from the expressions quoted, it is obvious that for all lattices likely to be used in practice the increases are negligibly small.

10. THE CASE OF $(p + 1)$ REPLICATIONS.

The analysis of arrangements involving $(p + 1)$ replications follows similar lines. Unadjusted estimates of the missing observations are found using the ordinary randomized block formula (Yates 1933) and intra row and column estimates from the appropriate form of formula (1) of Section 5 above. The latter estimates are substituted for the missing observations, and the experiment is analysed according to the method given by Yates (1940a), but, as in the case of $\frac{1}{2}(p + 1)$ replications, the mean squares for rows (eliminating treatments and columns) and for columns (eliminating treatments and rows) in this analysis are both exaggerated. The method of determining their correct reduced values is similar to that used above except that different formulae are needed for calculating the auxiliary estimates.

If V_i stands for the total yield of all replicates of a particular treatment i, $S(R_i)$ and $S(C_i)$ for the sums of all the row and column totals respectively containing treatment i, and G for the grand total, Yates (1940a) defines the quantities L'_i and M'_i by the equations

$$L'_i = pV_i - (p + 1)S(R_i) + G$$

$$M'_i = pV_i - (p + 1)S(C_i) + G.$$

The sums of squares for rows(eliminating treatments) and columns (eliminating treatments) are derived from these quantities.

Assume that one of the replicates of the ith treatment is missing and that the $(p - 1)$ other treatments in the same row as the missing observation are the jth, k^{th} .....

If V_i, G, L_j, L_k, ... now stand for the quantities defined above except for the missing observation, and T for the total of existing values in the square containing that observation, then the auxiliary estimate for the case where the columns are omitted from the analysis, is given by

$$a = \frac{1}{p^2(p - 1)} \left\{p^2V_i + (p + 1)T - G - (L_j + L_k + \dots)\right\}.$$

When the rows are omitted the auxiliary estimate is determined from a similar formula, substituting the quantities M for the quantities L.

The weighted estimates are determined by the procedure outlined above for the $\frac{1}{2}(p + 1)$ replications, but, owing to the fact that estimates of the varietal differences can be constructed from both the row and the column totals, the arithmetical details of the adjustment may be considerably more complicated.

11. REFERENCES.

Cornish, E.A. (1940). - The estimation of missing values in quasi-factorial designs. *Ann. Eugen.* 10: 137-143.

______(1943). - The recovery of the inter-block information in quasi-factorial designs with incomplete data. Coun. Sci. Ind. Res. (Aust.) Bull. 158.

Cox, G.M., Eckhardt, R.C., and Cochran, W.G. (1940). - The analysis of lattice and triple lattice experiments in corn varietal tests. Iowa Agr. Expt. Sta. Res. Bull.281.

Fisher, R.A., and Yates, F. (1938). - "Statistical Tables for Biological, Agricultural and Medical Research". (Oliver and Boyd : Edinburgh).

Stevens, W.L. (1939). - The completely orthogonalized Latin square. *Ann. Eugen.* 9: 82-93.

Yates, F. (1933). - The analysis of replicated experiments when the field results are incomplete. *Emp. J. Expt. Agric.* 1: 129-142.

______(1936a). - A new method of arranging variety trials involving a large number of varieties. *J. Agric. Sci.* 26: 424-455.

______(1936b). - Incomplete randomized blocks. *Ann. Eugen.* 7: 121-140.

______(1937). - A further note on the arrangement of varietal trials: Quasi-Latin squares. *Ibid.* 7: 319-332.

______(1939). - The recovery of inter-block information in variety trials arranged in three dimensional lattices. *Ibid.* 9: 136-156.

______(1940a). - Lattice Squares. *J. Agric. Sci.* 30: 672-687.

______(1940b). - The recovery of inter-block information in balanced incomplete block designs. *Ann. Eugen.* 10: 317-325.

Reprinted from the Australian Journal of Scientific Research, Series B, Biological Sciences, Vol. 2, No. 2, pp. 83-137, 1949

YIELD TRENDS IN THE WHEAT BELT OF SOUTH AUSTRALIA DURING 1896-1941

By E. A. Cornish*

[*Manuscript received January* 21, 1949]

Summary

The history of the wheat industry in South Australia is reviewed to provide a background for discussion of the forms of trend observed in yield. The period chosen for examination was 1896-1941, and the analysis extends to practically the entire wheat belt, the basic territorial unit used for assessing yield being the hundred, the mean area of which in South Australia is approximately 118 square miles. As a preliminary to the evaluation of the trends it was necessary to estimate and to eliminate the effects of variations in seasonal rainfall; the statistical technique used was that of partial regression, and reasons are given for the choice of rainfall variates.

The major soil groups under cultivation are described and mapped.

The elimination of phosphorus as a limiting factor in yield coincided with the beginning of the period under review, so that in classifying the forms of trend observed it was convenient to divide the hundreds into two groups, according to whether they were opened for cultivation before or after the advent of superphosphate.

The nitrogen status of the major wheat soils is discussed, and after consideration of relevant literature, it is concluded that the nitrogen required by the crop has been drawn almost entirely from soil reserves under the exploitative systems of cropping generally employed. The wheat belt is broadly divisible into three parts:

1. Sandy, stony, and mixed mallee soils and related types in which nitrogen becomes limiting after 20-40 years of cropping, and yields subsequently decline owing to exhaustion of the reserves.
2. Loamy mallee soils and red brown earths, where yields increase over the period 1896-1941, but at diminishing rates as nitrogen becomes limiting.
3. Sandy and loamy mallee and transitional mallee-solonetz soils, where yield increases linearly throughout, mainly because exploitative cropping has not been in progress long enough to make its influence apparent. These regions constitute only a small proportion of the total area.

The economic restoration and maintenance of the nitrogen status of the wheat soils are discussed briefly.

I. Historical Introduction

Wheat-farming is the major industry of South Australia. One feature of its expansion to this predominant position is illustrated in Figure 1, which shows the annual acreage of the crop from 1836 to 1941; four clearly defined phases

* Section of Mathematical Statistics, C.S.I.R.

are apparent, and for the purpose of tracing the history of the industry it is

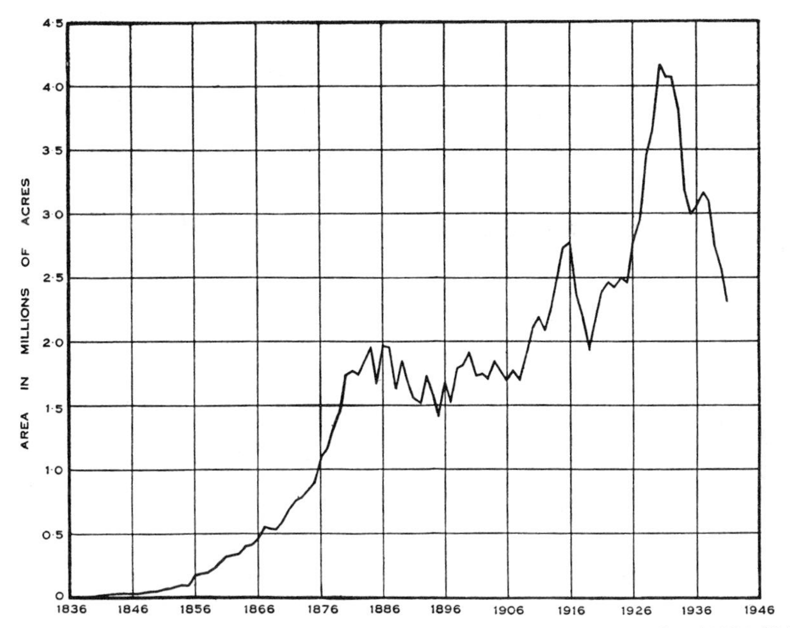

Fig. 1.—Annual acreage of wheat cultivated for grain in South Australia (1836-1941).

convenient to use them with the following end points: 1836-85, 1886-1908, 1909-32, 1933-41. This partitioning is made primarily on the basis of acreage and its temporal rate of change, but at the same time carries other factors of importance in the development. Reference to the effects of these factors is made in the notes which follow.

(a) 1836-85

The production of wheat was first attempted on the mainland in 1838, and although only European varieties were available, fair success was achieved because of the relatively favourable climate in the immediate vicinity of Adelaide. Experience soon demonstrated the correct time to sow the crop, and the area rapidly expanded to the point where production exceeded local demand.

During the decade 1841-50, three advances gave a great impetus to production:

1. The invention of the stripper in 1843; this machinery considerably reduced the labour costs of harvesting, and made possible expansion into greater areas.
2. The construction of a flour mill in 1843 stimulated production of wheat for export of flour to interstate markets.
3. The abolition of the British corn duties in 1847 encouraged production to meet an export trade with Britain.

In the early years, cropping was confined to the Mt. Lofty Ranges and their slopes to the east and west, both on account of their geographical situation and

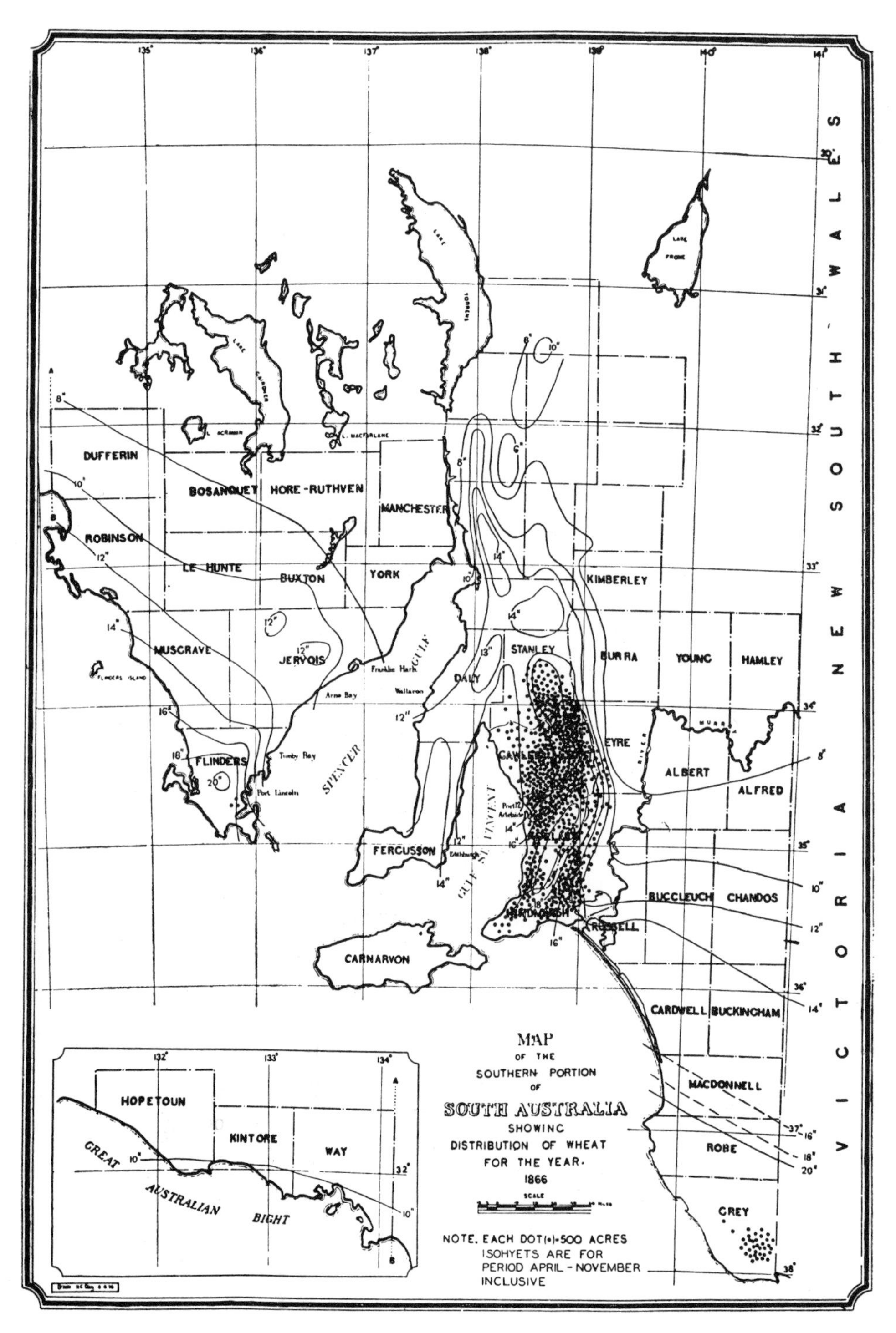
MAP
OF THE
SOUTHERN PORTION
OF
SOUTH AUSTRALIA
SHOWING
DISTRIBUTION OF WHEAT
FOR THE YEAR.
1866
SCALE
NOTE. EACH DOT(•)=500 ACRES
ISOHYETS ARE FOR
PERIOD APRIL - NOVEMBER
INCLUSIVE
NEW SOUTH WALES
VICTORIA
135°
136°
137°
138°
139°
140°
141°
DUFFERIN
BOSANQUET
HORE-RUTHVEN
MANCHESTER
ROBINSON
LE HUNTE
BUXTON
YORK
KIMBERLEY
MUSGRAVE
JERVOIS
STANLEY
BURRA
YOUNG
HAMLEY
DALY
EYRE
ALBERT
ALFRED
FLINDERS
FERGUSSON
BUCCLEUCH
CHANDOS
CARNARVON
CARDWELL
BUCKINGHAM
MACDONNELL
ROBE
GREY
SPENCER
GULF
GULF ST. VINCENT
Franklin Harb.
Wallaroo
Arno Bay
Tumby Bay
Port Lincoln
HOPETOUN
KINTORE
WAY
GREAT
AUSTRALIAN
BIGHT
132°
133°
134°

Fig. 2

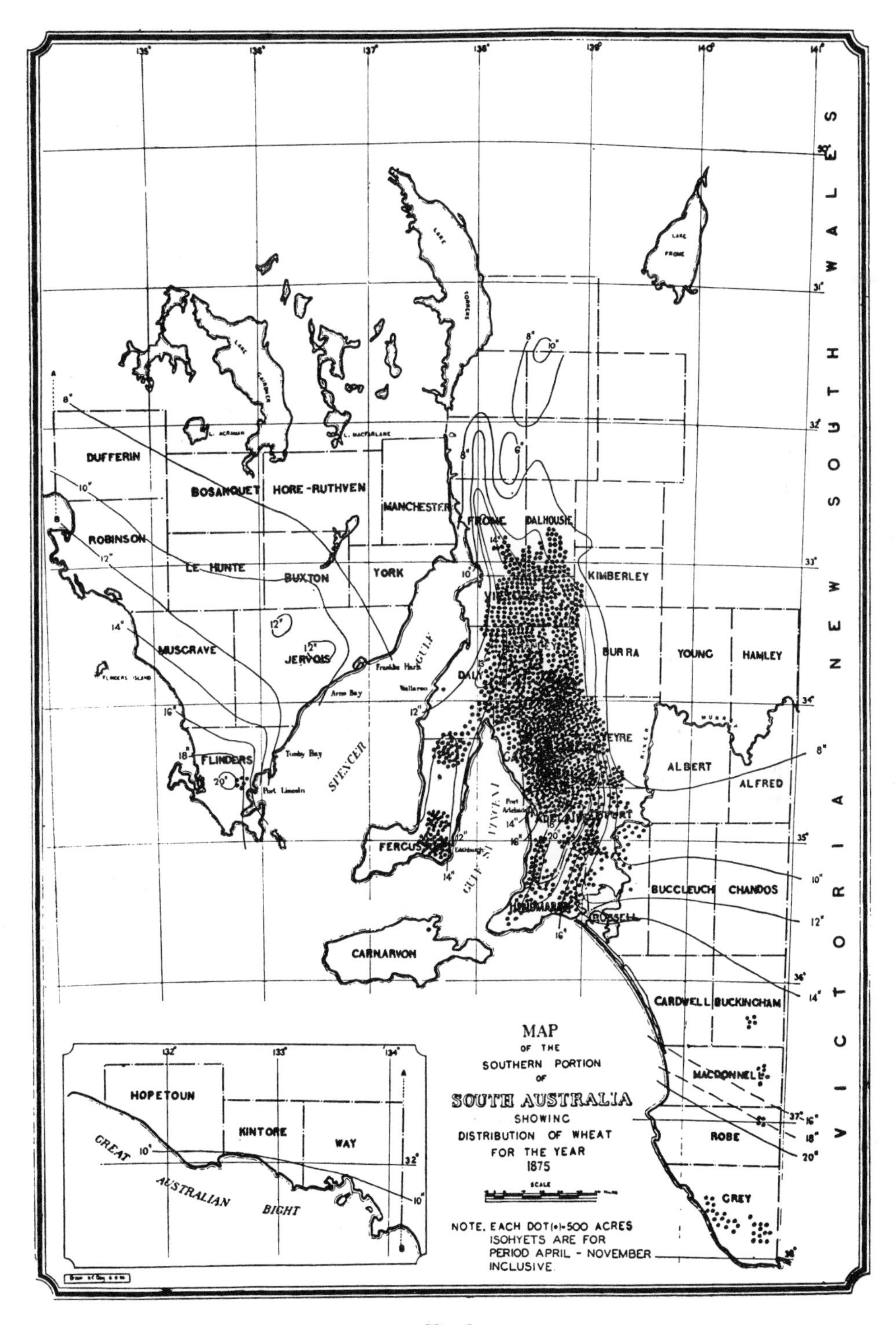
MAP
OF THE
SOUTHERN PORTION
OF
SOUTH AUSTRALIA
SHOWING
DISTRIBUTION OF WHEAT
FOR THE YEAR
1875
SCALE
NOTE. EACH DOT(•)=500 ACRES
ISOHYETS ARE FOR
PERIOD APRIL - NOVEMBER
INCLUSIVE
NEW SOUTH WALES
VICTORIA
135°
136°
137°
138°
139°
140°
141°
31°
32°
33°
34°
35°
36°
37°
38°
DUFFERIN
BOSANQUET
HORE-RUTHVEN
MANCHESTER
FROME
DALHOUSIE
ROBINSON
LE HUNTE
BUXTON
YORK
KIMBERLEY
MUSGRAVE
JERVOIS
BURRA
YOUNG
HAMLEY
FLINDERS
ALBERT
ALFRED
BUCCLEUCH
CHANDOS
RUSSELL
CARNARVON
CARDWELL
BUCKINGHAM
MACDONNELL
ROBE
GREY
SPENCER GULF
GULF ST VINCENT
Franklin Harb
Arno Bay
Wallaroo
Tumby Bay
Port Lincoln
Port Adelaide
HOPETOUN
KINTORE
WAY
GREAT AUSTRALIAN BIGHT
132°
133°
134°

Fig. 3

the readiness with which the arable parts could be brought into production. The combined effect of these factors is illustrated in Figure 2, which shows the distribution of the area under wheat in 1866, in relation to the isohyets of seasonal rainfall (total from April to November inclusive). Over practically all land thrown open for cultivation, the rainy season was followed by a long arid period extending throughout the summer, which precluded the use of rotation systems involving wheat that had been practised successfully in Europe. As a result, a new scheme was gradually developed, in which the land lay in fallow for periods of 9-12 months prior to sowing.

As the railway northward from Adelaide was extended, and subsidiary lines were constructed, expansion of the area followed, but cropping was still restricted mainly to regions with high rainfall. The distribution in 1875 is illustrated in Figure 3. It will be observed that this diagram shows the first signs of extension into areas with less than an average seasonal rainfall of 10 in.

Increasing demands of settlement next led to an attempt to convert large areas of mallee land to wheat-farming. The high costs and difficulties of clearing that were entailed would have retarded progress but for the invention, in 1876, of the multi-furrow stump-jump plough, followed later by the mallee roller. These two advances engendered a revolutionary change, by reducing costs and greatly facilitating the task of reclamation. Simultaneously with these developments, the railway network was carried further to the north, and considerable expansion followed in the most northerly counties. Figure 4 shows the distribution in 1884, and indicates clearly the extension in counties Blachford, Hanson, Newcastle, Granville, Frome, and Dalhousie, and the mallee districts of counties Daly and Fergusson, the area of crop sown in regions with a mean seasonal rainfall of less than 10 in. now assuming much larger proportions.

Toward the close of this period, two further, all-important, advances were made:

1. The selection, in 1881, of a variety known as Ward's Prolific, which was fairly resistant to rust, and yielded well under adverse conditions; it formed a base from which other selections were made, and constituted a major turning-point in varietal history.
2. During 1882-85, Custance demonstrated in the field at Roseworthy Agricultural College the value of superphosphate.

(*b*) *1886-1908*

Throughout this interval, the total area remained approximately constant, but the distribution of the crop was altered radically. In the first 10 years further expansion occurred in the northern districts, on Eyre Peninsula, and in counties Burra, Eyre, Sturt, and Albert. These increases were compensated by a reduction in acreage on Yorke Peninsula and in regions of higher rainfall to the north of Adelaide. The distribution in 1896 is given by Richardson (1936). The closing years brought further changes. Much of the far northern section proved unsuitable because of the limited and erratic rainfall, and a marked recession followed,

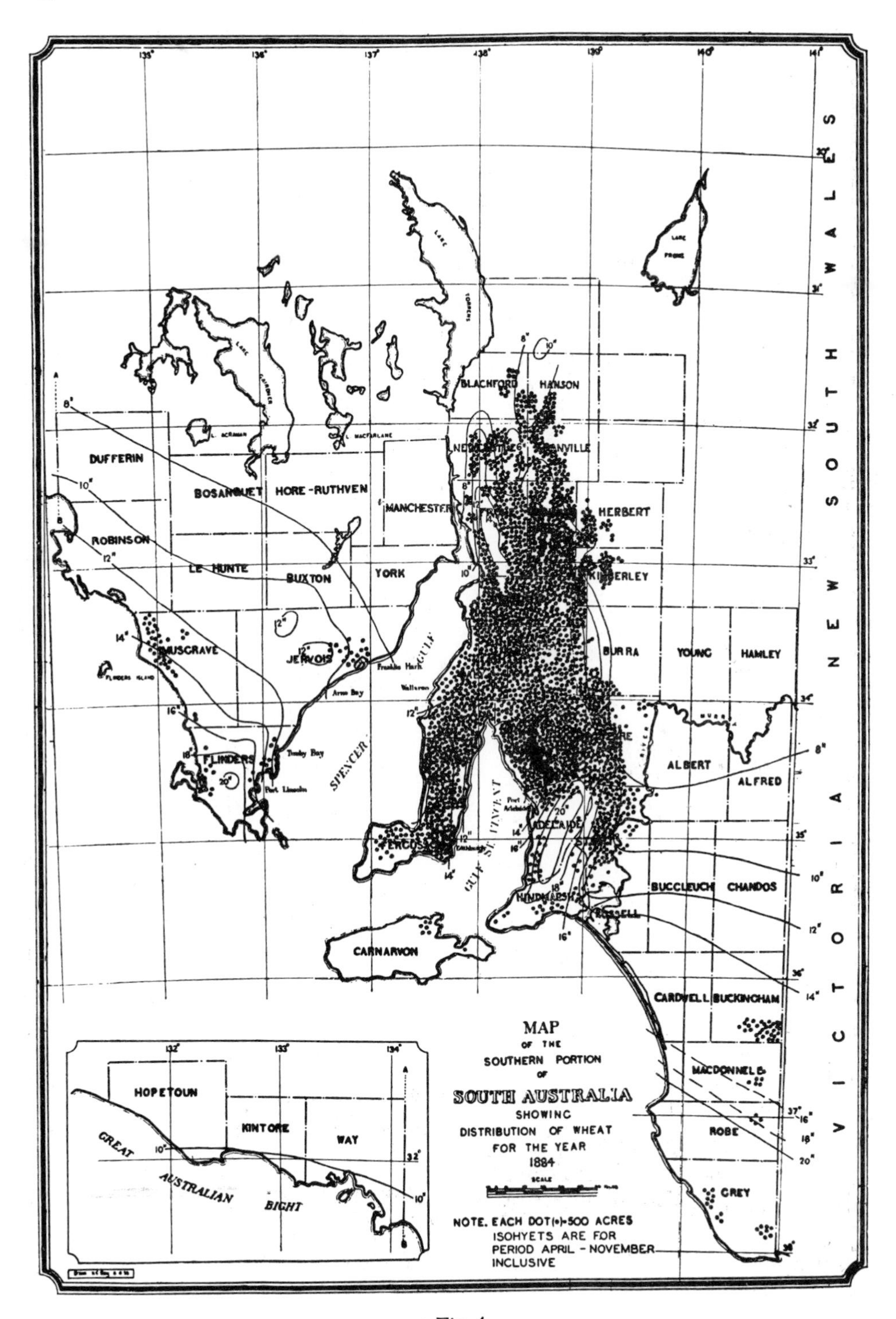
MAP
OF THE
SOUTHERN PORTION
OF
SOUTH AUSTRALIA
SHOWING
DISTRIBUTION OF WHEAT
FOR THE YEAR
1884
NOTE. EACH DOT (•)=500 ACRES
ISOHYETS ARE FOR
PERIOD APRIL - NOVEMBER
INCLUSIVE
NEW SOUTH WALES
VICTORIA
DUFFERIN
BOSANQUET
HORE-RUTHVEN
MANCHESTER
ROBINSON
LE HUNTE
BUXTON
YORK
MUSGRAVE
JERVOIS
FLINDERS
BLACKFORD
HANSON
HERBERT
BURRA
YOUNG
HAMLEY
ALBERT
ALFRED
ADELAIDE
BUCCLEUCH
CHANDOS
CARNARVON
CARDWELL
BUCKINGHAM
MACDONNELL
ROBE
GREY
HOPETOUN
KINTORE
WAY
GREAT AUSTRALIAN BIGHT
SPENCER
GULF
GULF ST. VINCENT

Fig. 4

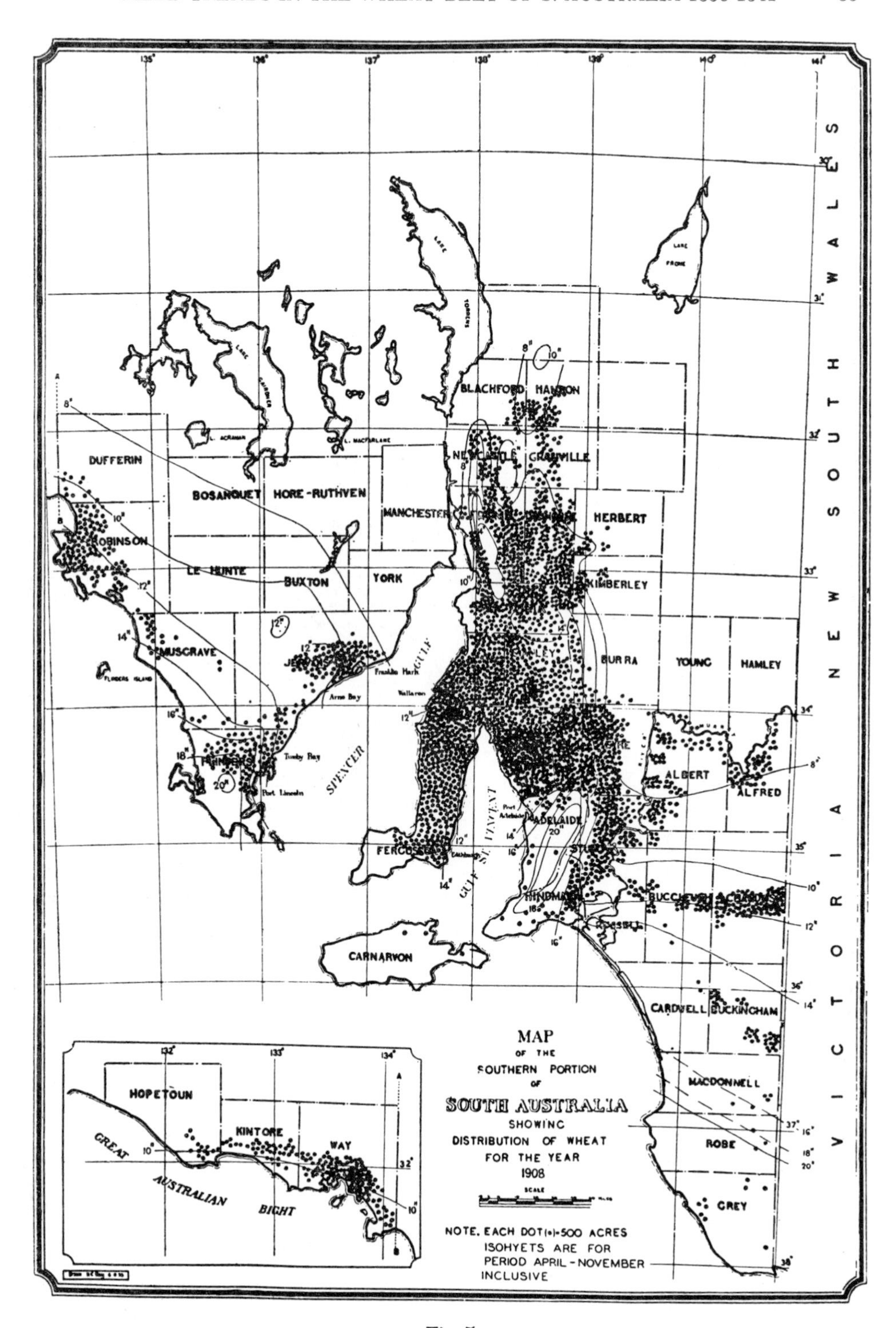

MAP
OF THE
SOUTHERN PORTION
OF
SOUTH AUSTRALIA
SHOWING
DISTRIBUTION OF WHEAT
FOR THE YEAR
1908
SCALE
NOTE. EACH DOT(•)=500 ACRES
ISOHYETS ARE FOR
PERIOD APRIL - NOVEMBER
INCLUSIVE
NEW SOUTH WALES
VICTORIA
DUFFERIN
ROBINSON
BOSANQUET
HORE-RUTHVEN
LE HUNTE
BUXTON
YORK
MANCHESTER
MUSGRAVE
FLINDERS
SPENCER
GULF
ST. VINCENT
FERGUSSON
CARNARVON
BLACKFORD
HANSON
NEWCASTLE
GRANVILLE
HERBERT
KIMBERLEY
BURRA
YOUNG
HAMLEY
ALBERT
ALFRED
ADELAIDE
HINDMARSH
CARDWELL
BUCKINGHAM
MACDONNELL
ROBE
GREY
Port Lincoln
Tumby Bay
Arno Bay
Franklin Harb.
Wallaroo
HOPETOUN
KINTORE
WAY
GREAT AUSTRALIAN BIGHT

Fig. 5

the northern boundary being withdrawn except for isolated localities. At the same time further settlement occurred on Eyre Peninsula in the vicinity of ports, and cropping was again intensified on Yorke Peninsula and in counties Daly, Gawler, and Light. Figure 5, which gives the distribution in 1908, shows also the developments on mallee land near the River Murray in counties Albert and Alfred, and the extension in counties Buccleuch and Chandos following construction of a railway through the latter region.

Finally, from 1900 onward, mean yields increased, thus giving the first tangible results of Lowrie's* efforts to popularize the use of superphosphate.

(*c*) *1909-32*

This period was marked by further withdrawals from the northern counties, but its outstanding feature is the second great surge of development, culminating in the peak of over 4 million acres during 1930-32 inclusive. This enormous expansion followed first from the extension of the railway system, during 1906-19, into the mallee areas of counties Albert and Alfred, and of Eyre Peninsula, and later from the stimulus given by the high prices offered for wheat, and repatriation policies following the 1914-18 war. The average distribution for the years 1924-5-6 is given by Richardson (loc. cit.), and that for 1930 in Figure 6.

Other important advances were a general adoption of the practice of applying superphosphate to the crop, the almost universal use of improved varieties that had been developed in the breeding programmes conducted in various parts of the Commonwealth, and the appearance of a close association between sheep- and wheat-farming.

(*d*) *1933-41*

A complete reversal of the trend in acreage characterizes this short sequence of years, and by 1941 the total area in the State had fallen to just over 2¼ million acres, the value attained about 1920. The rapidity with which this major readjustment occurred was due to the impact on the industry of the economic crisis of 1930 following immediately in the wake of four severe droughts during 1926-29. The balance was loaded most heavily against the wheat-growers of the marginal and submarginal regions, and comparison of the distribution in 1941 (Fig. 7) with that of 1930 demonstrates that by far the greater proportion of the reduction was made in the most recently settled tracts of mallee land. Actually, the exigencies of the economic situation only accelerated this reduction, for, as will appear, it was inevitable that the margin of cultivation would be withdrawn in these areas, just as it was in the most northerly parts after the expansive and optimistic settlement of the years 1880-96. This observation may even be extended. In the light of experience and from a consideration of the facts which have become available since 1941, it seems that additional large-scale adjustments should be made.

II. Origin and Scope of the Investigation

The problems associated with settlement of the marginal agricultural areas, which for a long time had been exercising the minds of governmental authorities.

* W. Lowrie succeeded J. D. Custance as Principal of Roseworthy Agricultural College.

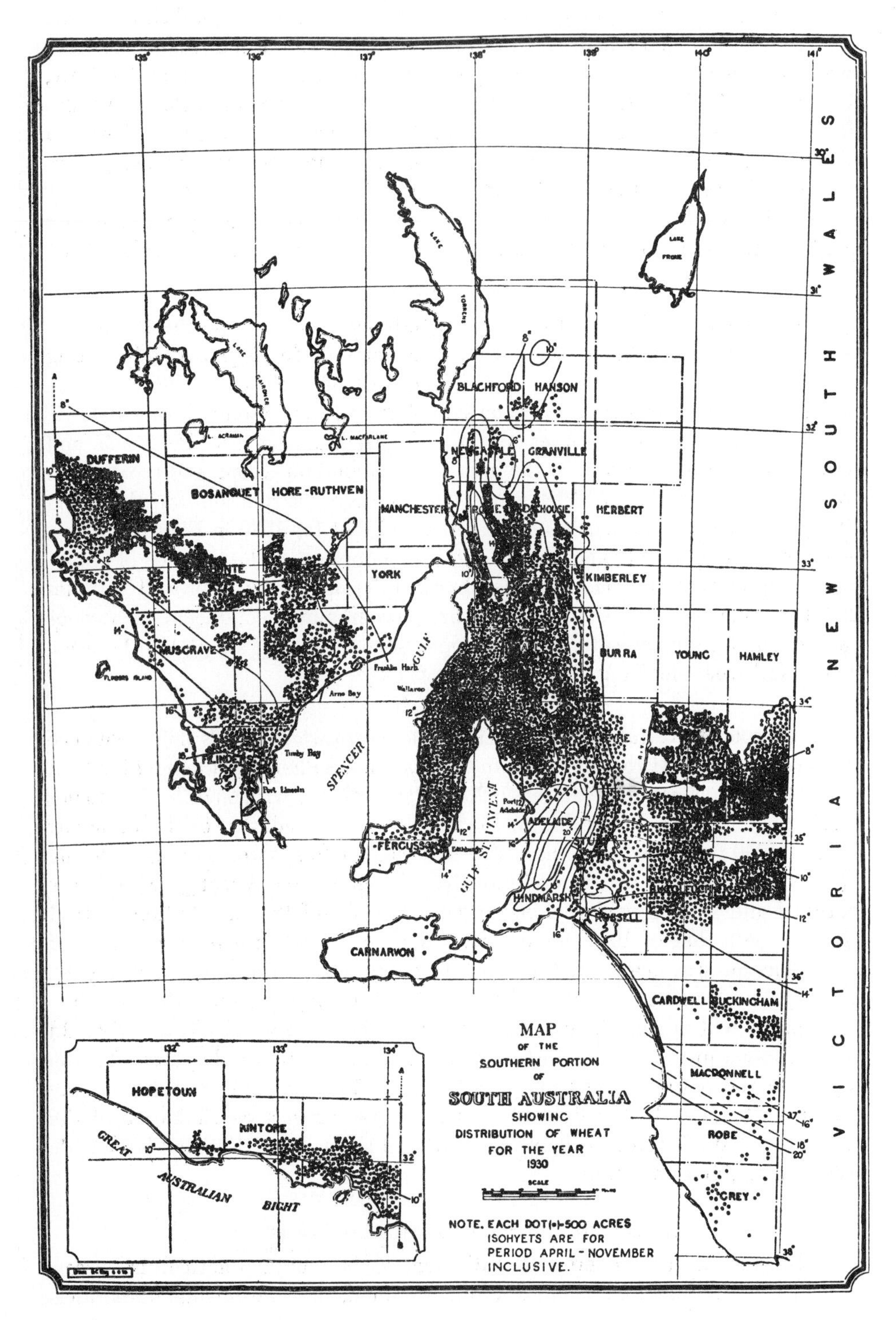

MAP
OF THE
SOUTHERN PORTION
OF
SOUTH AUSTRALIA
SHOWING
DISTRIBUTION OF WHEAT
FOR THE YEAR
1930
SCALE
NOTE. EACH DOT(•)=500 ACRES
ISOHYETS ARE FOR
PERIOD APRIL - NOVEMBER
INCLUSIVE.
NEW SOUTH WALES
VICTORIA
BLACHFORD
HANSON
NEWCASTLE
GRANVILLE
DUFFERIN
BOSANQUET
HORE-RUTHVEN
MANCHESTER
HERBERT
YORK
KIMBERLEY
MUSCRAVE
BURRA
YOUNG
HAMLEY
FLINDERS
Tumby Bay
Port Lincoln
SPENCER
GULF
Franklin Harb.
Arno Bay
Wallaroo
ST. VINCENT
GULF
FERGUSSON
ADELAIDE
HINDMARSH
RUSSELL
CARNARVON
CARDWELL
BUCKINGHAM
MACDONNELL
ROBE
GREY
HOPETOUN
KINTORE
WAY
GREAT
AUSTRALIAN
BIGHT

Fig. 6

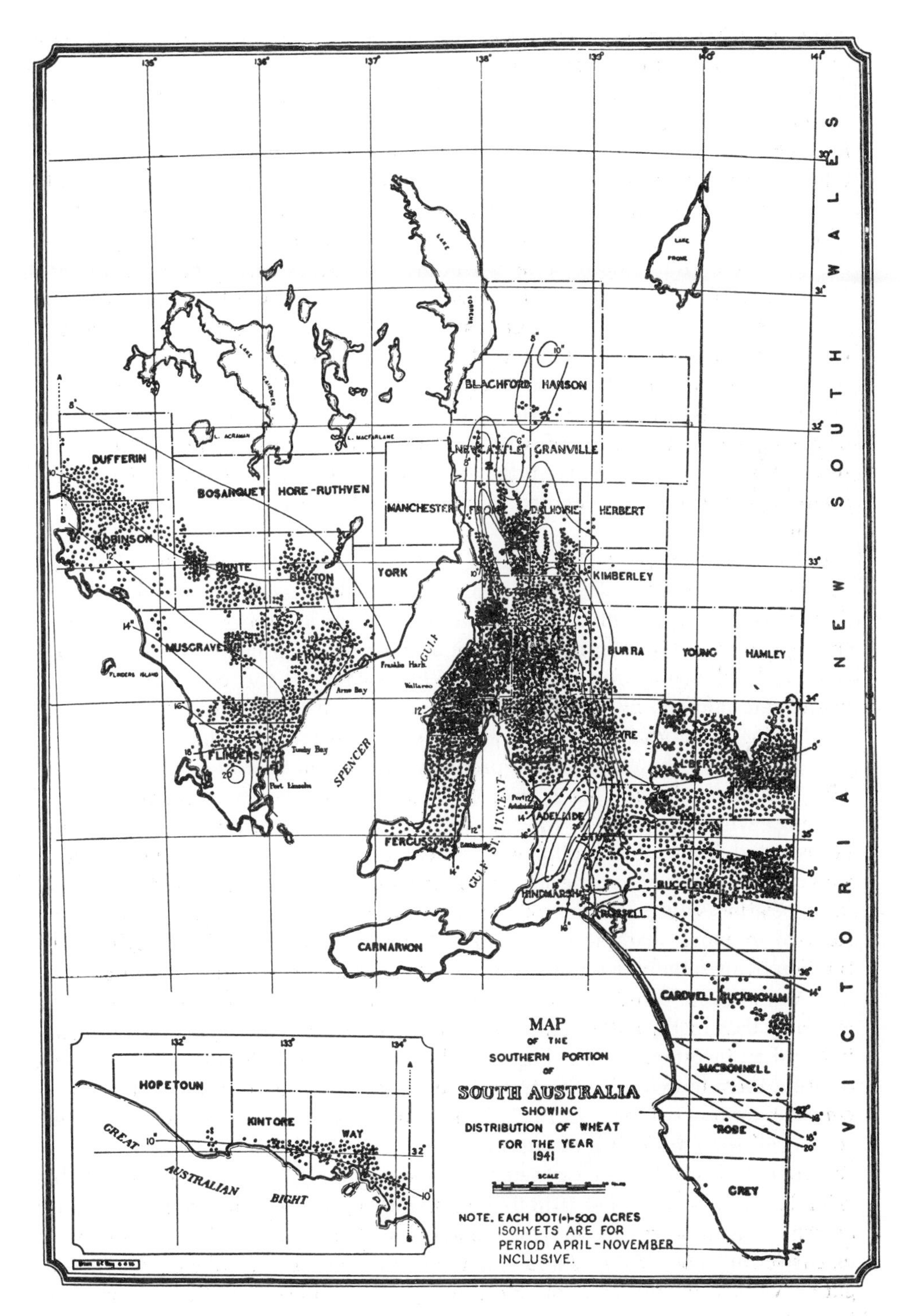
MAP
OF THE
SOUTHERN PORTION
OF
SOUTH AUSTRALIA
SHOWING
DISTRIBUTION OF WHEAT
FOR THE YEAR
1941
NOTE. EACH DOT(•)=500 ACRES
ISOHYETS ARE FOR
PERIOD APRIL-NOVEMBER
INCLUSIVE.
NEW SOUTH WALES
VICTORIA
DUFFERIN
BOSANQUET
HORE-RUTHVEN
MANCHESTER
FROME
DALHOUSIE
HERBERT
BLACHFORD
HANSON
NEWCASTLE
GRANVILLE
ROBINSON
BUXTON
YORK
KIMBERLEY
MUSGRAVE
JERVIS
BURRA
YOUNG
HAMLEY
FLINDERS
SPENCER
GULF
FERGUSSON
ADELAIDE
STURT
HINDMARSH
BUCCLEUCH
RUSSELL
CARNARVON
CARDWELL
BUCKINGHAM
MACDONNELL
ROBE
GREY
HOPETOUN
KINTORE
WAY
GREAT
AUSTRALIAN
BIGHT
GULF ST. VINCENT
FLINDERS ISLAND
Franklin Harb.
Arno Bay
Wallaroo
Tumby Bay
Port Lincoln

Fig. 7

were brought suddenly into focus by the economic situation, and several enquiries were instituted to study the position of the wheat industry in South Australia. The Agricultural Settlement Committee (1931) reported to the Government on agricultural policy, settlement, and development, and discussed *inter alia* the expansion and disabilities of the industry, the problems of drought relief, and measures for increasing production. The marginal areas were subjected to a particularly close examination, and two papers by Perkins (1934, 1936) again gave them prominence. In the second paper, Perkins segregated the profitable from unprofitable wheat-growing areas, using as a criterion a minimum mean yield of 6 bushels per acre for the 20 years 1915-34 inclusive, and made further constructive suggestions for rehabilitating and stabilizing agriculture in the districts he had designated as unprofitable and unsuitable for the production of wheat as the major crop.

In 1934, the Commonwealth set up a Royal Commission to investigate the economic position of the industry for the whole country. The Commission's second report (1935) covered the wide range of problems confronting the industry, and submitted detailed information relating to costs of production in various districts.

Another step was taken locally in 1939. The Marginal Lands Committee (1939) was formed to consider again the situation as it pertained to South Australia. This Committee defined marginal lands as "areas which have been subdivided into blocks intended principally for wheat-growing and which have been utilized mainly for that purpose, but owing to the combination of an inadequate rainfall and unsuitable land have proved to be unsuitable for wheat as a major operation," and surveyed the wheat belt of the State to determine the areas within the ambit of the problem and the number of settlers involved. The Committee found that a primary cause of failure was the small size of holding, which prevented the grower from carrying sufficient sheep, and necessitated frequent cropping with wheat to maintain income. The solution suggested was to increase the size of holding so that more sheep could be carried, in association with a longer rotation such as fallow, wheat, oats, and pasture for two or more years according to soil type and rainfall, the length of the rotation being largely determined by the rapidity of deterioration of the pasture.

At the same time, an independent and more detailed examination of the available data by the author indicated that the yields in these regions were declining, and an investigation was begun to determine the trends, with the objectives of using them to delineate the areas concerned and for correlation with relevant concomitant observations. In 1940, the remainder of the wheat belt was included, so that all the advantages of contrast could be gained from a comparison with the trends of yield from districts favoured with a higher and more reliable rainfall, the 25-year period 1913-37 inclusive being chosen for examination since at that time it gave the longest standard sequence of yields over the most extensive range of soil types and climatic conditions. The results of this investigation were communicated privately to the Rural Reconstruction Commission in 1944.

Subsequently, when more precise information on the major soil types became available, the scope of the enquiry was widened further by extending the analysis to a longer series of records.

III. Data and Analytical Method

(a) Yield Data

The smallest territorial unit for which consecutive yield data are available is the hundred, the mean area of which in South Australia is approximately 76,000 acres, or 118 square miles. This unit was adopted, and the yields, expressed as mean yields in bushels per acre *sown for grain*, were extracted from the South Australian Statistical Register for all hundreds with an average area of not less than 1,000 acres under crop. The yields used in compiling the records are not estimates, but are derived from returns submitted by the growers. Errors in the figures quoted (mainly printing errors) were eliminated by checking the means against total production and acreage.

In all, some 296 hundreds have been examined, and their distribution among the regional divisions of the State as given by the Regional Planning Committee (1946), is as follows:

Nuyts	48	Light	30	Flinders	42
Yorke	21	Sturt	12	Eyre	58
Pyap	19	Goyder	37	Fleurieu	1
Pinnaroo	24	Tatiara	3	Adelaide	1

In view of the possibility of future planning requirements, this method of reference was chosen rather than the older statistical subdivisions; an additional reason is provided by the fact that it has already been used in the report of the Pastoral and Marginal Agricultural Areas Inquiry Committee (1948). Figures 8 and 12 show the hundreds in relation to the regional boundaries.

In districts where cropping began prior to 1896, the analysis was confined to the period 1896-1941 inclusive, and where it commenced in 1896 or later only those hundreds which, by 1941, had a record of 20 years or more were selected. These limitations were imposed for three reasons:

1. All records of yield prior to 1896 are incomplete owing to the occurrence of two gaps, totalling 7 years, when no returns were taken, during 1885-88 and 1893-95, and sections of the records which exist do not give yields in sufficient detail for the purposes of the analysis.
2. In 1942, the Commonwealth Government, under war-time legislation, restricted the acreage to be sown to wheat.
3. The minimum of 20 years was chosen so that the trend in yield and the effects of seasonal rainfall could be accurately assessed.

(b) Rainfall Data

The finest subdivision of the year for which rainfall data are available in a form convenient for immediate use, is the calendar month. This unit was taken, and the data supplied by the Commonwealth Meteorological Bureau.

As so much was contingent upon making proper allowance for seasonal variations in yield, it was necessary to select the rainfall stations carefully. Owing to the circumstances that the area of each hundred is comparatively large, and that, in many instances, particularly those hundreds on the slopes of the Mt. Lofty and Flinders Ranges, there are considerable changes in altitude, the choice of rainfall stations would have been very difficult and of doubtful value, had recourse not been taken to two subsidiary sets of information:

1. A map showing the location of the area under crop in each hundred.
2. A map of isohyets of April-November rainfall.

On the whole, the rainfall records were remarkably complete, but there were, of course, the inevitable breaks in continuity. These were made good by substituting data from neighbouring stations, selected on the basis of the isohyets and the location of the crop. In the majority of hundreds, only one rainfall record was used, particularly in Nuyts, Eyre, Pyap, and Pinnaroo, where rainfall changes comparatively slowly with position; occasionally there were two, and more rarely three. In all, some 261 records were employed to eliminate the effects of seasonal rainfall.

(c) *Analytical Method*

In the examination of trends during the period 1913-37 the rainfall variates were determined by the following considerations.

(i) *The Nature and Paucity of the Yield and Rainfall Data.*—Since the yields in the greater proportion of hundreds were derived from very extensive areas, and the sequence was confined to a maximum of 25 years, these facts, coupled with the low density of rainfall stations and the enforced decision to employ monthly rainfall, precluded the possibility of making any detailed analysis of the rainfall. At the same time, however, it was recognized that some allowance would have to be made for the differential effects of rains in the several parts of the season.

(ii) *The Average Distribution of Seasonal Rainfall.*—Over the whole wheat belt the distribution of rainfall throughout the season is nearly symmetrical about a date in the interval mid-June to mid-July.

(iii) *The Form of the Regression Function.*—A previous analysis of yield data from Roseworthy College had shown that the regression of yield on rainfall, when expressed as a function of time, assumed the mathematically simple parabolic form, with a maximum in July. It was reasonable to assume that this form of relationship would be typical of most of the wheat-growing areas.*

(iv) *The Period during which Rainfall is Effective.*—It was shown by Trumble (1937) that the outer limits of the area at that time devoted to the cultivation of wheat corresponded roughly with the isochrone giving a season of five months' duration in which rainfall is effective, the average calendar interval being May-September. At the other extreme, the season ran to 7½ months in several isolated localities. The greatest proportion of the wheat belt, however, was situated in the zone between the 5- and

* This assumption will be substantiated in a paper in course of preparation.

7-month isochrones. In seeking a standard suitable over the whole wheat belt, April-November inclusive was chosen, as it gave a reasonable period prior to seeding in every district, and extended to harvest in all but a few places.

After taking account of these points, it appeared that the simplest adequate expressions which would effectively allow for variations in seasonal conditions would be the rains of the following subdivisions:

April and May
June, July, and August
September and October
November,

and these were taken as four rainfall variates.

Similar considerations apply in the analysis of the longer series of records, but the procedure has been modified by omitting November rainfall, since an exhaustive examination of the regressions of yield on this variate which had been obtained previously, showed that very little would be gained by its inclusion. With regard to the present investigation, the partial regressions on the remaining rainfall variates are by-products of the analysis, and their consideration has been relegated to another paper.

It is possible that rainfall of the preceding season and of the period December-March immediately prior to seeding, are correlated with yield, but such effects are also small when compared with those of the current season's rainfall to the end of October, and consequently they have been ignored. In any case, the crop records are not of sufficient length to account for them adequately.

At this juncture, it is important to mention two additional points. In the first place, only one meteorological element has been used to characterize the season, so that, in correlating it with yield, the regression obtained is the resultant of a number of components, one, the effect of the rainfall *per se*, and the remainder due to the direct and indirect effects of all other elements associated with the rainfall occurring in the several subdivisions under consideration. In the second place, it is obvious that definite optimal conditions must apply in various parts of the season, and the effects of the rainfall at all times are not strictly independent and additive. In the present enquiry, the quadratic terms, corresponding to such factors, have also been omitted, since in comparison with the linear terms they are of much less quantitative importance.

Four different functions of time have been employed to represent the temporal trends of yield. If y denotes yield, x_1, x_2, x_3 the rainfall variates in the order previously given, x_4 and x_5 time and its square, the four types of multiple regression were:

$$Y = \bar{y} + b_1(x_1 - \bar{x}_1) + b_2(x_2 - \bar{x}_2) + b_3(x_3 - \bar{x}_3) + b_4(x_4 - \bar{x}_4) \ldots \quad (1)$$

$$Y = \bar{y} + b_1(x_1 - \bar{x}_1) + b_2(x_2 - \bar{x}_2) + b_3(x_3 - \bar{x}_3) + b_4(x_4 - \bar{x}_4) + b_5(x_5 - \bar{x}_5) \ldots \quad (2)$$

$$Y = b_1(x_1 - \bar{x}_1) + b_2(x_2 - \bar{x}_2) + b_3(x_3 - \bar{x}_3) + b_4 x_4^{\,b_5} \ldots\ldots\ldots \quad (3)$$

$$Y = b_1(x_1 - \bar{x}_1) + b_2(x_2 - \bar{x}_2) + b_3(x_3 - \bar{x}_3) + x_4/(b_4 + b_5 x_4), \ldots \quad (4)$$

the bar over a symbol designating the arithmetic mean.

The coefficients were determined by the method of maximal likelihood (Fisher 1922), which, in the first two expressions, reduces directly to the standard technique of partial regression. In types (3) and (4) the logarithm of the likelihood is proportional to

$$L = S(y - Y)^2,$$

where the summation is taken over the n observations in the sample, and the equations which must be satisfied on maximizing L for variations in the coefficients are

$$\frac{\partial L}{\partial b_i} = 0, \qquad i = 1, 2, \ldots, 5.$$

Since these equations are non-linear in the parameters, the ordinary procedure of least squares is not directly applicable, but the solution may be obtained by iteration, taking as a starting-point any conveniently calculable approximate values.

If $\hat{b}_i$ and b'_i denote the maximal likelihood estimate and its approximation respectively, and

$$\hat{b}_i = b'_i + a_i,$$

the corrections to be applied are the solution of the equations

$$\left.\begin{array}{l}
\dfrac{\partial L}{\partial b_1} = \dfrac{\partial L}{\partial b_1} + a_1 \dfrac{\partial^2 L}{\partial b^2_1} + a_2 \dfrac{\partial^2 L}{\partial b_1 \partial b_2} + \ldots + a_5 \dfrac{\partial^2 L}{\partial b_1 \partial b_5} = 0 \\[2ex]
\dfrac{\partial L}{\partial b_2} = \dfrac{\partial L}{\partial b_2} + a_1 \dfrac{\partial^2 L}{\partial b_1 \partial b_2} + a_2 \dfrac{\partial^2 L}{\partial b^2_2} + \ldots + a_5 \dfrac{\partial^2 L}{\partial b_2 \partial b_5} = 0 \\[2ex]
\cdot \quad \cdot \quad \cdot \quad \cdot \quad \cdot \quad \cdot \quad \cdot \quad \cdot \quad \cdot \quad \cdot \\[2ex]
\dfrac{\partial L}{\partial b_5} = \dfrac{\partial L}{\partial b_5} + a_1 \dfrac{\partial^2 L}{\partial b_1 \partial b_5} + a_2 \dfrac{\partial^2 L}{\partial b_2 \partial b_5} + \ldots + a_5 \dfrac{\partial^2 L}{\partial b^2_5} = 0
\end{array}\right\} \quad \ldots \quad (5)$$

where in the set of derivatives on the extreme left, b_i is replaced by $\hat{b}_i$ after differentiation, and in the remainder by b'_i.

The solution is best obtained by inverting the matrix of second derivatives in equations (5), and if

$$[c_{ij}] \qquad i, j = 1, 2, \ldots, 5$$

denotes the inverse matrix, the a_i are given by

$$a_i = \mathop{S}_{j} c_{ij} \left(- \frac{\partial L}{\partial b_i} \right)_{b_i = b'_i} \qquad i, = 1, 2, \ldots, 5.$$

These corrections are added to the approximate values, and the whole process repeated if necessary.

The residual variance, s^2, is determined from the relation

$$s^2 = \frac{1}{(n-5)} S(y - Y)^2$$

since 5 adjustable parameters have been calculated from the data, and finally, the variance-covariance matrix of the $\hat{b}_i$ is computed by multiplying the elements of the inverse of the matrix

$$\left[\left(\frac{\partial^2 L}{\partial b_i \partial b_j} \right)_{\substack{b_i = \hat{b}_i \\ b_j = \hat{b}_j}} \right] \qquad i, j = 1, 2, \ldots, 5$$

by s^2.

The arithmetical detail of the process is illustrated with the calculations for the hundred of Crystal Brook, to the yields of which a multiple regression of type (**3**) was fitted. The steps in order are as follows:

1. Approximate values b'_1 (1.5208), b'_2 (2.0273), and b'_3 (0.6894) are found by fitting a multiple regression with x_1, x_2, x_3, and x_4 as independent variates, and the yields corrected for variations in seasonal rainfall.

2. Approximate values b'_4 (5.140) and b'_5 (0.2922) are then determined by taking the regression of log (corrected yield) on log (time).

3. After substituting these approximations, the numerical values of the elements in the matrix of coefficients and the quantities on the right-hand sides of equations (**5**) reduce to

		$\frac{\partial^2 L}{\partial b_i \partial b_j}$			$-\frac{\partial L}{\partial b_j}$
76.09	11.27	– 1.34	– 2.33	– 88.21	– 10.2270
11.27	158.18	40.32	– 2.71	– 104.74	– 12.7233
– 1.34	40.32	105.02	3.79	87.00	– 18.3782
– 2.33	– 2.71	3.79	276.54	4482.12	26.1761
– 88.21	– 104.74	87.00	4482.12	76774.51	425.7731

and the inverse of the matrix of second order derivatives is

		c_{ij}		
13.40377	– 1.03221	0.52849	– 2.25917	0.14528
– 1.03221	7.15029	– 2.78873	– 1.68294	0.10998
0.52849	– 2.78873	10.62995	1.45959	– 0.10045
– 2.25917	– 1.68294	1.45959	68.17862	– 3.98684
0.14528	0.10998	– 0.10045	– 3.98684	0.24621

each element being multiplied by 10^{-8}.

The correction term a_1, for example, is then

$$\{(-13.40377 \times 10.2270) + (1.03221 \times 12.7233) - (0.52849 \times 18.3782) - (2.25917 \times 26.1761) + (0.14528 \times 425.7731)\} \times 10^{-8} = -0.1309.$$

The corrections and second approximations are

a_1	a_2	a_3	a_4	a_5
– 0.1309	– 0.0264	– 0.1698	0.1049	– 0.0006

b_1	b_2	b_3	b_4	b_5
1.3899	2.0009	0.5196	5.2449	0.2916

4. Repetition of the process using the second approximations gives the quantities

		$\frac{\partial^2 L}{\partial b_i \partial b_j}$			$-\frac{\partial L}{\partial b_j}$
76.09	11.27	– 1.34	– 2.32	– 89.69	– 0.0049
11.27	158.18	40.32	– 2.70	– 106.50	0.0060
– 1.34	40.32	105.02	3.78	88.70	– 0.0018
– 2.32	– 2.70	3.78	275.48	4640.48	0.2139
– 89.69	– 106.50	88.70	4640.48	81162.64	– 4.8716

with inverse matrix

		c_{ij}			
13.43794	– 1.00674	0.50627	– 3.31375	0.20244	
– 1.00674	7.16927	– 2.80529	– 2.46979	0.15257	
0.50627	– 2.80529	10.64437	2.14881	– 0.13761	$\times 10^{-8}$
– 3.31375	– 2.46979	2.14881	100.43595	– 5.75168	
0.20244	0.15257	– 0.13761	– 5.75168	0.34175	

yielding the following second corrections and third approximations:

a_1	a_2	a_3	a_4	a_5	
– 0.0018	– 0.0012	0.0011	0.0495	– 0.0029	
b_1	b_2	b_3	b_4	b_5	. . . (6)
1.3881	1.9997	0.5207	5.2944	0.2887	

At this point the working could have been terminated, as the solution is sufficiently accurate, but the calculations are here taken through a further stage, using the third approximations, to show that the third corrections are negligible. The numerical values are as follows:

		$\frac{\partial^2 L}{\partial b_i \partial b_j}$			$-\frac{\partial L}{\partial b_j}$
76.09	11.27	– 1.34	– 2.28	– 89.01	0.0501
11.27	158.18	40.32	– 2.64	– 105.67	– 0.0122
– 1.34	40.32	105.02	3.73	89.14	0.0062
– 2.28	– 2.64	3.73	270.40	4593.47	0.0804
– 89.01	– 105.67	89.14	4593.47	81049.55	1.4668

		c_{ij}			
13.43509	– 1.00801	0.50610	– 3.28818	0.19924	
– 1.00801	7.16898	– 2.80602	– 2.47674	0.15169	
0.50610	– 2.80602	10.64593	2.19272	– 0.13908	$\times 10^{-8}$ . . . (7)
– 3.28818	– 2.47674	2.19272	101.36102	– 5.75387	
0.19924	0.15169	– 0.13908	– 5.75387	0.33901	

a_1	a_2	a_3	a_4	a_5
0.0007	– 0.0001	0.0001	– 0.0004	0.0000
b_1	b_2	b_3	b_4	b_5
1.3888	1.9996	0.5208	5.2940	0.2887

5. Taking the values in (6) as the solution, the residual sum of squares is 275.07, with 41 degrees of freedom ($n = 46$), giving a residual variance of 6.71,

from which the variances of the b_i are obtained by multiplying in turn by the diagonal elements of the matrix (7). The standard deviations and five values of t (Fisher 1946) in order are

Standard deviation	0.300	0.219	0.267	0.825	0.048
t	4.63	9.13	1.95	6.42	6.02

(41 degrees of freedom)

so that b_3 is the only insignificant coefficient.

After fitting the multiple regression (3), the sum of squares due to the regression formula can be very conveniently derived from the expression

$$b_1Sy(x_1 - \bar{x}_1) + b_2Sy(x_2 - \bar{x}_2) + b_3Sy(x_3 - \bar{x}_3) + b_4Syx_4^{\,b_5} \quad \ldots . \quad (8)$$

but for regressions of type (4) the formula cannot be written so simply, and takes the form

$$b^2{}_1S(x_1 - \bar{x}_1)^2 + b^2{}_2S(x_2 - \bar{x}_2)^2 + b^2{}_3S(x_3 - \bar{x}_3)^2 + 2b_1b_2S(x_1 - \bar{x}_1)(x_2 - \bar{x}_2)$$
$$+ 2b_1b_3S(x_1 - \bar{x}_1)(x_3 - \bar{x}_3) + 2b_2b_3S(x_2 - \bar{x}_2)(x_3 - \bar{x}_3)$$
$$+ 2Sy\left(\frac{x_4}{b_4 + b_5x_4}\right) - S\left(\frac{x_4}{b_4 + b_5x_4}\right)^2 \quad \ldots\ldots\ldots . \quad (9)$$

For all cases in which regressions of types (1) and (2) have been used, the multiple correlation coefficient has been determined, and thence the percentage of variance of yield, A, ascribable to the average effects of the rainfall variates and time, from the relation

$$A = 1 - \frac{(n-1)}{(n-p-1)}(1 - R^2) \quad \ldots\ldots\ldots\ldots \quad (10)$$

where R is the multiple correlation coefficient, n is the number of observations in the sample, and p the number of independent variates (Fisher 1924).

The sums of squares due to the regression formulae of types (3) and (4) contain the ordinary correction for the mean, $n\bar{y}^2$, and if this is deducted from both sums an index of multiple correlation may be defined, by analogy with the multiple correlation coefficient, as the ratio

$$R^2{}_{\prime} = \frac{-SY^2 + 2SyY - n\bar{y}^2}{S(y - \bar{y})^2} \quad \ldots\ldots \quad (11)$$

and the percentage of variance, corrected for positive bias in $R^2{}_{\prime}$, will be correspondingly

$$A_{\prime} = 1 - \frac{(n-1)}{(n-p-1)}(1 - R^2{}_{\prime}), \quad \ldots\ldots\ldots\ldots \quad (12)$$

p now designating the number of coefficients in the regression formula after deducting one for the mean. $R_{\prime}$ and $A_{\prime}$ have been determined for all hundreds in which they are appropriate.

IV. Soil Groups of the Wheat Belt

The geographical distribution of the soil groups in relation to the isohyets of seasonal rainfall and the boundaries of the hundreds, is illustrated in Figure 8. This map summarizes the best information available at the present time. It must be realized that the allocations have been made on a rather broad basis; within the boundaries as mapped, variations of type exist owing to geological

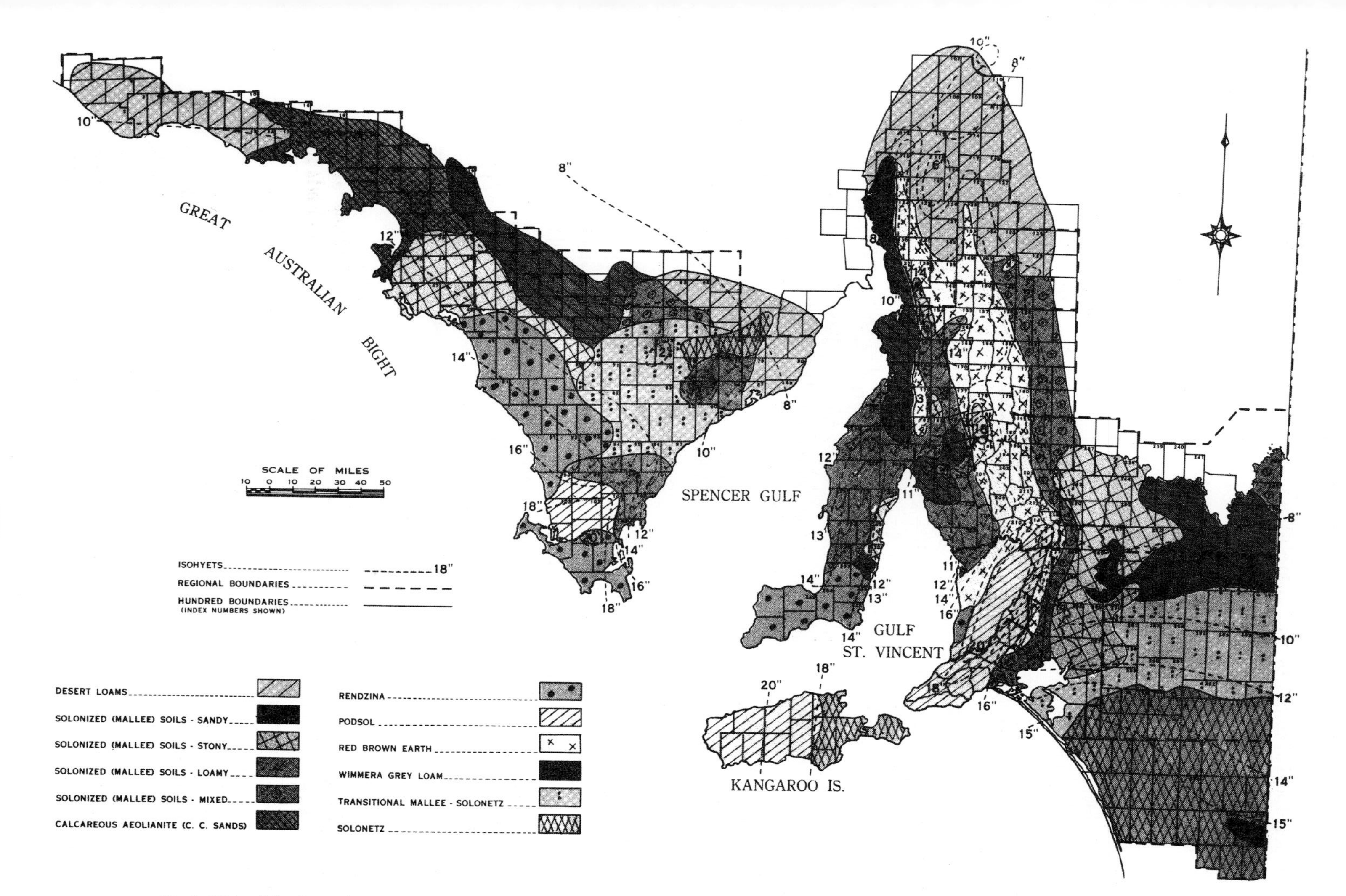

Fig. 8. Major Soils Zones in the Wheat Belt of South Australia (drawn from data supplied by the Division of Soils, C. S. I. R., and R. I. Herriot).

It should be noted that the method of presenting the zones in this map differs from the original, in which they were denoted by colours.

and climatic differentiation, and consequently, if the grouping is applied in fine detail it is liable to be misleading.

The descriptions of the soil groups are as follows:

(i) *Podsols.*—These are leached soils with a grey surface and generally a yellow or mottled clay subsoil. The surface soil is commonly of light texture, and acid in reaction; lime is absent from the profile except where the soil is residual on lime-bearing rocks, and even under these conditions it occurs only in the lowest horizon. The group as mapped includes the lateritic podsols of Kangaroo Island, the Mt. Lofty Ranges, and southern Eyre Peninsula, where ironstone gravel or massive laterite appears in the profile at or near the surface, Fertility of the podsols is low to moderate, but generally enhanced by humid climatic conditions.

(ii) *Red Brown Earths.*—This group has brown surface soils usually loamy in texture with red or red-brown structured clay subsoils. The surface is neutral to slightly acid in reaction, and lime occurs in the deeper subsoil. These are the most fertile of the soils used extensively for wheat-growing in South Australia, and are favoured by a fairly reliable winter rainfall. In the zone as mapped, there are very restricted occurrences of black earth or chernozem soils, the principal one being at Saddleworth.

(iii) *Desert Loams.*—In these brown soils of the arid areas there is usually a light to moderately heavy surface horizon over a clay subsoil in which occurs an accumulation of lime, and frequently, visible gypsum. The profile is alkaline throughout. The fertility of this group is very low and the various types erode rapidly under cultivation.

As mapped, the group includes:

1. The desert loams, which constitute the major portion of Flinders.
2. The types occurring at the western extremity of the wheat belt in the vicinity of Fowler's Bay. These are grey-brown loamy soils of varying depth which possess a high salinity at or near the surface in many parts, and have affinities to the desert loams.
3. The loamy mallee soils extending in an arc from Kimba to Cowell. These soils occur in a region of low rainfall, and hence are agriculturally similar to the desert loams and have accordingly been included with them.

Both the regions of the true desert loams and red brown earths are characterized by a high proportion of skeletal soils on the hills and ranges, too stony and shallow for cultivation.

(iv) *Rendzinas.*—Black soils, occasionally degraded to grey or grey-brown by saline conditions or leaching, lying residual on soft and/or clayey calcareous parent material. The surface is neutral to slightly alkaline in reaction. The soils are irregular in depth, of variable fertility, and occur over a moderately wide range of climatic conditions.

(v) *Solonized Soils.*—Soils of this type constitute the largest single group in the agricultural regions of South Australia. They are brown in colour and have been solonized. A prominent morphological feature is the large accession of calcium carbonate which was probably wind-borne.

The main group includes a subgroup originally described as mallee soils, and it is proposed to use this term here, since all occurrences in the wheat belt of the State carry the characteristic mallee vegetation associations (Wood 1937).

Texture is variable, and reaction alkaline. Different regions are dominated by (1) loamy soils; (2) stony soils where the development, exposure, and disintegration of a travertine limestone pan have resulted in a stony surface soil; (3) sandy soils, some of which have arisen by the accumulation of siliceous sands, stripped by aeolian action from the leached surface horizons of adjacent stony areas. Where all three or any two of these types occur together with no one predominant, the soils have been classed and mapped as mixed mallee.

A related class, described by the term calcareous aeolianite, has also been included in this group. These soils are consolidated calcareous sands which have *not* been solonized but in their original state carried mallee (*Eucalyptus oleosa*) interspersed with open spear grass (*Stipa* spp.) plains.

(vi) *Solonetz Soils.*—Former concentrations of salt have affected the profiles of these soils but they now contain relatively small amounts. The subsoil contains a marked proportion of exchangeable sodium. The soils are generally of light texture, particularly in the surface horizon, and in South Australia have developed on modified sand-ridge country. They are neutral to alkaline in the surface, and alkaline in the subsoil, and occur over a moderately wide range of climatic conditions. Fertility varies according to the character of the precursor soil and the stage of solonization, but is principally low.

(vii) *Transitional Mallee-Solonetz Soils.*—In the zone between the recognized brown solonized and solonetz soils, there occurs a substantial area featuring the characteristic mallee physiographic pattern of alternating dunes and flats, the former being solonetz, and the latter typical brown solonized soils; the proportion of solonetz to solonized soil increases on any line approaching the solonetz region. The rainfall of these areas is generally higher than that of the mallee, and cropping is confined mainly to the interdunal spaces.

(viii) *Wimmera Grey Soils.*—The profile, which is alkaline and may be partly solonized, is uniformly heavy, consisting of grey or brown clays, becoming mottled with depth; the grey clays may have a self-mulching surface. Lime and gypsum occur to some degree in the subsoil. The South Australian occurrence of this type is a small intrusion at the western extremity of the fertile soils of the Victorian Wimmera.

V. Summary of Results and Discussion

(*a*) *Accuracy of the Regression Formulae*

The complete tabulation of the results of this investigation has not been included, since it is too extensive for reproduction herein. Copies have, however, been filed, and are available for inspection.*

*At the libraries of the Council for Scientific and Industrial Research, 314 Albert Street, East Melbourne, and its Section of Mathematical Statistics, University of Adelaide.

The statistical significance, with minimum odds of 19:1, has been established for the trends in yield of practically all hundreds except those which have been classified as type III of Figure 11 (*vide infra*), and by appropriately combining the probabilities of both significant and insignificant coefficients, any doubts regarding the few exceptions are removed. Such tests are, however, really not required, since the close similarity of form in the trends of contiguous hundreds provides sufficient proof.

Table 1 sets out the distribution of the coefficients and indices of multiple correlation within arbitrary, but convenient, subdivisions of the range of seasonal rainfall (total of April-November inclusive) and the mean values of the percentage variance, the value 83 per cent., for example, being the mean of the 14 cases in which the coefficient or index was ⩾ 0.91. In 267 hundreds, 50 per cent. or more of the observed variation in yield is expressible in terms of the four types of function employed in the analysis.

TABLE 1

DISTRIBUTION OF COEFFICIENTS AND INDICES OF MULTIPLE CORRELATION WITHIN ARBITRARY CLASSES OF SEASONAL RAINFALL

Correlation	Rainfall (in.) < 8	8-10	10-12	12-14	14-16	16-18	> 18	Total	Per Cent. Variance (*A*)
⩾ 0.91	7	2	2	3				14	83
0.81-0.90	39	54	44	29	3			169	70
0.71-0.80	9	34	19	9	9	3	1	84	53
0.61-0.70		3	12	4	1	1		21	37
0.51-0.60		2	1	2		1		6	25
0.41-0.50				2				2	14
Total	55	95	78	49	13	5	1	296	

The relation which exists between the measures of correlation and seasonal rainfall reflects largely the strength of the relation between yield and rainfall, since, in the majority of hundreds, the regression of yield on time is the smaller contributor to the total variation ascribable to the regression formula. Considering the results as a whole, it can be claimed that a rainfall record provides a sufficiently accurate index of seasonal conditions in this environment.

Table 2 lists the distribution of the correlations within the several soil types of the wheat belt. As rainfall and soil type are, to some extent, related, these distributions are not completely independent of that in Table 1, but, nevertheless, they do indicate that the claim made above holds also on all principal soils. The grand total of 335 does not agree with that of Table 1 because 39 hundreds, comprised equally of two soil types, have been included under both in preparing the table.

(*b*) *The Types of Trend Which Occur*

Before proceeding further with the discussion, it will be helpful to amplify certain points mentioned in Section I, and introduce others, so that each individual sequence of yields can be viewed in its proper relation to the full cropping history of the corresponding district. With this objective, it is convenient to divide the hundreds into two categories: (i) those in which cropping began prior to 1896; and (ii) those in which cropping began during or after 1896.

TABLE 2

DISTRIBUTION OF COEFFICIENTS AND INDICES OF MULTIPLE CORRELATION WITHIN SOIL TYPES°

	Soil Type											
	Mallee Types											
Correlation	Sandy	Stony	Loamy	Mixed	Transitional	Calcareous Aeolianite	Desert Loam	Solonetz	Red Brown Earth	Rendzina	Podsol	Total
⩾ 0.91	2	4	3	3			3		2			17
0.81-0.90	39	10	27	19	6	15	28	1	34	6	1	186
0.71-0.80	16	8	9	5	15	7	15	2	13	2	5	97
0.61-0.70	2	3	4		9	1	1	1	1	3	1	26
0.51-0.60		1			2		1	1			1	6
0.41-0.50			1		1					1		3
Total	59	26	44	27	33	23	48	5	50	12	8	335

(i) *Hundreds in which Cropping Began prior to 1896.*—By 1896, districts in this class had records of yield varying up to 50 years in length, the duration of record being fairly strongly and positively correlated with seasonal rainfall. Since the land was virgin, yields obtained at the outset were reasonably high, and in some localities were probably maintained for varying short periods. Three principal systems of cropping came into use, which, in order of appearance, were wheat continuously, fallow-wheat, and fallow-wheat-oats, and no returns were made to the soil other than what was introduced in the fallows. These methods inevitably led to depletion of essential chemical elements, and, with phosphorus in particular, to almost complete exhaustion of soil reserves. With the consequent reduction of fertility, yields began declining at rates varying from district to district, depending upon initial fertility, date of settlement, rainfall, intensity of cropping, and other subsidiary factors, of which wind and water erosion were probably the most important. The available evidence indicates that during the last ten years of the century yields in practically all districts had reached a very low level.

During 1896-1910, the use of superphosphate had spread widely, and as is well known, there was an immediate and marked response in yield. The great changes in acreage on Yorke Peninsula and in counties Gawler, Light, and Daly between 1884 and 1908 are, in part, a reflection of these facts.† There seems little doubt that the area of wheat was reduced because yields had fallen to such

† Richardson's illustration of the distribution in 1896 reveals a very considerable reduction in the area of wheat for these parts of the State.

° Wimmera grey loam included with red brown earth.

a low level under the intense cropping (*vide* Figs. 3 and 4) of previous years: the response to superphosphate provided the stimulus for the re-intensification of cropping. But, apart from the increase in the phosphorus status of the soils, the advances made by the introduction of improved varieties and the closer association of sheep- and wheat-farming, the factors that had previously been responsible for the decline in yield were still operating, with the result that the rate of improvement was not generally maintained. In Figure 9, the courses of

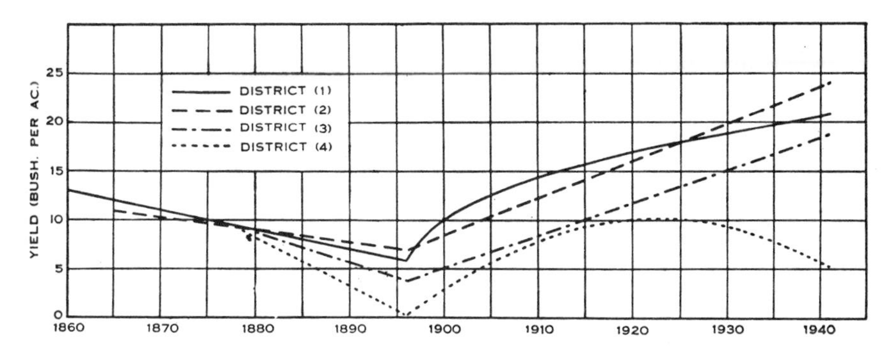

Fig. 9.—Reconstruction of the entire course of yield in four districts opened prior to 1896.

yield in four hundreds have been reconstructed, after making allowances for variations in seasonal rainfall, to illustrate some of these points. The following notes on each district are relevant:

District	Soil Type	Rainfall	Cropping Commenced
(1)	Red brown earth	14-16 in.	1850-1855
(2)	Red brown earth and loamy mallee	14-16 in.	1860-1865
(3)	Loamy mallee	12-14 in.	1880 (exact)
(4)	Mixed mallee	10-12 in.	1878 (exact)

The declines prior to 1896 are represented as linear functions of time, but it is probable that if complete data had been available they would have been found to be exponential. The differences between districts (1) and (2) during 1860-96 are consistent with the fact that the first possessed the more fertile soil, which had been cropped for some ten years longer. Districts (3) and (4) also fall appropriately into sequence according to their rainfall and the natural fertility of the soils. Their initial yields are considerably lower than, and the rates of decline exceed, those of (1) and (2).

After 1896, the response to superphosphate and other factors tending to increase yield, is manifest in the four curves, but their forms vary according to the general conditions in each area. A high rate of increase is present for varying short periods immediately after 1896 in all hundreds opened for settlement before that date, but has only been definitely established for curves similar to that of district (1) and one other type. In such cases the period of cropping before 1896 was generally long and intense. A second important factor which also partly accounts for the difference between districts (1) and (2) is that in the latter there has been a rapid increase during recent years in the number of stock, mainly sheep, associated with wheat-farming, accompanied by a lengthening of rotations, whereas in the former the general tendency has been to maintain the system based on short-term rotations.

District (4) demonstrates clearly the inability of these particular soils to withstand frequent cropping, since yield eventually declines again notwithstanding the advancements made in the past 40 years.

The six principal types of trend which occur are illustrated in Figure 10.

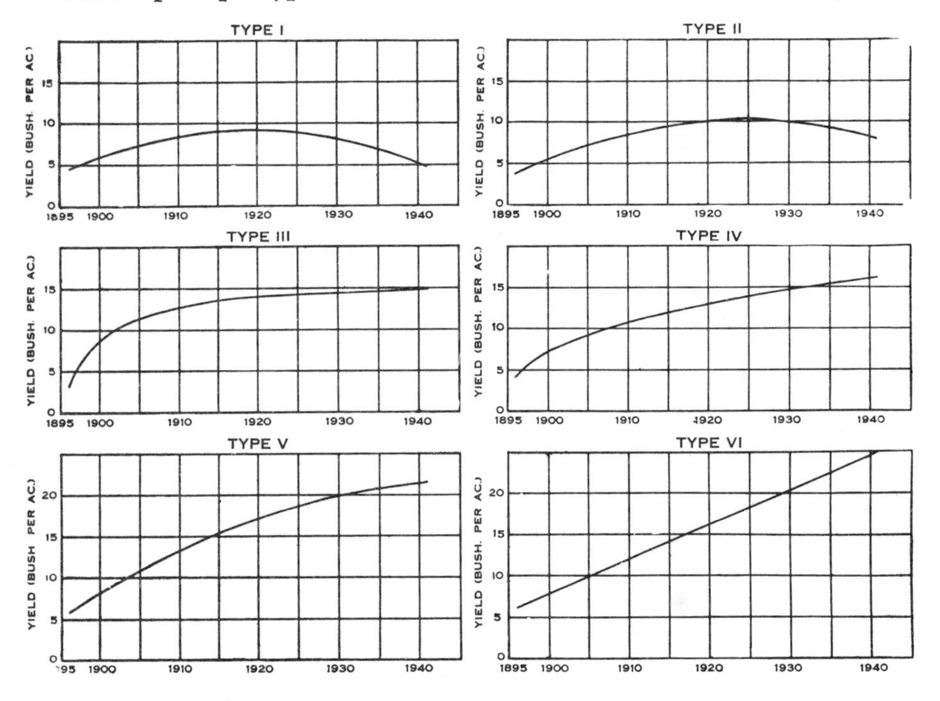

Fig. 10.—The principal types of yield trend in hundreds opened before 1896.

Types I and VI represent the extremes of the range, and there is a gradual transition from type to type within it.

(ii) *Hundreds in which Cropping Began during or after 1896*.—In this category, settlement coincided with the introduction of superphosphate, or occurred at a later date. The hundreds concerned are scattered over a wide range of soil and season, but the bulk of the group is located in the mallee areas of Nuyts, Eyre, Pinnaroo, and Pyap. In these regions, cropping methods in the early stages are exploitative through circumstances incidental to pioneering on land carrying mallee vegetation. Mallee stumps remaining in the ground after the first clearing repeatedly send out new growth, and the settler crops continuously for some years, burning the stubbles to destroy the shoots. The situation is also made more acute since the growers have only limited capital and credit resources, and this provides an additional reason for adopting such methods in an endeavour to improve the income of the first few years. During this phase, which takes on the average about five years to complete, the principal factor limiting yield, apart from rainfall, is probably a deficiency of available nitrogen. In the course of time, however, as fallowing and other measures for increasing production are introduced, yields would be expected to improve, but the subsequent course would vary according to soil, rainfall, and the practices adopted. The types which occur are illustrated in Figure 11. The range is not

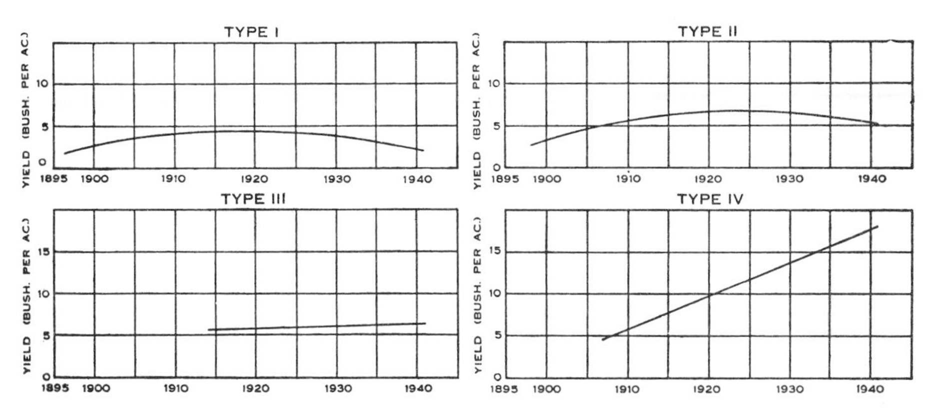

Fig. 11.—The principal types of yield trend in hundreds opened in 1896 or later.

as great as that of Figure 10, since the situation is not complicated by the effects of cropping prior to the advent of superphosphate. Type III applies to hundreds which have only been settled recently, the records being so short that the position with respect to trend is still indeterminate. The positive regressions on time are all statistically insignificant, and from this restricted point of view the sequences of yields are in accord with the hypothesis that the trends are zero, but consideration of all relevant information indicates clearly that hundreds classed in this group are repeating, some 10-20 years later, the first stage of either curve I or curve II. In this sense, the type III curve may be regarded as a subtype of I or II. Included in this group also are several cases in which the trend is practically zero. Hundreds with trends of type IV are easily distinguished, since they are almost entirely confined to two soil classes, and the regressions are strongly significant and of much greater magnitude than those of type III.

In general, both sequences of types are intimately related to improvement in the class of soil and in particular are arranged in order of increasing abundance of nitrogen; this point is elaborated further at various stages below in this section.

(*c*) *Survey of the Trends and Their Distribution*

Figure 12 depicts the geographical distribution of the trends. Diagrammatic representation has been simplified by assigning a specified type of trend to the whole of each hundred to avoid the complication that would inevitably ensue on a map of this scale if the actual location of the crop were given in each. Throughout the remainder of this section, frequent reference will be made to the curves of Figures 10 and 11, and to save repetition they will be quoted as, for example, I(10), III(11).

Index of Hundreds

*Nuyts**

1. Miller
2. Wookata
3. Sturdee
4. Caldwell
5. Nash
6. Magarey
7. Giles
8. Cohen
9. Burgoyne
10. Bagster
11. Kevin
12. Keith
13. Catt
14. Horn
15. Bartlett
16. O'Loughlin
17. Moule
18. Goode
19. Bonython
20. Wandana
21. Chillundie
22. Guthrie
23. Blacker
24. Wallanippie
25. Hague
26. Carawa
27. Petina
28. Wallala
29. Koolgera
30. Haslam
31. Perlubie
32. Walpuppie
33. Yantanabie
34. Finlayson
35. Tarlton
36. Cungena
37. Kaldoonera
38. Scott
39. Murray
40. Chandada
41. Karcultaby
42. Ripon
43. Forrest
44. Campbell
45. Inkster
46. Wrenfordsley
47. Rounsevell
48. Witera

Eyre

49. Wright
50. Condada
51. Carina
52. Minnipa
53. Yaninee
54. Pygery
55. Wudinna
56. Travers
57. Wallis
58. Palabie
59. Wannamana
60. Mamblin
61. Kappakoola
62. Warramboo
63. Cortlinye

Eyre (cont.)

64. Moseley
65. Solomon
66. Kelly
67. Colton
68. Talia
69. Barwell
70. McLachlan
71. Palkagee
72. Pascoe
73. James
74. Glynn
75. Smeaton
76. Campoona
77. Mangalo
78. Miltalie
79. Minbrie
80. Warren
81. Ward
82. Tooligie
83. Rudall
84. Yadnarie
85. Mann
86. Hawker
87. Playford
88. Wilton
89. Roberts
90. Boothby
91. Kiana
92. Mitchell
93. Shannon
94. Brooker
95. Moody
96. Butler
97. Dixson
98. Ulipa
99. Cummins
100. Stokes
101. Yaranyacka
102. Warrow
103. Mortlock
104. Koppio
105. Hutchison
106. Louth

Flinders

107. Woolyana
108. Wonoka
109. Arkaba
110. Warcowie
111. Adams
112. Wyacca
113. Kanyaka
114. Wirreanda
115. Yarrah
116. Boolcunda
117. Uroonda
118. Eurilpa
119. Pichi Richi
120. Palmer
121. Moockra
122. Yanyarrie
123. Bendleby
124. Woolundunga
125. Willochra

Flinders (cont.)

126. Coonatto
127. Pinda
128. Eurelia
129. Oladdie
130. Winninowie
131. Gregory
132. Willowie
133. Coomooroo
134. Walloway
135. Erskine
136. Cavenagh
137. Baroota
138. Wongyarra
139. Booleroo
140. Pekina
141. Black Rock Plain
142. Morgan
143. Coglin
144. Telowie
145. Appila
146. Yongala
147. Gumbowie
148. Parnaroo

Goyder

149. Tarcowie
150. Mannanarie
151. Pirie
152. Napperby
153. Howe
154. Booyoolie
155. Caltowie
156. Yangya
157. Belalie
158. Whyte
159. Terowie
160. Wandearah
161. Crystal Brook
162. Narridy
163. Bundaleer
164. Reynolds
165. Anne
166. Hallett
167. Mundoora
168. Red Hill
169. Koolunga
170. Yackamoorundie
171. Andrews
172. Ayers
173. Kingston
174. Wokurna
175. Barunga
176. Boucaut
177. Hart
178. Milne
179. Hanson
180. Kooringa
181. Baldina
182. Cameron
183. Everard
184. Blyth

Goyder (cont.)

185. Clare

Light

186. Goyder
187. Stow
188. Hall
189. Upper Wakefield
190. Stanley
191. Apoinga
192. Bright
193. Bundey
194. Saddleworth
195. Waterloo
196. English
197. Bower
198. Inkerman
199. Balaklava
200. Dalkey
201. Alma
202. Gilbert
203. Julia Creek
204. Neales
205. Dublin
206. Grace
207. Port Gawler
208. Mudlawirra
209. Light
210. Nuriootpa
211. Kapunda
212. Belvidere
213. Dutton
214. Moorooroo
215. Jellicoe

Yorke

216. Tickera
217. Wiltunga
218. Wallaroo
219. Kadina
220. Ninnes
221. Kulpara
222. Tiparra
223. Clinton
224. Kilkerran
225. Maitland
226. Cunningham
227. Wauraltee
228. Muloowurtie
229. Koolywurtie
230. Curramulka
231. Minlacowie
232. Ramsay
233. Para Wurlie
234. Moorowie
235. Dalrymple
236. Melville

Pyap

237. Beatty
238. Cadell
239. Markaranka
240. Pooginook
241. Parcoola
242. Murtho

Pyap (cont.)

243. Brownlow
244. Murbko
245. Waikerie
246. Holder
247. Moorook
248. Paringa
249. Gordon
250. Anna
251. Paisley
252. Bakara
253. Mantung
254. Pyap
255. Bookpurnong

Sturt

256. Bagot
257. Nildottie
258. Angas
259. Finniss
260. Ridley
261. Forster
262. Younghusband
263. Monarto
264. Mobilong
265. Burdett
266. Ettrick
267. Seymour

Adelaide

268. Tungkillo

Fleurieu

269. Freeling

Pinnaroo

270. Bowhill
271. Bandon
272. Chesson
273. Mindarie
274. Allen
275. Kekwick
276. McGorrery
277. Vincent
278. Wilson
279. McPherson
280. Peebinga
281. Hooper
282. Marmon Jabuk
283. Molineux
284. Cotton
285. Bews
286. Parilla
287. Pinnaroo
288. Sherlock
289. Roby
290. Peake
291. Price
292. Allenby
293. Livingston

Tatiara

294. Stirling
295. Wirrega
296. Tatiara

* Names of regions are printed in italics.

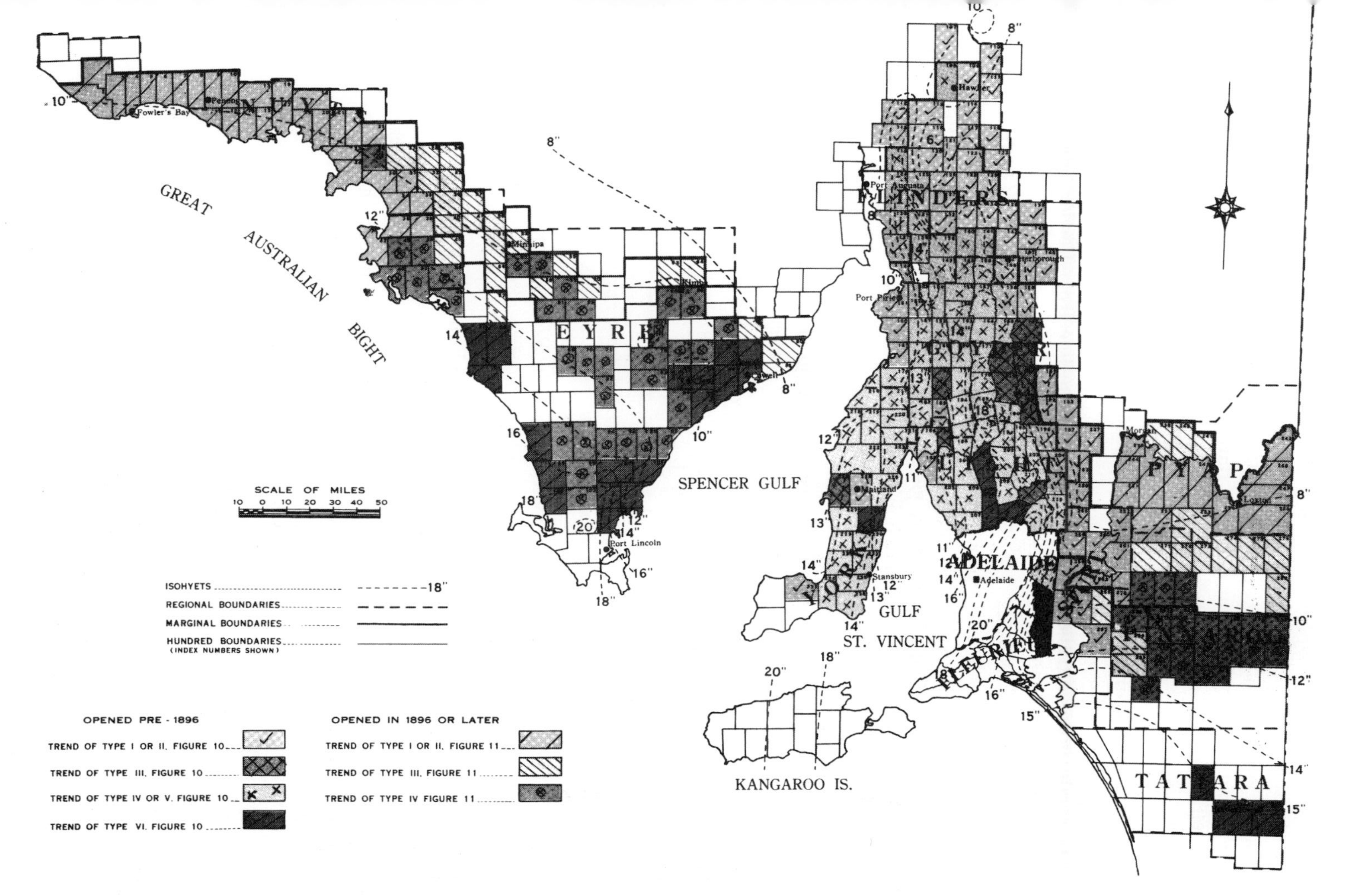

Fig. 12. Distribution of Yield Trends in the Wheat Belt of South Australia.

It should be noted that the method of presenting the trends in this map differs from the original, in which they were denoted by colours.

In surveying the whole wheat belt, reference to both maps is greatly facilitated by using a *primary* subdivision into the somewhat artificial* regions of the Regional Planning Committee (loc. cit.), rather than a natural one based on soil type and rainfall. The year or period during which settlement took place is listed with each group of hundreds.

Nuyts

1. Grey-brown loamy soils with affinities to the desert loams; rainfall approximately 10 in. throughout the group.

Generally speaking, the soils improve along the line from Fowler's Bay to a point just east of Penong.

a. Miller (1)†, Wookata (2), Sturdee (3), and Caldwell (4) (1896).—Although these hundreds were opened in 1896, analysis of the data could only be made from 1907, since all records prior to this date were included as one in the returns. During the first eleven years, yields may have increased slightly, but it is much more likely that they were only maintained. The curves show a steady decline from 1907, which is not typical of the group, possibly owing to a higher level of salinity in the soils and the fact that the areas under cultivation are too small to be representative. For convenience in mapping, they have been included

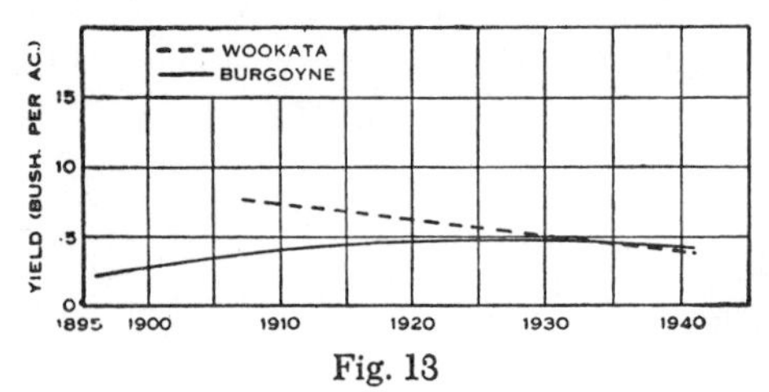

Fig. 13

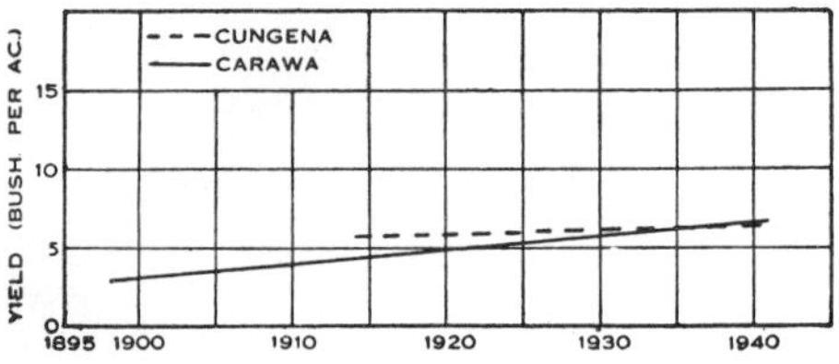

Fig. 14

as I(11). Figure 13 illustrates the smoothed trend in Wookata after variations due to seasonal rainfall have been eliminated, or as may be alternatively viewed, the course that yield would have followed if the rainfall variates had remained constant at their average values, throughout the period under examination.

b. Nash (5), Magarey (6), Giles (7), Cohen (8), Burgoyne (9), Bagster (10), Kevin (11), Keith (12), Catt (13), Horn (14), and Bartlett (15) (1890-1900).—For some 15-20 years yields slowly increased, but under the severe conditions this effect was transitory, and they subsequently declined, following courses I and II(11), according to local conditions. The curve for Burgoyne is also given in Figure 13. The influence of the slight improvement in the class of soil is shown by the sequence Wookata, Nash (the type I curve of Fig. 11), and Burgoyne.

2. Calcareous aeolianite; rainfall 9-10 in.

a. O'Loughlin (16), Moule (17), Goode (18), Bonython (19), Wandana (20), Chillundie (21), Guthrie (22), Blacker (23), Wallanippie (24), Carawa

* i.e. Artificial for present purposes.

† The numbers in brackets refer to the numbered hundreds on the maps (Figs. 8 and 12).

(26), Haslam (30), Perlubie (31), Finlayson (34), Ripon (42), Tarlton (35), Scott (38), and Murray (39) (1896).—The last three hundreds lie in the transitional zone between the calcareous sands and the adjoining stony mallee region; their trends resemble the remainder of the group, and consequently they have been included here. All yields show the expected increase, but as with the previous group, the effect is temporary, and they subsequently decline, the types being I and II(11) according to local conditions. The type II curve of Figure 11 is actually that of Perlubie. Carawa, with its type IV(11) curve, stands out as an exception. In this hundred, seasonal conditions are much better than indicated by the 10 in. isohyet because in the small section to which cropping has been restricted the rainfall is, on the average, nearly 11½ in. The maximum yield, attained in 1941, was, however, only about 7 bushels per acre.

b. Petina (27), Hague (25), Walpuppie (32), Cungena (36), and Wallala (28) (1907-17).—In these hundreds the records are too short to make a definite decision regarding trend, but the yields show small increases, and this observation agrees with the view that the subgroup is following, some 10-20 years later, the first phase of the course set by the older districts on this class of soil. The trends have been classified as type III(11).

Hague, which was opened in 1911, declines from the outset; only a small area is cultivated, and it has probably been exploited fully. In Figure 12 this hundred has been included under type I(11). Figure 14 shows the courses of yield in Carawa and Cungena.

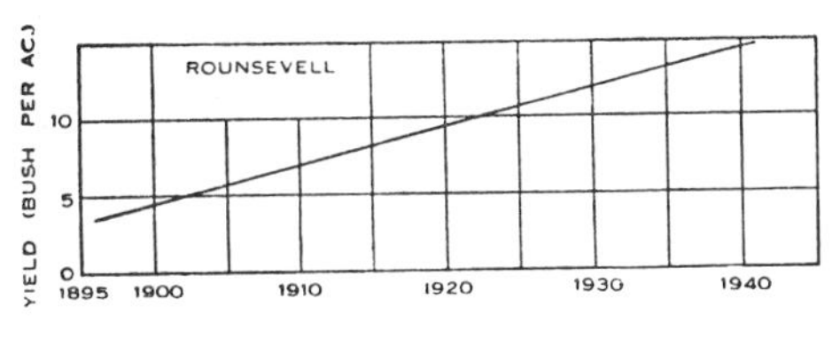

Fig. 15

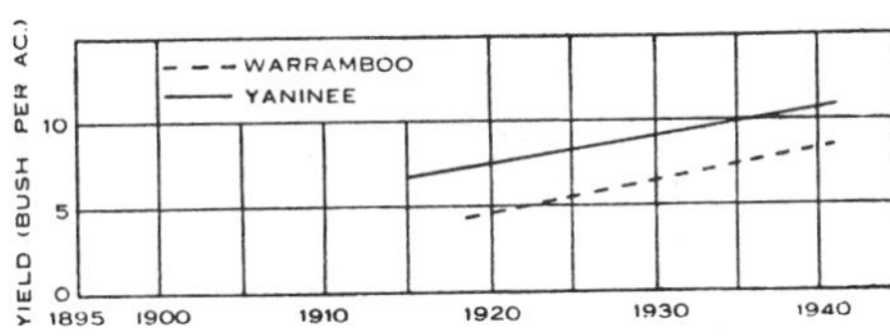

Fig. 16

3. Stony mallee; rainfall 10-13 in.

This group is mapped as stony mallee, but practically all the wheat is grown on scattered areas of sandy mallee which was derived from granitic parent rock.

a. Forrest (43), Wrenfordsley (46), Rounsevell (47), Witera (48), Campbell (44), and Wright (49) (1889-96).—With such a short period of light cropping prior to the advent of superphosphate, fertility was barely affected in these hundreds, and consequently no marked reaction occurs immediately after 1896. Yields steadily increase up to 1941, and all have been classified as type IV(11). These observations, coupled with the higher mean yields, provide a striking contrast with those made previously, and are directly ascribable to the more fertile class of soil and the additional 2-3 in. of seasonal rainfall. Figure 15 shows the course of yield in Rounsevell.

b. Travers (56) and Wallis (57) (from Eyre), Chandada (40), and Inkster (45) (1910-17).—Trends have not been definitely established, but the results obtained agree generally with the remainder of the group. They have been classified as type III(11).

Eyre

1. Sandy mallee; rainfall 9-11 in.

All hundreds were opened for settlement during 1914-19, and are divisible into two subgroups, depending on the nature of the soils.

a. Koolgera (29), Yantanabie (33), Kaldoonera (37), Karcultaby (41) (from Nuyts), Condada (50), Carina (51), Palabie (58), Wannamana (59), Mamblin (60), Kappakoola (61), and Warramboo (62).—The trends are in accordance with the expectation that yields will improve for some years, but owing to the short cropping history only three have been definitely established, namely, Wannamana, Kappakoola, and Warramboo. It will be noted that the latter two hundreds lie in close proximity to the 11 in. isohyet. This trio has been classified as type IV(11), and the remainder as type III(11).

b. Minnipa (52), Yaninee (53), Pygery (54), and Wudinna (55).—The distinguishing feature of the soils in these hundreds is their granitic origin, and the greater fertility is reflected immediately in the higher mean yields, which are similar to those of Rounsevell etc. The trends of Yaninee and Pygery have been established as type IV(11) but those of Minnipa and Wudinna are not so marked, and they have accordingly been grouped as type III(11). Yaninee and Warramboo are illustrated in Figure 16, which shows clearly the 2-3 bushel advantage in yield conferred by the better soil.

2. Transitional mallee-solonetz soils; rainfall 10-13 in.

a. Barwell (69), McLachlan (70), Tooligie (82), Palkagee (71), Rudall (83), Roberts (89), Dixson (97), Smeaton (75), Pascoe (72), Campoona (76), Solomon (65), and Kelly (66) (1907-21).—The last two hundreds contain, in addition, areas of mixed mallee soils. Even though the records are comparatively short, it has been possible to establish the reality of the improvement in yield, since the soils are more fertile and the seasonal conditions more favourable. The whole set is outstanding and has been classified as type IV(11). Tooligie (0.30 bushel per acre per annum) and Palkagee (0.42 bushel per acre per annum) have remarkably high increments.

b. Boothby (90), and Yadnarie (84) (1880-84).—The light and spasmodic cropping in these hundreds during the first twenty years did not materially reduce fertility, and consequently there is no marked reaction following the first applications of superphosphate. Yields increase steadily up to 1941, following the course of type VI(10). Figure 17 contrasts the curves of Tooligie and Boothby; the superiority of the former with respect to mean yield and trend is largely due to the advantages conferred by a difference of approximately 2 in. in the rainfall.

3. Loamy mallee soils merging into red brown earths.

The occurrences of these soils are in two separate localities within zones of different rainfall.

(i) Rainfall 9-12 in.

a. Miltalie (78), Minbrie (79), Mann (85), Hawker (86), and Playford (87) (1880-84).—The remarks here are similar to those given for Boothby and Yadnarie, all curves being of type VI(10). The result obtained for Playford is surprising, but it is readily understood when account is taken of the changes in acreage. The greatest area recorded was 8600 acres in 1913, and since then it has progressively decreased, at the same time concentrating in the north-west corner of the hundred on loamy mallee soils. Associated with this progressive retreat to the better areas is a gradual increase in mean yield, the effect of which is to reverse the curvature of the regression on time, the net result being that yield exhibits a small but real improvement, but the maximum yield was only about 6 bushels per acre.

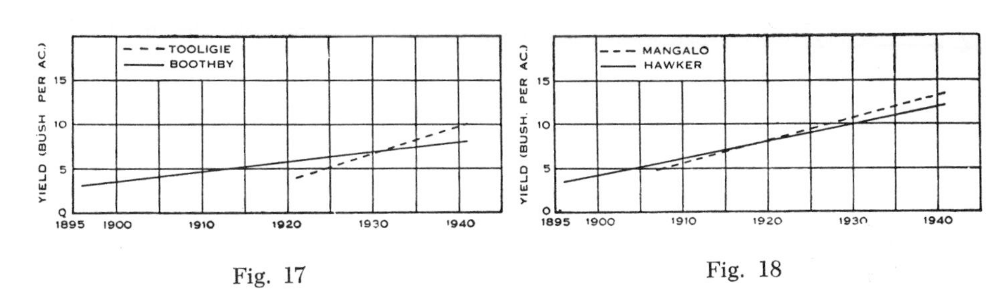

Fig. 17

Fig. 18

b. Mangalo (77) (1907).—The remarks here are similar to those given for Barwell etc., and the curve is of type IV(11).

Mangalo and Hawker, which are illustrated in Figure 18, resemble Tooligie and Boothby respectively. The loamy mallee soil is definitely superior to the transitional phase, but rainfall is the limiting factor.

(ii) Rainfall 12-16 in.

a. Stokes (100), Yaranyacka (101), Hutchison (105), and Louth (106) (1880).—From 1880 to 1896 only limited areas were sown to wheat, and to some extent cropping was also spasmodic. Very little deterioration occurred, so that the initial response to superphosphate is not strongly marked. All yields increase linearly, following curves of type VI(10).

b. Shannon (93), Brooker (94), Moody (95), and Butler (96) (1897-1907).—These hundreds all possess well-defined courses of type IV(11), Shannon, in particular, having the high annual increment of 0.40 bushel per acre. The curves of Hutchison and Shannon are given in Figure 19. Comparison of Figures 18 and 19 brings out clearly the superiority of the group in the zone of higher rainfall, where all rates of increase are greater both relatively and absolutely. This small section constitutes one of the most progressive portions of the wheat belt.

4. Podsols; rainfall 16-20 in.

a. Warrow (102), Ulipa (98), and Koppio (104) (1880).—The three curves are of type VI(10).

b. Cummins (99) and Mortlock (103) (1907).–These hundreds have been classed as type IV(11).

The series possesses an interesting feature, of which no trace is present in any other district. Neither subgroup shows an improvement in yield at the beginning of the several periods examined; in the three hundreds that were settled first, yield was maintained at 6-8 bushels per acre and did not increase

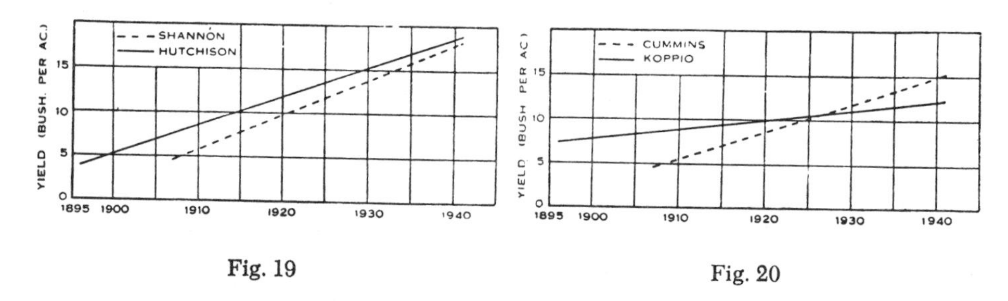

Fig. 19

Fig. 20

until some years after the introduction of superphosphate, and in the remaining two, the improvement likewise did not follow until after an initial stationary period. For the purposes of classifying and mapping the trends, this small departure from linearity was ignored. Figure 20 illustrates the courses in Koppio and Cummins.

5. Rendzinas; rainfall 12-16 in.

a. Colton (67), Talia (68), Ward (81), and Kiana (91) (1880).–As with other districts on Eyre Peninsula, in which early settlement occurred, initial cropping made only light demands on the soil, so that no marked changes occurred in these hundreds shortly after the first applications of superphosphate. From 1896, all yields increase steadily, following curves of type VI(10).

b. Mitchell (92) (1907).–In this area the curve is of type IV(11).

The group, as a whole, has a higher rainfall than the loamy mallee soils of the Cleve district, but the annual increments in yield, although quite significant, are only of the same order as those of the latter, i.e. about 0.10 bushel per acre per annum.

6. Loamy mallee soils agriculturally similar to the desert loams; rainfall 8-10 in.

Wilton (88), Glynn (74), Warren (80), Moseley (64), and Cortlinye (63) (1907-16).–This series has been classed as type III(11), since a definite decision regarding trend has not been obtained. Moseley and Cortlinye are the only important areas in the group.

7. Solonetz; rainfall 10 in.

The only hundred concerned is James (73), which was opened for settlement in 1910. The increase in yield is small (about 0.10 bushel per acre per annum) but definite, and has been included under type IV(11).

Flinders

1. Desert loams and mixed mallee types merging into desert loams; rainfall 6-10 in.

This group includes all hundreds in Flinders that are not listed under sections 2, 3, and 4 below (1875-84).

2. Sandy mallee; rainfall 10-12 in.

Winninowie (130), Baroota (137), and Telowie (144) (1875-84).

3. Red brown earth and mixed mallee; rainfall 10-12 in.

Black Rock Plain (141), and Yongala (146) (1875-84).

In the three groups, exploitative management of the first twenty years rapidly depleted soil reserves, yields declined, and incipient erosion developed. With the beginning of the new century came a full realization that in addition to the inferior types of soil, the settler had to contend with an extremely erratic seasonal rainfall. In consequence, the boundary of the area under cultivation was withdrawn during 1902-09, and cropping confined to scattered localities in the Flinders Ranges, which possessed arable pockets of somewhat better soils with a slightly higher rainfall. Superphosphate and the retreat to the more favourable sites were the principal factors responsible for the increase in yield

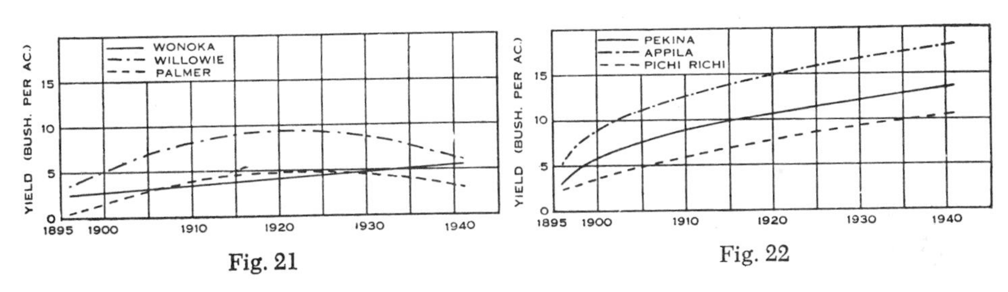

Fig. 21

Fig. 22

from 1896 to 1920, but this effect was transitory, since, with very few exceptions, even in these situations, the soils were incapable of withstanding frequent cropping. From about 1920, yields again declined, and apart from Wonoka, type V(10), and Black Rock Plain, type IV(10), the courses are of types I and II(10), depending upon local conditions. Wonoka and Black Rock Plain supply good illustrations of the effect on the trend in yield of a progressive reduction in acreage and concentration on better soils; in the latter almost the entire area occupied by the crop is red brown earth. In the majority of hundreds with type II curves the area of wheat was small and remained constant, the inference being that cultivation had been restricted to slightly better situations throughout. Palmer, Willowie, and Wonoka are contrasted in Figure 21; the type II curve of Figure 10 is that of Yongala.

4. Red brown earth; rainfall 10 in. to more than 14 in.

Pichi Richi (119), Gregory (131), Wongyarra (138), Booleroo (139), Appila (145), and Pekina (140) (1875).—After settlement, the area in each hundred increased to a point where it became stable. The initial exploitative cropping caused yields to decline, but they recovered after the introduction of superphosphate, and under the greatly improved conditions of soil and rainfall, continued to increase up to 1941, although at diminishing rates. Pichi Richi and Wongyarra are of type V(10), and the remainder type IV(10). In Pichi Richi the area of crop has remained practically constant since 1880, and actually only part of the hundred is classed as marginal land, but for convenience in mapping this has not been indicated in Figure 12. Figure 22 shows the courses in Pichi Richi, Pekina, and Appila, and illustrates the effect of an increase in seasonal rainfall from 10 in. to 14 in. The type V curve of Figure 10 is that of Wongyarra.

Goyder

1. Mixed mallee; rainfall 9-12 in.

Terowie (159), and Baldina (181) (1875-84).—The history of cropping in these hundreds runs exactly parallel with that of the hundreds with type I curves in Flinders; Baldina is type I(10), and Terowie type II(10), the latter being almost identical with the curve for Yongala in Figure 10.

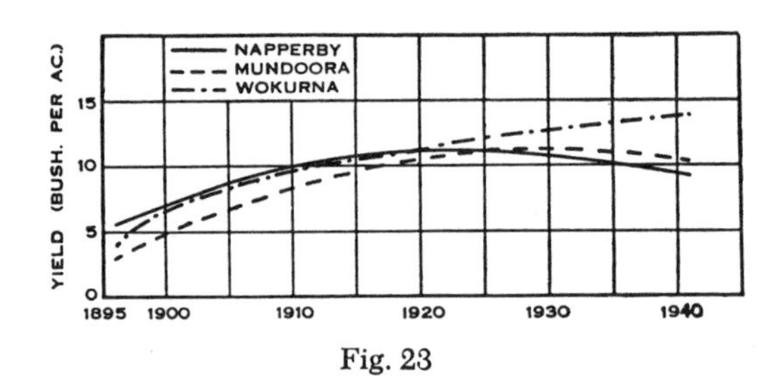

Fig. 23

2. Sandy mallee; rainfall 10-13 in.

Pirie (151), Wandearah (160), Napperby (152), Mundoora (167), and Wokurna (174) (1875-84).—In all districts the yields show the characteristic decline prior to 1896, and subsequent recovery which is transitory except in Wokurna. Rainfall increases from 10 in. to 13 in. in the order in which the hundreds are named, and the trends follow, in the sequence of types I, I, II, II, and IV(10), with improvement in the soils. The course in Pirie is the type I curve of Figure 10, Wandearah is closely similar, with a maximum of 11 bushels per acre, and Napperby, Mundoora, and Wokurna are given in Figure 23. Winninowie, Baroota, and Telowie (from Flinders) also fall properly into the sequence, their curves resembling that of Napperby.

3. Loamy mallee; rainfall 11-16 in.

Crystal Brook (161), Narridy (162), Yangya (156), Koolunga (169), Boucaut (176), Cameron (182), Everard (183), and Blyth (184) (1870-84).—All yields exhibit marked deterioration prior to 1896 and the subsequent recovery, but the curves vary according to the local conditions. Everard is of type I(10), and Boucaut type III(10). The soils of Everard are variable, and often poorer than the average grade of loamy mallee, while in Boucaut a small tract of inferior stony mallee occurs; in addition, both hundreds are endoreic zones in which problems, associated with the development of salinity in the surface soils, have arisen. The influence of the more fertile class of soil is apparent in the remaining hundreds, all of which are of type IV(10). Figure 24 shows the courses in

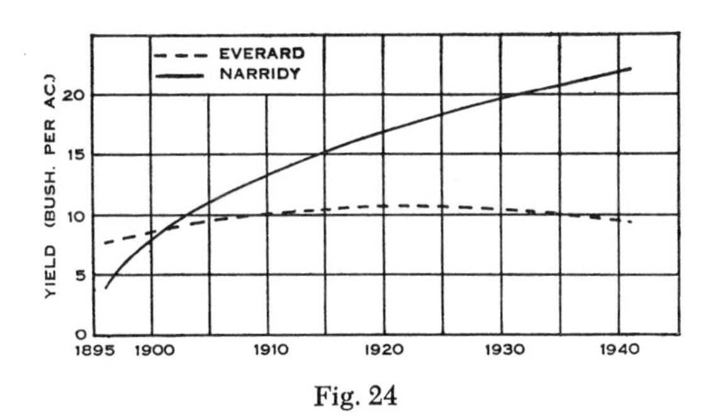

Fig. 24

Everard and Narridy, and the type III curve of Figure 10 is that of Boucaut.

4. Red brown earth; rainfall 12-18 in.

Tarcowie (149), and Mannanarie (150) (1870-75); Howe (153), Booyoolie (154), Caltowie (155), Belalie (157), Whyte (158), Bundaleer (163), Reynolds

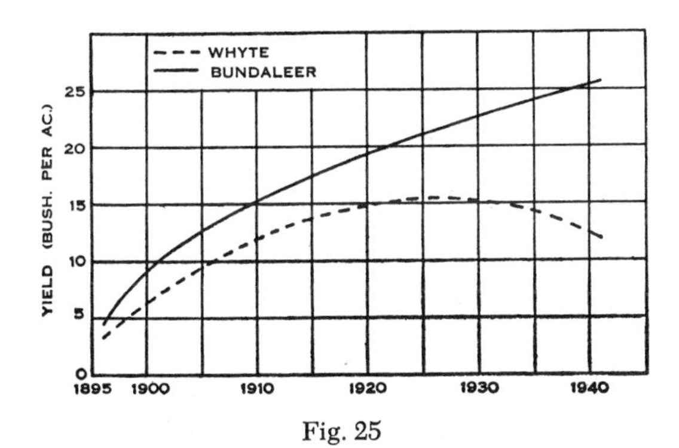

Fig. 25

(164), Anne (165), Red Hill (168), Barunga (175), Yackamoorundie (170), Hart (177), Ayers (172), and Andrews (171) (1866-75); Milne (178) and Hanson (179) (1860-70); Kingston (173), Kooringa (180), and Hallett (166) (1875-84); and Clare (185) (1860).—All trends display remarkably high rates of increase for some ten years after the initial applications of superphosphate, the general curve being type IV(10), but there are several exceptions. Whyte, type II(10), is comprised partly of mixed mallee soils, and just over half the area under wheat lies in a zone between the 10 in. and 12 in. isohyets. The acreage has been maintained throughout and this is the main factor underlying the contrast with Black Rock Plain. Ayers, Hanson, Hallett, Kingston, and Kooringa follow curves of type III(10). The best soils of these five hundreds are devoted principally to lucerne, and wheat is relegated to the lower slopes of the hills on shallow residual red brown earths, which are less fertile than the alluvial types of Belalie and Caltowie, for example. Hallett, Kingston, and Kooringa are extreme in this respect, while Ayers and Hanson carry an intergrade soil between the extremes. Whyte and Bundaleer are contrasted in Figure 25. Clare follows a course of type V(10).

Light

1. Mixed mallee

(i) Rainfall 6-10 in.

Bright (192), and Bundey (193) (1880); and Bower (197) (1884).—This trio forms part of the worst section of the wheat belt. The curves are type I(10), and the history of cropping is similar to that of hundreds in Flinders with the same form of trend.

(ii) Rainfall 10-14 in.

English (196), Neales (204), and Dutton (213) (1860-70).—In these cases the mean yields are considerably greater, and the curves are type II(10). Bower, Neales, and English are compared in Figure 26, which shows clearly the effect of the additional 4 in. seasonal rainfall. The course of yield in Dutton is almost identical with that of English.

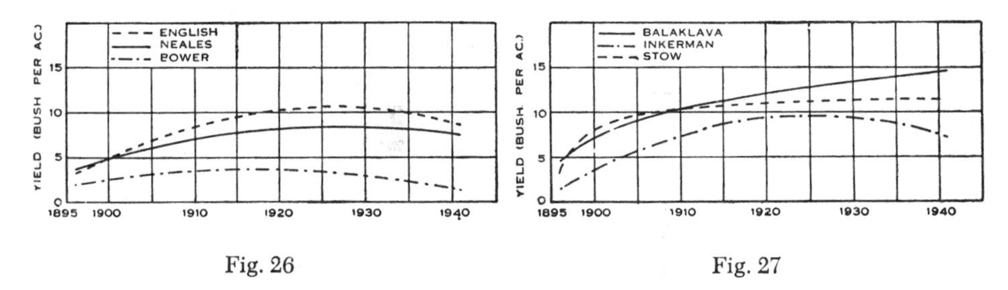

Fig. 26

Fig. 27

2. Sandy mallee; rainfall 10-14 in.

Inkerman (198), Balaklava (199), Hall (188), Goyder (186), and Stow (187) (1860-70).—In Inkerman, where conditions are similar to those of Mundoora, the curve is of type II(10) and falls properly into sequence with other members of the sandy mallee group immediately to the north. Problems associated with the development of salinity in the surface soils similar to those in Boucaut, also occur in Stow, and the resulting curve is of type III(10). The remaining hundreds are of type IV(10). Figure 27 gives the trends of Inkerman, Stow, and Balaklava.

3. Loamy mallee; rainfall 11-14 in.

Port Gawler (207), Dublin (205), Grace (206), Dalkey (200), and Mudlawirra (208) (1850-70).—The first four hundreds provide, in the order given, a sequence of improving variants of type IV(10), and Mudlawirra is type VI(10). In the latter, wheat was the dominant crop during the early years, reaching its peak in the decade 1870-80, when nearly 30,000 acres were cropped. The area was reduced to 10,000 acres by 1896, and since then has varied about this value. The general tendency has been to pass from wheat as the major enterprise to mixed farming, with an increase in the number of stock carried, accompanied by an increase in the length of rotations.

4. Red brown earth; rainfall 14-18 in.

Upper Wakefield (189), Stanley (190), Alma (201), Saddleworth (194), Gilbert (202), Waterloo (195), Julia Creek (203), Kapunda (211), Light (209), Belvidere (212), Nuriootpa (210), Moorooroo (214), and Jellicoe (215) (1850-66); and Apoinga (191) (1866-70).—Apoinga and Belvidere are of type III(10). The former is a member of the series Hallett etc., for the same reasons as given previously, while in the latter, vineyards occupy the best soils and wheat is cropped on intermediate grades of red brown earth resembling those of Ayers and Hanson. Alma and Nuriootpa follow courses of type VI(10), Moorooroo is of type V(10), and the remainder type IV(10). The position with

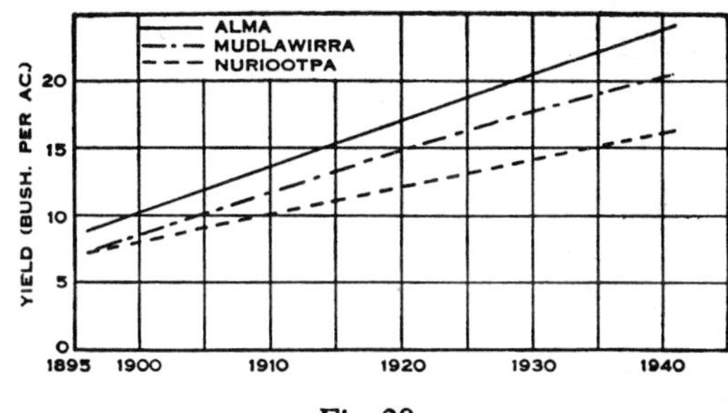

Fig. 28

respect to wheat as the major crop has undergone a radical change in Nuriootpa and Alma, more particularly in the former. In both hundreds the area decreased from its maximum, about 25,000 acres, attained during 1870-80, and by 1896 had fallen to 10,000 acres. With the appearance of superphosphate, the area increased to 15,000 acres in Alma, and remained approximately constant up to 1941. This hundred possesses particularly good soils, and with the increase in the number of stock carried there has been a trend to mixed farming under longer crop rotations. On the other hand, in Nuriootpa the area of crop was reduced still further, until it roughly stabilized at 5000 acres. During the period under review, the area devoted to vineyards has increased, and wheat has been replaced as the major crop, practically all of it having been produced under mixed-farming conditions.

Mudlawirra, Alma, and Nuriootpa are illustrated in Figure 28.

Yorke

1. Sandy mallee; rainfall 12-13 in.

Wiltunga (217), and Ninnes (220) (1875-84).—These two hundreds are situated at the southern extremity of the sandy mallee zone running parallel with the eastern coastline of Gulf St. Vincent. The curves are type IV(10) and follow in sequence after Wokurna in conformity with the higher rainfall and improvement in the soils.

2. Loamy mallee; rainfall 12-16 in.

Maitland (225) (1870-75); and Tickera (216), Wallaroo (218), Kadina (219), Tiparra (222), Kulpara (221), Clinton (223), Kilkerran (224), Wauraltee (227), Cunningham (226), Muloowurtie (228), Koolywurtie (229), and Curramulka (230) (1875-84).—In Kilkerran, salinity problems have developed as in Boucaut and Stow, and the yield follows a curve of type III(10). Muloowurtie is type VI(10), and although first settled during 1875-84, extensive development has only occurred in recent years, the effect of a progressive introduction of virgin land being to reverse the curvature of the regression on time, and thus maintain the rate of increase in yield. The results for Maitland, type II(10), with a maximum in 1938, and Curramulka, type V(10), in which the trend had nearly reached zero in 1941, are of far-reaching importance, and are directly ascribable to intensive cropping of cereals in the area. The remaining hundreds are type IV(10). Maitland, Tiparra, Curramulka, and Muloowurtie are illustrated in Figures 29 and 30.

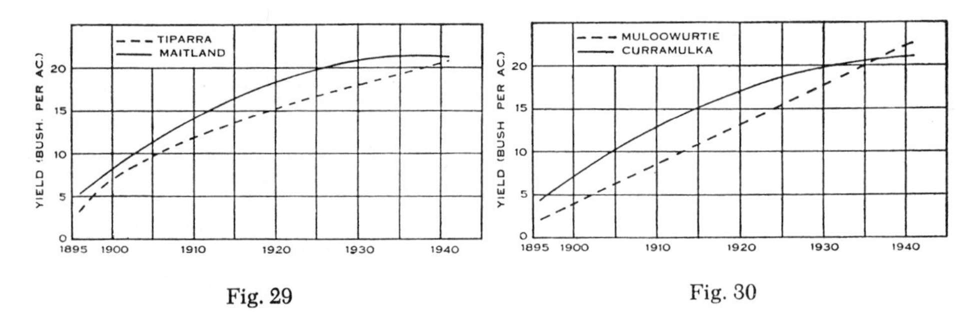

Fig. 29

Fig. 30

3. Rendzina; rainfall 14-16 in.

Minlacowie (231), Dalrymple (235), Melville (236), Moorowie (234), Para Wurlie (233), and Ramsay (232) (with some sandy mallee) (1870-80).—Moorowie and Ramsay are type IV(10), Para Wurlie is type II(10), and the remainder type V(10). The curve for Para Wurlie is an extreme variant of type II(10) in which the trend reached zero in 1940. Trace-element deficiencies are known to occur in this locality (*vide infra*).

As in southern Flinders and Goyder, the yields of all hundreds in Light and Yorke declined markedly during the period of cropping prior to 1896, and in general showed very high rates of recovery for some years after the initial applications of superphosphate. The group of four regions constitutes that part of the wheat belt which suffered most severely under the stringent conditions of cropping prior to 1896.

Sturt

1. Red brown earth; rainfall 12-14 in.

Monarto (263) and Tungkillo (268) (from Adelaide) (1860); and Freeling (269) (from Fleurieu) (1862).—In these hundreds there are occurrences of mixed mallee soils, but the bulk of the wheat is grown on red brown earth, and throughout the long history of cropping none of them has carried more than 3-5 thousand acres of wheat. The types are Monarto and Freeling VI(10), and Tungkillo V(10). Figure 31 illustrates Monarto.

2. Sandy mallee; rainfall 9 in.

Forster (261), the only hundred concerned, was opened in 1884. Yields show the characteristic decline prior to 1896, and the subsequent temporary recovery. The course of yield is type II(10), closely similar to the curve for Gordon in Figure 34.

3. Stony mallee; rainfall 8-12 in.

a. Ridley (260), Nildottie (257), Younghusband (262), Burdett (265), and Seymour (267) (1880).—The remarks here are similar to those made for Forster, all curves being of type II(10).

b. Ettrick (266) (1910).—The trend of yield during the shorter period of cropping in this hundred resembles the first phase of the remainder on this soil

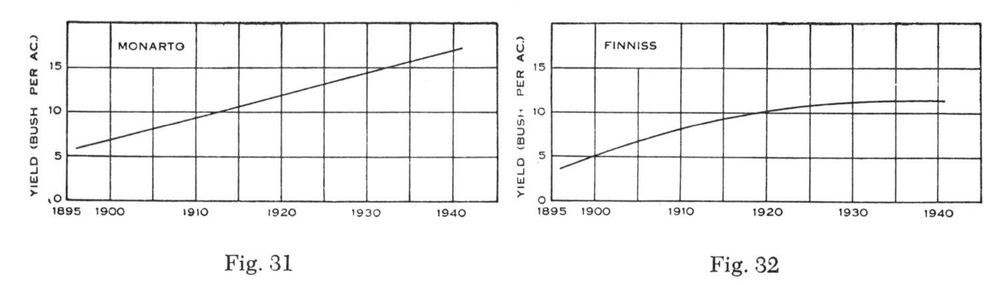

Fig. 31

Fig. 32

group, and has accordingly been classified as type III(11).

4. Mixed mallee; rainfall 8-12 in.

Bagot (256), Angas (258), Finniss (259), and Mobilong (264) (1862-75).—The courses of yield, which are of type II(10), are similar to those of English, Neales, and Dutton, on the same class of soil. Finniss is illustrated in Figure 32, and the curve for Mobilong is almost identical.

Pyap

1. Stony mallee; rainfall 6-10 in.

a. Anna (250), Brownlow (243), and Beatty (237) (1870-84).—These hundreds, together with Bagot from Sturt, Baldina from Goyder, and Bright, Bundey, and Bower from Light, comprise the worst section of the wheat belt. The three type I(10) curves are similar to that of Bower in Figure 26.

b. Cadell (238), Murbko (244), Paisley (251), and Bakara (252) (1896); and Mantung (253) (1909).—Yields of the first four hundreds show the expected increase, but as with all other mallee soils under similar conditions of rainfall, this effect is temporary. The curves are all of type II(11). Mantung, with its shorter cropping history, has improved slowly, following a course similar to the first phase of the curves in the older districts. The trend is not significant and has been classed as type III(11). The curves of Cadell and Mantung are given in Figure 33.

2. Sandy mallee; rainfall 6-8 in.

Waikerie (245) (partly stony), Holder (246), Moorook (247), Gordon (249), Pyap (254), and Bookpurnong (255) (1896); and Markaranka (239), Parcoola (241), and Pooginook (240) (1913).—The remarks here are similar to those of the previous section. Waikerie is type I(11), the next five are type II(11), and the last three type III(11). The curve for Gordon is given in Figure 34.

3. Mixed mallee; rainfall 6-8 in.

Murtho (242) and Paringa (248) (1896).—The curves are type I(11) and remarks similar to those above.

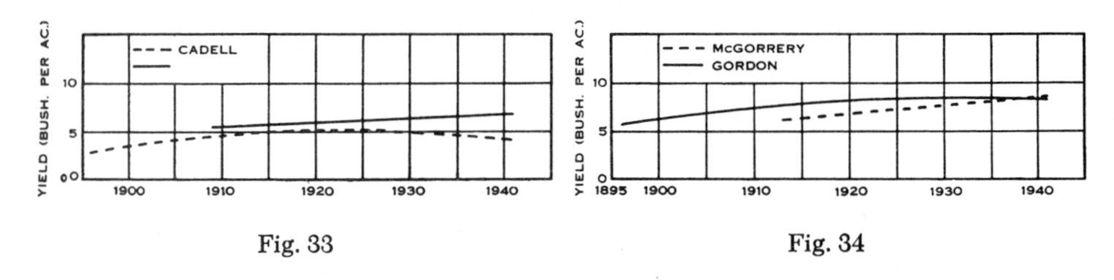

Fig. 33

Fig. 34

Throughout the region considerable reductions have been made in acreage since 1930, the nett effect of which has been to reduce the rate of decline in yield.

Pinnaroo

1. Sandy mallee; rainfall 8-10 in.

Bowhill (270) (with some stony and transitional mallee) (1896); and Bandon (271), Chesson (272), Mindarie (273), Allen (274), Kekwick (275), and McGorrery (276) (1907-19).—The curve for Bowhill is type II(11); and the remainder are of type III(11); remarks are similar to those above. McGorrery is shown in Figure 34.

2. Transitional mallee-solonetz; rainfall 10-12 in.

Vincent (277), Wilson (278), McPherson (279), Hooper (281), Marmon Jabuk (282), Molineux (283), Sherlock (288), Roby (289), Peake (290), Price (291), Cotton (284), Bews (285), Parilla (286), Pinnaroo (287), Peebinga (280), and Allenby (292) (1907-19).—The general curve is type IV(11) with the exception of Peebinga, Roby, and Sherlock, which are type III(11). This soil group is outstanding, and the contrast between it and neighbouring mallee types to the north is just as strongly marked as on Eyre Peninsula. The courses of yield in Pinnaroo and Roby are illustrated in Figure 35.

3. Solonetz; rainfall 14 in.

The only hundred concerned is Livingston (293), which is also of type IV(11) and was opened in 1907.

Tatiara

1. Solonetz; rainfall 15 in.

Stirling (294) (1889).—The curve is type VI(10), and the marked superiority of this hundred over Livingston is due to the fact that wheat is confined to a section in which the solonetzic soils are mixed with a rendzina-terra rossa complex of greater fertility.

2. Wimmera grey loam; rainfall 15 in.

Tatiara (296) and Wirrega (295) (1876 and 1877).—In both hundreds yields declined slightly under the cropping prior to 1896, but after the introduc-

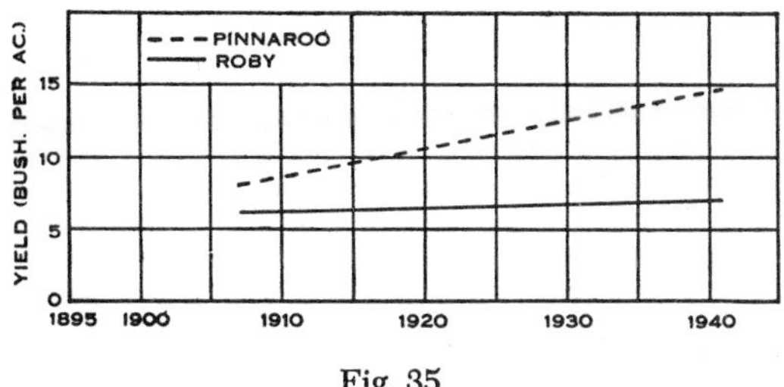

Fig. 35

tion of superphosphate increased steadily up to 1941, following curves of type VI(10). Stirling and Wirrega are contrasted in Figure 36, which shows the superiority of the Wimmera loam; the difference is even more marked when a comparison is made between Tatiara, the type VI curve of Figure 10, and Stirling.

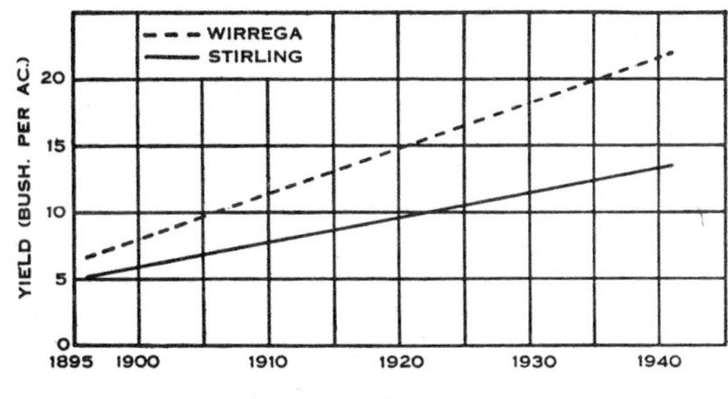

Fig. 36

The remarkable progress made in these districts agrees with observations made by Wadham and Wood (1939) on the Victorian counterpart of the soils.

(*d*) *Discussion of Trends*

It must be emphasized that this study is, by nature, only a broad survey. Obviously, with respect to any particular hundred there is, on the one hand, a set of factors tending to increase yields, and on the other, a set tending to decrease them. The observed trend is the resultant of their effects, and because of the nature of the data they are completely confounded and incapable of separate assessment; in fact, it is not even possible to enumerate either set exhaustively. But in viewing the matter broadly, some indication can be given of the major contributors in the two directions. Several instances have already been given in Section V(*c*).

On the positive side the principal factors are:

1. Maintenance of an adequate phosphorus and nitrogen supply.
2. Adoption of cultural practices suited to various types of soil.
3. Use of improved varieties adapted to the particular conditions of different localities.
4. Maintenance of the physical condition of the soil.

On the other hand, decreases can be unhesitatingly ascribed to a reduction of fertility consequent upon agricultural exploitation of the soils through over-cropping in short-term rotations. As pointed out above, this has proceeded so far in some areas that it outweighs the beneficial effects of recent advances, and results in a progressive decline. In other localities, although yield was still increasing in 1941, it would appear that farming practices in vogue were none the less exploitative, since rates of increase were being rapidly decelerated and yields were not as high as would be anticipated considering the nature of the soils, the seasonal rainfall, and the potentialities of the latest varieties at the time. Further reference is made to this point below.

The decline in fertility is attributable to (i) a cumulative deficiency of one or more essential elements, and (ii) a loss of soil structure. Each of these may exert its effects directly or indirectly.

(i) *Cumulative Deficiencies.*—Considering the soils of the wheat belt generally, no extensive quantitative evidence has hitherto been presented to show hat there is a cumulative deficiency of any element which is essential in the ıssimilatory processes of the wheat plant, such as occurred with phosphorus. The paramount importance of nitrogen in this connection was obscured until ıbout fifteen years ago by the fact that all experimental work had been con-lucted on fallowed land under conditions which were not conducive to substantial esponses from nitrogenous fertilizers (Richardson and Gurney 1933). The ıecessity of fallowing for securing high returns received general recognition arly in the development of the industry, and the practice was adopted widely. For many years its greatest advantage was considered to be conservation of water hat fell in the fallow season, but latterly this view has changed. Observations vhich were assembled and reviewed by Prescott (1933) showed that nitrate ccumulation is at least equally as important as moisture conservation and the xtermination of weeds. It is known that the production of nitrate in a fallow ; considerable (Prescott loc. cit.), and under optimal conditions is sufficient to ıeet the needs of heavy crops, i.e. 50-60 bushels per acre. Since, for example, 15 bushel crop removes approximately 20 lb. nitrogen per acre in its grain and raw, and to this must be added the amounts lost in other ways, the main-enance of fertility, provided no other factor is limiting, rests upon the replace-ıent of this nitrogen. Nitrogenous fertilizers are not used, so that unless natural ources exist to compensate the loss, it must be borne by the nitrogen reserves the soil.

Three contributors that may be considered in this connection are:

1. Non-symbiotic nitrogen fixation.
2. Symbiotic fixation with legumes.
3. Nitrogen introduced in rain water.

With respect to the first, the weight of evidence at the present time points to the fact that it is not of any consequence in the soils of the wheat belt. Thus Beck (1935), after examining South Australian soils, concluded that Lewcock's (1925) claim of the universal distribution of *Azotobacter* was not substantiated, and even in soils containing the organism the numbers present were so small that the amounts of nitrogen fixed would be negligible. Swaby (1939) working with Victorian soils found *Azotobacter* comparatively rare, and reached the same conclusion as Beck. On the other hand, *Clostridium butyricum* was much more widely distributed, and present in greater numbers, but environmental conditions in the wheat soils are not suitable for the development of this anaerobe to the point where it could fix appreciable amounts of nitrogen. Finally, Jensen (1939), in reviewing current work at the time on New South Wales soils, concluded that non-symbiotic fixation is only "a minor and mostly insignificant factor in maintaining the nitrogen content of soils under wheat cultivation." Recent observations made by Clarke and Marshall (1947) are also relevant in this connection; and at the same time demonstrate forcibly the outstanding weakness of the fallow-wheat rotation. These authors determined the reduction in the nitrogen content (expressed as a percentage of air-dry soil passing a 2 mm. sieve) of two slightly different red brown earths, after various periods of cultivation. The total declines in surface soils (top 4 in.) were 0.158 to 0.094 per cent. and 0.222 to 0.135 per cent. after 16 and 20 years, respectively, of cropping in a fallow-wheat rotation with superphosphate applied to the crop. The upper limit of each range is for virgin soil, and both reductions, of which the greater part occurred in the first five years, are statistically significant. Declines in subsurface soils were not significant. No legumes developed on either site, and consequently any nitrogen which may have been added to the soils came from other sources. Such contributions, if they existed, must, however, have been very small, since the ratio of carbon to nitrogen in these soils is approximately 12, thus making the nitrogen content a fairly accurate index of their organic matter status. It would appear then that the nitrogen withdrawn by successive crops was derived principally from the organic reserves of the soils.

Symbiotic fixation of nitrogen with legumes has undoubtedly been of some importance in certain areas of the wheat belt; further reference to this point is relegated to a later stage of the discussion.

Jensen (loc. cit.) quotes amounts of 3-4 lb. nitrogen per acre per annum in rain water, but these are trivial.

The soils of the South Australian mallee have been studied by Prescott and Piper (1932), and the red brown earths by Piper (1938). The percentage of nitrogen in the surface soils of the mallee group ranges from low (less than 0.10 per cent.) to moderate (0.10 to 0.30 per cent.) in passing from light textured sandy types to the heavier loams, with an average of 0.06 per cent., but texture and percentage nitrogen are not strongly associated; 87 per cent. of surface soils examined contained less than 0.10 per cent. nitrogen, and 83 per cent. of subsoils less than 0.05 per cent. In the red brown earths, the proportion of nitrogen varied from low to moderate, with an average of 0.10 per cent. for surface soils, 60 per cent. of which contained less than the average. The first 9 in. was taken in this group as surface soil, and this accounts for the fact that the average value lies at the lower extremity of the range for moderate amounts. The original reserves of nitrogen in the principal wheat soils are thus, at best, only moderate, and consequently it would not be surprising that an exploitative system of cropping has largely exhausted them.

Reference may be made to recent work on trace-element deficiencies, as this may also have a bearing on the nitrogen problem. The malady known as "coast disease" of sheep occurs in parts of South Australia, and has been shown to result from a dual deficiency of copper and cobalt in the fodder consumed by animals depastured in the affected areas (Marston, Lee, and McDonald, 1948*a*, 1948*b*). Other regions in which a deficiency of copper is not complicated by one of cobalt have been identified. On these much greater areas, the deficiency of copper varies in degree which is manifested by the animal in a range of symptoms, the first to appear being a lesion in the wool which is characteristic, specific, and easily discernible (Marston, Lee, and McDonald loc. cit.). The occurrence of these nutritional disorders in stock has led to intense work on the mineral requirements of oats and pasture plants, including legumes, in certain isolated areas (Riceman 1946, 1948), where essential trace elements, in particular copper and zinc, as well as phosphorus and nitrogen, are in extremely short supply. No widespread deficiencies have been observed with cereals, but observations made by Lee (personal communication, 1948) have shown that, except for the greater part of Flinders, copper deficiency, as indicated by the occurrence of lesions in wool, occurs throughout the portion of South Australia illustrated in Figures 8 and 12 and the extreme south-eastern section not included in the figures. The density of occurrence is greatest on the littoral calcareous dunes and adjacent areas, and is gradually reduced in passing from the mallee soils and genetically related types (Crocker 1946) to the red brown earths. As the requirements of the animal relative to those of the plant are enormous, it is much more sensitive and this would account for the observed facts. On the other hand, it is known that copper is essential in nitrification, and the possibility exists that in some areas after a period of cropping the originally low copper status of the soils has been reduced to a point where nitrifying organisms react to deficiencies which do not restrict the wheat plant. If this is so, it would tend to make nitrogen a limiting factor for wheat production; and of the main wheat soils, the mallee types are the most likely to be affected. Other elements are also required in nitrification, but particular reference is made only to copper in view of its known deficiencies.

There are also indications that deficiencies of certain elements are causing indirect effects by restricting symbiotic fixation of nitrogen. Trumble (personal communication, 1948), using various legumes as test plants, has obtained definite responses in yield to the elements potassium, copper, zinc, molybdenum, boron, and manganese at widely dispersed points of the wheat belt, either singly (potassium, molybdenum, zinc) or in certain combinations.

The extent to which these factors operate in limiting the nitrogen supply can, however, only be resolved by an extensive set of field trials.

(ii) *Loss of Soil Structure.*—Frequent cropping leads also to deterioration of soil structure and, indirectly through this, to losses of plant nutrients by erosion. The reduction in water-stable aggregates originally present in virgin soil by cultivation has been well established, and it will suffice to refer to the work of Clarke and Marshall (loc. cit.). Changes in water-stable aggregation of red brown earth surface soils, resulting from increasing periods of cultivation up to twenty years, were measured, and significant decreases found in all cases. The major part of the decline took place in the first five years of cultivation, and the total declines ranged from 42 to 69 per cent. of the aggregates in virgin grassland. The decrease results directly from the mechanical action of tillage, and indirectly through the decomposition of the organic matrix during fallowing. The effect on soils of heavy texture is to reduce their absorptive and retentive powers in coping with heavy rains; surfaces tend to set hard, causing uneven germination and making cultivation more difficult. Piper (loc. cit.) records such instances in the zone of the red brown earths. The most serious consequence, however, is that water erosion supervenes and may advance to its worst form — sheet erosion. Considerable areas in Flinders, Goyder, and Light have been eroded in varying degree (Rural Reconstruction Commission 1944), and the position in the hundred of Belalie may be taken as typical of a large proportion of the best agricultural land:

Degree of Erosion	Per Cent. of Surface Soil Remaining	Per Cent. of Total Arable Land
Slight	> 75	24
Moderate	25-75	69
Severe	< 25	7

On light textured sandy soils, removal of the original vegetal cover followed by frequent cropping has led to sand drift by aeolian action, the principal regions concerned in this respect being Sturt, Pyap, and northern Pinnaroo.

(iii) *General Relation between Form of Trend and Nitrogen Status of the Soils.*—The orderly geographical distribution of the various types of trend is striking, and accords generally with the view that after the deficiency of phosphorus has been overcome the progressively diminishing nitrogen status of the soils is a dominant factor influencing the form of the curves.

On sandy, stony, and mixed mallee soils, and related types that were opened in 1896 or several years later, the onset of a decline appears after 20-25 years of cropping with adequate supplies of superphosphate, and by 1941 yield in the majority of hundreds concerned had fallen to approximately the value it had attained at the time of settlement. In these soils, the proportion of nitrogen in their virgin state is low and the losses sustained by exploitation during the pioneering years and later cropping, together with those due to erosion where it has occurred, are great enough to exhaust the reserves in approximately 45 years.

Similar mallee types, and the desert loams which were settled between 1875 and 1896, had been largely abandoned after 10-20 years, and consequently are not represented among the curves of Figure 10. After the advent of superphosphate, the remaining small isolated areas of slightly better soils carried on for a further 20 years, when a second decline supervened and yield fell gradually to its value at the beginning of the century (hundreds with curves of type I(10)). In these soils exhaustion thus occurs after 55-65 years of cropping.

TABLE 3

FREQUENCY DISTRIBUTION OF SEASONAL RAINFALL AT SELECTED STATIONS

Station	Length of Record (years)	Rainfall (in.)						Mean Apr.-Nov.
		> 14	12-14	10-12	8-10	6-8	< 6	
Fowler's Bay	69	6	8	15	25	12	3	10.22
Ceduna	39	1	4	7	10	13	4	8.89
Petina	38	4	8	12	11	1	2	10.68
Cortlinye	26	3	4	3	8	7	1	9.66
Cowell	65	2	6	9	18	21	9	8.51
Hawker	64	6	6	10	16	11	15	9.25
Hammond	59	2	6	4	14	15	18	8.14
Orroroo	73	8	11	15	15	17	7	10.00
Terowie	64	7	9	11	17	14	6	9.96
Sandleton	57		4	4	20	9	20	7.54
Morgan	54		3	2	9	16	24	6.67
Loxton	37	2		5	9	9	12	8.01
Copeville	34		5	7	8	7	7	8.82

All diagrams given above, which refer to these districts, indicate that the maximum yields obtained under average rainfall are extremely low; considerable effort would have to be expended to increase them by 50 per cent. Without doubt, deterioration could be arrested and fertility restored in certain localities, but considering the areas generally, improvement would be of limited extent, as the steps taken must ultimately depend upon rainfall and its reliability. At the moment, research is proceeding on this vitally important feature of the climate, but no detailed quantitative data are available for the areas concerned. The approximate analysis of records from typical stations, set out in Table 3, illustrates the extremely hazardous nature of the seasonal rains, and requires no further comment, except perhaps to add that all stations are presented in the most favourable light, since the average effective rainfall at each is only slightly greater than 5 in. (*vide* Trumble loc. cit., map 2).

As pointed out above, the large tracts of these particular mallee soils that have been developed recently appear to be following courses similar to the districts with longer records, and although yields were still slowly increasing at the close of the period under review, it can be confidently anticipated that if cropping is continued under the old system, they will eventually decrease.

In an extensive and detailed report, the Pastoral and Marginal Agricultural Areas Inquiry Committee (loc. cit.) reviewed all enterprises within the marginal areas, and after examining the economics of wheat production both from the point of view of direct returns from wheat and that of indirect returns from sidelines, submitted a strong case for the abandonment of wheat-growing as the major operation, and hence mixed farming, in these regions. The Committee concluded that the worst portions of the marginal lands "within which it is impracticable to continue with any system of land use that involves even periodic cropping for grain, must of necessity be turned over to sheep grazing (with perhaps a few cattle) on an extensive scale, as no other proposition can provide the basis for security . . . " Apart from those hundreds which were omitted from consideration in the present analysis for reasons given in Section III(*a*), and ten minor exceptions, of which every one can be given adequate account, it will be observed in Figure 12 that the yields of all hundreds designated as marginal by the Committee follow courses of types I and II(10) or I, II, and III(11), and this result lends material support to their conclusion. The marginal boundaries as given in Figure 12 have been chosen to conform with the territorial unit which forms the basis of this investigation and should be regarded as indicating the approximate position of the line. Wannamana, Kappakoola, Warramboo (Eyre), and Vincent, Wilson, McPherson (Pinnaroo) constitute two doubtful areas. Their mean yields were low, but their rates of increase were quite definite and constant up to 1941, and consequently additional data must be examined before final judgment is passed. The remaining exceptions have been discussed in Section V(*c*).

Next in order are mallee soils situated in an intermediate zone of 10-14 in. seasonal rainfall. The hundreds concerned are Ripon, Scott, Murray (Nuyts), Telowie, Yongala (Flinders), Pirie, Napperby, Wandearah, Mundoora, Whyte, Terowie, Everard (Goyder), English, Neales, Dutton, Inkerman (Light), Finniss, Mobilong, Burdett, Ettrick, and Seymour (Sturt), which, except for Everard and Inkerman, are either contiguous or nearly so with the inner marginal boundary. In these areas the decline may not supervene until after 25-35 years of cropping with an ample supply of superphosphate, and except for Pirie its magnitude is not as great relatively as in the previous cases.

Thirdly, follow the heavier loamy mallee soils with a rainfall of 12 in. or more, and the red brown earths in Flinders, Goyder, Yorke, and Light. The dominant form of trend in this group of 78 hundreds is type IV(10). More significant observations are, however, that there are only six courses of type VI(10), each of which is characterized by special circumstances, and the presence of a type II(10) curve in Maitland, a district that has always been regarded as one of the best centres for cereal culture. The yields of barley in Maitland follow a similar course, thus confirming the result for wheat. It is of some moment also that the rate of increase in yield of the type V(10) curve of Curramulka, had very nearly reached zero by 1941. These observations, when coupled with the widespread advance of declining yields in the submarginal zone, provide a timely and salutary warning that the existing cropping systems must be altered radically to prevent further deterioration.

Since this group constitutes the heart of the industry, from which approximately 80 per cent. of the total harvest is gathered, it is worth while furnishing additional detail regarding the trends. Table 4 sets out the rates of increase in yield at four points of the period 1896-1941, together with the yield of 1941, adjusted for rainfall. It must be recognized that the production of the soil-

TABLE 4

RATES OF INCREASE OF YIELD AT SELECTED DATES IN THE PERIOD 1896-1941

Region* and Hundred	Rate of Increase of Yield (bush./ac./an.)				1941 Yield	
	1900	1911	1921	1941	(bush./ac.)	(bush./ac./in. seasonal rainfall)
Flinders						
Gregory	0.43	0.19	0.14	0.09	13.89	1.21
Wongyarra	0.59	0.44	0.32	0.06	21.37	1.31
Booleroo	0.53	0.25	0.18	0.12	16.58	1.38
Pekina	0.44	0.21	0.16	0.11	13.50	1.27
Black Rock Plain	0.40	0.21	0.16	0.11	12.18	1.27
Appila	0.57	0.26	0.19	0.13	18.22	1.34
Goyder						
Tarcowie	0.56	0.27	0.19	0.14	17.20	1.41
Mannanarie	0.50	0.22	0.16	0.11	15.81	1.39
Howe	0.61	0.31	0.23	0.16	18.37	1.35
Booyoolie	0.64	0.31	0.23	0.16	19.46	1.34
Caltowie	0.56	0.26	0.19	0.13	17.46	1.31
Yangya	0.65	0.30	0.22	0.15	20.44	1.55
Belalie	0.56	0.24	0.17	0.11	18.96	1.33
Crystal Brook	0.49	0.21	0.15	0.10	15.99	1.26
Narridy	0.73	0.39	0.30	0.22	22.14	1.73
Bundaleer	0.86	0.46	0.35	0.26	25.84	1.74
Reynolds	0.52	0.22	0.15	0.10	18.81	1.37
Anne	0.60	0.26	0.18	0.12	19.92	1.45
Hallett	0.96	0.28	0.14	0.06	17.30	1.32
Red Hill	0.50	0.22	0.16	0.10	16.58	1.33
Koolunga	0.57	0.27	0.20	0.14	17.54	1.42
Yackamoorundie	0.64	0.29	0.21	0.14	20.15	1.59
Andrews	0.67	0.31	0.22	0.15	21.35	1.61
Ayers	0.97	0.20	0.09	0.03	17.73	1.29
Kingston	0.96	0.25	0.12	0.05	17.19	1.27
Wokurna	0.44	0.20	0.14	0.10	13.96	1.15
Barunga	0.58	0.28	0.21	0.14	17.73	1.36
Boucaut	0.80	0.17	0.07	0.03	14.90	1.18
Hart	0.73	0.36	0.26	0.19	22.12	1.74
Milne	0.65	0.28	0.20	0.13	21.77	1.28
Hanson	1.09	0.30	0.15	0.06	19.49	1.31
Kooringa	1.03	0.31	0.15	0.06	18.60	1.28
Cameron	0.49	0.22	0.16	0.11	15.73	1.34
Blyth	0.47	0.19	0.13	0.09	16.58	1.21
Clare	0.46	0.36	0.27	0.09	21.09	1.03

* The name of the region is printed in italics.

TABLE 4 (*continued*)

Region* and Hundred	Rate of Increase of Yield (bush./ac./an.) 1900	1911	1921	1941	1941 Yield (bush./ac.)	1941 Yield (bush./ac./in. seasonal rainfall)
Light						
Goyder	0.48	0.24	0.17	0.12	14.50	1.32
Stow	0.55	0.09	0.04	0.01	11.67	0.97
Hall	0.60	0.28	0.20	0.14	18.66	1.39
Upper Wakefield	0.64	0.28	0.20	0.14	20.63	1.21
Stanley	0.70	0.32	0.23	0.16	22.34	1.32
Apoinga	0.97	0.33	0.17	0.07	18.05	1.25
Saddleworth	0.68	0.30	0.22	0.15	21.95	1.40
Waterloo	0.60	0.27	0.19	0.13	19.29	1.19
Balaklava	0.46	0.20	0.15	0.10	14.60	1.16
Dalkey	0.55	0.24	0.17	0.12	17.72	1.37
Alma	0.34				24.09	1.66
Gilbert	0.65	0.30	0.21	0.15	20.68	1.27
Julia Creek	0.51	0.22	0.16	0.10	17.15	1.15
Dublin	0.42	0.18	0.13	0.09	14.00	1.05
Grace	0.51	0.22	0.15	0.10	17.11	1.28
Port Gawler	0.39	0.17	0.12	0.08	13.05	1.01
Mudlawirra	0.30				20.41	1.42
Light	0.54	0.23	0.16	0.11	17.85	1.18
Nuriootpa	0.20				16.56	1.10
Kapunda	0.54	0.24	0.17	0.11	17.57	1.15
Belvidere	0.75	0.16	0.07	0.03	13.64	0.82
Moorooroo	0.20	0.19	0.18	0.17	15.30	0.93
Jellicoe	0.36	0.15	0.10	0.07	12.99	0.89
Yorke						
Tickera	0.49	0.26	0.19	0.14	14.83	1.32
Wiltunga	0.53	0.27	0.20	0.14	16.08	1.24
Kadina	0.52	0.25	0.18	0.13	16.12	1.25
Ninnes	0.59	0.30	0.23	0.16	17.78	1.43
Kulpara	0.56	0.26	0.19	0.13	17.63	1.37
Tiparra	0.68	0.38	0.30	0.22	20.61	1.55
Clinton	0.55	0.28	0.21	0.15	16.56	1.27
Kilkerran	0.98	0.28	0.14	0.05	17.60	1.44
Maitland	0.72	0.51	0.32	−0.06	21.45	1.28
Cunningham	0.63	0.35	0.27	0.20	18.98	1.52
Wauraltee	0.62	0.31	0.24	0.17	18.62	1.34
Muloowurtie	0.46				22.69	1.71
Koolywurtie	0.56	0.26	0.18	0.13	17.75	1.17
Curramulka	0.64	0.48	0.33	0.03	21.12	1.58
Minlacowie	0.44	0.35	0.27	0.11	19.79	1.38
Ramsay	0.52	0.25	0.18	0.13	16.06	1.09
Para Wurlie	0.65	0.48	0.32	0.00	20.11	1.33
Moorowie	0.60	0.29	0.22	0.15	18.20	1.20
Dalrymple	0.37	0.32	0.27	0.18	18.80	1.30
Melville	0.31	0.30	0.29	0.28	19.29	1.37

* The name of the region is printed in italics.

rainfall combination is limited, so that ultimately yield must slowly approach an asymptote. Comparison of the 1941 yields in Table 4, expressed as bushels per acre per inch seasonal rainfall, with the known potentialities of the most recent varieties at that time, which under favourable field conditions would yield 1½ to 2½ bushels per acre per inch, shows clearly that the retardation is due to causes other than those imposed by natural limitations. It will be observed that only 11 hundreds exceed the lower limit of 1½ bushels per acre per inch. Actually the figures given are slightly exaggerated, since no account has been taken of the amount of water conserved by fallowing. This depends on characteristics of the rainfall distribution, physical properties of the soil, and depth of penetration of the plant's root system, and in some places may exceed the equivalent of 4 in. of rain.

With these hundreds also, there can be no doubt that the nitrogen content of the soils is being depleted, as indicated by the progressive retardation in the course of yield, but their originally greater reserves and the regular appearance of improved varieties during the past 20-30 years have been major factors in preventing the occurrence of a decline in many of them. Under the better conditions of rainfall, continual use of superphosphate has encouraged the development of annual legumes, so that, in addition, symbiotic fixation of nitrogen has made appreciable contributions toward the maintenance of the nitrogen supply, but its effects cannot be assessed accurately without further information.

Finally, there are parts of southern Nuyts, Eyre, Pinnaroo, and Tatiara, the principal soils being sandy and loamy mallee, transitional mallee-solonetz, and Wimmera grey loam. In all hundreds the trends are either type VI(10) or type IV(11), these forms being directly attributable to the advances made in the past 40 years. The important point, however, is that the effects are manifested in this manner because a large proportion of the original nitrogen reserves still remains owing to the comparatively recent development of the areas. The Wimmera grey loam is a special case, noted for its high organic matter content, and constituting one of the most fertile wheat soils of Australia. The progress made is remarkable, despite the exploitative nature of the cropping system, and to date there is no conclusive evidence of physical deterioration. Of the remaining three soil types, the only one represented in places where cropping began prior to 1896 is loamy mallee. The marked contrast between the forms of trend in Figures 24 (Narridy), 29, 30 (Curramulka), and Figures 19, 30 (Muloowurtie) is largely an expression of this difference, the longer cropping history of the hundreds in the first group having led to a greater depletion of the nitrogen reserves. The experiences gained on the older areas provide a warning for the four soil types, particularly sandy mallee and transitional mallee-solonetz, as these are more likely to be affected first.

(iv) *Restoration and Maintenance of the Nitrogen Status.*—In these regions, with their comparatively reliable seasonal rains, the economic restoration and maintenance of the nitrogen supply can be effected by lengthening rotations to include several years of forage crops and/or temporary pasture which embody a legume and are capable of supporting maximal numbers of livestock. At the same time, this system periodically rebuilds soil structure.

A classical illustration is provided by the analysis of an accurate record of yield taken near Saddleworth. The property concerned passed into the hands of the present owners in 1897, after having been worked on a fallow-wheat rotation, probably for many years. In 1897 a three course rotation, fallow-wheat-pasture, was adopted and retained until 1924. The pasture phase of this rotation was replaced by oats in 1925, and shortly after, three longer rotations were introduced, namely,

fallow-wheat-oats-fallow-wheat-oats (grazed)-wheat (second fallow replaced occasionally by peas),

fallow-wheat-peas-wheat-oats-pasture,

fallow-wheat-oats-pasture,

in which the pasture contained burr clover as a constituent species. Sheep and cattle have been carried on the property since 1897, and superphosphate was first applied in 1902.

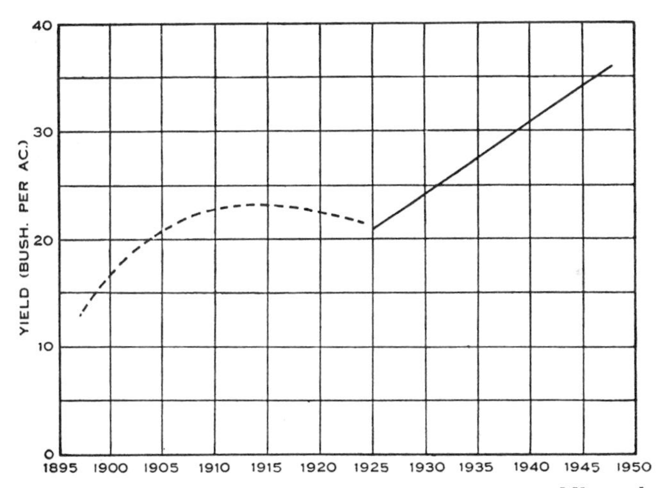

Fig. 37.—Analysis of wheat yield on a property near Saddleworth.

The sequence of yields was broken at 1924, and the two sections 1897-1924 and 1925-1948 analysed separately. Figure 37 illustrates the course of yield after allowance has been made for variations in the seasonal rainfall. The first phase shows a marked increase in yield due mainly to the use of superphosphate and improved varieties such as Federation and Yandilla King, followed by a slight fall over the 10 year period terminating in 1924, while the second shows a linear increase at the rate of $\frac{2}{3}$ bushel per acre per annum.

Cropping prior to 1897 had made phosphorus the limiting factor and reduced nitrogen reserves. As the phosphorus status was built up, consumption of nitrogen increased, and after 1915 it shows signs of becoming the limiting factor. Apart from re-arrangement of the rotations, the management has taken, since 1925, other important steps to improve the standard of farming. The total gain of 15 bushels per acre over the second phase is partly due to these factors, principally higher-yielding varieties, but the increase in fertility attributable to the new cropping systems cannot be doubted.

Owing to the extreme difficulty of securing long and accurate records such as this, it is impossible to obtain confirmation of these results over a range of large-scale field conditions. The only data suitable for analysis are the reported results of rotation trials that have been conducted in Victoria and South Australia.

The observations used by Forster (1939) were derived from experiments laid down by the Department of Agriculture, Victoria, at three widely separated centres, Longerenong, Werribee, and Rutherglen. Rainfall at the experimental sites ranges from 15 to 23 in., and the soils, in the order given, are the typical Wimmera grey loam, a basaltic alluvium, and a grey-red buckshot silty loam, none of which had been seriously depleted by cropping prior to the initiation of the trials. Superphosphate was the only manurial treatment applied to the plots. Yield of wheat from the rotation fallow-wheat, after adjustment for rainfall, was maintained at two centres, and increased at the third — Werribee. On the other hand, oat yields from the rotation fallow-wheat-oats declined markedly and significantly at the three places, indicating a definite reduction in fertility. The fact that wheat showed no evidence of this was probably attributable to the increased amounts of superphosphate and improved varieties that were incorporated from time to time in the trials. By differencing the yields of wheat from rotations involving one year's grazing and fallow only, Forster showed that yield from the former type of rotation was gaining significantly on the latter at Rutherglen and Werribee, but not at Longerenong. This would indicate that fertility was increased at the first two centres under the rotations used, since wheat yields were either maintained or increased at these places in the fallow-wheat rotation.

At Rutherglen, the responses to temporary sown pasture of the rotations was much greater than that obtained at Werribee and Longerenong, where the corresponding phase was a year's ley. Forster stated that other trials at Rutherglen in which pasture was ploughed in have shown a gain in fertility, thus demonstrating that with longer periods of pasture in the rotation, greater improvement in fertility can be expected.

Wark (1942), following the method outlined by Forster, analysed yield data from trials conducted at Booborowie, the Waite Agricultural Research Institute, and Roseworthy, in South Australia. The soil at the first two centres is a red brown earth, and at the third a loamy mallee, and seasonal rainfall is, in the order given, 14.4, 20.1, and 14.6 in. At the Waite Institute the trial was begun on virgin land, but the previous history of the experimental sites at the other two centres is not known definitely, though it is certain that cropping was conducted on each for some years.

The only data available at Roseworthy were yields derived from a fallow-wheat rotation under several manurial treatments, and in each, yield declined significantly, even in the presence of ample supplies of superphosphate.

At Booborowie, wheat yields were maintained in the fallow-wheat rotation, but yields of stubble-sown oats and barley, from fallow-wheat-oats and fallow-wheat-barley rotations, respectively, declined strongly and significantly, showing as with Forster's data, a reduction of fertility. By differencing yields of wheat from rotations incorporating either one year of natural pasture or two years of ryegrass pasture, and from the rotation fallow-wheat, it was found that yield was maintained. Inclusion of one year of pasture also materially reduced the rate at which stubble-sown oats was declining.

With the addition of seven years' observations to the data studied by Wark, the yields from four rotations at the Waite Institute have been re-examined. The period covered by the new analysis is 1925-47 inclusive, and the rotations concerned are

1. fallow-wheat,
2. fallow-wheat-oats,
3. fallow-wheat-barley,
4. fallow-wheat-oats-pasture (Wimmera ryegrass),

in which all crops received a dressing of 187 lb. per acre superphosphate. Wheat yields declined in the first three rotations, and increased in the last, but owing to the excessive annual variation, none of these results was statistically significant, even after making allowance for fluctuations in seasonal rainfall.

When differences between yields of wheat in the series were taken, the following results were obtained:

Difference of Wheat Yield	Regression Coefficient (bush./ac./an.)
fallow-wheat-oats-pasture – fallow-wheat	0.27
fallow-wheat-oats-pasture – fallow-wheat-oats	0.58
fallow-wheat-oats-pasture – fallow-wheat-barley	0.28

but, as before, after adjusting for rainfall, annual variation still masks other effects, and only the second coefficient is significant, though the other two have the sign which was expected. The mean yields of wheat in the four rotations over the period were

Rotation	1	2	3	4
Mean yield (bush./ac.)	36.2	35.8	37.5	44.9

from which the superiority of the four-course rotation is evident, since the 7-9 bushel differences are very significant.

Finally, pending a more detailed analysis, a preliminary examination has been made of a rotational trial embodying a modern experimental design. The five rotations are

1. fallow-wheat,
2. fallow-wheat-peas-pasture (Wimmera ryegrass),
3. fallow-wheat-subterranean clover-pasture (Wimmera ryegrass),
4. fallow-wheat-oats-pasture (Wimmera ryegrass)
5. fallow-wheat-barley-pasture (Wimmera ryegrass).

All crops, including the pasture, received superphosphate at 2 cwt. per acre during 1937-41, but this was reduced to 1 cwt. per acre in subsequent years, since the phosphate supply had been built up during an interval prior to the beginning of the experiment. Half of each plot sown to a forage crop, namely, peas, clover, oats, and barley, was harvested each year and the produce removed, while the other half was grazed *in situ.* For present purposes, it is sufficient to quote the mean yields at the termination of the eighth year. These are set out in Table 5.

TABLE 5

MEAN YIELDS OF ROTATION TRIAL AT WAITE AGRICULTURAL RESEARCH INSTITUTE

Rotation	1	2		3		4		5	
		Grazed	Harvested	Grazed	Harvested	Grazed	Harvested	Grazed	Harvested
Mean yields of wheat (bush./ac.)	31.87	43.49	41.59	37.41	34.32	39.77	39.14	37.74	36.19
		42.54		35.87		39.46		36.96	
Mean yields of forage crop total produce (air-dry) (cwt./ac.)		46.11	39.80	26.04	25.47	19.25	20.85	20.42	22.84
		42.96		25.76		20.05		21.63	
Mean yields of pasture (oven-dry) (cwt./ac.)		51.35	33.43	23.81	19.07	14.44	10.71	16.84	11.50
		42.39		21.44		12.57		14.17	

The yield of wheat from the fallow-wheat rotation is significantly lower than that of all four-course rotations, and there are significant differences among the latter, the most important being that the rotation including peas outyields the remainder. The pea crop is the outstanding forage, with a mean yield significantly greater than clover, oats, and barley. Two dry seasons, 1938 and 1940, greatly reduced the yield of subterranean clover as compared with peas, and this partly accounts for the large difference in the general means, but the mean yield of clover is significantly greater than that of barley or oats. Marked differences in the yield of temporary pasture in the various rotations have been evident throughout, and they have increased with time, all being significant. The effect of grazing is significant in the four cases.

Insufficient time has elapsed to show appreciable effects of the legumes on wheat yields in comparisons among the four-course rotations, but their superiority in other phases is clearly defined. The great advantage of peas compared with clover in this environment is also well established.

The high yields of wheat from the fallow-wheat rotation in this trial, 31.9 bushels per acre, and 36.2 bushels per acre in the one quoted above, are indicative of the ability of this particular soil type to withstand the strain of frequent cropping.

VI. Acknowledgments

The work described in this paper was carried out as part of the research programme of the Section of Mathematical Statistics, C.S.I.R., and arose as the natural outcome of an investigation begun in collaboration with Dr. I. F. Phipps and Mr. A. T. Pugsley at the Waite Agricultural Research Institute, University of Adelaide.

The salient features of the history of the industry have been drawn from "The Centenary History of South Australia," Royal Geographical Society of Australasia, South Australian Branch, in particular Chapter 10, by Dr. A. E. V. Richardson, and acknowledgment is made to the Society for permission to use the block from which Figures 2-7 were prepared.

The author is indebted to the Director of Lands, South Australia, for a map showing the location of the wheat crop; the Chief and officers, Division of Soils, C.S.I.R., for information concerning the soils, and for a map giving the boundaries of the main soil groups; Messrs. F. Coleman and Sons of Saddleworth, South Australia, for permission to use the record of yield from their property; Misses E. M. G. Goodale and P. M. Ohlsson, and Messrs. R. Birtwistle and G. G. Coote, Section of Mathematical Statistics, C.S.I.R., for assistance in the assembly and preparation of the data for analysis; and to Mr. M. C. Childs, Division of Biochemistry and General Nutrition, C.S.I.R., for preparation of the majority of the figures. Finally, special acknowledgment is made to colleagues at the Waite Agricultural Research Institute and the Division of Biochemistry and General Nutrition, C.S.I.R., for placing unpublished data at the author's disposal, and to Mr. R. I. Herriot, Soil Conservator, Department of Agriculture, South Australia, for ready cooperation in making available his unrivalled knowledge of the wheat soils.

VII. References

Agricultural Settlement Committee (1931).—Report. (Govt. Printer: Adelaide.)

Beck, A. B. (1935).—*Aust. J. Exp. Biol. Med. Sci.* **13**: 127-31.

Clarke, G. B., and Marshall, T. J. (1947).—*J. Coun. Sci. Industr. Res. Aust.* **20**: 162-75.

Crocker, R. L. (1946).—Coun. Sci. Industr. Res. Aust. Bull. No. 193, 56 pp.

Fisher, R. A. (1922).—*Philos. Trans.* A **222**: 309-68.

Fisher, R. A. (1924).—Ibid. B **213**: 89-142.

Fisher, R. A. (1946).—"Statistical Methods for Research Workers." 10th Ed. (Oliver and Boyd: Edinburgh.)

Forster, H. C. (1939).—*J. Dep. Agric. Vict.* **37**: 130-3.

Jensen, H. L. (1939).—*J. Aust. Inst. Agric. Sci.* **5**: 154-9.

LEWCOCK, H. K. (1925).—*Aust. J. Exp. Biol. Med. Sci.* **2**: 127-33.
MARGINAL LANDS COMMITTEE (1939).—Report. (Govt. Printer: Adelaide.)
MARSTON, H. R., LEE, H. J., and McDONALD, I. W. (1948*a*).—*J. Agric. Sci.* **38**: 216-21.
MARSTON, H. R., LEE, H. J., and McDONALD, I. W. (1948*b*).—Ibid. **38**: 222-8.
PASTORAL AND MARGINAL AGRICULTURAL AREAS INQUIRY COMMITTEE (1948).—Report. (Govt. Printer: Adelaide.)
PERKINS, A. J. (1934).—Dep. Agric. S. Aust. Bull. No. 298, 24 pp.
PERKINS, A. J. (1936).—*J. Dep. Agric. S. Aust.* **39**: 1199-222.
PIPER, C. S. (1938).—*Trans. Roy. Soc. S. Aust.* **62**: 53-100.
PRESCOTT, J. A. (1933).—Proc. 5th Pan-Pacif. Sci. Congr., pp. 2657-67.
PRESCOTT, J. A., and PIPER, C. S. (1932).—*Trans. Roy. Soc. S. Aust.* **56**: 118-47.
REGIONAL PLANNING COMMITTEE (1946).—Report on Regional Boundaries. (Govt. Printer: Adelaide.)
RICEMAN, D. S. (1946).—Thesis, University of Adelaide.
RICEMAN, D. S. (1948).—Coun. Sci. Industr. Res. Aust. Bull. No. 234, 45 pp.
RICHARDSON, A. E. V. (1936).—"The Centenary History of South Australia." (Roy. Geogr. Soc. A'sia., S. Aust. Branch: Adelaide.)
RICHARDSON, A. E. V., and GURNEY, H. C. (1933).—*Emp. J. Exp. Agric.* **1**: 193-205.
ROYAL COMMISSION ON THE WHEAT, FLOUR, AND BREAD INDUSTRIES (1935).—2nd Report. (Commonw. Govt. Printer: Canberra.)
RURAL RECONSTRUCTION COMMISSION (1944).—3rd Report. (Commonw. Govt. Printer: Canberra.)
SWABY, R. J. (1939).—*Aust. J. Exp. Biol. Med. Sci.* **17**: 401-23.
TRUMBLE, H. C. (1937).—*Trans. Roy. Soc. S. Aust.* **61**: 41-62.
WADHAM, S. M., and WOOD, G. L. (1939).—"Land Utilization in Australia." (Melbourne Univ. Press.)
WARK, D. C. (1942).—*Trans. Roy. Soc. S. Aust.* **66**: 133-41.
WOOD, J. G. (1937).—"The Vegetation of South Australia." (Govt. Printer: Adelaide.)

Reprinted from the Australian Journal of Scientific Research, Series B, Biological Sciences, Vol. 3, No. 2, pp. 178-218, 1950

THE INFLUENCE OF RAINFALL ON THE YIELD OF WHEAT IN SOUTH AUSTRALIA

By E. A. Cornish*

[*Manuscript received December* 20, 1949]

Summary

The relationship between the yield of wheat and seasonal rainfall in South Australia has been determined. The period chosen for examination was 1896-1941 and the analysis extends to practically the entire wheat belt of the State, all major soil groups and variants of climatic conditions within the area being represented. Yield was assessed using the hundred, which has an average area of approximately 118 square miles, as the basic territorial unit. Seasonal weather was represented by the rains in three subdivisions of the growing period of the crop, and these were taken as independent variates in an analysis by multiple regression. The regression of yield on rainfall, when considered as a function of time, takes a mathematically simple parabolic form with a maximum in winter and zeros in autumn and spring or early summer, and the coefficients obtained provide a clear demonstration of the sub-optimal character of average seasonal rainfall over the greater part of the season and almost throughout the wheat belt. The conclusions drawn from the general survey are substantiated by analyses of a limited number of exact records.

I. Introduction

In the Australian wheat belt the large fluctuations in yield are known to be dependent, to a very considerable extent, upon variations in the weather conditions prevailing in different years. Seasonal rainfall and its distribution in time constitute major determinants of yield, and associated factors, such as temperature and evaporation, also play their parts by modifying the duration of the season and determining the effective rainfall available to the crop. From accumulated experience, and recognition of the fact that the effects of any one meteorological element, for example rainfall, must vary throughout the season, the concept of an ideal distribution to meet the exigencies of critical periods during crop growth, has been developed. Apart, however, from such broad generalizations very little is known regarding the quantitative relationships of this phase of agricultural meteorology, notwithstanding the importance of such knowledge from both the scientific and the industrial points of view, including in the latter the direct and indirect value of forecasts in crop insurance and to trade and administration. This almost entire lack of accurate information is due principally to the complexity of the problem involved in the specification of the weather itself and the extreme paucity of reliable data.

Up to the present time the principal studies of Australian conditions have been contributed by Perkins and Spafford (1911, 1915), Richardson (1923),

* Section of Mathematical Statistics, C.S.I.R.O., University of Adelaide.

Perkins (1924), and Barkley (1927); reference is made again to the work of these authors in Section IV. With the advent of more powerful statistical techniques and accumulation of further data, the opportunity has been taken to carry out additional investigations. In a recent paper (Cornish 1949), the temporal trends of wheat yields in South Australia were discussed. The period chosen for examination was 1896-1941 inclusive and the analysis extended to practically the entire wheat belt of the State, all major soil groups and variants of climatic conditions within the area being represented. As a preliminary to the evaluation of the trends it was necessary to estimate and to eliminate the effects of variations in seasonal rainfall on yield, and the paramount importance of this step, with respect to both principle and accuracy, is illustrated by the results established. After making due allowance for changes in the seasons, it was possible to show that, in the maze of unordered data from this extensive and heterogeneous area, yields followed courses which were classifiable into two sequences, each containing a few simple types intimately related to the nitrogen status of the soils. Districts possessing the several types of trend were accurately delineated and conflicting opinions regarding the maintenance of fertility under the cropping systems in use were clarified, particularly for those sections known as marginal lands. As an immediate consequence, it is now possible from consideration of the trends and relevant observations on the soils and rainfall, including reliability of the latter, to decide upon the nature of remedial measures, their urgency and chances of success in individual districts.

In the paper cited no consideration was given to the effects of rainfall and one of the present main objectives is to take up this point. Since, for reasons which will be advanced later, the analysis must be regarded as only a general survey, a limited number of exact records obtained from experimental farms and private individuals have also been examined with a view to providing substantiation of the conclusions drawn from the general study.

II. The General Survey

(a) *Data and Analytical Method*

The data and analytical technique that were employed have already been described (Cornish loc. cit.) so that only the salient features need be repeated here.

(i) *Yield Data.*—The smallest territorial unit for which consecutive yield data are available is the hundred, the mean area of which in South Australia is approximately 76,000 acres, or 118 square miles. This unit was adopted and the yields, expressed as mean yields in bushels per acre *sown for grain* were extracted from the South Australian Statistical Register for all hundreds with an average area of not less than 1,000 acres under crop. The yields used in compiling the records are not estimates but are derived from returns submitted by the growers. Figure 1 shows the distribution, within the regional divisions of the State (Regional Planning Committee 1946), of the 296 hundreds that were examined.

In districts where cropping began prior to 1896, the analysis was confined to the period 1896-1941 inclusive and where it commenced in 1896 or later only those hundreds which by 1941 had records of 20 years or more were selected. These limitations were imposed for three reasons:

1. All records of yield prior to 1896 are incomplete owing to the occurrence of two gaps, totalling seven years, when no returns were taken during 1885-88 and 1893-95, and sections of the records which exist do not give yields in sufficient detail for the purposes of the analysis.
2. In 1942 the Commonwealth Government under war-time legislation, restricted the acreage to be sown to wheat.
3. The minimum of 20 years was chosen so that the trend in yield and the effects of seasonal rainfall could be accurately assessed.

(ii) *Rainfall Data.*—The finest subdivision of the year for which rainfall data were available in a form convenient for immediate use was the calendar month. This unit was taken and the data supplied by the Commonwealth Meteorological Bureau.

As so much was contingent upon making proper allowance for seasonal variations in yield, it was necessary to select the rainfall stations carefully. Owing to the circumstances that the area of each hundred is comparatively large, and that in many instances, particularly those hundreds on the slopes of the Mt. Lofty and Flinders Ranges, there are considerable changes in altitude, the choice of rainfall stations would have been very difficult and of doubtful value, had recourse not been taken to two subsidiary sets of information:

1. A map showing the location of the area under crop in each hundred.
2. A map of isohyets of April-November rainfall.

In the majority of hundreds only one rainfall record was used, particularly in the regions Nuyts, Eyre, Pyap, and Pinnaroo, where rainfall changes comparatively slowly with position; occasionally there were two and more rarely three. In all, some 261 records were used to estimate the effects of seasonal rainfall.

(iii) *The Choice of Rainfall Variates.*—The choice of rainfall variates was made after consideration of the following factors: (1) The nature and paucity of the yield and rainfall data; (2) the average distribution of seasonal rainfall; (3) the form of a previously determined regression function; (4) the period during which rainfall is effective; and (5) the general insignificance of the regressions of yield on November rainfall as determined in a previous analysis; and fell on the rains of the subdivisions:

April and May
June, July, and August
September and October,

as suitable variates in effectively allowing for variations in seasonal conditions.

It is possible that rainfall of the preceding season and of the period December-March immediately prior to seeding are correlated with yield but

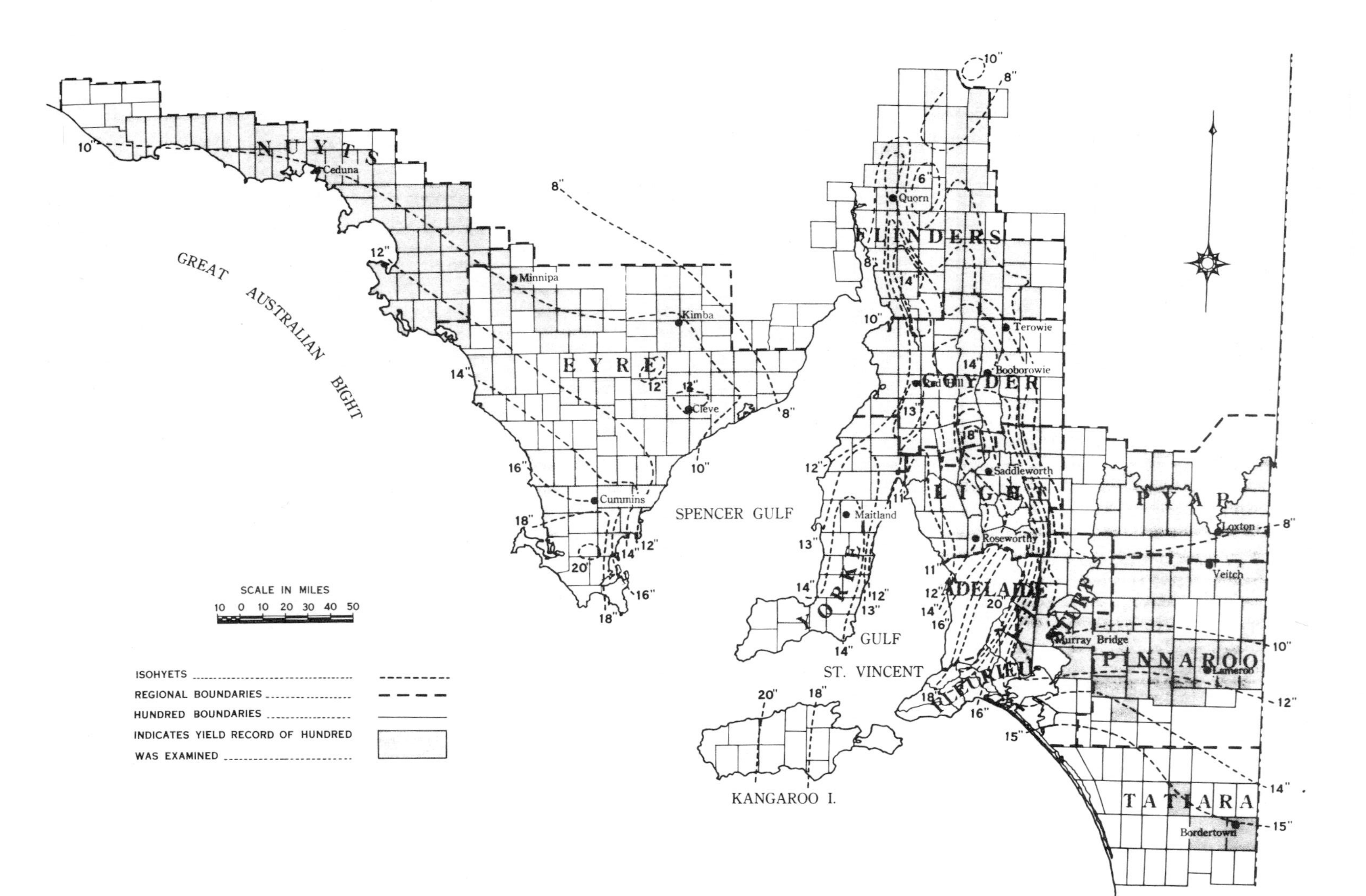

Fig. 1. – The wheat belt of South Australia.

such effects, if they exist, are small compared with those of the current season's rainfall to the end of October, and consequently they have been ignored. In any case the crop records are not of sufficient length to account for them adequately. Finally it is obvious that definite optimal conditions must apply in various parts of the season and the effects of rainfall at all times are not strictly independent and additive. In the present enquiry the quadratic terms corresponding to such factors have also been omitted, since in comparison with the linear terms they are of much less quantitative importance.

(iv) *The Statistical Technique.*—The statistical technique was that of multiple regression. Four different functions of time were employed to represent the temporal trends of yield, and if y denotes yield, x_1, x_2, x_3 the rainfall variates in the order given above, x_4 and x_5 time and its square, the types of multiple regression were

$$Y = \bar{y} + b_1(x_1 - \bar{x}_1) + b_2(x_2 - \bar{x}_2) + b_3(x_3 - \bar{x}_3) + b_4(x_4 - \bar{x}_4) \quad \ldots (1)$$

$$Y = \bar{y} + b_1(x_1 - \bar{x}_1) + b_2(x_2 - \bar{x}_2) + b_3(x_3 - \bar{x}_3) + b_4(x_4 - \bar{x}_4) + b_5(x_5 - \bar{x}_5) \quad \ldots (2)$$

$$Y = b_1(x_1 - \bar{x}_1) + b_2(x_2 - \bar{x}_2) + b_3(x_3 - \bar{x}_3) + b_4 x_4^{b_5} \quad \ldots (3)$$

$$Y = b_1(x_1 - \bar{x}_1) + b_2(x_2 - \bar{x}_2) + b_3(x_3 - \bar{x}_3) + x_4/(b_4 + b_5 x_4), \quad \ldots (4)$$

the bar over a symbol designating the arithmetic mean.

For all cases in which regressions of types (1) and (2) were used, the multiple correlation coefficient was determined and thence the percentage of variance of yield, A, ascribable to the average effects of the rainfall variates and time, from the relation

$$A = 1 - \frac{(n-1)}{(n-p-1)}(1 - R^2),$$

where R is the multiple correlation coefficient, n is the number of observations in the sample, and p the number of independent variates (Fisher 1924). Where regressions of types (3) and (4) were fitted an index of multiple correlation was calculated as the ratio

$$R^2_I = \frac{2SyY - SY^2 - n\bar{y}^2}{S(y - \bar{y})^2},$$

and the percentage of variance from the relation

$$A_I = 1 - \frac{(n-1)}{(n-p-1)}(1 - R^2_I),$$

p now designating the number of coefficients in the regression formula after deducting one for the mean.

(*b*) *Characteristics of the Season*

The most prominent feature of the season is the marked winter incidence of the rains. Approximately 70-80 per cent. of the annual precipitation occurs within the period April-October inclusive, the summer months being characterized by hot, dry atmospheric conditions, low rainfall, and high evaporation. Within the wheat belt, seasonal rains vary over the rather wide range 6 in. to 18 in. and their geographical distribution is illustrated by the isohyets of Figure 1.

Further details are presented in Table 1. Twelve stations, of which the locations are given in Figure 1, have been chosen as representative of the conditions prevailing in different parts of the wheat belt and the quantities tabled all relate to the rainfall variates used in the multiple regression equations. With all three variates the standard deviations per annum, in absolute terms, increase very closely in direct proportion to the means. Relative to the means April-May (breaking) rains and September-October (spring) rains are more variable than June-July-August (winter) rains; the first two show a general tendency to decrease as the mean increases, the effect being slightly more pronounced for April-May rains. Apart from the two most variable stations (Quorn and Loxton), the relative variability of the winter rains is comparatively constant. In all rainfall variates, the annual variation is very great, particularly when it is recalled that the variates are the total rains occurring in periods of 2 or 3 successive months; if shorter periods had been considered the relative variability would have been very much greater.

TABLE 1

CHARACTERISTICS OF SEASONAL RAINFALL

Station	Length of Record (years)	Rainfall Variates (in.)						Mean (Apr.-Oct.) (in.)	Intra-station Correlations		
		x_1		x_2		x_3					
		Mean	Stand. Dev.	Mean	Stand. Dev.	Mean	Stand. Dev.		x_1x_2	x_1x_3	x_2x_3
Ceduna	44	2.09	1.13	4.53	1.57	1.75	0.79	8.37	0.18	0.00	−0.02
Cleve	46	2.59	1.36	5.54	1.66	3.03	1.41	11.16	0.12	0.02	0.11
Minnipa	26	2.03	0.96	5.66	2.10	2.41	1.21	10.10	0.33	−0.03	0.13
Cummins	35	2.94	1.25	8.42	2.71	3.26	1.38	14.62	0.23	0.08	0.10
Terowie	46	1.84	1.09	4.43	1.63	2.55	1.28	8.82	0.07	0.01	0.33
Quorn	46	1.69	1.33	4.61	2.51	2.25	1.30	8.55	0.11	0.07	0.24
Red Hill	46	2.55	1.25	5.70	1.96	2.78	1.35	11.03	0.29	0.16	0.35
Loxton	46	1.58	1.03	3.23	1.43	2.24	1.41	7.05	−0.02	−0.16	0.20
Saddleworth	46	3.75	1.51	7.04	2.25	3.93	1.64	14.72	0.37	−0.01	0.19
Maitland	46	3.85	1.53	7.75	2.67	3.57	1.46	15.17	0.18	0.18	0.22
Murray Bridge	46	2.34	1.17	4.35	1.53	2.71	1.24	9.40	0.17	0.18	0.17
Bordertown	46	3.32	1.40	6.66	2.04	3.80	1.46	13.78	0.10	−0.06	0.21

Table 1 also gives the intra-station correlations between the three variates. Four of the 36 coefficients show evidence of weak covariation but a test of association, derived from the inverse hyperbolic transformations of the correlations, $z = \tanh^{-1} r$, gives $\chi^2 = 49.20$ with 36 degrees of freedom for which $0.05 < P < 0.10$, showing that the rains in the several parts of the season are practically independent, as judged from the data as a whole.

TABLE 2

INTER-STATION CORRELATIONS

	Cleve	Minnipa	Cummins	Terowie	Quorn	Red Hill	Loxton	Saddleworth	Maitland	Murray Bridge	Bordertown
Variate x_1											
Ceduna	0.42	0.72	0.48	0.56	0.62	0.60	0.52	0.54	0.50	0.52	0.32
Cleve		0.69	0.63	0.45	0.42	0.63	0.44	0.45	0.63	0.37	0.47
Minnipa			0.53	0.41	0.50	0.55	0.48	0.64	0.61	0.12 (n.s.)	0.51
Cummins				0.56	0.37	0.55	0.54	0.60	0.67	0.34	0.65
Terowie					0.73	0.74	0.85	0.74	0.57	0.52	0.53
Quorn						0.77	0.67	0.58	0.59	0.58	0.37
Red Hill							0.60	0.69	0.72	0.49	0.46
Loxton								0.67	0.47	0.54	0.63
Saddleworth									0.80	0.53	0.64
Maitland										0.56	0.61
Murray Bridge											0.41
Variate x_2											
Ceduna	0.58	0.74	0.63	0.61	0.52	0.60	0.46	0.57	0.62	0.59	0.51
Cleve		0.80	0.73	0.67	0.55	0.60	0.56	0.65	0.67	0.62	0.66
Minnipa			0.92	0.85	0.81	0.81	0.67	0.79	0.79	0.66	0.69
Cummins				0.85	0.79	0.83	0.71	0.80	0.85	0.74	0.80
Terowie					0.88	0.86	0.74	0.73	0.76	0.76	0.71
Quorn						0.83	0.77	0.68	0.73	0.64	0.62
Red Hill							0.81	0.80	0.88	0.72	0.65
Loxton								0.78	0.74	0.65	0.70
Saddleworth									0.81	0.63	0.79
Maitland										0.66	0.75
Murray Bridge											0.64
Variate x_3											
Ceduna	0.57	0.86	0.46	0.46	0.51	0.44	0.38	0.41	0.41	0.49	0.28 (n.s.)
Cleve		0.83	0.57	0.67	0.66	0.77	0.76	0.77	0.67	0.77	0.55
Minnipa			0.70	0.75	0.67	0.72	0.67	0.67	0.56	0.72	0.54
Cummins				0.65	0.47	0.71	0.52	0.76	0.67	0.68	0.66
Terowie					0.85	0.82	0.76	0.73	0.57	0.69	0.65
Quorn						0.75	0.72	0.54	0.42	0.52	0.42
Red Hill							0.72	0.85	0.79	0.70	0.69
Loxton								0.69	0.54	0.71	0.62
Saddleworth									0.88	0.75	0.83
Maitland										0.67	0.78
Murray Bridge											0.64

All coefficients are statistically significant (minimum odds 19 : 1) except the two indicated.

The inter-station correlations of Table 2 are in marked contrast, and indicate the very widespread nature of the disturbances producing the rains, despite the fact that the correlations decrease as inter-station distance increases. There is thus a definite tendency for the simultaneous occurrence of either very wet or very dry conditions over the whole wheat belt.

(c) *Discussion of Results*

The complete tabulation of the regressions of yield on rainfall has not been included since it is too extensive for reproduction herein; but all regressions are presented in summary tables and some of them graphically. Copies of the original data have, however, been filed and are available for inspection.*

Before proceeding to discussion of the results, it must be emphasized that only one meteorological element has been used to characterize the season, so that, in correlating it with yield, the regression obtained is the resultant of a number of components, one the direct effect of the rainfall *per se*, and the remainder due to the direct and indirect effects of all other elements to the extent to which they are associated with the rainfall occurring in the several subdivisions under consideration.

TABLE 3

DISTRIBUTION OF COEFFICIENTS AND INDICES OF MULTIPLE CORRELATION WITHIN ARBITRARY CLASSES OF SEASONAL RAINFALL

Correlation	Rainfall (in.)								Per Cent. Variance (A)
	<8	8-10	10-12	12-14	14-16	16-18	>18	Total	
$\geqslant 0.91$	7	2	2	3	—	—	—	14	83
0.81-0.90	39	54	44	29	3	—	—	169	70
0.71-0.80	9	34	19	9	9	3	1	84	53
0.61-0.70	—	3	12	4	1	1	—	21	37
0.51-0.60	—	2	1	2	—	1	—	6	25
0.41-0.50	—	—	—	2	—	—	—	2	14
Total	55	95	78	49	13	5	1	296	

Table 3 sets out the distribution of the coefficients and indices of multiple correlation within arbitrary but convenient subdivisions of the range of seasonal rainfall and the mean values of the percentage variance, the value 83 per cent., for example, being the mean of the 14 cases in which the coefficient or index was $\geqslant 0.91$. In 267 hundreds, 50 per cent. or more of the observed variation in yield is expressible in terms of the rainfall variates and time, a fact sufficient to excite remark when it is recalled that all other causes of variation without exception are included in the remaining 50 per cent. or less. The

* At the libraries of the Commonwealth Scientific and Industrial Research Organization, 314 Albert Street, East Melbourne, and its Section of Mathematical Statistics, University of Adelaide.

relation which exists between the measures of correlation and seasonal rainfall reflects largely the strength of the relation between yield and rainfall, since in the majority of hundreds the regression of yield on time is the smaller contributor to the total variation ascribable to the regression formula. Considering the results as a whole, it can be claimed that a rainfall record provides a sufficiently accurate index of seasonal conditions in this environment.

Table 4 lists the distribution of the correlations within the several soil types of the wheat belt.* As rainfall and soil type are to some extent related, these distributions are not completely independent of those in Table 3, but nevertheless they do indicate that the claim made above holds also on all principal soils. The grand total of 335 does not agree with that of Table 3 because 39 hundreds, comprised equally of two soil types, have been included under both in preparing the table.

TABLE 4

DISTRIBUTION OF COEFFICIENTS AND INDICES OF MULTIPLE CORRELATION WITHIN SOIL TYPES

	Soil Type											
	Mallee Types											
Correlation	Sandy	Stony	Loamy	Mixed	Transitional	Calcareous Aeolianite	Desert Loam	Solonetz	Red Brown Earth	Rendzina	Podsol	Total
⩾ 0.91	2	4	3	3	—	—	3	—	2	—	—	17
0.81-0.90	39	10	27	19	6	15	28	1	34	6	1	186
0.71-0.80	16	8	9	5	15	7	15	2	13	2	5	97
0.61-0.70	2	3	4	—	9	1	1	1	1	3	1	26
0.51-0.60	—	1	—	—	2	—	1	1	—	—	1	6
0.41-0.50	—	—	1	—	1	—	—	—	—	1	—	3
Total	59	26	44	27	33	23	48	5	50	12	8	335

This claim is illustrated from a complementary viewpoint by the frequency distributions of the hundreds and significant regressions in the double classification of soil type and rainfall in Table 5, in which allocations to the soil groups are the same as those used in the construction of Table 4. With the exception of the minor soil groups, solonetz, rendzina, and podsol, which represent a very small fraction of the total area, all soils possess high proportions of significant regressions over their respective ranges of rainfall. The coefficients b_2 are outstanding and nearly all are significant. Approximately 50 per cent. of each of b_1 and b_3 reach significance, but with these regressions the proportions fall rather more rapidly than with b_2 as the value of the corresponding rainfall variate increases.

* General descriptions of the soil types were given by the author (Cornish loc. cit.) and greater detail is obtainable from numerous publications of C.S.I.R.O.

Data for the principal soils are presented graphically in Figures 2-6 inclusive.

Red Brown Earths.—The three diagrams of Figure 2 show clearly that the magnitudes of the regressions decrease as the mean values of the rainfall variates increase, and as far as can be judged from the data these relationships are linear. Broadly speaking, hundreds classed in this soil group constitute a

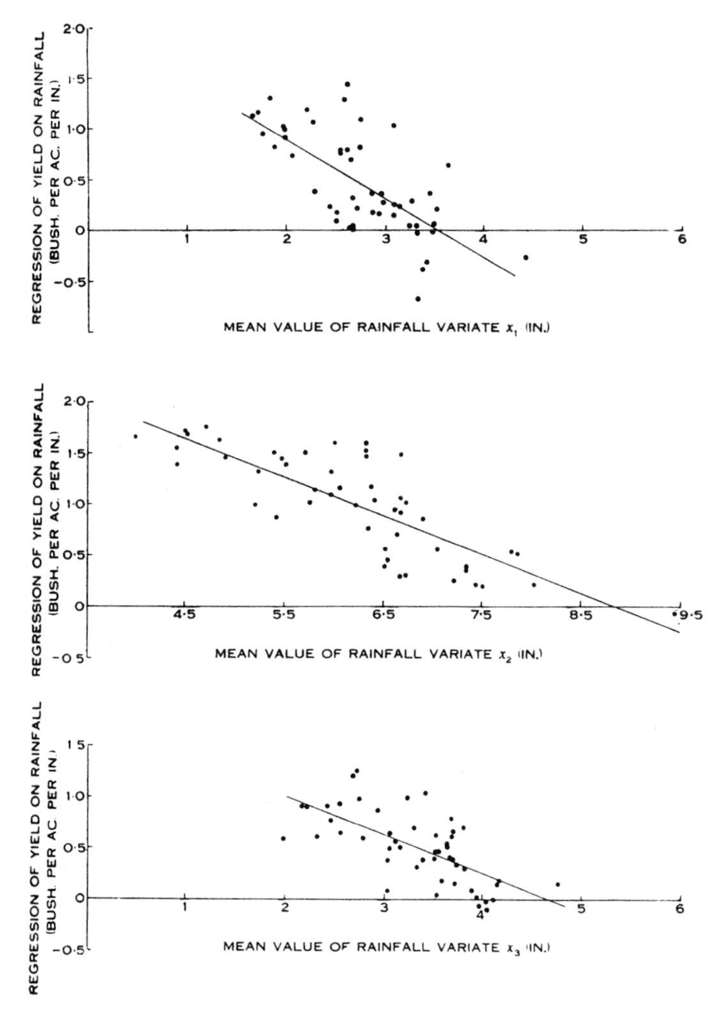

Fig. 2.—Regression coefficients of the red brown earths in relation to the corresponding rainfall variates.

homogeneous set with respect to the cropping methods in use and consequently the gradual fall in the regressions can be taken as evidence that the corresponding rainfall variates are approaching optimal values. Further reference is made to this point below.

Loamy Mallee.—Regression coefficients of the hundreds allocated to this soil type are plotted in Figure 3 and the diagrams indicate an association with rainfall. With respect to b_1 and b_2 in particular the relationships are weakened to some extent by a subdivision of the points which depends partly on a climatic differentiation and partly on a difference in standards of agriculture, according to regional location of the hundreds in (*a*) Eyre, and (*b*) Goyder, Yorke, and Light. To obtain rough approximations to the optimal rains, heterogeneity due to different standards and climate have been ignored and the lines have been fitted using each complete set of observations.

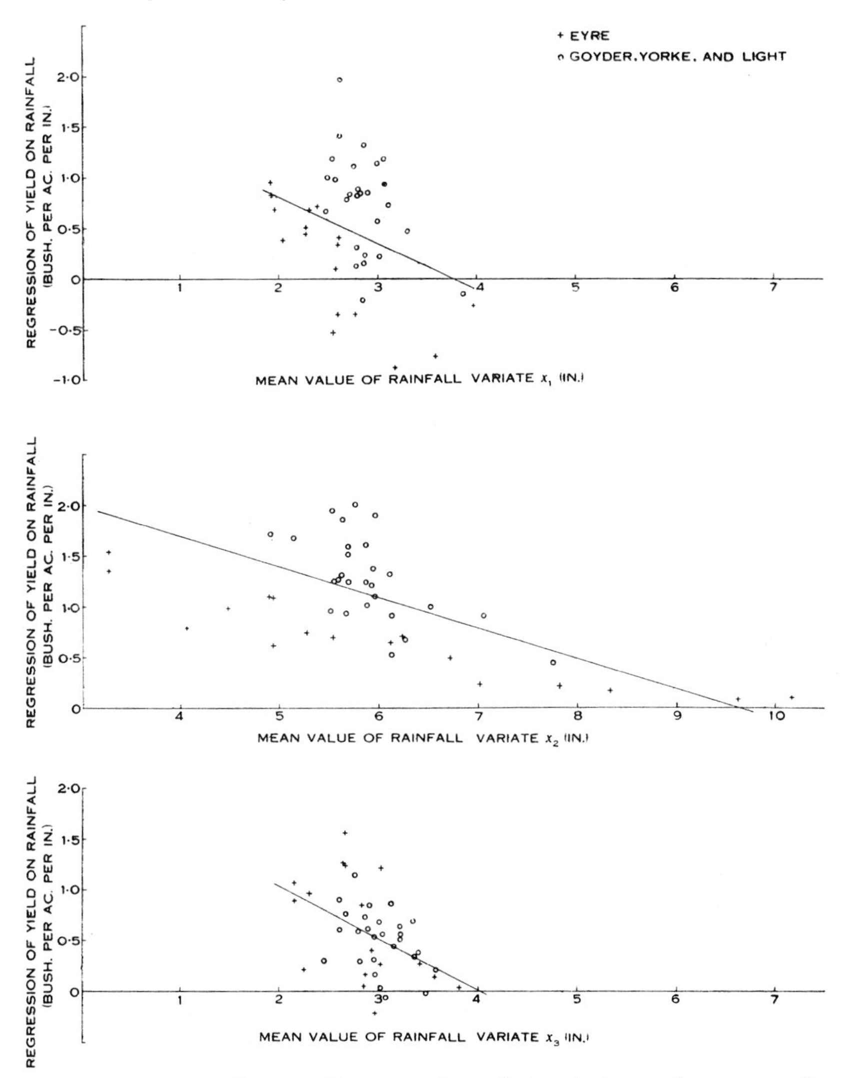

Fig. 3.—Regression coefficients of loamy mallee soils in relation to the corresponding rainfall variates.

Sandy Mallee.—Figure 4 shows the scatter diagrams of this soil type. The coefficients fall into two fairly distinct groups, the differentiation again being clearer for b_1 and b_2 than for b_3. In this case the regional subdivisions are (*a*) Nuyts and Eyre, and (*b*) Flinders, Goyder, Yorke, Light, Pyap, and Pinnaroo. There is no evidence that either b_1 or b_3 decreases as the corresponding rainfall variate increases; b_2 falls slightly but consideration of the data indicates that this is not due to the fact that rainfall is approaching an optimal value. Nearly all coefficients with values less than 1½ bushels per acre per inch and lying to the right of the diagram are from hundreds which possess low values of b_1.

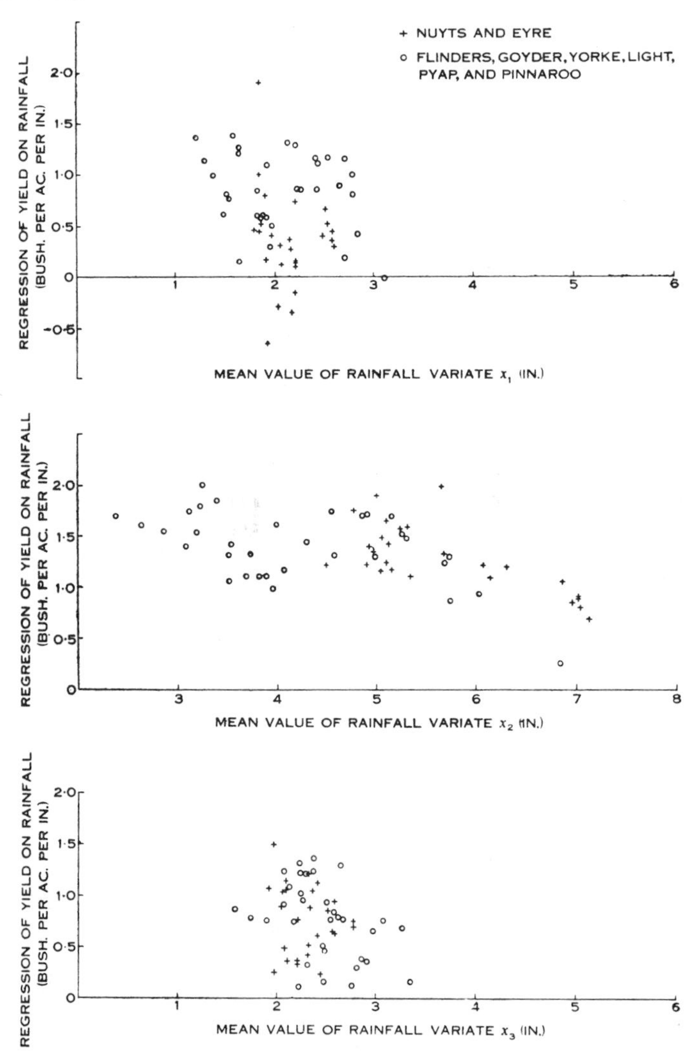

Fig. 4.—Regression coefficients of sandy mallee soils in relation to the corresponding rainfall variates.

Stony Mallee and Mixed Mallee.—In these two types the diagrams resemble those of the sandy mallee soils, but dispersion of the points is not so great and there is no evidence of grouping according to regional location of the hundreds.

Transitional Mallee-Solonetz.—This group occurs in two widely separated regions, Pinnaroo and Eyre, and it is clear in Figure 5 that no differentiation exists on this basis. The regressions are not related to rainfall.

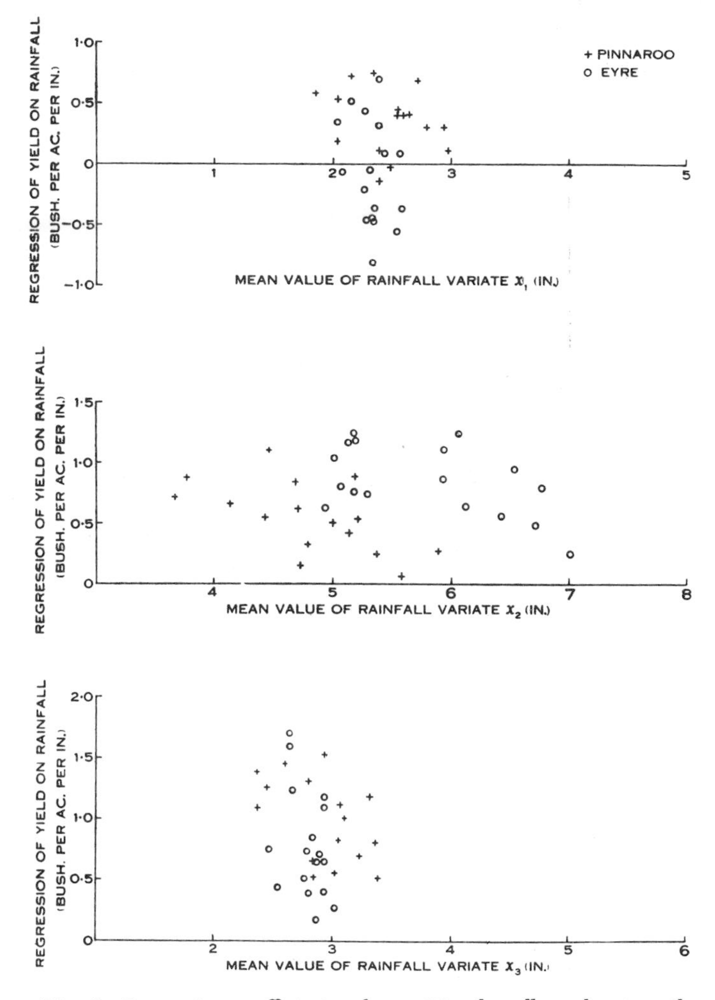

Fig. 5.—Regression coefficients of transitional mallee-solonetz soils in relation to the corresponding rainfall variates.

Calcareous Aeolianite.—The diagrams are similar to those in Figure 5.

Desert Loams.—Figure 6 illustrates the position in this soil type. The grouping of the coefficients according to location of the hundreds in (*a*) Nuyts, and (*b*) Flinders and Eyre is very distinct, but there is no sign that the rains approach optimal values.

The data of the red brown earths and loamy mallee soils have been fitted with weighted regression lines but it has been assumed that the coefficients within each set are independent. Obviously this condition is not fulfilled. Each set of coefficients is distributed approximately in a multivariate normal distribution with an unknown matrix of variances and covariances. As, however, the principal object of the analysis was to make an empirical fit of the lines, tests of significance being of secondary interest only, statistical dependence of the data is of no great consequence.

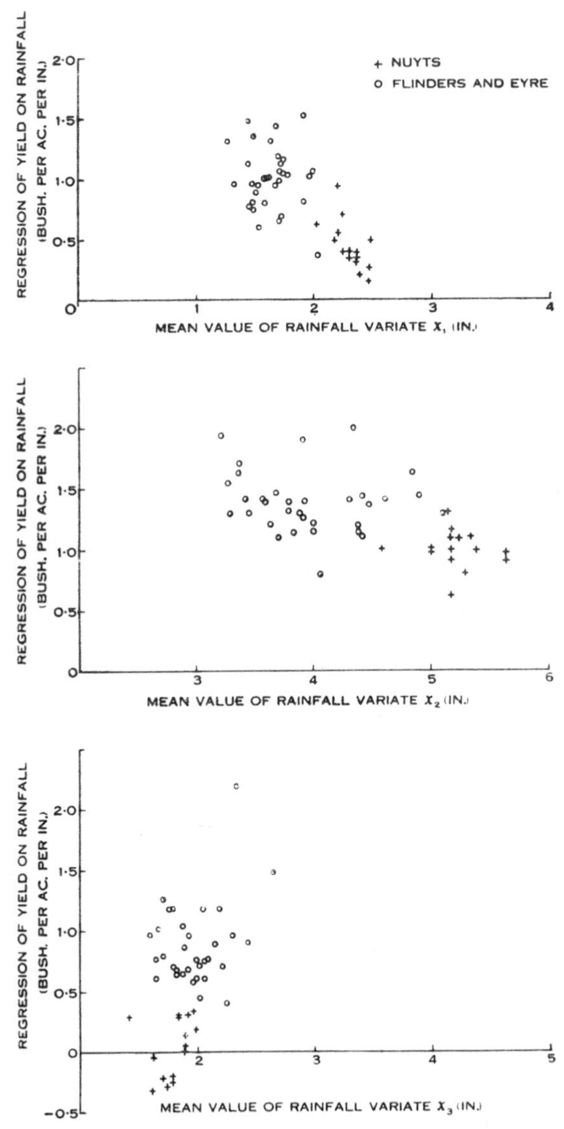

Fig. 6.—Regression coefficients of the desert loams in relation to the corresponding rainfall variates.

The frequency distributions of the significant regressions in Table 5 and the evident preponderance of high positive values, as indicated in the figures, provide a clear demonstration of the sub-optimal character of average seasonal rainfall over the greater part of the season and almost throughout the wheat belt.

Table 5

FREQUENCY DISTRIBUTIONS OF SIGNIFICANT REGRESSION COEFFICIENTS WITHIN SOIL TYPE — RAINFALL CLASSES

		Soil Type: Mallee Types								Soil Type														
Coefficient	Rainfall (in.)	Sandy		Stony		Loamy		Mixed		Transitional		Calcareous Aeolianite		Desert Loam		Solonetz		Red Brown Earth		Rendzina		Podsol		Total	
b_1	1-2	27	*13*	11	*11*	3	*2*	15	*15*	1	—	5	*1*	32	*31*	1	—	8	*8*	—	—	—	—	103	*81*
	2-3	31	*12*	15	*5*	32	*15*	12	*4*	32	*2*	18	*10*	16	*2*	—	—	25	*9*	7	*2*	1	—	189	*61*
	3-4	1	—	—	—	9	*4*	—	—	—	—	—	—	—	—	4	—	16	*1*	5	—	6	*3*	41	*8*
	4-5	—	—	—	—	—	—	—	—	—	—	—	—	—	—	—	—	1	—	—	—	1	—	2	—
Total		59	*25*	26	*16*	44	*21*	27	*19*	33	*2*	23	*11*	48	*33*	5	—	50	*18*	12	*2*	8	*3*	335	*150*
b_2	2-3	3	*3*	2	*2*	—	—	2	*2*	—	—	—	—	—	—	—	—	—	—	—	—	—	—	7	*7*
	3-4	15	*15*	8	*8*	2	*2*	5	*5*	2	*2*	—	—	21	*21*	1	*1*	1	*1*	—	—	—	—	55	*55*
	4-5	12	*12*	6	*6*	6	*6*	14	*14*	9	*7*	8	*8*	14	*14*	—	—	7	*7*	—	—	—	—	76	*74*
	5-6	18	*18*	4	*3*	21	*21*	5	*5*	15	*12*	14	*14*	13	*13*	1	—	12	*12*	—	—	—	—	103	*98*
	6-7	7	*6*	5	*4*	8	*8*	1	*1*	6	*5*	—	—	—	—	3	—	20	*17*	5	*2*	—	—	55	*43*
	7-8	4	*4*	1	*1*	4	*2*	—	—	1	—	1	*1*	—	—	—	—	8	*4*	5	*4*	1	—	25	*16*
	8-9	—	—	—	—	1	—	—	—	—	—	—	—	—	—	—	—	1	—	2	—	1	—	5	—
	9-10	—	—	—	—	1	—	—	—	—	—	—	—	—	—	—	—	1	—	—	—	4	—	6	—
	10-11	—	—	—	—	1	—	—	—	—	—	—	—	—	—	—	—	—	—	—	—	2	—	3	—
Total		59	*58*	26	*24*	44	*39*	27	*27*	33	*26*	23	*23*	48	*48*	5	*1*	50	*41*	12	*6*	8	—	335	*293*
b_3	1-2	6	*4*	3	*3*	—	—	2	*2*	—	—	13	*3*	34	*18*	—	—	1	—	—	—	—	—	59	*30*
	2-3	50	*29*	18	*12*	26	*11*	21	*18*	24	*14*	10	*7*	14	*11*	1	*1*	12	*9*	6	*1*	—	—	182	*113*
	3-4	3	*2*	5	*2*	18	*3*	4	*2*	9	*8*	—	—	—	—	4	—	31	*7*	6	—	8	—	88	*24*
	4-5	—	—	—	—	—	—	—	—	—	—	—	—	—	—	—	—	6	—	—	—	—	—	6	—
Total		59	*35*	26	*17*	44	*14*	27	*22*	33	*22*	23	*10*	48	*29*	5	*1*	50	*16*	12	*1*	8	—	335	*167*

Frequencies of significant coefficients are printed in italics.

The diagrams also show that the response to rainfall varies among the soil types as well as within the types to which reference has been made, but before making further comment it must be emphasized that this study is, by nature, only a broad survey. Obviously with respect to any particular hundred there is on the one hand a set of factors tending to increase the responses to rainfall and on the other a set tending to decrease them. The observed regression on any rainfall variate is the resultant of their effects, and because of the nature of the data they are completely confounded and incapable of separate assessment; in fact, it is not possible to enumerate either set exhaustively. In consequence the labour involved in providing greater detail than that presented by a series of mean values is not warranted, and it would be unreasonable to give more than a few general observations. It may be anticipated, however, that the accuracy of comparisons between the means will be enhanced since the regressions have been derived either from the data of identical sequences of seasons or from series having a large proportion of years in common.

TABLE 6

STANDARDIZED REGRESSIONS AND STANDARD DEVIATIONS OF RAINFALL VARIATES AND YIELD

Soil Type	b'_1	s_1	b'_2	s_2	b'_3	s_3	s_y
Sandy mallee (*a*)	0.47	1.10	1.12	1.64	0.52	1.29	2.04
(*b*)	0.15	1.09	1.07	2.05	0.39	1.21	2.45
Stony mallee	0.40	1.06	0.97	1.67	0.48	1.25	1.87
Loamy mallee (*a*)	0.35	1.30	0.92	1.99	0.24	1.32	2.72
(*b*)	0.07	1.26	0.47	2.07	0.30	1.29	2.53
Mixed mallee	0.37	1.10	0.96	1.68	0.58	1.28	2.27
Transitional mallee-solonetz	0.08	1.14	0.57	1.89	0.55	1.33	2.20
Calcareous aeolianite	0.32	1.08	1.13	1.85	0.32	0.96	1.96
Desert loam (*a*)	0.56	1.22	1.20	1.95	0.47	1.09	2.15
(*b*)	0.25	1.26	1.00	1.97	0.03	0.90	1.99
Red brown earth	0.21	1.37	0.70	2.16	0.26	1.53	3.07

The subdivisions of sandy mallee, loamy mallee, and desert loam correspond to those given in the text.

In absolute terms, the residual standard deviation of yield and the standard deviations of the rainfall variates vary considerably, and consequently to render the regression coefficients directly comparable, it is necessary to standardize them by multiplying each by the ratio

$$\frac{\text{standard deviation of rainfall}}{\text{standard deviation of yield}}.$$

Average values of these standardized regressions, denoted by b'_i, are set out in Table 6 together with the corresponding average standard deviations of yield, s_y, and the rainfall variates, s_i, to facilitate reconversion of the regressions to their original dimensions. The figures given are averages of all hundreds classified in the several soil types listed in the table.

With respect to any particular regression the contrasts are now substantially reduced, but several still persist. Transitional mallee-solonetz behaves differently from all other soils and the differences between the subdivisions of the sandy mallee and desert loam types stand out clearly. Seasonal rains vary in the neighbourhood of optimal values throughout considerable areas of the red brown earths and loamy mallee soils and consequently the means of these two groups are lower than those of many of the remaining soil classes. Within the latter set, the means of all three coefficients are lower than might be expected on the basis of a comparison with the corresponding quantities from those parts of the red brown earths and loamy mallees with a comparable rainfall, the differences observed being attributable to a number of factors which may be integrated appropriately in the term "standard of wheat farming." In varying degree, the yields of all mallee districts have been affected by the lower standard of the practices adopted. The majority of hundreds in these areas were opened for cultivation after 1896, and consequently the pioneering phase represents an increasingly larger proportion of the period under review as the date of settlement advances from 1896. Cropping methods in the early stages are exploitative through circumstances incidental to pioneering on land carrying mallee vegetation. Mallee stumps remaining in the ground after the first clearing repeatedly send out new growth and the settler crops continuously for some years, burning the stubbles to destroy the shoots. During this phase, which may take up to 10 years to complete, the principal factor limiting yield, apart from rainfall, is probably a deficiency of nitrogen. In the course of time, yields improve as fallowing and other measures for increasing production are introduced, but the methods of cropping employed do not reach the same standard as in districts with a rather more reliable rainfall. Under such conditions the response of the crop to rainfall is naturally not as great as it would otherwise be.

The most significant observation is, however, that the figures quoted establish the greater effectiveness of winter rains as compared with autumn and spring rains; on all soils, apart from transitional mallee-solonetz, b'_2 is 2-2½ times as great as either b'_1 or b'_3. In the past considerable emphasis has been laid upon the importance of spring rains, but it is clear from the sign, magnitude, and duration of the response that the ultimate success of the harvest depends upon favourable rains throughout the season, and in particular during the winter months. Seeding operations, when properly conducted, commence after the opening rains and range from mid-April to the end of June, depending upon date of break in the season and location within the wheat belt. The mean time of seeding can be taken as the last week of May and it is significant that rain above average for some weeks prior to sowing should have a marked influence on yield. Early rains facilitate preparations for seeding and coincide with soil temperatures that are still high enough to promote germination and early development, two factors of material value in a growing season of short duration. The characteristic feature of autumnal weather over a large proportion of the South Australian wheat belt is that suitable conditions for seeding operations, rapid germination, and early growth, are confined to a very limited period. Earlier seeding is hazardous and later seeding almost invariably leads to unsatisfactory development which can only be successful in the event of a protracted and otherwise favourable season.

As time advances from April 1, average effective rainfall increases as a result of the combined action of increasing rainfall and associated atmospheric conditions of rising humidity, decreasing temperature, and decreasing solar radiation. The increasing response to winter rains is due principally to an indirect effect following upon accumulation of water reserves for use in later stages of the life cycle, but there exists also a direct effect in stimulating surface growth since temperatures do not fall to the point where development is arrested completely.

From mid-August, growth and transpiration accelerate rapidly as each atmospheric condition traverses the second part of the course in its periodic movement. The fall in response to rainfall in the latter portion of the season can reasonably be ascribed to the interplay of several factors. In the first place, average effective rainfall is rapidly reduced as the season advances and the regression shows, in reality, the benefit from only the available moisture. Secondly, rain during September, October, and November is often preceded by strong north winds which directly damage the crop through shattering, lodging, etc., and finally, if suitable conditions prevail after rain, they are conducive to the development of disease, for example rust.

The regularity and simplicity of the changes in magnitude of the coefficients, taken in their natural order, suggest that the regression of yield on rainfall, considered as a function of time, assumes a form of simple character, which could, in fact, be represented accurately by a parabola having a maximum in winter and zeros in autumn and spring or early summer. Later sections will demonstrate that this actually holds true over a large proportion of the wheat belt. The shape of the curve would be expected to vary with soil, standard of cropping, stage of development, and so on but the effects of these factors are subordinate to the dominating influence of rainfall. In regions of lowest rainfall, the curve has its greatest maximum, but as the mean rainfall increases, the winter maximum is reduced in value, accompanied by a general flattening of the curve. An extremely simplified diagrammatic representation is given in Figure 7, where the regression functions, in order, correspond to districts with increasing mean rainfall.

The change in the form of these curves, as the mean rainfall increases, is readily explicable when consideration is given to the concept of optimal rains. Thus, corresponding to any particular time t during the season, there will be an optimal rainfall, and the relation between yield and rainfall at t would be expected to follow a curve rising to, and falling from, the yield at the optimum, in a manner resembling that given in Figure 8; as drawn the curve is a parabola but there are no theoretical grounds to suggest that the true relationship would possess this particular mathematical form. The four points marked on the abscissa in Figure 8 correspond to the mean rainfall at t in the four districts of Figure 7 and constitute a sequence of values approaching optimal rain. Tangents at the points where the four ordinates intersect the curve represent a series of diminishing positive regression coefficients corresponding to the ordinates at t of the four curves in Figure 7.

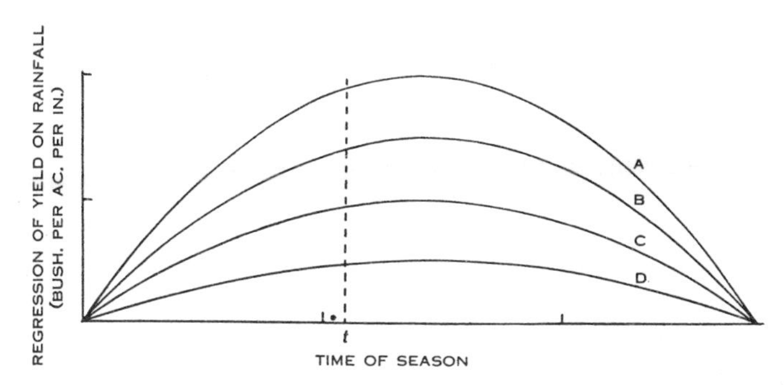

Fig. 7.—Diagrammatic representation of regression functions in four districts, A, B, C, and D in order of increasing rainfall.

It was expected originally that direct evidence of well-defined optimal conditions would be obtained for individual hundreds throughout a substantial proportion of the wheat belt, but as mentioned previously, an exhaustive examination failed to reveal, in any instance, corresponding quadratic terms of quantitative importance, the greater part of the effect of rainfall being expressible

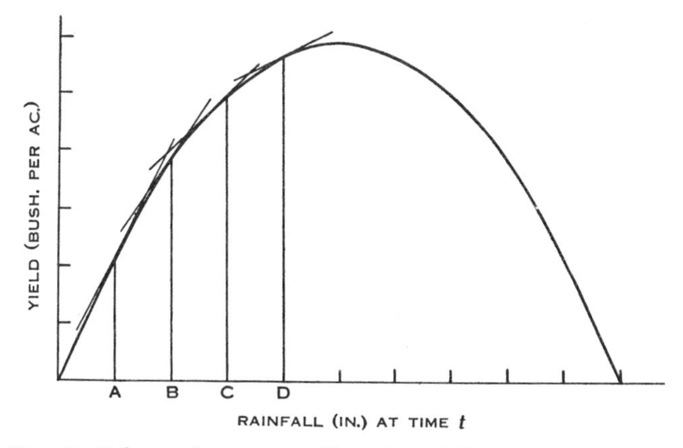

Fig. 8.—Relation between yield and rainfall at time t, showing regression of yield on rainfall at t, in the four districts of Figure 7.

in terms of linear relations in the variates employed herein. The explanation of this feature lies simply in the fact that the rains actually received in the majority of seasons in the period under review are distributed over a range with an upper limit which is in defect of the optimum.

It is, however, possible with the red brown earths and loamy mallee types to make an indirect assessment of optimal rains by taking in Figures 2 and 3 the values where the fitted regression lines intersect the axes of the rainfall variates, and the relationship between yield and rainfall may be inferred from the solution of the elementary differential equation

$$dy/dx = a + bx,$$

where y denotes yield, x any rainfall variate, and a and b the constants of the appropriate regression line in the figures. Integration gives

$$y = k + ax + \tfrac{1}{2}bx^2,$$

and since the regressions are approximately linear, a parabola describes, with sufficient accuracy, the relation between yield and rainfall for the three subdivisions of the season used in the analysis. This convenient procedure has been adopted since it does not seem justifiable to press refinement too far owing to the nature of the data and the assumptions that have already been made. The equations to the curves for the red brown earths are as follows:

Rains of the Period	Equation	Optimal Rain (in.)
April-May	$y = k + 2.07x_1 - 0.29x_1^2$	3.6
June-July-August	$y = k + 3.39x_2 - 0.19x_2^2$	8.8
September-October	$y = k + 1.77x_3 - 0.19x_3^2$	4.7

so that optimal seasonal rains on this soil type amount to approximately 17 in. appropriately distributed. Ignoring regional differentiation, loamy mallee soils give the following results:

April-May	$y = k + 1.73x_1 - 0.23x_1^2$	3.8
June-July-August	$y = k + 2.90x_2 - 0.15x_2^2$	9.6
September-October	$y = k + 2.06x_3 - 0.26x_3^2$	4.0

yielding an optimal seasonal rain of nearly 17½ in., which, considering the crude nature of the estimation, agrees reasonably well with what would be expected in comparison with the red brown earths. The constant of integration must, of course, remain unknown in the absence of any knowledge regarding boundary conditions.

III. Examination of Individual Records

It is clear that the regressions for the yields of whole hundreds have been calculated on an extremely broad basis and for this reason an attempt has been made to substantiate the general results by examination of data obtained under more uniform conditions. Owing to the extreme difficulty of securing long and accurate records only a limited number were available; the locations of the places whence they were derived are given in Figure 1.

(a) *Roseworthy*

(i) *Yield and Rainfall Data.*—The yield data were taken from the records of the so-called permanent manurial experiment on wheat at Roseworthy Agricultural College. This trial was initiated in 1905 and designed on similar lines to the classical experiment on Broadbalk at Rothamsted, but differed from the latter in two important features. At Rothamsted the manurial treatments were applied to single plots, but at Roseworthy the treatments were applied to pairs of plots, which for several treatments were replicated. Previous experience had demonstrated that it was essential to fallow and in order to accomplish this and simultaneously maintain the continuity of the record, each treatment was applied to two adjacent plots which were alternately cropped and fallowed. The trial was continued until 1930, by which time, notwithstanding the fallowing, the yields had declined to extremely low values. In 1931 all plots were fallowed but poor yields were obtained again in 1932, and for this and other reasons the trial was terminated.

In the original plan the experimental field comprised 122 acres divided into 61 two-acre plots. In 1911 a further 10 plots were added, making the area 142 acres. Over the major portion of the field the soil, which originally carried scattered groups of Murray box (*Eucalyptus largiflorens*), consists of a fairly uniform medium to heavy chocolate loam overlying a yellow-red clay subsoil and may be regarded as typically good wheat land. The only irregularity of note is due to small areas of travertine limestone with their covering of loose, light soil, characteristic of mallee country; soils of this type carry mallee (*E. dumosa* and *E. oleosa*) in their natural state. Each plot consisted of a narrow strip about 4/5 chain in breadtn and some 24-25 chains long. In 1905 the plot area was 1.97 acres but in subsequent years this was slightly reduced.

Owing to the numerous modifications and changes made in the original plan, the yields from only eight pairs of plots for the 26 seasons 1905-30 inclusive are suitable for analysis but two of these have been omitted since the manurial treatments are not now of any practical importance. The numbers and manurial treatments of the selected plots are given below:

Plot Number	Treatment
26 and 27	No manure
52 and 53	No manure
28 and 29	Superphosphate 2 cwt./ac.
58 and 59	Superphosphate 2 cwt./ac.
30 and 31	Superphosphate 2 cwt./ac., sodium nitrate 1 cwt./ac. applied at seeding
32 and 33	Superphosphate 2 cwt./ac., sodium nitrate 1 cwt./ac. applied in spring.

In 1909 sodium nitrate was not applied to plots 32-33 and the recorded changes in variety are as follows:

1905 Gluyas, 1906 King's Early, 1907 Gluyas, 1908 King's Early, 1909-22 King's White, 1924-27 Gluyas, 1928 Ford, 1929-30 Gluford.

The rainfall data were taken from the meteorological records at the College; rainfall was measured with a standard 8 in. gauge on a site approximately one mile from the experimental plots.

In 1923 excessive autumnal and early winter rains made normal seeding operations impracticable with the result that no crop yield is available for this year. Allowance for this fact has been made in the correlational analysis.

(ii) *Preliminary Analysis.*—It has been pointed out that yields were obtained from adjacent plots sown in alternate years. Inspection of the records showed clearly that the mean yields of the members of each selected pair were not significantly different but it was not obvious that the trends of these yields also differed only by errors of random sampling. Before proceeding with the main analysis it was deemed advisable, therefore, to subject the data to preliminary tests in order to assess the effect of the alternate fallows on the continuity of the records. The series of yields involving plots scheduled for sowing in 1923 were completed by inserting the average of the three preceding and three succeeding yields and the linear regression on time of each individual sequence determined. In all cases, the mean yields, linear regressions, and residual variances were not significantly different and consequently it was considered legitimate to combine the two series to form a single continuous record for each pair of plots.

(iii) *Analysis of Yield Data.*—On the assumption that the succession of seasons had been fortuitous, and this is justifiable, the linear regression of yield on time was determined and the total variation in yield partitioned into two components, the first ascribable to the regression and the second to annual fluctuations of yield about the fitted line.

Large and significant declines in yield were observed under all manurial treatments, and from the evidence available, they may be attributed primarily to increasing infestation of the plots with weeds, particularly wild oats (*Avena fatua*). Since the six unmanured plots averaged approximately 20 bushels per acre in 1905, there seems no doubt that, from the standpoint of local practice at the time, the experiment was initiated at a stage where the field was in a high state of fertility and it is possible that the fall in yield was due in part to depletion of this fertility, but it is certain that this factor is negligible in comparison with the predominant damaging effect of the weeds.

(iv) *Correlation of Yield and Rainfall.*—Fluctuations in yield about the course of the decline may be attributed to random variations in the seasons, and for the purpose of correlating them with the weather, the method developed by Fisher (loc. cit.) has been followed. Seasonal conditions have been represented conventionally by rainfall of the period March 1 to December 14, an interval of sufficient length to embrace all seasons in which rainfall is effective for some weeks prior to the average date of seeding (approximately May 30) and all seasons in which harvesting was delayed until early December.

The rainfall of this period in each year was divided into 48 six-day totals and to each set of values was fitted a series of orthogonal polynomial functions of the third degree in time, thus furnishing four constants, denoted by a', b', c', and d' with which to express the quantity and distribution of rainfall in each season. As the linear component of secular change had been eliminated from each sequence of yields it was necessary to treat the rainfall distribution values in a similar manner. Actually each series was fitted initially with a curve of the second degree in time. This precaution was taken because a previous analysis had demonstrated the presence of well-defined changes in the distribution of seasonal rainfall at Adelaide, a neighbouring observing station. The parabolic terms were, however, all insignificant and were not retained. In making allowance for the fact that no yields were available in 1923, the polynomial value of each rainfall distribution constant was calculated for that year and the difference, observed value – polynomial value, found in each case.

The quantities a', b', c', and d', which are proportional to the polynomial terms, were taken as independent variates in terms of which yields were to be expressed in the form of a multiple regression relationship. The standard technique is used to determine the partial regression coefficients, but before proceeding to the solution of the normal equations, the matrix of sums of squares and products of the independent variates must be corrected by making allowance for secular trend and omission of the year 1923.

With rainfall measured in the unit 10^{-3} in., the matrix of corrected sums of squares and products is

	a'	b'	c'	d'
a'	171396	– 4398	– 14289	– 7934
b'	– 4398	22179	– 691	– 282
c'	– 14289	– 691	12423	– 5579
d'	– 7934	– 282	– 5579	11858

and the inverse of this matrix is

7.618731	2.103194	14.189972	11.823744
2.103194	45.823220	7.720513	6.129333
14.189972	7.720513	128.808584	70.280276
11.823744	6.129333	70.280276	125.453847

each value being multiplied by 10^{-6}.

TABLE 7

CORRECTED SUMS OF PRODUCTS (YIELD AND RAINFALL DISTRIBUTION CONSTANTS)

Dimensions (bush.in.)

Plots	a'	b'	c'	d'
26-27	7.31530	– 0.63527	– 1.97764	0.71288
52-53	6.41082	– 0.89363	– 1.65307	0.45876
28-29	9.42050	– 0.42372	– 2.17046	0.75947
58-59	4.79836	– 0.34972	– 2.07570	1.16497
30-31	9.87960	– 0.17377	– 2.54519	1.07318
32-33	9.91132	0.00528	– 2.78624	1.22241

After correction for secular trend and omission of the data for 1923, the sums of products of yield and the rainfall distribution constants are given in Table 7, yield being measured in bushels per acre and rainfall in inches, and with the same units the partial regressions of yield on the rainfall variates are as given in Table 8.

It is of interest at this stage to calculate how much of the annual variability in yield is expressible in terms of the quantity and distribution of seasonal rainfall. For example, with respect to plots 26-27 the sum of squares ascribable to the regression on all four variates is 502.61, but if a' had been the only measure of rainfall, the regression of yield on it would have been

$$7.31530/0.171396 = 42.68 \text{ bushels per acre per inch}$$

and the sum of squares due to this regression

$$7.31530^2/0.171396 = 312.22.$$

TABLE 8

PARTIAL REGRESSIONS OF YIELD ON RAINFALL DISTRIBUTION CONSTANTS

Dimensions (bush.in.$^{-1}$)

Plots	a'	b'	c'	d'
26-27	34.7635	− 24.6234	− 105.7361	33.0453
52-53	28.9300	− 37.4166	− 96.6184	11.6969
28-29	49.0620	− 11.7053	− 95.7936	51.5256
58-59	20.1420	− 14.8184	− 120.1057	54.8595
30-31	51.4774	− 0.2562	− 113.5696	71.5066
32-33	50.4396	7.0685	− 132.2979	74.7594

The difference between these two sums of squares is the amount attributable to the regression on the distribution constants b', c', and d'. This series of calculations is summarized in Table 9 in an analysis of the annual variation of yield, plots 26-27.

TABLE 9

ANALYSIS OF VARIANCE (ANNUAL VARIATION IN YIELD, PLOTS 26-27)

Variation Due to	Deg. Freed.	Sum Sq.	Mean Sq.	Variance Ratio
Regression on a', b', c', d'	4	502.61	125.65	14.87
Regression on a' alone	1	312.22	312.22	36.95
Difference (Regression on b', c', d')	3	190.39	63.46	7.51
Deviations from regression	19	160.55	8.45	
Total	23	663.16		

The regression on a' alone is highly significant and thus gives decisive information regarding the effect of quantity of rain, namely that the yield is increased on the average by

$$42.68/48 = 0.89 \text{ bushels per acre}$$

per inch of rain per six-day period.

The mean square for regression on b', c', and d' is also strongly significant, showing that the distribution of rainfall throughout the season has a marked effect on yield. The multiple correlation is $R = 0.87$ and the percentage of variance, A, ascribable to the average effect of the rain is 71 per cent.

Table 10 summarizes these analyses for the selected plots.

The regressions of yield on quantity of rain alone are

26–27	52–53	28–29	58–59	30–31	32–33
42.68	37.40	54.96	28.00	57.64	57.83

Comparison of these values with the partial regressions on a' given above in

TABLE 10

SUMMARY OF ANALYSES OF VARIANCE

Plot Number	Sums of Squares: Regression Formula (4 deg. freed.)	Deviations from Regression (19 deg. freed.)	Total (23 deg. freed.)	R	Per Cent. Variance (A)
26-27	502.61	160.55	663.16	0.87	71
52-53	383.99	239.66	623.65	0.78	53
28-29	714.20	286.24	1000.44	0.84	65
58-59	415.04	473.36	888.40	0.68	36
30-31	874.42	337.48	1211.90	0.85	66
32-33	959.96	284.66	1244.62	0.88	72
				Mean	61

Table 8, show that the latter are all smaller in magnitude and indicates roughly the degree to which the estimated average effect of the rainfall is modified when seasonal distribution is taken into account.

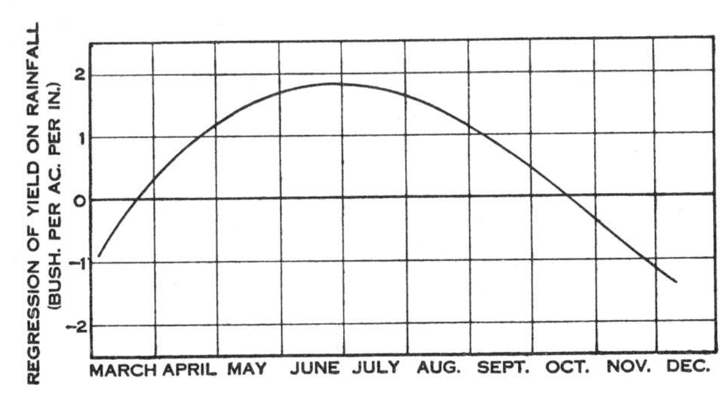

Fig. 9.—Regression function at Roseworthy (no manure treatment).

(v) *The Regression Functions.*—In order to construct regression functions for the various treatments, showing the estimated average effect of an additional inch of rain at any time in the season from March 1 to mid-December, the partial regressions of each set are multiplied by factors of the form

$$\frac{(2p)!}{(p!)^2\ 48.49\ldots(48+p)},$$

where p takes successively the values 0, 1, 2, and 3. Substituting for p in this order the numerical values are 1/48, 1/1176, 1/19600, and 1/299880 and the four coefficients obtained from the partial regressions of plots 26-27, for example, are 0.7242, -0.02094, -0.005395, and 0.0001102. These coefficients are then multiplied in order by the polynomials tabulated by Fisher and Yates (1948) for the series $n = 48$, to give the sequence of points of the regression function. Figures 9, 10, and 11 show the course of this function throughout

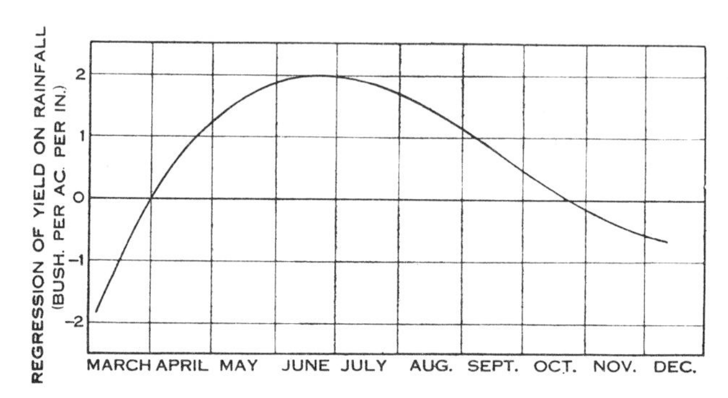

Fig. 10.—Regression function at Roseworthy (superphosphate 2 cwt. per acre).

the selected period. Only the average curves of the duplicates are shown in Figures 9 and 10, and in Figure 11 the curve is the average of two treatments since the individual functions are almost identical.

The three curves follow similar courses of a predominantly parabolic form but do differ in certain points. Similarity of form indicates the dominant role of seasonal conditions in contrast to the secondary influence that manurial treatment has in modifying the response of the crop, even in cases where the differences between the curves are significant. It must be remembered, however, that the sequence of years may be too short, in view of the extremely erratic nature of seasonal conditions, to establish the differences accurately.

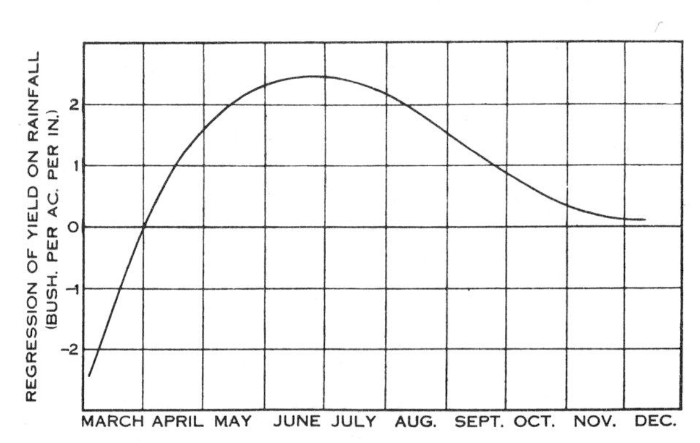

Fig. 11.—Regression function at Roseworthy (superphosphate 2 cwt. per acre and sodium nitrate 1 cwt. per acre).

The detrimental effect of excess rain in the initial stages, exhibited by all curves, is rapidly reduced, and by April 1 is transformed to a beneficial effect. The extent of the damage is greatest on plots 30-31 and 32-33, followed by

28-29 and 58-59, 26-27 and 52-53. From April the advantage to the crop increases under all treatments, reaching a maximum during the last 10 days of June; the greatest response is, as would be expected, on plots 30-31 and 32-33 (2.49), followed by 28-29 and 58-59 (1.98), 26-27 and 52-53 (1.78). Thereafter the response gradually diminishes and with the exception of plots 30-31 and 32-33 assumes negative values during the latter part of October and maintains them until mid-December. These observations are in the reverse order of those at the beginning of the season, but no special significance can be attached to the decreases noted in Figures 9 and 10 after the end of the second week in November, since from this date onward an increasing proportion of years occurs in which rains are too late to have any influence on the crop.

The presence of such pronounced decreases due to March rains was unexpected since on general grounds and from the evidence available it might be anticipated that the effects would be negligible. Several possible explanations of this observation immediately suggest themselves: (1) A spurious effect introduced through correlation of March rains with those of other parts of the season; (2) a chance correlation existing in the series of years under review; (3) rust epidemics associated with heavy rains of late summer (*vide* Cass Smith and Millington 1944); or (4) distortion at the ends of the range due to the inherent disadvantage of polynomial approximation; but none of them withstands close examination. The agreement of all plots and the magnitude of the initial negative regressions are, however, indicative of a genuine effect and this view is supported, though not with completely independent testimony, by Prescott's observation (1933) that, at Roseworthy, the correlation between yield of wheat and rains of the period January-March inclusive, immediately prior to seeding was -0.45. The probability that such a result should be obtained by chance from uncorrelated material is approximately 1 in 50. Prescott examined data from three other South Australian stations, Minnipa, Booborowie, and Veitch, the correlations being respectively -0.22, -0.23, and -0.27, and remarked that the series of four negative coefficients was suggestive of a harmful effect of these rains. Further consideration does, however, render untenable any suggestion that such a relation holds widely *under edaphic and climatic conditions similar to those of the stations mentioned.* In the first place, when the four coefficients are taken individually only that for Roseworthy is significant and, when taken conjointly, the probability of a chance result is about 1 in 10; secondly, data from stations similarly situated provide no evidence that the relation exists. It thus appears that some particular circumstance is associated with the locality. Since Roseworthy College is nearly centrally situated in the hundred of Mudlawirra its rainfall may be taken as representative of that area, and consequently if the feature observed is not confined to the College farm itself, it should appear on examination of the yields from the hundred. To test this point yields of the College and the hundred, for the same sequence of years 1905-30 inclusive, were correlated with the distribution constants a', b', c', and d' and two regression functions were constructed. The first few polynomial values are:

Date	Roseworthy College Regression Coefficient (bush./ac./in.)	Hundred of Mudlawirra Regression Coefficient (bush./ac./in.)
March 3	−1.77	−0.26
9	−1.39	0.01
15	−1.04	0.25
21	−0.72	0.46
27	−0.43	0.66
April 2	−0.16	0.83
9	0.09	0.97

thus establishing, beyond all doubt, the extremely localized character of the phenomenon.

The effect observed is attributable to a temporary depletion of the reserves of available plant nutrients, particularly nitrates, caused by a chain of circumstances depending ultimately upon the high degree of weed infestation, which, incidentally, was not confined merely to the experimental plots, but extended throughout the College area.

Early seasonal rains stimulate germination and rapid development of weeds which immediately draw on the nitrate supply generated in the fallows. Although this organic material is returned to the soil during preparations for seeding, its nitrogen would not become available again until decomposition occurred, and moreover, in the various stages of the latter process, in particular the initial stages, soil nitrates suffer a further reduction as microorganisms draw on the supply to synthesize microbial protein. The influence on the regression function of nitrate deficiency caused in this way is transient since it is rapidly counterbalanced by opposing beneficial effects of later rains and nitrogen gradually becomes available. The crop is, however, deprived of a certain proportion of the nitrogen as some of the weed population survives to complete the life cycle. Two features of the analysis support this explanation of the negative March regressions. In the first place, the order of magnitude of the damaging effect of excessive March rains is the same as the order in decrement of yield which, as stated above, is itself due principally to infestation by weeds, and secondly, there is a response to nitrogen as indicated by contrasting the curve of Figure 11 with those of Figures 9 and 10. The general management of the farm was subjected to radical changes in 1932-33 so that further confirmation may be sought by examining data of the period 1933-48 inclusive. Only yields for the whole farm are available, and their correlation with March rainfall of the 16 years is −0.06, an insignificant value. The paucity of the data does not warrant any greater detail in the analysis.

One other point should be mentioned. It is known that heavy autumnal rains are very effective leaching agents (Prescott and Piper 1930). Penman (personal communication 1949) has pointed out that this is due to the structural condition of aggregated soil particles following the long arid summer season, and that a substantial proportion of nitrate leached to the subsoil may disappear, probably by denitrification and subsequent dissipation of nitrogen in gaseous form. At Roseworthy, leaching, especially in the surface layers, may thus tend to accentuate the nitrate deficiency, but under the prevailing climatic conditions, it could not reasonably be regarded as a major contributor. In *all* the data herein examined, the only locality that shows definite evidence of nitrate deficiency, caused by leaching, and *sufficiently acute to give rise to negative regressions on rainfall,* is a small area of podsolic soils situated in southern Eyre Peninsula (*vide* Cornish loc. cit., Fig. 8).

(vi) *Significant Differences between Regression Functions.*—At this juncture it is perhaps worth while to give, at some length, a description of the significance test of the difference between two curves as a preliminary to contrasting the results obtained under the several treatments. Actually, the test can be made in two ways, depending on whether replicates of the treatments exist or not. If there is no replication, the test is made by comparing the variance ascribable to the rainfall curve fitted to the differences between the two treatments with deviations of the differences from the fitted curve. On the other hand, if replicates of the treatment exist, a rainfall curve is fitted to each replicate and the difference between the average curves of the treatments is then tested against the differences between rainfall curves of replicates of the same treatment. Obviously in this case, the first test can also be applied in comparing the average curves of the two treatments. The two tests are not necessarily equivalent, but the cogency of this statement may not be apparent until further details of each test, and the appropriate interpretation of the two results are given.

First test.—To illustrate this test consider the duplicate sets of yields from plots 26-27 and 52-53. In applying the test, a rainfall curve is fitted to the differences between the deviations of the observed yields in the two series from their respective linear trends in a manner similar to that used in finding the rainfall curves for individual plots. The same curve can obviously be obtained by taking, at all points, the difference between the curves already fitted to the two series of yields. This provides a very expeditious means for finding the actual curve but, in order to make the test of significance conveniently, the regression coefficients are required. It is not, however, necessary to start *ab initio.* Since the regression coefficients and product sums of yield with the rainfall distribution constants are available for each plot, it can be shown that the quantities required in making the significance test can be obtained by taking differences. Thus from Tables 7 and 8, the product sums and regression coefficients, together with their differences, are as follows:

	Plots		
Products of Yield with	26-27	52-53	Difference
a'	7.31530	6.41082	0.90448
b'	−0.63527	−0.89363	0.25836
c'	−1.97764	−1.65307	−0.32457
d'	0.71288	0.45876	0.25412

Regression on

a'	34.7635	28.9300	5.8335
b'	−24.6234	−37.4166	12.7932
c'	−105.7361	−96.6184	−9.1177
d'	33.0453	11.6969	21.3484

and the sum of the squares ascribable to the regression of the differences between yield on the rainfall distribution constants is thus

$$5.8335 \times 0.90448 + \ldots\ldots + 21.3484 \times 0.25412 = 16.97.$$

The total sum of squares of the differences is now required. If y stands for an observed yield, Y for the predicted yield according to the linear trend, and subscripts 1 and 2 are added to distinguish the replicates, this sum is

$$S\{(y_1 - Y_1) - (y_2 - Y_2)\}^2 = S(y_1 - Y_1)^2 + S(y_2 - Y_2)^2 - 2S(y_1 - Y_1)(y_2 - Y_2).$$

The two sums of squares on the right-hand side are already available and their actual values are, from Table 10, 663.16 and 623.65 respectively. The sum of products is obtained in a similar manner to those given in Table 7, making due allowance for secular trend and omission of the year 1923. The corrected sum is 536.32 and consequently the sum of the squares of differences is 214.17. This series of operations can then be summarized in the analysis of variance of Table 11. Obviously, no significant difference exists between the two curves on the basis of this test.

TABLE 11

ANALYSIS OF VARIANCE (DIFFERENCE BETWEEN RAINFALL CURVES, PLOTS 26-27 AND 52-53)

Variation due to	Deg. Freed.	Sum Sq.	Mean Sq.
Regression of difference between replicate curves on a', b', c', d'	4	16.97	4.24
Regression on a' alone	1	4.77	4.77
Difference (regression on b', c', d')	3	12.20	4.06
Deviations from regression	19	197.20	10.38
Total	23	214.17	

A similar test of the difference between the rainfall curves of plots 28-29 and 58-59 showed that the regression on a' alone was greater in the former replicate; the difference is significant at the 5 per cent. point but in other respects the curves may be regarded as equivalent. Re-examination of the original data did not disclose any peculiar circumstance relating to plots 58-59 which might reasonably account for the discrepancy between the regressions.

Second test.—Since there is no evidence to show that the autumnal and spring dressings of sodium nitrate differ in their effects, the series 30-31 and 32-33 were taken as replicates and the second test may be illustrated by comparing their average curve with the average of the duplicates 26-27 and 52-53. The appropriate product sums and regression coefficients from Tables 7 and 8 respectively are first averaged and then differences between the averages are taken. The figures are as follows:

	Average of Plots		
Product of Yield with	30-31 and 32-33	26-27 and 52-53	Difference
a'	9.89546	6.86306	3.03240
b'	– 0.08425	– 0.76445	0.68020
c'	– 2.66571	– 1.81536	– 0.85035
d'	1.14779	0.58582	0.56197
Regression on			
a'	50.9585	31.8468	19.1117
b'	3.4062	– 31.0200	34.4262
c'	– 122.9338	– 101.1772	– 21.7566
d'	73.1330	22.3711	50.7619

The sum of squares ascribable to the regression of the difference on the rainfall distribution values is thus 128.40.

The total sum of squares of the differences between the average deviations is now required. If, as before, y and Y stand for observed and predicted yields respectively, subscripts 1 and 2 distinguish the replicates and superscript dashes are added to denote manurial treatment, the sum of squares (superphosphate and sodium nitrate – no manure) is

$$\begin{aligned} &S\{\tfrac{1}{2}[(y'_1 - Y'_1) + (y'_2 - Y'_2)] - \tfrac{1}{2}[(y''_1 - Y''_1) + (y''_2 - Y''_2)]\}^2 \\ &= \tfrac{1}{4}\{S(y'_1 - Y'_1)^2 + S(y'_2 - Y'_2)^2 + S(y''_1 - Y''_1)^2 + S(y''_2 - Y''_2)^2 \\ &+ 2[S(y'_1 - Y'_1)(y'_2 - Y'_2) + S(y''_1 - Y''_1)(y''_2 - Y''_2)] \\ &- 2[S(y'_1 - Y'_1)(y''_1 - Y''_1) + S(y'_1 - Y'_1)(y''_2 - Y''_2) \\ &+ S(y'_2 - Y'_2)(y''_1 - Y''_1) + S(y'_2 - Y'_2)(y''_2 - Y''_2)]\}. \end{aligned}$$

Only the last four product sums need be calculated since the remaining quantities are available from previous tests. Numerically the sum of squares is 280.47 and the analysis of variance is given in Table 12.

In conducting the second test, a comparison is made between the mean square 32.10 with 4 degrees of freedom and the mean square 2.91 with 8 degrees of freedom; in this illustration the difference is strongly significant, $0.001 < P < 0.01$.

The analysis may be carried further by partitioning the sum 128.40 into two of its components

(i) regression on a'
(ii) regression on b', c', and d',

and comparing the mean square of each with its corresponding components in the sum 23.25 with 8 degrees of freedom. This test shows that the regressions on quantity of rain and its distribution are individually significant, $0.01 < P < 0.05$, in each case, and consequently there is a real difference between the average curves of plots 30-31, 32-33 and 26-27, 52-53.

After comparing the two tests, it is apparent that only in the second is account taken of soil heterogeneity. Consequently, any significant difference demonstrated by the first test is not necessarily attributable to differential treatment as it may be due to inherent differences between the plots. Apart from this, another important difference exists between the two tests. Suppose the treatments to be compared are replicated and the average regression of each set of replicates is taken. Consider one of the treatments. If the sequence of years under examination can be regarded as a random sample of rainfall values, and is of sufficient length to sample adequately the range of rainfall variation, the expectation is that the average regression, obtained from any other sample containing the same number of seasons, will only differ from that calculated on the first sample by more than twice the standard deviation of the first regression (as computed from the deviations from the average regression line) in 5 per cent. of trials. The two regressions may or may not be

TABLE 12

ANALYSIS OF VARIANCE (DIFFERENCE BETWEEN AVERAGE CURVES, PLOTS 30-31 AND 32-33 LESS PLOTS 26-27 AND 52-53)

Variation due to	Deg. Freed.		Sum Sq.		Mean Sq.	Variance Ratio
Total	23		280.47			
Regression of difference between average curves on a', b', c', d'	4		128.40		32.10	11.03
Deviations from regression	19		152.07		8.00	
Regression of difference between replicate curves on a', b', c', d'						
(i) Plots 26-27 and 52-53	4	8	16.97	23.25	2.91	
(ii) Plots 30-31 and 32-33	4		6.28			
Deviations from regression						
(i) Plots 26-27 and 52-53	19	38	197.20	234.23	6.16	
(ii) Plots 30-31 and 32-33	19		37.03			

really different, but the important point is that their difference is unlikely to be significant using sampling errors calculated by the first test. As replicates exist, a separate regression can be fitted to each and the significance of the average regression may be tested by comparison with the variation among regressions of individual replicates. Since the experience of all replicates is based upon an identical sequence of seasons, it may be anticipated that the random errors associated with the replicates are on a smaller scale, and if the effects due to soil heterogeneity (between replicates) are small or non-existent, the average regression calculated from the first sample of seasons would probably be significant. The same remark applies equally well to the average regression based on the second set of seasons. Thus the average regressions may be established as individually significant, but it is also possible that their difference is significant, or that they exhibit opposing effects due to rainfall, especially if the two samples of seasons differ markedly. The smaller error in the second test cannot thus be relied upon to provide an accurate guide as to the nature of the regression obtained from a different sample of seasons, and consequently conclusions based on the second test are valid only for the particular set of seasons which have actually been experienced. If soil heterogeneity makes no contribution to the difference between the regressions, the first test is the more exacting when the tests are not equivalent because it accounts for all sources of variation of which account is taken in the second test apart from soil heterogeneity. This point is well illustrated in the example used above in the second test of significance. Soil heterogeneity does not affect the duplicates and, as shown, the difference between the average regressions is strongly significant ($0.001 < P < 0.01$) on the second test. On the other hand, comparison of the mean square 32.10 with the mean square 8.00 (19 degrees of freedom) by the first test, gives a variance ratio of 4.01 for which $0.01 < P < 0.05$.

Comparative tests among the three treatments gave the following results:

Superphosphate plus sodium nitrate	*v.*	No manure	Significant differences between the regressions on both quantity and seasonal distribution of the rains.
Superphosphate plus sodium nitrate	*v.*	Superphosphate	Significant difference between regressions on quantity of rain only.
Superphosphate	*v.*	No manure	No significant differences.

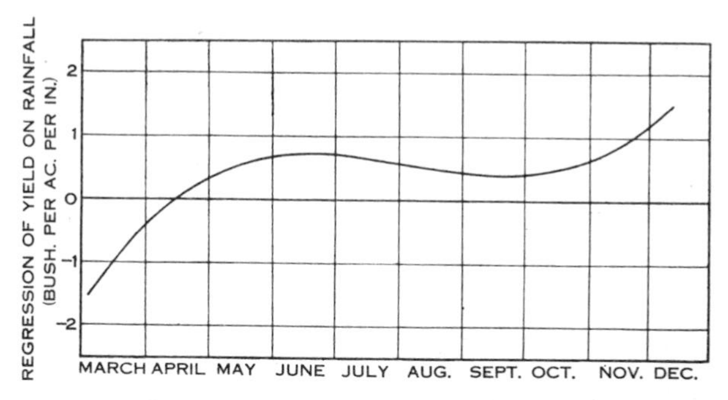

Fig. 12.—Difference between regression functions of Figures 9 and 11.

The first of these comparisons provides the greatest contrast and the regression function representing it is illustrated in Figure 12. The two salient features are the heavier damage due to excess March rain and the greater response to winter rains on plots 30-31 and 32-33. The former has been covered in the discussion above, while the latter follows from the response of the crop to nitrogen and phosphorus supplied in the fertilizer. The treatment replaces some of the nitrogen rendered unavailable and may even assist in the liberation of a certain proportion of it by facilitating decomposition of the organic material ploughed into the soil. The reaction must occur in this way since on well-prepared fallows nitrification generates sufficient nitrate to meet the demands of cropping so that normally there is no response to dressings of nitrogenous fertilizers.

The significant difference on quantity of rain in the second comparison follows also from the stimulus given by the additional nitrate.

The negative result of the third comparison was unexpected since it was anticipated that a difference in favour of superphosphate would be obtained in accordance with the established observations that soils of the wheat belt are deficient in available phosphorus and that phosphatic fertilizer induces early development of the wheat plant. The response of the crop to seasonal rains must necessarily be intimately related to the fertility of the soil, but the dominating effect of rainfall, coupled with the fact that the sample of years is too small to allow the disturbing influence of erratic seasons to be smoothed out, make it impossible to establish any but major differences. The solution of the main problem is complicated also by the competitive effect of the weeds since the shortage of nitrogen on the plots treated with phosphate alone no doubt influences the response to phosphorus. This factor is probably not operative on the plots treated with both superphosphate and sodium nitrate after application of the latter.

(*b*) *Minnipa*

The yield data were taken from records of the State experimental station at Minnipa for the period 1916-44 inclusive. Farming commenced in 1916 on virginal sandy mallee soils of granitic origin, which in their original state carried mallee eucalypts (*E. gracilis, E. dumosa, and E. oleosa*). The property was operated as an experimental station until 1930 and subsequently on a basis of share-farming. All wheat crops have been grown with superphosphate on mixed areas of fallow, stubble, and new land, the total area in any

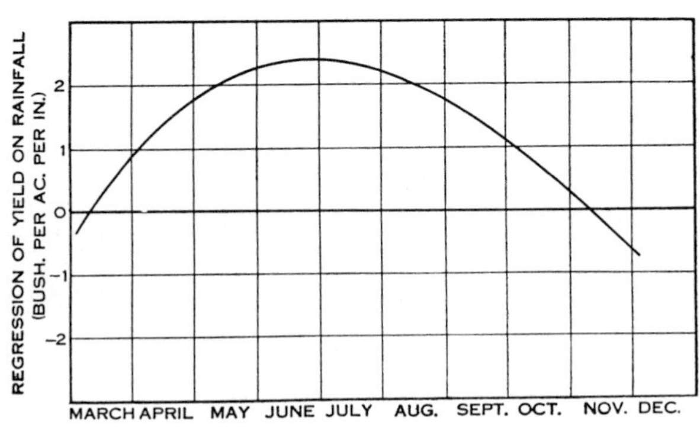

Fig. 13.—Regression function at Minnipa.

year ranging from 100 to 1,000 acres. Rainfall data were recorded in the township of Minnipa adjacent to the farm. The method of analysis was similar to that used for Roseworthy, seasonal weather being represented by rains of the period March 1-December 8 to which a polynomial of the third degree was fitted. The regression function is illustrated in Figure 13.

(c) *Kimba*

This record was obtained from a large private property on which approximately 600 acres of wheat were harvested annually, and extends over the period 1917-43 inclusive. The farm was established in 1916-17 on virginal sandy mallee which in its native state carried *Callitris* spp. in addition to typical mallee. All crops have been supplied with superphosphate and in the early years were sown on fallow, stubble, and new land, but latterly in 4 course rotations. Rainfall observations were taken in Kimba nearby. Seasonal rains were computed for the period March 1-December 8 and the analysis was similar to that used for Roseworthy. The regression function is given in Figure 14.

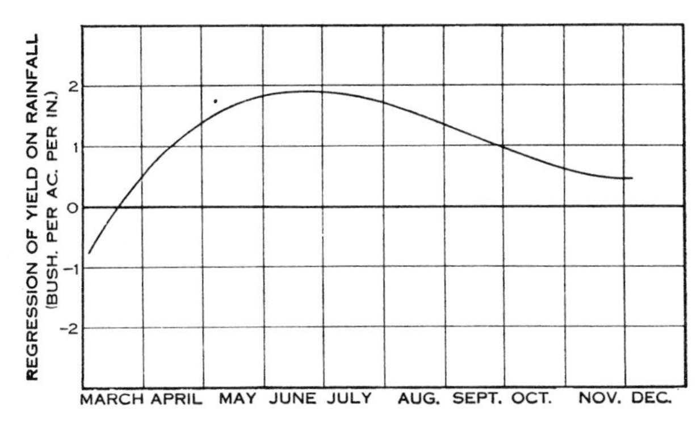

Fig. 14.—Regression function at Kimba.

(d) *Veitch*

The yield data were recorded at the State experimental farm, Veitch, for the years 1909-30 inclusive. Farming was commenced in 1909 on virginal sandy mallee soils which carried mallee eucalypts with only a sparse bushy undergrowth indicative of the low rainfall, and all wheat crops have been grown with superphosphate on mixed areas of fallow, stubble, and new land, the total area in any year ranging from 100 to nearly 1,000 acres. Rainfall observations were taken at Veitch's Well on a site adjoining the farm. As the data were recorded only by calendar months, seasonal conditions could not be considered in detail and recourse was therefore taken to an analysis by multiple regression in which rains of the periods April-May, June-July-August, and September-October were taken as variates. The partial regressions in bushels per acre per inch are

April-May	1.89 ± 0.486	
June-July-August	2.11 ± 0.264	18 degrees of freedom
September-October	1.54 ± 0.354	

(e) *Booborowie*

This State experimental farm formed part of North Booborowie station which for many years had carried sheep and was repurchased by the State for closer settlement in 1911. The yields were obtained from areas ranging from

100 to 400 acres during the period 1912-30 inclusive. The soil is a typical red brown earth which in its native state carried bluegum (*Eucalyptus leucoxylon*) on the flats and she-oak (*Casuarina stricta*) on the rises with a dense covering of *Lomandra*, Wallaby grass (*Danthonia* spp.), and Spear grass (*Stipa* spp.). All wheat crops were grown with superphosphate on fallow. Rainfall observations were taken on the property and as the data were available only in calendar months, the analysis was similar to that used for Veitch. The partial regressions in bushels per acre per inch are

April-May	1.10 ± 0.375	15 degrees of freedom
June-July-August	1.80 ± 0.255	
September-October	0.78 ± 0.330	

(*f*) *Saddleworth*

This record of yield data, which extends over the period 1897-1943, was taken from a large private holding near Saddleworth. The property is comprised almost equally of two soil types, a black earth (chernozem) and a type intermediate in grade between a red brown earth and a podsol. Wheat has been sown with superphosphate on areas averaging 500 acres annually in various crop rotations involving 3-7 courses. Rainfall observations were taken in the nearby township of Saddleworth and the season has been represented by rains of the period March 1-December 31, the analysis being similar to that used for Roseworthy except that each season's data were fitted with a polynomial of the 5th degree. The regression function is given in Figure 15.

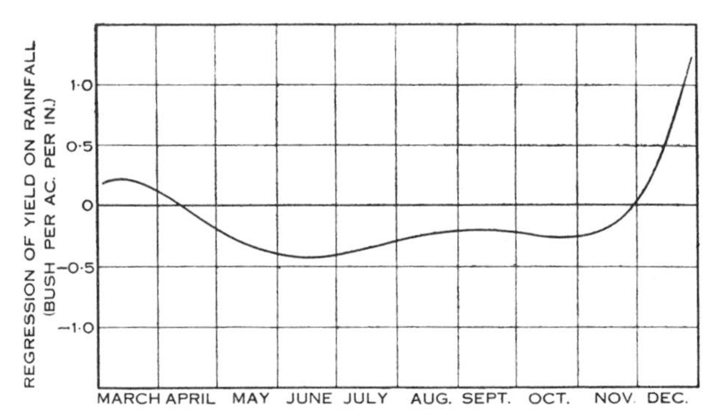

Fig. 15.—Regression function at Saddleworth.

(*g*) *Lameroo*

The private property from which this record for the years 1911-46 inclusive was taken, is situated near Lameroo and was established in 1910-11 on virginal transitional mallee-solonetz soil carrying in its native state mallee (*E. incrassata*) and broombush (*Melaleuca uncinata*). At this particular site the soils tend toward complete solonization. Wheat crops have all been supplied with superphosphate and have been sown on mixed areas, ranging from 100 to 450 acres, of fallow and stubble and latterly in rotations. Rainfall observations were taken on the property. The season was represented by rains of the period March 1-December 20, the analysis being similar to that used for Roseworthy. The regression function is illustrated in Figure 16.

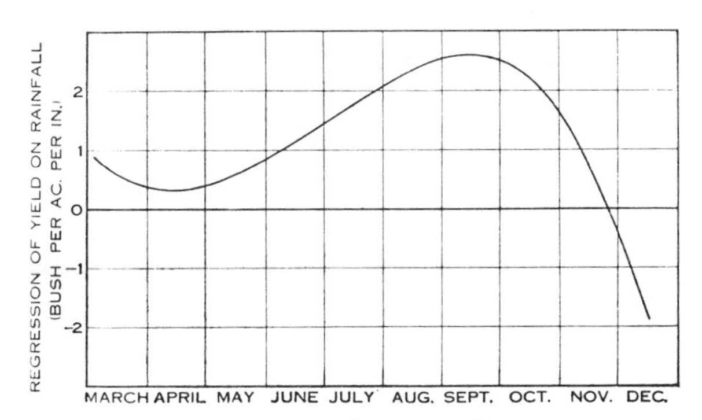

Fig. 16.—Regression function at Lameroo.

(*h*) *General Remarks*

In the analyses of the data listed in Sections III(*b*) to III(*g*) above, secular variations in yield and/or rainfall, wherever they occurred, have been eliminated before correlating the variables.

Considering the soils of the wheat belt broadly, the more lightly textured mallee types may be represented by Minnipa, Kimba, and Veitch, the heavier loamy mallee types by Roseworthy, transitional mallee-solonetz by Lameroo, red brown earths in zones of less than 15 in. seasonal rainfall by Booborowie, and red brown earths with their associated soils in zones of more than 15 in. seasonal rainfall by Saddleworth (*vide* Cornish loc. cit., Fig. 8).

The results obtained from these individual records substantiate the observations and general conclusions given in Section II(*c*), and vindicate the argument used in the choice of the rainfall variates and analytical technique. Regressions on both quantity and seasonal distribution of rainfall are significant at Kimba, Minnipa, and Lameroo, but not at Saddleworth. After taking the average curve of the three treatments at Roseworthy to represent that locality, comparative tests among the regression functions yielded significant differences in all cases except the contrast between Minnipa and Kimba. It is, however, certain that, if the particular circumstances noted at Roseworthy had not applied, the regression function would not have differed significantly from either Minnipa or Kimba. The partial regressions obtained at Booborowie indicate that the regression function of the red brown earths also assumes a parabolic form.

Average values of the ordinates for the periods April-May, June-July-August, September-October, taken from the regression functions of Figures 9 to 11, 13, 14, and 16, together with means of the components of seasonal rainfall, are set out in Table 13. These regressions, and the corresponding quantities from Veitch and Booborowie, are in close agreement with the coefficients obtained for hundreds on the same broad soil types in which the standard of wheat farming attains a reasonable level. Average figures for the hundreds are naturally somewhat lower since they have been derived from yields obtained under general industrial conditions and in certain areas are influenced also by the fact that seasonal rains approach their optimal values.

TABLE 13

AVERAGE REGRESSIONS AND COMPONENTS OF SEASONAL RAINFALL FROM SELECTED STATIONS

Station	Regressions (bush./ac./in.)			R	Components of Seasonal Rainfall (in.)			Total
	Apr.-May	June-July-Aug.	Sept.-Oct.		Apr.-May	June-July-Aug.	Sept.-Oct.	
Roseworthy Fig. 9	1.1	1.6	0.5					
Fig. 10	1.1	1.8	0.6					
Fig. 11	1.4	2.2	1.0					
Roseworthy mean	1.2	1.9	0.7		3.11	6.30	3.85	13.26
Minnipa	1.7	2.2	1.2	0.85	2.03	5.66	2.41	10.10
Kimba	1.3	1.8	1.0	0.79	2.00	5.10	2.65	9.75
Mean	1.5	2.0	1.1					
Veitch	1.9	2.1	1.5	0.93	1.85	3.60	2.63	8.08
Booborowie	1.1	1.8	0.8	0.91	2.78	6.81	4.00	13.59
Saddleworth	—	—	—	—	3.80	7.25	3.93	14.98
Lameroo	0.5	1.7	2.4	0.78	2.81	5.28	3.36	11.45

The curve for Lameroo contrasts strongly with the remainder, and confirms the observation made above that transitional mallee-solonetz soils behave differently from other soil types within the same zone of rainfall, but the cause of this is at present obscure. At Saddleworth, average rainfall is very nearly optimal at all stages of the season. Since harvesting normally takes place in late December, the rains of November should be included in the seasonal total; this gives an amount of 16.15 in. which would point to the fact that the estimate of 17 in. for red brown earths and associated types given in Section II(*c*) is too high, even after allowing for errors of estimation. This view may be tested by examination of the records from the Waite Agricultural Research Institute. The soil at this station is a typical red brown earth and the mean values of the components of seasonal rainfall are April-May 5.03 in., June-July-August 9.14 in., September-October 4.61 in., and November 1.47 in. It is known (Cornish unpublished data) that April-May rains are considerably in excess of the optimum; June-July-August rains are moderately so; while September-October and November rains are close to their optimal values in this locality. As at Saddleworth, November rains must be included in the seasonal total, and consequently, in districts where harvesting normally occurs during the latter half of December, it appears that the estimate of 8.8 in. for winter rains is between 1 and 2 in. too high, but that the remaining two estimates are fairly accurate. On the other hand, in regions where the crop reaches maturity between mid-November and mid-December, it is probable that the figure given is substantially correct. The estimated optimal seasonal total for the heavier mallee soils stands in proper relation to that of the red brown earths since effective rainfall is lower in the climatic zone of the former soil types.

The values of the quantity *A* are Minnipa 67 per cent., Kimba 54 per cent., Veitch 83 per cent., Booborowie 79 per cent., and Lameroo 55 per cent., which in combination with the figures from Roseworthy, show that approximately 65 per cent. of the annual variance of yield is ascribable to the average effects of the rainfall, thus strongly supporting the contention made above that rainfall is a sufficiently reliable index of seasonal weather in the wheat belt of South Australia. The result also indicates that in the near future it should be possible to bring fairly accurate forecasts within reach of the State and the industrial farmer when the effects of rainfall have been more accurately assessed by taking account of the information available in additional data.

IV. Comparison with Previous Work

Perkins and Spafford (loc. cit.) in two early reports on the work of the permanent experiment field at Roseworthy summarized their conclusions after studying the effects on yield of seasons favourable and unfavourable to wheat Their examination was confined to the data accumulated during the first 10 years of the experimental work, but in framing their reports, they no doubt drew upon experience gained from cropping in the district over a longer term. The summary of their conclusions from the second report is as follows:

"In the Roseworthy district, and those similarly situated, 15 in. to 16 in. of rain well distributed over the growing period of cereals—April to November—render possible yields of 28 bush. to 30 bush. of wheat or 3 to 3¼ tons of hay.

"In the same districts the most favourable rain distribution for maximum crops is represented by good early seeding rains aggregating 4 in. to 5 in. in April and May, followed by 4 in. to 5 in. in June and July, and by 6 in. to 7 in. between August 1 and October 31. November rains are, as a rule, without significance in seasons that are otherwise favourable.

"In these districts the seasonal factors exercising most powerfully a depressing influence on yields, whether of grain or of hay, are as follows, in order of relative importance: (*a*) Unduly late seeding rains; (*b*) a shortage of seeding rains followed by a dry winter; (*c*) a shortage of spring rains following on an unfavourable winter; (*d*) unseasonably sudden hot weather accompanied by northerly winds in October or even in November; and (*e*) moist warm weather in October and November leading to attacks of red rust." In both reports considerable emphasis was laid upon the importance of seeding and winter rains. The conclusions arrived at here, using exact statistical methods and much more data reaffirm their findings with greater precision, and in fact extend them to a major proportion of the wheat belt.

In a general survey of the relation between yield and seasonal rainfall for the principal districts of South Australia, Perkins (loc. cit.) discussed, *inter alia,* the relative effectiveness of April-November rains and summarized the position thus:

"In each district for every additional inch above April-November rainfall requisite to economic crops, given satisfactory distribution, we may anticipate increases in yield above the minimum of

1½ bushels in the Upper North
1 bushel in the Lower North
½ bushel in the Central District."

The districts to which Perkins referred are three of the former statistical subdivisions of the State. Upper North corresponds almost identically with Flinders (*vide* Fig. 1); Lower North to Goyder, northern Yorke and hundreds on the northern boundary of Light; and Central is comprised mainly of the remaining parts of Yorke and Light. The responses per inch of rainfall may then be compared with the regression coefficients plotted in Figures 2 to 6. Thus Perkins's three estimates, in the order given above, correspond approximately to the average values of the following groups:

1. Desert loams of group (a) in the three parts of Figure 6.
2. Red brown earths that have $x_1 < 2¼$ in., $x_2 < 5½$ in., and $x_3 < 2¾$ in. in Figure 2.

1. Red brown earths for which 2¼ in. $< x_1 <$ 3 in., 5½ in. $< x_2 <$ 6½ in., and 2¾ in. $< x_3 <$ 3½ in. in Figure 2.
2. Loamy mallee soils of group (b) that have $b_1 > 0.9$, $b_2 > 1.5$, and $b_3 > 0.7$ in Figure 3.

1. Red brown earths that have $x_1 > 3$ in., $x_2 > 6½$ in., and $x_3 > 3½$ in. in Figure 2.
2. Loamy mallee soils of group (b) that have $b_1 < 0.9$, $b_2 < 1.5$, and $b_3 < 0.7$ in Figure 3.

Perkins also gave a table showing the ideal distribution of so-called "useful" rain corresponding to requirements of the bulk of South Australian wheat-growing areas. The figures given are substantially the same as those quoted from his previous paper (Perkins and Spafford loc. cit.).

In reporting the results of an extensive research on the water requirements of farm crops under Victorian conditions, Richardson (loc. cit.) gave transpiration ratios of 395 for dry matter and 1067 for grain as averages of winter-grown wheat. Taking these figures as representing an average of seasons, he computed that, for moderate rainfall, each inch of rain transpired by the crop is capable of producing approximately 3½ bushels per acre and then compared actual production with this hypothetical figure. For both Victoria and South Australia, seasonal rainfall was taken as that occurring during the period April-October inclusive and, in order to provide an average, representative of the main areas in each State, the mean of a number of selected stations was taken. In both States a very close relationship between State yield and average composite seasonal rainfall was found. The period under review embraced the years 1892-1921 inclusive and on partitioning it into three decennial periods, the following ratios were found:

Period	Average Yield (bush./ac.)		Composite Seasonal Rainfall (in.)		Bush./ac./in. of Seasonal Rain	
	Vict.	S. Aust.	Vict.	S. Aust.	Vict.	S. Aust.
1892-1901	7.65	4.73	11.49	11.55	0.67	0.41
1902-1911	10.50	9.62	11.63	12.18	0.90	0.79
1912-1921	12.62	10.47	11.59	11.18	1.09	0.94
Mean	10.26	8.27	11.59	11.64	0.89	0.71

Since the average rainfall in the several periods remains practically constant, the increases in the ratios show the extent to which wheat farming had progressed toward more effective utilization of soil moisture. Richardson observed that the transpiration rate increases exponentially during the early stages of growth and reaches a maximum during October which, with winter-grown cereals, follows closely after flowering, and usually coincides with stringent atmospheric conditions. The fact that approximately 40 per cent. of the total water used by the crop is transpired during October suggested that spring rains prior to, and during, that month should exercise a pronounced effect on the ultimate yield.

Following on Richardson's work, Barkley (loc. cit.) investigated the relationship between yield and monthly rainfall for the Wimmera district of Victoria, and found certain strong correlations. Thus he states:

"The outstanding feature of the rainfall effect in Australia is the definite limits of the critical seasons. In Victoria, for instance, eighty per cent. of the variations in the wheat harvest can be ascribed to the rainfall of August and September, leaving only twenty per cent. of the departures from an average yield to be accounted for by (*a*) moisture stored in the fallow, (*b*) May to July rains, and (*c*) the October rainfall. Of these subsidiary influences outside the critical spring months the falls in June are most important, and provide the earliest clue to the opening of the season."

He noted that the response in yield to increasing rainfall was logarithmic, the spring rains of August and September adding to the yields in the proportions of 8.22, 3.65, 2.03, 1.55, 1.01, and 0.98 bushels per acre for each successive inch up to the sixth. As already stated no such direct evidence of an approach to the optimum could be found in the data herein examined.

Barkley laid great stress upon the existence of these so-called critical periods and nominated them for other parts of the Australian wheat belt. For example, June was given as the critical month on the west coast of South Australia (Nuyts and Eyre of Fig. 1) and the Darling Downs (Queensland), and August in the Lower North of South Australia. If Barkley's statements are accepted the rains of June alone, and August alone would be expected to exercise strong control over yields in the respective districts mentioned, but it is clear that this is not the case. Whatever may be expected on physiological grounds on the basis of such work as that of Richardson, the results of the present investigation show that, in the environment to which they are applicable, the influence of rainfall on yield is strongly marked throughout practically the entire growing season and is not confined to a few relatively short intervals of time.

V. Ackowledgments

The author is indebted to Messrs. Cant Bros. of Kimba, F. Coleman and Sons of Saddleworth, and Mr. A. J. A. Koch of Lameroo for placing their records at his disposal and for permission to use them in publication; to Miss E. M. G. Goodale, Miss P. M. Ohlsson, Mr. R. Birtwistle, and Mr. G. G. Coote, Section of Mathematical Statistics, C.S.I.R.O., for assistance in the assembly and preparation of the data for analysis; and to Mr. M. C. Childs, Division of Biochemistry and General Nutrition, C.S.I.R.O., for preparation of the diagrams.

VI. References

Barkley, H. (1927).—*Wheat and Grain Review* 7: 8-11.

Cass Smith, W. P., and Millington, A. J. (1944).—*J. Dep. Agric. W. Aust.* **21**: 1-16.

Cornish, E. A. (1949).—*Aust. J. Sci. Res.* B **2**: 83-137.

Fisher, R. A. (1924).—*Philos. Trans.* B **213**: 89-142.

Fisher, R. A., and Yates, F. (1948).—"Statistical Tables for Biological, Agricultural, and Medical Research." 3rd Ed. (Oliver and Boyd: Edinburgh.)

Perkins, A. J. (1924).—*Rep. Aust. Ass. Adv. Sci.*, Adelaide, **17**: 122-37.

Perkins, A. J., and Spafford, W. J. (1911).—*J. Dep. Agric. S. Aust.* **14**: 959-67; 1030-37; 1141-54. **15**: 10-25; 112-9.

Perkins, A. J., and Spafford, W. J. (1915).—*J. Dep. Agric. S. Aust.* **17**: 1250-68; **18**: 13-25; 484-96; 563-75; 648-52.

Prescott, J. A. (1933).—Proc. 5th Pan-Pacif. Sci. Congr., pp. 2657-67.

Prescott, J. A., and Piper, G. R. (1930).—*J. Agric. Sci.* **20**: 517-31.

Regional Planning Committee (1946).—Report on Regional Boundaries. (Govt. Printer, Adelaide.)

Richardson, A. E. V. (1923).—*J. Dep. Agric. Vict.* **21**: 193-212; 257-84; 321-39; 385-404; 449-81.

Reprinted from the Australian Journal of Applied Science, Vol.4, No. 4, pp. 489-507 1953.

THE FREQUENCY DISTRIBUTION OF THE SPECTROGRAPHIC (D.C. ARC) ERROR

By A. C. OERTEL* and E. A. CORNISH†

(*Manuscript received August* 20, 1953)

Summary

A sample of 1161 errors, derived from 259 sets of triplicate and 64 sets of sextuplicate measurements of line intensity, has been used to determine the frequency distribution of the spectrographic error which occurs in the D.C. arc excitation of samples of soil and plant ash. It has been found that the error is, in the greater part at least, a proportional one, and when expressed on a logarithmic scale, its frequency distribution, so far as can be ascertained from these data, is normal. It follows that measures and tests of significance based upon the hypothesis of normality are applicable in spectrographic (D.C. arc) investigations. Additional data are being accumulated to provide the means for making more comprehensive tests of the nature of the frequency distribution.

I. INTRODUCTION

The theory of tests of significance, with particular reference to small samples, pioneered by "Student" (1908), and later developed and extended considerably by Fisher (1925 and subsequent editions), is based upon the assumption of normality in the frequency distribution of the data under examination. The value of these researches is greatly enhanced by the fact that even if the original distribution is not exactly normal, those of the mean and other statistics calculable from the observations usually tend to normality as the size of the sample is increased. The tests may therefore be applied legitimately to data for which there is not sufficient evidence to show that the original distribution is normal, but for which there is ample reason to suppose that it does not belong to the exceptional class of distributions in which sample statistics do not tend to normality.

The application of modern statistical measures and tests of significance to spectrographic observations determined in this Laboratory has never given any indication that the procedures could not be validly applied, but as frequently only small numbers of observations are available, it appeared desirable to check this point. Apart from this, an investigation was worth while since, so far as is known, the frequency distribution of the spectrographic error has not been determined. The data employed herein were derived from routine analytical measurements which have been replicated only a few times, and the method to be followed in such circumstances is probably of interest to the experimentalist.

* Division of Soils, C.S.I.R.O., Adelaide.

† Section of Mathematical Statistics, C.S.I.R.O., Adelaide.

II. The Samples of Errors

The samples of errors were derived from measurements of line intensity, not of the ratio of line intensities; and the measurements were selected from analytical results obtained under the conditions used in this Laboratory (Oertel 1950*a*, 1950*b*). The following restrictions were necessarily imposed on the selection. The measurements of a line intensity must have been made at least in triplicate, because duplicate measurements automatically give a perfectly symmetrical distribution of errors. Each sample must contain not less than 100 errors, so that it may be considered as a passably representative sample. The samples should cover elements and basic compositions which behave differently in the source of excitation, so that the conclusions are likely to be generally valid. Each sample of errors must be derived from results obtained by one operator in one series of analyses, so that the sample may be expected to be homogeneous. Five samples of errors were obtained from:

A and *B*.—Triplicated determinations of copper and manganese, respectively, in 63 samples of soil, arced with an equal part by weight of powdered graphite.

C and *D*.—Triplicated determinations of, respectively, copper in 68, and molybdenum in 65, samples of plant ash, arced alone.

E.—Sextuplicated determinations of copper, manganese, molybdenum, and tin, in 16 samples of modified plant ash (eight arced alone, and eight with an equal part by weight of powdered graphite).

The experimental values of the line intensities were subject to errors caused by:

(i) Sampling fluctuations which occurred when the 10 mg sample for excitation was taken.

(ii) Variations in the amount of sample actually excited in the D.C. arc.

(iii) Fluctuations in the current and other parameters of the source of excitation.

(iv) Variations in the photographic materials and processes.

(v) Densitometric errors.

On the assumption that the experimental results had been carefully obtained, it is almost certain that all errors, except a few of category (iv), were purely random errors. Those errors which are classified as constant in the theory of errors, and systematic errors (which become constant in the hands of a skilled observer) are outside the scope of the present investigation.

III. The Proportional Nature of the Error

The random error may be any one of the following five types:

(i) Independent of the magnitude of measured line intensity, the mean value of the error being constant for changes in the latter.

(ii) Dependent on the magnitude of line intensity, the mean value varying directly with the latter, that is, a proportional error.
(iii) Dependent on the magnitude of line intensity, the mean value varying in a more complex manner with the latter.
(iv) A composite error of types (i) and (ii).
(v) A composite error of types (i) and (iii).

Errors of categories (i), (ii), and (iii) of Section II are very likely to be purely proportional. Some of those of category (iv) may vary from

TABLE 1

STANDARD DEVIATIONS s AND MEANS m FROM TRIPLICATED MEASUREMENTS
SAMPLE D: NATURAL NUMBERS

s	m	s	m	s	m
0.0346	0.18	0.1997	1.18	0.3863	5.90
0.0625	0.22	0.1212	1.27	0.8089	6.57
0.0819	0.24	0.0625	1.31	1.6820	6.79
0.0520	0.25	0.1836	1.37	0.5511	6.88
0.0608	0.31	0.2905	1.63	0.4330	7.34
0.0361	0.42	0.3832	1.77	0.1700	7.50
0.0458	0.51	0.1493	2.01	0.8487	7.73
0.0964	0.52	0.3659	2.26	0.2848	7.91
0.0721	0.53	0.3751	2.26	0.9526	8.20
0.0557	0.54	0.1609	2.39	1.2150	8.29
0.1114	0.55	0.4341	2.56	0.6127	8.33
0.0529	0.56	0.1500	2.57	0.6338	8.53
0.1127	0.59	0.3858	2.60	0.3012	8.58
0.0346	0.60	0.2663	2.60	1.0670	8.99
0.0964	0.61	0.1929	2.61	0.6245	10.60
0.0100	0.71	0.1300	2.79	1.2880	12.53
0.0608	0.72	0.4059	2.96	1.6640	13.90
0.1552	0.83	0.1732	3.45	0.9491	14.83
0.1217	0.89	0.9427	4.10	1.3400	17.97
0.0458	0.92	0.4387	4.42	1.8560	19.07
0.1513	1.06	0.2234	5.25	1.4560	29.97
0.1510	1.14	0.4503	5.59		

constancy to dependence on a complicated function of the measured line intensity. However, these are not likely to be large, and the remainder of the errors in this category are almost certainly proportional. It has been shown (Oertel, unpublished data) with a sample of 25 readings at each of nine levels of density, that the densitometric error is, in essence, proportional.

If the error as a whole is proportional, then the standard deviation at any level of intensity will be, on the average, a constant fraction of the magnitude of the intensity. An estimate of the standard deviation can be obtained from any replicated measurement, and the mean of the readings gives an estimate of the magnitude of the intensity. The values for the corresponding pairs of estimates recorded in Table 1 are for the sample of measurements D.

The equation for the regression of the standard deviation s on the mean m represents the best linear relationship between the two. It was found to be, for sample D,

$$s = 0.0699\,\mathrm{m} + 0.0996.$$

The value for the coefficient of regression, 0.0699, was quite significantly different from zero ($P < 0.001$). Therefore the value for s depended directly on that for m. It was not possible to make a valid test of the reality of the constant, 0.0996; but it appeared that it might be real.

It was found for each of the five samples of measurements that the value for the standard deviation was directly proportional to the magnitude of the measured line intensity. Except for sample D, there was no evidence at all of a constant component in the standard deviation. Taking into consideration the details given in Section II, it may be concluded that the random error which occurs in the D.C. arc excitation of samples of soil and plant ash is essentially a proportional error.

IV. Preparation of the Frequency Histogram

The actual value of a proportional error is difficult to interpret, but if the data are transformed to a logarithmic scale, the error is constant in that scale, its value being independent of the magnitude of the measured line intensity. The value for the logarithmic error, on its own, is a complete measure of the error.

The five samples of observations were converted to logarithms, and all subsequent calculations were made on these quantities. To make the data comparable, they were first reduced to standard measure by taking deviations from the mean of each set of replicates, and expressing them in terms of the appropriate standard deviation as the unit of measurement. For example, the standard deviation of the measurements on copper (sample A) was derived from the analysis of variance between and within the 63 triplicate sets, and was therefore based on 126 degrees of freedom. Similarly, the remaining standard deviations of samples B, C, D, and E were based, respectively, on 126, 136, 130, and 80 degrees of freedom. For convenient presentation in tabular form, the standardized errors have been arranged in order of magnitude, and grouped into classes, taking one-tenth of the appropriate standard deviation as the unit for the class interval. The frequency distributions so formed are given in Table 2.

V. Determination of the Frequency Distribution

To examine the hypothesis of normality, a preliminary test was conducted on the standardized variates. The standard deviations were presumed to be known with exactitude, and on this assumption χ^2, given by the relation

$$\chi^2 = \Sigma(x - \bar{x})^2/\text{variance},$$

was computed for each set of triplicates and sextuplicates, the summation being taken over the observations in a set of replicates. If the original

TABLE 2

FREQUENCY TABLE FOR TRIPLICATES AND SEXTUPLICATES

Centre of Class Interval	Positive Deviations					Negative Deviations				
	Triplets				Sextets	Triplets				Sextets
	A	*B*	*C*	*D*	*E*	*A*	*B*	*C*	*D*	*E*
0.05	11	14	8	9	22	10	8	8	14	10
0.15	8	7	13	5	26	8	11	10	9	16
0.25	9	3	8	9	11	10	6	10	9	19
0.35	12	13	12	12	24	8	8	8	5	18
0.45	5	8	15	8	17	7	10	8	11	12
0.55	9	5	7	12	9	2	6	9	6	10
0.65	7	7	4	4	8	9	5	9	10	12
0.75	7	9	6	6	13	6	6	10	4	18
0.85	8	10	7	6	14	10	6	4	9	13
0.95	4	7	4	3	12	6	3	4	3	6
1.05	2	2	6	5	5	4	4	4	4	4
1.15	4	1	2	3	8	2	3	4	1	6
1.25	1	1		1	9	1	2	2	6	10
1.35	1	3	1	4	7	2	2	2	2	5
1.45	3	3	2	1	3	3	2	3	1	3
1.55	1		1	1	1	1	3	1	1	5
1.65		2	3	1	2	1	2		1	7
1.75		2		3	2		1	1	1	1
1.85			1		1					1
1.95	1	1	2		1	1			1	
2.05	1				2					
2.15			1	1	2		1	1	1	2
2.25			1		1	1				
2.35					1		2		1	
2.45								1		1
2.55				1		1		1		1
2.65	2									1
—										
3.65										1
—										
4.15					1					

logarithmic data are normally distributed, then χ^2 so obtained from triplicates is distributed in the well-known χ^2-distribution with 2 degrees of freedom. Table 3 shows the distribution of the 259 values, and the test of agreement with expectation. The fit is good ($\chi^2 = 5.749$ with 9 degrees of freedom and $0.70 < P < 0.80$). The total of all 259 values is 517.86, giving a mean of 1.999 also agreeing closely with its expectation of two.

TABLE 3

FREQUENCY DISTRIBUTION OF χ^2 FROM TRIPLICATES: SAMPLES *A*, *B*, *C*, AND *D*

P	χ^2	Observed Frequency $x + m$	Expected Frequency m	x^2/m
		3	2.59	
0.99	0.0201			
		1	2.59	0.672
0.98	0.0404			
		6	7.77	
0.95	0.103			
		16	12.95	0.718
0.90	0.211			
		30	25.90	0.649
0.80	0.446			
		24	25.90	0.139
0.70	0.713			
		59	51.80	1.001
0.50	1.386			
		51	51.80	0.012
0.30	2.408			
		19	25.90	1.838
0.20	3.219			
		23	25.90	0.325
0.10	4.605			
		12	12.95	0.070
0.05	5.991			
		4	7.77	
0.02	7.824			
		4	2.59	0.325
0.01	9.210			
		7	2.59	
Total		259	259	5.749*

(The first three and the last three rows of observed and expected frequencies are bracketed together, with a single x^2/m value for each group: 0.672 and 0.325.)

* $\chi^2 = 5.749$ with 9 degrees of freedom, $0.70 < P < 0.80$; $\Sigma\chi^2 = 517.86$; and mean value of $\chi^2 = 1.999$.

In a similar manner, the 64 values of χ^2 from the sextuplicate sets should be distributed in the χ^2-distribution with 5 degrees of freedom. The relevant details are given in Table 4.

In neither case is the test exact, since the estimated variances have been substituted as parametric values in the formula given above, but nevertheless there is a strong indication that the logarithmic data exhibit variability which could be expected with normally distributed quantities. It may be mentioned that as a check on the test of the triplicates, visual interpolation was made in the tables of variance ratio distribution (Merrington and Thompson 1943), and so far as could be judged, the slight changes in the values defining the several classes of Table 3 would not have altered the distribution of the observed frequencies.

Owing to the particular nature of the original data, certain modifications were necessary before further tests of normality were made. These modifications are dependent upon the statistical properties of deviations

TABLE 4

FREQUENCY DISTRIBUTION OF χ^2 FROM SEXTUPLICATES: SAMPLE E

P	χ^2	Observed Frequency $x + m$	Expected Frequency m	x^2/m
		1	0.64	
0.99	0.554			
		0	0.64	
0.98	0.752			
		1	1.92	
0.95	1.145			2.408
		7	3.20	
0.90	1.610			
		9	6.40	
0.80	2.343			
		8	6.40	
0.70	3.000			
		13	12.80	0.003
0.50	4.351			
		8	12.80	1.800
0.30	6.064			
		7	6.40	
0.20	7.289			
		4	6.40	
0.10	9.236			
		1	3.20	
0.05	11.070			0.252
		2	1.92	
0.02	13.388			
		1	0.64	
0.01	15.086			
		2	0.64	
Total		64	64	4.463*

* $\chi^2 = 4.463$ with 3 degrees of freedom, $0.20 < P < 0.30$; $\Sigma\chi^2 = 319.39$; and mean value of $\chi^2 = 4.990$.

from the mean in samples from normal distributions, which, for the purposes of the present paper, may be summarized in the following theorem:

If $x_1\, x_2, \ldots\, x_n$ is a random sample of n observations from a normal distribution with a mean ξ and variance σ^2, and

$$\bar{x} = \Sigma x/n$$

is the sample mean, then any $n - r$ deviates $(r \neq 0)$, say $x_1 - \bar{x}$, $x_2 - \bar{x}, \ldots x_{n-r} - \bar{x}$ are distributed in a multivariate normal distribution of order $n - r$, independently of $\bar{x}$. The deviates each have a mean equal to zero, variance equal to $(n - 1)\sigma^2/n$, and the correlation between any pair is $-1/(n - 1)$.

Allowance for the change in variance was made by multiplying the standardized deviates by the factor $\sqrt{3/2}$ in the triplicates and by the factor $\sqrt{6/5}$ in the sextuplicates, and difficulties associated with the correlations were met by taking random selections from each triplicate and sextuplicate set.

(a) *First Approximate Test*

A sample of 323 observations was obtained by selecting one observation at random from each set of replicates. Assuming, as before, that the variances are known exactly, such a set of observations should, if the hypothesis of normality is true, constitute a sample from a normal distribution with a mean equal to zero and variance equal to unity. Table 5 summarizes the tests made on the three random selections taken from the data. The first shows no evidence of departure from normality, but the second and third are suspect. At first sight, it seems difficult to reconcile these results with that obtained previously, but it must be remembered that the preliminary test related only to agreement with respect to variability of the observations. The contrast given in Table 5 is much more comprehensive since it tests the agreement over the full range of the observed data. The fact that an exact test, following these lines, is available renders this approximate test redundant, but it has been included here because it provides the only means of combining all the data to build the sample up to the maximum number of independent observations (323) obtainable from the original replicate sets.

(b) *Second Approximate Test*

Pairs of observations were chosen at random from the sets of replicates, and for the data from the triplicates, correlation tables were constructed. If the original data are normally distributed, then the 259 pairs so obtained from the triplicates should constitute a sample from a normal bivariate distribution in which the correlation is $-1/2$.

The actual correlations r in the three samples taken, together with their inverse hyperbolic transformations z are:

Sample	r	z	S.D. of z
1	−0.4475	−0.4818	
2	−0.4693	−0.5092	0.0625
3	−0.5243	−0.5825	
Mean	−0.4811	−0.5245	0.0361
Expectation	−0.5000	−0.5493	

TABLE 5

COMPARISON OF THE FREQUENCY DISTRIBUTION OF THE STANDARDIZED LOGARITHMIC SPECTROGRAPHIC ERROR WITH THE NORMAL DISTRIBUTION HAVING ZERO MEAN AND UNIT STANDARD DEVIATION

Class Limits in Terms of Standard Normal Deviate	Expected Frequency m	Random Selection					
		I		II		III	
		Observed Frequency $x+m$	x^2/m	Observed Frequency $x+m$	x^2/m	Observed Frequency $x+m$	x^2/m
	0.16	0		1		0	
<—3.2905							
	1.45	5		6		8	
—3.2905-(—2.5758)							
	1.62	0	1.071	2	0.210	1	2.895
—2.5758-(—2.3263)							
	4.85	4		3		5	
—2.3263-(—1.9600)							
	8.08	3		6		9	
—1.9600-(—1.6449)							
	16.15	21	1.457	21	1.457	20	0.918
—1.6449-(—1.2816)							
	16.15	7	5.184	14	0.286	9	3.165
—1.2816-(—1.0364)							
	16.15	17	0.045	24	3.816	17	0.045
—1.0364-(—0.8416)							
	16.15	19	0.503	22	2.119	13	0.614
—0.8416-(—0.6745)							
	16.15	11	1.642	15	0.082	6	6.379
—0.6745-(—0.5244)							
	16.15	16	0.001	20	0.918	19	0.503
—0.5244-(—0.3853)							
	16.15	13	0.614	13	0.614	8	4.113
—0.3853-(—0.2533)							
	16.15	18	0.212	22	2.119	17	0.045
—0.2533-(—0.1257)							
	32.30	28	0.572	26	1.229	41	2.343

Class Limits in Terms of Standard Normal Deviate	Expected Frequency m	Random Selection I: Observed Frequency $x+m$	I: x^2/m	II: Observed Frequency $x+m$	II: x^2/m	III: Observed Frequency $x+m$	III: x^2/m
—0.1257-0.1257	16.15	22	2.119	15	0.082	21	1.457
0.1257-0.2533	16.15	12	1.066	9	3.165	17	0.045
0.2533-0.3853	16.15	25	4.850	26	6.008	26	6.008
0.3853-0.5244	16.15	20	0.918	10	2.342	17	0.045
0.5244-0.6745	16.15	18	0.212	16	0.001	11	1.642
0.6745-0.8416	16.15	21	1.457	17	0.045	16	0.001
0.8416-1.0364	16.15	17	0.045	19	0.503	17	0.045
1.0364-1.2816	16.15	11	1.642	5	7.698	14	0.286
1.2816-1.6449	8.08	7		4		4	
1.6449-1.9600	4.85	5		6		4	
1.9600-2.3263	1.62	1	0.083	0	1.648	1	1.648
2.3263-2.5758	1.45	1		1		2	
2.5758-3.2905 and >3.2905	0.16	1		0		0	
Total	323.02	323	23.693	323	34.342	323	32.197*

* χ^2: random selection I = 23.693 with 18 degrees of freedom, $0.10 < P < 0.20$; II = 34.342 with 18 degrees of freedom, $0.01 < P < 0.02$; III = 32.197 with 18 degrees of freedom, $0.02 < P < 0.05$.

The mean value of z is -0.5245, to which corresponds a value -0.4811 for the correlation, insignificantly different from the hypothetical value $-1/2$. Table 6 summarizes the test of agreement between observed and expected frequencies in the three samples.

No attempt was made to test the bivariate frequencies from the sextuplicate sets with their expectations, as the total number of observations (64) was too small. The correlations were as follows:

Sample	r	z	S.D. of z
1	-0.2238	-0.2277	
2	-0.1321	-0.1329	0.1280
3	-0.1469	-0.1480	
Mean	-0.1679	-0.1695	0.0739
Expectation	-0.2000	-0.2026	

the value obtained by combining the three estimates again agreeing, within the limits of sampling, with its expectation $-1/5$.

Further tests on the partial correlations, up to the third order, could be made with these data, but were not considered worth while in view of the small number of observations.

(*c*) *Exact Tests*

Exact tests are conducted by making due allowance for the fact that the variances have been estimated. This is accomplished by taking a standardized deviate as the statistic t,* instead of a normal deviate, and testing the agreement of the observed distribution with that of t having appropriate degrees of freedom. This test was first made using data from all the replicated sets. Under these conditions, samples A and B could be combined to give 126 observations from the t-distribution with 126 degrees of freedom, but samples C, D, and E had to be treated individually, since they yielded, respectively:

Observations from t-distribution	68	65	64
Degrees of freedom	136	130	80

As before, three random selections were made, and in all cases the observed distributions agreed with expectation within the limits of variation attributable to random sampling.

The triplicates of samples A, B, C, and D were then used to obtain a larger set of observations. From sample C, five triplicate sets, and from sample D, two triplicate sets were selected at random and discarded;

* Strictly speaking, these quantities are not values of the variate t, because the standard deviations used in the process of standardization were derived, for convenience, from all the replicate sets in each of the samples, A, B, C, and D.

TABLE 6

COMPARISON OF THE BIVARIATE FREQUENCY DISTRIBUTION OF THE STANDARDIZED LOGARITHMIC SPECTROGRAPHIC ERROR WITH THE BIVARIATE NORMAL DISTRIBUTION HAVING ZERO MEANS, UNIT STANDARD DEVIATIONS, AND CORRELATION COEFFICIENT $-\frac{1}{2}$

Class Limits in Terms of Standard Normal Deviate for Both Variates	Expected Frequency m	Random Selection I			
		Observed* Frequency $x+m$	x^2/m	Observed† Frequency $x+m$	x^2/m
0-0.1633	17.73	19	0.091	18	0.004
0.1633-0.3266	11.96	13	0.090	7	2.057
0.3266-0.4899	7.11	12		10	
0.4899-0.6532	3.70	3		4	
0.6532-0.8165	1.68	1	0.475	3	0.925
0.8165-0.9798	0.66	—		—	
>0.9798	0.32	—		—	
0-0.1633	19.59	20‡	0.009	19§	0.018
0.1633-0.3266	17.02	16	0.061	15	0.240
0.3266-0.4899	14.09	16	0.259	15	0.059
0.4899-0.6532	11.11	14	0.752	6	2.350
0.6532-0.8165	8.33	9	0.036	11	0.969
0.8165-0.9798	5.95	6		7	
0.9798-1.1431	4.04	1		6	
1.1431-1.3064	2.61	1		2	
1.3064-1.4697	1.60	2		—	
1.4697-1.6330	0.94	1	2.590	—	0.001
1.6330-1.7963	0.52	—		—	
1.7963-1.9596	0.27	—		1	
1.9596-2.1229	0.14	—		1	
>2.1229	0.07	—		—	

χ^2: random selection I = 10.986 with 17 degrees of freedom, $0.80 < P < 0.90$

* Both variates + ve. † Both variates — ve.

‡ First variate + ve and second variate — ve to end of column.

§ First variate — ve and second variate + ve to end of column.

TABLE 6 (*Continued*)

Class Limits in Terms of Standard Normal Deviate for Both Variates	Expected Frequency m	Random Selection II: Observed* Frequency $x + m$	x^2/m	Observed† Frequency $x + m$	x^2/m
0-0.1633	17.73	15	0.420	17	0.030
0.1633-0.3266	11.96	15	0.773	10	0.321
0.3266-0.4899	7.11	10		3	
0.4899-0.6532	3.70	1		—	
0.6532-0.8165	1.68	2	0.021	1	4.143
0.8165-0.9798	0.66	1		2	
>0.9798	0.32	—		—	
0-0.1633	19.59	26‡	2.097	19§	0.018
0.1633-0.3266	17.02	12	1.481	16	0.061
0.3266-0.4899	14.09	27	11.289	20	2.479
0.4899-0.6532	11.11	14	0.752	8	0.871
0.6532-0.8165	8.33	5	0.753	7	0.115
0.8165-0.9798	5.95	6		6	
0.9798-1.1431	4.04	3		2	
1.1431-1.3064	2.61	2		1	
1.3064-1.4697	1.60	5		2	
1.4697-1.6330	0.94	1	0.077	—	2.590
1.6330-1.7963	0.52	—		—	
1.7963-1.9596	0.27	—		—	
1.9596-2.1229	0.14	—		—	
>2.1229	0.07	—		—	

χ^2: random selection II = 28.831 with 17 degrees of freedom, $0.02 < P < 0.05$

* Both variates + ve. † Both variates — ve.
‡ First variate + ve and second variate — ve to end of column.
§ First variate — ve and second variate + ve to end of column.

TABLE 6 (*Continued*)

Class Limits in Terms of Standard Normal Deviate for Both Variates	Expected Frequency m	Random Selection III			
		Observed* Frequency $x+m$	x^2/m	Observed† Frequency $x+m$	x^2/m
0-0.1633	17.73	14	0.785	17	0.030
0.1633-0.3266	11.96	8	1.311	9	0.733
0.3266-0.4899	7.11 }	10 }		7 }	
0.4899-0.6532	3.70 }	4 }		— }	
0.6532-0.8165	1.68 }	1 }	0.475	2 }	1.483
0.8165-0.9798	0.66 }	1 }		— }	
>0.9798	0.32 }	— }		— }	
0-0.1633	19.59	24‡	0.993	20§	0.009
0.1633-0.3266	17.02	18	0.056	17	0.000
0.3266-0.4899	14.09	18	1.085	18	1.085
0.4899-0.6532	11.11	15	1.362	10	0.111
0.6532-0.8165	8.33 }	10 }	0.005	10 }	0.207
0.8165-0.9798	5.95 }	4 }		6 }	
0.9798-1.1431	4.04 }	2 }		3 }	
1.1431-1.3064	2.61 }	— }		1 }	
1.3064-1.4697	1.60 }	5 }		2 }	
1.4697-1.6330	0.94 }	1 }	0.001	— }	1.677
1.6330-1.7963	0.52 }	— }		— }	
1.7963-1.9596	0.27 }	— }		— }	
1.9596-2.1229	0.14 }	1 }		— }	
>2.1229	0.07 }	1 }		— }	

χ^2: random selection III = 11.408 with 17 degrees of freedom; $0.80 < P < 0.90$

* Both variates + ve. † Both variates — ve.

‡ First variate + ve and second variate — ve to end of column.

§ First variate — ve and second variate + ve to end of column.

TABLE 7

COMPARISON OF THE FREQUENCY DISTRIBUTION OF THE STANDARDIZED LOGARITHMIC SPECTROGRAPHIC ERROR WITH THE t-DISTRIBUTION OF 126 DEGREES OF FREEDOM

Class Limits for t	Expected Frequency m	Random Selection I Observed Frequency $x+m$	I x^2/m	II Observed Frequency $x+m$	II x^2/m	III Observed Frequency $x+m$	III x^2/m
	0.13	0		0		0	
<—3.3694							
	1.13	3		4		5	
3.3694-(—2.6154)							
	1.26	1	0.537	2	0.457	2	2.314
—2.6154-(—2.3563)							
	3.78	3		3		5	
—2.3563-(—1.9790)							
	6.30	3		6		6	
—1.9790-(—1.6570)							
	12.60	16	0.917	16	0.917	18	2.314
—1.6570-(—1.2883)							
	12.60	4	5.870	11	0.203	7	2.489
—1.2883-(—1.0407)							
	12.60	13	0.013	16	0.917	13	0.013
—1.0407-(—0.8445)							
	12.60	18	2.314	16	0.917	12	0.029
—0.8445-(—0.6764)							
	12.60	7	2.489	10	0.537	5	4.584
—0.6764-(—0.5257)							
	12.60	11	0.203	18	2.314	14	0.156
—0.5257-(—0.3862)							
	12.60	10	0.537	11	0.203	6	3.457
—0.3862-(—0.2539)							
	12.60	13	0.013	19	3.251	11	0.203
—0.2539-(—0.1259)							
	25.20	25	0.002	22	0.406	32	1.835

TABLE 7 (*Continued*)

Class Limits for t	Expected Frequency m	Random Selection I: Observed Frequency $x+m$	I: x^2/m	II: Observed Frequency $x+m$	II: x^2/m	III: Observed Frequency $x+m$	III: x^2/m
—0.1259-0.1259	12.60	12	0.029	10	0.537	14	0.156
0.1259-0.2539	12.60	12	0.029	5	4.584	14	0.156
0.2539-0.3862	12.60	22	7.013	17	1.537	22	7.013
0.3862-0.5257	12.60	17	1.537	9	1.029	14	0.156
0.5257-0.6764	12.60	13	0.013	12	0.029	10	0.537
0.6764-0.8445	12.60	16	0.917	15	0.457	11	0.203
0.8445-1.0407	12.60	16	0.917	16	0.917	11	0.203
1.0407-1.2883	12.60	8	1.679	5	4.584	9	1.029
1.2883-1.6570	6.30	5		2		4	
1.6570-1.9790	3.78	3		6		4	
1.9790-2.3563	1.26	0	1.029	0	1.029	1	0.203
2.3563-2.6154	1.13	1		1		2	
2.6154-3.3694, >3.3694	0.13	0		0		0	
Total	252.00	252	26.058	252	24.825	252	27.050

χ^2: random selection I = 26.058 with 18 degrees of freedom, $0.05 < P < 0.10$; II = 24.825 with 18 degrees of freedom, $0.10 < P < 0.20$; III = 27.050 with 18 degrees of freedom, $0.05 < P < 0.10$

two new estimates of variance, now on 126 degrees of freedom, were obtained, and a completely new set of standardized deviates calculated. Random selections from samples C and D could then be amalgamated with those from samples A and B to give 252 observations from the t-distribution with 126 degrees of freedom. Table 7 shows the comparison of observed and expected frequencies. The values of χ^2 are high, but not beyond the limits usually accepted as due to sampling.

The k-statistics up to the sixth order were computed for these three sampling distributions, and it is of some interest to compare the values obtained with the known values of the first six cumulants of the t-distribution with 126 degrees of freedom. The contrast is made in Table 8.

VI. The Frequency Distribution of the Coefficient of Variation

Although all errors derived from the samples of triplicated and sextuplicated measurements were also converted into what may be conveniently termed coefficients of variation (natural number deviation as fraction of natural number mean), the frequency distribution of this variate was not examined in the same detail. A coefficient of variation has one advantage only, namely, that its practical significance is readily appreciated. Its statistical properties are complicated by the fact that it is the ratio of two variates, and its use in statistical tests is inconvenient and time-consuming.

In general, it was found that the properties of this distribution did not differ appreciably from those of the logarithmic standard deviation.

VII. Conclusion

For some years following publication of the first edition of *Statistical Methods for Research Workers* (Fisher loc. cit.), doubts respecting the validity of the tests given in this book to non-normal data arose in many quarters. From the outset, Fisher maintained that he had never known difficulty to arise from imperfect normality of variation, and demonstrated (Fisher 1942) how randomization, which is necessary for the validity of any test of significance, affords the means of verifying arithmetically the appropriateness of the t- and z-tests for any particular series of observations. Further contributions to work along these lines, which have been made by Pearson (1931), Rider (1931), Baker (1932), Eden and Yates (1933), Pitman (1937*a*, 1937*b*, 1938), Davies and Cornish (unpublished data), and others, have served to reaffirm Fisher's original statements. The general conclusion to be drawn from this work is that the t- and z-distributions are remarkably insensitive to the departures from normality usually encountered in practice, and this, coupled with the fact that the observations reported herein are not numerous and the degree of replication is so small, makes a statement regarding the normality of the

TABLE 8

CUMULANTS OF THE t-DISTRIBUTION AND k-STATISTICS OF RANDOM SAMPLES

Cumulant	General Formulae for ν Degrees of Freedom	Value for $\nu = 126$	k-Statistic	Random Selection			Standard Deviation
				I	II	III	
κ_1	0	0	k_1	0.03485	—0.1382	—0.06481	0.0635
κ_2	$\nu/(\nu-2)$	1.0161	k_2	0.9311	1.0181	1.1259	0.0918
κ_3	0	0	k_3	—0.2861	—0.1388	—0.2733	0.1649
κ_4	$6\nu^2/(\nu-2)^2(\nu-4)$	0.0508	k_4	0.4260	0.2546	0.8493	0.3480
κ_5	0	0	k_5	0.6794	—0.1756	1.5218	0.8388
κ_6	$240\nu^3/(\nu-2)^3(\nu-4)(\nu-6)$	0.0172	k_6	—0.3167	—1.4927	—4.1997	2.2781

distribution of the logarithmic spectrographic error tentative only. At the same time, the results indicate that statistical measures and tests of significance can be confidently used with this type of observation. The complete set of data was remarkably homogeneous, although drawn from a variety of sources, and, consequently, the conclusions are more reliable than would otherwise be expected.

VIII. References

BAKER, G. A. (1932).—*Ann. Math. Statist.* **3**: 1.

EDEN, T., and YATES, F. (1933).—*J. Agric. Sci.* **23**: 6.

FISHER, R. A. (1925).—"Statistical Methods for Research Workers." (Oliver and Boyd: Edinburgh.)

FISHER, R. A. (1942).—"The Design of Experiments." (Oliver and Boyd: Edinburgh.)

MERRINGTON, M., and THOMPSON, C. M. (1943).—*Biometrika* **33**: 74.

OERTEL, A. C. (1950*a*).—*Aust. J. Appl. Sci.* 1: 152.

OERTEL, A. C. (1950*b*).—*Aust. J. Appl. Sci.* 1: 259.

PEARSON, E. S. (1931).—*Biometrika* **23**: 114.

PITMAN, E. J. G. (1937*a*).—*Suppl. J. R. Statist. Soc.* **4**: 119.

PITMAN, E. J. G. (1937*b*).—*Suppl. J. R. Statist. Soc.* **4**: 225.

PITMAN, E. J. G. (1938).—*Biometrika* **29**: 322.

RIDER, P. R. (1931).—*Ann. Math. Statist.* 2: 48.

"STUDENT" (1908).—*Biometrika* **6**: 1.

Reprinted from the Australian Journal of Physics, Volume 7, Number 2, pp. 334–346, 1954

ON THE SECULAR VARIATION OF RAINFALL AT ADELAIDE

By E. A. Cornish*

[*Manuscript received January 25, 1954*]

Summary

A detailed analysis of the rainfall of Adelaide has established that periodic changes occur in the incidence and duration of the winter rains. These changes have a period and amplitude of approximately 23 years and 30 days respectively, and superimposed on them is a long-term trend which is manifested by protraction of the latter half of the season, spring rains now occurring about 3 weeks later than they did just over 100 years ago. The total quantity of rain precipitated has shown no statistically significant changes.

I. Introduction

The daily rainfall observations recorded at Adelaide during the 95 years 1839–1933 inclusive have been subjected to detailed analysis, and the results reported (Cornish 1936). This investigation pointed to the occurrence of periodic changes in the incidence and duration of the winter rains, with a period of approximately 23 years and amplitude of 30 days, but no statistically significant changes were demonstrable in the amount of rain precipitated. The periodic trend was very clearly defined from 1839 to about 1912, but thereafter the observations showed some evidence of a breakdown in the law of change. With the accumulation of further data, the opportunity has been taken to re-examine this point and to revise the analysis.

II. Data

On January 1, 1839, Sir George Kingston established a daily rainfall record at Adelaide on a site approximately 500 yd from the present position of the Observatory. This record was continued until November 1879. From May 1860, readings have been taken at the Observatory, so that over 19 years the two sets of observations were concurrent. During this interval, the average annual difference between the gauges was only 0·26 in. (Kingston's being the greater), and, considering the close proximity of the sites, it may be assumed that the two series in combination give a continuous and practically uniform record of the Adelaide rainfall. No definite statement could be found regarding the diameter of the gauge employed by Kingston, nor, in fact, the size of the gauge used in the early days of the existence of the Observatory. It is fairly certain, however, that no radical departure could have been made from the standard 8-in. gauge which has been in use since 1870.

The analysis previously reported has now been extended to include the rainfall observations of 1934–1950 inclusive.

* Section of Mathematical Statistics, C.S.I.R.O., Adelaide.

III. Analysis

The basic data were obtained by a method devised by Fisher (1924). The rainfall of each year was divided into 61 six-day totals, and to each of these annual sequences a series of orthogonal polynomial functions of the fifth degree in time was fitted, thus furnishing six quantities with which the amount and distribution of rain in each year could be represented. These distribution constants, designated a', b', . . ., f', have been tabled (Cornish loc. cit.) for the years 1839–1933 inclusive, and supplementary data for the period 1934–1950

Table 1

RAINFALL DISTRIBUTION VALUES FOR 1934–1950

Unit : 10^{-3} in.

Year	a'	b'	c'	d'	e'	f'
1934	332	+96	—27	—47	—16	+ 3
1935	384	+23	—64	— 6	+17	— 5
1936	317	+44	— 1	+ 1	+31	—11
1937	378	+15	—30	— 7	+18	+ 1
1938	316	—46	—49	+26	+ 5	— 1
1939	381	— 6	—48	— 9	— 1	—12
1940	265	+ 9	—23	— 2	+16	—16
1941	370	—13	—26	—42	— 1	+27
1942	417	+ 6	—94	— 9	+27	— 9
1943	291	—27	—27	— 5	— 1	+13
1944	281	+20	—18	+33	+ 2	—18
1945	293	+64	— 9	—26	— 4	— 2
1946	370	—23	—11	+17	+10	+13
1947	359	+41	—47	— 3	+ 6	+16
1948	351	+33	—51	+15	—14	—23
1949	299	+43	—29	—29	—27	+ 1
1950	263	+29	—52	—10	+11	— 9
Mean 1839–1950	345·35	+14·80	—56·07	— 2·72	+15·67	— 3·28

inclusive are given in Table 1. For convenience of presentation in tabular form, the unit is 10^{-3} in. These quantities are actually proportional to the coefficients of the polynomial terms, and to obtain the latter they must be divided by factors of the form

$$\{(r!)^2 60.59 \ . \ . \ . \ (61-r)\}/(2r+1)!,$$

where r is the degree of the term fitted.

The first constant a' represents the average rainfall in 10^{-3} in. per 6-day period, b' is proportional to the linear term of the seasonal sequence, c' is proportional to the parabolic term, and so on, taking more complex features of the seasonal distribution into account.

In Figures 1, 2, and 3, the courses of the secular changes in a', b', and c' respectively have been depicted by plotting running 10-year means of each.

The diagram of a' shows quite marked changes in the amount of rain precipitated, with two short periods of high rainfall centred at 1850 and 1920, when the mean annual rainfall was approximately 24 in., separated by a long interval in which the mean fell to about 20 in.

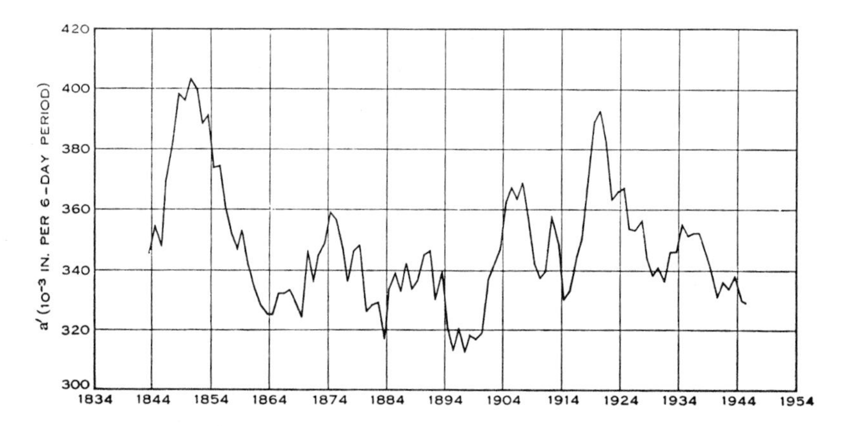

Fig. 1.—Ten-year means of the distribution value a'.

Changes in the course of b' are also prominent, and even more complex than those observed with a', the diagram showing clearly that the swing in the mean is again becoming regular, following the departure observed between 1910 and 1930. As will appear later, the effects of this disturbance are not so serious as

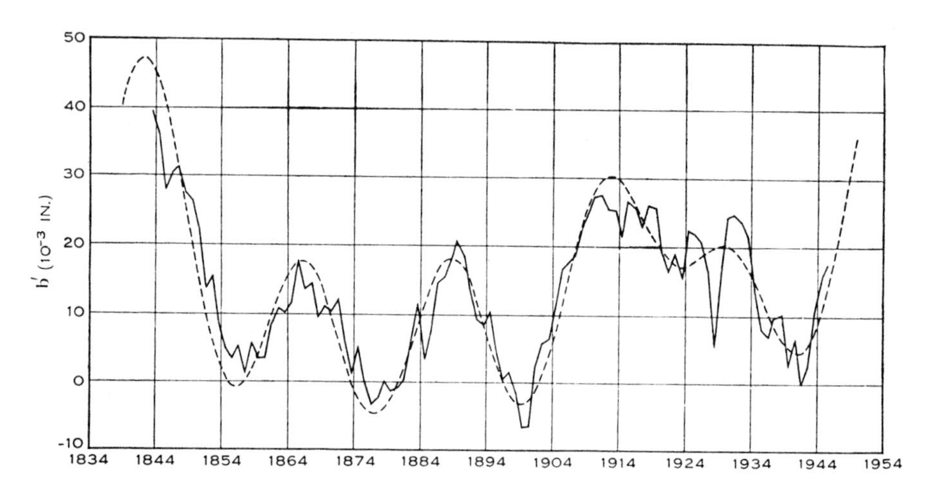

Fig. 2.—Ten-year means and harmonic curve of the distribution value b'.

the figure might indicate. By comparison, c' follows a simple course with a distinct and apparently uniform increase, the mean having increased from -67 in 1839 to -45 in 1950 (both values approximate).

The secular changes observed in the six distribution constants were then examined in more detail by fitting each series of 112 values with an orthogonal polynomial of the fifth degree in time, and the significance of the polynomials re-tested on all the available data. The procedure adopted in this step may be illustrated by the calculations relating to c'. The following values were obtained, the unit being 10^{-3} in. :

Mean	$-56\cdot07143$	x'_1	$-593\cdot40$	$(x'_1)^2$	$352{,}123\cdot56$
Coefficient of 1st degree	$3\cdot67983$	x'_2	$68\cdot06$	$(x'_2)^2$	$4{,}632\cdot16$
,, ,, 2nd ,,	$1\cdot76860$	x'_3	$42\cdot99$	$(x'_3)^2$	$1{,}848\cdot14$
,, ,, 3rd ,,	$0\cdot25608$	x'_4	$7\cdot56$	$(x'_4)^2$	$57\cdot15$
,, ,, 4th ,,	$1\cdot03557$	x'_5	$35\cdot95$	$(x'_5)^2$	$1{,}292\cdot40$
,, ,, 5th ,,	$-1\cdot07097$	x'_6	$-42\cdot98$	$(x'_6)^2$	$1{,}847\cdot28$

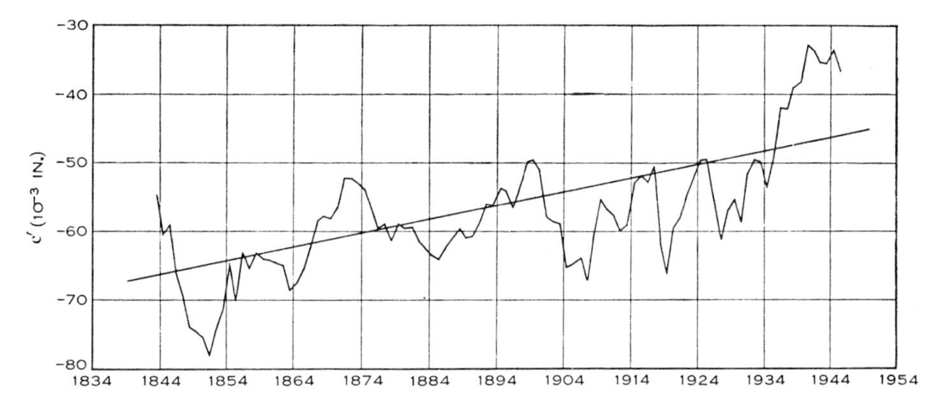

Fig. 3.—Ten-year means and linear regression of the distribution value c'.

The quantities $x'_1, \ldots, x'_6$ are orthogonal and normal linear functions of the 112 values of c', obtained from the corresponding quantities in the first column by multiplying the coefficient of the term of degree r by

$$\left\{\frac{(2r+1).112.113. \ldots (112+r)}{111.110. \ldots (112-r)}\right\}^{\frac{1}{2}},$$

and $x'_2, \ldots, x'_6$ represent the several components of secular change in the c' sequence. The last column gives the squares of these functions. The total variation of the c' values from their mean may be divided into two portions :

(1) a sum of squares associated with regression on time, and due to a comparatively simple temporal trend predominating over the random fluctuations,

(2) the remaining sum of squares which may be attributed to random annual variation,

and the first of these components may be partitioned further to give the individual contributions of the terms of the several degrees to the total for regression. The analysis of variance given in Table 2 establishes the strong significance ($P<0\cdot01$) of the linear term, thus confirming the feature noted in Figure 3.

Table 3 provides a summary analysis of all six distribution constants, the unit being 10^{-3} in.

No significant changes have occurred in a', d', e', and f', which thus merely fluctuate with large standard deviations about their respective mean values. This confirms the previous findings with respect to a', d', and e'. With f', however, the original analysis indicated an apparently significant parabolic

TABLE 2

ANALYSIS OF VARIANCE OF THE c' SERIES

Variation Due to	Degrees of Freedom	Sum of Squares	Mean Square	Variance Ratio
Linear component	1	4,632·16	4,632·16	6·94
Quadratic ,, ...	1	1,848·14	1,848·14	n.s.
Cubic ,, ...	1	57·15	57·15	n.s.
Quartic ,, ...	1	1,292·40	1,292·40	n.s.
Quintic ,, ...	1	1,847·28	1,847·28	n.s.
Deviations from regression ..	106	70,714·30	667·12	
Total	111	80,391·43		

component, but this result had been treated with caution since the f' series had shown a considerable departure from normality. It is of some interest to see that, with more extensive data, this term now degenerates to non-significance.

The significance of the linear term in the course of c' is a new feature, and the remaining significant effects, namely, the quadratic and cubic components

TABLE 3

SUMMARY POLYNOMIAL ANALYSIS OF RAINFALL DISTRIBUTION VALUES

Unit: 10^{-3} in.

	a'	b'	c'	d'	e'	f'
Mean	345·35	14·80	—56·07	— 2·72	15·67	— 3·28
x_2'	— 45·22	— 7·49	68·06	—16·98	—30·87	15·94
x_3'	17·46	62·70	42·99	—12·63	—23·20	22·72
x_4'	— 72·95	—64·85	7·56	14·91	—32·91	—27·55
x_5'	—116·24	36·96	35·95	3·02	—21·84	— 6·49
x_6'	65·14	19·22	—42·98	11·88	1·20	—16·77
Standard residual ..	68·36	29·88	25·83	18·91	17·55	14·10
t (106 deg. freedom)		5·24	22·97	1·52	9·45	2·46

of the changes in b', confirm the conclusion drawn earlier. The analysis of the b' sequence was actually carried as far as the term of the ninth degree, but since coefficients of terms beyond the fifth degree are quite insignificant they have not been quoted.

Oscillations in the course of b' not associated with the polynomial were conspicuous, and it thus appeared desirable to test whether a harmonic series would take account of them. The analysis was taken as far as the fifth harmonic yielding the series

$$14\cdot80+6\cdot72\cos\theta-5\cdot00\sin\theta+2\cdot46\cos2\theta+4\cdot68\sin2\theta+6\cdot96\cos3\theta$$
$$+5\cdot02\sin3\theta+4\cdot89\cos4\theta-0\cdot12\sin4\theta+3\cdot96\cos5\theta+10\cdot03\sin5\theta,$$

where $\theta=0$ in 1839. The significance of this series is due to the fifth harmonic. The fitted curve is given in Figure 2, and evidently follows the oscillations not represented by the polynomial. For this reason, the harmonic series was used in place of the polynomial fitted to b' in subsequent statistical tests.

An attempt was made to provide a more precise test of the significance of the harmonic series fitted to the b' sequence and the polynomials fitted to the remaining distribution constants, by subjecting the whole of the original data, comprising 6832(112 $\times$ 61) six-day totals of rainfall to one comprehensive analysis of variance. With rainfall measured in inches, the primary subdivision of the total variation is as follows :

Variation Due to	Degrees of Freedom	Sum of Squares
Between 6-day means ..	60	159·3083
Between years	111	31·7711
Residual	6660	1412·7703
Total	6831	1603·8497

In the further subdivision of these portions of the total variation, to show the contributions of the harmonic and the polynomials, the individual sums of squares must be placed on a comparable basis. This is accomplished by normalizing the distribution constant corresponding to the original polynomial of degree r using the factor

$$\left\{\frac{(2r+1).61.62\ldots(61+r)}{60.59\ldots(61-r)}\right\}^{\frac{1}{2}},$$

or, more simply, by multiplying the appropriate sum of squares in the analysis, by the square of the above quantity. This test confirmed the significances obtained above, but, in the presence of such excessive annual variation, is still not sensitive enough to detect any other secular effects.

The last line of Table 3 gives the results of tests of significance on the mean values of the constants, with the exception of a'. Apart from the mean value of d', the remainder are significant and thus may be regarded as permanent characteristics of the season. The average seasonal distribution, represented by these mean values, is given in Figure 4, where it is contrasted with the means of the 6-day totals (on a per-day basis) and the distribution obtained by fitting the average 6-day totals with a harmonic series which takes the form

$$5\cdot77-3\cdot38\cos\theta-0\cdot384\sin\theta+0\cdot290\cos2\theta-0\cdot636\sin2\theta,$$

where $\theta=0$ at the mid point of the first 6-day interval in the year. This curve is identical with what would have been obtained if harmonics of this form had been fitted to the 61 values of each year, and averaged over the 112 years. Further discussion of the polynomial and harmonic representations is given below.

IV. Representation of the Season

When the original investigation was reported (Cornish loc. cit.), Whipple (1936) criticized the representation of individual years of rainfall and the average annual seasonal distribution by means of polynomials. The polynomials were designed to represent individual seasons, and are appropriate for this purpose because they do not impose the condition of periodicity on the weather, which does not, in fact, repeat itself year by year. On the other hand, the *average* rainfall sequence, being periodic, is properly represented by a harmonic curve.

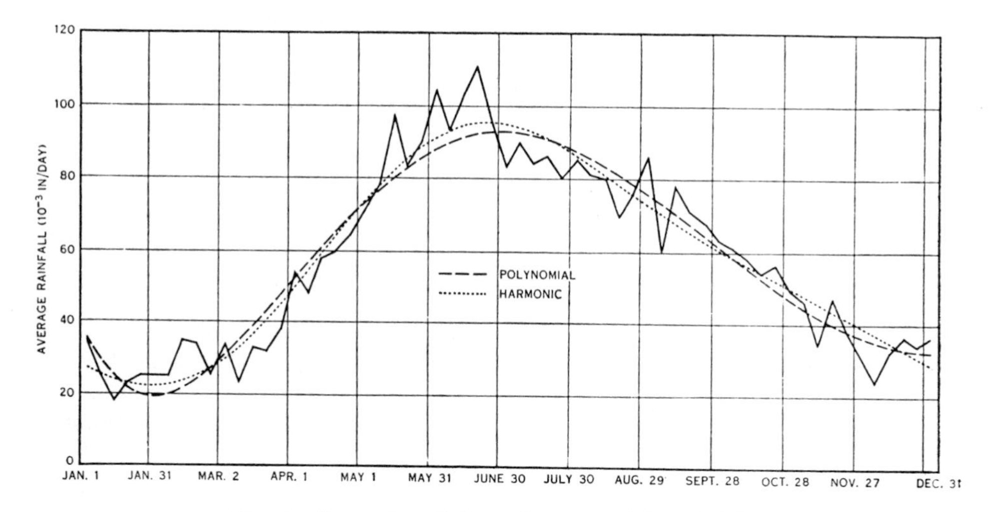

Fig. 4.—Seasonal variation of average daily rainfall.

As polynomials had been used for the annual observations, it was a natural step to show how they represented the average season, but at no stage was it claimed that they were designed for this purpose, or that their average was superior to a harmonic series. Whipple stated " the 12 monthly totals would give twice as much information about the distribution of rain through the year as six terms of the series of polynomials ", but gave no reason why they should do so when fitted with a harmonic series.

Instead, however, of using mean monthly rainfall, advantage may be taken of the finer detail supplied by the 61 six-day totals, since the superiority of either method of representation will be demonstrated more clearly with these totals than with the monthly means. The primary subdivision of the sum of squares given in Section III forms the starting point. The contribution made by the average polynomial is 145·8501, and the sum due to the harmonic is 148·2759, so that, after making allowance for the secular changes in the distribution constants, the two curves can be properly contrasted in the following analysis of variance :

Variation	Polynomial Curve			Variation	Harmonic Curve		
	Degrees of Freedom	Sum of Squares	Mean Square		Degrees of Freedom	Sum of Squares	Mean Square
Polynomial ..	5	145·8501	29·1700	Harmonic ..	4	148·2759	37·0690
Residual ..	55	13·4582	0·2447	Residual ..	56	11·0324	0·1970
Between 6-day means ..	60	159·3083			60	159·3083	
Residual ..	6630		0·2114				

Neither mean square, 0·2447 and 0·1970, differs significantly from 0·2114, and consequently both curves represent the average sequence well. The harmonic gives slightly the better representation, since its residual variance is smaller (and it absorbs one less degree of freedom), but comparison of the respective mean squares does not substantiate Whipple's estimate. The amount of information is the inverse of the variance, and the figures are 4·09 units for the polynomial and 5·08 units for the harmonic, the latter being only 25 per cent. greater. It must be emphasized again that the test has been made merely to examine the force of Whipple's statement.

TABLE 4

TEST OF NORMALITY OF RAINFALL DISTRIBUTION VALUES

Measure of Departure	a'	b'	c'	d'	e'	f'	Standard Deviation
g_1	0·2873	0·1666	0·0400	0·0781	0·2398	0·3900	0·2284
g_2	—0·3400	—0·1270	—0·2474	0·3485	0·4626	2·6497	0·4531

V. FREQUENCY DISTRIBUTION OF a', b', . . ., f'

All tests of significance employed are based upon the assumption of normality of the distribution constants a', b', . . ., f'. Table 4 summarizes a test of normality, and shows the measures of departure with standard deviations appropriate to a normal distribution. These tests confirm the results previously given, f' again being the only quantity to show a significant departure from normality ; the positive value of g_2 indicates a symmetrical departure such that the apex and two tails of the distribution are increased at the expense of the shoulders.

VI. CORRELATIONS OF THE DISTRIBUTION CONSTANTS

It has been observed that in the South Australian environment intra-station correlations of rainfall for subdivisions of the season are very weak, and in this connexion it is of some interest to compare these observations with correlations obtained from the extensive Adelaide record. The distribution constants have been calculated from uncorrelated functions of time, so that if rainfall at any time of the season is correlated with rainfall at any other time, the relation should appear on correlating them. After making allowance for secular changes, the correlations were determined and their inverse hyperbolic transformations $z = \tanh^{-1} r$, where r is the correlation coefficient, are presented in Table 5. The variate z is very nearly normally distributed, each value in the table having a standard deviation of 0·0958. Judging the data as a whole, the covariation is undoubtedly real ($\chi^2 = 81{\cdot}35$ with 15 degrees of freedom), but obviously from the individual values of z the correlations are all very weak, a result which confirms the observations made for stations distributed throughout the winter rainfall zone in South Australia.

TABLE 5

INTER-CORRELATIONS OF RAINFALL DISTRIBUTION VALUES

Values of $z = \tanh^{-1} r$: standard deviation 0·0958

	a'	b'	c'	d'	e'
b'	0·02536				
c'	—0·5565	0·02709			
d'	0·006209	—0·2769	0·06659		
e'	0·2420	0·1539	—0·2536	—0·03694	
f'	—0·2665	0·1121	0·2785	—0·06205	—0·2030

VII. CHANGES IN THE RAINFALL SEQUENCE

To express in simpler form the changes in the rainfall sequence represented by the secular trends of b' and c', the data were next presented from another standpoint. Examination of the monthly records disclosed that the main fluctuation seemed to concern the date of incidence of the winter rains, and it therefore appeared desirable to ascertain to what extent the oscillation in this date and the mean date of attainment of the yearly maximum of rainfall would account for the disturbances observed. In addition, there was definite evidence that spring and early summer rains had advanced toward the end of the year.

Consider the total rainfall of any two successive periods of 183 days. If the date dividing them falls in the spring, the second will generally contain the smaller quantity of rain, and vice versa if the day of division falls in the autumn. When daily differences of such totals are taken, a sequence of values should be obtained which changes sign regularly twice per rainfall year, first in the winter from negative to positive, and secondly in the summer from positive to negative.

These differences were determined for each day of the year in the 112 years of observations. With the exception of several years, the change of sign in winter and summer was defined very clearly. In exceptional years, the differences alternated in sign for short periods of varying lengths before definitely adopting the opposite sign, and in such cases the day of zero difference was obtained by smoothing the series with successive 10-day means.

The date in the winter at which the preceding 6 months had received as much rain as the 6 months following may be regarded as an empirical median of the rainfall sequence, and, in the original investigation of the changes in the mean value of b', two quartiles were located, one on either side of the median, between each of which and the median, one-quarter of the rainfall of the year surrounding the median date, had fallen. To these dates, three others corresponding to the octiles $\frac{3}{8}$, $\frac{5}{8}$, and $\frac{7}{8}$ have been added. These additional figures

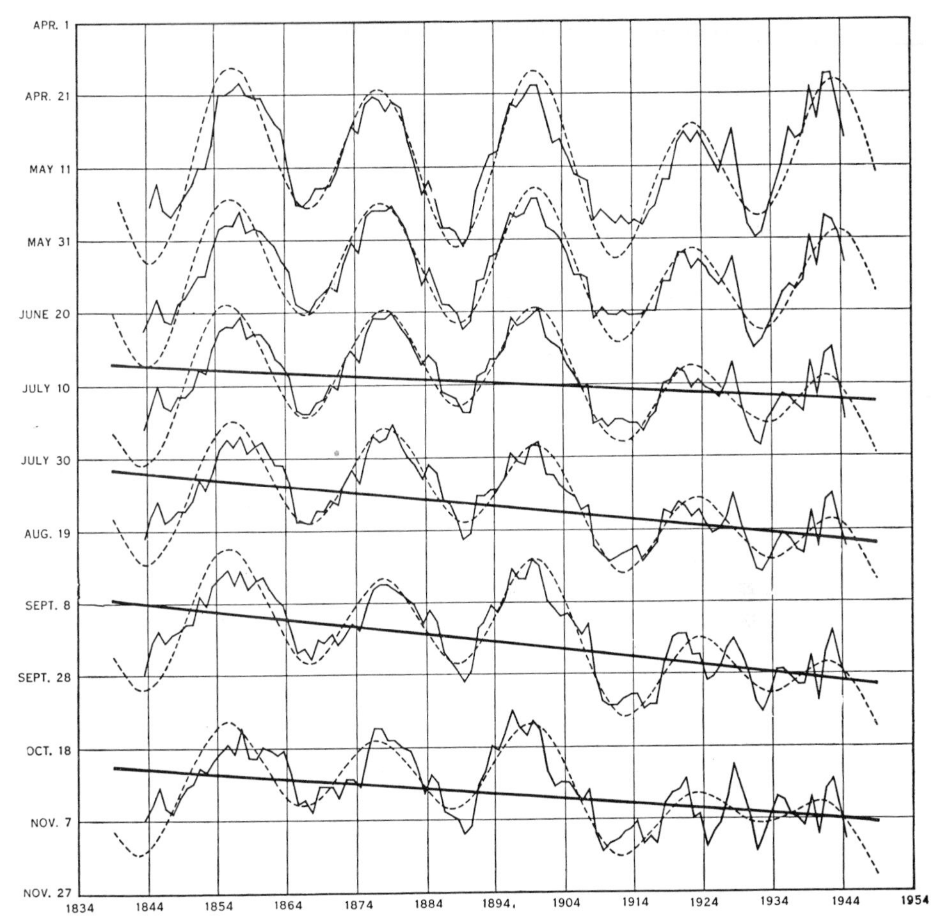

Fig. 5.—Ten-year means and harmonic curves of the octiles of the rainfall year, with linear components in the dates from the median onward.

serve to provide more detail of the changes expressed by b', but their principal use is to illustrate more clearly the linear trend represented by c'. The whole series is given in Figure 5 by plotting 10-year means of the several dates, and superimposing the harmonic curves *fitted to the original data.*

The six graphs all agree with the course of b'. The periods closely approximate each other, and the movements are in phase, but the displacements are in opposite directions, which must necessarily follow from the physical nature of the coefficient b'. The abnormal behaviour of the latter between 1910 and 1930 is reflected in all the curves of Figure 5, though not nearly to the same degree in the first quartile and third octile.* The reason for this is that the curves of Figure 5 are based on the rainfall year, which is a natural unit, whereas b' is dependent upon rain falling in the calendar year. There is definite evidence also from the curves of the first quartile and third octile that the course of b' is again becoming regular.

The changes expressed by b' are quite distinct from those of c', and represent a regular oscillation, of which the period and amplitude are approximately 23 years and 30 days respectively, in the dates of incidence of the winter rains and attainment of the yearly maximum of rainfall. This oscillation has been confined principally to a portion of the seasonal distribution, for the date of the minimum has remained practically constant throughout the 112 years, the section exhibiting greatest movement being that extending from April to November.

The amplitude of cycles occurring after 1910 is gradually reduced in the curves from the median date onward, and in the seventh octile has been almost entirely eliminated. This is undoubtedly due to the progressive change represented by c'; evidently some considerable time had to lapse before the cumulative effect became dominant, because the first three oscillations are clearly defined. The linear components in the curves from the median onward are all strongly significant, and are given in Figure 5. As the diagrams indicate, the changes expressed by c' have been confined to the latter half of the season, and amount to protraction of the later rains, the median, fifth octile, third quartile, and seventh octile having advanced by 11, 21, 24, and 17 days respectively in the period under review. The record is too short to permit any definite statement regarding this remarkable feature of the data, but it probably constitutes portion of a long-term oscillation, of which all that can be said at the moment is that the period (if it is periodic) is not less than 225 years.

The mutual agreement of the curves substantiates the view that a shift of date, regardless of quantity of seasonal precipitation, provides an adequate description of the changes in progress. In absolute terms, neither the amplitude of the cyclic movement, nor the total extension of the season, is great, but relative to a season of short duration such as occurs over a large proportion of the winter rainfall zone in South Australia, they may well assume considerable significance, particularly from the standpoint of agriculture.

VIII. Concluding Remarks

Whipple (loc. cit.) was the first to draw attention, in publication, to the extraordinary coincidence of this oscillation in the rainy season, with the alternation, from cycle to cycle, of the magnetic polarity of sunspots in both solar hemispheres, at sunspot minima.

* It will be observed that the median and quartiles of Figure 5 do not agree with the curves given previously (Cornish loc. cit.). When the original calculations were conducted, an error was made in finding the median date about 1900, which was automatically transmitted to the two quartiles and became progressively worse. This has now been rectified.

The latter phenomenon, now well known from the work of Hale and others, was first observed in 1912 (see, e.g. Chapman and Bartels 1940). Sunspots generally occur in pairs, the two spots being of opposite polarity. During any 11-year spot cycle, the polarities of the leading spots are usually the same for all pairs on the same side of the solar equator, but of opposite polarity on the other side. At the commencement of each new cycle there is a reversal of polarity, between leading and following spots in the pairs, and between northern and southern solar hemispheres. The two types of 11-year cycle, designated as *P* and *E* cycles, occurred as follows :

P Cycles	*E* Cycles
1843–1856	1856–1867
1867–1878	1878–1889
1889–1901	1901–1913
1913–1923	1923–1933
1933–1944	1944–

During a *P* cycle, the polarity of the leading spot of a pair in the northern hemisphere of the Sun is the same as that of the magnetic pole in the southern hemisphere of the Earth. Comparison of these intervals with those obtainable from Figure 5 shows how closely seasons having progressively early winter rains correspond with the *P* cycles, and seasons having progressively late rains correspond with the *E* cycles.

The oscillations in the Adelaide records must be due to secular changes in the latitudinal paths of anticyclones (with their attendant cyclones) across southern Australia, superimposed on the normal seasonal variation. Kidson (1925) demonstrated the pronounced seasonal variation in the latitude of the mean monthly tracks, and attempted to relate latitudinal departures from normal, at various longitudes, ranging from 120 °E. to 170 °E., to variations during the sunspot cycle. He made no differentiation of the cycles, as indicated above, so that the opposing changes noted herein have probably tended to annul each other, at 138 °E., the approximate longitude of Adelaide. In the absence of Kidson's original data, and later observations in convenient form, it is not possible to examine this point further.

The presence of a progressive change in the season at Adelaide agrees with a general observation made recently by Deacon (1953). Deacon's evidence was derived from mean differences in summer temperature, summer and winter precipitation, and barometric pressure between the two epochs 1880–1910 and 1910–1940, for stations in south-eastern Australia, and he concluded that a major climatic change has been operating progressively since 1880 (the earliest record he used), contemporaneously with changes occurring in the northern hemisphere.

The gradual advance in the latter half of the season at Adelaide also agrees with an increase in late spring and early summer rainfall at other stations as found by Mason (personal communication 1953).

IX. Acknowledgments

Grateful acknowledgment is made to the Divisional Meteorologist for South Australia, who placed the Adelaide records at the author's disposal, and to Mr. M. C. Childs, of the Division of Biochemistry and General Nutrition, C.S.I.R.O., for preparation of the diagrams.

X. References

Chapman, S., and Bartels, J. (1940).—" Geomagnetism." (Clarendon Press : Oxford.)
Cornish, E. A. (1936).—*Quart. J. R. Met. Soc.* **62** : 481.
Deacon, E. L. (1953).—*Aust. J. Phys.* **6** : 209.
Fisher, R. A. (1924).—*Phil. Trans.* B **213** : 89.
Kidson, E. (1925).—Bur. Met. Aust. Bull. No. 17.
Whipple, F. J. W. (1936).—*Quart. J. R. Met. Soc.* **62** : 492.

AUSTRALASIAN MEDICAL PUBLISHING CO. LTD.
SEAMER AND ARUNDEL STS., GLEBE, SYDNEY

Reprinted from the Australian Journal of Physics, Volume 7, Number 4, pp. 531–542. 1954

THE MULTIVARIATE *t*-DISTRIBUTION ASSOCIATED WITH A SET OF NORMAL SAMPLE DEVIATES

By E. A. Cornish*

[*Manuscript received July* 29, 1954]

Summary

This paper gives a short account of the more important properties of the multivariate *t*-distribution, which arises in association with a set of normal sample deviates.

I. Introduction

The multivariate *t*-distribution described herein was encountered in the course of a recent investigation of the frequency distribution of the spectrographic error which occurs in the D.C. arc excitation of samples of soil and plant ash (Oertel and Cornish 1953). The original observations consisted of triplicate and sextuplicate determinations of copper, manganese, molybdenum, and tin in samples containing varying amounts of each of these elements, and it was found that the variance of measured line intensity was proportional to the square of the mean intensity. A logarithmic transformation stabilized the variance, which, for any individual element, could then be estimated from all the samples involving that element, and by subsequent standardization all the values of the new metric could be placed on a comparable basis. The standardized variates appeared to be normally distributed, and it became desirable, for various reasons, to test this point.

In providing tests of normality, two general lines of approach have been followed :

(1) a normal distribution is fitted to the sample data, and the χ^2 test of goodness of fit is applied ;

(2) certain functions of the sample moments are calculated, and the significance of their departure from expectations based on the assumption of normality is examined,

but neither of these procedures, as ordinarily used, was suitable owing to the particular nature of the data. For this reason, and as the situation is likely to arise frequently in practice, other means were sought to make the necessary tests. A very limited number of exact tests is available (Fisher 1946), and reasonably accurate approximate tests have been devised by Pearson and Welch (1937) using as a base either a standard test of type (2) above or the work of Geary (1935, 1936). To meet the situation, other exact tests for small samples

* Division of Mathematical Statistics, C.S.I.R.O., Adelaide.

are required to supplement those given by Fisher, and consequently this particular t-distribution has been studied because it offered distinct possibilities in this connexion.

II. Distribution Function of Normal Sample Deviates

Suppose $x_1, x_2, \ldots, x_n$ is a random sample of observations from a normal distribution specified by a mean ξ and variance σ^2.

Let

$$\bar{x} = \sum_{i=1}^{n} x_i/n$$

be the sample mean, and

$$y_i = x_i - \bar{x}$$

be the deviates from the sample mean. The distribution function of one deviate is a classical result; Fisher (1920) gave the distribution of two deviates, and Irwin (1929) established the general non-singular distribution of $(n-1)$ deviates, which, without loss of generality, may be assumed to be the first $(n-1)$. In conformity with the notation used herein, the general distribution may be stated in the following manner:

If $x_1, x_2, \ldots, x_n$ are distributed in a multivariate normal distribution with variance-covariance matrix $\sigma^2\mathbf{I}_n$,* then $y_1, y_2, \ldots, y_{n-1}$ and $\bar{x}$ are distributed in a non-singular multivariate normal distribution with variance-covariance matrix

$$\mathbf{V} = \sigma^2 \begin{bmatrix} (n-1)/n & -1/n & \cdot & \cdot & \cdot & -1/n & 0 \\ -1/n & (n-1)/n & \cdot & \cdot & \cdot & -1/n & 0 \\ \cdot & \cdot & \cdot & \cdot & \cdot & \cdot & \cdot \\ \cdot & \cdot & \cdot & \cdot & \cdot & \cdot & \cdot \\ -1/n & -1/n & \cdot & \cdot & \cdot & (n-1)/n & 0 \\ 0 & 0 & \cdot & \cdot & \cdot & 0 & 1/n \end{bmatrix}$$

of order $n \times n$.

The deviates $y_1, y_2, \ldots, y_{n-1}$ are thus distributed independently of $\bar{x}$. Alternatively, $y_1, y_2, \ldots, y_n$ are distributed in a singular multivariate normal distribution with variance-covariance matrix

$$\sigma^2 \begin{bmatrix} (n-1)/n & -1/n & \cdot & \cdot & \cdot & -1/n \\ -1/n & (n-1)/n & \cdot & \cdot & \cdot & -1/n \\ & \cdot & \cdot & \cdot & \cdot & \cdot \\ \cdot & \cdot & \cdot & \cdot & \cdot & \cdot \\ -1/n & -1/n & \cdot & \cdot & \cdot & (n-1)/n \end{bmatrix}$$

of order $n \times n$ and rank $(n-1)$.

III. The Multivariate t-Distribution

Let $\mathbf{B}$ denote the leading submatrix of order $(n-1) \times (n-1)$ in the matrix $\mathbf{V}$ of Section II. If s^2 is an estimate of σ^2, based on ν degrees of freedom and

* The symbol $\mathbf{I}$ will designate the unit matrix, and the attached subscript will indicate its order.

distributed independently of $y_1, y_2, \ldots, y_{n-1}$, the distribution function of the y_i and s may be written

$$\frac{|\mathbf{B}|^{-\frac{1}{2}}}{(2\pi)^{(n-1)/2}} \frac{\nu^{\nu/2}}{2^{(\nu-2)/2}\sigma^\nu\Gamma\nu/2} \int \ldots \int e^{-\mathbf{y}'\mathbf{B}^{-1}\mathbf{y}/2} s^{\nu-1} e^{-\nu s^2/2\sigma^2} ds\,d\mathbf{y},^*$$

the multiple integral being taken over the region defined by the inequalities

$$-\infty \leqslant y_i \leqslant Y_i, \quad i=1,2,\ldots,n-1,$$
$$0 \leqslant s \leqslant S.$$

$\mathbf{B}^{-1}$ is the reciprocal matrix of $\mathbf{B}$, and $\mathbf{y}'$ is the row vector $[y_1\, y_2 \ldots y_{n-1}]$.

Make the non-singular transformation indicated by the matrix equation

$$\mathbf{t}=\mathbf{Q}\mathbf{y},$$

where $\mathbf{Q}$ is a diagonal matrix, whose diagonal elements are each equal to

$$\frac{1}{s}\sqrt{\frac{n}{n-1}}.$$

The jacobian is $|\mathbf{Q}|^{-1}$, and the distribution function becomes

$$\frac{\nu^{\nu/2}}{(2\pi)^{(n-1)/2}2^{(\nu-2)/2}\sigma^\nu\Gamma\nu/2} \int \ldots \int |\mathbf{QBQ}'|^{-\frac{1}{2}} e^{-\mathbf{t}'(\mathbf{QBQ}')^{-1}\mathbf{t}/2} e^{-\nu s^2/2\sigma^2} s^{\nu-1} d\mathbf{t}\,ds,$$

in which the domain of integration is defined by

$$-\infty \leqslant t_i \leqslant T_i, \quad i=1,2,\ldots,n-1,$$
$$0 \leqslant s \leqslant S.$$

Since

$$\mathbf{QBQ}' = \frac{\sigma^2}{s^2} \begin{bmatrix} 1 & -1/(n-1) & \cdot & \cdot & \cdot & -1/(n-1) \\ -1/(n-1) & 1 & \cdot & \cdot & \cdot & -1/(n-1) \\ \cdot & \cdot & \cdot & \cdot & \cdot & \cdot \\ \cdot & \cdot & \cdot & \cdot & \cdot & \cdot \\ -1/(n-1) & -1/(n-1) & \cdot & \cdot & \cdot & 1 \end{bmatrix}$$

$$= \frac{\sigma^2}{s^2}\mathbf{R} \text{ say,}$$

it follows that

$$|\mathbf{QBQ}'|^{-\frac{1}{2}} = \left(\frac{s}{\sigma}\right)^{n-1} |\mathbf{R}|^{-\frac{1}{2}},$$

and integrating for s from 0 to ∞, the multivariate distribution function of $t_1, t_2, \ldots, t_{n-1}$ is

$$\frac{\Gamma(\nu+n-1)/2\,|\mathbf{R}|^{-\frac{1}{2}}}{(\pi\nu)^{(n-1)/2}\Gamma\nu/2} \int \ldots \int (1+\mathbf{t}'\mathbf{R}^{-1}\mathbf{t}/\nu)^{-(\nu+n-1)/2} d\mathbf{t}, \quad \ldots. \quad (\mathbf{1})$$

the integral being taken over the region defined by the inequalities

$$-\infty \leqslant t_i \leqslant T_i, \quad i=1,2,\ldots,n-1.$$

The limiting form of the distribution function (**1**), as $\nu \rightarrow \infty$, is that of a multivariate normal distribution with variance-covariance matrix $\mathbf{R}$.

* The notation $\Gamma m/2$ denotes $\Gamma(m/2)$ throughout.

IV. Properties of the Distribution

(a) *Mean Values*

The vector of mean values, $E(\mathbf{t})$, is obviously null, and consequently the variance-covariance matrix is

$$E(\mathbf{tt}')=E\left(\frac{n}{s^2(n-1)}\mathbf{yy}'\right)$$

$$=E\left(\frac{1}{s^2}\right)E\left(\frac{n}{n-1}\mathbf{yy}'\right)$$

$$=\frac{\nu}{\sigma^2(\nu-2)}\cdot\sigma^2\mathbf{R}$$

$$=\frac{\nu}{\nu-2}\mathbf{R}. \qquad \ldots\ldots\ldots\ldots\ldots\ldots\ldots\ldots \quad (2)$$

(b) *Distribution of Linear Functions*

Suppose

$$\mathbf{x}=\mathbf{Ht}$$

are any $p\leqslant(n-1)$ linearly independent linear functions of the t_i. The distribution of these functions can be derived directly but is more conveniently obtained from the corresponding result for normally distributed variates, namely, that if $y_1, y_2, \ldots, y_{n-1}$ are distributed in a multivariate normal distribution with variance-covariance matrix $\mathbf{B}$, then the variates

$$\mathbf{z}=\mathbf{Hy}$$

are distributed in a multivariate normal distribution with variance-covariance matrix $\mathbf{HBH}'$.

The distribution function of the z_i is thus

$$\frac{|\,\mathbf{HBH}'\,|^{-\frac{1}{2}}}{(2\pi)^{p/2}}\int\ldots\int e^{-\mathbf{z}'(\mathbf{HBH}')^{-1}\mathbf{z}/2}d\mathbf{z}$$

over the region defined by the inequalities

$$z_i\leqslant Z_i, \qquad i=1,2,\ldots,p,$$

and consequently, the distribution function of

$$\mathbf{x}=\mathbf{Ht}=\mathbf{HQy}=\mathbf{Qz}$$

and s is

$$\frac{\nu^{\frac{1}{2}\nu}}{(2\pi)^{p/2}2^{(\nu-2)/2}\sigma^{\nu}\Gamma\nu/2}\int\ldots\int|\,\mathbf{QHBH}'\mathbf{Q}'\,|^{-\frac{1}{2}}e^{-\mathbf{x}'(\mathbf{QHBH}'\mathbf{Q}')^{-1}\mathbf{x}/2}e^{-\nu s^2/2\sigma^2}s^{\nu-1}ds\,d\mathbf{z}$$

over the region defined by

$$x_i\leqslant X_i, \qquad i=1,2,\ldots,p,$$
$$s\leqslant S.$$

Since

$$\mathbf{QHBH}'\mathbf{Q}'=\frac{\sigma^2}{s^2}\mathbf{HRH}',$$

integration for s from 0 to ∞ leaves the distribution of $x_1, x_2, \ldots, x_p$ in the form

$$\frac{\Gamma(\nu+p)/2 \mid \mathbf{HRH}' \mid^{-\frac{1}{2}}}{(\pi\nu)^{p/2}\Gamma\nu/2}\int \ldots \int \{1+\mathbf{x}'(\mathbf{HRH}')^{-1}\mathbf{x}/\nu\}^{-(\nu+p)/2}d\mathbf{x}. \quad .. \; (3)$$

over the region defined by

$$x_i \leqslant X_i, \qquad i=1,2,\ldots,p,$$

that is, a multivariate t-distribution of order p characterized by the matrix $(\mathbf{HRH}')^{-1}$.

(c) Marginal Distribution of $t_1, t_2, \ldots, t_r$

The marginal distribution of $t_1, t_2, \ldots, t_r$, $r<(n-1)$, follows from the previous result by taking

$$\mathbf{H}=\begin{bmatrix} \mathbf{I}_r & \cdot \\ \cdot & \cdot \end{bmatrix},$$

and is thus a multivariate t-distribution of order r, characterized by the matrix $\mathbf{R}_1^{-1}$, $\mathbf{R}_1$ being the leading submatrix of order $r \times r$ in $\mathbf{R}$. The variance-covariance matrix is $\{\nu/(\nu-2)\}\mathbf{R}_1$, and consequently, the variances and covariances of the marginal distribution are identical with their values in the original distribution. The limiting form of this distribution as $\nu \to \infty$ is the marginal multivariate normal distribution with variance-covariance matrix $\mathbf{R}_1$.

(d) Conditional Distribution of $t_1, t_2, \ldots, t_r$

To find the conditional distribution of $t_1, t_2, \ldots, t_r$ when $t_{r+1}, \ldots, t_{n-1}$ are given specified values, first partition the matrix $\mathbf{R}^{-1}$ so that it takes the form

$$\begin{bmatrix} \mathbf{R}_1 & \mathbf{R}_3 \\ \mathbf{R}_3' & \mathbf{R}_2 \end{bmatrix}$$

where the submatrices $\mathbf{R}_1$, $\mathbf{R}_2$, and $\mathbf{R}_3$ respectively are of orders $r \times r$, $\{n-(r+1)\} \times \{n-(r+1)\}$, and $r \times \{n-(r+1)\}$ and the row vector $\mathbf{t}'$ so that

$$\mathbf{t}'=[t_1 \ldots t_r \mid t_{r+1} \ldots t_{n-1}],$$

which may be written

$$[\mathbf{t}_1' \quad \mathbf{t}_2'].$$

Since

$$\begin{bmatrix} \mathbf{I}_r & \cdot \\ -\mathbf{R}_3'\mathbf{R}_1^{-1} & \mathbf{I}_{n-(r+1)} \end{bmatrix}\begin{bmatrix} \mathbf{R}_1 & \mathbf{R}_3 \\ \mathbf{R}_3' & \mathbf{R}_2 \end{bmatrix}\begin{bmatrix} \mathbf{I}_r & -\mathbf{R}_1^{-1}\mathbf{R}_3 \\ \cdot & \mathbf{I}_{n-(r+1)} \end{bmatrix}=\begin{bmatrix} \mathbf{R}_1 & \cdot \\ \cdot & \mathbf{R}_2-\mathbf{R}_3'\mathbf{R}_1^{-1}\mathbf{R}_3 \end{bmatrix}, \quad \ldots\ldots (4)$$

on taking reciprocals

$$(\mathbf{R}_2-\mathbf{R}_3'\mathbf{R}_1^{-1}\mathbf{R}_3)^{-1}=\mathbf{R}_2,$$

where $\mathbf{R}_2$ is the submatrix in $\mathbf{R}$ corresponding to $\mathbf{R}_2$ in $\mathbf{R}^{-1}$.

The marginal distribution of $t_{r+1}, t_{r+2}, \ldots, t_{n-1}$ may thus be written

$$\frac{\Gamma\{\nu+n-(r+1)\}/2 \mid \mathbf{R}_2-\mathbf{R}_3'\mathbf{R}_1^{-1}\mathbf{R}_3 \mid^{\frac{1}{2}}}{(\pi\nu)^{\{n-(r+1)\}/2}\Gamma\nu/2}\{1+\mathbf{t}_2'(\mathbf{R}_2-\mathbf{R}_3'\mathbf{R}_1^{-1}\mathbf{R}_3)\mathbf{t}_2/\nu\}^{-\{\nu+n-(r+1)\}/2}d\mathbf{t}_2.$$

Express the quadratic form $\mathbf{t}'\mathbf{R}^{-1}\mathbf{t}$ as

$$\mathbf{t}'\mathbf{R}^{-1}\mathbf{t}=\mathbf{t}_1'\mathbf{R}_1\mathbf{t}_1+2\mathbf{t}_2'\mathbf{R}_3'\mathbf{t}_1+\mathbf{t}_2'\mathbf{R}_3'\mathbf{R}_1^{-1}\mathbf{R}_3\mathbf{t}_2+\mathbf{t}_2'(\mathbf{R}_2-\mathbf{R}_3'\mathbf{R}_1^{-1}\mathbf{R}_3)\mathbf{t}_2,$$

and take as the fixed set of values for the variables in $\mathbf{t}_2$, the elements of a vector $\mathbf{a}$. Using the determinantal relation found by taking determinants of both sides of (4), the conditional distribution function of $\mathbf{t}_1$ is

$$\frac{\Gamma(\nu+n-1)/2\,|\,\mathbf{R}_1\,|^{\frac{1}{2}}}{\pi\nu)^{r/2}\Gamma\{\nu+n-(r+1)\}/2}\{1+\mathbf{a}'(\mathbf{R}_2-\mathbf{R}_3'\mathbf{R}_1^{-1}\mathbf{R}_3)\mathbf{a}/\nu\}^{\{\nu+n-(r+1)\}/2}$$

$$\times\int\cdots\int\left\{1+\frac{(\mathbf{t}_1+\mathbf{R}_1^{-1}\mathbf{R}_3\mathbf{a})'\mathbf{R}_1(\mathbf{t}_1+\mathbf{R}_1^{-1}\mathbf{R}_3\mathbf{a})+\mathbf{a}'(\mathbf{R}_2-\mathbf{R}_3'\mathbf{R}_1^{-1}\mathbf{R}_3)\mathbf{a}}{\nu}\right\}^{-(\nu+n-1)/2}d\mathbf{t}_1, \qquad \text{(5)}$$

integration being taken over the domain specified by

$$t_i \leqslant T_i, \qquad i=1,2,\ldots,r.$$

The distribution function (5) is dependent upon the particular set of values chosen for the variables in the vector $\mathbf{t}_2$, and its limiting form, as $\nu\to\infty$, is the conditional multivariate normal distribution of order r, with vector of means $-\mathbf{R}_1^{-1}\mathbf{R}_3\mathbf{a}$, and variance-covariance matrix $\mathbf{R}_1^{-1}$.

The mean value of $\mathbf{t}_1$ is

$$E(\mathbf{t}_1)=\frac{\Gamma(\nu+n-1)/2\,|\,\mathbf{R}_1\,|^{\frac{1}{2}}}{(\pi\nu)^{r/2}\Gamma\{\nu+n-(r+1)\}/2}\{1+\mathbf{a}'(\mathbf{R}_2-\mathbf{R}_3'\mathbf{R}_1^{-1}\mathbf{R}_3)\mathbf{a}/\nu\}^{\{\nu+n-(r+1)\}/2}$$

$$\times\int_{-\infty}^{\infty}\cdots\int\mathbf{t}_1\left\{1+\frac{(\mathbf{t}_1+\mathbf{R}_1^{-1}\mathbf{R}_3\mathbf{a})'\mathbf{R}_1(\mathbf{t}_1+\mathbf{R}_1^{-1}\mathbf{R}_3\mathbf{a})+\mathbf{a}'(\mathbf{R}_2-\mathbf{R}_3'\mathbf{R}_1^{-1}\mathbf{R}_3\mathbf{a})}{\nu}\right\}^{-(\nu+n-1)/2}d\mathbf{t}_1.$$

To evaluate this integral, first make the change of variable

$$\mathbf{t}_1+\mathbf{R}_1^{-1}\mathbf{R}_3\mathbf{a}=\mathbf{z}\,;$$

the jacobian is 1, and, omitting the constant factor, the integral becomes

$$\int_{-\infty}^{\infty}\cdots\int(\mathbf{z}-\mathbf{R}_1^{-1}\mathbf{R}_3\mathbf{a})\left\{1+\frac{\mathbf{z}'\mathbf{R}_1\mathbf{z}+\mathbf{a}'(\mathbf{R}_2-\mathbf{R}_3'\mathbf{R}_1^{-1}\mathbf{R}_3)\mathbf{a}}{\nu}\right\}^{-(\nu+n-1)/2}d\mathbf{z}$$

$$=-\mathbf{R}_1^{-1}\mathbf{R}_3\mathbf{a}\int_{-\infty}^{\infty}\cdots\int\left\{1+\frac{\mathbf{z}'\mathbf{R}_1\mathbf{z}+\mathbf{a}'(\mathbf{R}_2-\mathbf{R}_3'\mathbf{R}_1^{-1}\mathbf{R}_3)\mathbf{a}}{\nu}\right\}^{-(\nu+n-1)/2}d\mathbf{z}.$$

Since $\mathbf{R}_1$ is a real, positive definite symmetric matrix, the quadratic form $\mathbf{z}'\mathbf{R}_1\mathbf{z}$ may be reduced to a sum of squares, after which the integration is easily performed, and finally yields

$$E(\mathbf{t}_1)=-\mathbf{R}_1^{-1}\mathbf{R}_3\mathbf{a}, \qquad \text{(6)}$$

which is a linear function of the elements of the vector $\mathbf{a}$. The regression of $\mathbf{t}_1$ on $\mathbf{t}_2$ thus exists and is linear.

The variance-covariance matrix is then

$$E(\mathbf{t}_1+\mathbf{R}_1^{-1}\mathbf{R}_3\mathbf{a})(\mathbf{t}_1+\mathbf{R}_1^{-1}\mathbf{R}_3\mathbf{a})'=\frac{\Gamma(\nu+n-1)/2\mid\mathbf{R}_1\mid^{\frac{1}{2}}}{(\pi\nu)^{r/2}\Gamma\{\nu+n-(r+1)\}/2}\{1+\mathbf{a}'(\mathbf{R}_2-\mathbf{R}_3'\mathbf{R}_1^{-1}\mathbf{R}_3)\mathbf{a}/\nu\}^{\{\nu+n-(r+1)\}/2}$$

$$\times\int_{-\infty}^{\infty}\cdots\int(\mathbf{t}_1+\mathbf{R}_1^{-1}\mathbf{R}_3\mathbf{a})(\mathbf{t}_1+\mathbf{R}_1^{-1}\mathbf{R}_3\mathbf{a})'$$

$$\left\{1+\frac{(\mathbf{t}_1+\mathbf{R}_1^{-1}\mathbf{R}_3\mathbf{a})'\mathbf{R}_1(\mathbf{t}_1+\mathbf{R}_1^{-1}\mathbf{R}_3\mathbf{a})+\mathbf{a}'(\mathbf{R}_2-\mathbf{R}_3'\mathbf{R}_1^{-1}\mathbf{R}_3)\mathbf{a}}{\nu}\right\}^{-(\nu+n-1)/2}d\mathbf{t}_1,$$

and this integral is also readily evaluated after making the congruent transformation

$$\mathbf{t}_1+\mathbf{R}_1^{-1}\mathbf{R}_3\mathbf{a}=\mathbf{P}\mathbf{z}$$

such that $\mathbf{P}'\mathbf{R}_1\mathbf{P}=\mathbf{I}_r$.

The variance-covariance matrix is

$$\frac{\nu+\mathbf{a}'(\mathbf{R}_2-\mathbf{R}_3'\mathbf{R}_1^{-1}\mathbf{R}_3)\mathbf{a}}{\nu+n-(r+3)}\mathbf{R}_1^{-1}, \quad \ldots\ldots\ldots\ldots\ldots \quad (7)$$

which depends upon the particular values chosen for the variables in the vector $\mathbf{t}_2$.

The limiting form of the matrix (7), as $\nu\rightarrow\infty$, is $\mathbf{R}_1^{-1}$, which is the variance-covariance matrix of the limiting conditional multivariate normal distribution, independent, as it should be, of the values of the fixed variates.

(e) *Distribution Function of* $\mathbf{t}'\mathbf{R}^{-1}\mathbf{t}$

(i) From the transformation of Section III

$$\mathbf{t}=\mathbf{Q}\mathbf{y},$$

and so

$$\mathbf{t}'\mathbf{R}^{-1}\mathbf{t}=\mathbf{y}'\mathbf{Q}'\mathbf{R}^{-1}\mathbf{Q}\mathbf{y}$$

$$=\sum_{i=1}^{n} y_i^2/s^2.$$

Consequently

$$\mathbf{t}'\mathbf{R}^{-1}\mathbf{t}/(n-1)$$

is distributed as e^{2z} with $(n-1)$ and ν degrees of freedom.

(ii) The distribution of $\mathbf{t}'\mathbf{R}^{-1}\mathbf{t}$ may now be examined when the variables are subject to the linearly independent linear homogeneous conditions represented by the matrix equation

$$\mathbf{S}\mathbf{t}=\mathbf{0},$$

where $\mathbf{S}$ is of order $p\times(n-1)$ and rank $p<(n-1)$.

Construct the matrix $\mathbf{H}$ such that

$$\mathbf{H}\mathbf{R}\mathbf{H}'=\mathbf{I}_{n-p-1},$$

and

$$\mathbf{H}\mathbf{R}\mathbf{S}'=\mathbf{0},$$

and make the non-singular transformation

$$\mathbf{x}=\mathbf{P}\mathbf{t}=\begin{bmatrix}\mathbf{H}\\ \mathbf{S}\end{bmatrix}\mathbf{t}.$$

The distribution function of $\mathbf{t}'\mathbf{R}^{-1}\mathbf{t}$ thus becomes

$$\frac{|\,\mathbf{PRP}'\,|^{-\frac{1}{2}}\Gamma(\nu+n-1)/2}{(\pi\nu)^{(n-1)/2}\Gamma\nu/2}\int\cdots\int\{1+\mathbf{x}'(\mathbf{PRP}')^{-1}\mathbf{x}/\nu\}^{-(\nu+n-1)/2}d\mathbf{x},$$

where the domain of integration is defined by

$$\mathbf{x}'(\mathbf{PRP}')^{-1}\mathbf{x}\leqslant Q, \text{ say.}$$

Now impose the conditions

$$\mathbf{St}=\mathbf{0}.$$

This makes

$$x_{n-p}=x_{n-p+1}=\cdots=x_{n-1}=0, \qquad \ldots\ldots\ldots\ldots (8)$$

and, using the results of Section IV (d), the conditional distribution of $x_1, x_2, \ldots, x_{n-p-1}$ when (8) holds is

$$\frac{\Gamma(\nu+n-1)/2}{(\pi\nu)^{(n-p-1)/2}\Gamma(\nu+p)/2}\int\cdots\int(1+\mathbf{x}_1'\mathbf{x}_1/\nu)^{-(\nu+n-1)/2}d\mathbf{x}_1,$$

where $\mathbf{x}_1'$ is the vector $[x_1, x_2, \ldots, x_{n-p-1}]$ and the integral is taken over the region defined by

$$\mathbf{x}_1'\mathbf{x}_1\leqslant Q.$$

A spherical polar transformation in $(n-p-1)$ dimensions, followed by a change of variable which makes the square of the radius vector equal to $\frac{\nu}{\nu+p}\chi^2$ reduces this integral to the form

$$\frac{\Gamma(\nu+n-1)/2}{(\nu+p)^{(n-p-1)/2}\Gamma(n-p-1)/2\;\Gamma(\nu+p)/2}\int_0^Q(\chi^2)^{(n-p-3)/2}\{1+\chi^2/(\nu+p)\}^{-(\nu+n-1)/2}d(\chi^2),$$

$$\ldots\ldots\ldots\ldots\ldots\ldots (9)$$

which is equivalent to Fisher's z-distribution with $(n-p-1)$ and $(\nu+p)$ degrees of freedom. The transfer of p degrees of freedom from one set of degrees of freedom to the other is a consequence of the fact that the conditional distribution is dependent upon the values of the fixed variates.

The distinction between the distribution (9) and that of $\mathbf{t}'\mathbf{R}^{-1}\mathbf{t}=\sum_{i=1}^{n} y_i^2/s^2$ when the y_i are subject to the restrictions $\mathbf{Sy}=\mathbf{0}$, should be noted. The latter distribution is, of course, Fisher's z-distribution with degrees of freedom $(n-p-1)$ and ν. Since s is essentially positive, either set of restrictive conditions then implies the other, and at first sight it might seem that the two distributions are identical.

(f) *Distribution Function of* $\mathbf{t}'\mathbf{At}$

The distribution function of the quadratic form $\mathbf{t}'\mathbf{At}$, of rank $r\leqslant(n-1)$, is given by

$$\frac{\Gamma(\nu+n-1)/2\,|\,\mathbf{R}\,|^{-\frac{1}{2}}}{(\pi\nu)^{(n-1)/2}\Gamma\nu/2}\int\cdots\int(1+\mathbf{t}'\mathbf{R}^{-1}\mathbf{t}/\nu)^{-(\nu+n-1)/2}d\mathbf{t},$$

where the domain of integration is defined by

$$\mathbf{t'At} \leqslant Q.$$

Make the congruent transformation

$$\mathbf{t} = \mathbf{Hx},$$

the matrix $\mathbf{H}$ being chosen so that

$$\mathbf{HR}^{-1}\mathbf{H'} = \mathbf{I}_{n-1},$$

and

$$\mathbf{HAH'} = \mathbf{\Lambda},$$

where Λ is a diagonal matrix whose diagonal elements are the roots of the equation

$$|\lambda \mathbf{R}^{-1} - \mathbf{A}| = 0,$$

or, alternatively, the latent roots of the matrix $\mathbf{RA}$.

The jacobian is $|\mathbf{R}|^{\frac{1}{2}}$, so that the distribution function becomes

$$\frac{\Gamma(\nu+n-1)/2}{(\pi\nu)^{(n-1)/2}\Gamma\nu/2}\int \ldots \int (1+\mathbf{x'x}/\nu)^{-(\nu+n-1)/2}d\mathbf{x},$$

the integral being taken over the region defined by

$$\mathbf{x'\Lambda x} \leqslant Q,$$

or

$$\sum_{i=1}^{r} \lambda_i x_i^2 \leqslant Q,$$

where $\lambda_1, \lambda_2, \ldots, \lambda_r$ are the non-zero latent roots of $\mathbf{RA}$.

After integrating for $x_{r+1}, x_{r+2}, \ldots, x_{n-1}$, the distribution function becomes

$$\frac{\Gamma(\nu+r)/2}{(\pi\nu)^{r/2}\Gamma\nu/2}\int \ldots \int (1+\mathbf{x}_1'\mathbf{x}_1/\nu)^{-(\nu+r)/2}d\mathbf{x}_1, \quad \ldots\ldots \quad (10)$$

where $\mathbf{x}_1'$ is the vector $[x_1\, x_2 \ldots x_r]$, and the domain of integration is defined by

$$\sum_{i=1}^{r} \lambda_i x_i^2 > Q.$$

Consequently, the necessary and sufficient condition that the distribution function (**10**) is equivalent to the z-distribution with degrees of freedom r and ν, is that the non-zero latent roots of the matrix $\mathbf{RA}$ are all equal to unity.

V. Parametric Values of the Multivariate Distribution

In further discussion of the conditional distribution and of regression and correlation among the t-variates, consideration is given in particular to the case $r=1$, since this value is of the greatest importance in practice.

(a) *Conditional Distribution*

The determinant

$$|\mathbf{R}| = n^{n-2}/(n-1)^{n-1},$$

and, if R_{ij} denotes the co-factor of the (ij)th element in $|\mathbf{R}|$, then

$$R_{ii} = 2n^{n-3}/(n-1)^{n-2},$$

$$R_{ij} = n^{n-3}/(n-1)^{n-2}, \quad i \neq j,$$

and hence

$$\mathbf{R}^{-1}=\begin{bmatrix} 2(n-1)/n & (n-1)/n & \cdot & \cdot & \cdot & (n-1)/n \\ (n-1)/n & 2(n-1)/n & \cdot & \cdot & \cdot & (n-1)/n \\ \cdot & \cdot & \cdot & \cdot & \cdot & \cdot \\ \cdot & \cdot & \cdot & \cdot & \cdot & \cdot \\ (n-1)/n & (n-1)/n & \cdot & \cdot & \cdot & 2(n-1)/n \end{bmatrix}.$$

Taking $r=1$, the relation

$$\mathbf{R}^{-1}=\begin{bmatrix} \mathbf{R}_1 & \mathbf{R}_3 \\ \mathbf{R}_3' & \mathbf{R}_2 \end{bmatrix}$$

gives

$$\mathbf{R}_1=2(n-1)/n,$$

and

$$\mathbf{R}_3=[(n-1)/n \ (n-1)/n \ . \ . \ . \ (n-1)/n],$$

a row vector of order $(n-2)$.

The conditional mean value of the variate t_1 is then

$$-\mathbf{R}_1^{-1}\mathbf{R}_3\mathbf{a}=-\tfrac{1}{2}\sum_{j=2}^{n-1} a_j.$$

Moreover,

$$\mathbf{R}_2-\mathbf{R}_3'\mathbf{R}_1^{-1}\mathbf{R}_3=\begin{bmatrix} 3(n-1)/2n & (n-1)/2n & \cdot & \cdot & \cdot & (n-1)/2n \\ (n-1)/2n & 3(n-1)/2n & \cdot & \cdot & \cdot & (n-1)/2n \\ \cdot & \cdot & \cdot & \cdot & \cdot & \cdot \\ \cdot & \cdot & \cdot & \cdot & \cdot & \cdot \\ (n-1)/2n & (n-1)/2n & \cdot & \cdot & \cdot & 3(n-1)/2n \end{bmatrix},$$

and hence

$$\nu+\mathbf{a}'(\mathbf{R}_2-\mathbf{R}_3'\mathbf{R}_1^{-1}\mathbf{R}_3)\mathbf{a}=\nu+\sum_j 3(n-1)a_j^2/2n+2\sum_{j<k}(n-1)a_ja_k/2n.$$

The conditional variance of the variate t_1 thus becomes

$$\frac{\nu+3(n-1)\sum_j a_j^2/2n+(n-1)\sum_{j<k} a_ja_k/n}{\nu+n-4}\,\frac{n}{2(n-1)},$$

and the average value of this quantity, for all possible values of $a_2, a_3, . \ . \ ., a_{n-1}$, is

$$\frac{\nu+\dfrac{3(n-1)}{2n}(n-2)\dfrac{\nu}{\nu-2}-\dfrac{n-1}{n}\tfrac{1}{2}(n-3)(n-2)\dfrac{\nu}{(\nu-2)(n-1)}}{\nu+n-4}\,\frac{n}{2(n-1)}=\frac{\nu}{\nu-2}\,\frac{n}{2(n-1)}.$$

(*b*) *Regression and Correlation*

From (**2**), the variance-covariance matrix of the determining variates $t_2, t_3, . \ . \ ., t_{n-1}$ is

$$\frac{\nu}{\nu-2}\mathbf{R}_2,$$

of which the determinant is

$$\frac{2n^{n-3}}{(n-1)^{n-2}}\left(\frac{\nu}{\nu-2}\right)^{n-2},$$

and, consequently, the reciprocal matrix

$$\left[\frac{\nu}{\nu-2}\mathbf{R}_2\right]^{-1}=\begin{bmatrix} \frac{3(n-1)}{2n}\frac{\nu-2}{\nu} & \frac{n-1}{2n}\frac{\nu-2}{\nu} & \cdots & \frac{n-1}{2n}\frac{\nu-2}{\nu} \\ \frac{n-1}{2n}\frac{\nu-2}{\nu} & \frac{3(n-1)}{2n}\frac{\nu-2}{\nu} & \cdots & \frac{n-1}{2n}\frac{\nu-2}{\nu} \\ \vdots & \vdots & & \vdots \\ \frac{n-1}{2n}\frac{\nu-2}{\nu} & \frac{n-1}{2n}\frac{\nu-2}{\nu} & \cdots & \frac{3(n-1)}{2n}\frac{\nu-2}{\nu} \end{bmatrix}.$$

Also from (**2**), the covariance of t_1 with each of $t_2, t_3, \ldots, t_{n-1}$ is

$$-\nu/(\nu-2)(n-1),$$

so that the vector of regression coefficients in the multiple regression of t_1 on $t_2, t_3, \ldots, t_{n-1}$ is

$$\left[\frac{\nu}{\nu-2}\mathbf{R}_2\right]^{-1}\begin{bmatrix} \frac{-\nu}{(\nu-2)(n-1)} \\ \frac{-\nu}{(\nu-2)(n-1)} \\ \vdots \\ \frac{-\nu}{(\nu-2)(n-1)} \end{bmatrix}=\begin{bmatrix} -\frac{1}{2} \\ -\frac{1}{2} \\ \vdots \\ -\frac{1}{2} \end{bmatrix},$$

in agreement with the coefficients of the linear function for the conditional mean value of t_1.

The residual variance of t_1 with respect to $t_2, t_3, \ldots, t_{n-1}$ is

$$\frac{\left|\frac{\nu}{\nu-2}\mathbf{R}\right|}{\left|\frac{\nu}{\nu-2}\mathbf{R}_2\right|}=\frac{\nu}{\nu-2}\,\frac{n}{2(n-1)},$$

independent of the values of the determining variates, and equal to the average conditional variance given above.

The square of the multiple correlation of t_1 with $t_2, t_3, \ldots, t_{n-1}$ is

$$1-\frac{\left|\frac{\nu}{\nu-2}\mathbf{R}\right|}{\frac{\nu}{\nu-2}\left|\frac{\nu}{\nu-2}\mathbf{R}_2\right|}=(n-2)/2(n-1),$$

and the partial correlation of t_1 with t_j $(j=2,3,\ldots,n-1)$ is

$$\frac{-\left(\frac{\nu}{\nu-2}\right)^{n-2}R_{1j}}{\sqrt{\left(\frac{\nu}{\nu-2}\right)^{n-2}R_{11}\left(\frac{\nu}{\nu-2}\right)^{n-2}R_{jj}}}=-\tfrac{1}{2}.$$

VI. References

FISHER, R. A. (1920).—*Mon. Not. R. Astr. Soc.* **80**: 758.
FISHER, R. A. (1946).—" Statistical Methods for Research Workers." 10th Ed. (Oliver and Boyd: Edinburgh.)
GEARY, R. C. (1935).—*Biometrika* **27**: 310.
GEARY, R. C. (1936).—*Biometrika* **28**: 295.
IRWIN, J. O. (1929).—*J. R. Statist. Soc.* **92**: 580.
OERTEL, A. C., and CORNISH, E. A. (1953).—*Aust. J. Appl. Sci.* **4**: 489.
PEARSON, E. S., and WELCH, B. L. (1937).—*J. R. Statist. Soc. Suppl.* **4**: 94.

AUSTRALASIAN MEDICAL PUBLISHING CO. LTD.
SEAMER AND ARUNDEL STS., GLEBE, SYDNEY

Reprinted from the Australian Journal of Physics, Volume 8, Number 2, pp. 193–199, 1955

THE SAMPLING DISTRIBUTIONS OF STATISTICS DERIVED FROM THE MULTIVARIATE t-DISTRIBUTION

By E. A. Cornish*

[*Manuscript received November* 30, 1954]

Summary

The sampling distributions of the more important statistical derivates from the multivariate t-distribution are established.

I. Introduction

A multivariate generalization of Student's t-distribution has been considered by Dunnett and Sobel (1954) in connexion with certain multiple decision problems concerned with the ranking, according to their mean values, of normal populations having a common unknown variance. These authors studied the probability integral of the bivariate population in detail, giving exact and asymptotic expressions, and tables for certain special cases. The particular applications to which they refer have been discussed by Bechhofer, Dunnett, and Sobel (1954). The same general distribution was derived, and its principal properties established independently by the author (Cornish 1954), when considering the pretreatment to be given to certain types of replicated experimental observations, before applying tests of normality. This distribution possesses properties which make it suitable as a basis for exact tests of significance in various problems, and Dunnett and Sobel have taken the first step towards its use in practice by providing tables of the probability integral. In this paper, we shall be concerned with sampling distributions of statistics derived from the multivariate t-distribution. The general sampling distribution of the means and sums of squares and products of the variates is first established, and from it, the sampling distributions of the more important statistical derivates are obtained.

II. General Distribution of Means and Sums of Squares and Products

Suppose $x_1, x_2, \ldots, x_p$ are distributed in a non-singular multivariate normal distribution having a null vector of means and variance-covariance matrix $\sigma^2\mathbf{R}$. In the class of cases to be considered, the symmetric correlation matrix $\mathbf{R}=[\rho_{ij}]$ is known, but the variance σ^2 is unknown. If s^2 is an estimate of σ^2, based on ν degrees of freedom and distributed independently of $x_1, x_2, \ldots, x_p$, then, as indicated above, it has been shown that the variates

$$t_i=x_i/s, \quad i=1, 2, \ldots, p, \quad \ldots\ldots\ldots\ldots\ (1)$$

* Division of Mathematical Statistics, C.S.I.R.O., Adelaide.

have the distribution

$$\frac{|\mathbf{R}|^{-\frac{1}{2}}\Gamma\frac{1}{2}(\nu+p)}{(\pi\nu)^{\frac{1}{2}p}\Gamma\frac{1}{2}\nu}(1+\mathbf{t}'\mathbf{R}^{-1}\mathbf{t}/\nu)^{-\frac{1}{2}(\nu+p)}d\mathbf{t}, \quad \ldots\ldots\ldots\ldots (2)$$

where $\mathbf{t}'$ is the row vector $[t_1\, t_2 \ldots t_p]$.

If n samples are taken from the multivariate normal distribution yielding the variate values

$$x_{ik}, \qquad i=1, 2, \ldots, p, \quad k=1, 2, \ldots, n,$$

the probability of obtaining the corresponding sets of t-variates, defined by

$$t_{ik}=x_{ik}/s, \quad \ldots\ldots\ldots\ldots\ldots\ldots (3)$$

is

$$\left\{\frac{|\mathbf{R}|^{-\frac{1}{2}}\Gamma\frac{1}{2}(\nu+p)}{(\pi\nu)^{\frac{1}{2}p}\Gamma\frac{1}{2}\nu}\right\}^n \prod_{k=1}^{n}(1+\mathbf{t}_k'\mathbf{R}^{-1}\mathbf{t}_k/\nu)^{-\frac{1}{2}(\nu+p)}d\mathbf{t}_k. \quad \ldots\ldots (4)$$

The means and sums of squares and products of deviations of these t-variates are defined respectively by the relations

$$\left.\begin{aligned} \bar{t}_i &= \frac{1}{n}\sum_{k=1}^{n} t_{ik}, & i&=1, 2, \ldots, p, \\ T_{ij} &= \sum_{k=1}^{n}(t_{ik}-\bar{t}_i)(t_{jk}-\bar{t}_j), & i, j&=1, 2, \ldots, p. \end{aligned}\right\} \quad \ldots\ldots (5)$$

Formally, the distribution function of the new variates given by the relations (5) can be derived from the distribution (4), but it is much more simply obtained from the known corresponding result for normally distributed quantities.

The sample means of the normal variates are defined by

$$\bar{x}_i=\frac{1}{n}\sum_{k=1}^{n} x_{ik}, \qquad i=1, 2, \ldots, p.$$

Denoting the matrix

$$[x_{ik}-\bar{x}_i], \qquad i=1, 2, \ldots, p, \quad k=1, 2, \ldots, n,$$

by $\mathbf{X}$, and writing

$$\mathbf{XX}'=\mathbf{C}=[C_{ij}],$$

the distribution of the means and sums of squares and products (Wishart 1928) is

$$\frac{\left|\frac{\sigma^2}{n}\mathbf{R}\right|^{-\frac{1}{2}}}{(2\pi)^{\frac{1}{2}p}}\exp\left\{-\tfrac{1}{2}\bar{\mathbf{x}}'\left(\frac{\sigma^2}{n}\mathbf{R}\right)^{-1}\bar{\mathbf{x}}\right\}d\bar{\mathbf{x}}$$

$$\times\frac{|\sigma^2\mathbf{R}|^{-\frac{1}{2}(n-1)}|\mathbf{C}|^{\frac{1}{2}(n-p-2)}\exp\left\{-\frac{1}{2\sigma^2}\operatorname{Tr}(\mathbf{R}^{-1}\mathbf{C})\right\}}{2^{\frac{1}{2}p(n-1)}\pi^{\frac{1}{4}p(p-1)}\prod_{i=1}^{p}\Gamma\frac{1}{2}(n-i)}\prod_{i\leqslant j}dC_{ij}, \quad \ldots (6)$$

where $\bar{\mathbf{x}}'$ is the row vector $[\bar{x}_1\, \bar{x}_2 \ldots \bar{x}_p]$.

From (3) and (5)

$$\left.\begin{aligned} \bar{t}_i &= \bar{x}_i/s, \\ T_{ij} &= C_{ij}/s^2, \end{aligned}\right\} \quad \ldots\ldots\ldots\ldots\ldots\ldots\ldots \quad (7)$$

and the jacobian of (7) is $s^{p(p+2)}$. Substitute in (6), using the relations (7), multiply by the distribution of s and integrate for s from 0 to ∞, and the distribution of the $\bar{t}_i$ and T_{ij} takes the form

$$\frac{\left|\frac{1}{n}\mathbf{R}\right|^{-\frac{1}{2}}\Gamma\tfrac{1}{2}(\nu+pn)}{(\pi\nu)^{\frac{1}{2}p}\Gamma\tfrac{1}{2}\nu} \; \frac{|\mathbf{R}|^{-\frac{1}{2}(n-1)}\,|\mathbf{T}|^{\frac{1}{2}(n-p-2)}}{\pi^{\frac{1}{4}p(p-1)}\nu^{\frac{1}{2}p(n-1)}\prod_{i=1}^{p}\Gamma\tfrac{1}{2}(n-i)}$$

$$\times \frac{d\bar{\mathbf{t}}\prod_{i\leqslant j} dT_{ij}}{\left\{1+\dfrac{\bar{\mathbf{t}}'\left(\frac{1}{n}\mathbf{R}\right)^{-1}\bar{\mathbf{t}}+\mathrm{Tr}\,(\mathbf{R}^{-1}\mathbf{T})}{\nu}\right\}^{\frac{1}{2}(\nu+pn)}}, \quad \ldots\ldots\ldots\ldots \quad (8)$$

where $\mathbf{T}$ is the matrix $[T_{ij}]$. The limiting form of the distribution (8), as $\nu\to\infty$, is the Wishart distribution.

III. Marginal Distribution of the $\bar{t}_i$

To obtain the marginal distribution of the means, integration of (8) with respect to the T_{ij} will be made following a method similar to that used by Cramer (1946, Section 29.5). Since $\mathbf{R}^{-1}$ is positive definite, it may be reduced by an orthogonal matrix $\mathbf{H}$, so that

$$\mathbf{H}\mathbf{R}^{-1}\mathbf{H}' = \mathbf{\Lambda},$$

where $\mathbf{\Lambda}$ is diagonal, its diagonal elements λ_i being the latent roots of $\mathbf{R}^{-1}$. The same transformation applied to the matrix $\mathbf{T}$ yields

$$\mathbf{H}\mathbf{T}\mathbf{H}' = \mathbf{Y}, \text{ say}, \quad \ldots\ldots\ldots\ldots\ldots\ldots \quad (9)$$

transforming the $\frac{1}{2}p(p+1)$ variates in $\mathbf{T}$ to $\frac{1}{2}p(p+1)$ new variates in $\mathbf{Y}$. The relation (9) represents a linear transformation of the variables whose determinant (the jacobian required) is a power of the determinant $|\mathbf{H}|$ (MacDuffee 1943; James 1954). As $\mathbf{H}$ is orthogonal, the jacobian is unity, and, omitting the constant, (8) becomes

$$\frac{|\mathbf{Y}|^{\frac{1}{2}(n-p-2)} d\bar{\mathbf{t}}\prod_{i\leqslant j} dY_{ij}}{\left\{1+\dfrac{\bar{\mathbf{t}}'\left(\frac{1}{n}\mathbf{R}\right)^{-1}\bar{\mathbf{t}}+\mathrm{Tr}\,(\mathbf{\Lambda}\mathbf{Y})}{\nu}\right\}^{\frac{1}{2}(\nu+pn)}}. \quad \ldots\ldots\ldots \quad (10)$$

Next, for $i\neq j$, let

$$Y_{ij} = z_{ij}\sqrt{Y_{ii}Y_{jj}}, \quad \ldots\ldots\ldots\ldots\ldots\ldots\ldots \quad (11)$$

and denote the diagonal matrix, whose ith diagonal element is $\sqrt{Y_{ii}}$, by $\mathbf{D}$.

The jacobian of the transformation (**11**) is $(Y_{11}Y_{22}\dots Y_{pp})^{\frac{1}{2}(p-1)}$ and

$$|\mathbf{Y}|=|\mathbf{DZD}|=Y_{11}Y_{22}\dots Y_{pp}|\mathbf{Z}|,$$

where

$$\mathbf{Z}=\begin{bmatrix} 1 & z_{12} & \cdot & \cdot & \cdot & z_{1p} \\ z_{21} & 1 & \cdot & \cdot & \cdot & z_{2p} \\ \cdot & \cdot & \cdot & \cdot & \cdot & \cdot \\ z_{p1} & z_{p2} & \cdot & \cdot & \cdot & 1 \end{bmatrix}.$$

The expression (**10**) thus changes to

$$\frac{(Y_{11}Y_{22}\dots Y_{pp})^{\frac{1}{2}(n-3)}|\mathbf{Z}|^{\frac{1}{2}(n-p-2)}\mathrm{d}\bar{\mathbf{t}}\prod_{i=1}^{p}\mathrm{d}Y_{ii}\prod_{i<j}\mathrm{d}z_{ij}}{\left\{1+\dfrac{\bar{\mathbf{t}}'\left(\frac{1}{n}\mathbf{R}\right)^{-1}\bar{\mathbf{t}}+\mathrm{Tr}\,(\mathbf{\Lambda Y})}{\nu}\right\}^{\frac{1}{2}(\nu+pn)}}. \quad \dots (\mathbf{12})$$

Integration of (**12**) over the domain where **Z** is positive definite yields the factor

$$\frac{\pi^{\frac{1}{4}p(p-1)}}{\{\Gamma\frac{1}{2}(n-1)\}^p}\prod_{i=1}^{p}\Gamma\tfrac{1}{2}(n-i),$$

and integration over the range 0 to ∞ for each of the Y_{ii} gives

$$\frac{\nu^{\frac{1}{2}p(n-1)}\{\Gamma\frac{1}{2}(n-1)\}^p\Gamma\frac{1}{2}(\nu+p)}{|\mathbf{\Lambda}|^{\frac{1}{2}(n-1)}\Gamma\frac{1}{2}(\nu+pn)},$$

leaving the marginal distribution of the $\bar{t}_i$ in the form

$$\frac{\left|\frac{1}{n}\mathbf{R}\right|^{-\frac{1}{2}}\Gamma\frac{1}{2}(\nu+p)}{(\pi\nu)^{\frac{1}{2}p}\Gamma\frac{1}{2}\nu}\cdot\frac{\mathrm{d}\bar{\mathbf{t}}}{\left\{1+\bar{\mathbf{t}}'\left(\frac{1}{n}\mathbf{R}\right)^{-1}\bar{\mathbf{t}}/\nu\right\}^{\frac{1}{2}(\nu+p)}}, \quad \dots\dots (\mathbf{13})$$

which is the multivariate t-distribution, characterized by the matrix $\{(1/n)\mathbf{R}\}^{-1}$. The variance-covariance matrix of the $\bar{t}_i$ is thus $(1/n)\{\nu/(\nu-2)\}\mathbf{R}$, and the marginal distribution of any variate $\bar{t}_i$ is

$$\frac{n^{\frac{1}{2}}\Gamma\frac{1}{2}(\nu+1)}{(\pi\nu)^{\frac{1}{2}}\Gamma\frac{1}{2}\nu}(1+n\bar{t}_i^2/\nu)^{-\frac{1}{2}(\nu+1)}\mathrm{d}\bar{t}_i$$

(Cornish loc. cit.). The variate $\bar{t}_i\sqrt{n}$ is then distributed in Student's distribution with ν degrees of freedom.

IV. Marginal Distribution of the T_{ij}

Since $\left(\frac{1}{n}\mathbf{R}\right)^{-1}$ is a positive definite symmetric matrix, the distribution (8) may be integrated easily over the range $-\infty$ to ∞ for each of the variables in $\bar{\mathbf{t}}$, yielding the numerical factor

$$\frac{\left|\frac{1}{n}\mathbf{R}\right|^{\frac{1}{2}}(\pi\nu)^{\frac{1}{2}p}\Gamma\frac{1}{2}[\nu+p(n-1)]}{\Gamma\frac{1}{2}(\nu+pn)},$$

and hence the marginal distribution of the T_{ij} is

$$\frac{|\mathbf{R}|^{-\frac{1}{2}(n-1)}\Gamma\frac{1}{2}[\nu+p(n-1)]}{\pi^{\frac{1}{4}p(p-1)}\nu^{\frac{1}{2}p(n-1)}\Gamma\frac{1}{2}\nu\prod_{i=1}^{p}\Gamma\frac{1}{2}(n-i)}\cdot\frac{|\mathbf{T}|^{\frac{1}{2}(n-p-2)}\prod_{i\leqslant j}\mathrm{d}T_{ij}}{\{1+\mathrm{Tr}(\mathbf{R}^{-1}\mathbf{T})/\nu\}^{\frac{1}{2}[\nu+p(n-1)]}}. \quad \ldots (\mathbf{14})$$

The substitutions $(n-1)s_i^2=T_{ii}$ and $(n-1)w_{ij}=T_{ij}$ in (**14**) then give the simultaneous marginal distribution of the variances and covariances.

From (**14**) the marginal distribution of T_{ii} is

$$\frac{\Gamma\frac{1}{2}(\nu+n-1)}{\nu^{\frac{1}{2}(n-1)}\Gamma\frac{1}{2}\nu\Gamma\frac{1}{2}(n-1)}\cdot\frac{T_{ii}^{\frac{1}{2}(n-3)}\mathrm{d}T_{ii}}{(1+T_{ii}/\nu)^{\frac{1}{2}(\nu+n-1)}}, \quad \ldots\ldots\ldots (\mathbf{15})$$

which is equivalent to the distribution of Fisher's z with $(n-1)$ and ν degrees of freedom. The χ^2 distribution, as the distribution of the sum of the squares of independent normal variates in standard measure, is thus replaced by the distribution of z as the sum of the squares of uncorrelated but dependent t-variates.

Writing $T_{ii}=(n-1)s_i^2$ in (**15**) will give the marginal distribution of the variance, and from this the distribution of the average conditional variance of any admissible order (partial variance in Bartlett's (1933) terminology) can be obtained by appropriately adjusting the degrees of freedom involving the sample number n.

V. Marginal Distributions of Correlation and Regression Coefficients

Since regression and correlation coefficients, both total and partial, and the multiple correlation coefficient are essentially ratios of quantities involving the t-variates, in which the standardizing variate, the estimated standard deviation, cancels out, their distributions are identical with those of their counterparts relating to normally distributed quantities. These distributions may, however, be obtained directly, and as an illustration we give the derivation of the distribution of the regression coefficient, from the marginal distribution (**14**) when $p=2$.

With this value of p, (**14**) reduces to

$$\frac{\Gamma\frac{1}{2}[\nu+2(n-1)]}{\pi^{\frac{1}{2}}\nu^{(n-1)}(1-\rho^2)^{\frac{1}{2}(n-1)}\Gamma\frac{1}{2}\nu\Gamma\frac{1}{2}(n-1)\Gamma\frac{1}{2}(n-2)}$$

$$\times\frac{\begin{vmatrix}T_{11} & T_{12}\\ T_{12} & T_{22}\end{vmatrix}^{\frac{1}{2}(n-4)}\mathrm{d}T_{11}\mathrm{d}T_{12}\mathrm{d}T_{22}}{\left\{1+\frac{1}{\nu(1-\rho^2)}(T_{11}-2\rho T_{12}+T_{22})\right\}^{\frac{1}{2}[\nu+2(n-1)]}}. \quad \ldots\ldots (\mathbf{16})$$

Substitute

$$T_{12}=b_{21}T_{11}$$

in (**16**), and take the integral with respect to T_{11} and T_{22}. Hence

$$\iint\frac{T_{11}{}^{\frac{1}{2}(n-2)}(T_{22}-b_{21}T_{11})^{\frac{1}{2}(n-4)}\mathrm{d}T_{11}\mathrm{d}T_{22}\mathrm{d}b_{21}}{\left[1+\frac{1}{\nu(1-\rho^2)}\{T_{22}+T_{11}(1-2\rho b_{21})\}\right]^{\frac{1}{2}[\nu+2(n-1)]}}$$

has to be taken over the region defined by the inequality

$$T_{22}-b_{21}^2T_{11}\geqslant 0.$$

Change the variable to $x=T_{22}-b_{21}^2T_{11}$, and integrate with respect to both x and T_{11} from 0 to ∞, and the distribution of b_{21} takes the form

$$\frac{(1-\rho^2)^{\frac{1}{2}(n-1)}\Gamma\frac{1}{2}n}{\pi^{\frac{1}{2}}\Gamma\frac{1}{2}(n-1)}(1-2\rho b_{21}+b_{21}^2)^{-\frac{1}{2}n}\mathrm{d}b_{21},$$

as first found for normally distributed variates by Pearson (1926) and Romanovsky (1926).

VI. Marginal Distribution of the Covariance

As the starting point for the derivation of the marginal distribution of the covariance, we use the result for normally distributed quantities (Pearson, Jeffery, and Elderton 1929; Wishart and Bartlett 1932). With the notation of Section II, and taking $p=2$, the distribution of C_{12} is

$$\frac{(1-\rho^2)^{\frac{1}{2}(n-3)}\exp\{\rho C_{12}/\sigma^2(1-\rho)^2\}}{\pi^{\frac{1}{2}}\sigma^2 2^{\frac{1}{2}(n-2)}\Gamma\frac{1}{2}(n-1)}\left\{\frac{|C_{12}|}{\sigma^2(1-\rho^2)}\right\}^{\frac{1}{2}(n-2)}K_{\frac{1}{2}(n-2)}\left\{\frac{|C_{12}|}{\sigma^2(1-\rho^2)}\right\}\mathrm{d}C_{12},$$

$$\ldots\ldots\ldots\ldots\ldots\ldots (\mathbf{17})$$

where the vertical bars now designate the modulus, and $K_m(x)$ is the Bessel function of the second kind with imaginary argument.

For a fixed value of s^2, make the substitution $C_{12}=s^2T_{12}$ in (**17**), multiply by the distribution of s^2, and integrate with respect to s^2 from 0 to ∞, obtaining

$$\frac{\nu^{\frac{1}{2}\nu}(1-\rho^2)^{\frac{1}{2}(n-3)}\mathrm{d}T_{12}}{\pi^{\frac{1}{2}}2^{\frac{1}{2}(\nu+n-2)}\sigma^{\nu+2}\Gamma\frac{1}{2}\nu\Gamma\frac{1}{2}(n-1)}\int_0^\infty(s^2)^{\frac{1}{2}\nu}\exp\left\{-\frac{s^2}{2\sigma^2}\left(\nu-\frac{2\rho T_{12}}{1-\rho^2}\right)\right\}$$

$$\times\left\{\frac{s^2|T_{12}|}{\sigma^2(1-\rho^2)}\right\}^{\frac{1}{2}(n-2)}K_{\frac{1}{2}(n-2)}\left\{\frac{s^2|T_{12}|}{\sigma^2(1-\rho^2)}\right\}\mathrm{d}(s^2).$$

$$\ldots\ldots\ldots\ldots\ldots\ldots (\mathbf{18})$$

A second substitution

$$y=\frac{s^2\,|\,T_{12}\,|}{\sigma^2(1-\rho^2)}$$

reduces the integral in this expression to

$$\int_0^\infty y^{\frac{1}{2}(\nu+n-2)}\exp\left\{-\frac{1-\rho^2}{2\,|\,T_{12}\,|}\left(\nu-\frac{2\rho T_{12}}{1-\rho^2}\right)y\right\}K_{\frac{1}{2}(n-2)}(y)\mathrm{d}y,$$

and this may be evaluated using the result given by Watson (1922, p. 388 (7)), yielding

$$\frac{(\frac{1}{2}\pi)^{\frac{1}{2}}\Gamma\frac{1}{2}(\nu+2)\Gamma\frac{1}{2}[\nu+2(n-1)]P_{\frac{1}{2}(n-3)}^{-\frac{1}{2}(\nu+n-1)}(\cosh\alpha)}{(\sinh\alpha)^{\frac{1}{2}(\nu+n-1)}}, \quad \ldots\ldots \quad (\mathbf{19})$$

where $\cosh\alpha=\nu(1-\rho^2)-2\rho T_{12}/(2\,|\,T_{12}\,|)$ and $P_\mu^\lambda(x)$ is the associated Legendre function of the first kind.

Multiplying (**19**) by the constant from (**18**) and substituting for the hyperbolic functions reduces the distribution of T_{12} to the form

$$\frac{\nu^{\frac{1}{2}(\nu+2)}\Gamma\frac{1}{2}[\nu+2(n-1)]}{2\Gamma\frac{1}{2}(n-1)}\cdot\frac{(1-\rho^2)^{\frac{1}{2}(\nu+n-1)}\,|\,T_{12}\,|^{\frac{1}{2}(n-3)}}{[\{\nu(1-\rho^2)-2\rho T_{12}\}^2-4\,|\,T_{12}\,|^2]^{\frac{1}{2}(\nu+n-1)}}$$

$$\times P_{\frac{1}{2}(n-3)}^{-\frac{1}{2}(\nu+n-1)}\left\{\frac{\nu(1-\rho^2)-2\rho T_{12}}{2\,|\,T_{12}\,|}\right\}\mathrm{d}T_{12}. \quad \ldots\ldots\ldots\ldots \quad (\mathbf{20})$$

The change of variable $T_{12}=(n-1)w_{12}$ will give the marginal distribution of the covariance, and from this the distribution of the partial covariance of any admissible order can be obtained by appropriately adjusting the degrees of freedom involving the sample number n.

VII. References

Bartlett, M. S. (1933).—*Proc. Roy. Soc. Edinb.* **53** (3) : 260.

Bechhofer, R. E., Dunnett, C. W., and Sobel, M. (1954).—*Biometrika* **41** : 170.

Cornish, E. A. (1954).—*Aust. J. Phys.* **7** : 531.

Cramer, H. (1946).—" Mathematical Methods of Statistics." (Princeton Univ. Press.)

Dunnett, C. W., and Sobel, M. (1954).—*Biometrika* **41** : 153.

James, A. T. (1954).—*Ann. Math. Statist.* **25** : 40.

MacDuffee, C. C. (1943).—" Vectors and Matrices." (Math. Assoc. Amer.)

Pearson, K. (1926).—*Proc. Roy. Soc.* A **122** : 1.

Pearson, K., Jeffery, G. B., and Elderton, E. M. (1929).—*Biometrika* **21** : 164.

Romanovsky, V. (1926).—*Bull. Acad. Sci. U.R.S.S.* [6] **10** : 643.

Watson, G. N. (1922).—" Theory of Bessel Functions." (Cambridge Univ. Press.)

Wishart, J. (1928).—*Biometrika* **20A** : 32.

Wishart, J., and Bartlett, M. S. (1932).—*Proc. Camb. Phil. Soc.* **28** : 455.

AUSTRALASIAN MEDICAL PUBLISHING CO. LTD.
SEAMER AND ARUNDEL STS., GLEBE, SYDNEY

Commonwealth Scientific and Industrial Research Organization Australia.
Division of Mathematical Statistics Technical Paper No. 3, Melbourne, 1956.

THE RECOVERY OF INTERBLOCK INFORMATION IN QUASI-FACTORIAL DESIGNS WITH INCOMPLETE DATA

3. BALANCED INCOMPLETE BLOCKS

By E. A. Cornish*

(*Manuscript received October* 10, 1955)

Summary

The exact analysis for combining the weighted interblock and intrablock information in a single estimate of a missing observation is given for balanced incomplete-block, unbalanced lattice, and lattice-square designs.

I. Introduction

The quasi-factorial and balanced incomplete-block designs were introduced by Yates (1936*a*, 1936*b*, 1937), and designed for the purpose of eliminating heterogeneity in experimental material to a greater degree than is possible by the use of ordinary randomized blocks. In the original papers, only the complete elimination of group (block, row, or column) differences was considered, but it was pointed out that as the intergroup comparisons contained information on the treatments, some loss of information would inevitably result from the methods of analysis advocated. It was shown that the amount of information lost depends upon the extent to which the group means differ, and that in the limiting case, where the intergroup and intragroup comparisons are of equal accuracy, this loss amounts to $1-E$, where E is the efficiency factor of the particular design. Since Yates's pioneering research, Harshbarger has examined certain types of rectangular lattice, and the Indian school, principally through the work of Bose, Nair, and Rao, has been responsible for partially balanced incomplete-block designs and their generalizations. In this paper, we shall confine attention to the designs introduced by Yates.

The means for recovering intergroup information have been discussed by Yates (1939, 1940*a*, 1940*b*) and Cox, Eckhardt, and Cochran (1940). The methods of analysis are quite straightforward provided the data are complete, but if the designs have been impaired by the loss or unreliability of one or more observations, the analysis requires modification. An approximate method of recovering intergroup information has been described for square, triple, and cubic lattices (Cornish 1943), and has been extended to the lattice-square designs (Cornish 1944), but it could be used only with the limited number of special cases of balanced incomplete-block designs which can be arranged in groups of blocks containing complete replications of the treatments. In this paper, the exact procedure is given for combining the weighted interblock and intrablock information in a single estimate of a missing observation, thus covering the analysis of all balanced incomplete-block designs, and the exact analysis is extended to square, triple, and cubic lattices, and the lattice squares.

* Division of Mathematical Statistics, C.S.I.R.O., Adelaide.

II. Estimation of Treatment Effects

(a) *Intrablock Estimates*

(i) *Designs in Blocks.*—Consider a general design in which t treatments are to be tested in b blocks containing $k<t$ units each, such that the jth treatment is replicated r_j times, and the ith and jth treatments occur together in λ_{ij} blocks.

The linear model is

$$y_{qj} = \mu+\tau_j+\beta_q+\epsilon_{qj}, \quad \ldots\ldots\ldots\ldots\ldots\ldots (1)$$

where μ is a constant representing the general mean, τ_j is constant for all units in the jth treatment, β_q is constant for all units in the qth block, the ϵ_{qj} are the intrablock residuals assumed independently and normally distributed with mean $= 0$ and variance $= \sigma^2$, and the τ_j and β_q are subject to the conditions

$$\Sigma r_j\tau_j = \Sigma\beta_q = 0. \quad \ldots\ldots\ldots\ldots\ldots\ldots (2)$$

Application of the method of maximal likelihood leads directly to the operation of minimizing

$$\sum_j\sum_q(y_{qj}-\mu-\tau_j-\beta_q)^2, \quad \ldots\ldots\ldots\ldots\ldots\ldots (3)$$

subject to the conditions (**2**), thus yielding the normal equations

$$\left.\begin{aligned} bkm &= G \\ k(m+b_q)+\sum_q t_j &= B_q \qquad q = 1, 2, \ldots, b \\ r_j(m+t_j)+\sum_j b_q &= T_j \qquad j = 1, 2, \ldots, t, \end{aligned}\right\} \quad \ldots\ldots (4)$$

from which the estimates m, b_q, and t_j of μ, β_q, and τ_j respectively are obtained. G is the grand total of all observations, and B_q and T_j are totals of the qth block and jth treatment.

The intrablock estimate of the treatment effect is then given by the relation

$$\frac{r_j(k-1)}{k}t_j-\frac{1}{k}\Sigma\lambda_{jl}t_l = T_j-\frac{1}{k}\Sigma B = Q_j, \quad \ldots\ldots\ldots\ldots (5)$$

where the first summation is over all treatments occurring with treatment j, and the second is over all blocks containing treatment j.

1. *Balanced Incomplete Blocks.*—In these designs

$$r_j = r \qquad \text{for } j = 1, 2, \ldots, t$$
$$\lambda_{ij} = \lambda \qquad \text{for } i, j = 1, 2, \ldots, t,$$

so that from (**5**), the intrablock estimate of the treatment mean is

$$t_j+m = \frac{k(t-1)}{rt(k-1)}Q_j+m = \frac{1}{rE}Q_j+m,$$

where E, the efficiency factor for the design, is $t(k-1)/k(t-1)$.

2. *Square Lattice.*—In this series $t = k^2$, and it is assumed that the two groupings of the treatments in the lattice are each replicated r times. Assigning a double subscript to denote a treatment,

$$\begin{aligned} r_{ij} &= 2r && \text{for } i, j = 1, 2, \ldots, k \\ \lambda_{ij\cdot uv} &= r && \text{for } i = u \text{ or } j = v \\ &= 0 && \text{otherwise,} \end{aligned}$$

and using the groupings previously employed (Cornish 1943), the intrablock estimate of a treatment mean is

$$t_{uv}+m = \frac{1}{2r}T_{uv}-\frac{1}{2kr}(Y_{u\cdot}-X_{u\cdot})-\frac{1}{2kr}(X_{\cdot v}-Y_{\cdot v}).$$

3. *Triple Lattice.*—In the structure of these designs, $t = k^2$, and there are three groupings of the treatments in the lattice which are assumed to be each replicated r times. Assigning a triple subscript to denote treatment,

$$\begin{aligned} r_{ijl} &= 3r && \text{for } i, j, l = 1, 2, \ldots, k \\ \lambda_{ijl\cdot uvw} &= r && \text{for } i = u \text{ or } j = v \text{ or } l = w \\ &= 0 && \text{otherwise,} \end{aligned}$$

so that the intrablock estimate of a treatment mean is

$$t_{uvw}+m = \frac{1}{3r}T_{uvw}-\frac{1}{6rk}(2X_{u\cdot\cdot}-Y_{u\cdot\cdot}-Z_{u\cdot\cdot})-\frac{1}{6rk}(2Y_{\cdot v\cdot}-X_{\cdot v\cdot}-Z_{\cdot v\cdot})$$
$$-\frac{1}{6rk}(2Z_{\cdot\cdot w}-X_{\cdot\cdot w}-Y_{\cdot\cdot w}).$$

4. *Cubic Lattice.*—In these designs, $t = k^3$, and there are three groupings of the treatments in the lattice which are assumed to be each replicated r times. Assigning a triple subscript to denote treatment,

$$\begin{aligned} r_{ijl} &= 3r && \text{for } i, j, l = 1, 2, \ldots, k \\ \lambda_{ijl\cdot uvw} &= r && \text{for } \begin{matrix} i = u \\ j = v \end{matrix} \text{ or } \begin{matrix} i = u \\ l = w \end{matrix} \text{ or } \begin{matrix} j = v \\ l = w \end{matrix} \\ &= 0 && \text{otherwise,} \end{aligned}$$

so that the intrablock estimate of a treatment mean is

$$t_{uvw}+m = \frac{1}{3r}T_{uvw}-\frac{1}{6rk^2}(Y_{u\cdot\cdot}+Z_{u\cdot\cdot}-2X_{u\cdot\cdot})-\frac{1}{6rk^2}(X_{\cdot v\cdot}+Z_{\cdot v\cdot}-2Y_{\cdot v\cdot})$$
$$-\frac{1}{6rk^2}(X_{\cdot\cdot w}+Y_{\cdot\cdot w}-2Z_{\cdot\cdot w})-\frac{1}{6rk}(2Z_{uv\cdot}-X_{uv\cdot}-Y_{uv\cdot})$$
$$-\frac{1}{6rk}(2Y_{u\cdot w}-X_{u\cdot w}-Z_{u\cdot w})-\frac{1}{6rk}(2X_{\cdot vw}-Y_{\cdot vw}-Z_{\cdot vw}).$$

(ii) *Designs in Latin Squares.*—1. *Lattice Squares.* By virtue of the double restriction of the Latin square, assignable variability must be eliminated in two ways, and consequently the analysis of these designs cannot be subsumed under the general treatment in subsection (i) above. The analysis is made separately, and in conformity with Yates's notation (1940*a*), the intragroup estimate of a treatment mean is, for example, in designs arranged in $\frac{1}{2}(k+1)$ squares,

$$t+m=\frac{2}{k(k-1)}\left[\sum_{i=1}^{\frac{1}{2}(k+1)}({}_sP_{iv}-{}_iP_{iv})+\sum_{i'=1}^{\frac{1}{2}(k+1)}({}_sP_{i'v}-{}_{i'}P_{i'v})-\frac{k-1}{k+1}G\right].$$

(*b*) *Weighted Interblock and Intrablock Estimates*

(i) *Designs in Blocks.*—To illustrate the basic analysis with recovery of interblock information, consider the analysis of a balanced incomplete-block design.

The linear model

$$y_{qj}=\mu+\tau_j+\beta_q+\epsilon_{qj}, \qquad \ldots\ldots\ldots\ldots\ldots\ldots(6)$$

is similar to that in (**1**), μ, τ_j, and ϵ_{qj} having the same meanings, but the β_q are now taken to be normally and independently distributed, with zero mean and variance σ_β^2, and also independent of the ϵ_{qj}. From the nature of the assumptions in this model, observations in the same block are positively correlated, so that the probability of obtaining the observed set is

$$\frac{|\mathbf{A}|^{\frac{1}{2}}}{(2\pi)^{\frac{1}{2}bk}}\exp\{-\tfrac{1}{2}(\mathbf{y}-\boldsymbol{\mu}-\boldsymbol{\tau})'\mathbf{A}(\mathbf{y}-\boldsymbol{\mu}-\boldsymbol{\tau})\}d\mathbf{y},$$

where $\mathbf{A}$ is the $bk\times bk$ matrix

$$\begin{bmatrix}\mathbf{B} & & & & \\ & \mathbf{B} & & & \\ & & \cdot & & \\ & & & \cdot & \\ & & & & \cdot \\ & & & & & \mathbf{B}\end{bmatrix},$$

$\mathbf{B}$ is the $k\times k$ matrix

$$\begin{bmatrix}\frac{(k-1)\sigma_\beta^2+\sigma^2}{\sigma^2(k\sigma_\beta^2+\sigma^2)} & \frac{-\sigma_\beta^2}{\sigma^2(k\sigma_\beta^2+\sigma^2)} & \cdots & \frac{-\sigma_\beta^2}{\sigma^2(k\sigma_\beta^2+\sigma^2)}\\ \frac{-\sigma_\beta^2}{\sigma^2(k\sigma_\beta^2+\sigma^2)} & \frac{(k-1)\sigma_\beta^2+\sigma^2}{\sigma^2(k\sigma_\beta^2+\sigma^2)} & \cdots & \frac{-\sigma_\beta^2}{\sigma^2(k\sigma_\beta^2+\sigma^2)}\\ \cdots & \cdots & \cdots & \cdots\\ \frac{-\sigma_\beta^2}{\sigma^2(k\sigma_\beta^2+\sigma^2)} & \frac{-\sigma_\beta^2}{\sigma^2(k\sigma_\beta^2+\sigma^2)} & \cdots & \frac{(k-1)\sigma_\beta^2+\sigma^2}{\sigma^2(k\sigma_\beta^2+\sigma^2)}\end{bmatrix},$$

and $(\mathbf{y}-\boldsymbol{\mu}-\boldsymbol{\tau})$ is a column vector of order bk.

Writing

$$\mathbf{y}_q'=[y_{q1}y_{q2}\ \cdot\ \cdot\ \cdot\ y_{qk}]$$

for the row vector of observations in the qth block and $\boldsymbol{\tau}_q'$ for the corresponding vector of treatment constants,

$$(\mathbf{y}-\boldsymbol{\mu}-\boldsymbol{\tau})'\mathbf{A}(\mathbf{y}-\boldsymbol{\mu}-\boldsymbol{\tau})=\sum_q(\mathbf{y}_q-\boldsymbol{\mu}-\boldsymbol{\tau}_q)'\mathbf{B}(\mathbf{y}_q-\boldsymbol{\mu}-\boldsymbol{\tau}_q).$$

Make the orthogonal transformation

$$\mathbf{x} = \mathbf{C}(\mathbf{y}-\boldsymbol{\mu}-\boldsymbol{\tau}),$$

where the $bk \times bk$ matrix $\mathbf{C}$ is

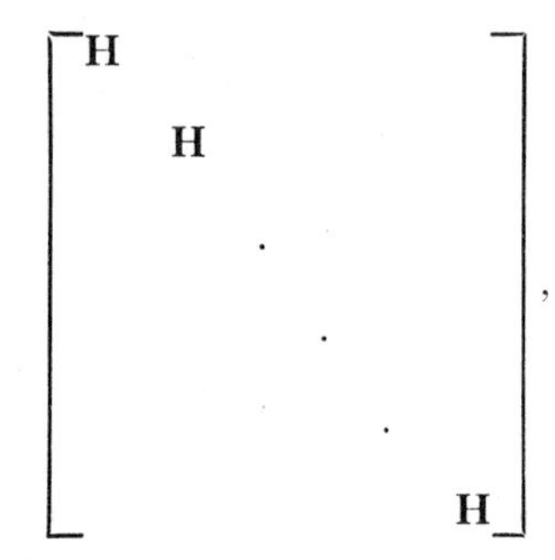

and $\mathbf{H}$ is an orthogonal $k \times k$ matrix having

$$\frac{1}{\sqrt{k}}, \frac{1}{\sqrt{k}}, \ldots, \frac{1}{\sqrt{k}}$$

for the elements of its last row.

Then

$$\mathbf{x}_q = \mathbf{H}(\mathbf{y}_q-\boldsymbol{\mu}-\boldsymbol{\tau}_q)$$

and, as $\mathbf{H}$ is orthogonal, the variance-covariance matrix of the elements of $\mathbf{x}_q$ is

$$\mathbf{HB}^{-1}\mathbf{H}' = \begin{bmatrix} \sigma^2 & & & & & \\ & \sigma^2 & & & & \\ & & \cdot & & & \\ & & & \cdot & & \\ & & & & \cdot & \\ & & & & \sigma^2 & \\ & & & & & k\sigma_\beta^2+\sigma^2 \end{bmatrix}$$

so that

$$\sum_q \mathbf{x}_q'(\mathbf{HBH}')\mathbf{x}_q = \frac{1}{\sigma^2}\sum_q \sum_{j=1}^{k-1} x_{qj}^2 + \frac{1}{k\sigma_\beta^2+\sigma^2}\sum_q x_{qk}^2$$

$$= \frac{1}{\sigma^2}\sum_q \sum_{j=1}^{k} \{y_{qj}-\bar{y}_{q\cdot}-(\tau_{qj}-\bar{\tau}_{q\cdot})\}^2 + \frac{k}{k\sigma_\beta^2+\sigma^2}\sum_q \{\bar{y}_{q\cdot}-\mu-\bar{\tau}_{q\cdot}\}^2,$$

where $\bar{y}_{q\cdot}$ is the mean of the observations in the qth block, and $\bar{\tau}_{q\cdot}$ is the mean of the τ_j occurring in that block.

Writing $\omega = 1/\sigma^2$ and $\omega' = 1/(k\sigma_\beta^2+\sigma^2)$, the probability of obtaining the observed set becomes

$$\frac{|\mathbf{A}|^{\frac{1}{2}}}{(2\pi)^{\frac{1}{2}bk}}\exp[-\tfrac{1}{2}(\omega\Sigma\Sigma\{y_{qj}-\bar{y}_{q\cdot}-(\tau_{qj}-\bar{\tau}_{q\cdot})\}^2+k\omega'\Sigma\{\bar{y}_{q\cdot}-\mu-\bar{\tau}_{q\cdot}\}^2)]d\mathbf{y},$$

and assuming ω and ω' are known, the maximal likelihood estimates are found by minimizing the weighted sum of squares

$$\omega\Sigma\Sigma\{y_{qj}-\bar{y}_{q\cdot}-(\tau_{qj}-\bar{\tau}_{q\cdot})\}^2+k\omega'\Sigma\{\bar{y}_{q\cdot}-\mu-\bar{\tau}_{q\cdot}\}^2. \quad \ldots\ldots(7)$$

Adverting to the general case in II(a)(i) above, and minimizing the result corresponding to (7), the normal equations become

$$bkm = G$$

$$r_j\left(\frac{k-1}{k}\omega+\frac{\omega'}{k}\right)t_j-\Sigma\left(\frac{\lambda_{jl}}{k}\omega-\frac{\lambda_{jl}}{k}\omega'\right)t_l = \omega(T_j-\Sigma B/k)+\omega'(\Sigma B/k-r_jG/bk).$$

$$\ldots\ldots\ldots\ldots(8)$$

Substituting

$$R_j = r_j\left(\omega+\frac{\omega'}{k-1}\right)$$

$$\Lambda_{jl} = \lambda_{jl}(\omega-\omega')$$

$$Q_j' = \Sigma B/k-r_jG/bk$$

$$\omega Q_j+\omega'Q_j' = P_j,$$

the equation for t_j assumes the form

$$R_j\left(\frac{k-1}{k}\right)t_j-\frac{1}{k}\Sigma\Lambda_{jl}t_l = P_j, \quad \ldots\ldots\ldots\ldots\ldots\ldots(9)$$

which is similar in form to (**5**).

Making the necessary substitutions, as in the previous section, the weighted interblock and intrablock estimates of the treatment means are as follows.

1. *Balanced Incomplete Blocks.*

$$t_j+m = \frac{T_j}{r}+\frac{(\omega-\omega')\{(t-k)T_j-(t-1)\Sigma B+(k-1)G\}}{r\{t(k-1)\omega+(t-k)\omega'\}}.$$

2. *Square Lattice.*

$$t_{uv}+m = \frac{1}{2r}T_{uv}-\frac{(\omega-\omega')}{2rk(\omega+\omega')}(Y_{u\cdot}-X_{u\cdot})-\frac{(\omega-\omega')}{2rk(\omega+\omega')}(X_{\cdot v}-Y_{\cdot v}).$$

3. *Triple Lattice.*

$$t_{uvw}+m = \frac{1}{3r}T_{uvw}-\frac{\omega-\omega'}{3rk(2\omega+\omega')}\{(2X_{u\cdot\cdot}-Y_{u\cdot\cdot}-Z_{u\cdot\cdot})+(2Y_{\cdot v\cdot}-X_{\cdot v\cdot}-Z_{\cdot v\cdot})$$
$$+(2Z_{\cdot\cdot w}-X_{\cdot\cdot w}-Y_{\cdot\cdot w})\}.$$

4. *Cubic Lattice.*

$$t_{uvw}+m = \frac{1}{3r}T_{uvw}-\frac{(\omega-\omega')^2}{3rk^2(\omega+2\omega')(2\omega+\omega')}\{(Y_{u\cdot\cdot}+Z_{u\cdot\cdot}-2X_{u\cdot\cdot})$$
$$+(X_{\cdot v\cdot}+Z_{\cdot v\cdot}-2Y_{\cdot v\cdot})+(X_{\cdot\cdot w}+Y_{\cdot\cdot w}-2Z_{\cdot\cdot w})\}$$
$$-\frac{\omega-\omega'}{3rk(2\omega+\omega')}\{(2Z_{uv\cdot}-X_{uv\cdot}-Y_{uv\cdot})+(2Y_{u\cdot w}-X_{u\cdot w}-Z_{u\cdot w})$$
$$+(2X_{\cdot vw}-Y_{\cdot vw}-Z_{\cdot vw})\}.$$

(ii) *Designs in Latin Squares.*—1. *Lattice Squares.* For the same reasons as given above, designs of this class require separate treatment, and again following Yates's notation, the weighted estimate of a treatment mean, with $\frac{1}{2}(k+1)$ replications, is

$$t+m = \frac{2T}{(k+1)} + \frac{2(\omega_i-\omega_r)}{k(k+1)\{\omega_r+\frac{1}{2}(k-1)\omega_i\}}\left\{\sum_{i=1}^{\frac{1}{2}(k+1)} {}_sP_{iv} - \tfrac{1}{2}(k+1)_iP_{iv}\right\}$$

$$+\frac{2(\omega_i-\omega_c)}{k(k+1)\{\omega_c+\frac{1}{2}(k-1)\omega_i\}}\left\{\sum_{i'=1}^{\frac{1}{2}(k+1)} {}_sP_{i'v} - \tfrac{1}{2}(k+1)_{i'}P_{i'v}\right\}.$$

III. Estimation of Missing Observations

As in cases previously discussed, the problem of analysis in these designs with incomplete data can best be approached through the estimation of values for the missing observations.

(*a*) *Balanced Incomplete Blocks*

(i) *Designs not Arranged in Replications.*—When the data are complete, the minimal value of the weighted sum of squares (**7**) is

$$\omega\Sigma\Sigma\{y_{qj}-\bar{y}_{q\cdot}-(t_{qj}-\bar{t}_{q\cdot})\}^2+k\omega'\Sigma\{\bar{y}_{q\cdot}-m-\bar{t}_{q\cdot}\}^2, \quad \ldots\ldots(\mathbf{10})$$

where

$$m = G/bk$$

$$t_j = \frac{k(t-1)}{rs}P_j,$$

and in the latter,

$$s = t(k-1)\omega+(t-k)\omega'.$$

Substituting for $\bar{y}_{q\cdot}$ and m, (**10**) may be reduced to the convenient form

$$\omega\Sigma\Sigma y_{qj}^2-(\omega-\omega')\Sigma B^2/k-\omega'G^2/bk-\Sigma t_jP_j\,. \quad \ldots\ldots\ldots(\mathbf{11})$$

Now suppose that an observation, corresponding to treatment j in the qth block, is missing. Substitute a for the unknown value in (**11**), and minimize the expression for variations in a. The only terms of (**11**) involving a are

$$\omega a^2-(\omega-\omega')(B_q+a)^2/k-\left\{\omega(T_{qj}+a)-(\omega-\omega')\frac{(B_q+a+\sum_j B)}{k}\right\}^2\frac{k(t-1)}{rs}$$

$$-\Sigma\left\{\omega T_{qi}-(\omega-\omega')\frac{(B_q+a+\sum_i B)}{k}\right\}^2\frac{k(t-1)}{rs}+\frac{\omega'(G+a)^2}{bk}\left\{\frac{k(t-1)\omega'}{s}-1\right\},$$

where T_{qj}, B_q, and G now stand for totals of existing observations, and the value of a which minimizes this expression is

$$a = \frac{b}{rU}[(\omega-\omega')\{VB_q-(t-1)W\sum_j B+(\omega-\omega')(t-1)\Sigma\sum_i B+k(k-1)\omega'G\}$$

$$+k(t-1)\omega\{WT_{qj}-(\omega-\omega')\Sigma T_{qi}\}], \quad \ldots(\mathbf{12})$$

where

$$U = t(k-1)\{b(k-1)-(t-1)\}\omega^2+t(k-1)(2b-r-1)\omega\omega'+(t-k)(b-t)\omega'^2,$$
$$V = rs-k(t-1)\omega',$$
$$W = (k-1)\omega+\omega'.$$

(ii) *Designs Arranged in Complete Replications.*—When the designs can be arranged in groups of blocks, each group containing a *single* replicate of all treatments, allowance must be made in the analysis for the elimination of variability between groups. The linear model in (**6**) is changed by inclusion of a term γ_l which is constant for all units in group l. In consequence, the maximal likelihood estimates have to be found by minimizing the weighted sum of squares

$$\omega\Sigma\Sigma\{y_{qj}-\bar{y}_{q\cdot}-(\tau_{qj}-\bar{\tau}_{q\cdot})\}^2+k\omega'\Sigma\{\bar{y}_{q\cdot}-\mu-\gamma_l-\bar{\tau}_{q\cdot}\}^2, \quad \ldots.(\mathbf{13})$$

subject to the additional restriction

$$\Sigma\gamma_l = 0.$$

As each treatment occurs once only in each group of blocks, the normal equations remain unaltered except for the addition of the equation

$$t(g_l+m) = G_l,$$

from which g_l, the estimate of γ_l, is determined. The minimal value of the sum of squares (**13**) now becomes

$$\omega\Sigma\Sigma y^2-(\omega-\omega')\Sigma B^2/k-\omega'\Sigma G_l^2/t-\Sigma t_jP_j, \quad \ldots\ldots..(\mathbf{14})$$

and substituting a for the missing observation yields the solution

$$a = [bt(\omega-\omega')VB_q+\omega bkt(t-1)WT_{qj}-\omega bkt(t-1)(\omega-\omega')\Sigma T_{qi}-bt(t-1)(\omega-\omega')W\sum_j B$$
$$+bt(t-1)(\omega-\omega')^2\Sigma\sum_i B-krt\omega'^2(t-1)G+bkrs\omega' G_l]$$
$$\div[bt(k-1)\{rt(k-1)-k(t-1)\}\omega^2+2btr(k-1)(t-k)\omega\omega'+r(t-k)\{b(t-k)-t(t-1)\}\omega'^2],$$
$$\ldots\ldots\ldots.(\mathbf{15})$$

which minimizes (**14**).

(b) *Other Designs Arranged in Complete Replications*

(i) *Designs in Blocks.*—In the arrangements listed below, the blocks can always be subdivided into groups, each group containing a single replication of all treatments. The formulae for a single missing value are as follows.

1. *Square Lattice.*—The observation is missing from treatment uv in block q of the lth replication of the X grouping of treatments.

$$a = [\{2rk(\omega-\omega')B_q+k^2\omega T_{uv}+2r\omega' X_{..}^{(l)}\}(\omega+\omega')-k\omega(\omega-\omega')(X_{.v}+Y_{.v})$$
$$+k\omega(\omega-\omega')(X_{u.}-Y_{u.})-2k\omega'(\omega-\omega')X_{.v}-(\omega-\omega')^2(X_{..}-Y_{..})-\omega'(\omega+\omega')(X_{..}+Y_{..})]$$
$$\div[(k-1)\{\omega^2(2rk-k-1)+\omega\omega'(2rk+2r-k+1)+2\omega'^2(r-1)\}]. \quad \ldots..(\mathbf{16})$$

2. *Triple Lattice.* The observation is missing from treatment uvw in block q of the lth replication of the X grouping of treatments.

$$a = [\{3rk(\omega-\omega')B_q+k^2\omega T_{uvw}+3r\omega' X_{...}^{(l)}\}(2\omega+\omega')$$
$$-k\omega(\omega-\omega')\{2(X_{u..}+Y_{u..}+Z_{u..})+(2Y_{.v.}-X_{.v.}-Z_{.v.})+(2Z_{..w}-X_{..w}-Y_{..w})\}$$
$$-3k\omega'(\omega-\omega')X_{u..}+\omega(\omega-4\omega')(X_{...}+Y_{...}+Z_{...})-3(\omega-\omega')^2X_{...}]$$
$$\div[(k-1)\{2\omega^2(3rk-k-1)+\omega\omega'(3rk+6r-k+2)+3(r-1)\omega'^2\}]. \quad \ldots..(\mathbf{17})$$

3. *Cubic Lattice.*—The observation is missing from treatment uvw in block q of the lth replication of the X grouping of treatments.

$$
\begin{aligned}
a = {} & [3rk^2(\omega-\omega')B_q+k^3\omega T_{uvw}+3r\omega' X_{...}^{(l)}-\omega(X_{...}+Y_{...}+Z_{...}) \\
& +\frac{3k(\omega-\omega')}{2}(X_{.v.}-Z_{.v.}+X_{..w}-Y_{..w})+\frac{3(\omega-\omega')}{2}(Y_{...}+Z_{...}) \\
& -3k^2(\omega-\omega')X_{.vw}+k^2\omega\mu\{2(2X_{.vw}-Y_{.vw}-Z_{.vw})-(2Y_{u \cdot w}-X_{u \cdot w}-Z_{u \cdot w}) \\
& -(2Z_{uv.}-X_{uv.}-Y_{uv.})-\frac{2}{k}(2X_{.v.}-Y_{.v.}-Z_{.v.})-\frac{2}{k}(2X_{..w}-Y_{..w}-Z_{..w}) \\
& +\frac{1}{k}(2Y_{u..}-X_{u..}-Z_{u..})+\frac{1}{k}(2Y_{..w}-X_{..w}-Z_{..w})+\frac{1}{k}(2Z_{u..}-X_{u..}-Y_{u..}) \\
& +\frac{1}{k}(2Z_{.v.}-X_{.v.}-Y_{.v.})+\frac{3}{k^2}(2X_{...}-Y_{...}-Z_{...})\}+\frac{k\omega\lambda}{2}\{2(2X_{u..}-Y_{u..}-Z_{u..}) \\
& -(2Y_{.v.}-X_{.v.}-Z_{.v.})-(2Z_{..w}-X_{..w}-Y_{..w})-\frac{3}{k}(2X_{...}-Y_{...}-Z_{...})\}] \\
& \div[(k-1)[(3rk^2-k^2+2k-1)\omega-6(k-1)\omega\mu-3\omega\lambda+3\omega'\{r(k+1)-k\}]], \quad \textbf{(18)}
\end{aligned}
$$

where

$$
\lambda = \frac{\omega-\omega'}{\omega+2\omega'} \quad \text{and} \quad \mu = \frac{\omega-\omega'}{2\omega+\omega'}.
$$

In all cases, formulae for missing values in the remaining groups are obtained by following similar rules to those given previously (Cornish 1943).

(ii) *Designs in Latin Squares.*—(1) *Lattice Squares.* The observation is missing from a treatment in the ith row and jth column of the lth square.

$$
\begin{aligned}
a = {} & [k^2T+\{\tfrac{1}{2}(k+1)(\lambda+\mu)-1\}\{G-\tfrac{1}{2}(k+1)S_l\}-k\lambda\{\tfrac{1}{2}(k-1)L_{iv}-\Sigma L_{mv}\} \\
& -k\mu\{\tfrac{1}{2}(k-1)M_{j'v}-\Sigma M_{n'v}\}] \\
& \div[\tfrac{1}{2}(k-1)^2(k+1)\{1-(\lambda+\mu)/2\}], \quad \ldots\ldots\ldots\ldots \textbf{(19)}
\end{aligned}
$$

where

$$
\lambda = \frac{\omega_i-\omega_r}{\omega_r+\frac{1}{2}(k-1)\omega_i}, \qquad \mu = \frac{\omega_i-\omega_c}{\omega_c+\frac{1}{2}(k-1)\omega_i},
$$

and S_l is the total of existing values in the lth square.

Note that in all cases, when there is no recoverable interblock information, i.e. $\omega' = 0$ (or in the lattice squares $\omega_r' = \omega_c' = 0$), these formulae reduce to the corresponding forms given previously (Cornish 1943, 1944) for the intrablock or fully adjusted estimates, and when the recoverable interblock information is a maximum, i.e. $\omega' = \omega$ (or $\omega_r' = \omega_c' = \omega_i$), they reduce to the ordinary randomized block formula given by Yates (1933), except for balanced incomplete-block designs which cannot be arranged in complete replications.

IV. Estimation of Weights

(a) *Balanced Incomplete Blocks*

(i) *Designs not Arranged in Complete Replications.*—If the intrablock estimates are included with the existing observations, the analysis of the completed set may

TABLE 1

DATA REQUIRED FOR ANALYSIS OF THE COMPLETED SET (*Example* 1, *Sect.* IV)

Treatment	a	b	c	d	e	f	g	h	i	Total
T	41·9	39·1	41·3	43·6	41·7	38·3	28·6	42·8	37·0	354·3
ΣB	152·9	155·6	167·6	154·4	159·2	161·9	137·5	175·2	152·9	1417·2
W	49·2	13·6	−71·4	45·7	−2·2	−40·8	105·9	−124·7	24·7	0

be made, the mean squares for blocks (eliminating treatments) and residual providing the basis for the estimation of ω' and ω. The intrablock error variance is correctly estimated by dividing the residual sum of squares by the corresponding number of degrees of freedom, reduced by the number of missing observations, but the interblock variance requires further consideration before it can be used. It can be shown that in the analysis of the completed set of observations this variance is always too large, and denoting the intrablock estimate of a missing value in a treatment containing one such value by a, and the treatment total *including* this value by T_a, with corresponding quantities b_1, b_2, and T_b for a treatment with two missing values, and so on, the necessary reduction in the sum of squares is given by

$$\frac{1}{r(r-1)}\underset{1}{S}(T_a-ra)^2+\frac{1}{2r(r-2)}\underset{2}{S}\{2T_b-r(b_1+b_2)\}^2+\tfrac{1}{2}\underset{2}{S}(b_1-b_2)^2+\ .\ .\ .\ , \qquad \text{(20)}$$

TABLE 2

ANALYSIS OF COMPLETED SET (*Example* 1, *Sect.* IV)

Variation Due To	Degrees of Freedom	Sum of Squares	Mean Square
Treatment component	8	35·9597	
Remainder	9	155·4522	
Blocks (eliminating treatments)	17	191·4119	
Treatment (ignoring blocks)	8	21·0050	
Residual	43	112·5618	2·6177
Total	68	324·9787	

where $\underset{1}{S}$, $\underset{2}{S}$, . . . indicate summation over all treatments containing one, two, . . . missing values.

The following examples illustrate the procedure.

Example 1

The observations have been taken from the book by Fisher and Yates (1953) and have already been used in illustration (Cornish 1940). The values for treatments a, f, and i in blocks 3, 6, and 2 have been assumed to be missing. The intrablock estimates, in order, are 5·0, 8·4, and 5·1, and substituting these values gives the requisite additional data summarized in Table 1; $t = 9$, $b = 18$, $k = 4$, $r = 8$, $\lambda = 3$, and $E = 27/32$.

The analysis of variance for the completed set is given in Table 2. The corrected sum of squares for blocks (eliminating treatments) is 176·1751, so that the appropriate mean square for estimating ω' is 10·3632.

TABLE 3

DATA REQUIRED FOR CALCULATION OF INTRABLOCK ESTIMATES AND ANALYSIS OF COMPLETED SET

(*Example* 2, *Sect.* IV)

Variety	Total	Block	Total	Varieties
4	$61·1+a_1$	9	$111·6+a_1$	4, 8, 10, 19, 21
6	$109·6+a_2$	15	$110·8+a_2$	5, 6, 16, 18, 21
10	$149·2+a_3$	11	$100·6+a_3$	1, 7, 10, 16, 17

Variety	1	2	3	4	5	6	7	8	9	10	11	12	13	14
T	102·9	120·4	122·7	75·5	140·8	138·7	120·6	186·2	190·9	190·0	184·4	188·4	188·3	193·1
ΣB	677·9	691·0	693·9	530·0 656·0	727·4	556·1 696·0	711·1	768·2	731·8	589·8 732·6	726·0	756·6	738·4	741·6
W	22·0	40·0	18·8	21·6	−361·6	232·8	−358·8	−451·2	352·0	321·6	364·0	−184·0	178·4	191·2

Variety	15	16	17	18	19	20	21
T	129·0	102·4	148·8	192·9	87·1	72·7	107·6
ΣB	691·1	688·9	738·5	750·5	650·4	668·9	680·2
W	175·6	−206·0	−455·6	10·0	319·2	−281·2	51·2

With a complete set of observations, the expectation of the mean square for blocks (eliminating treatments) is

$$\frac{bk-t}{b-1}\sigma_\beta^2+\sigma^2,$$

as given by Yates (1940*b*). If p observations are missing, it can be shown that this expectation is reduced to

$$\frac{(bk-t-p)}{b-1}\sigma_\beta^2+\sigma^2, \quad \ldots\ldots\ldots\ldots\ldots\ldots (\mathbf{21})$$

so that, in the example above,

$$\omega = 0{\cdot}3820 \text{ and } \omega' = 0{\cdot}08775.$$

Example 2

These observations have been taken from a varietal trial of wheat at the Waite Institute (Waite Agricultural Research Institute 1943). One plot yield from each of varieties 4, 6, and 10 in blocks 9, 15, and 11 respectively, with actual values 13·0, 30·0, and 45·8, has been assumed to be missing. The data required to find the intrablock estimates, and complete the analysis, are given in Table 3; $t = 21$, $b = 21$, $k = 5$, $r = 5$, $\lambda = 1$, and $E = 21/25$.

TABLE 4

ANALYSIS OF COMPLETED SET (*Example* 2, *Sect.* IV)

Variation Due To	Degrees of Freedom	Sum of Squares	Mean Square
Blocks (eliminating varieties)	20	215·2926	
Varieties (ignoring blocks)	20	7197·7693	
Residual	61	395·0394	6·4761
Total	101	7808·1013	

The intrablock estimates are $a_1 = 14{\cdot}4$, $a_2 = 29{\cdot}1$, and $a_3 = 40{\cdot}8$, and the analysis of the completed set is presented in Table 4. The reduction in the sum of squares for blocks (eliminating varieties) calculated from (**20**) is 12·7240, giving a mean square of 10·1284, and from (**21**) the expectation of the reduced block variance is $(81/20)\sigma_\beta^2+\sigma^2$, so that the weights are $\omega = 0{\cdot}1544$ and $\omega' = 0{\cdot}09103$..

(ii) *Designs Arranged in Replications*.—As in the previous subsection, intrablock estimates are substituted for the missing values, and the analysis of the completed set is made. When only one value is missing, the necessary reduction in the sum of squares for blocks (eliminating treatments) may be found by the application of a simple formula; thus, if a is the estimate, and T_a, R_a, and G_a are the treatment, group, and grand totals respectively, including the estimate, the reduction is given by

$$(tT_a+rR_a-G_a-rt_a)^2/rt(r-1)(t-1). \quad \ldots\ldots\ldots (\mathbf{22})$$

On the other hand, if more than one observation is missing, the ease with which the reduction may be obtained depends upon the distribution of the missing values in the design, and Wilkinson (personal communication) has shown that, under certain conditions, which incidentally should cover the majority of cases, formulae as simple as (**22**) may be applied. Otherwise, the procedure outlined previously (Cornish 1943) must be followed.

The analysis is illustrated below in example 3, using data selected from example 2 to construct the balanced incomplete-block design, combinatorial solution number 6, from Fisher and Yates (1953).

Table 5

DATA REQUIRED FOR CALCULATION OF INTRABLOCK ESTIMATES AND ANALYSIS OF COMPLETED SET

(*Example* 3, *Sect.* IV)

Variety	Total	Block	Total	Varieties
2	$129{\cdot}4+a_1$	1	$63{\cdot}4+a_1$	1, 2, 3, 4
5	$178{\cdot}1+a_2$	4	$86{\cdot}4+a_2$	3, 4, 5, 6
7	$232{\cdot}2+a_3$	10	$96{\cdot}6+a_3$	3, 4, 7, 8

Variety	1	2	3	4	5	6	7	8
T	103·8	150·9	166·7	184·8	207·8	262·9	274·6	267·5
ΣB	756·5	693·5 778·4	801·8	812·9	689·3 805·4	828·3	709·3 848·3	844·4
W	−23·3	11·8	−88·8	−94·1	50·4	110·5	17·3	16·2

Example 3

One plot yield from each of varieties 2, 5, and 7, in blocks 1, 4, and 10 respectively, with actual values 18·3, 30·0, and 42·0, has been assumed to be missing. The data required to find the intrablock estimates, and complete the analysis, are given in Table 5; $t = 8$, $b = 14$, $k = 4$, $r = 7$, $\lambda = 3$, and $E = 6/7$.

The intrablock estimates are 21·5, 29·7, and 42·4, and the analysis of the completed set is presented in Table 6. In this example, where the missing observations are in different varieties and different groups, the reduction in the sum of squares for blocks (eliminating varieties) is easily obtained. Under these conditions, the residuals (the quantity in the numerator of (**22**)) corresponding to the intrablock estimates are first required, and denoting them by $x_1, x_2, \ldots, x_p$, the general formula for the amount to be deducted is

$$\frac{rt}{(rt-r-t)}\left\{\sum_{i=1}^{p} x_i^2-\left(\sum_{i=1}^{p} x_i\right)^2/(rt-r-t+p)\right\}, \qquad \ldots\ldots(\mathbf{23})$$

which, in this case with $x_1 = 2{\cdot}1036$, $x_2 = 0{\cdot}5500$, and $x_3 = 1{\cdot}6696$, yields 9·6844, making the mean square for blocks 8·8722.

With a complete set of observations, the expectation of the mean square for blocks (eliminating treatments) is

$$\frac{bk-v-k(r-1)}{b-r}\sigma_\beta^2+\sigma^2, \qquad \ldots\ldots\ldots\ldots\ldots\ldots(\mathbf{24})$$

TABLE 6

ANALYSIS OF COMPLETED SET (*Example* 3, *Sect.* IV)

Variation Due To	Degrees of Freedom	Sum of Squares	Mean Square
Groups	6	76·3111	
Varietal component	7	48·7121	
Remainder	6	76·3111	
Blocks (eliminating varieties)	13	125·0232	
Varieties (ignoring blocks)	7	3871·2450	
Residual	26	123·5943	4·7536
Total	52	4196·1736	

but if p values are missing, this expectation takes the form

$$\frac{bk-v-p-k'(r-1)}{b-r}\sigma_\beta^2+\sigma^2,$$

where the multiple k' is less than k, and can be found exactly only with considerable labour. The coefficient of σ_β^2 in this expectation must, however, lie between

$$\frac{bk-v-k(r-1)}{b-r} \text{ and } \frac{bk-v-p-k(r-1)}{b-r},$$

and a very good approximation to k' can be found by assigning weight 1 to all units in complete blocks, and weight $(k-p_i)/k$ to all existing units in a block with p_i missing values. Using this rule, the approximation to k' in the above example is 3·625, so that the weighting coefficients become $\omega = 0{\cdot}2104$ and $\omega' = 0{\cdot}1030$.

(b) *Unbalanced Lattices and Lattice Squares*

In both these classes of design, the expectation of the mean square for blocks (rows, columns), eliminating treatments, is also difficult to obtain because of the presence of one or more additional classifications, but since the expectations may be expressed in the same form as (**24**), the method of approximation given above may be used. In the unbalanced lattices the same weighting procedure is applied, but in the lattice square designs with $\frac{1}{2}(k+1)$ replications the rule requires modification. Taking rows, for example, all units in complete rows are allotted weight 1, and all existing units in a row with p_i missing values, weight $(k-p_i)/k$, provided that in both

TABLE 7

DATA REQUIRED FOR ANALYSIS OF THE COMPLETED SET (*Example* 4, *Sect.* IV)

Square I		Square II		Square III	
L_1	5·36	L_1	−28·64	L_1	90·07
L_4	−125·53	L_5	64·27	L_2	−21·43
M_1	−50·94	M_5	52·23	M_1	95·17
M_2	−57·83			M_5	37·37
Total	−268·17		120·03		148·14
	1709·07		1579·67		1570·30

sy = 183·37, cx = 219·67, grand total 4859·04.

cases the columns are otherwise intact, but if another value were missing from a second row, the weight of the existing unit at the intersection of the corresponding column and the first row would be halved.

TABLE 8

ANALYSIS OF COMPLETED SET (*Example* 4, *Sect.* IV)

Variation Due To	Degrees of Freedom	Sum of Squares	Mean Square
Squares	2	481·1903	
Rows (eliminating treatments)	12	1010·0837	
Columns (eliminating treatments)	12	353·9223	
Treatments (ignoring rows and columns)	24	7204·9323	
Residual	22	389·1029	17·6865
Total	72	9439·2315	

Example 4

The observations have been taken from the paper by Yates (1940*a*), and have already been used in illustration (Cornish 1944). Two values have been assumed missing: that of treatment sy in the first square (actual value 61·2) and that of treatment cx in the second square (actual value 74·0). The intrarow and column estimates are a_1 = 67·07 and a_2 = 69·27 respectively. These yield the supplementary data in Table 7, which are required for the analysis of the completed set given in Table 8.

The reduced values of the sums of squares for rows (eliminating treatments) and columns (eliminating treatments), which are most conveniently found by calculating two auxiliary sets of estimates, as indicated (Cornish 1944), are respectively 1008·8281 and 329·6393, giving mean squares 84·0690 and 27·4699, of which the approximate expectations are $3{\cdot}207\sigma_r^2+\sigma^2$ and $3{\cdot}207\sigma_c^2+\sigma^2$, determined by applying the rule given above. We then have

$$\omega_i = 0{\cdot}05654, \quad \omega_r = 0{\cdot}008251, \quad \omega_c = 0{\cdot}03036,$$
$$\lambda = 0{\cdot}3980 \quad \text{and} \quad \mu = 0{\cdot}1825.$$

V. Calculation of Adjusted Treatment Means

After the weighting coefficients ω and ω' have been found, the formulae of Section III may be applied to determine estimates, combining the weighted inter-block and intrablock information, which may be subsequently used to calculate the adjusted treatment means. The application of the formulae is illustrated with the data of the examples used above.

Table 9

DATA REQUIRED FOR CALCULATION OF ADJUSTED TREATMENT TOTALS (*Example* 1, *Sect.* V)

Treatment	a	b	c	d	e	f	g	h	i
T	42·0	39·1	41·3	43·6	41·7	37·1	28·6	42·8	36·9
ΣB	151·8	155·5	167·7	153·2	159·2	160·7	137·4	174·0	152·9
W	54·9	10·8	−75·8	51·7	−5·8	−40·8	103·1	−118·7	20·6
Y	43·5	39·4	39·2	45·0	41·5	36·0	31·4	39·6	37·5

$$\mu = \frac{\omega-\omega'}{\omega t(k-1)+\omega'(t-k)} = 0{\cdot}02737.$$

Example 1

The intrablock estimates $a_1 = 5{\cdot}0$, $a_2 = 8{\cdot}4$, and $a_3 = 5{\cdot}1$ are taken as first approximations, and (**12**) is applied successively to obtain improved values. Since $\omega = 0{\cdot}3820$ and $\omega' = 0{\cdot}08775$,

$$s = 27\omega+5\omega' = 10{\cdot}7530, \quad U = 1242\omega^2+729\omega\omega'+45\omega'^2 = 206{\cdot}0307,$$
$$V = 8s-32\omega' = 83{\cdot}2162, \quad W = 3\omega+\omega' = 1{\cdot}2338, \ \omega-\omega' = 0{\cdot}2943,$$

and the second approximation to a_3 is

$$\frac{18}{8\times 206{\cdot}0307}[0{\cdot}2943(83{\cdot}2162\times 14{\cdot}8-8\times 1{\cdot}2338\times 133{\cdot}0$$
$$+0{\cdot}2943\times 8\times 395{\cdot}3+4\times 3\times 0{\cdot}08775\times 349{\cdot}2)$$
$$+4\times 8\times 0{\cdot}3820(1{\cdot}2338\times 31{\cdot}9-0{\cdot}2943\times 106{\cdot}0)]$$
$$= 5{\cdot}0.$$

The final values are $a_1 = 5{\cdot}1$, $a_2 = 7{\cdot}2$, $a_3 = 5{\cdot}0$, and the adjusted treatment totals are given in Table 9.

Example 2

We take the intrablock estimates $a_1 = 14{\cdot}4$, $a_2 = 29{\cdot}1$, and $a_3 = 40{\cdot}8$ as the first approximations, and apply (**12**) successively to obtain improved values. We have $\omega = 0{\cdot}1544$, $\omega' = 0{\cdot}09103$, so that

$$\begin{aligned} s &= 14{\cdot}4269 \\ U &= 170{\cdot}6822 \\ V &= 63{\cdot}0316 \\ W &= 0{\cdot}7087 \\ \omega-\omega' &= 0{\cdot}06338, \end{aligned}$$

and the second approximation to a_3 is

$$\frac{21}{5\times170{\cdot}6822}[0{\cdot}06338(63{\cdot}0316\times100{\cdot}6-20\times0{\cdot}7087\times591{\cdot}2$$

$$+0{\cdot}06338\times20\times2250{\cdot}8+5\times4\times0{\cdot}09103\times2942{\cdot}6)$$

$$+5\times20\times0{\cdot}1544(0{\cdot}7087\times149{\cdot}2-0{\cdot}06338\times474{\cdot}7)]$$

$$= 38{\cdot}4.$$

The final values are $a_1 = 15{\cdot}0$, $a_2 = 27{\cdot}9$, $a_3 = 38{\cdot}4$.

Example 3

We take the intrablock estimates $a_1 = 21{\cdot}5$, $a_2 = 29{\cdot}7$, $a_3 = 42{\cdot}4$ as first approximations and apply (**15**) successively to improve the values. We have $\omega = 0{\cdot}2104$, $\omega' = 0{\cdot}1030$, and thus

$$\begin{aligned} s &= 5{\cdot}4607 \\ V &= 35{\cdot}3422 \\ W &= 0{\cdot}7341, \end{aligned}$$

the denominator of (**15**) reducing in this case to 2489·2897, so that the second approximation to a_1 is

$$[425{\cdot}2029\times63{\cdot}4+484{\cdot}2743\times129{\cdot}4-70{\cdot}8672\times455{\cdot}3-61{\cdot}8205\times693{\cdot}5$$
$$+9{\cdot}0466\times2116{\cdot}5-16{\cdot}6188\times1597{\cdot}5+220{\cdot}3734\times192{\cdot}5]/2489{\cdot}2897$$
$$= 19{\cdot}9.$$

The final values are $a_1 = 19{\cdot}9$, $a_2 = 29{\cdot}3$, $a_3 = 41{\cdot}1$.

Example 4

We take the intrarow and intracolumn estimates $a_1 = 67{\cdot}07$ and $a_2 = 69{\cdot}27$ as first approximations, and apply (**19**) successively to improve the values. Since $\lambda+\mu = 0{\cdot}5805$, $\frac{1}{2}(k+1)(\lambda+\mu) = 1{\cdot}7415$, $(k-1)^2(k+1)\{2-(\lambda+\mu)\} = 136{\cdot}2720$, the second approximation to a_2 is

$$\frac{4}{136{\cdot}2720}[25\times150{\cdot}4+0{\cdot}7415(4789{\cdot}77-3\times1510{\cdot}4)$$
$$-1{\cdot}9900(-2\times40{\cdot}5+586{\cdot}3)-0{\cdot}9125(2\times40{\cdot}37+459{\cdot}8)]$$
$$= 72{\cdot}00.$$

The final values are $a_1 = 63{\cdot}12$ and $a_2 = 72{\cdot}13$, which are substituted to determine the best weighted estimates of the treatment means.

Example 5

To illustrate the calculations for the square lattice, the data from Cox, Eckhardt, and Cochran (1940) which have already been employed to show the details of the approximate procedure mentioned in Section I may be used. The approximation to the expectation of the mean square for blocks (eliminating treatments) is $6 \cdot 6973\sigma_\beta^2+\sigma^2$, so that from the analysis of the completed set (Cornish loc. cit.)

$$\omega = 0 \cdot 2365, \quad \omega' = 0 \cdot 01883, \quad \omega-\omega' = 0 \cdot 2177, \quad \omega+\omega' = 0 \cdot 2553,$$
$$\text{and} \quad \lambda = (\omega-\omega')/(\omega+\omega') = 0 \cdot 8525.$$

Now take the intrablock estimates $a_1 = 27 \cdot 91$ and $a_2 = 29 \cdot 76$ as the first approximations, and successively apply (**16**) to improve the values.

The second approximation to a_2 is then

$$[(0 \cdot 2177 \times 36 \times 233 \cdot 9+0 \cdot 2365 \times 81 \times 87 \cdot 7+0 \cdot 01883 \times 4 \times 2593 \cdot 8)$$
$$\times 0 \cdot 2553-9 \times 0 \cdot 2365 \times 0 \cdot 2177(477 \cdot 6+551 \cdot 6)+9 \times 0 \cdot 2365 \times 0 \cdot 2177$$
$$(499 \cdot 2-538 \cdot 41)-18 \times 0 \cdot 01883 \times 0 \cdot 2177 \times 477 \cdot 6-0 \cdot 2177^2$$
$$(4767 \cdot 3-5049 \cdot 21)-0 \cdot 01883 \times 0 \cdot 2553 \times 9816 \cdot 51]$$
$$\div[8(26 \times 0 \cdot 2365^2+32 \times 0 \cdot 2365 \times 0 \cdot 01883+2 \times 0 \cdot 01883^2)]$$
$$= 29 \cdot 95.$$

The final values are $a_1 = 27 \cdot 77$, $a_2 = 29 \cdot 95$, which give the following totals required for adjusting the varietal means:

$$T_{1.} = 1006 \cdot 17 \qquad T_{2.} = 1059 \cdot 15 \qquad T_{.1} = 1067 \cdot 42$$
$$Y_{1.} = 485 \cdot 6 \qquad Y_{2.} = 507 \cdot 55 \qquad X_{.1} = 538 \cdot 27.$$

The corresponding adjustments are

$$0 \cdot 8281 \qquad 1 \cdot 0431 \qquad -0 \cdot 2159,$$

and these quantities may be compared with the true values (Cox, Eckhardt, and Cochran 1940) or the approximations given by Cornish (1943).

VI. References

CORNISH, E. A. (1940).—*Ann. Eugen., Lond.* **10**: 112.

CORNISH, E. A. (1943).—Coun. Sci. Industr. Res. Aust. Bull. No. 158.

CORNISH, E. A. (1944).—Coun. Sci. Industr. Res. Aust. Bull. No. 175.

COX, G. M., ECKHARDT, R. C., and COCHRAN, W. G. (1940).—Res. Bull. Iowa Agric. Exp. Sta. No. 281.

FISHER, R. A., and YATES, F. (1953).—"Statistical Tables for Biological, Agricultural and Medical Research." (Oliver & Boyd: London.)

WAITE AGRICULTURAL RESEARCH INSTITUTE (1943).—Report, 1941–1942.

YATES, F. (1933).—*Emp. J. Exp. Agric.* **1**: 129.

YATES, F. (1936*a*).—*J. Agric. Sci.* **26**: 424.

YATES, F. (1936*b*).—*Ann. Eugen., Lond.* **7**: 121.

YATES, F. (1937).—*Ann. Eugen., Lond.* **7**: 319.

YATES, F. (1939).—*Ann. Eugen., Lond.* **9**: 136.

YATES, F. (1940*a*).—*J. Agric. Sci.* **30**: 672.

YATES, F. (1940*b*).—*Ann. Eugen., Lond.* **10**: 317.

Reprinted from *Biometrics*
Vol. 13, No. 1, pp. 19-27, March, 1957

AN APPLICATION OF THE KRONECKER PRODUCT OF MATRICES IN MULTIPLE REGRESSION

E. A. Cornish
Commonwealth Scientific and Industrial Research Organization, Division of Mathematical Statistics, Adelaide, Australia

Introduction

In a recent climatological investigation, the author had occasion to compute the regression of monthly rainfall on the altitude, latitude and longitude of certain observing stations in South Australia. The original data comprised the monthly rainfall at each of fifty stations for the years 1888–1949 inclusive, and regression formulae were calculated for both mean monthly rainfall and monthly totals in individual years.

For any particular month, the analysis of variance of the 3100 (50×62) rainfall totals then assumed the form shown in Table 1. Designating the determining variates altitude, latitude and longitude by x_1, x_2 and x_3 respectively, with corresponding means $\bar{x}_1$, $\bar{x}_2$ and $\bar{x}_3$, and regression coefficients b_1, b_2 and b_3, and writing

$$\mathbf{X} = [x_{ij} - \bar{x}_i] \qquad \begin{array}{l} i = 1, 2, 3 \\ j = 1, 2, \cdots, 50 \end{array}$$

for the matrix of deviations from means,

$$\mathbf{b}_s = \begin{bmatrix} b_{1s} \\ b_{2s} \\ b_{3s} \end{bmatrix}$$

TABLE 1

ANALYSIS OF VARIANCE

Variation due to		Degrees of freedom	
Between years		61	
Between stations	average regression	49	3
	average deviations from regression		46
Years × stations	regressions × years	2989	183
	deviations from regressions within years		2806
Total		3099	

for the vector of regression coefficients in the sth year, and

$$\bar{\mathbf{b}} = \begin{bmatrix} \bar{b}_1 \\ \bar{b}_2 \\ \bar{b}_3 \end{bmatrix}$$

for the vector of average regression coefficients, the sum of squares for the 62 regression formulae, with 186 degrees of freedom is

$$\sum_{s=1}^{62} \mathbf{b}_s'(\mathbf{XX}')\mathbf{b}_s ,$$

and the sum of squares ascribable to the average regression is

$$62\bar{\mathbf{b}}'(\mathbf{XX}')\bar{\mathbf{b}}.$$

The difference between these two quantities

$$\sum_{s=1}^{62} \mathbf{b}_s'(\mathbf{XX}')\mathbf{b}_s - 62\bar{\mathbf{b}}'(\mathbf{XX}')\bar{\mathbf{b}}$$

is the sum of squares for the interaction of regression coefficients with years. The relevant point here is that the corresponding mean square was strongly significant in all months, and thus for the purposes of the enquiry it became important to find assignable causes of this variability.

With this objective, the regression of the $b_{1s}(s = 1, 2, \cdots, 62)$ on two further variates, mean monthly rainfall and time, was calculated; this was done also for the b_{2s} and b_{3s}, thus introducing in all, 6 new regression coefficients. Obviously these coefficients are linear functions of the original rainfall observations, but it is not immediately clear how other points regarding them are resolved; for example, how is their component sum of squares calculated and what constitutes their variance-covariance matrix?

The procedure to be followed in problems of this type has been given in certain special cases; for example, Yates ([1935,] pp. 209, 210) outlined the calculation of the component sums of squares in the interaction of two fertilizer treatments which had been varied quantitatively as an essential feature of a particular experimental design. The full implications of the statistical methods given in this and other cases are not, however, directly apparent, since the variates used were orthogonal both within and between sets, as in the example quoted, or the analyses were simple because there was only one variate in one or both sets.

General case

Denote the dependent variate by y, and the independent variates by $x_1, x_2, \cdots, x_p$. Assume there are n values of each, and that the observations are taken m times on the variate y.

Let

$$\mathbf{X} = [x_{ij} - \bar{x}_i] \qquad \begin{array}{l} i = 1, 2, \cdots, p \\ j = 1, 2, \cdots, n \end{array}$$

$$\mathbf{Y} = [y_{kl} - \bar{y}_l] \qquad \begin{array}{l} k = 1, 2, \cdots, n \\ l = 1, 2, \cdots, m \end{array}$$

and

$$\mathbf{B} = [b_{rs}] \qquad \begin{array}{l} r = 1, 2, \cdots, p \\ s = 1, 2, \cdots, m. \end{array}$$

The m sets of normal equations corresponding to the m regressions of y on $x_1, x_2, \cdots, x_p$ may then be written as

$$(\mathbf{XX}')\mathbf{B} = \mathbf{XY}. \tag{1}$$

Let the second set of determining variates be z_1, z_2, $\cdots$, z_q,

$$\mathbf{Z} = [z_{uv} - \bar{z}_u], \qquad \begin{array}{l} u = 1, 2, \cdots, q \\ v = 1, 2, \cdots, m \end{array}$$

and the second set of regression coefficients be

$$\boldsymbol{\beta} = [\beta_{gh}] \qquad \begin{array}{l} g = 1, 2, \cdots, q \\ h = 1, 2, \cdots, p. \end{array}$$

The normal equations involving the regression coefficients β_{gh} of the b_{rs} on the variates z_1, z_2, $\cdots$, z_q are then

$$(\mathbf{ZZ}')\boldsymbol{\beta} = \mathbf{ZB}'. \tag{2}$$

From (1)

$$\mathbf{B} = (\mathbf{XX}')^{-1}\mathbf{XY},$$

so that from (2)

$$(\mathbf{ZZ}')\boldsymbol{\beta} = \mathbf{ZY}'\mathbf{X}'(\mathbf{XX}')^{-1}$$

or

$$(\mathbf{ZZ}')\boldsymbol{\beta}(\mathbf{XX}') = \mathbf{ZY}'\mathbf{X}'. \tag{3}$$

After carrying out the matrix multiplications indicated by (3), and equating corresponding elements of the matrix products, a set of pq equations is obtained, from which the pq elements in $\boldsymbol{\beta}$ may be calculated.

Taking $p = 3$, $q = 2$, for example, and letting $X_{\delta\epsilon}$ and $Z_{\mu\nu}$ *now stand for the general elements in* ($\mathbf{XX}'$) *and* ($\mathbf{ZZ}'$) *respectively*, the left hand members of the equations (3) may be set out as the product

$$\begin{bmatrix} X_{11}Z_{11} & X_{11}Z_{12} & X_{12}Z_{11} & X_{12}Z_{12} & X_{13}Z_{11} & X_{13}Z_{12} \\ X_{11}Z_{12} & X_{11}Z_{22} & X_{12}Z_{12} & X_{12}Z_{22} & X_{13}Z_{12} & X_{13}Z_{22} \\ X_{12}Z_{11} & X_{12}Z_{12} & X_{22}Z_{11} & X_{22}Z_{12} & X_{23}Z_{11} & X_{23}Z_{12} \\ X_{12}Z_{12} & X_{12}Z_{22} & X_{22}Z_{12} & X_{22}Z_{22} & X_{23}Z_{12} & X_{23}Z_{22} \\ X_{13}Z_{11} & X_{13}Z_{12} & X_{23}Z_{11} & X_{23}Z_{12} & X_{33}Z_{11} & X_{33}Z_{12} \\ X_{13}Z_{12} & X_{13}Z_{22} & X_{23}Z_{12} & X_{23}Z_{22} & X_{33}Z_{12} & X_{33}Z_{22} \end{bmatrix} \begin{bmatrix} \beta_{11} \\ \beta_{21} \\ \beta_{12} \\ \beta_{22} \\ \beta_{13} \\ \beta_{23} \end{bmatrix}$$

The 6×6 matrix here given may also be presented more concisely in the form

$$\begin{bmatrix} X_{11}(\mathbf{ZZ}') & X_{12}(\mathbf{ZZ}') & X_{13}(\mathbf{ZZ}') \\ X_{12}(\mathbf{ZZ}') & X_{22}(\mathbf{ZZ}') & X_{23}(\mathbf{ZZ}') \\ X_{13}(\mathbf{ZZ}') & X_{23}(\mathbf{ZZ}') & X_{33}(\mathbf{ZZ}') \end{bmatrix};$$

it is, in fact, a special type of product of $(\mathbf{XX}')$ and $(\mathbf{ZZ}')$ designated the Kronecker (or direct) product (see, for example, Murnaghan [1938], pp. 68–70) and is frequently written

$$(\mathbf{XX}') \times (\mathbf{ZZ}').$$

Correspondingly, the coefficients of the y_{kl} in the right hand members of (3) form a matrix which is the Kronecker product

$$\mathbf{X} \times \mathbf{Z}.$$

The argument holds true for general values p and q, so that if the elements of $\boldsymbol{\beta}$ are written out as a column vector, taking the columns of $\boldsymbol{\beta}$ in order, and the elements of $\mathbf{Y}$ are also written as a column vector taking the rows of $\mathbf{Y}$ in order, and these vectors are designated $\boldsymbol{\beta}_1$ and $\mathbf{Y}_1$ respectively, the equations (3) may be set out in the alternate form

$$\{(\mathbf{XX}') \times (\mathbf{ZZ}')\}\boldsymbol{\beta}_1 = (\mathbf{X} \times \mathbf{Z})\mathbf{Y}_1 \tag{4}$$

of which the solution is

$$\begin{aligned} \boldsymbol{\beta}_1 &= \{(\mathbf{XX}') \times (\mathbf{ZZ}')\}^{-1}(\mathbf{X} \times \mathbf{Z})\mathbf{Y}_1 \\ &= \{(\mathbf{XX}')^{-1} \times (\mathbf{ZZ}')^{-1}\}(\mathbf{X} \times \mathbf{Z})\mathbf{Y}_1 \end{aligned} \tag{5}$$

using the property that the reciprocal matrix of a Kronecker product is the Kronecker product of the reciprocals.

The set of normal equations (4) is thus what would have been derived, by the standard least squares method of fitting, from the original nm values of the y_{kl} and the nm values of each of the pq variates

$$(x_1 - \bar{x}_1)(z_1 - \bar{z}_1),\ (x_1 - \bar{x}_1)(z_2 - \bar{z}_2),\ \cdots,\ (x_1 - \bar{x}_1)(z_q - \bar{z}_q),$$
$$(x_2 - \bar{x}_2)(z_1 - \bar{z}_1),\ (x_2 - \bar{x}_2)(z_2 - \bar{z}_2),\ \cdots,\ (x_2 - \bar{x}_2)(z_q - \bar{z}_q),$$
$$\cdots,\ (x_p - \bar{x}_p)(z_1 - \bar{z}_1),\ (x_p - \bar{x}_p)(z_2 - \bar{z}_2),\ \cdots,\ (x_p - \bar{x}_p)(z_q - \bar{z}_q),$$

which may be regarded as having been generated by taking the Kronecker product of the two sets

$$(x_1 - \bar{x}_1),\ (x_2 - \bar{x}_2),\ \cdots,\ (x_p - \bar{x}_p)$$

and

$$(z_1 - \bar{z}_1),\ (z_2 - \bar{z}_2),\ \cdots,\ (z_q - \bar{z}_q).$$

All subsequent operations thus reduce to standard procedure; for example, the sum of squares ascribable to the coefficients β_{gh} is

$$\boldsymbol{\beta}_1'\{(\mathbf{XX}') \times (\mathbf{ZZ}')\}\boldsymbol{\beta}_1 , \qquad (6)$$

or in its alternate form

$$\boldsymbol{\beta}_1'(\mathbf{X} \times \mathbf{Z})\mathbf{Y}_1 , \qquad (7)$$

and the variance-covariance matrix of the β_{gh} is

$$\sigma^2\{(\mathbf{XX}') \times (\mathbf{ZZ}')\}^{-1} = \sigma^2\{(\mathbf{XX}')^{-1} \times (\mathbf{ZZ}')^{-1}\}, \qquad (8)$$

where σ^2 is the residual variance of the y_{kl} .

If both members of (3) are transposed, the relation

$$(\mathbf{XX}')\boldsymbol{\beta}'(\mathbf{ZZ}') = \mathbf{XYZ}' \qquad (9)$$

is obtained. This corresponds to a reversal in the order of fitting the variates x and z, and consequently, if the order of the elements in the vectors $\boldsymbol{\beta}_1$ and $\mathbf{Y}_1$ is adjusted accordingly to give vectors $\boldsymbol{\beta}_2$ and $\mathbf{Y}_2$, the counterpart of (4) is

$$\{(\mathbf{ZZ}') \times (\mathbf{XX}')\}\boldsymbol{\beta}_2 = (\mathbf{Z} \times \mathbf{X})\mathbf{Y}_2 , \qquad (10)$$

a set of normal equations identical with what would have been obtained by taking the regression of the y_{kl} on the variates

$$(z_1 - \bar{z}_1)(x_1 - \bar{x}_1), (z_1 - \bar{z}_1)(x_2 - \bar{x}_2), \cdots , (z_1 - \bar{z}_1)(x_p - \bar{x}_p),$$
$$(z_2 - \bar{z}_2)(x_1 - \bar{x}_1), (z_2 - \bar{z}_2)(x_2 - \bar{x}_2), \cdots , (z_2 - \bar{z}_2)(x_p - \bar{x}_p),$$
$$\cdots , (z_q - \bar{z}_q)(x_1 - \bar{x}_1), (z_q - \bar{z}_q)(x_2 - \bar{x}_2), \cdots , (z_q - \bar{z}_q)(x_p - \bar{x}_p).$$

Note also that the equations (10) can be derived from (3), and the equations (4) can be derived from (9), in accordance with the fact that the matrices $(\mathbf{XX}') \times (\mathbf{ZZ}')$ and $(\mathbf{ZZ}') \times (\mathbf{XX}')$ can be converted into each other by a suitable permutation of rows and columns, and similarly for $\mathbf{X} \times \mathbf{Z}$ and $\mathbf{Z} \times \mathbf{X}$.

The final result is thus independent of the manner in which the calculations are made, but in general, either two-stage method will involve less computing than the standard procedure on pq variates.

Example

In the example of the Introduction, the units of measurement were inches for rainfall, 100′ for altitude, 10^{-1} degree for both latitude and longitude. With these units and the variates of the first set in the order altitude, latitude, longitude, and those of the second set in the order mean rainfall and time,

$$(\mathbf{XX}') = \begin{bmatrix} 1138.265050 & -161.151320 & 215.304620 \\ -161.151320 & 534.296632 & -125.495288 \\ 215.304630 & -125.495288 & 199.183242 \end{bmatrix},$$

$$(\mathbf{XX}')^{-1} = \begin{bmatrix} 0.00110607 & 0.00006195 & -0.00115655 \\ 0.00006195 & 0.00220017 & 0.00131925 \\ -0.00115655 & 0.00131925 & 0.00710186 \end{bmatrix},$$

$$(\mathbf{ZZ}') = \begin{bmatrix} 175.2654 & -722.3850 \\ -722.3850 & 19855.5000 \end{bmatrix},$$

$$(\mathbf{ZZ}')^{-1} = \begin{bmatrix} 0.00671215 & 0.00024420 \\ 0.00024420 & 0.00005925 \end{bmatrix},$$

$$\mathbf{ZB}' = \begin{bmatrix} 13.1641 & 3.5660 & -14.6490 \\ -63.1084 & -4.2488 & 35.5231 \end{bmatrix}$$

so that

$$\boldsymbol{\beta} = (\mathbf{ZZ}')^{-1}\mathbf{ZB}'$$

$$= \begin{bmatrix} 0.00671215 & 0.00024420 \\ 0.00024420 & 0.00005925 \end{bmatrix} \begin{bmatrix} 13.1641 & 3.5660 & -14.6490 \\ -63.1084 & -4.2488 & 35.5231 \end{bmatrix}$$

$$= \begin{bmatrix} 0.072948 & 0.022898 & -0.089651 \\ -0.000524 & 0.000619 & -0.001473 \end{bmatrix}.$$

From these we obtain

$$
\mathbf{XX}' \times \mathbf{ZZ}' = \begin{bmatrix} 199498.479 & -822265.598 & -28244.251 & 116413.296 & 37735.452 & -155532.835 \\ -822265.598 & 22600821.700 & 116413.296 & -3199740.034 & -155532.835 & 4274981.081 \\ -28244.251 & 116413.296 & 93643.713 & -385967.873 & -21994.982 & 90655.914 \\ 116413.296 & -3199740.034 & -385967.873 & 10608726.777 & 90655.914 & -2491771.691 \\ 37735.452 & -155532.835 & -21994.982 & 90655.914 & 34909.931 & -143886.986 \\ -155532.835 & 4274981.081 & 90655.914 & -2491771.691 & -143886.986 & 3954882.862 \end{bmatrix},
$$

in which, for example, the 2×2 submatrix

$$
\begin{bmatrix} 199498.479 & -822265.598 \\ -822265.598 & 22600821.700 \end{bmatrix} = 1138.265050 \begin{bmatrix} 175.2654 & -722.3850 \\ -722.3850 & 19855.5000 \end{bmatrix},
$$

$$
(\mathbf{XX}')^{-1} \times (\mathbf{ZZ}')^{-1} = \begin{bmatrix} 0.0^{5}742408 & 0.0^{6}270104 & 0.0^{6}415846 & 0.0^{7}151293 & -0.0^{5}776296 & -0.0^{6}282433 \\ 0.0^{6}270104 & 0.0^{7}655327 & 0.0^{7}151293 & 0.0^{8}367069 & -0.0^{6}282433 & -0.0^{7}685240 \\ 0.0^{6}415846 & 0.0^{7}151293 & 0.0^{4}147679 & 0.0^{6}537286 & 0.0^{5}885499 & 0.0^{6}322163 \\ 0.0^{7}151293 & 0.0^{8}367069 & 0.0^{6}537286 & 0.0^{6}130357 & 0.0^{6}322163 & 0.0^{7}781634 \\ -0.0^{5}776296 & -0.0^{6}282433 & 0.0^{5}885499 & 0.0^{6}322163 & 0.0^{4}476687 & 0.0^{5}173429 \\ -0.0^{6}282433 & -0.0^{7}685240 & 0.0^{6}322163 & 0.0^{7}781634 & 0.0^{5}173429 & 0.0^{6}420774 \end{bmatrix},
$$

and

$$
\boldsymbol{\beta}_1' = [0.072948 \quad -0.000524 \quad 0.022898 \quad 0.000619 \quad -0.089651 \quad -0.001473].
$$

The sum of squares ascribable to the regression coefficients in $\boldsymbol{\beta}_1$ is then

$$\boldsymbol{\beta}_1'\{(\mathbf{XX}') \times (\mathbf{ZZ}')\}\boldsymbol{\beta}_1 = 950.06,$$

and the analysis of variance for the month of June takes the form shown in Table 2.

TABLE 2

Analysis of Variance of June Regression Coefficients

Variation due to	Degrees of freedom	Sum of squares
Between years	61	8765.99
Average regression	3	4135.11
Average deviations from average regression	46	896.49
Between stations	49	5031.60
Regression on mean rainfall and time	6	950.06
Remainder	177	415.58
Regressions × years	183	1365.64
Deviations from regressions within years	2806	1337.28
Years × stations	2989	2702.92
Total	3099	16500.51

REFERENCES

Murnaghan, F. D. [1938]. *The theory of group representations.* Johns Hopkins Press, Baltimore.

Yates, F. [1935]. Complex experiments. *Suppl. J. Roy. Stat. Soc. 2*: 181–223.

Commonwealth Scientific and Industrial Research Organization, Australia. Division of Mathematical Statistics Technical Paper No. 4, Melbourne, 1958.

THE CORRELATION OF MONTHLY RAINFALL WITH POSITION AND ALTITUDE OF OBSERVING STATIONS IN SOUTH AUSTRALIA

By G. G. Coote* and E. A. Cornish†

[*Manuscript received February 25, 1957*]

Summary

Linear partial regression coefficients of mean monthly rainfall (1888–1949) on the latitude, longitude, and altitude of 50 stations in South Australia have been determined. Linear prediction in the manner here represented is very accurate, so the analysis was extended to a further 47 stations; with 97 stations, the residual standard deviation is reduced to approximately 1·5 per cent. of the mean monthly rainfall. The results have been applied in the construction of monthly isohyetal maps of the region concerned.

I. Introduction

The extent to which different parts of a country experience similar deviations from their average weather sequences is of paramount importance in several aspects of climatology.

Firstly, a decision must frequently be made as to whether records of a meteorological element, e.g. rainfall, taken simultaneously at a number of observing stations for relatively short periods of time, can be used to replace observations taken at one station over a long time interval. This question often arises in recently developed areas where reliable records are either unavailable or very rare, and the importance of any decision reached is due to the fact that its nature determines the manner in which forecasts of agricultural production from weather data are made, or the manner in which the characteristic climatic features of a given area are assessed. Forecasts are of both direct and indirect value in crop insurance, trade, and administration, and the more important climatic features are required for the purpose of developing agriculture, and for use in problems of engineering and soil conservation. Thus, again taking rainfall as an illustration, suppose that the observations at four stations within a region are completely independent, then 25 years' data at each point will provide as much information on, say, rainfall variability as a record of 100 years at a single station. If, on the other hand, rainfalls at the four stations are correlated, the gain in information resulting from inclusion of additional stations will be reduced by an amount depending upon the magnitude of the correlation, and in the extreme case of perfect correlation, the inclusion of additional stations will provide no more information than that obtainable from the record of one station.

Secondly, and this applies particularly in regions where observing stations are only sparsely distributed, it has become increasingly important to interpolate for meteorological variates. Various isopleths are required, and, considering local demands only, those most urgently needed are isohyets. The accuracy with which rainfall, or any other meteorological variate, may be estimated at a given point

* Division of Mathematical Statistics, C.S.I.R.O., Homebush, N.S.W.
† Division of Mathematical Statistics, C.S.I.R.O., Adelaide.

from known observations at neighbouring points, must evidently depend upon the density of observing stations and their inter-correlations, and, knowing the latter, it is possible to state how densely located the stations should be in order to provide estimates with any assigned degree of accuracy.

Thirdly, the reliance which can be placed on all meteorological estimates, or predictions, is dependent upon the distribution of variance between localized and widespread fluctuations in the weather, and the only way of investigating this distribution is to determine empirically the manner in which the variance is partitioned.

Attention has been directed first to monthly rainfall, and the problem of determining both the correlations and the distribution of variance has been approached in the following ways:

(i) by relating rainfall to the position and altitude of the observing stations within a region of relatively limited area,

(ii) by direct correlation of rainfall at stations dispersed over a much greater area, and determination of the relation to inter-station distance.

The object of this paper is to present an account of an investigation of type (i), with particular reference to the problem of interpolating monthly rainfall.

II. Data

The monthly rainfall observations were extracted from the records of 50 stations for the years 1888–1949 inclusive, situated in the region bounded by latitudes 34° 24′ and 35° 40′ S. and longitudes 138° 6′ and 139° 18′ E. Map 1 depicts the area and shows the individual stations and their locations. The sites chosen represent the maximum number obtainable with the longest standard record at the time the investigation was planned. Rainfall records were provided either by the Commonwealth Meteorological Bureau or private individuals, positions of the stations were made available by the Department of Lands, South Australia, and altitudes were obtained from maps compiled by the Royal Australian Survey Corps or by direct instrumental determination.

This area was chosen partly for the reasons given above, partly because it contains an important region devoted to intensive agriculture, and a substantial proportion of it also constitutes the watershed of the Adelaide water supply.

III. Climatic and Topographic Features

The season is marked by the strong winter incidence of its rains. Approximately 70–80 per cent. of the annual precipitation occurs within the period April–October inclusive, the summer months being characterized by hot, dry atmospheric conditions, low rainfall, and high evaporation.

The area chosen for examination includes portion of the Mount Lofty Ranges, which themselves form part of a major meridional horst region, bounded on the west by Gulf St. Vincent, and on the east by the basin of the River Murray. A latitudinal profile shows the main topographical features of the area to be the range, with its highest point at 2334 ft, constituting a well-defined central plateau bounded by fault escarpments, and plains bordering the range on the west and east. In general, the western fault escarpment is the more abrupt, leading to steeper gradients on the western aspect.

The range lies at an angle of approximately 30° to the prevailing direction of advance of the fronts, and frontal rains comprise the major component of the annual rainfall, the remainder being convectional rains occurring mainly in winter, and general rains usually over very large areas due to convergence in the northerly stream ahead of the advancing fronts. Purely orographic rains are negligible, but topography plays a major part, by inducing local intensification of precipitation arising from activity on the fronts and from other causes.

IV. Regression on Determining Variates

In the analysis of the data, distinction must be made between the monthly means and the rainfall totals of individual months in different years; the latter are affected not only by changes in rainfall over the region as a whole, but also by local irregularities depending upon topography, aspect of the site, and so on. Within any climatic zone, the averages may be anticipated to be more stable, and thus capable of representation to a greater degree of accuracy. The cogency of this remark is emphasized by the results obtained in the two cases. Therefore, Section IV has been divided into two parts, and the two types of data treated separately.

The statistical technique was that of multiple regression, first suggested by Fisher (1925), later advocated by Irwin (1929), and used by Hopkins (1938*a*, 1938*b*). Taking monthly rainfall (averages and individual totals) as the dependent variate y, and the altitude x_1, latitude x_2, and longitude x_3 of the observing stations as the determining variates, the regression function assumed the form

$$Y = \bar{y}+b_1(x_1-\bar{x}_1)+b_2(x_2-\bar{x}_2)+b_3(x_3-\bar{x}_3),$$

the bar over a symbol designating the arithmetic mean. The units of measurement were inches for rainfall, 10^{-1} degrees for both latitude and longitude, and 100 ft for altitude. At Adelaide, degrees of latitude and longitude are very nearly equivalent to 69 and 57 statute miles, so that the units used herein represent approximately 7 and 6 miles respectively.

Since the same set of stations was used throughout both analyses, the matrix of coefficients in the normal equations was constant and took the following form, the order of the variates being altitude, latitude, and longitude:

$$\begin{bmatrix} 1138\cdot265050 & -161\cdot151320 & 215\cdot304630 \\ -161\cdot151320 & 534\cdot296632 & -125\cdot495288 \\ 215\cdot304630 & -125\cdot495288 & 199\cdot183242 \end{bmatrix},$$

with an inverse

$$\begin{bmatrix} 0\cdot0^{2}1106065067 & 0\cdot0^{4}6195415008 & -0\cdot0^{2}1156553000 \\ 0\cdot0^{4}6195415008 & 0\cdot0^{2}2200169761 & 0\cdot0^{2}1319247141 \\ -0\cdot0^{2}1156553000 & 0\cdot0^{2}1319247141 & 0\cdot0^{2}7101855063 \end{bmatrix}.$$

The sums of products of mean monthly rainfall and the determining variates are listed in Table 1.

TABLE 1

SUMS OF PRODUCTS: RAINFALL WITH ALTITUDE, LATITUDE, AND LONGITUDE

	January	February	March	April	May	June
Altitude (in. $\times$ 10^2 ft)	31·83162	29·93149	34·38988	94·22567	139·56235	208·50783
Latitude (in. $\times$ 10^{-1} deg)	8·205528	7·798336	15·237372	27·649888	44·941840	57·833912
Longitude (in. $\times$ 10^{-1} deg)	−1·475252	−1·192574	−5·558948	−10·379442	−19·447610	−20·519058
	July	**August**	**September**	**October**	**November**	**December**
Altitude (in. $\times$ 10^2 ft)	195·78685	203·90867	158·72253	107·16198	58·25993	42·67211
Latitude (in. $\times$ 10^{-1} deg)	65·438920	32·912768	36·981512	17·599352	10·802952	3·427764
Longitude (in. $\times$ 10^{-1} deg)	−15·372030	−3·461462	−0·987658	−1·194668	−1·101518	0·892474

(a) *Mean Monthly Rainfall*

The regression coefficients obtained are presented in Table 2 with their standard deviations, the positive directions for the measurement of latitude and longitude being taken as south and east respectively. All coefficients are strongly significant ($P<0\cdot001$ in all cases). As would be expected from the nature, position, and orientation of the area, rainfall increases with altitude and latitude, and decreases with longitude, but the absolute value of the regression on longitude relative to the regression on latitude for all months is greater than might at first be anticipated. In addition to the normal decrease in rainfall with increasing distance from the coast, the longitudinal coefficient contains components attributable respectively to imperfections in the allowance made by the variate altitude, as it does not account completely for the reduction in rainfall resulting from subsidence of the air mass as the fronts stream down the leeward slopes of the range, and to the greater increase in rainfall on a windward than on a leeward slope.

TABLE 2

PARTIAL REGRESSIONS OF MEAN MONTHLY RAINFALL ON ALTITUDE, LATITUDE, AND LONGITUDE

Month	Partial Regression on		
	Altitude ($in/10^2$ ft)	Latitude ($in/10^{-1}$ deg)	Longitude ($in/10^{-1}$ deg)
January	0·0374 ± 0·00280	0·0181 ± 0·00395	−0·0365 ± 0·00710
February	0·0350 ± 0·00268	0·0174 ± 0·00378	−0·0328 ± 0·00679
March	0·0454 ± 0·00391	0·0283 ± 0·00551	−0·0592 ± 0·00990
April	0·118 ± 0·00851	0·0530 ± 0·0120	−0·146 ± 0·0216
May	0·180 ± 0·0122	0·0819 ± 0·0172	−0·240 ± 0·0308
June	0·258 ± 0·0186	0·113 ± 0·0263	−0·311 ± 0·0472
July	0·238 ± 0·0153	0·136 ± 0·0216	−0·249 ± 0·0388
August	0·232 ± 0·0160	0·0805 ± 0·0226	−0·217 ± 0·0406
September	0·179 ± 0·0120	0·0899 ± 0·0169	−0·142 ± 0·0303
October	0·121 ± 0·00845	0·0438 ± 0·0119	−0·109 ± 0·0214
November	0·0664 ± 0·00480	0·0259 ± 0·00676	−0·0610 ± 0·0122
December	0·0464 ± 0·00391	0·0114 ± 0·00551	−0·0385 ± 0·00990
Total	1·556	0·699	−1·642

On the assumption that the rains of different months are independent (and this is essentially true), the totals given in Table 2 indicate that the average annual rainfall increases by approximately $1\frac{1}{2}$ in/100 ft altitude, and by 0·7 in/7 miles south, and decreases by 1·6 in/6 miles east.

The three sets of coefficients exhibit very regular oscillations, which are illustrated in Figure 1 with harmonic curves of the form

$$B = \bar{b}+\alpha \sin(\theta+\phi_1)+\beta \sin(2\theta+\phi_2)$$

fitted to the observations of each set. Judging from the data for altitude and longitude, the temporal change is not a simple annual oscillation, as the second harmonic is also significant in both cases. With latitude, however, the second coefficient is not significant, but it has been retained since the curve is closely similar in form to that of the altitudinal regressions. The constants of the three curves are given in Table 3, the phase angles agreeing within the limits of sampling with the phase angles of the harmonic fitted to the rainfall distribution.

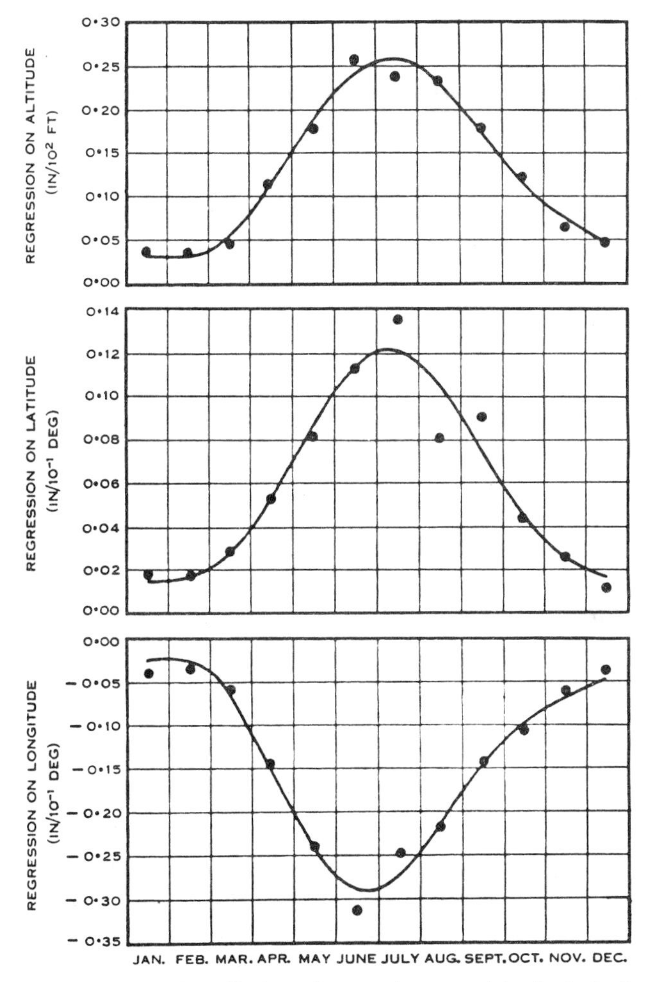

Fig. 1.—Annual oscillation of regressions on altitude, latitude, and longitude.

The significance of the harmonic terms, and the goodness of fit of the curves, may be tested using all the data on each regression coefficient. Since, in addition to the monthly means, the regressions were determined for all months in all years, there are $12 \times 62 = 744$ regression coefficients corresponding to each determining variate, and the total variation of each set of these regressions from their respective means may be partitioned into several components. As the analysis presents a feature of interest, the details for altitude are given in Table 4. The subdivision of the total variation must be made in the manner indicated, since the change in

TABLE 3

CONSTANTS OF HARMONICS FITTED TO SEQUENCES OF REGRESSION COEFFICIENTS

Regression on	$\bar{b}$	a	ϕ_1	β	ϕ_2
Altitude	0·130	0·113	88° 55′	0·0169	120° 20′
Latitude	0·0583	0·0533	94° 25′	0·00944	103° 48′
Longitude	−0·137	0·126	102° 32′	0·0316	155° 8′

magnitude of the coefficients, as the season advances, carries with it differential variability. Assuming that the variances of individual months have been accurately

TABLE 4

ANALYSIS OF VARIANCE: REGRESSION COEFFICIENTS ON ALTITUDE

Variation Due to	Degrees of Freedom		Sum of Squares		Mean Square
Harmonic curve	4		485·3360		121·3340
Deviations	7		6·1689		0·8813
Between months		11		491·5049	
January	61		14·8775		0·2439
February	61		10·9077		0·1788
March	61		13·6958		0·2245
April	61		58·4053		0·9575
May	61		99·8847		1·6375
June	61		115·9275		1·9005
July	61		58·6713		0·9618
August	61		63·0682		1·0339
September	61		65·0235		1·0660
October	61		37·3885		0·6129
November	61		14·8559		0·2435
December	61		12·3054		0·2017
Within months		732		565·0113	
Total		743		1056·5162	

determined on 61 degrees of freedom, an approximate test of goodness of fit of the harmonic curve may be made by calculating

$$\chi^2 = \sum_{j=1}^{12} 62(\bar{b}_{1j} - B_{1j})^2/s_j^2,$$

with 7 degrees of freedom. B_{1j} is the mean value of the coefficient computed from the harmonic, $\bar{b}_{1j}$ is the observed mean, and $s_j{}^2$ is the variance of the jth month in the analysis of variance in Table 4. The regressions on altitude give $\chi^2 = 11 \cdot 031$ for which $0 \cdot 10 < P < 0 \cdot 20$, so that the harmonic fits the observations well. For latitude and longitude, the values of χ^2 are $12 \cdot 611$ and $8 \cdot 460$ respectively.

The observed seasonal changes represent a direct expression of the dependence of the regression coefficients on mean rainfall. Seasonal incidence of the rains yields marked differences in mean rainfall; consequently, it may be anticipated that the

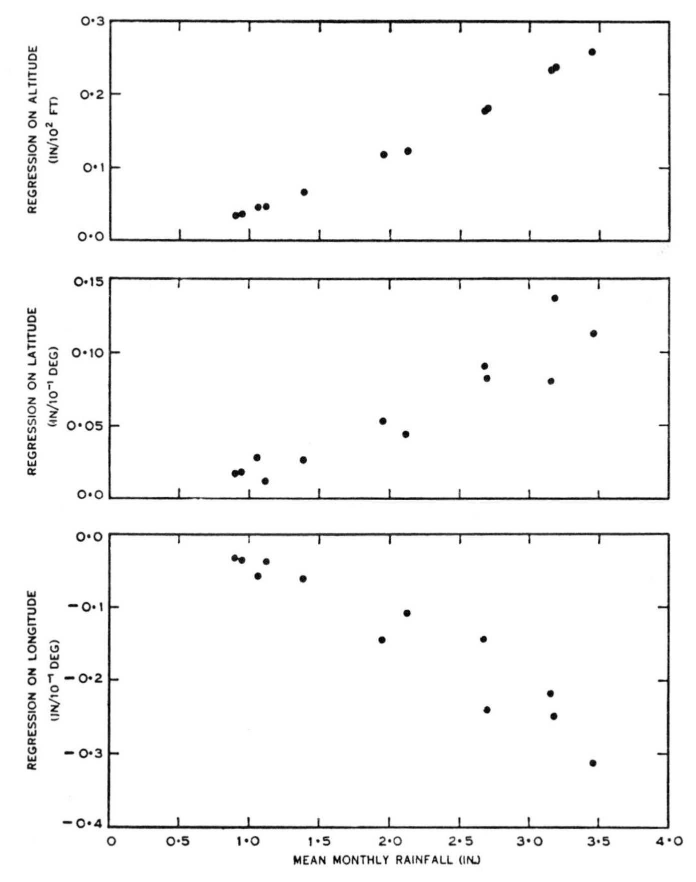

Fig. 2.—Regressions on altitude, latitude, and longitude related to monthly rainfall.

effects of the three determining variates would be closely related to the mean monthly rainfall, and hence that the comparison of the regressions on any particular determining variate for different months would be more appropriate in the proportionate than in the absolute effects. The observations of Figure 1 are illustrated from this new standpoint in Figure 2. As the curvature in each part of the diagram indicates, the rate of change of the coefficients is slightly more rapid than demanded by proportionality, but the quadratic term in the regressions which have been fitted to these data is significant for altitude only. When the regression coefficients are expressed as percentages of the corresponding mean monthly rainfall, the sevenfold increase in the absolute values of the coefficients is reduced to a multiple of 2. This residual departure from proportionality is due to the fact that the topographic effect produces greater intensification of the rains arising from the several types of activity during the winter when conditions for rain are more favourable than in other seasons.

TABLE 5

NORMALIZED VECTORS OF REGRESSION COEFFICIENTS

Month	Altitude	Latitude	Longitude
January	3·17	1·53	−3·09
February	3·16	1·58	−2·97
March	2·98	1·86	−3·88
April	3·14	1·41	−3·89
May	3·11	1·42	−4·16
June	3·16	1·39	−3·80
July	3·09	1·76	−3·24
August	3·26	1·13	−3·05
September	3·17	1·59	−2·51
October	3·25	1·18	−2·93
November	3·23	1·26	−2·97
December	3·29	0·81	−2·73

The set of regression coefficients for each month may be regarded as a vector, and an alternate means of placing the coefficients on a comparable basis is by normalization of these vectors. This could, in effect, be done by multiplying each regression coefficient by the ratio

$$\frac{\text{standard deviation of determining variate}}{\text{standard deviation of rainfall}},$$

but this procedure would, strictly speaking, apply only when the determining variates, and hence the regression coefficients, are uncorrelated. In the present circumstances, where the coefficients, although linearly independent, are correlated, it is more appropriate to use the square root of the sum of squares for regression as the normalizing factor. The values so obtained are given in Table 5. On this basis, it is obvious that the relative effect for any given coefficient is practically constant over all months, this observation applying particularly for altitude.

The relations illustrated in Figure 2 hold also for the regression coefficients calculated from individual monthly totals in different years. Figure 3 shows the regressions on altitude in relation to the mean June rainfall of the 50 stations in each of the 62 years. Corresponding diagrams for the remaining months exhibit similar features, but they stand out more clearly for altitude than either latitude or longitude, and for the winter months as compared with the summer months. The curves of Figure 2 summarize all the information contained in diagrams such

TABLE 6

SUMMARY OF ANALYSES OF VARIANCE: MEAN MONTHLY RAINFALL

Month	Multiple Correlation	% Variance Due to Regression	% Residual Variance	Total Variance (in^2)	Residual Variance (in^2)	Mean Rainfall (in.)	Residual S.D. (as % of mean)	S.D. of Mean (as % of mean)
January	0·90	79·68	20·32	0·0351	0·0071	0·95	8·8	1·26
February	0·90	79·17	20·83	0·0310	0·0065	0·90	9·0	1·22
March	0·89	77·08	22·92	0·0604	0·0138	1·06	11·0	1·60
April	0·91	81·24	18·76	0·3491	0·0655	1·95	13·1	1·85
May	0·92	83·44	16·56	0·8077	0·1338	2·70	13·6	1·93
June	0·91	81·07	18·93	1·6553	0·3134	3·47	16·1	2·28
July	0·93	85·00	15·00	1·4108	0·2116	3·18	14·5	2·04
August	0·91	81·46	18·54	1·2509	0·2320	3·16	15·3	2·15
September	0·92	83·25	16·75	0·7720	0·1294	2·69	13·4	1·90
October	0·91	81·20	18·80	0·3436	0·0646	2·13	11·9	1·69
November	0·90	80·26	19·74	0·1056	0·0208	1·39	10·4	1·44
December	0·87	74·12	25·88	0·0535	0·0138	1·12	10·4	1·52

as Figure 3, and the process of averaging smooths out irregularities, thus giving the definition of curvature observable in Figure 2.

Linear prediction in the manner here represented is very accurate, considering that the residual standard deviation includes, in addition to errors of observation, any systematic deviations from the linear formulae, as, for example, non-linear deviations or other quadratic terms, and irregularities arising from local topographic features or exposure of the stations. Table 6 gives the multiple correlation coefficients and other data relevant to the specification of the accuracy of the prediction formulae. Approximately 80 per cent. of the inter-station variance in the averages of 62 years' data is explicable in terms of the variates used, leaving about 20 per cent. to be ascribed to other factors, such as those mentioned above.

In absolute terms, the residual variance increases with the mean, and also when considered relative to the mean; but relative to the total variance it decreases in the winter months. The variables used thus account for changes in the rainfall more satisfactorily in the winter than in the summer, but the difference is not great.

(b) *Monthly Totals in Individual Years*

Since the regression coefficients were determined for all months in all years, the further analysis of the data can be made in either of two ways. Considering only the regression coefficients themselves, the analysis follows the lines given in the detail of Table 4 for the regressions on altitude, but it can be also extended to the individual monthly rainfall totals. For any particular month there are $62 \times 50 = 3100$ totals, and the primary analysis of these data takes the form:

Variation due to	Degrees of Freedom
Between years	61
Between stations	49
Years $\times$ stations	2989
Total	3099

Table 7 shows this primary subdivision for all months.

Taking account of the regressions obtained in individual years, and also those from the analysis of the mean monthly rainfall over the whole period, the above analysis can be extended to show the contribution of these items to the several sums of squares. Table 8 gives the analysis for June.

The variation between stations, averaged over all years, is divisible into two parts:

(i) a portion attributable to the average regression on the three determining variates (3 degrees of freedom),

(ii) residual deviations from the average regression, due to local variations which have been averaged over all years (46 degrees of freedom).

Correspondingly, the interaction of years and stations yields two parts:

(i) the variation between regression coefficients from year to year (183 degrees of freedom),

(ii) residual deviations from the individual regressions, due to local variations within individual years (2806 degrees of freedom),

and the latter mean square is the appropriate term for testing the interaction of regression coefficients with years. As would be anticipated from the relationships exhibited in diagrams such as Figure 3, the interaction is strongly significant in all months. The coefficients in each of the three series fluctuate widely from year to year in a manner depending principally on the nature of the variations in rainfall,

TABLE 7

ANALYSES OF VARIANCE: INDIVIDUAL MONTHLY RAINFALL TOTALS

Variation Due to	Sum of Squares					
	January	February	March	April	May	June
Between years (61 degrees of freedom)	2550·82	2670·38	2810·89	6852·07	6775·85	8765·99
Between stations (49 degrees of freedom)	106·56	95·32	181·98	1060·12	2454·72	5031·60
Residual (2989 degrees of freedom)	780·62	658·29	721·08	1534·41	2275·28	2702·92
Variation Due to	**July**	**August**	**September**	**October**	**November**	**December**
Between years (61 degrees of freedom)	4433·05	4756·42	5041·10	3869·42	2723·77	2251·63
Between stations (49 degrees of freedom)	4285·49	3801·98	2346·13	1043·08	320·86	162·08
Residual (2989 degrees of freedom)	1614·31	1641·56	1490·81	1104·66	562·75	618·85

and this accounts for a substantial proportion of the sum of squares with 183 degrees of freedom. This point was established by taking the regression of each series of regression coefficients on mean monthly rainfall; as indicated in Table 8, fluctuations in the mean rainfall of June account for as much as $\frac{2}{3}$ of the variation ascribable to the interaction. This proportion is greatest in the winter months, varying from approximately $\frac{1}{2}$ to $\frac{2}{3}$ for April to October inclusive, but in the remaining months gradually falls to its lowest value, 13 per cent., in January. The remaining component, with 180 degrees of freedom, is also strongly significant in all months, but has not been investigated beyond establishing that it is not due to any secular changes in the climate of the region. The calculation of the appropriate sums of squares in the analysis of variance presents certain points of interest, as it employs the properties of the Kronecker product of matrices; the procedure has been discussed by Cornish (1957) in a recent paper.

Adverting to the variation between stations, it should be pointed out that the residual variance with 46 degrees of freedom is not strictly appropriate for testing the significance of the average regression, since the former does not contain the component representing the interaction of regression coefficients and years.

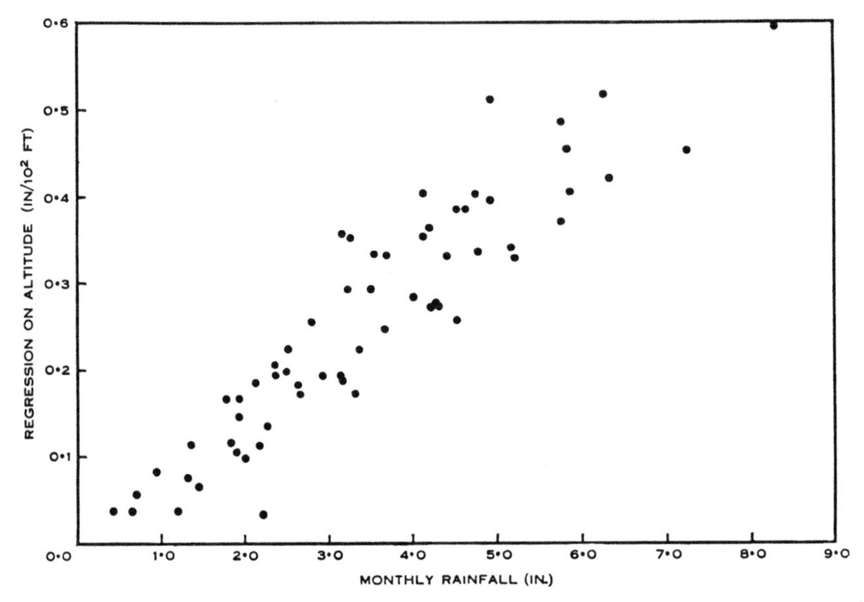

Fig. 3.—Regressions on altitude related to individual annual means (June rainfall).

Table 8

ANALYSIS OF VARIANCE: JUNE RAINFALL

Variation Due to	Degrees of Freedom			Sum of Squares			Mean Square
Between years			61			8765·99	
Average regression		3			4135·11		1378·37
Average deviations from average regression		46			896·50		19·49
Between stations			49			5031·61	
Regression on mean rainfall	3			922·79			307·60
Remainder	180			442·85			2·46
Regressions × years		183			1365·64		7·46
Deviations from regressions within years		2806			1337·27		0·48
Years × stations			2989			2702·91	
Total			3099			16500·51	

However, tests are not really necessary, as the contribution made by the average regression is so large. The complete set of components of variability for all months is presented in Table 9.

TABLE 9

MEAN SQUARES AND THEIR COMPONENTS

Variation Due to	Degrees of Freedom	Expectation of Mean Square	January	February	March	April	May	June
Average regression	3	$62\sigma_4^2+62\sigma_3^2+\sigma_2^2+\sigma_1^2$	28·7969	25·2505	47·9880	291·3008	690·7337	1378·3695
Deviations	46	$62\sigma_3^2+\sigma_1^2$	0·4384	0·4255	0·8265	4·0483	8·3157	19·4889
Regressions × years	183	$\sigma_2^2+\sigma_1^2$	1·7733	1·4943	1·6786	4·3804	6·6521	7·4625
Deviations	2806	σ_1^2	0·1625	0·1371	0·1475	0·2612	0·3770	0·4766
Components:								
Deviations from regressions within years		σ_1^2	0·1625	0·1371	0·1475	0·2612	0·3770	0·4766
Regressions × years		σ_2^2	1·6108	1·3572	1·5311	4·1192	6·2751	6·9859
Average deviations from average regression		σ_3^2	0·0045	0·0047	0·0110	0·0611	0·1280	0·3067
Average regression		σ_4^2	0·4314	0·3785	0·7360	4·5667	10·9055	21·8048

Variation Due to	Degrees of Freedom	Expectation of Mean Square	July	August	September	October	November	December
Average regression	3	$62\sigma_4^2+62\sigma_3^2+\sigma_2^2+\sigma_1^2$	1227·4574	1046·1632	658·7376	286·5991	87·1017	40·9965
Deviations	46	$62\sigma_3^2+\sigma_1^2$	13·1112	14·4237	8·0417	3·9843	1·2946	0·8499
Regressions × years	183	$\sigma_2^2+\sigma_1^2$	4·3378	4·4630	4·3389	2·8789	1·1873	1·2975
Deviations	2806	σ_1^2	0·2924	0·2940	0·2483	0·2059	0·1231	0·1359
Components:								
Deviations from regressions within years		σ_1^2	0·2924	0·2940	0·2483	0·2059	0·1231	0·1359
Regressions × years		σ_2^2	4·0454	4·1690	4·0906	2·6730	1·0642	1·1616
Average deviations from average regression		σ_3^2	0·2068	0·2279	0·1257	0·0609	0·0189	0·0115
Average regression		σ_4^2	19·5210	16·5737	10·4291	4·5152	1·3668	0·6288

The most important comparison in analyses such as given in Table 8, is that of the mean square residual between stations (46 degrees of freedom) with the mean square residual of the interaction of stations and years (2806 degrees of freedom). For all months the former is significantly greater. Consequently, although the regression formula is sufficiently accurate for the estimation of mean monthly rainfall, it is inadequate for prediction of individual monthly totals. The residual between stations still contains a large component of variation ascribable to factors not represented by the variates used, and until the regression formula can be improved, estimation of individual totals cannot be made satisfactorily. The residuals at the stations have been computed, and form the basic data of a further investigation which will be presented elsewhere.

V. Correlations of Rainfall

The information relating to the distribution of variance as given in Tables 7 and 9 may be summarized and reviewed generally in terms of the correlations between the rainfall at the several stations. The intra-class correlations of monthly rainfall for years and stations are obtained from the mean squares of the analyses given in Table 7. If v_y, v_s, and v are the mean squares between years, between stations, and the residual (stations $\times$ years) respectively, the correlations are

$$\text{intra-year } r_y = v_y/(v_y+v), \qquad \text{intra-station } r_s = v_s/(v_s+v).$$

The correlations are given in Table 10, with their inverse hyperbolic transformations $z = \tanh^{-1} r$, and the latter are illustrated in their temporal order in Figure 4; these correlations are all significant, since the mean squares on which they are based are all significant.

The intra-year correlation provides a measure of the distribution of variance between the localized and widespread components in the fluctuations of rainfall in the region from year to year. For example, if in any year June rainfall is above the general average taken over all stations and years, a high correlation indicates that all stations tend to agree in having June rainfall above average, that is, the primary rain-bearing disturbances tend to affect the region as a whole. As Table 10 shows, and a test of significance verifies, the correlation is approximately 0·75 in all months, so that the distribution of variance between the two components is practically constant, irrespective of the season in which the month occurs. This approximate constancy may, however, be a consequence of the length of the interval over which rainfall is measured, namely, the month which could be of sufficient length to allow compensatory effects. A shorter interval, for example, a week, might well show that the distribution of variance has a well-defined annual periodicity. This point is at present under investigation on an extensive scale.

The extent to which the stations are related is not, however, as high as might at first be anticipated, but it must be remembered that in the relatively small area under consideration, irregular local effects constitute an increased proportion of the total variance of rainfall.

On the other hand, the intra-station correlation measures the distribution of variance between the disturbances operating sporadically and those which occur much more regularly from year to year at a particular station. Thus, if at any site, June rainfall is above the general average, a high correlation indicates that the rainfall of this month tends to be above the general average in all years. The marked seasonal oscillation in the intra-station correlations contrasts strongly with the constant intra-year correlation, and directly reflects the characteristics of a winter rainfall. In the summer months, the rains at any particular station vary considerably from year to year, as they arise from sporadic causes, but in the winter

TABLE 10

INTRA-YEAR AND INTRA-STATION CORRELATIONS

Correlations	January	February	March	April	May	June
Years r_y	0·76	0·80	0·79	0·81	0·74	0·76
$z = \tanh^{-1} r_y$	1·00	1·09	1·08	1·14	0·96	1·00
Stations r_s:						
(a)	0·11	0·11	0·19	0·40	0·51	0·64
$z = \tanh^{-1} r_s$	0·11	0·11	0·19	0·42	0·56	0·77
(b)	0·027	0·033	0·069	0·19	0·25	0·39
$z = \tanh^{-1} r_s$	0·027	0·033	0·069	0·19	0·26	0·41

Correlations	July	August	September	October	November	December
Years r_y	0·73	0·74	0·77	0·77	0·83	0·78
$z = \tanh^{-1} r_y$	0·92	0·95	1·01	1·03	1·17	1·05
Stations r_s:						
(a)	0·72	0·69	0·61	0·48	0·35	0·19
$z = \tanh^{-1} r_s$	0·91	0·85	0·70	0·52	0·37	0·20
(b)	0·41	0·44	0·34	0·23	0·13	0·078
$z = \tanh^{-1} r_s$	0·44	0·47	0·35	0·23	0·13	0·078

the rains are much more reliable, since they originate from causes which tend to be repeated from year to year; at the same time, the correlations for the winter months are not high, thus indicating the moderately large variability of the winter rains in this environment.

In these considerations, the mean square between stations with 49 degrees of freedom has been used in calculating the correlations, so that the latter include the component of variation ascribable to the position and altitude of the stations. For the purpose of assessing the general agreement from year to year at any station, this component should be included, as position and altitude are among the characteristic features of a station. It is, however, instructive to recalculate the correlations after eliminating the component due to regression, and the values so found are also given in Table 10 and Figure 4. Naturally, the correlations are much reduced, but the series still exhibits a marked seasonal oscillation. In conformity with the principal observation of the last paragraph of Section IV, these correlations are all significant, and judging from their sign, the variations from year to year at a particular station still tend to be alike under the influence of factors not included in the regression.

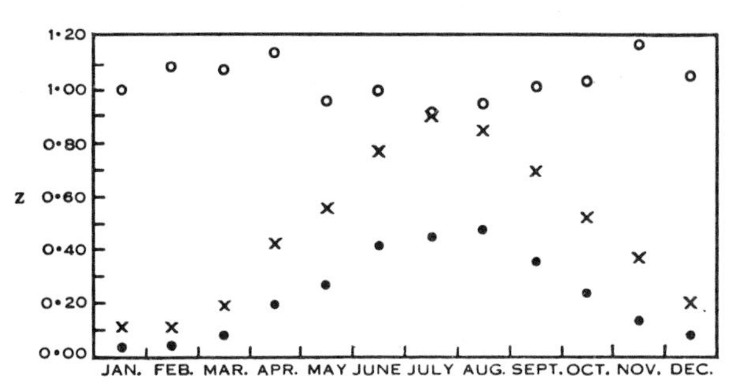

Fig. 4.—Intra-year and intra-station correlations.
○ Intra-year correlation.
× Intra-station correlation.
● Intra-station correlation (after eliminating effects of position and altitude).

VI. Construction of Isohyets

The analysis has demonstrated that mean monthly rainfall can be predicted with sufficient accuracy in terms of the position and altitude of the observing stations, and may thus be used to provide the basic data for interpolation, and hence for the construction of isohyets. For this purpose, the data from 47 additional stations were combined with those of the 50 stations used previously, and the multiple regressions were redetermined for all months. The dependent variate, mean monthly rainfall, was based on varying numbers of years, but differences in weight were ignored in fitting the regression formulae. The regression coefficients so obtained did not differ significantly from those in Table 2, and were therefore used in the new series of calculations. Table 11 summarizes the information on residual variance from the analyses of the set of 97 stations. In absolute terms, the residual variance increases with the mean, and also when considered relative to the mean, though in the latter the increase is not as strongly marked as in the analyses of the set of 50 stations. The residual variance, relative to the total variance, decreases in the winter months, the seasonal change being more clearly defined than in the analyses of the set of 50 stations, and the variables used accounting for changes in the rainfall much more satisfactorily in the winter than in the summer.

If the residual variances are accepted as sufficiently accurate estimates of the variance for each month, after allowing for the position and altitude of the stations, then 100 stations in the area reduce the residual standard deviation to approximately 1·5 per cent. of the mean.

Using the same units as previously, the regression of May, for example, became

$$Y = 2 \cdot 90 + 0 \cdot 170 x_1 + 0 \cdot 0641 x_2 - 0 \cdot 244 x_3.$$

By assigning a value to the rainfall corresponding to the isohyet required, and a succession of values to altitude, this equation can be reduced to a series of linear relations in latitude and longitude, each of which corresponds to a straight line on a map of the region. The intersections of these lines with the orographical contours provide a set of points on the required isohyet, which may then be readily constructed.

Maps 2–13 inclusive show the geographical distribution of the isohyets for each month. In Maps 3 and 7, sea-level isohyets have been superimposed to illustrate the profound influence of altitude; in all diagrams it is clear how closely the isohyets follow the contours. By taking sections drawn to scale from figures such as Maps 3 and 7, most instructive diagrams can be prepared, showing the normal sea-level decrease in rainfall as distance from the coast increases, against the topographical background.

Table 11

Analyses of the set of 97 stations: data on residual variance

Month	Residual Variance	Mean Rainfall	Residual S.D. (as % of mean)	S.D. of Mean (as % of mean)
January	0·0266	0·97	16·49	1·67
February	0·0310	0·97	18·56	1·88
March	0·0261	1·03	15·53	1·58
April	0·0836	1·91	15·18	1·54
May	0·1799	2·78	15·11	1·53
June	0·3829	3·37	18·40	1·87
July	0·1912	3·18	13·84	1·41
August	0·2123	3·13	14·70	1·49
September	0·1543	2·67	14·61	1·48
October	0·0688	2·10	12·38	1·26
November	0·0278	1·39	12·23	1·24
December	0·0199	1·11	12·61	1·28

VII. Acknowledgments

The authors are indebted to Mr. M. C. Childs of the Division of Biochemistry and General Nutrition, C.S.I.R.O., for the preparation of the diagrams.

VIII. References

Cornish, E. A. (1957).—*Biometrics* **13**: 19–27.

Fisher, R. A. (1925).—"Statistical Methods for Research Workers." (Oliver & Boyd: Edinburgh.)

Hopkins, J. W. (1938*a*).—*Canad. J. Res.* C **16**: 16–26.

Hopkins, J. W. (1938*b*).—*Canad. J. Res.* C **16**: 214–24.

Irwin, J. O. (1929).—Crop forecasting and the use of meteorological data in its improvement. Conf. Empire Meteorologists, Agric. Section.

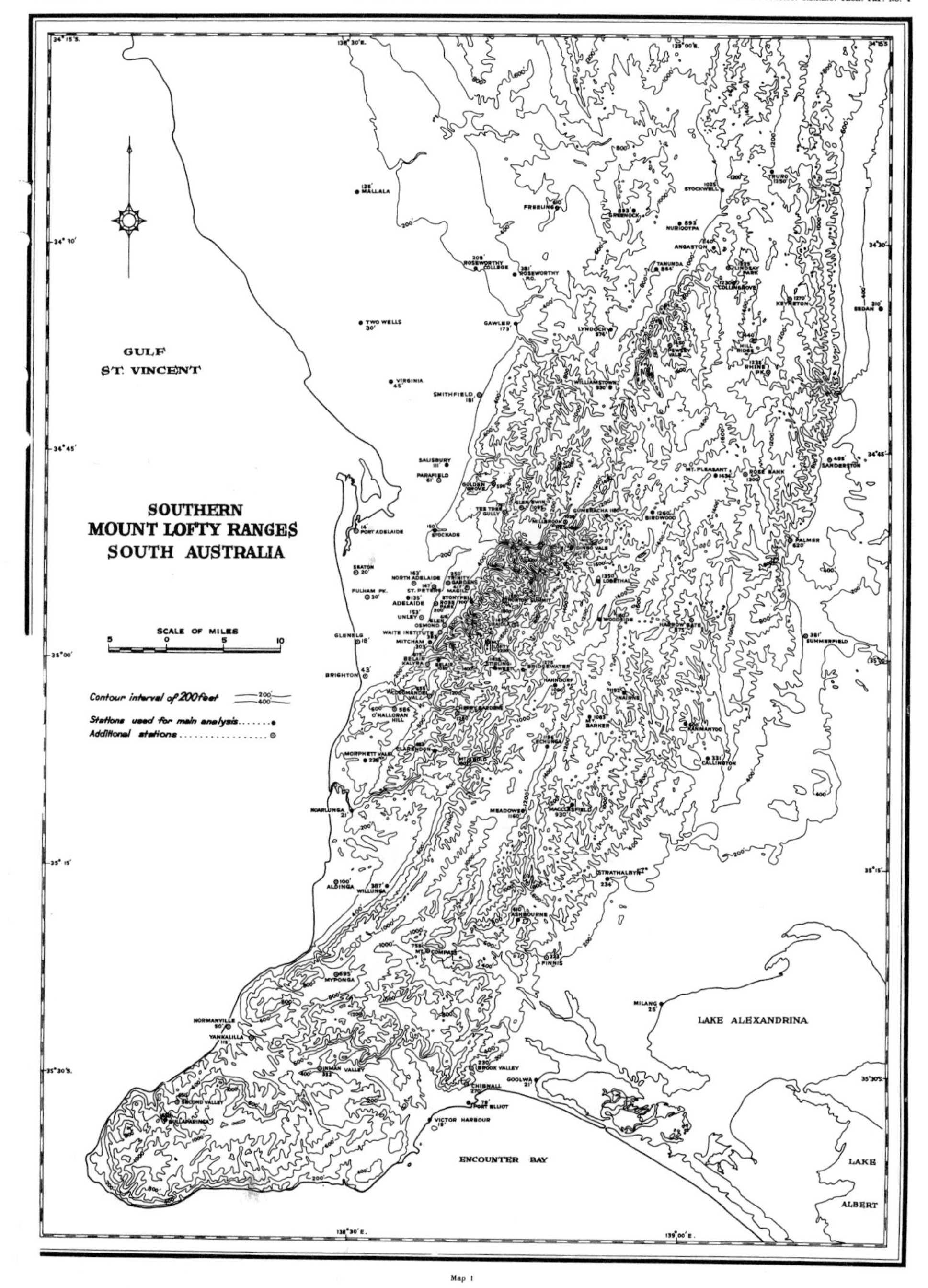
SOUTHERN
MOUNT LOFTY RANGES
SOUTH AUSTRALIA
GULF
ST. VINCENT
SCALE OF MILES
5 0 5 10
Contour interval of 200 feet
Stations used for main analysis
Additional stations
LAKE ALEXANDRINA
ENCOUNTER BAY
LAKE
ALBERT
138° 30' E.
139° 00' E.

Map 1

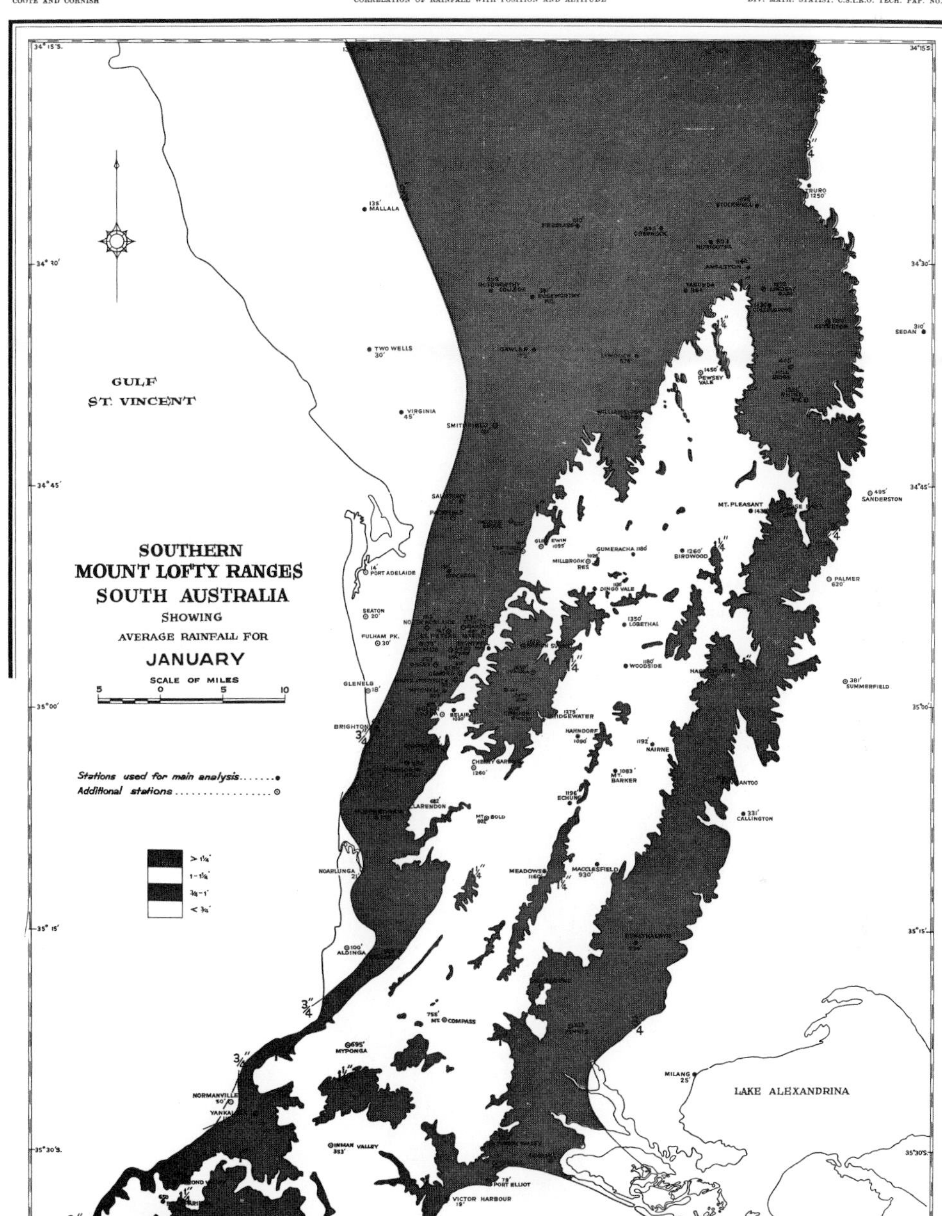

Map 2

Any imperfections in the presentation of this map are due to the present difficulty of reproduction. It should be noted that in the original the regions were denoted by colours.

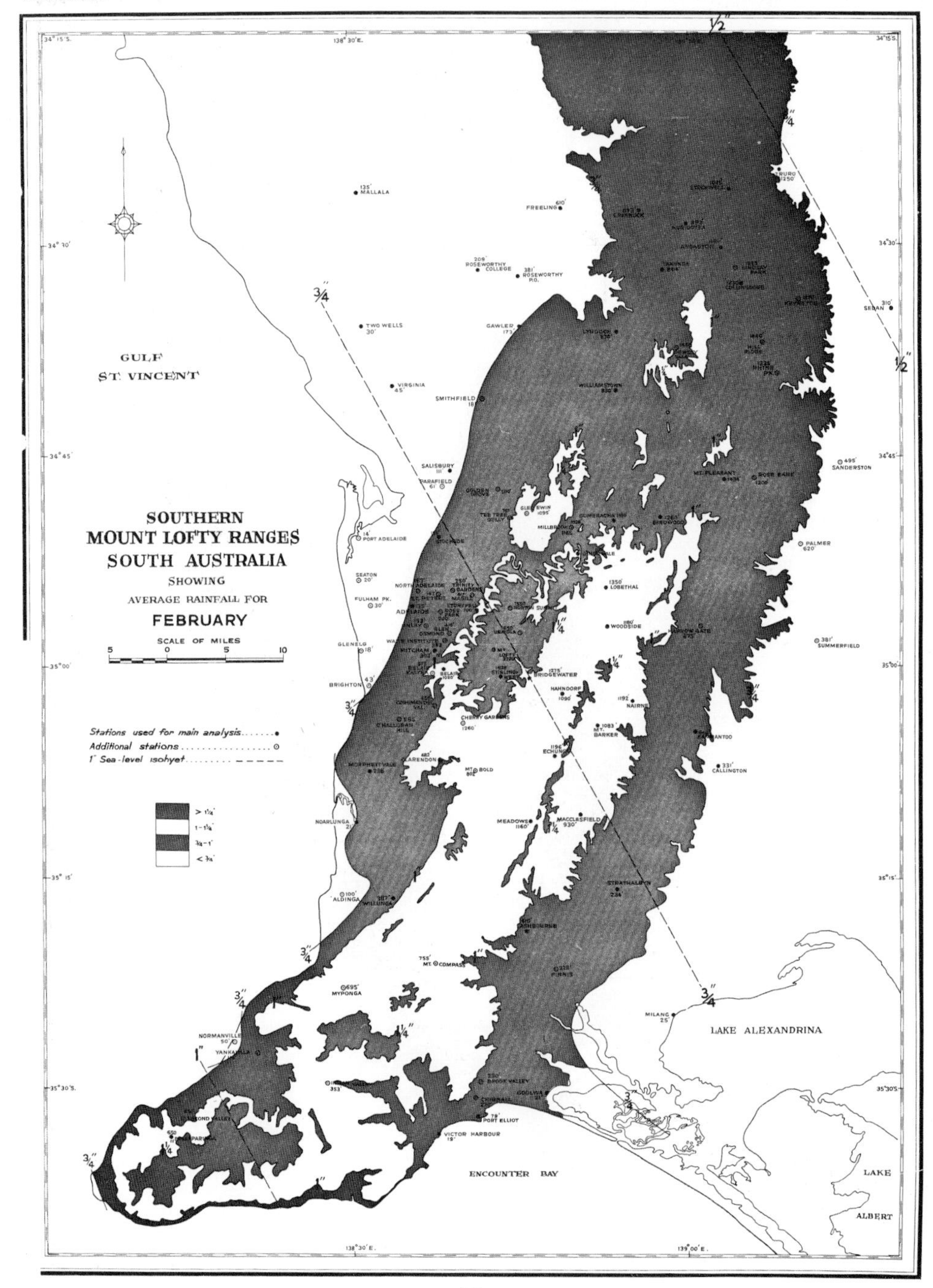

Map 3

Any imperfections in the presentation of this map are due to the present difficulty of reproduction. It should be noted that in the original the regions were denoted by colours.

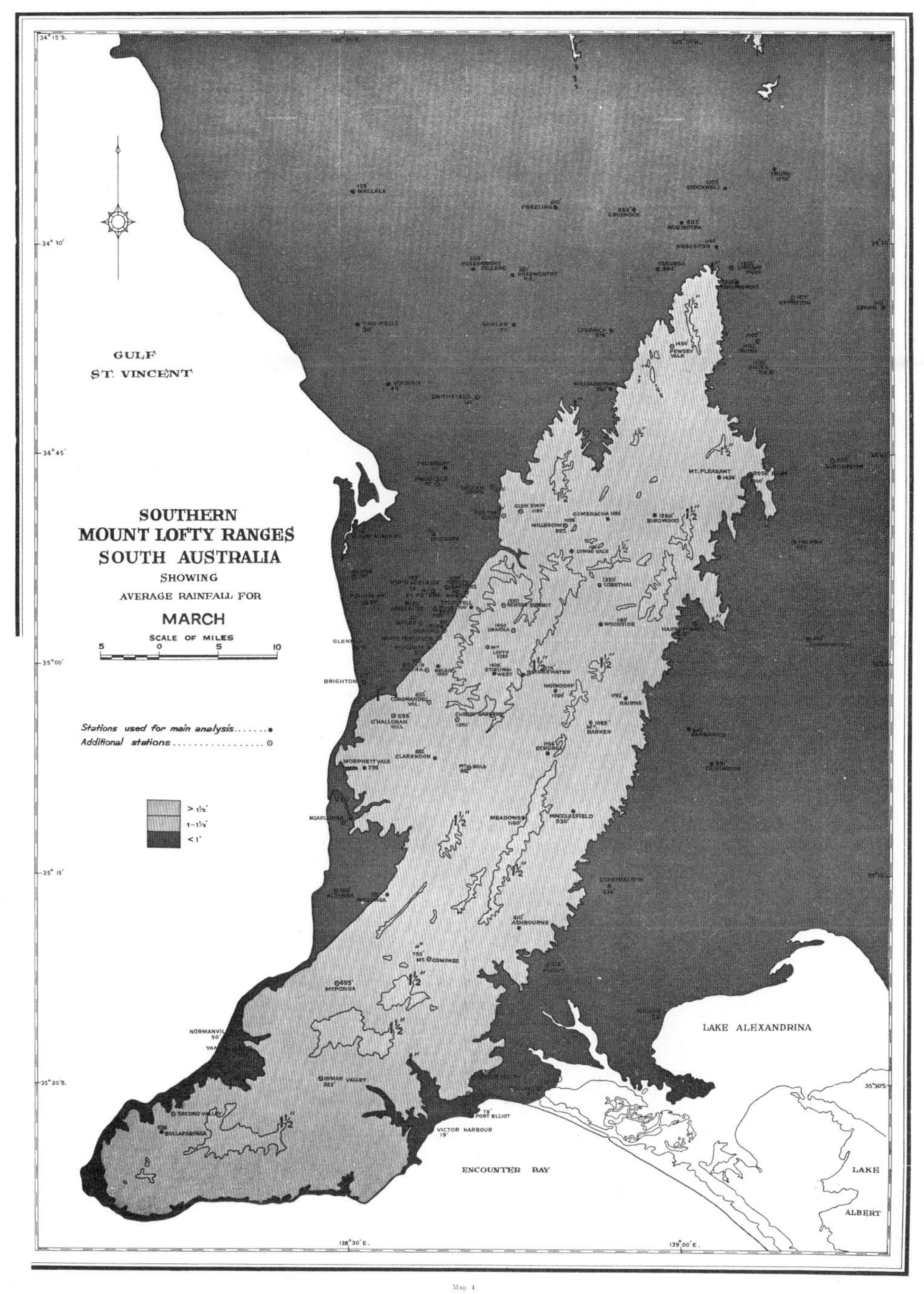

Map 4

Any imperfections in the presentation of this map are due to the present difficulty of reproduction. It should be noted that in the original the regions were denoted by colours.

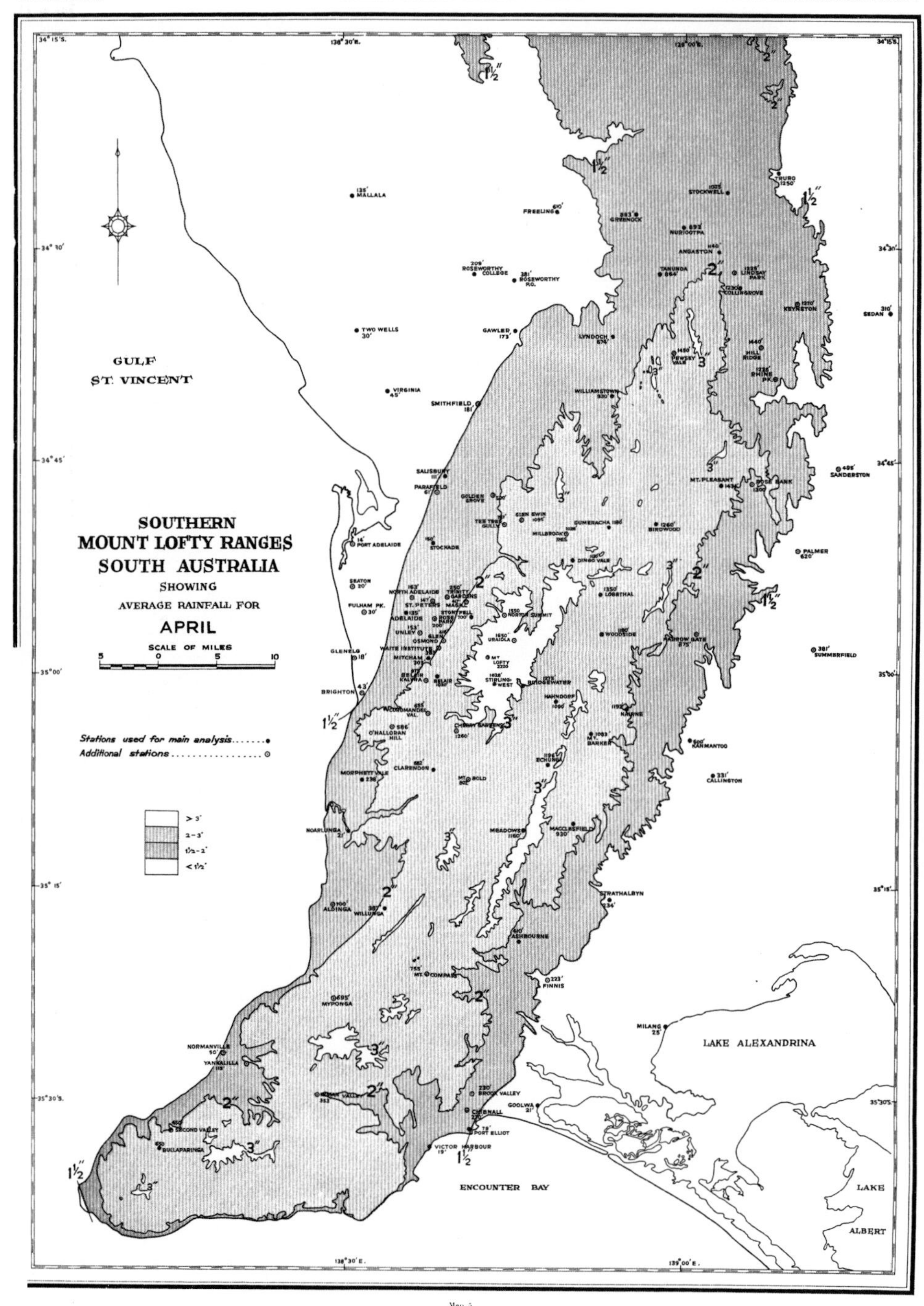

Map 5

Any imperfections in the presentation of this map are due to the present difficulty of reproduction. It should be noted that in the original the regions were denoted by colours.

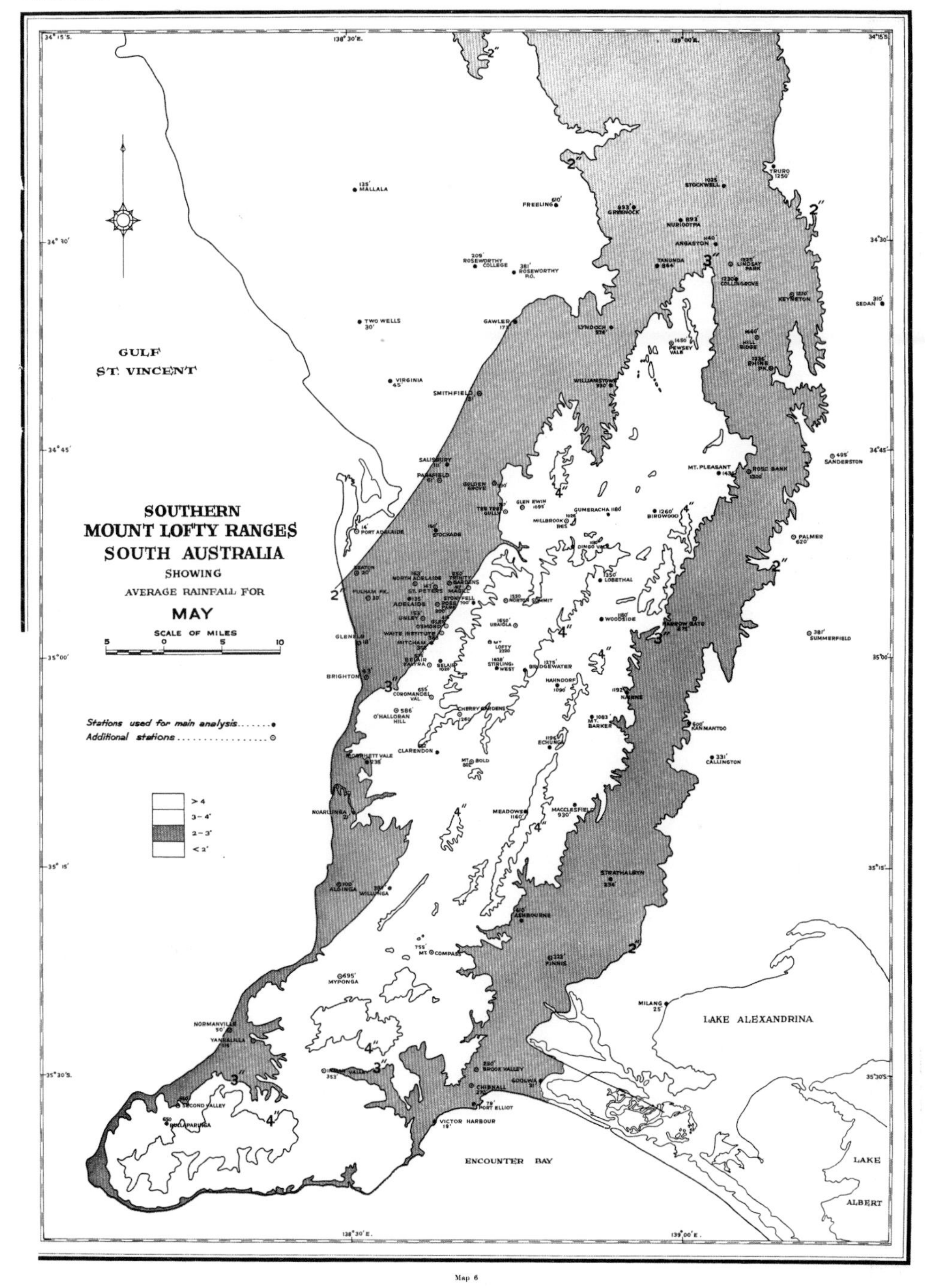

Map 6

Any imperfections in the presentation of this map are due to the present difficulty of reproduction. It should be noted that in the original the regions were denoted by colours.

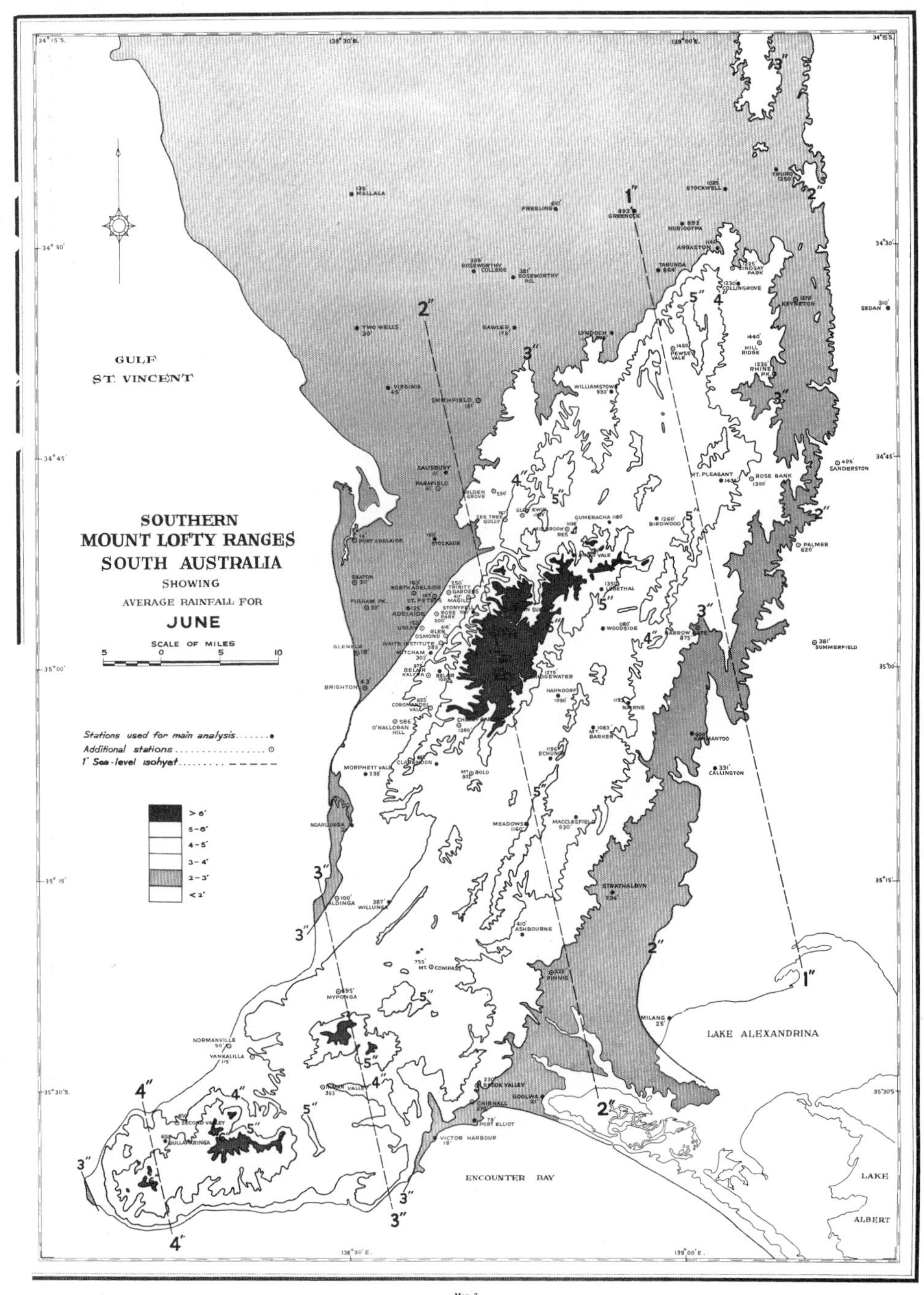

Map 7

Any imperfections in the presentation of this map are due to the present difficulty of reproduction. It should be noted that in the original the regions were denoted by colours.

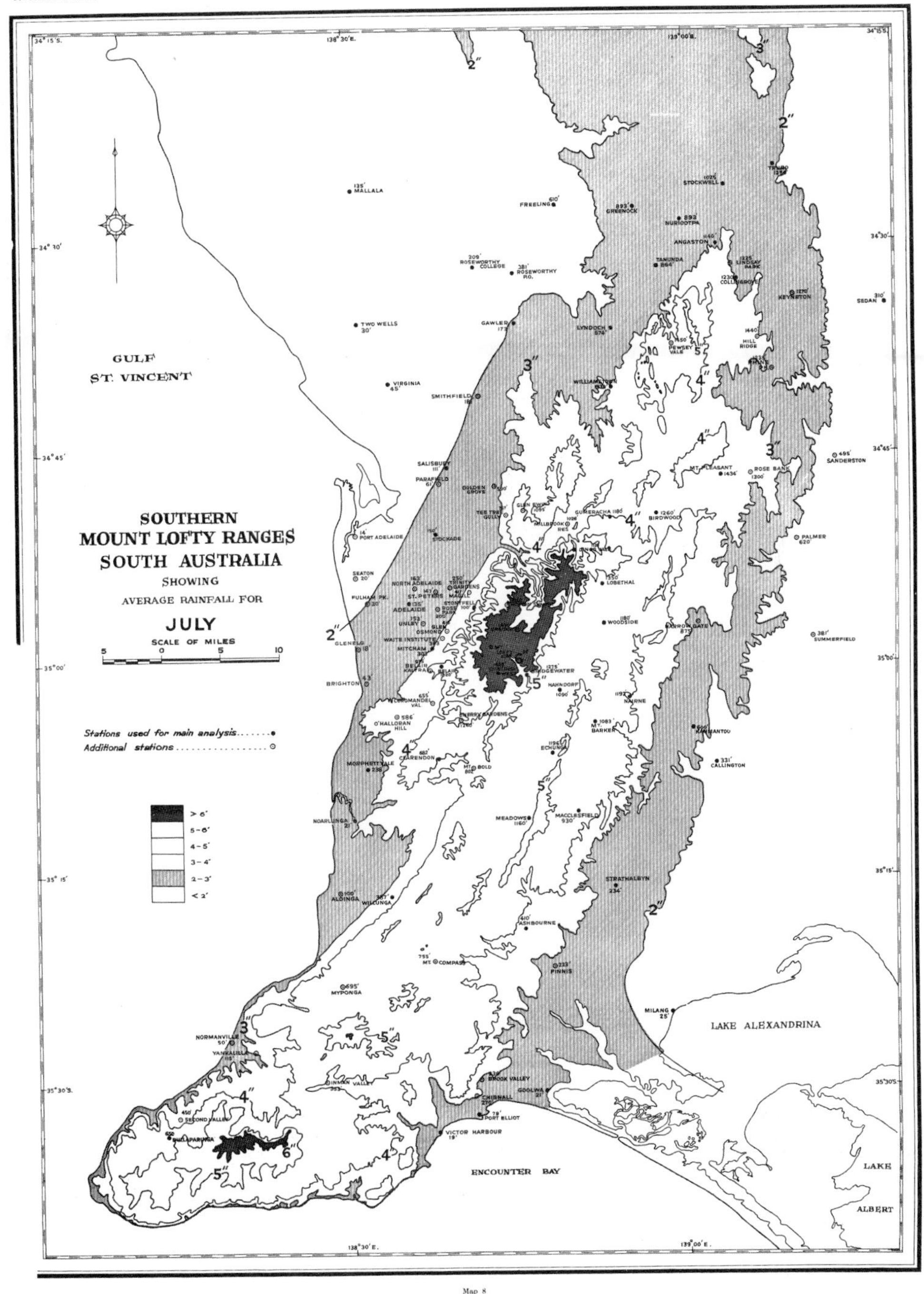

Map 8

Any imperfections in the presentation of this map are due to the present difficulty of reproduction. It should be noted that in the original the regions were denoted by colours.

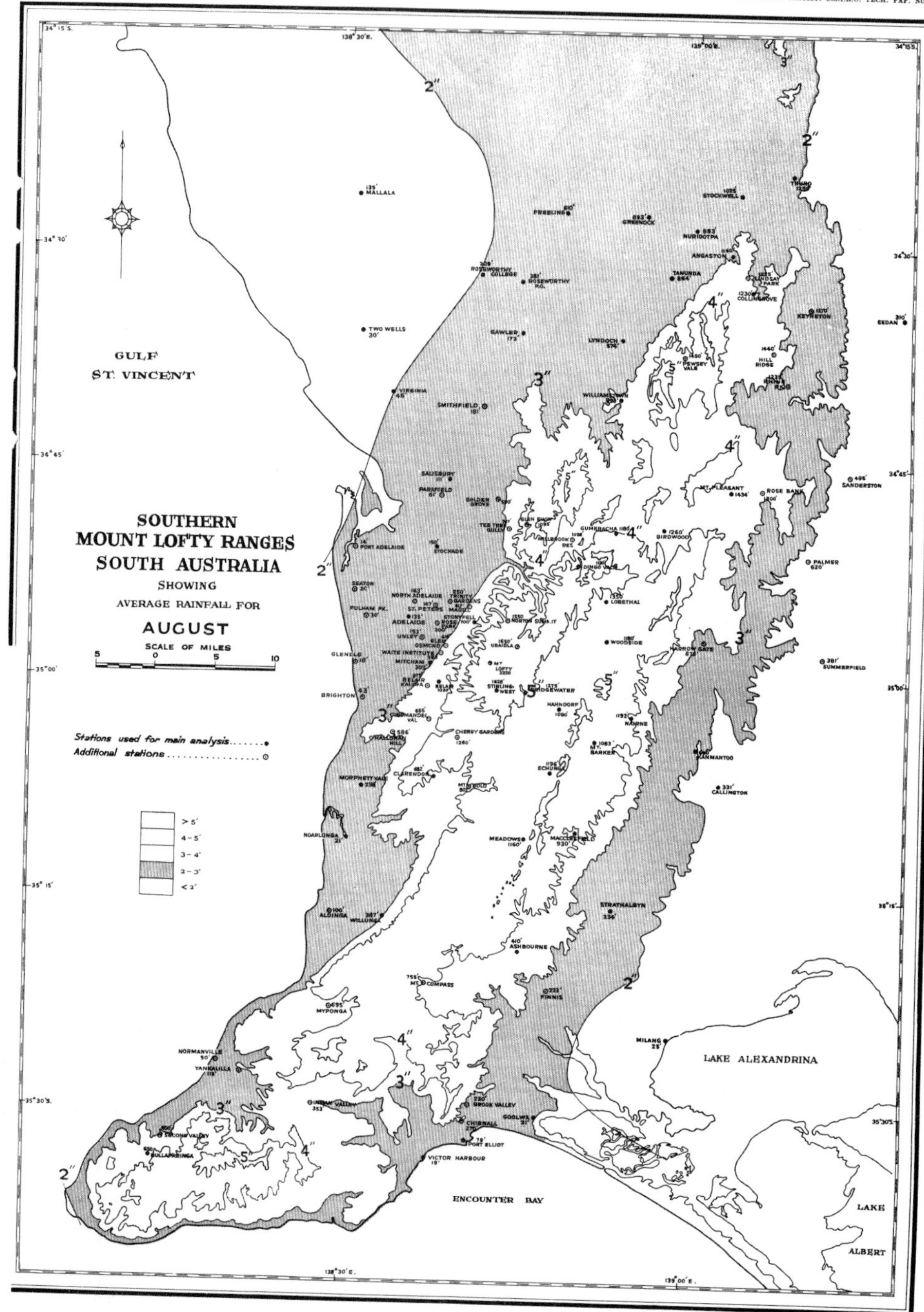

Map 9

Any imperfections in the presentation of this map are due to the present difficulty of reproduction. It should be noted that in the original the regions were denoted by colours.

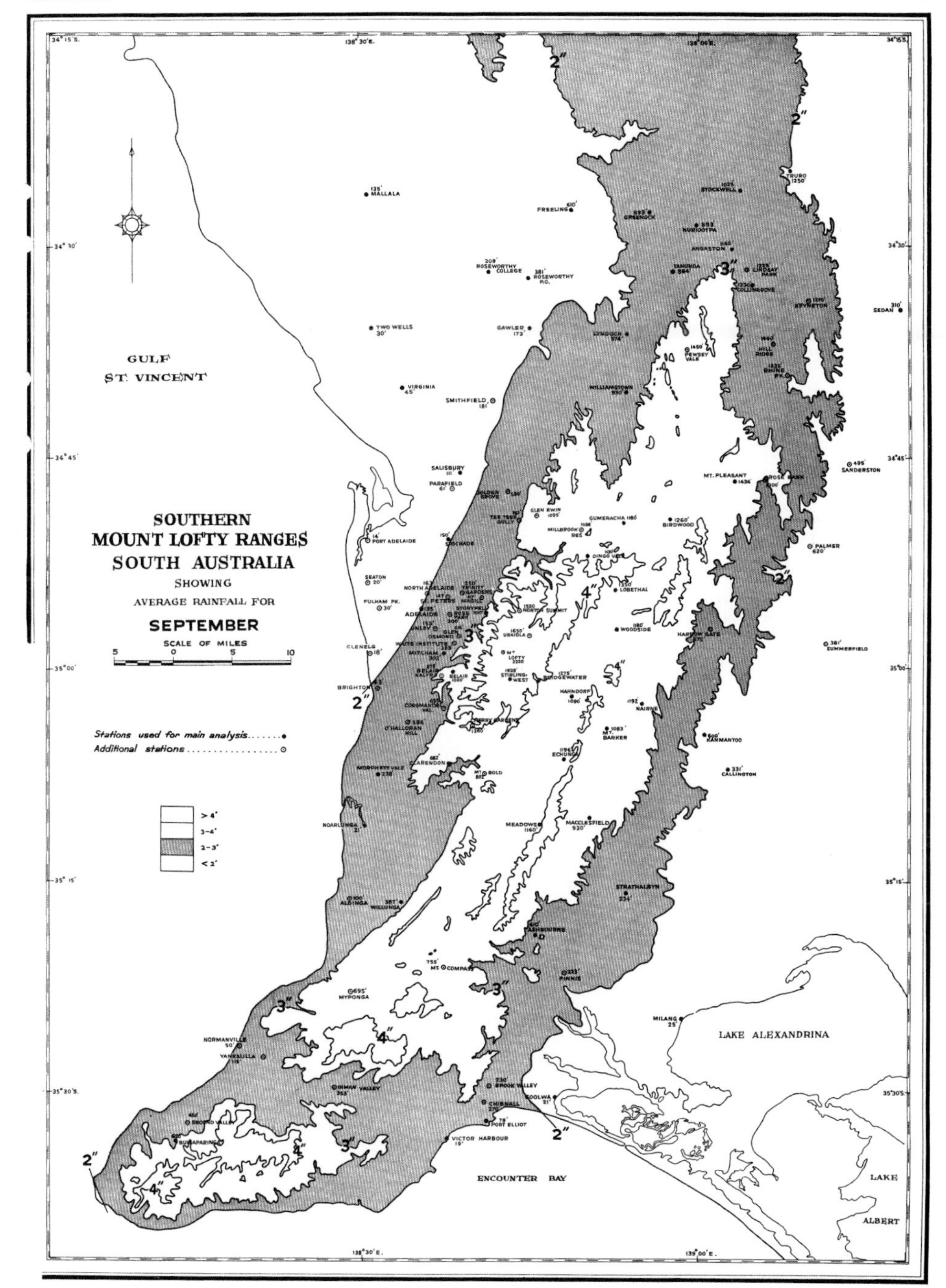

Map 10

Any imperfections in the presentation of this map are due to the present difficulty of reproduction. It should be noted that in the original the regions were denoted by colours.

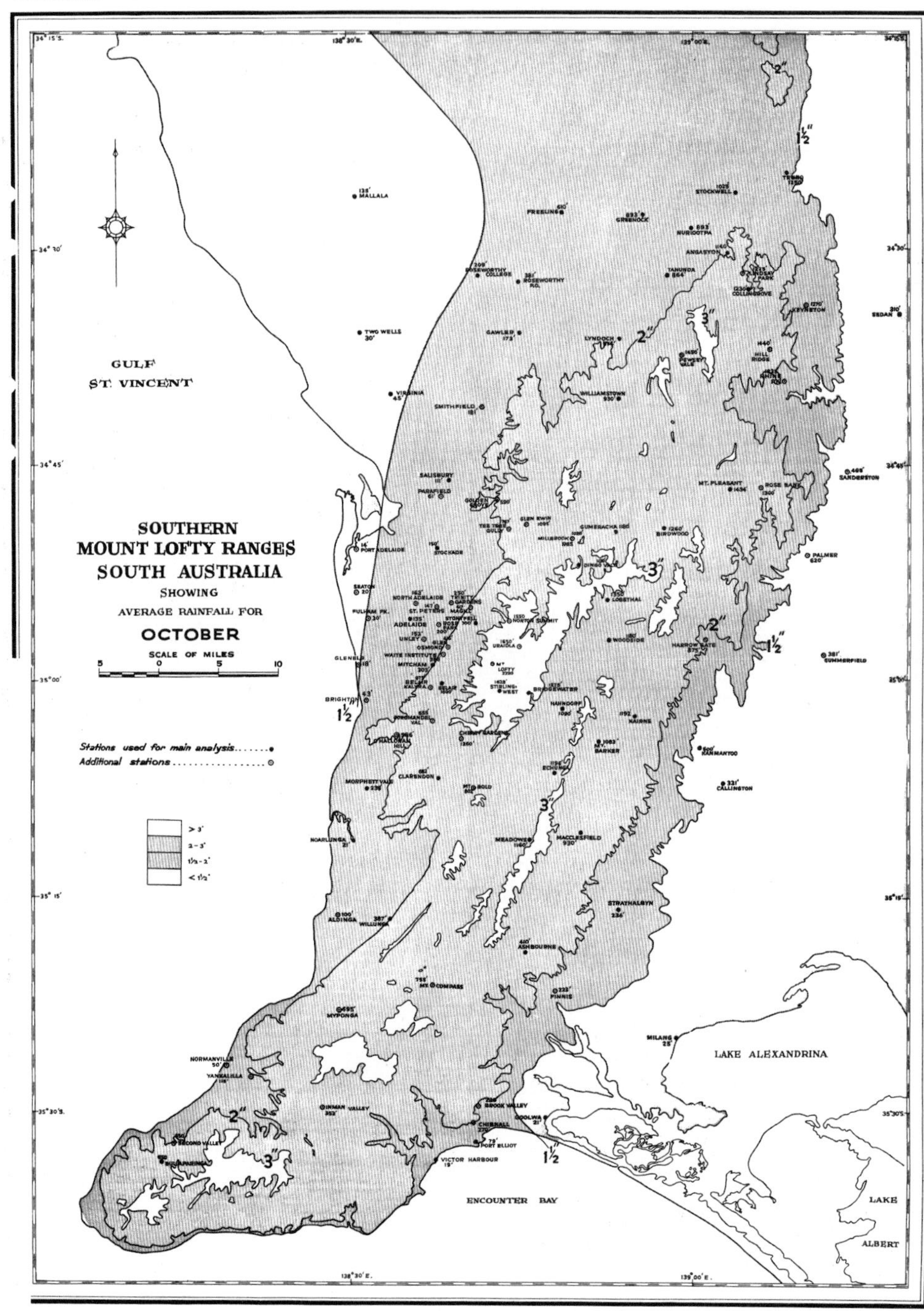

Map 11

Any imperfections in the presentation of this map are due to the present difficulty of reproduction. It should be noted that in the original the regions were denoted by colours.

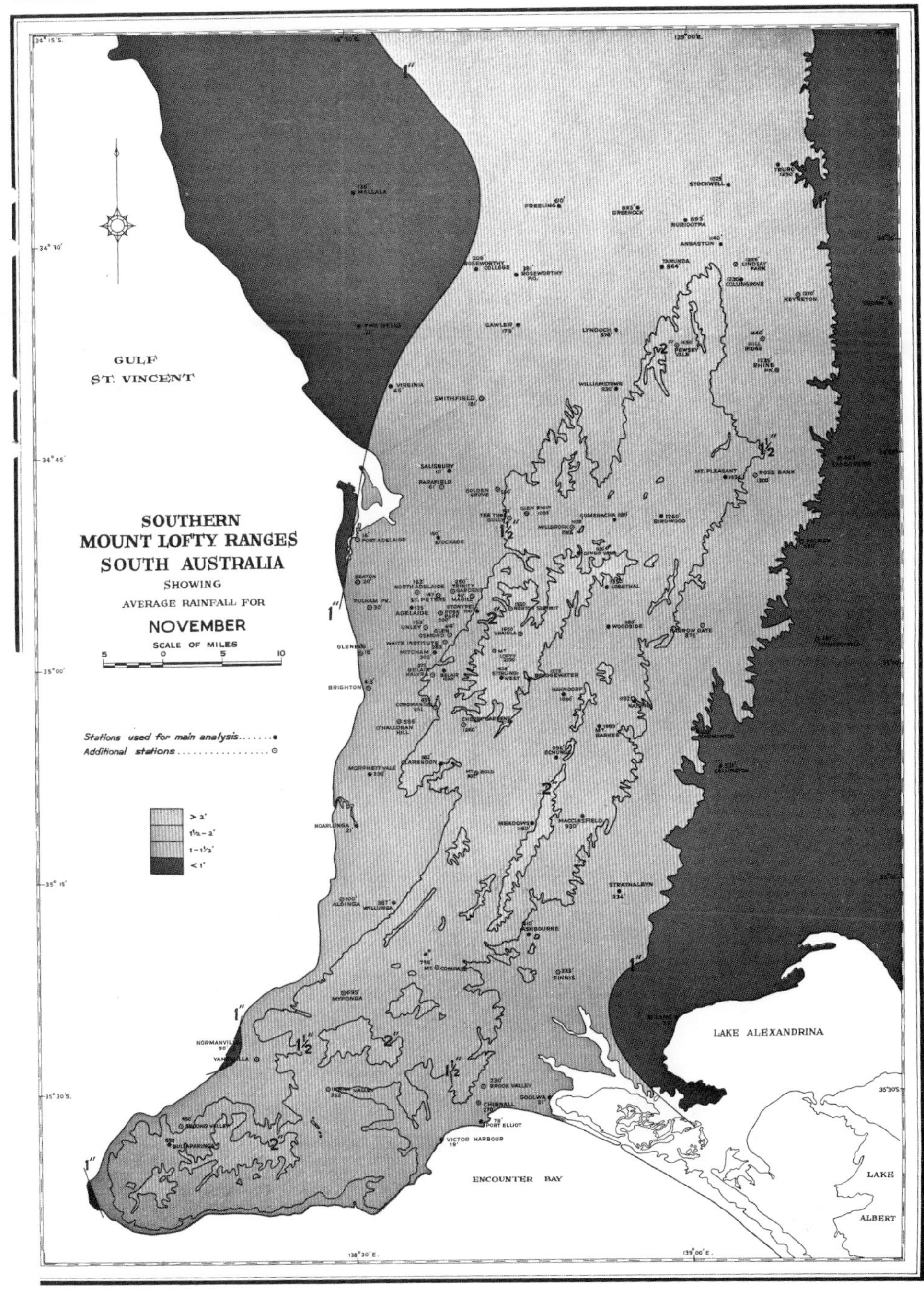

Map 12

Any imperfections in the presentation of this map are due to the present difficulty of reproduction. It should be noted that in the original the regions were denoted by colours.

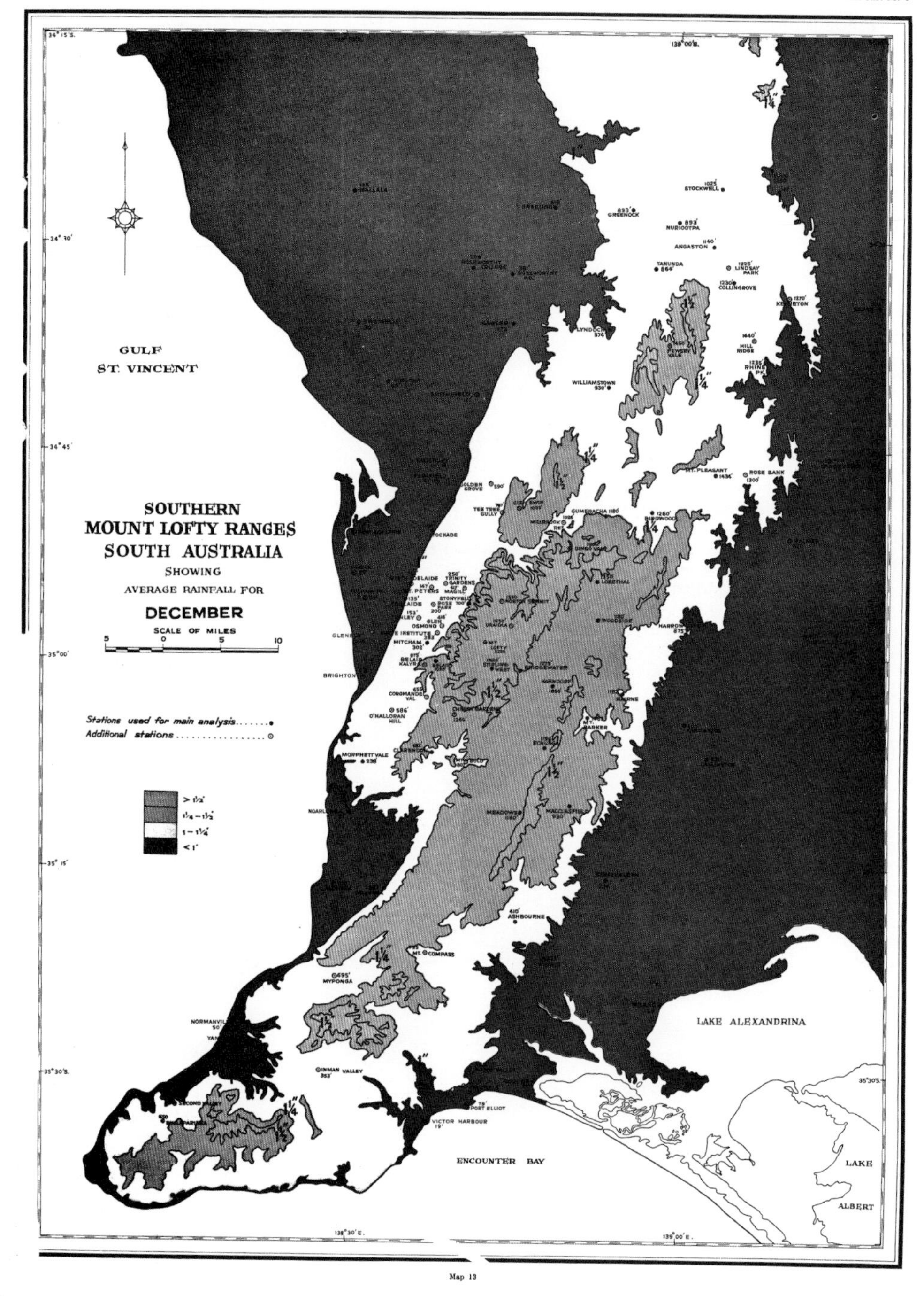

Map 13

Any imperfections in the presentation of this map are due to the present difficulty of reproduction. It should be noted that in the original the regions were denoted by colours.

Commonwealth Scientific and Industrial Research Organization, Australia. Division of Mathematical Statistics Technical Paper No. 5, Melbourne 1958.

INTER-STATION CORRELATIONS OF MONTHLY RAINFALL IN SOUTH AUSTRALIA

By N. S. STENHOUSE* and E. A. CORNISH†

[*Manuscript received September* 16, 1957]

Summary

Inter-station correlations of monthly rainfall data have been calculated for 25 stations in south-eastern South Australia from records taken over the period 1860–1947.

Inter-station distances and bearings account for a substantial proportion of the variation observed among the transformed correlations. The relation with distance and bearing is simple, leading to a simple spatial distribution of the correlations, the axis of maximal correlation having an orientation approximating that of the fronts as they move eastward over the region under investigation.

A method is outlined for determining an empirical spectral distribution of variance over distance.

I. INTRODUCTION

The extent to which different parts of a country experience similar deviations from their average weather sequences is of paramount importance in several aspects of climatology. Firstly, a decision must frequently be made whether records of a meteorological element, e.g. rainfall, taken simultaneously at a number of observing stations for relatively short periods of time, can be used to replace observations taken at one station over a long time-interval. This question often arises in recently developed areas where reliable records are either unavailable or very rare, and the importance of any decision reached is due to the fact that its nature determines the manner in which forecasts of agricultural production from weather data are made, or the manner in which the characteristic climatic features of a given area are assessed. Forecasts are of both direct and indirect value in crop insurance, trade, and administration; and the more important climatic features are required for the purpose of developing agriculture and for use in problems of engineering and soil conservation. Thus, again taking rainfall as an illustration, suppose that the observations at four stations within a region are completely independent, then 25 years' data at each point will provide as much information on, say, rainfall variability as a record of 100 years at a single station. If, on the other hand, rainfalls at the four stations are correlated, the gain in information resulting from inclusion of additional stations will be reduced by an amount depending upon the magnitude of the correlation, and in the extreme case of perfect correlation, the inclusion of additional stations will provide no more information than that obtainable from the record of one station.

Secondly—and this applies particularly in regions where observing stations are only sparsely distributed—it has become increasingly important to interpolate for meteorological variates. Various isopleths are required and considering local

* Division of Mathematical Statistics, C.S.I.R.O., University Grounds, Nedlands, W.A.
† Division of Mathematical Statistics, C.S.I.R.O., Adelaide.

demands only, those most urgently needed are isohyets. The accuracy with which rainfall, or any other meteorological variate, may be estimated, at a given point, from known observations at neighbouring points, must evidently depend upon the density of observing stations and their inter-correlations, and, knowing the latter, it is possible to state how densely located the stations should be to provide estimates with any assigned degree of accuracy.

Thirdly, the reliance which can be placed on all meteorological estimates, or predictions, is dependent upon the distribution of variance between localized and widespread fluctuations in the weather, and the only way of investigating this distribution is to determine empirically the manner in which the variance is partitioned.

Attention has been directed first to monthly rainfall, and the problem of determining both the correlations and the distribution of variance has been approached in two ways:

(i) by relating rainfall to the position and altitude of the observing stations within a region of relatively limited area;

(ii) by direct correlation of rainfall at stations dispersed over a much greater area, and determination of the relation to inter-station distance.

An investigation of type (i), with particular reference to the problem of interpolating monthly rainfall, has been reported (Coote and Cornish 1958), and the technique subsequently improved (Evans and Cornish, unpublished data). Our present object is to submit results from an inquiry of type (ii).

In a previous paper (Cornish 1950), seasonal characteristics in the wheat belt of South Australia were described in part by using inter-station and intra-station correlations of rainfall. For practical purposes, the latter were negligible (incidentally facilitating the interpretation of the multiple regressions) but, in strong contrast, the inter-station correlations were very high, and despite the fact that they diminished with inter-station distance, they demonstrated clearly the widespread nature of the major rain-bearing disturbances. This investigation embraced only winter rainfall in the broad subdivisions April–May, June–July–August, and September–October, so the decision was made to extend the study to rains throughout the year, using the calendar month as the unit of time, and a selection of stations over a wide range of topographical conditions in south-eastern South Australia.

II. Data

The monthly rainfall observations were extracted from the records of 25 stations taken during the period 1860–1947, the names and positions of the stations being given in each of the Figures 4–6. The number of years for which simultaneous observations were obtainable was not constant for all pairs of stations, but differences in the weights of the correlations were small enough to be neglected in the subsequent analysis. Rainfall records were provided by the Commonwealth Meteorological Bureau, and the positions of the stations, with inter-station distances and bearings, were made available by the Department of Lands, South Australia.

III. Climatic and Topographic Features

The principal topographical features of the region are the meridional horsts of the Mount Lofty and Flinders Ranges, situated in the north-west sector on a line running from Cape Jervis to a point beyond the most northerly station (Hawker), and the plains of the south-eastern section. The whole area lies in the winter rainfall zone of southern Australia, the season being marked by the strong winter incidence of its rains. Approximately 70–80 per cent. of the annual precipitation occurs within the period April–October inclusive, the summer months being characterized by hot, dry atmospheric conditions, low rainfall, and high evaporation. The winter rainfall zone itself lies in the path of the migratory anticyclones with their attendant

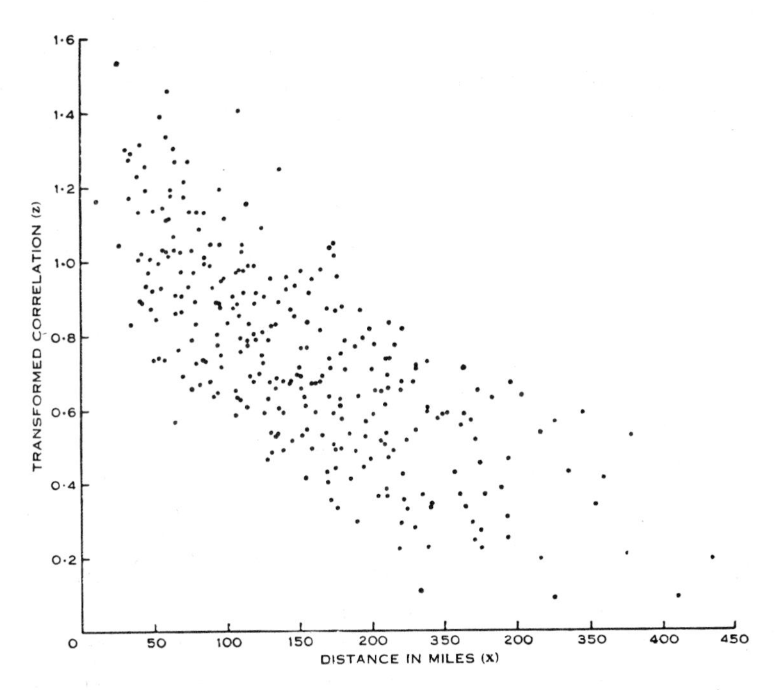

Fig. 1.—Relationship of transformed correlation and inter-station distance (November).

cyclones, as they move eastward. The major component of the annual rainfall is due to activity on the associated cold fronts, the remainder being convectional rains occurring mainly in winter, and general rains usually over large areas due to convergence in the northerly or north-westerly stream ahead of the advancing fronts.

Throughout the year, the orientation of the fronts is approximately NW.–SE., so it may be anticipated that the rains at stations along this line will be more closely correlated than rains at stations along any other directed line, or, expressed alternatively, the axis of maximal correlation will be approximately NW.–SE.

IV. Analysis

The primary data were obtained by directly correlating the monthly rainfall of all pairs of stations, yielding 300 correlation coefficients (r) for each month, which, for subsequent stages of the analysis, were replaced by their inverse hyperbolic transformations $z = \tanh^{-1} r$. The advantages of using this transformation have been pointed out and emphasized by Fisher (1946), and in the present circumstances it must be approximately normally distributed.

On general grounds, the correlations should all be positive in the region chosen for examination, and inspection of the data shows clearly the inverse relationship of z with increasing distance. Figure 1 (p. 5) illustrates this relationship for November, the remaining months being similar in form. With two stations close together, r approaches $+1$ and z tends to infinity, and for stations a great distance apart r and z both approach zero. The relation between z and distance is therefore very likely representable by a form in which z varies as some inverse power of the distance, being asymptotic to the axis of z as distance tends to zero, and asymptotic to the axis of distance as the latter tends to infinity. Over the range of distances at our disposal, the relationship should be reasonably well represented by the parabola

$$Z = k + b_1 x + b_2 x^2,$$

where x is the distance in miles. This method of representation has been chosen for

Table 1

ADJUSTED MEAN VALUES OF THE CORRELATIONS

z \ r	Mt. Gambier 1	Kingston 2	Bordertown 3	Meningie 4	Cape Borda 5	Pinnaroo 6	Murray Bridge 7	Yorketown 8	Adelaide 9	Maitland 10	Renmark 11	Morgan 12
1		0·8088	0·7086	0·6284	0·6041	0·5365	0·5267	0·5670	0·5604	0·5739	0·3336	0·2986
2	1·1235		0·7388	0·7255	0·6624	0·5659	0·5666	0·6550	0·6578	0·6232	0·4107	0·3880
3	0·8843	0·9478		0·7236	0·5993	0·7235	0·6542	0·6332	0·6904	0·6537	0·5565	0·5012
4	0·7388	0·9191	0·9153		0·6045	0·6596	0·7411	0·7184	0·7578	0·7147	0·4983	0·4916
5	0·6995	0·7971	0·6920	0·7002		0·4482	0·4977	0·6820	0·6380	0·6320	0·3702	0·3930
6	0·5993	0·6415	0·9151	0·7922	0·4825		0·7450	0·6178	0·6581	0·6662	0·7707	0·7168
7	0·5855	0·6425	0·7827	0·9529	0·5463	0·9617		0·6763	0·7174	0·7000	0·6029	0·6408
8	0·6431	0·7841	0·7468	0·9043	0·8328	0·7214	0·8223		0·7629	0·8078	0·4916	0·5000
9	0·6335	0·7890	0·8487	0·9909	0·7548	0·7894	0·9022	1·0030		0·8054	0·5239	0·5308
10	0·6533	0·7302	0·7817	0·8968	0·7448	0·8039	0·8672	1·1206	1·1137		0·5218	0·5535
11	0·3469	0·4365	0·6278	0·5471	0·3886	1·0221	0·6977	0·5381	0·5816	0·5787		0·7798
12	0·3080	0·4094	0·5509	0·5381	0·4154	0·9011	0·7595	0·5493	0·5913	0·6235	1·0448	
13	0·6246	0·7623	0·9615	0·9675	0·7137	0·9386	1·0180	0·8782	1·0297	1·0407	0·7491	0·7287
14	0·5656	0·6502	0·8417	0·8054	0·6997	0·8768	0·9270	0·8382	0·9056	1·0530	0·7633	0·7718
15	0·4959	0·5737	0·7618	0·6951	0·5411	0·8561	0·7793	0·6641	0·7961	0·8030	0·8695	0·9365
16	0·5077	0·6280	0·7137	0·6991	0·5907	0·8960	0·8108	0·7486	0·8177	0·8813	0·8024	0·7936
17	0·4071	0·5075	0·5742	0·6145	0·5521	0·8119	0·7429	0·6712	0·7236	0·7919	0·8148	0·8627
18	0·4494	0·5149	0·6203	0·6456	0·5759	0·8052	0·7547	0·6948	0·7440	0·8269	0·7810	0·7922
19	0·2115	0·3120	0·3853	0·4041	0·3508	0·6350	0·5564	0·4859	0·4519	0·5538	0·6969	0·7779
20	0·3010	0·4070	0·5122	0·4866	0·4249	0·7157	0·5956	0·5617	0·5599	0·6200	0·8155	0·8329
21	0·2250	0·3261	0·3995	0·3656	0·3346	0·6795	0·4598	0·4601	0·4493	0·4879	0·7161	0·7143
22	0·7795	0·9147	0·6933	0·8334	0·9064	0·6071	0·6486	0·9932	0·7930	0·8158	0·3957	0·4329
23	0·9252	1·0149	1·2256	1·1601	0·6694	0·8799	0·8663	0·8271	0·9980	0·8721	0·5632	0·5414
24	0·5245	0·6137	0·7122	0·7466	0·6555	0·8908	0·8268	0·8995	0·9477	1·1203	0·6966	0·8268
25	0·6559	0·6571	0·7936	0·7823	0·5727	1·1506	1·0042	0·7211	0·8263	0·7637	0·9457	0·9728

convenience in this preliminary survey. Examination of the data and the results indicates that this formula is sufficiently accurate for distances greater than approximately 50 miles. In only one month does the fitted curve have a minimal value in the range of distances herein employed, and then only near the end of the range at a distance greater than 400 miles.

As a consequence of the observations made in Section III, the correlations in the neighbourhood of a given point will not fall off isotropically, and from consideration of the dominant type of rain-bearing disturbance, it may be expected that allowance for the direction of displacement could be made effectively by modifying the regression function to take the form

$$Z = k + b_1x + b_2x^2 + b_3(x\sin 2\theta) + b_4(x\cos 2\theta), \quad \ldots\ldots\ldots(1)$$

where θ is the bearing of one station on another, measured clockwise from the north. There is no evidence to suggest that, for the monthly data herein used, a more elaborate specification of direction is required. Inclusion of distance in the additional variates definitely improves the fit in all months.

Table 1 gives the mean values of z, with their corresponding correlation coefficients, for the 300 pairs of stations, averaged over the twelve months of the year, and corrected for bias using the formula $-\frac{1}{2}r/(n-1)$ (Fisher 1946); the inverse relationship with distance and the effects of direction of displacement are apparent.

Table 1 (*Continued*)

Adjusted Mean Values of the Correlations

Kapunda 13	Clare 14	Hallett 15	Gladstone 16	Port Pirie 17	Melrose 18	Port Augusta 19	Orroroo 20	Hawker 21	Cape Willoughby 22	Tintinara 23	Bute 24	Rosalind 25	
0·5543	0·5121	0·4589	0·4681	0·3860	0·4214	0·2084	0·2922	0·2213	0·6524	0·7283	0·4812	0·5757	1
0·6424	0·5718	0·5181	0·5566	0·4680	0·4737	0·3022	0·3860	0·3150	0·7234	0·7678	0·5467	0·5765	2
0·7450	0·6867	0·6421	0·6130	0·5185	0·5513	0·3673	0·4716	0·3796	0·6001	0·8413	0·6121	0·6604	3
0·7476	0·6670	0·6013	0·6038	0·5473	0·5687	0·3835	0·4515	0·3501	0·6823	0·8210	0·6331	0·6540	4
0·6130	0·6042	0·4938	0·5304	0·5021	0·5197	0·3371	0·4010	0·3226	0·7194	0·5846	0·5754	0·5174	5
0·7346	0·7048	0·6943	0·7143	0·6706	0·6669	0·5615	0·6142	0·5912	0·5420	0·7064	0·7118	0·8180	6
0·7691	0·7292	0·6523	0·6700	0·6308	0·6379	0·5053	0·5339	0·4299	0·5707	0·6995	0·6788	0·7634	7
0·7055	0·6848	0·5811	0·6343	0·5858	0·6011	0·4509	0·5093	0·4302	0·7587	0·6789	0·7161	0·6176	8
0·7738	0·7190	0·6618	0·6738	0·6191	0·6315	0·4235	0·5079	0·4213	0·6601	0·7608	0·7387	0·6785	9
0·7782	0·7830	0·6657	0·7071	0·6595	0·6788	0·5034	0·5511	0·4525	0·6728	0·7025	0·8077	0·6432	10
0·6346	0·6430	0·7011	0·6653	0·6722	0·6533	0·6024	0·6726	0·6145	0·3763	0·5104	0·6022	0·7378	11
0·6223	0·6479	0·7336	0·6604	0·6977	0·6596	0·6515	0·6820	0·6134	0·4077	0·4941	0·6788	0·7499	12
	0·8345	0·7674	0·7529	0·6860	0·7060	0·4822	0·5952	0·4726	0·6407	0·7454	0·7922	0·7557	13
1·2026		0·7953	0·7915	0·7324	0·7457	0·5084	0·6199	0·5243	0·6016	0·6698	0·8171	0·6940	14
1·0140	1·0857		0·8218	0·7278	0·7659	0·6364	0·7521	0·6142	0·5249	0·6173	0·7392	0·7044	15
0·9795	1·0754	1·1623		0·8256	0·8370	0·6689	0·7486	0·6292	0·5801	0·6273	0·7721	0·6922	16
0·8404	0·9339	0·9240	1·1741		0·8414	0·7686	0·7362	0·6536	0·5197	0·5209	0·7678	0·6765	17
0·8792	0·9632	1·0103	1·2109	1·2259		0·7249	0·8190	0·7264	0·5369	0·5542	0·7352	0·6617	18
0·5258	0·5606	0·7522	0·8087	1·0168	0·9180		0·7912	0·7735	0·3644	0·3768	0·6232	0·5487	19
0·6857	0·7249	0·9777	0·9698	0·9422	1·1538	1·0745		0·8239	0·4275	0·4532	0·6443	0·6062	20
0·5135	0·5822	0·7157	0·7401	0·7816	0·9210	1·0289	1·1689		0·3230	0·3517	0·5417	0·5126	21
0·7593	0·6957	0·5831	0·6625	0·5759	0·5998	0·3819	0·4568	0·3350		0·6572	0·6132	0·5471	22
0·9624	0·8103	0·7207	0·7370	0·5776	0·6245	0·3963	0·4888	0·3674	0·7879		0·6680	0·6958	23
1·0772	1·1480	0·9487	1·0256	1·0148	0·9401	0·7302	0·7656	0·6066	0·7141	0·8072		0·7209	24
0·9860	0·8556	0·8759	0·8521	0·8226	0·7959	0·6165	0·7028	0·5662	0·6143	0·8590	0·9095		25

These results substantiate those obtained previously (Cornish 1950), and in fact extend them to all months of the year.

Table 2, on the other hand, gives the means for the several months, averaged over the 300 pairs of stations, and also corrected for bias. Compared with the correlations obtained in another earlier study (Coote and Cornish 1958) the present values are much lower, but it must be remembered that the average distance between stations is now considerably greater. As a check on this point a selection of pairs of stations, at distances approximating those employed previously, yielded mean values of comparable magnitude. This is shown clearly by the distributions illustrated in Figure 2.

As with previous data, there is no clear evidence of an annual oscillation in the monthly means, but it is possible that the month is too long an interval over which the rain is measured to allow a seasonal trend to express itself; for example,

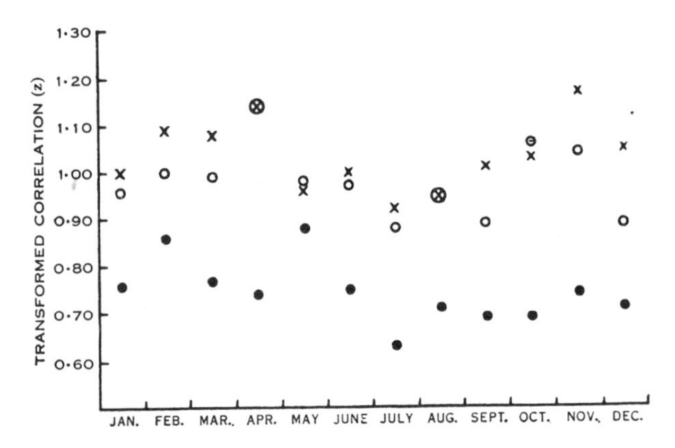

Fig. 2.—Monthly means of transformed correlations. ●, means of 300 correlations; ○, means of selected correlations; ×, correlations from Coote and Cornish (1958).

the long interval may well allow the occurrence of compensatory effects. A more detailed and extensive study now in progress gives definite evidence of an annual oscillation in some of the correlations.

The coefficients of the regression formulae are presented in Table 3 with their standard deviations. The linear terms are all negative and strongly significant ($P<0{\cdot}001$) in all months, while the quadratic terms are all positive and significant in seven of the twelve cases. Among the coefficients of the displacement terms, those of $\sin 2\theta$ are all negative and significant, whereas those of $\cos 2\theta$ are all negative except for February, and only seven are significant. There is some slight evidence of a seasonal trend in the coefficients of the linear term on distance and the displacement term involving the sine of the angle of bearing. Tests have been made on the differences between months, but the inconsistencies in each annual sequence make it impossible to draw any definite conclusion. Figures 3(a), 3(b), and 3(c) show each sequence graphically (p. 12).

TABLE 2

SUMMARY OF ANALYSES OF VARIANCE: TRANSFORMED CORRELATIONS

Approximate theoretical residual variance: 0·01879

Month	Multiple Correlation	% Variance Due to Regression	% Residual Variance	Total Variance	Residual Variance	Mean z	Residual S.D. as % of Mean	Mean Correlation r
Jan.	0·76	0·5658	0·4342	0·0775	0·0337	0·7673	23·46	0·64
Feb.	0·71	0·4950	0·5050	0·0629	0·0318	0·8685	20·73	0·70
Mar.	0·73	0·5232	0·4768	0·0706	0·0337	0·7726	23·30	0·65
Apr.	0·70	0·4847	0·5153	0·0876	0·0451	0·7494	28·02	0·63
May	0·65	0·4143	0·5857	0·0438	0·0257	0·8847	18·09	0·71
June	0·68	0·4513	0·5487	0·0671	0·0368	0·7583	25·06	0·64
July	0·69	0·4711	0·5289	0·0586	0·0310	0·6382	28·20	0·56
Aug.	0·70	0·4855	0·5145	0·0552	0·0284	0·7200	23·61	0·61
Sept.	0·77	0·5858	0·4142	0·0664	0·0275	0·6986	24·33	0·60
Oct.	0·88	0·7719	0·2281	0·0857	0·0196	0·6925	20·22	0·60
Nov.	0·81	0·6509	0·3491	0·0763	0·0266	0·7492	21·36	0·63
Dec.	0·73	0·5262	0·4738	0·0490	0·0232	0·7144	21·00	0·61

TABLE 3

PARTIAL REGRESSIONS OF THE TRANSFORMED CORRELATION ON DISTANCE AND DIRECTION OF DISPLACEMENT

Month	b_1	b_2	b_3	b_4	Axis of Correlation	
					Max.	Min.
Jan.	$-0{\cdot}0^{2}4453^{\dagger} \pm 0{\cdot}0^{3}4719$	$0{\cdot}0^{5}5013 \pm 0{\cdot}0^{5}1310$	$-0{\cdot}0^{2}1179 \pm 0{\cdot}0^{3}1108$	$-0{\cdot}0^{3}3600 \pm 0{\cdot}0^{3}1152$	126°31′	36°31′
Feb.	$-0{\cdot}0^{2}3429 \pm 0{\cdot}0^{3}4584$	$0{\cdot}0^{5}2374^{*} \pm 0{\cdot}0^{5}1272$	$-0{\cdot}0^{3}7191 \pm 0{\cdot}0^{3}1077$	$0{\cdot}0^{3}1134^{*} \pm 0{\cdot}0^{3}1119$	139°29′	49°29′
Mar.	$-0{\cdot}0^{2}3086 \pm 0{\cdot}0^{3}4719$	$0{\cdot}0^{5}1180^{*} \pm 0{\cdot}0^{5}1310$	$-0{\cdot}0^{3}8169 \pm 0{\cdot}0^{3}1108$	$-0{\cdot}0^{3}1417^{*} \pm 0{\cdot}0^{3}1152$	130° 5′	40° 5′
Apr.	$-0{\cdot}0^{2}4005 \pm 0{\cdot}0^{3}5459$	$0{\cdot}0^{5}4175 \pm 0{\cdot}0^{5}1515$	$-0{\cdot}0^{3}7299 \pm 0{\cdot}0^{3}1282$	$-0{\cdot}0^{3}5003 \pm 0{\cdot}0^{3}1333$	117°47′	27°47′
May	$-0{\cdot}0^{2}2474 \pm 0{\cdot}0^{3}4121$	$0{\cdot}0^{5}1919^{*} \pm 0{\cdot}0^{5}1144$	$-0{\cdot}0^{3}5232 \pm 0{\cdot}0^{4}9680$	$-0{\cdot}0^{3}1860^{*} \pm 0{\cdot}0^{3}1006$	125°13′	35°13′
June	$-0{\cdot}0^{2}3385 \pm 0{\cdot}0^{3}4931$	$0{\cdot}0^{5}3745 \pm 0{\cdot}0^{5}1369$	$-0{\cdot}0^{3}6568 \pm 0{\cdot}0^{3}1158$	$-0{\cdot}0^{3}5091 \pm 0{\cdot}0^{3}1204$	116° 7′	26° 7′
July	$-0{\cdot}0^{2}3310 \pm 0{\cdot}0^{3}4526$	$0{\cdot}0^{5}3154 \pm 0{\cdot}0^{5}1256$	$-0{\cdot}0^{3}3533 \pm 0{\cdot}0^{3}1063$	$-0{\cdot}0^{3}1225^{*} \pm 0{\cdot}0^{3}1105$	125°27′	35°27′
Aug.	$-0{\cdot}0^{2}2566 \pm 0{\cdot}0^{3}4332$	$0{\cdot}0^{5}2354^{*} \pm 0{\cdot}0^{5}1203$	$-0{\cdot}0^{3}5082 \pm 0{\cdot}0^{3}1018$	$-0{\cdot}0^{3}6611 \pm 0{\cdot}0^{3}1058$	108°47′	18°47′
Sept.	$-0{\cdot}0^{2}2698 \pm 0{\cdot}0^{3}4263$	$0{\cdot}0^{5}1360^{*} \pm 0{\cdot}0^{5}1183$	$-0{\cdot}0^{3}7673 \pm 0{\cdot}0^{3}1001$	$-0{\cdot}0^{3}7049 \pm 0{\cdot}0^{3}1041$	113°43′	23°43′
Oct.	$-0{\cdot}0^{2}4777 \pm 0{\cdot}0^{3}3599$	$0{\cdot}0^{5}3668 \pm 0{\cdot}0^{6}9990$	$-0{\cdot}0^{2}1020 \pm 0{\cdot}0^{4}8454$	$-0{\cdot}0^{3}2560 \pm 0{\cdot}0^{4}8787$	127°58′	37°58′
Nov.	$-0{\cdot}0^{2}4520 \pm 0{\cdot}0^{3}4192$	$0{\cdot}0^{5}5253 \pm 0{\cdot}0^{5}1164$	$-0{\cdot}0^{3}5886 \pm 0{\cdot}0^{4}9848$	$-0{\cdot}0^{3}5398 \pm 0{\cdot}0^{3}1024$	113°44′	23°44′
Dec.	$-0{\cdot}0^{2}3536 \pm 0{\cdot}0^{3}3915$	$0{\cdot}0^{5}3936 \pm 0{\cdot}0^{5}1087$	$-0{\cdot}0^{3}7887 \pm 0{\cdot}0^{4}9197$	$-0{\cdot}0^{3}1807^{*} \pm 0{\cdot}0^{4}9560$	128°33′	38°33′

* Not significant

† $0^{2} \equiv {\cdot}00$, etc.

Table 2 completes the summary of the analyses, and gives the multiple correlation coefficients and other data relevant to the specification of the accuracy of the prediction formulae. The regressions on distance provide by far the more important contribution, both relatively and absolutely, to the total variation ascribable to the regression formula. It is of some consequence that as much as from 40 to 80 per cent. of the variance of the correlations (measured on the scale of the transform) is explicable in terms of the variates used, considering that the residual standard deviation includes, in addition to errors of observation, any systematic deviations from the formulae and irregularities arising from local topographical features or exposure of the stations. The absolute value of the residual standard deviation is not constant, but there is no regular sequence in the change from month to month; when considered relative to the mean, there is a more clearly defined trend, the winter months, with the exception of May, being slightly higher. Moreover, if the residual variance is considered relative to the total variance, the winter months give the higher percentages. The variables used thus account for changes in the correlation more satisfactorily in the summer, but the difference is not great.

In sampling from a bivariate normal distribution, the variance of the transform z is, for all practical purposes, dependent on the sample size and not on the correlations, so that an accurate estimate of the average theoretical variance may be obtained from the harmonic mean of the numbers $(n_{ij}-3)$, n_{ij} being the number of pairs of observations for stations i and j. This average value for all months is 0·0188, and the table shows that only in October does the residual variance from the fitted regression approach the theoretical value. The discrepancies indicate that the sampling errors are higher than expected from a bivariate normal distribution or that the regression formulae are inadequate, or that both causes are operating. The distribution of monthly rainfall, as already stated, is not normal, and consequently the first cause is present, but the extent to which it influences the result is not known, so that, without considerable extension of the work, the contribution arising from the error of the regression formula cannot be determined. In all tests of significance, the actual residual variances have been employed. The residuals at the stations have been computed and closely examined, but there was no regularity in their spatial distribution, nor was there any marked discrepancy, for any individual station, between the observed and predicted values of the transformed correlation.

Maximizing the right-hand side of equation (**1**) for variations in θ yields the solution

$$\theta = \tfrac{1}{2}\arctan(b_3/b_4),$$

for the bearing of the axis of maximal correlation, and because of the manner in which the regression formula has been fitted, the axis of minimal correlation will lie at right angles. The bearings of the axes are given in Table 3, and those of the axis of maximal correlation are illustrated in Figure 3(d).

The maps of Figures 4–6 have been constructed from the regression formula (**1**) by assigning a range of values of Z from 0·6 to 1·2 in steps of 0·05, and a range of values of θ from 0° to 175° by steps of 5°, and solving the resultant quadratic equations for distance. The isopleths, for which the term *isohomeotrope* is introduced, were then found by interpolating in this basic framework. Adelaide has been chosen as the

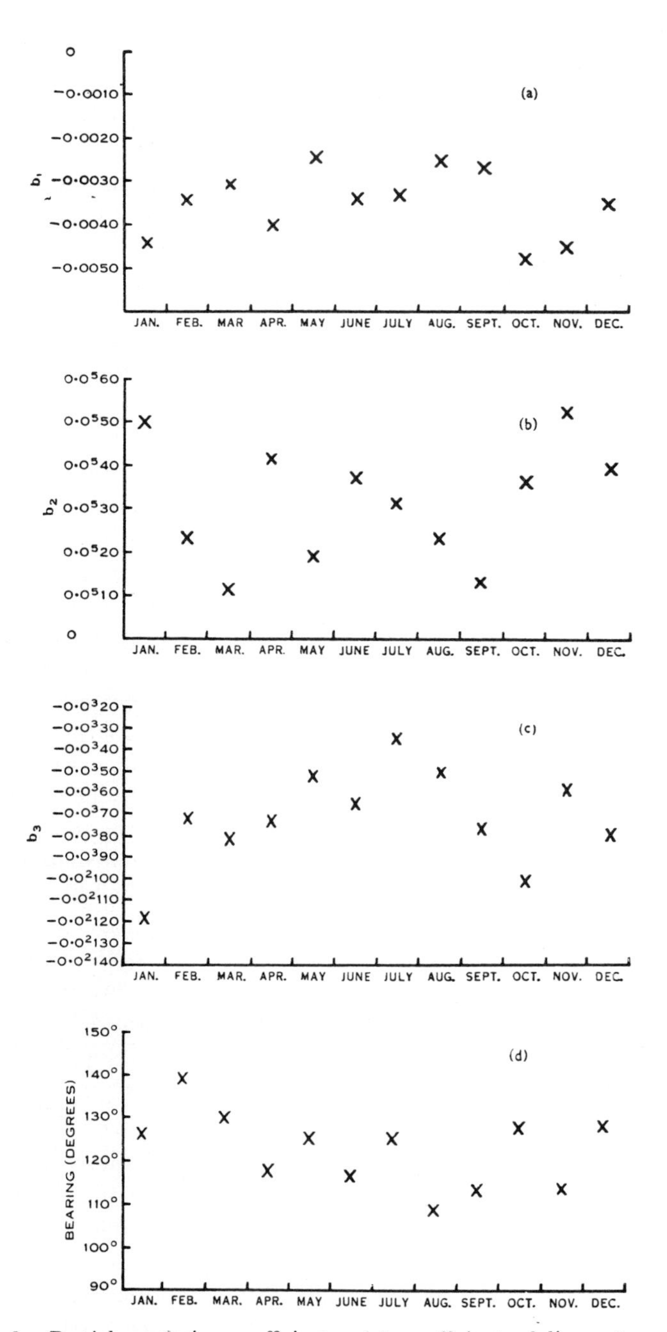

Fig. 3.—Partial regression coefficients. (*a*), coefficient of linear term on distance; (*b*), coefficient of quadratic term on distance; (*c*), coefficient of ($x\sin 2\theta$); (*d*), axis of maximal correlation.

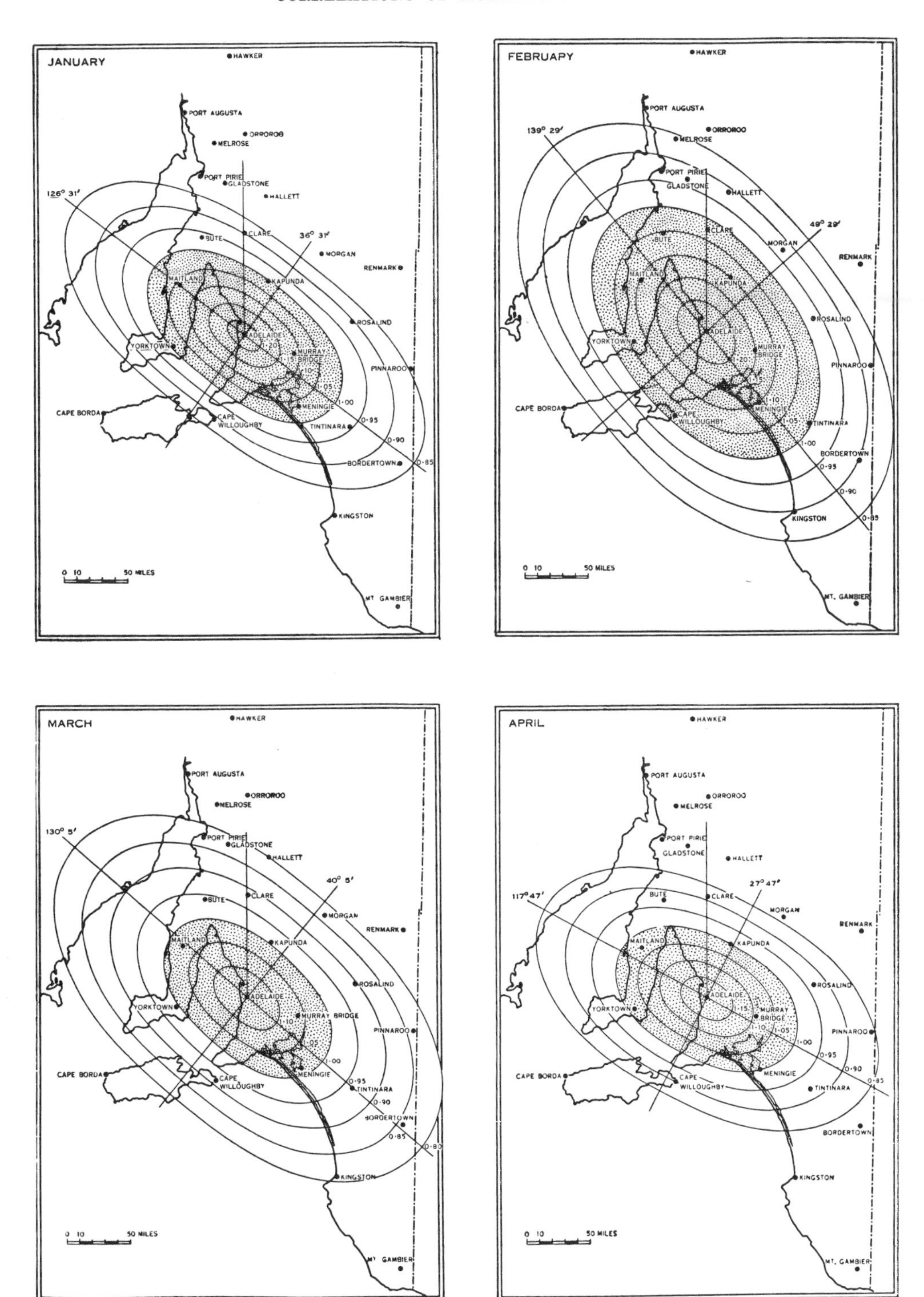

Fig. 4.—Isohomeotropes for January to April.

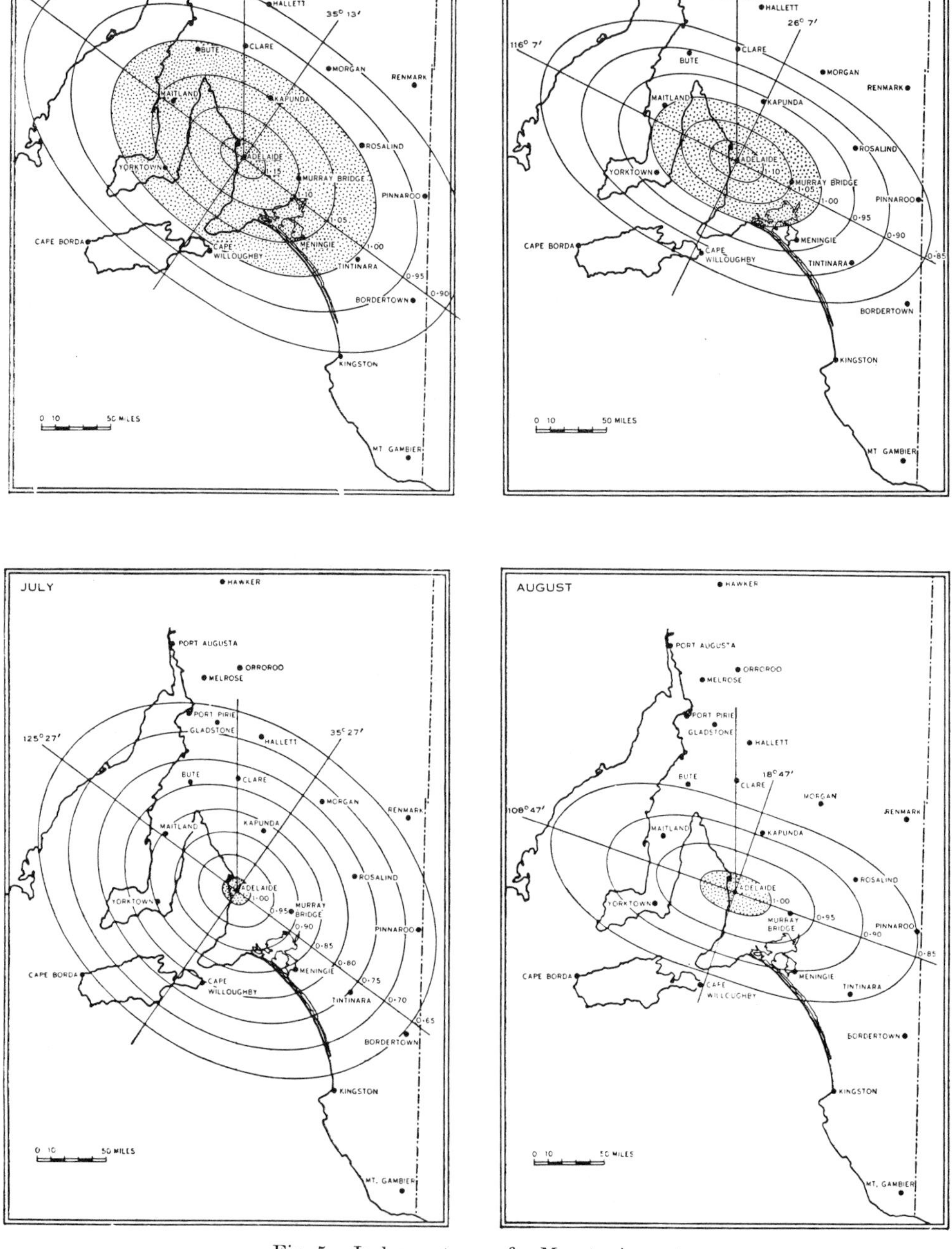

Fig. 5.—Isohomeotropes for May to August.

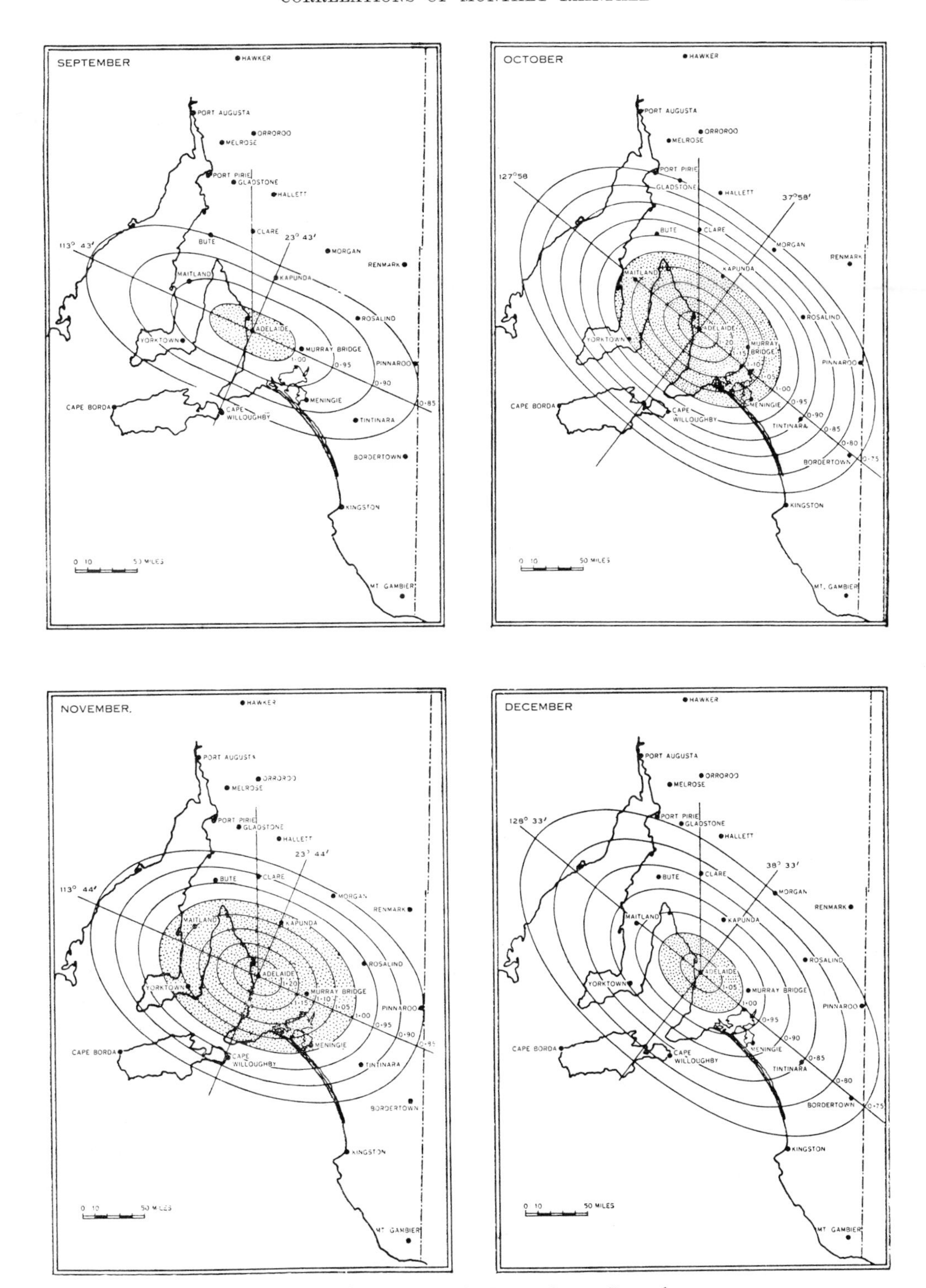

Fig. 6.—Isohomeotropes for September to December.

reference point for constructing these diagrams because of its approximate central position, but it must be emphasized that any other point in the region could have been chosen, the effect of another choice being merely to translate the figure to a new origin. The data only warrant the estimation of the average effects of distance and direction of displacement, and for this reason have been pooled.

The axes of maximal and minimal correlation have been inserted in each figure, and, in addition, the N.–S. line through Adelaide. The shaded area on each map is the zone where $Z \geqslant 1 \cdot 00$, corresponding to a correlation coefficient of 0·76. In June, August, and September the direction of the axis of maximal correlation is such that the number of isopleths included has been limited, to avoid excessive extrapolation.

Despite the limitations of the data, the diagrams present features of a graded sequence. Against the background of the map, the seasonal change of the axis of maximal correlation from NW. in midsummer to WNW. in late winter is shown more clearly, May, July, and October appearing to be the discrepant months.

The correlations tend to be higher and to persist over greater distances in summer than in winter, the exceptional months with respect to the latter being May and December. The contrasts are brought out clearly by the shaded zones in each diagram. At first sight it might be anticipated that the higher correlations would occur in the winter and also be maintained over distances, but it must be remembered that deviations below the mean rainfall contribute equally to the correlations, and that fine weather prevails over extensive areas in summer.

In general, the curvature of the isopleths is reduced practically to zero in the neighbourhood of the axis of minimal correlation, and in some months even shows a slight and curious reversal which is probably fortuitous.

The simplicity of the changes in the correlation with distance and direction of displacement is reflected in the form of the figures. The definite indication is that simple laws relate these quantities over very considerable areas, which will provide sufficiently exact knowledge of the accuracy of monthly rainfall estimates, based on a limited number of stations.

As mentioned above, the region under investigation lies in the eastward latitudinal paths of the migratory anticyclones. There is a well-defined annual oscillation in these paths associated with the sequence of seasons during the year (see, for example, Kidson 1925), which leads to a definite annual oscillation in the orientation of the frontal lines over the area. In summer, the centres of the anticyclones have a mean track at a greater latitude than that of Adelaide giving a NW.–SE. orientation of the frontal line, but in the winter the centres of the anticyclones have a mean track at a latitude less than that of Adelaide, yielding a frontal line with an orientation N. of NW. The synoptic situations in the immediate vicinity of Adelaide corresponding to these extremes are represented diagrammatically in Figure 7.

There are four primary factors concerned with the formation of rain in this region, and the general effects they exert on the correlations may now be considered in order.

(i) *Frontal Activity.*—The two main features of relevance in this case are as follows:

(1) slope of the frontal surface which depends upon the temperature difference at the interface of the two air masses, slope varying inversely as the temperature difference, and rainfall varying as the slope;

(2) volume of air uplifted, which varies as the pressure gradient, the resultant rainfall varying as the volume of air.

This assumes, of course, that the properties of the air mass ahead of the front remain constant. In the region concerned, the slope of the front and the pressure gradient decrease northward along the front.

The extent to which these factors operate to produce similar deviations of rainfall is impressed on the correlations of all months, as is apparent from the isopleths given in the maps.

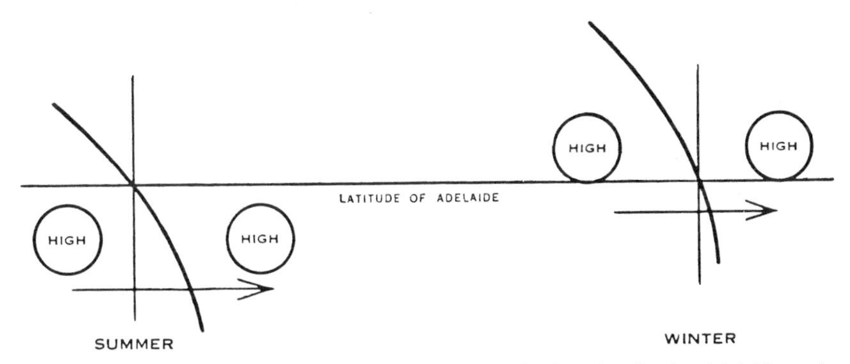

Fig. 7.—Diagrammatic representation of the synoptic situation in the Adelaide region during summer and winter.

On the other hand, an important factor, acting along the line of the front, which tends to reduce the correlations, is that the temperature differences between the air masses, and hence the slope of the front, vary irregularly as the front moves inland.

With the eastward movement of the front, points whose join lies at right angles to the front belong to different frontal sections, thus deriving their rain from different sets of causes, and consequently are subjected to a reduction in correlation in this direction. The spatial relationships are illustrated in Figure 8 (p. 18).

(ii) *Latitudinal Convergence.*—In South Australia, rains from this source are due to convergence in a north or north-westerly stream, resulting in uplift of the air mass. Convergence usually occurs on the rear side of an anticyclone, ahead of the advancing front; widespread cloud forms at middle level, and occasionally general rains fall over the whole region, the effect of topography being practically obliterated. These rains tend to maintain the correlations over large distances, irrespective of direction of displacement.

(iii) *Convection.*—Convectional rains occur mainly in winter, and do not conform to any pattern in the spatial distribution of the rainfall, which is generally very localized. The general effect is to reduce the correlations in all directions.

(iv) *Topography*.—In South Australia purely orographic rainfall does not occur, but the effect of topography is extremely important in modifying the results following from frontal uplift. The effects of topography on frontal disturbances and the distribution of convectional rains associated with these disturbances will raise the mean rainfall by different amounts in different parts of the area. Between pairs of stations, the correlations will be greater or less, according to whether both stations are similarly situated or not—for example, whether they are situated in the ranges, where topography intensifies frontal rains, or on the plains, where convectional rains augment frontal rains. The major elevated land mass and the plain regions run almost north and south, so that topographic control would tend to impose a N.–S. axis in the pattern of the correlations.

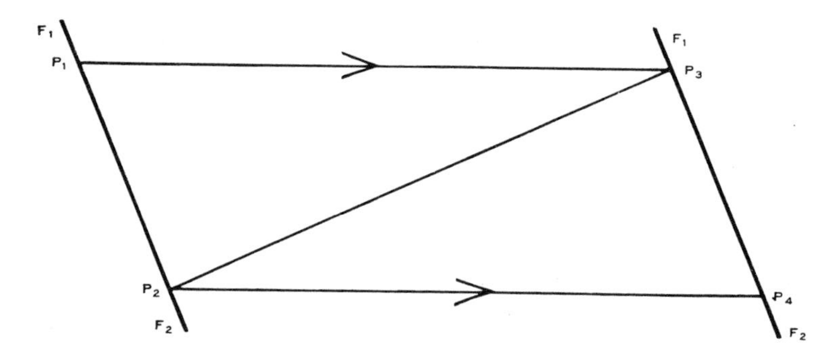

Fig. 8.—The relation between points P whose join is at right angles to the front, and frontal sections F.

The effects of topography, convergence, and convection are dominated by associations along the line of the fronts, as shown by the diagrams, but at the same time it is clear that they modify the influence of frontal factors by decreasing the eccentricity of the approximate elliptical form of the isopleths. An important and cogent observation is that the bearings of the axes of maximal correlation show a general tendency to swing to the west of NW. in the winter months, in fact to a direction 20–30° west of the orientation of the fronts. If topography, for example, were exercising a strong influence, the reverse should hold.

The trend in the axis of maximal correlation indicates that associations arise from translation of the fronts, in addition to those engendered by the fronts themselves. Two features of the movement of the pressure systems and the fronts account for this change of direction. Firstly, a wave frequently forms on the frontal surface and travels southward along the front. This motion, in conjunction with the eastward movement of the front, gives to the centre of the wave a resultant course represented diagrammatically in Figure 9, which, from the speed of the wave and that of the front, would in general have an orientation west of NW. Pressure gradients are steepest at the centre of the wave, promoting greater uplift and therefore generally more rain. Secondly, the two high-pressure systems in advance of, and behind, the advancing front, extend into the atmosphere, but with increasing altitude lag behind, and at the same time are further to the north, than the centres at the surface. The stream of the trailing edge of the leading high is elevated by the front to yield rain, and its prevailing direction is W. of NW. The associations originating in these ways combine with those along the line of the front to give the observed effect.

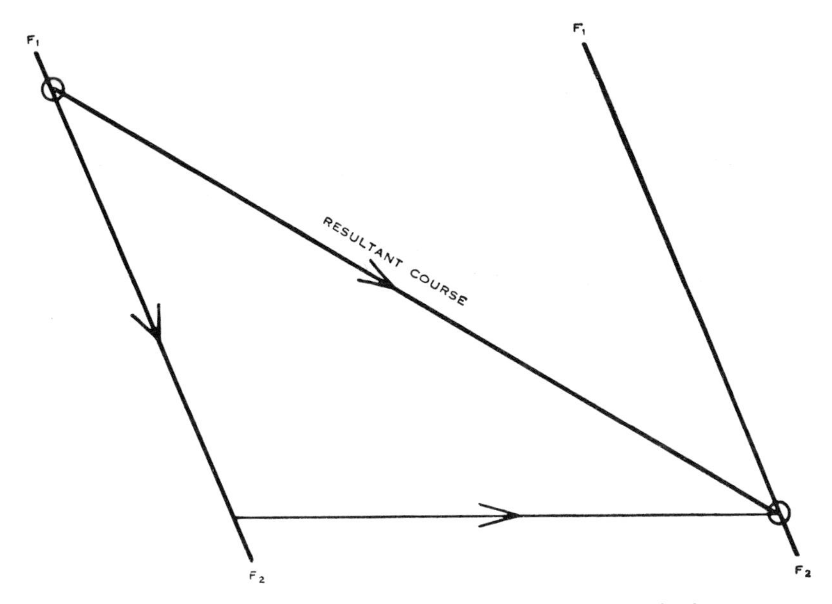

Fig. 9.—The resultant course of a wave which forms on the front.

V. Empirical Spectral Distribution of Variance

As indicated in Section I, assessment of inter-station correlations of rainfall is required to place those aspects of meteorological physics which concern localization of meteorological effects, upon an experimental and quantitative basis. Thus it is possible that the greater part of the differences between one season and another is due to very widespread rain-bearing disturbances; in such circumstances, and as observed with the present data, high correlations exist even between distant stations. On the other hand, a substantial proportion of the variation from season to season may be ascribed to causes which have extremely localized consequences. It is of some moment to indicate how this distribution of variance may be investigated empirically.

The general analysis outlined above was not designed specifically for investigating this point, as the form of the chosen regression function was selected for convenience in fitting, and while it is sufficiently accurate for distances greater than approximately 50 miles, it is not accurate enough for smaller distances, and is certainly not of the proper functional form to represent the relation with distance in the neighbourhood of zero.

To eliminate the effects of direction of displacement, two groups have been selected for each month from the 300 possible pairs, by taking

(i) pairs of stations whose join lies in a direction within $\pm 15°$ of the axis of maximal correlation,

(ii) pairs of stations whose join lies in a direction within $\pm 15°$ of the axis of minimal correlation,

and in each group the regression

$$Z = ax^{-b} \quad \ldots\ldots\ldots\ldots\ldots\ldots\ldots\ldots(2)$$

has been fitted to represent the relation with distance in a form appropriate for very small distances.

Consider a fixed point P and a variable point Q in the region. The square of the correlation between P and Q measures the proportion of the variance of the rainfall at Q which is associated with the variability of rainfall at P, and this association is a consequence of rain-bearing disturbances which, so far as these two sites are concerned, are as extensive or more extensive than the distance between them.

If v be the variance of rainfall at Q, for some particular time of the year t, and $v_x dx$ is the proportion of the variance at Q associated with the variability at P, distant x miles, then

$$\int_0^\infty v_x dx = 1.$$

From the relation (**2**) above, the empirically determined proportion of the variance associated with distances $\geqslant x$ is

$$r_x^2 = \tanh^2 z_x = \tanh^2(ax^{-b}),$$

so that

$$\int_x^\infty v_x dx = \tanh^2(ax^{-b}),$$

or alternatively, from the proportion of variance associated with distances $\leqslant x$,

$$\int_0^x v_x dx = \text{sech}^2(ax^{-b}).$$

From either of these integrals, a distribution, which may be conveniently designated an empirical spectral distribution of variance, is obtained in the form

$$2abx^{-(b+1)}\,\text{sech}^2(ax^{-b})\tanh(ax^{-b})\,dx.$$

In accordance with the observed facts regarding direction of displacement from a fixed point, there will be such a distribution of variance over distance for each direction.

Using the coefficients from (**2**), approximations to the two extremes of these distributions, along the axes of maximal and minimal correlation, may be obtained. The crudity of these approximations arises from three principal causes:

(i) the limitations of the calendar month for estimating the correlations,
(ii) the paucity of observations satisfying the requirements set out above,
(iii) the rather wide tolerances of $\pm 15°$ about the axes.

The few general observations which follow are thus made with the reservation that they may well need revision when more accurate assessments of the distributions have been obtained from the detailed investigation now in progress.

Figure 10 presents the distributions obtained for January, the extreme asymmetry being typical of the distributions for all months. The maximum of the

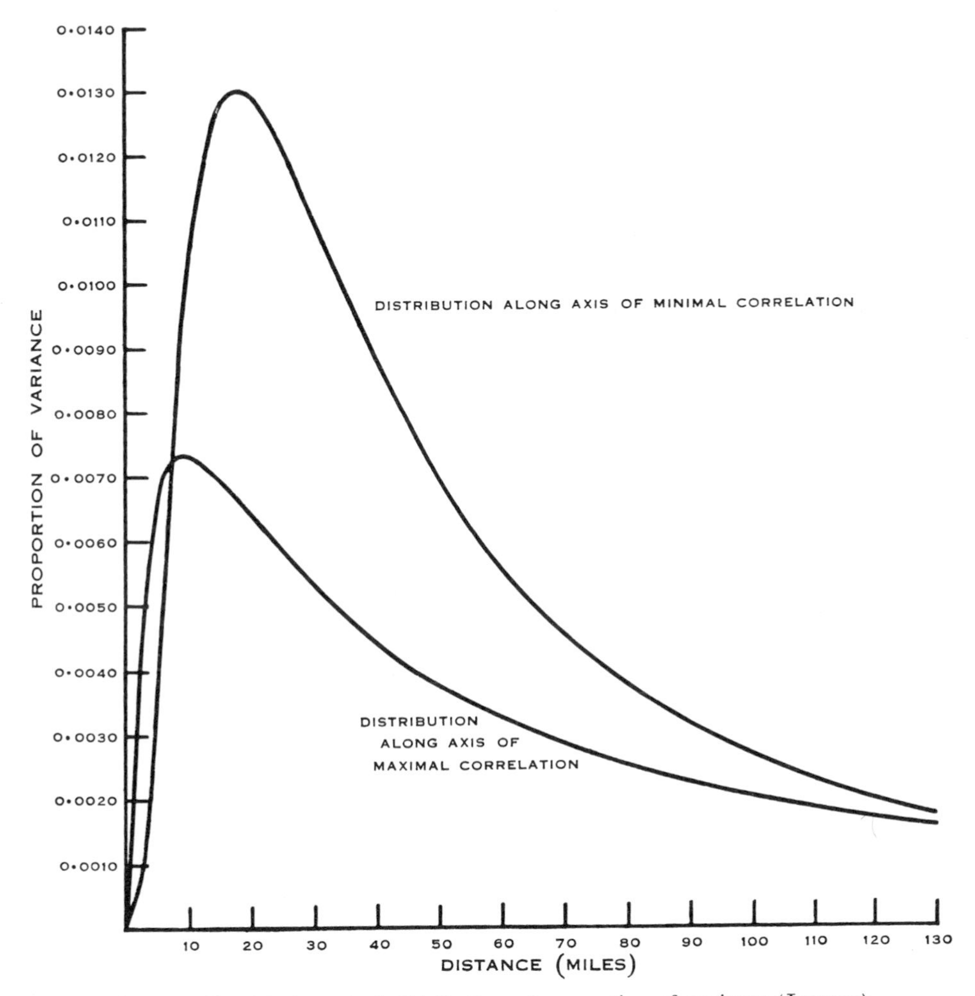

Fig. 10.—Empirical spectral distribution of proportion of variance (January).

distribution along the axis of maximal correlation is always attained in a shorter distance than that of the distribution along the axis of minimal correlation; for most of the former, the maximum is attained in just under 10 miles, whereas for the latter, distances range from 15 to 40 miles, the majority being 15–20 miles. There is no consistency in the relative magnitude of the two maxima; in fact, in a number of months the positions are reversed, but in all cases, the cumulative loss of associated variability along the axis of minimal correlation is either in excess of, or rapidly overtakes (within 20 miles) and remains greater than, the corresponding quantity of the distribution along the axis of maximal correlation. At 200 miles, the respective average cumulative losses are 80 per cent. and 55 per cent.

VI. Acknowledgments

The authors are indebted to Mr. B. Mason, of the Commonwealth Meteorological Bureau, for helpful discussions, and to Mr. M. C. Childs, of the Division of Biochemistry and General Nutrition, C.S.I.R.O., for the preparation of the diagrams.

VII. References

Coote, G. G., and Cornish, E. A. (1958).—C.S.I.R.O. Divn. Math. Stats. Tech. Paper No. 4.
Cornish, E. A. (1950).—*Aust. J. Sci. Res.* B **3**: 178–218.
Fisher, R. A. (1946).—"Statistical Methods for Research Workers." 10th Ed. (Oliver and Boyd: Edinburgh.)
Kidson, E. (1925).—Aust. Bur. Meteor. Bull. No. 17.

Reprinted from the Australian Journal of Statistics, Vol. 2, No. 1, Apr. 1960.

FIDUCIAL LIMITS FOR PARAMETERS IN COMPOUND HYPOTHESES

E. A. CORNISH
C.S.I.R.O. Division of Mathematical Statistics, Adelaide

1. Introductory. In a series of papers, dating from 1930, Fisher has strongly advocated the fiducial argument as a tool of scientific inference. For the class of cases to which it is appropriate, this type of argument leads to rigorous probability statements about the parameters of the population, from which the observational record constitutes a random sample, without the assumption of any knowledge whatever, respecting their probability distributions *a priori*.

We may, for exanple, consider a set of N observations $x_1, x_2, \ldots, x_N$ taken from a normal distribution having a mean ξ and variance σ^2 and calculate

$$\bar{x}=\frac{1}{N}\Sigma(x),$$

$$s^2=\frac{1}{N(N-1)}\Sigma(x-\bar{x})^2,$$

then Student's statistic t, defined by

$$t=(\bar{x}-\xi)/s,$$

is distributed in the well-known form

$$\frac{\Gamma\frac{1}{2}(n+1)}{\sqrt{\pi n}\,\Gamma\frac{1}{2}n}\left(1+\frac{t^2}{n}\right)^{-\frac{1}{2}(n+1)}dt, \tag{1}$$

which depends only on the parameter $n=N-1$. The distribution (1), and the relation,

$$\xi=\bar{x}-st,$$

together establish the fiducial distribution of ξ, based solely on the observed sample, in the form,

$$\frac{\Gamma\frac{1}{2}(n+1)}{\sqrt{\pi n}\,\Gamma\frac{1}{2}n}\left\{1+\frac{(\bar{x}-\xi)^2}{ns^2}\right\}^{-\frac{1}{2}(n+1)}\frac{d\xi}{s}. \tag{2}$$

Fiducial probabilities which may be required in practice are obtainable by integrating (2) over appropriate ranges of ξ, but clearly, are already available from tables of the probability integral of (1).

In recent years, there has been a tendency to consider the more general type of problem to which the z test is applicable, and in which the analysis of variance distinguishes a number of component parts due to different causes of variation, thus leading to a wide choice of hypotheses to be tested. On this point, however, it should be observed that, in the majority of fields of application, knowledge is so lacking that it is preferable to design an experiment and select a system of treatments so as to provide uniquely a series of tests for hypotheses chosen in advance. This is especially true when the results of the tests have to be interpreted for the purposes of examining a scientific theory, designing future experiments, or in formulating practical

Received for publication February 24, 1960.

advice, and theoretical independence of the quantities under test is an obvious advantage for such purposes.

Not infrequently, however, occasions arise where it is appropriate, or desirable, to examine a compound hypothesis in which a number of contrasts, independent or otherwise, are to be examined conjointly. This type of problem has been considered in recent years by various workers including Fisher (1935–1951), Tukey (1953), Scheffé (1953), and Dunnett (1955).

Our purpose is to generalize (2) to give the simultaneous fiducial distribution of any number of parameters, based on the joint distribution of a set of t-statistics, which are not only dependent through the common estimate of the variance, but also correlated because of the nature of the enquiry under consideration. Then, following Fisher (*loc. cit.* section 64) we indicate how a comprehensive test may be conducted so as to specify the aggregate of compound hypotheses which are contradicted by experimental or observational data at any assigned level of significance.

2. The General Fiducial Distribution. Suppose $\mathbf{x}$ is a column vector with elements $x_1, x_2, \ldots, x_N$ which are normally and independently distributed with means $\xi_1, \xi_2, \ldots, \xi_N$ and common variance σ^2, and let

$$\mathbf{y}=\mathbf{Hx}=(h_{ij})\mathbf{x} \qquad i=1, 2, \ldots, p; \; j=1, 2, \ldots, N$$

be any $p<N$, linearly independent, linear functions of the x_i, then the distribution of the y_i is well known to be

$$\frac{|\sigma^2\mathbf{HH}'|^{-\frac{1}{2}}}{(2\pi)^{\frac{1}{2}p}} e^{-\frac{1}{2}(\mathbf{y}-\boldsymbol{\alpha})'(\sigma^2\mathbf{HH}')^{-1}(\mathbf{y}-\boldsymbol{\alpha})} d\mathbf{y}$$

with a vector of means $\boldsymbol{\alpha}$.

If s^2 is an estimate of σ^2, on n degrees of freedom, distributed independently of the x_i and therefore independently of the y_i, and

$$\mathbf{t}=\mathbf{Q}(\mathbf{y}-\boldsymbol{\alpha})/s, \tag{3}$$

where $\mathbf{Q}$ is a diagonal matrix whose i^{th} diagonal element is $(\mathbf{h}_i\mathbf{h}_i')^{-\frac{1}{2}}$, $\mathbf{h}_i$ being the ith row vector of $\mathbf{H}$, it has been shown (Cornish 1954, Dunnett and Sobel 1954) that the distribution of the t_i takes the form

$$\frac{\Gamma\frac{1}{2}(n+p)}{(\pi n)^{\frac{1}{2}p}\Gamma\frac{1}{2}n}\left(1+\frac{\mathbf{t}'\mathbf{R}^{-1}\mathbf{t}}{n}\right)^{-\frac{1}{2}(n+p)} \frac{d\mathbf{t}}{|\mathbf{R}|^{\frac{1}{2}}}, \tag{4}$$

where $\mathbf{R}=\mathbf{Q}(\mathbf{HH}^1)\mathbf{Q}$ is the non-singular, symmetric correlation matrix of the y_i, the correlation ρ_{ij} being equal to

$$\frac{\mathbf{h}_i\mathbf{h}_j'}{(\mathbf{h}_i\mathbf{h}_i'\cdot\mathbf{h}_j\mathbf{h}_j')^{\frac{1}{2}}}.$$

The elements of $\mathbf{H}$ are known constants, satisfying the conditions

$$\sum_{j=1}^{N} h_{i.}=0 \quad \text{for } i=1, 2, \ldots, p.$$

Using (3) and the relation,

$$\mathbf{t}'\mathbf{R}^{-1}\mathbf{t}=(\mathbf{y}-\boldsymbol{\alpha})'(s^2\mathbf{HH}')^{-1}(\mathbf{y}-\boldsymbol{\alpha}), \tag{5}$$

the simultaneous fiducial distribution of the parameters α_i, is immediately obtainable from (4) in the form

$$\frac{\Gamma\frac{1}{2}(n+p)}{(\pi n)^{\frac{1}{2}p}\Gamma\frac{1}{2}n}\left\{1+\frac{(\mathbf{y}-\boldsymbol{\alpha})'(s^2\mathbf{HH}')^{-1}(\mathbf{y}-\boldsymbol{\alpha})}{n}\right\}^{-\frac{1}{2}(n+p)} \frac{d\boldsymbol{\alpha}}{|s^2\mathbf{HH}'|^{\frac{1}{2}}}, \tag{6}$$

which is a direct generalization of the fiducial distribution (2). Any fiducial probability required may be obtained by integrating (6) over appropriate ranges of the α_i. For example, if $\boldsymbol{\lambda}$ is a vector whose elements are constants, the fiducial probability that the parameters satisfy the inequality,

$$\frac{\boldsymbol{\lambda}'(\mathbf{y}-\boldsymbol{\alpha})}{\{s^2\boldsymbol{\lambda}'(\mathbf{HH}')\boldsymbol{\lambda}\}^{\frac{1}{2}}} \leqslant t_1$$

is given by

$$\frac{\Gamma\frac{1}{2}(n+1)}{\sqrt{\pi n}\,\Gamma\frac{1}{2}n}\int_{-\infty}^{t_1}\frac{dt}{\left(1+\frac{t^2}{n}\right)^{\frac{1}{2}(n+1)}}.$$

The fiducial distribution has been developed by Fisher (1954) under rather more general conditions than presented here. Thus Fisher gives the distribution when the variance-covariance matrix, involving p different variances (of p variates) and $\frac{1}{2}p(p-1)$ covariances, has to be estimated from the data.

For the particular type of problem in which we wish to examine the possible variations of any hypothesis which specifies all parameters simultaneously while maintaining an assigned level of significance, a single measure of the simultaneous discrepancies between the several observed quantities and their hypothetical values, is required. It is now standard practice to use such a measure for testing a null hypothesis by the z test in an analysis of variance. Thus, if the observations $x_1, x_2, \ldots, x_N$ had yielded treatment means $\bar{x}_1, \bar{x}_2, \ldots, \bar{x}_{p+1}$ involving, respectively, $r_1, r_2, \ldots, r_{p+1}$ replicates, the basic measure for the test of the null hypothesis $\xi_i=\xi$ for $i=1, 2, \ldots, p+1$ is the sum of squares

$$\sum_{i=1}^{p+1} r_i(\bar{x}_i-\bar{x})^2,$$

with p degrees of freedom, $\bar{x}$ denoting the general mean. On the other hand, if the linear functions y_i had been chosen to represent the p degrees of freedom for treatment, the corresponding null hypothesis would have been that the vector $\mathbf{y}$ is null, and the sum of squares would have then assumed the alternative form

$$\mathbf{y}'(\mathbf{HH}')^{-1}\mathbf{y}.$$

Adverting to the case on hand, the appropriate measure of the simultaneous discrepancies between observation and hypothesis is Hotelling's (1931) generalization of Student's t-statistic, with the particular definition

$$T^2=(\mathbf{y}-\boldsymbol{\alpha})'(s^2\mathbf{HH}')^{-1}(\mathbf{y}-\boldsymbol{\alpha}),$$

and the required fiducial probability corresponding to a given value, T_1^2, is then obtained by integrating (6) over the domain defined by

$$(\mathbf{y}-\boldsymbol{\alpha})'(s^2\mathbf{HH}')^{-1}(\mathbf{y}-\boldsymbol{\alpha})\leqslant T_1^2. \tag{7}$$

From the distribution (4), it has been shown (Cornish 1954) that the quadratic form

$$\mathbf{t}'\mathbf{R}^{-1}\mathbf{t}/p$$

is distributed as e^{2z} with p and n degrees of freedom. Using this fact, and the relation (5), fiducial probabilities associated with regions of the type defined by (7) are obtainable directly from the distribution of z with parameters $n_1=p$ and $n_2=n$. At this juncture we may distinguish three important special cases which may occur in practice :

(*a*) The rows of $\mathbf{H}$ are orthogonal, but not normalized, in which case $\mathbf{HH}'=\mathbf{\Lambda}$, a diagonal matrix with diagonal elements λ_i, $i=1, 2, \ldots, p$. The distribution (4) reduces to

$$\frac{\Gamma\frac{1}{2}(n+p)}{(\pi n)^{\frac{1}{2}p}\Gamma\frac{1}{2}n}\left(1+\frac{\mathbf{t}'\mathbf{t}}{n}\right)^{-\frac{1}{2}(n+p)}d\mathbf{t}, \tag{8}$$

and the simultaneous fiducial distribution assumes the form

$$\frac{\Gamma\frac{1}{2}(n+p)}{(\pi n)^{\frac{1}{2}p}\Gamma\frac{1}{2}n}\left\{1+\frac{(\mathbf{y}-\boldsymbol{\alpha})'(s^2\mathbf{\Lambda})^{-1}(\mathbf{y}-\boldsymbol{\alpha})}{n}\right\}^{-\frac{1}{2}(n+p)}\frac{d\boldsymbol{\alpha}}{|s^2\mathbf{\Lambda}|^{\frac{1}{2}}}. \tag{9}$$

(*b*) The rows of $\mathbf{H}$ are orthogonal, each having the same norm ν, in which case $\mathbf{HH}'=\nu\mathbf{I}$. The distribution of the t_i is identical with (8) and the fiducial distribution is

$$\frac{\Gamma\frac{1}{2}(n+p)}{(\pi n)^{\frac{1}{2}p}\Gamma\frac{1}{2}n}\left\{1+\frac{(\mathbf{y}-\boldsymbol{\alpha})'(s^2\nu\mathbf{I})^{-1}(\mathbf{y}-\boldsymbol{\alpha})}{n}\right\}^{-\frac{1}{2}(n+p)}\frac{d\boldsymbol{\alpha}}{|s^2\nu\mathbf{I}|^{\frac{1}{2}}}. \tag{10}$$

(*c*) The rows of $\mathbf{H}$ are orthogonal and normalized, in which case $\mathbf{HH}'=\mathbf{I}$. The distribution of the t_i, as before, is identical with (8) and the fiducial distribution is

$$\frac{\Gamma\frac{1}{2}(n+p)}{(\pi n)^{\frac{1}{2}p}\Gamma\frac{1}{2}n}\left\{1+\frac{(\mathbf{y}-\boldsymbol{\alpha})'(\mathbf{y}-\boldsymbol{\alpha})}{n}\right\}^{-\frac{1}{2}(n+p)}\frac{d\boldsymbol{\alpha}}{s^p}. \tag{11}$$

Since $\mathbf{HH}'$ is positive definite and symmetric, the general case may be reduced to any one of the three forms (*a*), (*b*) or (*c*), and equally either (*a*) or (*b*) may be reduced to (*c*). The course to be followed depends upon the nature of the enquiry ; orthogonal comparisons, for example, may not provide just what is wanted.

3. Illustrative Examples. (i) The illustration used by Fisher (1935–1951) is the particular case (*a*) above. The original data comprised 36 yields of potatoes from a 6×6 Latin square involving 6 treatments in a 2×3 factorial arrangement of nitrogenous and phosphatic fertilizers. The sum of squares for treatments in the analysis of variance was partitioned into 5 components ascribable to linear contrasts among the original data corresponding to the main effects and interactions of the fertilizers. Denoting the treatment combinations by n_ip_j ($i=0, 1$; $j=0, 1, 2$) and taking them in order, and the linear contrasts in the order N, P_1, P_2, NP_1, NP_2, the matrix $\mathbf{H}$, with orthogonal, but not normalized rows, is

n_0p_0	n_0p_1	n_0p_2	n_1p_0	n_1p_1	n_1p_2
−1	−1	−1	1	1	1
−1	·	1	−1	·	1
1	−2	1	1	−2	1
1	·	−1	−1	·	1
−1	2	−1	1	−2	1

each column being repeated 6 times, corresponding to the six replicates of each treatment.

Then

$$\mathbf{y}=\begin{bmatrix} N \\ P_1 \\ P_2 \\ NP_1 \\ NP_2 \end{bmatrix}=\begin{bmatrix} 1667 \\ 1977 \\ 381 \\ 383 \\ 19 \end{bmatrix}, \quad \mathbf{HH}'=\mathbf{\Lambda}=\begin{bmatrix} 36 & \cdot & \cdot & \cdot & \cdot \\ \cdot & 24 & \cdot & \cdot & \cdot \\ \cdot & \cdot & 72 & \cdot & \cdot \\ \cdot & \cdot & \cdot & 24 & \cdot \\ \cdot & \cdot & \cdot & \cdot & 72 \end{bmatrix},$$

and since $s^2=1527$, on 20 degrees of freedom,

$$(\mathbf{y}-\boldsymbol{\alpha})'(s^2\boldsymbol{\Lambda})^{-1}(\mathbf{y}-\boldsymbol{\alpha})=$$

$$\frac{1}{1527}\left\{\frac{(1667-\alpha_1)^2}{36}+\frac{(1977-\alpha_2)^2}{24}+\frac{(381-\alpha_3)^2}{72}+\frac{(383-\alpha_4)^2}{24}+\frac{(19-\alpha_5)^2}{72}\right\}.$$

Then, for example, the 1% test of significance for a hypothesis specifying all the values of $\alpha_1, \alpha_2, \ldots, \alpha_5$ will reject any hypothesis for which the value of the above expression, when divided by 5, exceeds 4·10, and will accept all hypotheses for which it is less than 4·10, the tabular value of e^{2z} for $n_1=5$ and $n_2=20$, at the 1% point.

Since $\sqrt{5\times1527\times4\cdot10}=177$, all acceptable values of the parameters $\alpha_1, \alpha_2, \ldots, \alpha_5$ constitute the interior and boundary of the ellipsoid centred at the point (1667, 1977, 381, 383, 19) with semi-axes (in order) $177\sqrt{36}$, $177\sqrt{24}$, $177\sqrt{72}$, $177\sqrt{24}$, $177\sqrt{72}$.

(ii) Bliss (1958) fitted the first term of a Fourier series to observations on the mean hourly standing potentials in an elm tree and obtained the following relation describing the diurnal changes

$$Y=-66\cdot54+2\cdot2017\cos\theta+5\cdot0279\sin\theta,$$

in which θ takes the values 0, $\pi/12$, $2\pi/12, \ldots, 23\pi/12$.
Since

$$\sum_{m=1}^{24}\cos^2\frac{(m-1)\pi}{12}=\sum_{m=1}^{24}\sin^2\frac{(m-1)\pi}{12}=12,$$

H is an orthogonal two-rowed matrix

$$\frac{1}{12}\begin{bmatrix}\cos 0 & \cos\pi/12 & \cos 2\pi/12 \ldots \cos 23\pi/12\\ \sin 0 & \sin\pi/12 & \sin 2\pi/12 \ldots \sin 23\pi/12\end{bmatrix},$$

the norm of each row being $\nu=\frac{1}{12}$, so that this example is the particular case (*b*) above, with

$$\mathbf{HH}'=\frac{1}{12}\ldots\ldots\mathbf{I},\ (\mathbf{HH}')^{-1}=12\mathbf{I}\text{ and }\mathbf{y}=\begin{bmatrix}2\cdot2017\\5\cdot0279\end{bmatrix}.$$

Bliss gives $s^2=21\cdot8544$ on (15·50) degrees of freedom, and $e^{2z}=3\cdot6572$ at the 5% point for $n_1=2$ and $n_2=15\cdot50$. Introducing the factor 8 to allow for the 8-fold replication of the observations on standing potential

$$(\mathbf{y}-\boldsymbol{\alpha})'(s^2\nu\mathbf{I})^{-1}(\mathbf{y}-\boldsymbol{\alpha})=\frac{96}{21\cdot8544}\{(2\cdot2017-\alpha_1)^2+(5\cdot0279-\alpha_2)^2\},$$

and the 5% test of significance for a hypothesis specifying the values of α_1 and α_2 will accept any hypothesis for which

$$(2\cdot2017-\alpha_1)^2+(5\cdot0279-\alpha_2)^2\leqslant2\times21\cdot8544\times3\cdot6572/96=1\cdot6651.$$

At this level of fiducial probability, all acceptable values of the parameters fall within, or on the circle, centred at the point (2·2017, 5·0279), with radius $\sqrt{1\cdot6651}=1\cdot29$.

(iii) Dunnett (1955) has presented measurements on the breaking strength of a fabric treated by three different chemical processes and a standard method of manufacture. There were triplicate determinations on each of the four methods, and the object of the experiment was to compare the breaking strength of each process with the standard. The mean breaking strengths, in pounds, were

Standard	*Process* 1	*Process* 2	*Process* 3
50	61	52	45

The linear contrasts among the observations are therefore $61-50=11$, $52-50=2$, and $45-50=-5$, and these are the elements of the vector $\mathbf{y}$, obtained from the vector of original observations by the matrix

$$\mathbf{H}=\begin{bmatrix} -1/3 & -1/3 & -1/3 & 1/3 & 1/3 & 1/3 & \cdot & \cdot & \cdot & \cdot & \cdot & \cdot \\ -1/3 & -1/3 & -1/3 & \cdot & \cdot & \cdot & 1/3 & 1/3 & 1/3 & \cdot & \cdot & \cdot \\ -1/3 & -1/3 & -1/3 & \cdot & \cdot & \cdot & \cdot & \cdot & \cdot & 1/3 & 1/3 & 1/3 \end{bmatrix}.$$

We then have

$$\mathbf{HH}'=\frac{1}{3}\begin{bmatrix} 2 & 1 & 1 \\ 1 & 2 & 1 \\ 1 & 1 & 2 \end{bmatrix} \text{ and } (\mathbf{HH}')^{-1}=\frac{3}{4}\begin{bmatrix} 3 & -1 & -1 \\ -1 & 3 & -1 \\ -1 & -1 & 3 \end{bmatrix}.$$

Dunnett gives the estimate of variance, $s^2=19$, on 8 degrees of freedom, so that

$$(\mathbf{y}-\boldsymbol{\alpha})'(s^2\mathbf{HH}')^{-1}(\mathbf{y}-\boldsymbol{\alpha})=\frac{1}{19}\Big\{\frac{9}{4}[(11-\alpha_1)^2+(2-\alpha_2)^2+(-5-\alpha_3)^2]$$

$$-\frac{6}{4}[(11-\alpha_1)(2-\alpha_2)+(11-\alpha_1)(-5-\alpha_3)+(2-\alpha_2)(-5-\alpha_3)\Big\},$$

and the 5% test of significance for a hypothesis specifying all the values of α_1, α_2 and α_3 will accept any hypothesis for which

$$\frac{9}{4}\{(11-\alpha_1)^2+(2-\alpha_2)^2+(-5-\alpha_3)^2\}$$

$$-\frac{6}{4}\{(11-\alpha_1)(2-\alpha_2)+(11-\alpha_1)(-5-\alpha_3) \tag{12}$$

$$+(2-\alpha_2)(-5-\alpha_3)\}\leqslant 3\times 19\times 4\cdot 07=232,$$

where $4\cdot 07$ is the 5% value of e^{2z} for $n_1=3$ and $n_2=8$.

At this level of fiducial probability all acceptable values of the parameters fall within, or on the ellipsoid, defined by (12) and centred at the point (11, 2, −5).

(iv) Dunnett (*loc. cit.*) has given blood count measurements on three groups of animals, one of which served as a control while the other two were treated with different drugs A and B.

The counts (millions of cells per cubic millimetre) were:

	Control	*Drug A*	*Drug B*
Mean	8·25	8·90	10·88
Number of replicates ..	6	4	5 .

The linear contrasts among the observations are therefore $8\cdot 90-8\cdot 25=0\cdot 65$ and $10\cdot 88-8\cdot 25=2\cdot 63$, and these are the elements of the vector $\mathbf{y}$, obtained from the original observations by the matrix

$$\mathbf{H}=\begin{bmatrix} -1/6 & -1/6 & -1/6 & -1/6 & -1/6 & -1/6 & 1/4 & 1/4 & 1/4 & 1/4 & \cdot & \cdot & \cdot & \cdot & \cdot \\ -1/6 & -1/6 & -1/6 & -1/6 & -1/6 & -1/6 & \cdot & \cdot & \cdot & \cdot & 1/5 & 1/5 & 1/5 & 1/5 & 1/5 \end{bmatrix}.$$

Then

$$\mathbf{HH}'=\begin{bmatrix} 5/12 & 1/6 \\ 1/6 & 11/30 \end{bmatrix} \text{ and } (\mathbf{HH}')^{-1}=\begin{bmatrix} 44/15 & -4/3 \\ -4/3 & 10/3 \end{bmatrix}.$$

Dunnett gives $s^2=1\cdot3805$ on 12 degrees of freedom, so that the 5% test of significance for a hypothesis specifying the values of α_1 and α_2 will accept any hypothesis for which

$$(13)\quad \frac{44}{15}(0\cdot65-\alpha_1)^2+\frac{10}{3}(2\cdot63-\alpha_2)^2-\frac{8}{3}(0\cdot65-\alpha_1)(2\cdot63-\alpha_2)$$
$$\leqslant 2\times1\cdot3805\times3\cdot88=10\cdot72$$

where $3\cdot88$ is the 5% value of e^{2z} for $n_1=2$ and $n_2=12$.

At this level of probability all acceptable values of the parameters fall within, or on, the ellipse defined by (13) and centred at the point $(0\cdot65,\ 2\cdot63)$, *vide* Figure 1.

The procedure in the general case has been illustrated by applying it to examples of the type used by Dunnett for comparing treatments with the standard or control, but it should be emphasized that the method is applicable to any set of p contrasts among the treatments for which the matrix $\mathbf{H}$ is known and of rank p.

4. Comparison with Other Methods. Dunnett (*loc. cit.*) has outlined a method for testing a compound hypothesis for the particular case where a number of treatments are compared with a control or standard. A set of experimental observations has yielded $p+1$ means $\bar{x}_0, \bar{x}_1, \bar{x}_2, \ldots, \bar{x}_p$ from a control and p other treatments respectively, together with an independent estimate of the variance, s^2, based on n degrees of freedom. Dunnett determined confidence limits for each of the differences $\bar{x}_i-\bar{x}_0$, and presented tables for conducting compound one-sided and two-sided tests at each of two values of the specified joint confidence coefficient ($0\cdot95$ and $0\cdot99$), for values of $p=1(1)9$ and a comprehensive range of values n. He pointed out that, except for certain special cases, only approximate solutions are available until either the multivariate t-distribution has been tabulated, or appropriate tables have been constructed from extended tables of the multivariate normal distribution.

In this connection, the important feature of the procedure outlined in section 3 and illustrated in examples (iii) and (iv) is that it provides an *exact* solution when the matrix $\mathbf{H}$ is known, and this will always be true of the class of cases with which we are concerned. The only possible limitation to its use would be the extent to which the z distribution has been tabled, but with the wealth of tabular material now available this limitation is quite unimportant.

Figure 1 shows the ellipse comprising the 95% fiducial region, defined by (13) in example (iv) of the previous section, contrasted with the corresponding confidence regions computed according to the procedures advocated by Dunnett (inner rectangle), and Scheffé (outer rectangle).

Dunnett's confidence region was obtained by a method which is equivalent to integrating (6) with $p=2$, $n=12$ and $\rho_{12}=\frac{1}{2}$, instead of $\sqrt{\frac{2}{11}}\doteqdot0\cdot43$, over the domain defined by

$$(14)\qquad \begin{aligned} y_1-ds_1 &\leqslant \alpha_1 \leqslant y_1+ds_1 \\ y_2-ds_2 &\leqslant \alpha_2 \leqslant y_2+ds_2 \end{aligned}$$

where s_1 and s_2 are the estimated standard deviations of y_1 and y_2 respectively and d is a multiple corresponding to a probability of $0\cdot95$. Ignoring the slight inaccuracy in the boundaries of this region, arising from the difference in the correlations, the value of the integral taken over the domain is 0·95. The contributions to the integral, which should have come from the four zones A, B, C and D (and which are excluded by Dunnett) are exactly compensated by the contributions from the regions E, F, G and H, adjacent to the respective ends

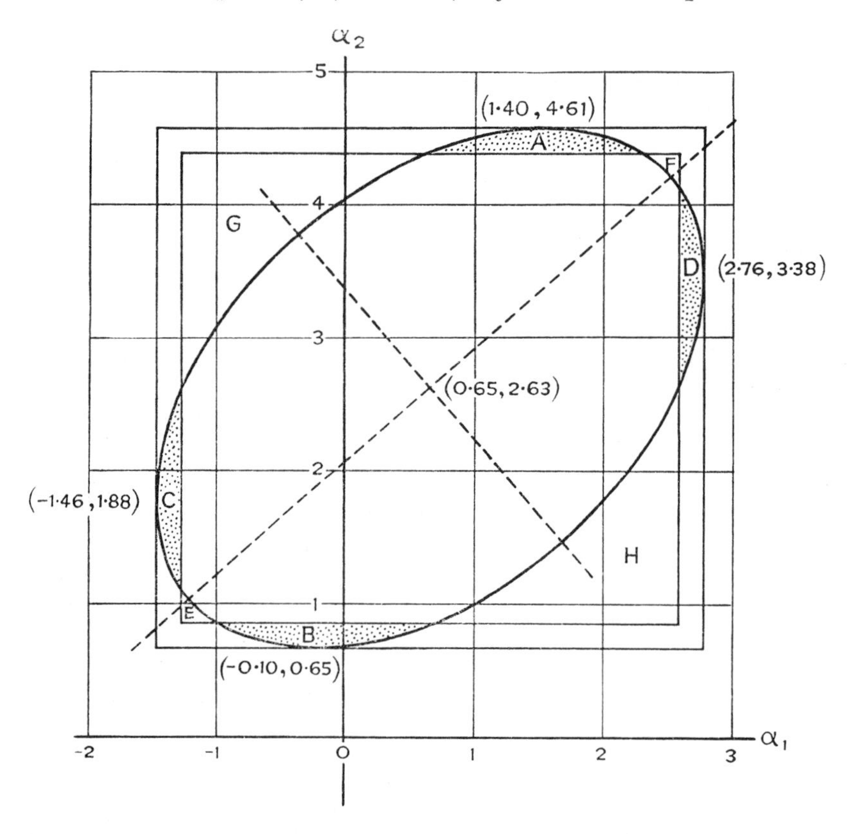

Figure 1. 95% fiducial region and corresponding confidence regions of Dunnett and Scheffé, for example (iv)

of the major and minor axes of the ellipse. The diagram emphasizes the fact that, when determining that region of the parameter space which agrees with the observed data, *in a test of a compound hypothesis* at any assigned level of fiducial probability, we must integrate over a domain throughout which the *joint* discrepancies are properly represented. The limits for either parameter are conditional on the particular value of the other parameter, and domains of the type defined by (14) do not properly account for this fact.

Scheffé's procedure, on the other hand, locates the domain defined by (13) and projects it on to the axes to obtain unconditional limits for the parameters. Although the limits so obtained correctly represent the extremes, each pair corresponds only to particular values of the other parameter, and their use in delineating the boundaries of the fiducial region ignores information supplied by the simultaneous distribution of the parameters. The rectangle having the projections as its sides constitutes Scheffé's confidence region, and the integral taken over it yields a probability $>0 \cdot 95$.

References

Bliss, C. I. (1958). "Periodic Regression in Biology and Climatology." *Conn. Agric. Expt. Stat. Bull.* 615.

Cornish, E. A. (1954). "The Multivariate t-Distribution Associated with a Set of Normal Sample Deviates", *Aust. J. Phys.*, 7, 531–542.

Dunnett, C. W. (1955). "A Multiple Comparison Procedure for Comparing Several Treatments with a Control." *J. Amer. Statist. Assoc.*, 50, 1096–1121.

Dunnett, C. W., and Sobel, M. (1954). "A Bivariate Generalisation of Student's t-Distribution, with Tables for Certain Special Cases." *Biometrika*, 41, 153–169.

Fisher, R. A. (1935–1951). "The Design of Experiments." Oliver and Boyd, Edinburgh.

Fisher, R. A. (1954). *J. Roy. Statist. Soc.* (Series B), 16, 212–213.

Hotelling, H. (1931). "The Generalisation of Student's Ratio." *Ann. Math. Statist.*, 2, 360–378.

Scheffé, H. (1953). "A Method for Judging all Contrasts in the Analysis of Variance." *Biometrika*, 40, 87–104.

Tukey, J. W. (1953). "The Problem of Multiple Comparisons." Unpublished notes. Princeton University.

NEWS AND NOTES

The American Mathematical Society announces that it is about to commence publication of a new journal, *Soviet Mathematics—Doklady*, being translations of the articles in *Doklady Akademii Nauk, S.S.S.R.*, dealing with pure mathematics.

Mr. B. L. Adkins has been appointed Lecturer in Statistics in the University of Melbourne. He is B.Sc.(Q'land) and has had considerable experience in forestry statistics.

Professor P. A. P. Moran, of the Australian National University, is spending a sabbatical year at Oxford at the Department of Social Medicine. His book, "The Theory of Storage", has recently been published as a Methuen Monograph on Applied Probability and Statistics.

Applications are invited for two positions of Progammer in the Adolph Basser Computing Laboratory of the School of Physics from graduates in Science or Engineering. Those appointed will work on general programs for Silliac and on Monte Carlo calculations concerned with cosmic ray shower series.

The Statistical Society of New South Wales announces a symposium on Statistics in Industrial Management to be held on 30th and 31st May, 1960, at the University of Sydney. Topics will include measurement and standardization, sampling industry, controlled experiment in industry and the training of statisticians for industry.

Copies of the *Bibliography of Research in the Social Sciences in Australia 1954-1957* are still available and may be obtained from the Secretary, Social Science Research Council of Australia, Academy of Science Building, Canberra City, A.C.T. Price (including postage): Australia, 5/6; United Kingdom 5/6 stg.; U.S.A. & Canada, $1.

AUSTRALASIAN MEDICAL PUBLISHING CO. LTD.
SEAMER AND ARUNDEL STS., GLEBE, SYDNEY

Reprinted from "Technometrics"
Vol. 2, No. 2, pp. 209–226, May, 1960

The Percentile Points of Distributions Having Known Cumulants

SIR RONALD A. FISHER
AND
E. A. CORNISH

In an earlier study of the uses of moments and cumulants in the specification of statistical distributions, the authors developed explicit asymptotic expansions, expressing any desired percentile point of such distributions in terms of known cumulants.

The general formulae are now presented as far as the sixth adjustment, based on the eighth cumulant, and also numerical tables showing the coefficients of all terms for ten chosen levels of significance over the range 0.5 to 0.0005 (single tail), together with the first five Hermite polynomials and tables for the common tests of significance, χ^2, t and z, at the same levels.

I. INTRODUCTORY

In 1937, in a study of the uses of moments and cumulants in the specification of statistical distributions, the authors (1937)* were led to develop explicit asymptotic expansions, expressing any desired percentile point of such distributions in terms of known cumulants. The general formulae were presented so far as the fourth adjustment, based on the sixth cumulant, and also numerical tables showing the coefficients of all terms for nine chosen levels of significance over the range 0.25 to 0.0005 (single tail), together with the first five Hermite polynomials at the same levels.

As an illustrative example, the cumulants of the z distribution were expressed in terms of $1/n_1$ and $1/n_2$, the reciprocals of the two numbers of degrees of freedom, and the rapid convergence at the 5% point exhibited for the case $n_1 = 24$ and $n_2 = 60$. In the intervening period, the formulae have frequently been found useful, either for calculations of higher accuracy in the case of functions already tabulated, or for values outside their range, or especially with tables of multiple entry to supply intermediate values more accurate than can be obtained by interpolation (Fisher 1941, Goldberg and Levine 1946), or for cases where no tables existed (Johnson and Welch 1939).

II. THE METHOD OF EXPANSION

The several steps in the method of expansion are set out in Sections 7 and 8 of our previous paper. Here a brief outline of these steps will be sufficient, as the principal purpose of this note is to extend the formulae and tables to the sixth corrective term, so widening the range of useful application.

If the element of frequency in the distribution of a variate ξ is $f(\xi)d\xi$, the

* Attention is drawn to this reference, as it has been repeatedly misquoted.

effect of the operator

$$\exp\left\{\frac{a_r}{r!}\left(-\frac{d}{d\xi}\right)^r\right\},$$

when acting on the frequency function, is to increase the rth cumulant by a_r, but leaves the distribution otherwise unchanged. This important operational property of the cumulants is basic to the method, the essential steps of which are as follows:

(i) If the cumulants $\kappa_1, \kappa_2, \cdots$ of the distribution of ξ are expressible in power series of the reciprocal of some number n, the frequency element may be represented as

$$\exp\left\{-av^{1/2}\frac{d}{d\xi}+\frac{1}{2}bv\frac{d^2}{d\xi^2}-\frac{1}{6}cv^{3/2}\frac{d^3}{d\xi^3}+\frac{1}{24}dv^2\frac{d^4}{d\xi^4}-\frac{1}{120}ev^{5/2}\frac{d^5}{d\xi^5}+\frac{1}{720}fv^3\frac{d^6}{d\xi^6}-\cdots\right\}\frac{1}{\sqrt{2\pi v}}e^{-(\xi-m)^2/2v}\,d\xi \tag{1}$$

where the coefficients a and c are of order $n^{-1/2}$, b and d of order n^{-1}, e of order $n^{-3/2}$, f of order n^{-2} $\cdots$, and are related, respectively, to κ_1 and κ_3, κ_2 and $\kappa_4, \kappa_5, \kappa_6, \cdots$, and m and v are the mean and variance of a normal distribution chosen for convenience.

(ii) Expanding the operator and integrating, the frequency less than $m + \xi v^{1/2}$ may be expressed in terms of the corresponding normal probability integral and a series of adjustments of decreasing order of magnitude.

(iii) If x is the normal deviate at some chosen level of probability, and ξ the corresponding deviate of the distribution under consideration, the difference $\xi - x$ may be found by equating the expression for the probability that the variate has a value less than ξ to

$$\frac{1}{\sqrt{2\pi}}\int_{-\infty}^{\xi}e^{-\xi^2/2}\,d\xi-\frac{1}{\sqrt{2\pi}}e^{-\xi^2/2}\left\{(\xi-x)-\frac{1}{2}(\xi-x)^2\xi+\frac{1}{6}(\xi-x)^3(\xi^2-1)-\frac{1}{24}(\xi-x)^4(\xi^3-3\xi)+\cdots\right\} \tag{2}$$

in which the coefficients are the Hermite polynomials. By considering the terms of each order of magnitude in succession, we may develop an expansion for $\xi - x$ in terms of successive polynomials in ξ.

(iv) The expansion for $\xi - x$ is converted to a much more useful expansion in terms of x, the values of which are known in advance, so obtaining the percentile deviate ξ explicitly in terms of the normal deviate.

The adjustments to the normal deviate having the required probability integral are set out in tabular form below. Adjustments V and VI are new with this paper.

Table I gives the numerical values of the first seven Hermite polynomials, over the same range of percentiles as given previously, and Table II gives the numerical values of the polynomials in the several adjustment terms.

Table I—*Hermite Polynomials*

p	x	$x^2 - 1$	$x^3 - 3x$	$x^4 - 6x^2 + 3$	$x^5 - 10x^3 + 15x$	$x^6 - 15x^4 + 45x^2 - 15$	$x^7 - 21x^5 + 105x^3 - 105x$
.5	0	−1.0000 0000 0000	0	3.0000 0000 0000	0	−15.0000 0000 0000	0
.25	0.6744 8975 0296	−0.5450 6357 6746	−1.7166 1929 6367	0.4773 4860 9677	7.1884 4393 0012	2.4617 8870 2987	−41.4702 1233 2512
.10	1.2815 5156 5545	0.6423 7441 5150	−1.7398 6719 3688	−4.1568 5277 1360	1.6322 4759 7878	22.8760 7332 1217	19.5233 8199 1054
.05	1.6448 5362 6951	1.7055 4345 4095	−0.4843 3791 7511	−5.9132 9534 2574	−7.7891 5362 1425	16.7544 5912 7787	74.2935 5459 2500
.025	1.9599 6398 4540	2.8414 5882 0694	1.6492 2898 3034	−5.2919 4705 3076	−16.9689 4156 4258	−6.7987 7905 6330	88.4882 8729 6296
.01	2.3263 4787 4041	4.4118 9443 1054	5.6109 0548 2094	−0.1827 6525 3449	−22.8687 9748 7187	−52.2869 5214 8942	15.5751 4495 1354
.005	2.5758 2930 3549	5.6348 9660 1021	9.3628 7318 0281	7.2124 7330 0116	−18.8733 9264 3621	−84.6770 0432 9402	−104.8731 5322 6683
.0025	2.8070 3376 8344	6.8794 3857 6622	13.6967 4885 5139	17.8089 2082 3031	−4.7967 5329 2545	−102.5092 5258 5743	−258.9664 1382 0596
.001	3.0902 3230 6168	8.5495 3570 6083	20.2395 8682 9338	36.8964 1796 5259	33.0601 5546 0763	−82.3185 2937 4518	−452.7443 1163 3935
.0005	3.2905 2673 1492	9.8275 6617 0663	25.7568 1572 7087	55.2707 9215 6104	78.8427 5615 2042	−16.9197 6407 7727	−528.7314 7290 0548

Table II—*Numerical values of the polynomials in the general formula*

	p									
	.5	.25	.10	.05	.025	.01	.005	.0025	.001	.0005
c	−0.16667	−0.09084	0.10706	0.28426	0.47358	0.73532	0.93915	1.14657	1.42492	1.63793
b	0	0.33724	0.64078	0.82243	0.97998	1.16317	1.28791	1.40352	1.54512	1.64526
d	0	−0.07153	−0.07249	−0.02018	0.06872	0.23379	0.39012	0.57070	0.84332	1.07320
c^2	0	0.07663	0.06106	−0.01878	−0.14607	−0.37634	−0.59171	−0.83890	−1.21026	−1.52234
bc	0.16667	0.09084	−0.10706	−0.28426	−0.47358	−0.73532	−0.93915	−1.14657	−1.42492	−1.63793
e	0.02500	0.00398	−0.03464	−0.04928	−0.04410	−0.00152	0.06010	0.14841	0.30747	0.46059
cd	−0.08333	0.00282	0.14644	0.17532	0.10210	−0.17621	−0.53531	−1.02868	−1.89358	−2.71243
c^3	0.05247	−0.01428	−0.11629	−0.11899	−0.02937	0.25195	0.59757	1.06301	1.86790	2.62337
b^2	0	−0.08431	−0.16019	−0.20561	−0.24500	−0.29079	−0.32198	−0.35088	−0.38628	−0.41132
bd	0	0.10729	0.10874	0.03027	−0.10308	−0.35068	−0.58518	−0.85605	−1.26497	−1.60980
bc^2	0	−0.19158	−0.15265	−0.04696	0.36517	0.94084	1.47928	2.09726	3.02565	3.80584
f	0	0.00998	0.00227	−0.01082	−0.02357	−0.03176	−0.02621	−0.00666	0.04592	0.10950
ce	0	−0.05126	0.01086	0.09462	0.16106	0.16058	0.05366	−0.17498	−0.70466	−1.30531
d^2	0	−0.03285	0.00776	0.05985	0.09659	0.07888	−0.01226	−0.19116	−0.59062	−1.03555
c^2d	0	0.14764	−0.10858	−0.39517	−0.55856	−0.32621	0.35696	1.60445	4.29316	7.23307
c^4	0	−0.06898	0.09585	0.25623	0.31624	0.07286	−0.46535	−1.39199	−3.32716	−5.40702
b^2c	−0.16667	−0.09084	0.10706	0.28426	0.47358	0.73532	0.93915	1.14657	1.42492	1.63793
be	−0.05000	−0.00796	0.06928	0.09855	0.08820	0.00305	−0.12021	−0.29682	−0.61494	−0.92118
bcd	0.25000	−0.00846	−0.43931	−0.52597	−0.30631	0.52864	1.60592	3.08604	5.68074	8.13729
bc^3	−0.20988	0.05714	0.46515	0.47598	0.11748	−1.00781	−2.39028	−4.25205	−7.47158	−10.49350
g	−0.00298	0.00049	0.00454	0.00332	−0.00135	−0.01037	−0.01680	−0.02034	−0.01633	−0.00336
cf	0.02083	−0.00791	−0.03371	−0.01141	0.04024	0.12188	0.16262	0.15484	0.01974	−0.21672
de	0.03333	−0.01168	−0.04871	−0.01011	0.07080	0.18338	0.21571	0.14718	−0.18946	−0.70228
c^2e	−0.08333	0.04892	0.12788	−0.03219	−0.31878	−0.67431	−0.71959	−0.37377	1.00984	3.01445
cd^2	−0.11111	0.05854	0.14979	−0.06496	−0.42767	−0.82209	−0.76354	−0.11687	2.10109	5.17256
c^3d	0.23457	−0.17412	−0.28497	0.38896	1.39396	2.27932	1.73496	−0.69355	−8.16706	−18.12350
c^5	−0.09091	0.08374	0.07657	−0.28485	−0.76311	−1.05359	−0.53986	1.05630	5.56273	11.36150

Table II—*Continued*

	p									
	.5	.25	.10	.05	.025	.01	.005	.0025	.001	.0005
b^3	0	0.04216	0.08010	0.10280	0.12250	0.14540	0.16099	0.17544	0.19314	0.20566
b^2d	0	−1.13411	−0.13593	−0.03784	0.12885	0.43835	0.73147	1.07006	1.58122	2.01225
b^2c^2	0	0.33526	0.26714	−0.08217	−0.63904	−1.64648	−2.58873	−3.67020	−5.29489	−6.66023
bf	0	−0.02496	−0.00567	0.02705	0.05892	0.07941	0.06553	0.01666	−0.11479	−0.27376
bce	0	0.17941	−0.03802	−0.33116	−0.56370	−0.56204	−0.18780	0.61244	2.46632	4.56859
bd^2	0	0.11498	−0.02716	−0.20949	−0.33807	−0.27606	0.04292	0.66905	2.06717	3.62442
bc^2d	0	−0.66436	0.48861	1.77828	2.51350	1.46795	−1.60634	−7.22002	−19.31920	−32.54883
bc^4	0	0.37938	−0.52718	−1.40927	−1.73933	−0.40075	2.55940	7.65594	18.29937	29.73862
h	0	−0.00103	0.00048	0.00184	0.00219	0.00039	−0.00260	−0.00642	−0.01123	−0.01311
cg	0	0.00933	−0.00937	−0.02175	−0.01828	0.01651	0.06034	0.10752	0.14686	0.13160
df	0	0.01700	−0.01657	−0.03576	−0.02362	0.04640	0.12406	0.19389	0.20794	0.09578
e^2	0	0.01023	−0.01020	−0.02156	−0.01387	0.02788	0.07209	0.10787	0.09883	0.00803
c^2f	0	−0.04600	0.07442	0.12464	0.04811	−0.25595	−0.56524	−0.81931	−0.79570	−0.24501
cde	0	−0.14366	0.22074	0.34077	0.05220	−0.91516	−1.77648	−2.26940	−1.31805	1.58689
d^3	0	−0.03120	0.04431	0.06346	−0.00731	−0.21997	−0.39141	−0.45485	−0.12354	0.67346
c^3e	0	0.15345	−0.35265	−0.42236	0.15607	1.76170	3.04573	3.56562	1.33265	−4.18956
c^2d^2	0	0.30132	−0.63747	−0.68830	0.52403	3.60694	5.77533	5.99761	−0.47341	−13.97518
c^4d	0	−0.36979	1.09609	0.79893	−1.77731	−7.45279	−10.76740	−9.59341	6.35417	36.18663
c^6	0	0.10100	−0.40194	−0.12422	1.07691	3.38425	4.37679	3.02544	−5.79725	−20.92219

III. Algebraic Forms of the Expansion for Common Tests of Significance

In general applications of the formulae given above, the coefficients a, b, c, $\cdots$ can be evaluated numerically, and their values substituted, but to facilitate the use of the formulae with those statistics in frequent use, the substitution has been made algebraically to give the general form of the expansion in terms of the parameters of the distributions concerned.

Adjustment	Coefficient	Polynomial in x	Divisor
I	a	1	1
	c	$x^2 - 1$	6
II	b	x	2
	d	$x^3 - 3x$	24
	c^2	$-(2x^3 - 5x)$	36
III	bc	$-(x^2 - 1)$	6
	e	$x^4 - 6x^2 + 3$	120
	cd	$-(x^4 - 5x^2 + 2)$	24
	c^3	$12x^4 - 53x^2 + 17$	324
IV	b^2	$-x$	8
	bd	$-(x^3 - 3x)$	16
	bc^2	$5(2x^3 - 5x)$	72
	f	$x^5 - 10x^3 + 15x$	720
	ce	$-(2x^5 - 17x^3 + 21x)$	180
	d^2	$-(3x^5 - 24x^3 + 29x)$	384
	c^2d	$14x^5 - 103x^3 + 107x$	288
	c^4	$-(252x^5 - 1688x + 1511x)$	7776
V	b^2c	$x^2 - 1$	6
	be	$-(x^4 - 6x^2 + 3)$	60
	bcd	$x^4 - 5x^2 + 2$	8
	bc^3	$-(12x^4 - 53x^2 + 17)$	81
	g	$x^6 - 15x^4 + 45x^2 - 15$	5040
	cf	$-(x^6 - 13x^4 + 33x^2 - 9)$	432
	de	$-(x^6 - 12x^4 + 29x^2 - 8)$	240
	c^2e	$16x^6 - 181x^4 + 393x^2 - 90$	1080
	cd^2	$12x^6 - 129x^4 + 271x^2 - 64$	576
	c^3d	$-(80x^6 - 803x^4 + 1513x^2 - 304)$	1296
	c^5	$960x^6 - 8937x^4 + 15062x^2 - 2651$	29160

Adjustment	Coefficient	Polynomial in x	Divisor
VI	b^3	x	16
	b^2d	$5(x^3 - 3x)$	64
	b^2c^2	$-35(2x^3 - 5x)$	288
	bf	$-(x^5 - 10x^3 + 15x)$	288
	bce	$7(2x^5 - 17x^3 + 21x)$	360
	bd^2	$7(3x^5 - 24x^3 + 29x)$	768
	bc^2d	$-(14x^5 - 103x^3 + 107x)$	64
	bc^4	$11(252x^5 - 1688x^3 + 1511x)$	15552
	h	$x^7 - 21x^5 + 105x^3 - 105x$	40320
	cg	$-(2x^7 - 37x^5 + 160x^3 - 135x)$	5040
	df	$-(x^7 - 17x^5 + 69x^3 - 57x)$	1152
	e^2	$-(2x^7 - 33x^5 + 132x^3 - 108x)$	3600
	c^2f	$18x^7 - 293x^5 + 1100x^3 - 795x$	5184
	cde	$18x^7 - 273x^5 + 974x^3 - 695x$	1440
	d^3	$9x^7 - 131x^5 + 451x^3 - 321x$	3072
	c^3e	$-(396x^7 - 5708x^5 + 18755x^3 - 11811x)$	19440
	c^2d^2	$-(594x^7 - 8193x^5 + 26006x^3 - 16367x)$	13824
	c^4d	$5148x^7 - 67004x^5 + 195259x^3 - 109553x$	62208
	c^6	$-(154440x^7 - 1887684x^5 + 5033714x^3 - 2542637x)$	4199040

(a) χ^2 *distribution*

If n is the number of degrees of freedom

$$\begin{aligned} \chi^2 = n & + \sqrt{n}\,(x\sqrt{2}) + \tfrac{2}{3}(x^2 - 1) + \frac{1}{\sqrt{n}}\left(\frac{x^3 - 7x}{9\sqrt{2}}\right) - \frac{1}{n}\left(\frac{6x^4 + 14x^2 - 32}{405}\right) \\ & + \frac{1}{n\sqrt{n}}\left(\frac{9x^5 + 256x^3 - 433x}{4860\sqrt{2}}\right) + \frac{1}{n^2}\left(\frac{12x^6 - 243x^4 - 923x^2 + 1472}{25515}\right) \\ & - \frac{1}{n^2\sqrt{n}}\left(\frac{3753x^7 + 4353x^5 - 289517x^3 - 289717x}{9185400\sqrt{2}}\right). \end{aligned} \tag{3a}$$

(b) *t distribution*

If n is the number of degrees of freedom

$$
\begin{aligned}
t = {} & x \\
& + \frac{1}{n}\left(\frac{x^3 + x}{4}\right) \\
& + \frac{1}{n^2}\left(\frac{5x^5 + 16x^3 + 3x}{96}\right) \\
& + \frac{1}{n^3}\left(\frac{3x^7 + 19x^5 + 17x^3 - 15x}{384}\right) \\
& + \frac{1}{n^4}\left(\frac{79x^9 + 776x^7 + 1482x^5 - 1920x^3 - 945x}{92160}\right) \\
& + \frac{1}{n^5}\left(\frac{27x^{11} + 339x^9 + 930x^7 - 1782x^5 - 765x^3 + 17955x}{368640}\right).
\end{aligned} \tag{3b}
$$

The number of terms and the orders of magnitude differ in this expansion because it has been derived from the original expansion of Student's integral in powers of n^{-1} as given by Fisher (1926).

(c) *z distribution*

If n_1 and n_2 are the degrees of freedom, and

$$\sigma = \frac{1}{n_1} + \frac{1}{n_2} \quad \delta = \frac{1}{n_1} - \frac{1}{n_2},$$

then

$$
\begin{aligned}
z = {} & \sqrt{\frac{\sigma}{2}}\,(x) \\
& - \delta\left(\frac{x^2 + 2}{6}\right) \\
& + \sqrt{\frac{\sigma}{2}}\left\{\sigma\left(\frac{x^3 + 3x}{24}\right) + \frac{\delta^2}{\sigma}\left(\frac{x^3 + 11x}{72}\right)\right\} \\
& - \left\{\delta\sigma\left(\frac{x^4 + 9x^2 + 8}{120}\right) - \frac{\delta^3}{\sigma}\left(\frac{3x^4 + 7x^2 - 16}{3240}\right)\right\} \\
& + \sqrt{\frac{\sigma}{2}}\left\{\sigma^2\left(\frac{x^5 + 20x^3 + 15x}{1920}\right) + \delta^2\left(\frac{x^5 + 44x^3 + 183x}{2880}\right)\right. \\
& \left. + \frac{\delta^4}{\sigma^2}\left(\frac{9x^5 - 284x^3 - 1513x}{155520}\right)\right\} + \left\{\delta\sigma^2\left(\frac{4x^6 - 25x^4 - 177x^2 + 192}{20160}\right)\right. \\
& \left. + \delta^3\left(\frac{4x^6 + 101x^4 + 117x^2 - 480}{90720}\right) - \frac{\delta^5}{\sigma^2}\left(\frac{12x^6 + 513x^4 + 841x^2 - 2560}{1632960}\right)\right\}
\end{aligned} \tag{3c}
$$

$$- \sqrt{\frac{\sigma}{2}} \left\{ \sigma^3 \left(\frac{x^7 + 7x^5 + 7x^3 + 105x}{21504} \right) \right.$$

$$+ \delta\sigma^2 \left(\frac{801x^7 + 10511x^5 + 30151x^3 + 62241x}{4838400} \right)$$

$$- \frac{\delta^4}{\sigma} \left(\frac{477x^7 + 4507x^5 - 82933x^3 - 264363x}{43545600} \right)$$

$$\left. + \frac{\delta^6}{\sigma^3} \left(\frac{3753x^7 + 55383x^5 - 368897x^3 - 1213927x}{1175731200} \right) \right\}.$$

As an illustration of the accuracy, we may use the example given previously. When $n_1 = 24$ and $n_2 = 60$, the 5% value of z is 0.26534844, and the asymptotic expansion yields the following values:

Order of magnitude	Successive terms	Successive totals	Successive errors
0	0.2809 1224	0.2809 1224	0.0155 6380
1	— 196 0643	2613 0581	— 40 4263
2	44 6851	2657 7432	4 2588
3	— 4 8004	2652 9428	— 5416
4	5645	2653 5073	229
5	— 154	2653 4919	75
6	— 102	2653 4817	— 27

The numerical values of the polynomials in x occurring in the above formulae are given in Table III.

IV. Examples of the Types of Problem to Which the Expansion Has Been Applied

(a) *The asymptotic approach to Behrens' integral*

Fisher (1926) developed the ordinate and integral of Student's distribution in a series of powers of n^{-1}, giving the polynomial coefficients so far as the fifth adjustment. The purpose of this expansion was to supply sufficiently accurate values of the probabilities corresponding to any values of t for values of n beyond the range which it was proposed to tabulate.

With Behrens' extension of Student's test there were even stronger reasons for using a similar method. The direct calculations carried out by Sukhatme (1938) are very much more laborious than those needed for Student's integral. At any single level of significance, various values are required for three parameters, provided by the two numbers of degrees of freedom of the two samples, and the estimated ratio of the variances of the two means. For functions of many variables, there is a great advantage in the use of explicit formulae in which the several variables may be substituted, and there is much to be gained by extending the use of such formulae over regions too extensive for complete tabulation. Finally, it should be noted that the logical situation in which we would prefer to rely on the separate estimates of variances from the two samples rather than on any process of pooling these estimates, is of more frequent occurrence with large samples than with small, and is particularly applicable to cases,

Table III—*Numerical values of the polynomials in the expansions for χ^2, t and z.*

(a) χ^2

p									
.5	.25	.10	.05	.025	.01	.005	.0025	.001	.0005
0	0.9538726	1.8123876	2.3261743	2.7718076	3.2899527	3.6427727	3.9697452	4.3702484	4.6535075
−0.666667	−0.363376	0.428250	1.137029	1.894306	2.941263	3.756598	4.586292	5.699690	6.551711
0	−0.346842	−0.539450	−0.554981	−0.486382	−0.290266	−0.073888	0.193953	0.619006	0.989534
0.07901	0.06022	−0.01772	−0.12296	−0.27240	−0.54197	−0.80252	−1.11315	−1.60211	−2.03211
0	−0.0309	0.0022	0.0779	0.1948	0.4116	0.6228	0.8752	1.2735	1.6249
0.0577	0.0393	−0.0253	−0.1006	−0.1952	−0.3425	−0.4642	−0.5886	−0.7467	−0.8535
0	0.012	0.073	0.122	0.170	0.203	0.183	0.100	−0.145	−0.469

The values in the columns are in the same order as the polynomials in expansion 3a. Sufficient figures have been retained to ensure accuracy in the fourth decimal place for $n > 30$, except for $x\sqrt{2}$ which should be taken more accurately for $n > 1600$.

(b) t

p									
.5	.25	.10	.05	.025	.01	.005	.0025	.001	.0005
0	0.24533	0.84658	1.52377	2.37227	3.72907	4.91655	6.23122	8.15013	9.72973
0	0.0795	0.5709	1.4202	2.8225	5.7197	8.8348	12.8509	19.6925	26.1330
0	−0.005	0.259	0.983	2.556	6.719	12.144	20.221	36.154	53.169
0	0	0.1	0.4	1.6	5.6	12.1	23.2	48.6	79.4

The values in the columns are in the same order as the polynomials in expansion 3b. Sufficient figures have been retained to ensure accuracy in the fourth decimal place for $n > 30$.

TABLE III—*Continued*

(c) z

p									
.5	.25	.10	.05	.025	.01	.005	.0025	.001	.0005
0	0.67448975	1.28155157	1.64485363	1.95996398	2.32634787	2.57582930	2.80703377	3.09023231	3.29052673
0.33333333	0.40915607	0.60706240	0.78425724	0.97357647	1.23531574	1.43914943	1.64657310	1.92492262	2.13792770
0	0.0970966	0.2478934	0.3910327	0.5587089	0.8153747	1.0340770	1.2724563	1.6158742	1.8958323
0	0.1073089	0.2250258	0.3131057	0.4040101	0.5302747	0.6308956	0.7360447	0.8819839	0.9975582
0.0666667	0.1025116	0.2123230	0.3305821	0.4777495	0.7166304	0.9311327	1.1750042	1.5428288	1.8557024
−0.004938	−0.003764	0.001108	0.007685	0.017025	0.033873	0.050157	0.069572	0.100132	0.127007
0	0.008539	0.033737	0.065478	0.108805	0.184807	0.257207	0.343093	0.478317	0.597758
0	0.047595	0.114789	0.176687	0.249610	0.363825	0.464148	0.576788	0.745061	0.887356
0	−0.00711	−0.01611	−0.02343	−0.03114	−0.04168	−0.04971	−0.05761	−0.06765	−0.07475
0.00952	0.00529	−0.00736	−0.01938	−0.03126	−0.04286	−0.04537	−0.03958	−0.01462	0.02094
−0.00529	−0.00447	0	0.00722	0.01859	0.04128	0.06515	0.09556	0.14695	0.19516
0	0	0	0	0	0	0.0178	0.0256	0.0384	0.0502
0	0.00344	0.00833	0.01491	0.02660	0.05478	0.09004	0.14149	0.24158	0.34748
0	0.0109	0.0381	0.0804	0.1534	0.3174	0.5105	0.7799	1.2814	1.7939
0	0	0	0	0	0	0	0	0	0
0	0	0	0	0	0	0	0	0	0

The values in the columns are in the same order as the polynomials in expansion 3c. Sufficient figures have been retained to ensure accuracy in the sixth decimal place for $n_1 > 24$ and $n_2 > 60$.

such as arise in Physics and Astronomy, in which we wish to compare estimates of the value of the same quantity (a) from relatively ample data of low intrinsic accuracy, and (b) from a small series of observations of relatively high precision. When, as often happens, the estimates of precision of the means obtained in these two ways are of the same order of magnitude, the only satisfactory test is that based on Behrens' solution. The asymptotic expansion is particularly suitable for evaluating the percentiles for this special application. There were thus four manifest advantages of the asymptotic approach to Behrens' integral (Fisher 1941):

(i) a check on Sukhatme's values, obtained by a completely independent method, and applicable at least for the higher values of n_1 and n_2,

(ii) greater accuracy than could be obtained for percentiles from Sukhatme's table for values of n_1 and n_2 greater than 12,

(iii) a wider range of levels of significance in the region to which the asymptotic expansion is applicable,

(iv) the theoretical guidance offered by the algebraic form of the leading terms of the expansion.

(b) *The fiducial distribution of the binomial parameter*, p

When discussing the application of the fiducial argument to discontinuous observations, Fisher (1959) found that the mean of the fiducial distribution of p, the parameter of the binomial distribution, for given observational frequencies a, b out of N, was

$$\bar{p} = \frac{a}{N} + \frac{b-a}{2N^2} - \frac{3(b-a)}{2N^3} + \frac{15(b-a)}{2N^4} - \cdots \quad (4)$$

if χ were taken to be normally distributed.

On the other hand, the mean of the Bayesian distribution *a posteriori*, using the Bayesian probability *a priori*

$$\frac{1}{\pi\sqrt{pq}}\,dp,$$

was

$$\frac{a+\frac{1}{2}}{N+1} = \frac{a}{N} + \frac{b-a}{2N^2} - \frac{b-a}{2N^3} + \frac{b-a}{2N^4} - \cdots. \quad (5)$$

Asymptotic agreement between these means appears when allowance is made for the effects of departure from normality in the binomial distribution, which are appreciable in the expression (4).

Direct application of the asymptotic expansion, using the six adjustment terms gave the following expansion for the binomial variate in terms of the normal deviate x

$$a = pN + x\sqrt{Npq} + \tfrac{1}{6}(q-p)(x^2-1) + \frac{1}{72\sqrt{Npq}}\{-x^3 + x - pq(2x^3 - 14x)\} + \cdots$$

which, after inversion, gave*

$$p = \frac{a}{N} - \frac{x\sqrt{ab}}{N^{3/2}} + \frac{(b-a)(2x^2+1)}{6N^{\ 2}} + \left\{\frac{(-2N^2+26ab)x^3+(-7N^2+34ab)x}{72N^{5/2}\sqrt{ab}}\right\} + \cdots \tag{6}$$

whence, substituting its average value for each power of x, the mean of the fiducial distribution is

$$\bar{p} = \frac{a}{N} + \frac{b-a}{2N^2} - \frac{b-a}{2N^3} + \frac{b-a}{2N^4} - \cdots \tag{7}$$

agreeing so far as the fourth term with (5).

The expansion (6) also provides a ready means for comparing the two distributions with respect to other properties. For example, although the means are in agreement, the asymptotic fiducial distribution has the higher variance.

See also the alternative treatment in Fisher (1957).

(c) *Quantitative inheritance*

Panse (1940) has described a statistical technique for the study of quantitative inheritance, in which genetic models, based on data from the F_2 and F_3 generations, are used to represent the constitution of particular characters. The statistical consequences in the population, corresponding to these models were assessed, using the cumulant function of the joint distribution of the F_2 phenotypic value, the mean of the F_3 progeny, and the genotypic variance of F_3 progeny. These functions provided the data for expressing an attribute of the F_3 progeny in terms of the F_2 phenotypic values, and thus the effects of selection in the F_2 phenotype on the mean value of the F_3 progeny could be determined by integration over the F_2 distribution. When the intensity of selection was assigned, the limits of integration were calculable from the asymptotic expansion of the deviate.

V. A Class of Distributions with a Finite Condensation at Zero

The Poisson Series, a discontinuous distribution of positive integers, is well known to have the simple series of cumulants

$$\kappa_r = m,$$

for all values of r. Correspondingly, the cumulative function is

$$K = m(e^{it} - 1),$$

and the characteristic function

$$M = \exp\{m(e^{it} - 1)\}.$$

* For the remaining terms see Fisher (1959)

It is less well known that if, for all values of r,

$$\kappa_r = r!m,$$

or, if a scaling factor be introduced,

$$\kappa_r = r!a^r m,$$

the distribution, derived from

$$K = \frac{mait}{1 - ait}$$

and

$$M = e^{-m}e^{m/(1-ait)}$$

$$= e^{-m}\sum_0^\infty \frac{m^n}{n!}(1 - ait)^{-n}$$

is continuous over the range of positive values, with a finite condensation at zero. For $(1 - ait)^{-n}$ is the characteristic function of the Eulerian distribution

$$\frac{1}{(n-1)!}\left(\frac{X}{a}\right)^{n-1} e^{-X/a}\frac{dX}{a}$$

or of

$$\chi^2 = \frac{2X}{a},$$

for $2n$ degrees of freedom, or of the sum of n random variables, each distributed as

$$e^{-x/a}\frac{dx}{a}, \; x \geq 0, \text{ a positive}$$

Hence the distribution is that of the sum of a number of such variables, when the number is distributed in a Poisson Series of parameter m. The variate is, therefore, never negative, but is zero with finite frequency

$$e^{-m}.$$

Over the range of positive values, the distribution is continuous, and can be expressed as

$$\sqrt{\frac{m}{xa}}\, e^{-m-(x/a)} I_1\left(2\sqrt{\frac{xm}{a}}\right) dx$$

where I_1 is a Bessel function, specified by

$$I_1(u) = \tfrac{1}{2}u + \frac{1}{2^2.4}u^3 + \frac{1}{2^2.4^2.6}u^5 + \cdots$$

The distribution was first recognized (Bennett 1952, Fisher 1954) as characteristic of that of the length of germ plasm still heterogenic at an advanced stage of inbreeding, but its intrinsic incorporation of a finite condensation at zero makes it appropriate to a number of natural phenomena, a good illustration being the rainfall of an arid region. A continuous model for rainfall is unsatisfactory for such regions. For many localities it is preferable to use a model ascribing the total rainfall for a given period, for example, a month or a year, to a number of showers, the number being a random sample from a Poisson series with parameter m, the rainfall of the showers having positive values only, representable by the Eulerian distribution

$$\frac{1}{p!}\, x^p e^{-x}\, dx$$

where p can be small, as in the previous example where it is actually 0, but must be > -1.

The advantage of this type of distribution, relevant for the purposes of the water engineer, is that there is a finite probability, namely e^{-m}, of no rain, whereas a continuous distribution would make this probability zero, contrary to experience.

For the asymptotic expansion, when m is sufficiently large, we may take the exact values for the mean (m) and variance ($2m$) and for the measures of non-normality

$$c = \frac{3}{\sqrt{2m}} \qquad f = \frac{90}{m^2}$$

$$d = \frac{6}{m} \qquad g = \frac{315\sqrt{2}}{m^2\sqrt{m}}$$

$$e = \frac{15\sqrt{2}}{m\sqrt{m}} \qquad h = \frac{2520}{m^3}.$$

These yield the six adjustments, to the normal deviate x

$$\text{I}\ \frac{x^2 - 1}{\sqrt{8m}} \qquad \text{IV}\ -\frac{4x^3 - x}{384m^2}$$

$$\text{II}\ \frac{-x}{8m} \qquad \text{V}\ \frac{3x^4 + 2x^2 - 11}{480m^2\sqrt{2m}}$$

$$\text{III}\ \frac{x^2 - 1}{24m\sqrt{2m}} \qquad \text{VI}\ -\frac{96x^5 + 164x^3 - 767x}{46080m^3}.$$

The coefficients of Table IV give a rather comprehensive tabulation of the distribution, when m is sufficiently large for the accuracy required, and the levels of significance of interest. These large values of m would be troublesome to use in direct evaluation. For sufficiently small values of m, however, the probability that the variate x exceeds any limit X may be evaluated as

$$P = e^{-m}e^{-X} \sum_{i=0}^{\infty} \frac{X^i}{i!} \sum_{j>i} \frac{m^j}{j!}$$

which may be recognized as the probability that a random Poisson variate with parameter m shall exceed a random Poisson variate with parameter X.

Table IV—*Coefficients of powers of* $(2m)^{-1/2}$

p	x	$\frac{x^2-1}{2}$	$-\frac{x}{4}$	$\frac{x^2-1}{12}$	$-\frac{4x^3-x}{96}$	$\frac{3x^4+2x^2-11}{120}$	$-\frac{96x^5+164x^3-767x}{5760}$
.5	0	−0.50000	0	−0.08333	0	−0.09167	0
.25	0.67449	−0.27253	−0.16862	−0.04542	−0.00576	−0.07891	−0.09486
.10	1.28155	0.32119	−0.32039	0.05353	−0.07435	0.00314	0.05311
.05	1.64485	0.85277	−0.41121	0.14213	−0.16829	0.13642	−0.10835
.025	1.95996	1.42073	−0.48999	0.23679	−0.29330	0.34128	0.43543
.01	2.32635	2.20595	−0.58159	0.36766	−0.50035	0.73075	1.18428
.005	2.57583	2.81745	−0.64396	0.46957	−0.68527	1.11946	2.03348
0025	2.80703	3.43972	−0.70176	0.57329	−0.89234	1.59180	3.16056
.001	3.09023	4.27477	−0.77256	0.71246	−1.19741	2.34733	−5.12555
.0005	3.29053	4.91378	−0.82263	0.81896	−1.45024	3.01970	7.00573

References

J. H. Bennett (1952). Junctions in inbreeding. *Genetica 26:* 392–406.

E. A. Cornish and R. A. Fisher (1937). Moments and cumulants in the specification of distributions. *Rev. de l'Inst. int. de stat. 5:* 307–22.

R. A. Fisher (1926). Expansion of "Student's" integral in powers of n^{-1}. *Metron 5:* 109–12.

R. A. Fisher (1941). The asymptotic approach to Behrens' integral, with further tables for the d test of significance. *Ann. eugen. 11:* 141–72.

R. A. Fisher (1954). A fuller theory of "junctions" in inbreeding. *Heredity 8:* 187–97.

R. A. Fisher (1957). The underworld of probability. *Sankhya 18:* 201–10.

R. A. Fisher (1959). "Statistical Methods and Scientific Inference." 2nd ed. Oliver and Boyd, Edin.

H. Goldberg and H. Levine (1946). Approximate formulas for the percentage points and normalization of t and χ^2. *Ann. math. statist. 17:* 216–25.

N. L. Johnson and B. L. Welch (~~1309~~ 1939). Applications of the non-central t-distribution. *Biometrika 31:* 362-89.

V. G. Panse (1940). A statistical study of quantitative inheritance. *Ann. eugen. 10:* 76–105.

P. V. Sukhatme (1938). On Fisher and Behrens' test of significance for the difference in means of two normal samples. *Sankhya 4:* 39–48.

Commonwealth Scientific and Industrial Research Organization, Australia. Division of Mathematical Statistics Technical Paper No. 6, Melbourne, 1960.

THE SIMULTANEOUS FIDUCIAL DISTRIBUTION OF THE PARAMETERS OF A NORMAL BIVARIATE DISTRIBUTION WITH EQUAL VARIANCES

By E. A. Cornish*

[*Manuscript received July* 6, 1960]

Summary

The simultaneous distribution of the four statistical derivates (the two means, the common variance, and the correlation coefficient) in samples from a bivariate normal distribution with equal variances is first obtained and then used to construct the simultaneous fiducial distribution of the four parameters.

I. Introduction

The extension of the notion of fiducial probability to that of the fiducial distribution of a parameter is an important and useful development in inductive inference. Thus, for example, if the fiducial distribution of a parameter θ is known, it may be employed to determine the fiducial distribution of any function, $f(\theta)$, which is uniquely defined by θ.

A natural development is to extend the concept to derive the simultaneous distribution of two or more parameters, but such a distribution does not, in general, follow by simple generalization of the argument used to obtain the distribution of a single parameter. In the particular case of the mean, ξ, and variance, σ^2, of a normal distribution Fisher (1935, 1941) has developed the simultaneous fiducial distribution of these parameters, and has completed the logic of the demonstration (Fisher 1959); by integrating with respect to the parameters individually, the two appropriate marginal fiducial distributions are verified. The simultaneous distribution may also be used to determine the fiducial distribution of any function of the parameters $f(\xi,\sigma)$, in particular, the function $\xi+a\sigma$, which is of considerable importance in practice (Fisher 1931, 1941, 1959). More recently Fisher (1959) has derived, by an extension of the process used for the mean and variance of a single variate, the fiducial distribution of the parameters in a normal bivariate population with means ξ_1 and ξ_2, and variance–covariance matrix

$$\begin{bmatrix} \sigma_1^2 & \rho\sigma_1\sigma_2 \\ \rho\sigma_1\sigma_2 & \sigma_2^2 \end{bmatrix}.$$

The primary object of this paper is to derive the corresponding distribution for the four parameters of a normal bivariate distribution with means ξ_1 and ξ_2, and variance–covariance matrix

$$\sigma^2\begin{bmatrix} 1 & \rho \\ \rho & 1 \end{bmatrix}.$$

For this purpose we require the sampling distribution of the statistics which are to be used as estimates of the parameters, firstly, to demonstrate their sufficiency, and secondly, to

* Division of Mathematical Statistics, C.S.I.R.O., Adelaide.

provide the basic material for constructing the fiducial distribution from its components. The marginal distribution of r, the statistic which estimates the correlation ρ, was given by de Lury (1938), and is implied in the solution due to Pitman (1939) of a problem first proposed by Finney (1938). The sampling distribution we require is most easily established using the artifice introduced by Pitman.

II. The Distribution of the Sample Statistics

Suppose the variates x_1 and x_2 have a normal bivariate distribution with means ξ_1 and ξ_2, respectively and variance–covariance matrix

$$\sigma^2\begin{bmatrix}1 & \rho\\ \rho & 1\end{bmatrix},$$

and that x_{1j}, x_{2j}, $j = 1, 2, \ldots, N$, is a sample of N pairs from this distribution. The probability of obtaining this sample is

$$\{2\pi\sigma^2(1-\rho^2)^{\frac{1}{2}}\}^{-N}\exp\left[-\frac{1}{2\sigma^2(1-\rho^2)}\sum_j\{(x_{1j}-\xi_1)^2-2\rho(x_{1j}-\xi_1)(x_{2j}-\xi_2)+(x_{2j}-\xi_2)^2\}\right] \times \prod_{i,j} dx_{ij}. \quad \ldots.(1)$$

The four statistics defined by the relations:

$$\left.\begin{aligned}\bar{x}_1 &= \frac{1}{N}\Sigma(x_{1j}), \qquad \bar{x}_2 = \frac{1}{N}\Sigma(x_{2j}),\\ s^2 &= \{\Sigma(x_{1j}-\bar{x}_1)^2+\Sigma(x_{2j}-\bar{x}_2)^2\}/2(N-1),\\ r &= 2\Sigma(x_{1j}-\bar{x}_1)(x_{2j}-\bar{x}_2)/\{\Sigma(x_{1j}-\bar{x}_1)^2+\Sigma(x_{2j}-\bar{x}_2)^2\},\end{aligned}\right\} \quad \ldots..(2)$$

are calculated from the sample observations, and their sampling distribution is required.

If two new variates y_1 and y_2, defined by

$$y_1 = x_1+x_2$$

$$y_2 = x_1-x_2$$

are taken, they will be normally and independently distributed with means $\xi_1+\xi_2$ and $\xi_1-\xi_2$, respectively, and variance–covariance matrix

$$2\sigma^2\begin{bmatrix}1+\rho & \cdot\\ \cdot & 1-\rho\end{bmatrix}.$$

From the N values of

$$\left.\begin{aligned}y_{1j} &= x_{1j}+x_{2j},\\ y_{2j} &= x_{1j}-x_{2j},\end{aligned}\right\} \quad \ldots\ldots\ldots\ldots\ldots\ldots\ldots.(3)$$

we then determine five new statistics

$$\left.\begin{aligned} \bar{y}_1 &= \frac{1}{N}\Sigma(y_{1j}), & \bar{y}_2 &= \frac{1}{N}\Sigma(y_{2j}), \\ s_1^2 &= \Sigma(y_{1j}-\bar{y}_1)^2/(N-1), & s_2^2 &= \Sigma(y_{2j}-\bar{y}_2)^2/(N-1), \\ r_{12} &= \Sigma(y_{1j}-\bar{y}_1)(y_{2j}-\bar{y}_2)/\{\Sigma(y_{1j}-\bar{y}_1)^2 . \Sigma(y_{2j}-\bar{y}_2)^2\}^{\frac{1}{2}}, \end{aligned}\right\} \quad \ldots.(4)$$

which are distributed in the well-known form

$$\frac{N}{4\pi\sigma^2(1-\rho^2)^{\frac{1}{2}}}\exp\left\{-\frac{N}{2(1-\rho^2)}\left[\frac{\{\bar{y}_1-(\xi_1+\xi_2)\}^2}{2\sigma^2(1+\rho)}+\frac{\{\bar{y}_2-(\xi_1-\xi_2)\}^2}{2\sigma^2(1-\rho)}\right]\right\}d\bar{y}_1 d\bar{y}_2$$

$$\times \frac{(N-1)^{N-1}}{2^{2(N-2)}\sigma^{2(N-1)}(1-\rho^2)^{\frac{1}{2}(N-1)}\pi^{\frac{1}{2}}\Gamma\frac{1}{2}(N-1) . \Gamma\frac{1}{2}(N-2)} s_1^{N-2} s_2^{N-2}(1-r_{12}^2)^{\frac{1}{2}(N-4)}$$

$$\times \exp\left[-\frac{N-1}{2}\left\{\frac{s_1^2}{2\sigma^2(1+\rho)}+\frac{s_2^2}{2\sigma^2(1-\rho)}\right\}\right]ds_1 ds_2 dr_{12}. \quad \ldots\ldots\ldots\ldots\ldots\ldots(5)$$

Integration for the redundant variable r_{12} yields

$$\frac{\Gamma\frac{1}{2} . \Gamma\frac{1}{2}(N-2)}{\Gamma\frac{1}{2}(N-1)}.$$

Using (**3**) and the facts that

$$\begin{aligned} s_1^2 &= \Sigma(y_{1j}-\bar{y}_1)/(N-1) \\ &= \{\Sigma(x_{1j}-\bar{x}_1)^2+\Sigma(x_{2j}-\bar{x}_2)^2+2\Sigma(x_{1j}-\bar{x}_1)(x_{2j}-\bar{x}_2)\}/(N-1) \\ &= 2s^2(1+r), \\ s_2^2 &= \Sigma(y_{2j}-\bar{y}_2)^2/(N-1) \\ &= \{\Sigma(x_{1j}-\bar{x}_1)^2+\Sigma(x_{2j}-\bar{x}_2)^2-2\Sigma(x_{1j}-\bar{x}_1)(x_{2j}-\bar{x}_2)\}/(N-1) \\ &= 2s^2(1-r), \end{aligned}$$

make the change of variables

$$\begin{aligned} s_1 &= s\sqrt{[2(1+r)]}, & \bar{y}_1 &= \bar{x}_1+\bar{x}_2, \\ s_2 &= s\sqrt{[2(1-r)]}, & \bar{y}_2 &= \bar{x}_1-\bar{x}_2. \end{aligned}$$

The jacobian is

$$\frac{4s}{\sqrt{(1-r^2)}},$$

and the distribution (**5**) becomes

$$\frac{N}{2\pi\sigma^2(1-\rho^2)^{\frac{1}{2}}}\exp\left[-\frac{N}{2\sigma^2(1-\rho^2)}\{(\bar{x}_1-\xi_1)^2-2\rho(\bar{x}_1-\xi_1)(\bar{x}_2-\xi_2)+(\bar{x}_2-\xi_2)^2\}\right]\mathrm{d}\bar{x}_1\mathrm{d}\bar{x}_2$$

$$\times\frac{(N-1)^{N-1}}{2^{N-3}(1-\rho^2)^{\frac{1}{2}(N-1)}\{\Gamma\frac{1}{2}(N-1)\}^2}\left(\frac{s}{\sigma}\right)^{2N-3}(1-r^2)^{\frac{1}{2}(N-3)}\exp\left[-\left\{\frac{(N-1)s^2}{\sigma^2(1-\rho^2)}(1-\rho r)\right\}\right]\frac{\mathrm{d}s}{\sigma}\mathrm{d}r, \qquad \text{..........(6)}$$

showing that $\bar{x}_1$ and $\bar{x}_2$ are distributed in a bivariate normal distribution, with means ξ_1 and ξ_2, and variance–covariance matrix

$$\frac{\sigma^2}{N}\begin{bmatrix}1 & \rho\\ \rho & 1\end{bmatrix},$$

independently of the distribution of s and r.

It is clear from the form of (**6**) that the likelihood function of the parameters obtainable from the observed sample, is expressible as the product

$$L(\bar{x}_1\bar{x}_2|\xi_1\xi_2\sigma^2\rho).L(rs|\sigma^2\rho).L(\text{independent of the parameters}),$$

thus making the maximal likelihood estimates $\bar{x}_1$, $\bar{x}_2$, s, and r jointly sufficient, from which it follows that we may proceed to find the simultaneous fiducial distribution of the parameters.

III. The Fiducial Distribution of the Parameters

Taking the second factor of (**6**), involving s and r, and integrating for s from 0 to ∞, leaves the marginal distribution of r in the form

$$\frac{\Gamma\frac{1}{2}N.(1-\rho^2)^{\frac{1}{2}(N-1)}}{\pi^{\frac{1}{2}}.\Gamma\frac{1}{2}(N-1)}\frac{(1-r^2)^{\frac{1}{2}(N-3)}}{(1-\rho r)^{N-1}}\mathrm{d}r, \qquad \text{............(7)}$$

as given by de Lury (1938).

This distribution is dependent only on ρ and forms the starting point in the development of the simultaneous fiducial distribution of the parameters. The probability that $r \leqslant r_1$ is

$$F(r_1\rho)=\frac{\Gamma\frac{1}{2}N.(1-\rho^2)^{\frac{1}{2}(N-1)}}{\Gamma\frac{1}{2}.\Gamma\frac{1}{2}(N-1)}\int_{-1}^{r_1}\frac{(1-r^2)^{\frac{1}{2}(N-3)}}{(1-\rho r)^{N-1}}dr,$$

and differentiating with respect to ρ,

$$-\frac{\partial}{\partial\rho}F(r_1\rho)=\frac{-\Gamma\frac{1}{2}N}{\Gamma\frac{1}{2}.\Gamma\frac{1}{2}(N-1)}\frac{\partial}{\partial\rho}\int_{-1}^{r_1}\frac{(1-\rho^2)^{\frac{1}{2}(N-1)}(1-r^2)^{\frac{1}{2}(N-3)}}{(1-\rho r)^{N-1}}dr$$

$$=\frac{\Gamma\frac{1}{2}N.(1-\rho^2)^{\frac{1}{2}(N-3)}}{\Gamma\frac{1}{2}.\Gamma\frac{1}{2}(N-1)}(N-1)\int_{-1}^{r_1}\frac{(\rho-r)(1-r^2)^{\frac{1}{2}(N-3)}}{(1-\rho r)^{N}}dr$$

$$=\frac{\Gamma\frac{1}{2}N.(1-\rho^2)^{\frac{1}{2}(N-3)}}{\Gamma\frac{1}{2}.\Gamma\frac{1}{2}(N-1)}\frac{(1-r_1^2)^{\frac{1}{2}(N-1)}}{(1-\rho r_1)^{N-1}},$$

so that the marginal distribution of ρ, for a given value of r, is

$$\frac{\Gamma\frac{1}{2}N\,.\,(1-r^2)^{\frac{1}{2}(N-1)}}{\Gamma\frac{1}{2}\,.\,\Gamma\frac{1}{2}(N-1)}\;\frac{(1-\rho^2)^{\frac{1}{2}(N-3)}}{(1-\rho r)^{N-1}}\,d\rho. \qquad \ldots\ldots\ldots\ldots(8)$$

For any given values of r and ρ, the conditional distribution of s is, from (**7**) and the second factor of (**6**),

$$\frac{2}{\Gamma(N-1)}\left(\frac{N-1}{1-\rho^2}\right)^{N-1}(1-\rho r)^{N-1}\left(\frac{s}{\sigma}\right)^{2N-3}\exp\left[-\left\{\frac{(N-1)s^2}{\sigma^2(1-\rho^2)}(1-\rho r)\right\}\right]\frac{ds}{\sigma},$$

and the probability that $s \leqslant s_1$, for given values of r and ρ, is

$$G(s_1\sigma) = \frac{2}{\Gamma(N-1)}\left(\frac{N-1}{1-\rho^2}\right)^{N-1}(1-\rho r)^{N-1}\int_0^{s_1}\left(\frac{s}{\sigma}\right)^{2N-3}\exp\left[-\left\{\frac{(N-1)s^2}{\sigma^2(1-\rho^2)}(1-\rho r)\right\}\right]\frac{ds}{\sigma}.$$

Differentiating with respect to σ,

$$-\frac{\partial}{\partial\sigma}G(s_1\sigma) = \frac{-2}{\Gamma(N-1)}\left(\frac{N-1}{1-\rho^2}\right)^{N-1}(1-\rho r)^{N-1}$$

$$\times\frac{\partial}{\partial\sigma}\int_0^{s_1}\left(\frac{s}{\sigma}\right)^{2N-3}\exp\left[-\left\{\frac{(N-1)s^2}{\sigma^2(1-\rho^2)}(1-\rho r)\right\}\right]\frac{ds}{\sigma}$$

$$= \frac{2}{\Gamma(N-1)}\left(\frac{N-1}{1-\rho^2}\right)^{N-1}(1-\rho r)^{N-1}\int_0^{s_1}\frac{2(N-1)s^{2N-3}}{\sigma^{2N-1}}$$

$$\times\left\{1-\frac{s^2(1-\rho r)}{\sigma^2(1-\rho^2)}\right\}\exp\left[-\left\{\frac{(N-1)s^2}{\sigma^2(1-\rho^2)}(1-\rho r)\right\}\right]ds$$

$$= \frac{2}{\Gamma(N-1)}\left(\frac{N-1}{1-\rho^2}\right)^{N-1}\frac{(1-\rho r)^{N-1}}{\sigma}\left(\frac{s_1}{\sigma}\right)^{2(N-1)}\exp\left[-\left\{\frac{(N-1)s_1^2}{\sigma^2(1-\rho^2)}(1-\rho r)\right\}\right],$$

so that the fiducial distribution of σ for given s, conditional on the values of ρ and r, is

$$\frac{2}{\Gamma(N-1)}\left(\frac{N-1}{1-\rho^2}\right)^{N-1}(1-\rho r)^{N-1}\left(\frac{s}{\sigma}\right)^{2(N-1)}\exp\left[-\left\{\frac{(N-1)s^2}{\sigma^2(1-\rho^2)}(1-\rho r)\right\}\right]\frac{d\sigma}{\sigma}.$$

$$\ldots\ldots\ldots\ldots(9)$$

The simultaneous fiducial distribution of ρ and σ is then obtained by mutiplying the conditional distribution (**9**) by the marginal distribution (**8**) yielding

$$\frac{(N-1)^{N-1}}{2^{N-3}\{\Gamma\frac{1}{2}(N-1)\}^2(1-\rho^2)^{\frac{1}{2}(N+1)}}\left(\frac{s}{\sigma}\right)^{2(N-1)}(1-r^2)^{\frac{1}{2}(N-1)}\exp\left[-\left\{\frac{(N-1)s^2}{\sigma^2(1-\rho^2)}(1-\rho r)\right\}\right]\frac{d\sigma}{\sigma}\,d\rho.$$

$$\ldots\ldots\ldots\ldots(10)$$

As $\bar{x}_1$ and $\bar{x}_2$ are distributed independently of s and r, the fiducial distribution of the means, for known values of σ and ρ, is

$$\frac{N}{2\pi\sigma^2(1-\rho^2)^{\frac{1}{2}}}\exp\left[-\frac{N}{2\sigma^2(1-\rho^2)}\left\{(\xi_1-\bar{x}_1)^2-2\rho(\xi_1-\bar{x}_1)(\xi_2-\bar{x}_2)+(\xi_2-\bar{x}_2)^2\right\}\right]d\xi_1\,d\xi_2, \quad \ldots\ldots\ldots\ldots(\mathbf{11})$$

and the simultaneous distribution of the four parameters is then the product of the fiducial distributions (**10**) and (**11**).

IV. References

Finney, D. J. (1938).—*Biometrika* **30**: 190–2.

Fisher, R. A. (1931).—"British Association Mathematical Tables." Vol. 1. pp. XXVI–XXXV. (British Association: London.)

Fisher, R. A. (1935).—*Ann. Eugen. Lond.* **6**: 391–8.

Fisher, R. A. (1941).—*Ann. Eugen. Lond.* **11**: 141–72.

Fisher, R. A. (1959).—"Statistical Methods and Scientific Inference." 2nd Ed. (Oliver & Boyd: Edinburgh.)

de Lury, D. B. (1938).—*Ann. Math. Statist.* **9**: 149–51.

Pitman, E. J. G. (1939).—*Biometrika* **31**: 9–12.

Commonwealth Scientific and Industrial Research Organization, Australia. Division of Mathematical Statistics Technical Paper No. 8, Melbourne, 1961.

THE SIMULTANEOUS FIDUCIAL DISTRIBUTION OF THE LOCATION PARAMETERS IN A MULTIVARIATE NORMAL DISTRIBUTION

By E. A. Cornish*

[*Manuscript received November* 8, 1960]

Summary

The precise form of the simultaneous fiducial distribution of the location parameters in a multivariate normal distribution, first conjectured by Fisher, is established by direct proof. The application of this distribution in the practical testing of compound hypotheses respecting the values of the parameters is illustrated by examples, and attention is directed to the inappropriateness of Hotelling's distribution for this purpose.

I. Introduction

In a series of papers, dating from 1930, Fisher has strongly advocated the fiducial argument as a tool of scientific inference. For the class of cases to which it is appropriate, this type of argument leads to rigorous probability statements about the parameters of the population, from which the observational record constitutes a random sample, without the assumption of any knowledge whatever, respecting their probability distributions *a priori*.

We may, for example, consider a set of N observations $x_1, x_2, \ldots, x_N$ taken from a normal distribution having a mean ξ and variance σ^2, and calculate

$$\bar{x} = \frac{1}{N}\Sigma(x),$$

$$s^2 = \frac{1}{N(N-1)}\Sigma(x-\bar{x})^2,$$

then Student's statistic t, defined by

$$t = (\bar{x}-\xi)/s,$$

is distributed in the well-known form

$$\frac{\Gamma\frac{1}{2}(n+1)}{(\pi n)^{\frac{1}{2}}\Gamma\frac{1}{2}n}\left(1+\frac{t^2}{n}\right)^{-\frac{1}{2}(n+1)}dt, \tag{1}$$

which depends only on the parameter $n = N-1$. The distribution (1), and the relation

$$\xi = \bar{x}-st$$

together establish the fiducial distribution of ξ, based solely on the observed sample, in the form

$$\frac{\Gamma\frac{1}{2}(n+1)}{(\pi n)^{\frac{1}{2}}\Gamma\frac{1}{2}n}\left\{1+\frac{(\xi-\bar{x})^2}{ns^2}\right\}^{-\frac{1}{2}(n+1)}\frac{d\xi}{s}. \tag{2}$$

* Division of Mathematical Statistics, C.S.I.R.O., Adelaide.

Fiducial probabilities which may be required in practice are obtainable by integrating (2) over appropriate ranges of ξ, but clearly, are already available from tables of the probability integral of (1).

The distribution (2) has recently been generalized (Cornish 1960) to give the simultaneous fiducial distribution of any number of parameters, based on the joint distribution of a set of t-statistics, which are not only dependent through a common estimate of variance, but also correlated because of the nature of the enquiry under consideration. Thus if $\mathbf{x}$ is a column vector of normally and independently distributed variates $x_1, x_2, \ldots, x_N$ with means $\xi_1, \xi_2, \ldots, \xi_N$ and common variance σ^2, and

$$\mathbf{y} = \mathbf{H}\mathbf{x}$$

are any $p<N$, linearly independent, linear functions of the x_i, then the y_i are distributed in a multivariate normal distribution with a variance-covariance matrix $\sigma^2\mathbf{H}\mathbf{H}'$ and a vector of means $\boldsymbol{\eta} = \mathbf{H}\boldsymbol{\xi}$.

If, now, s^2 is an estimate of σ^2, on n degrees of freedom, distributed independently of the x_i and therefore independently of the y_i, the simultaneous fiducial distribution of the parameters η_i is

$$\frac{\Gamma\frac{1}{2}(n+p)}{(\pi n)^{\frac{1}{2}p}\Gamma\frac{1}{2}n}\left\{1+\frac{(\boldsymbol{\eta}-\mathbf{y})'(s^2\mathbf{H}\mathbf{H}')^{-1}(\boldsymbol{\eta}-\mathbf{y})}{n}\right\}^{-\frac{1}{2}(n+p)}\frac{d\boldsymbol{\eta}}{|s^2\mathbf{H}\mathbf{H}'|^{\frac{1}{2}}}, \tag{3}$$

from which any fiducial probability required may be obtained by integrating over appropriate ranges of the η_i.

Fisher (1954) has conjectured a rather more general distribution than (3). Thus, if $z_1, z_2, \ldots, z_p$ are distributed in a multivariate normal distribution with means $\zeta_1, \zeta_2, \ldots, \zeta_p$ and variance-covariance matrix $\boldsymbol{\Sigma}$, of which there is a non-singular estimate $\mathbf{S}$, based on n degrees of freedom and distributed independently of the z_i, Fisher's generalization takes the form

$$\frac{\Gamma\frac{1}{2}(n+p)}{(\pi n)^{\frac{1}{2}p}\Gamma\frac{1}{2}n}\left\{1+\frac{(\boldsymbol{\zeta}-\mathbf{z})'\mathbf{S}^{-1}(\boldsymbol{\zeta}-\mathbf{z})}{n}\right\}^{-\frac{1}{2}(n+p)}\frac{d\boldsymbol{\zeta}}{|\mathbf{S}|^{\frac{1}{2}}}. \tag{4}$$

In a personal communication Fisher has indicated that G. A. Barnard has probably verified this result using the properties of the characteristic function. Our objects are to establish the distribution (4) by alternate and perhaps more direct means, and to illustrate its use in practice.

II. The Multivariate Fiducial Distribution

Suppose $\mathbf{x}$ is a column vector whose elements $x_1, x_2, \ldots, x_p$ are distributed in a multivariate normal distribution with a vector of means $\boldsymbol{\xi}' = [\xi_1\ \xi_2 \ldots \xi_p]$ and variance-covariance matrix $\boldsymbol{\Sigma}$. We assume that there is available a non-singular estimate $\mathbf{S} = [s_{ij}]$, $i, j = 1, 2, \ldots, p$, of $\boldsymbol{\Sigma}$ on n degrees of freedom, distributed independently of the x_i.

To establish the distribution we shall use the well-known result that if $\boldsymbol{\lambda}$ is a non-null column vector with p *arbitrary*, real, non-stochastic elements, the observable variate $\boldsymbol{\lambda}'\mathbf{x}$ is normally distributed with a mean $\boldsymbol{\lambda}'\boldsymbol{\xi}$ and variance $\boldsymbol{\lambda}'\boldsymbol{\Sigma}\boldsymbol{\lambda}$ and consequently

$$\frac{\boldsymbol{\lambda}'(\boldsymbol{\xi}-\mathbf{x})}{(\boldsymbol{\lambda}'\mathbf{S}\boldsymbol{\lambda})^{\frac{1}{2}}} = t$$

is distributed in Student's distribution with n degrees of freedom, independently of $\boldsymbol{\lambda}'\boldsymbol{\xi}$ and $\boldsymbol{\lambda}'\boldsymbol{\Sigma}\boldsymbol{\lambda}$. As this distribution is independent of the unknown parameters, values of t_P can be computed such that, for any random sample of any normal distribution

$$\Pr(t \leqslant t_P) = P$$

for all values of P from 0 to 1. After substituting for t in this expression, we may make the following statement, namely,

$$\Pr\{\boldsymbol{\lambda}'\boldsymbol{\xi} \leqslant \boldsymbol{\lambda}'\mathbf{x}+(\boldsymbol{\lambda}'\mathbf{S}\boldsymbol{\lambda})^{\frac{1}{2}}t_P\} = P,$$

in which the right-hand side of the inequality contains only known quantities. This is a definite and objective statement respecting the values of the parameters ξ_i, which supplies information that may be used to determine the form of their fiducial distribution, namely, that the total frequency in the simultaneous distribution, lying on one side of the hyperplane

$$\boldsymbol{\lambda}'\boldsymbol{\xi} = \boldsymbol{\lambda}'\mathbf{x}+(\boldsymbol{\lambda}'\mathbf{S}\boldsymbol{\lambda})^{\frac{1}{2}}t_P,$$

is P, corresponding to the value t_P, or the probability is P that the point with coordinates $\xi_1, \xi_2, \ldots, \xi_p$ lies in the specified region of the parameter space. In Figure 1, which illustrates geometrically the case $p = 2$, the probability is P that a point, with coordinates ξ_1, ξ_2, lies in the shaded region bounded by the line

$$\lambda_1\xi_1+\lambda_2\xi_2 = \lambda_1x_1+\lambda_2x_2+(\lambda_1^2s_{11}+2\lambda_1\lambda_2s_{12}+\lambda_2^2s_{22})^{\frac{1}{2}}t_P,$$

where s_{11}, s_{12}, and s_{22} are elements of the matrix $\mathbf{S}$ of order 2×2.

We assume that the fiducial distribution is non-singular (*vide*, for example, Cramér 1946, section 22.5), and no restriction is introduced by assuming also that the frequency density takes the form

$$f[1+\phi\{(\xi_1-x_1), (\xi_2-x_2), \ldots, (\xi_p-x_p)\}].$$

As the variables of the fiducial distribution are statistically dependent through the common variance–covariance matrix $\mathbf{S}$, the function f is not expressible as a product such as

$$\prod_i f_i(\xi_i-x_i)$$

having p factors.

Using the established fact above, the integral

$$\int \ldots \int f\,[1+\phi\{(\xi_1-x_1), (\xi_2-x_2), \ldots, (\xi_p-x_p)\}]\mathrm{d}\xi_1\mathrm{d}\xi_2 \ldots \mathrm{d}\xi_p, \tag{5}$$

when taken over the domain defined by

$$\frac{\boldsymbol{\lambda}'(\boldsymbol{\xi}-\mathbf{x})}{(\boldsymbol{\lambda}'\mathbf{S}\boldsymbol{\lambda})^{\frac{1}{2}}} = t_1 \leqslant t_P,$$

is

$$\frac{\Gamma\frac{1}{2}(n+1)}{(\pi n)^{\frac{1}{2}}\Gamma\frac{1}{2}n}\int_{-\infty}^{t_P}\left(1+\frac{t_1^2}{n}\right)^{-\frac{1}{2}(n+1)} \mathrm{d}t_1. \tag{6}$$

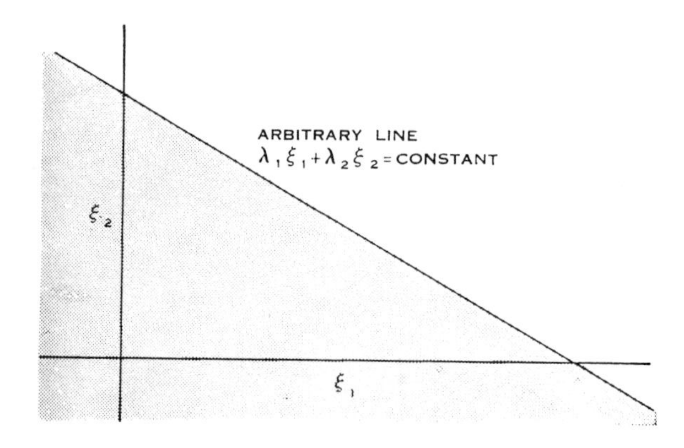

Fig. 1.—Illustrating the subdivision of a two-dimensional parameter space by an arbitrary line.

The function f must involve the elements of $\mathbf{S}$ as parameters and to evaluate the integral (5) we may, since $\mathbf{S}$ is positive definite and symmetric and the vector $\boldsymbol{\lambda}$ is arbitrary, make the real, non-singular, linear transformation

$$\mathbf{t} = \mathbf{H}(\boldsymbol{\xi}-\mathbf{x}),$$

with $\mathbf{H}$ chosen so that

$$\mathbf{HSH}' = \mathbf{I},$$

at the same time making one of the elements of $\mathbf{t}$, t_1 say, a Student t-statistic. For example, with $p = 3$ we may use

$$t_1 = \frac{1}{s_{11}^{\frac{1}{2}}}(\xi_1-x_1),$$

$$t_2 = \frac{-s_{12}/s_{11}}{\{(s_{11}s_{22}-s_{12}^2)/s_{11}\}^{\frac{1}{2}}}(\xi_1-x_1) + \frac{1}{\{(s_{11}s_{22}-s_{12}^2)/s_{11}\}^{\frac{1}{2}}}(\xi_2-x_2),$$

$$t_3 = \frac{-s_{13}/s_{11}}{\{|\mathbf{S}|/(s_{11}s_{22}-s_{12}^2)\}^{\frac{1}{2}}}(\xi_1-x_1)$$

$$-\frac{\{(s_{11}s_{23}-s_{12}s_{13})/s_{11}\}/\{(s_{11}s_{22}-s_{12}^2)/s_{11}\}}{\{|\mathbf{S}|/(s_{11}s_{22}-s_{12}^2)\}^{\frac{1}{2}}}(\xi_2-x_2)$$

$$+\frac{1}{\{|\mathbf{S}|/(s_{11}s_{22}-s_{12}^2)\}^{\frac{1}{2}}}(\xi_3-x_3).$$

As $\mathbf{H}$ is non-singular,

$$(\boldsymbol{\xi}-\mathbf{x}) = \mathbf{H}^{-1}\mathbf{t}, \tag{7}$$

of which the jacobian is

$$|\mathbf{H}^{-1}| = |\mathbf{S}|^{\frac{1}{2}},$$

so that the integral (5) becomes

$$\int \ldots \int f\,[1+\phi\{(t_1, t_2, \ldots, t_p)\}]|\mathbf{S}|^{\frac{1}{2}}\,\mathrm{d}t_1\mathrm{d}t_2\ldots\mathrm{d}t_p \tag{8}$$

over the region defined by

$$t_1 \leqslant t_P.$$

We now have to trace, in reverse order, the steps by which the result (6) was obtained from the integral (8).

Consider the Weyl fractional integral of order $\frac{1}{2}$,

$$\frac{1}{\Gamma\frac{1}{2}}\int_y^\infty f(x)(x-y)^{-\frac{1}{2}}\,\mathrm{d}x,$$

and the relation

$$\frac{1}{\Gamma\frac{1}{2}}\int_y^\infty f(x)(x-y)^{-\frac{1}{2}}\,\mathrm{d}x = h(y), \tag{9}$$

(Erdélyi *et al.* 1954). If a function $f(x)$ exists such that (9) holds, when $h(y)$ is given, then $f(x)$ is *uniquely* determined (Kober 1940).* In order to use this fact we first convert the integrand of (6) to a form more convenient for our purposes. Thus we have

$$\frac{\Gamma\frac{1}{2}(n+1)}{(\pi n)^{\frac{1}{2}}\Gamma\frac{1}{2}n}\left(1+\frac{t_1^2}{n}\right)^{-\frac{1}{2}(n+1)} = \frac{n^{\frac{1}{2}(n+2)}}{2\Gamma\frac{1}{2}}\left\{\frac{\Gamma\frac{1}{2}(n+1)}{\Gamma\frac{1}{2}(n+2)}(n+t_1^2)^{-\frac{1}{2}(n+1)}\right\},$$

and the factor within the brackets plays the role of the known function $h(y)$. This was obtained by integrating some function of t_1 and t_2 with respect to t_2 over the range $-\infty$ to ∞. We now assume that the integrand was the function $f(t_1^2+t_2^2)$ and consequently

$$\frac{1}{\Gamma\frac{1}{2}}\int_{-\infty}^{\infty} f(t_1^2+t_2^2)\mathrm{d}t_2 = \frac{2}{\Gamma\frac{1}{2}}\int_0^\infty f(t_1^2+t_2^2)\mathrm{d}t_2 = \frac{1}{\Gamma\frac{1}{2}}\int_y^\infty f(x)(x-y)^{-\frac{1}{2}}\,\mathrm{d}x$$

$$= \frac{\Gamma\frac{1}{2}(n+1)}{\Gamma\frac{1}{2}(n+2)}(n+y)^{-\frac{1}{2}(n+1)},$$

* I am indebted to Professor E. S. Barnes for this reference.

with $y = t_1^2$ and $x = t_1^2+t_2^2$. The function of x which satisfies this relation is

$$f(x) = (n+x)^{-\frac{1}{2}(n+2)};$$

hence we have, after inserting the constant multiplier,

$$\frac{n^{\frac{1}{2}(n+2)}}{2(\Gamma\frac{1}{2})^2}\int_y^\infty \frac{(x-y)^{-\frac{1}{2}}}{(n+x)^{\frac{1}{2}(n+2)}}\,dx = \frac{1}{(\Gamma\frac{1}{2})^2}\int_0^\infty \frac{dt_2}{\{1+(t_1^2+t_2^2)/n\}^{\frac{1}{2}(n+2)}},$$

and, in virtue of the property of uniqueness, the function

$$\left(1+\frac{t_1^2+t_2^2}{n}\right)^{-\frac{1}{2}(n+2)}$$

is the only one which could have given the known result.

We now take the function

$$\frac{\Gamma\frac{1}{2}(n+2)}{\Gamma\frac{1}{2}(n+3)}\left(1+\frac{t_1^2+t_2^2}{n}\right)^{-\frac{1}{2}(n+2)}$$

and again apply the Weyl transform. Proceeding in this way, we find after $p-1$ steps that the integral must have been

$$\frac{1}{(\Gamma\frac{1}{2})^p}\,\frac{2^{p-2}}{n^{\frac{1}{2}(p-2)}}\,\frac{\Gamma\frac{1}{2}(n+p)}{\Gamma\frac{1}{2}(n+2)}\int_0^\infty dt_2\int_0^\infty dt_3 \ldots \int_0^\infty dt_{p-1}\int_0^\infty \frac{dt_p}{(1+\mathbf{t}'\mathbf{t}/n)^{\frac{1}{2}(n+p)}}$$

$$= \frac{\Gamma\frac{1}{2}(n+p)}{(\pi n)^{\frac{1}{2}p}\Gamma\frac{1}{2}n}\int_{-\infty}^\infty \ldots \int_{-\infty}^\infty \left(1+\frac{\mathbf{t}'\mathbf{t}}{n}\right)^{-\frac{1}{2}(n+p)} dt_2\, dt_3 \ldots dt_p. \tag{10}$$

The real function

$$\phi\{(\xi_1-x_1), (\xi_2-x_2), \ldots, (\xi_p-x_p)\}$$

is thus uniquely transformed to the real positive definite, quadratic form

$$\frac{1}{n}\mathbf{t}'\mathbf{t}$$

of rank p, by the real, non-singular, linear transformation (7), and consequently ϕ must be the real, positive definite, quadratic form

$$\frac{1}{n}(\boldsymbol{\xi}-\mathbf{x})'\mathbf{H}'\mathbf{H}(\boldsymbol{\xi}-\mathbf{x}) = \frac{1}{n}(\boldsymbol{\xi}-\mathbf{x})'\mathbf{S}^{-1}(\boldsymbol{\xi}-\mathbf{x})$$

of rank p.

The simultaneous fiducial distribution of the parameters is thus

$$\frac{\Gamma\frac{1}{2}(n+p)}{(\pi n)^{\frac{1}{2}p}\Gamma\frac{1}{2}n}\left\{1+\frac{(\boldsymbol{\xi}-\mathbf{x})'\mathbf{S}^{-1}(\boldsymbol{\xi}-\mathbf{x})}{n}\right\}^{-\frac{1}{2}(n+p)}\frac{d\boldsymbol{\xi}}{|\mathbf{S}|^{\frac{1}{2}}}, \tag{11}$$

from which any fiducial probability required may be obtained by integrating over appropriate ranges of the ξ_i.

The marginal fiducial distribution of ξ_i derived from (11) takes the form

$$\frac{\Gamma\frac{1}{2}(n+1)}{(\pi n)^{\frac{1}{2}}\Gamma\frac{1}{2}n}\int_{-\infty}^{x_i+s_{ii}^{\frac{1}{2}}t_P}\left\{1+\frac{(\xi_i-x_i)^2}{ns_{ii}}\right\}^{-\frac{1}{2}(n+1)}\frac{d\xi_i}{s_{ii}^{\frac{1}{2}}},$$

in conformity with (2).

As with the distribution previously discussed (Cornish 1960), the form of the fiducial distribution (11) is particularly appropriate for testing compound hypotheses respecting the elements of the vector of means $\boldsymbol{\xi}$. For this type of problem in which we wish to examine the possible variations of any hypothesis which specifies all parameters simultaneously, while maintaining an assigned level of significance, a single measure of the simultaneous discrepancies between the several observed quantities and their hypothetical values, is required. The appropriate measure is Hotelling's (1931) generalization of Student's t-statistic

$$T^2=(\boldsymbol{\xi}-\mathbf{x})'\,\mathbf{S}^{-1}(\boldsymbol{\xi}-\mathbf{x}),$$

in which $\mathbf{S}$ is the observed variance–covariance matrix of the sample values x_i.

The required fiducial probability, corresponding to a given value, T_1^2, is obtained by integrating (11) over the domain defined by the inequality

$$\frac{1}{p}(\boldsymbol{\xi}-\mathbf{x})'\mathbf{S}^{-1}(\boldsymbol{\xi}-\mathbf{x})\leqslant T_1^2. \tag{12}$$

The transformation (7) reduces this integral to

$$\frac{\Gamma\frac{1}{2}(n+p)}{(\pi n)^{\frac{1}{2}p}\Gamma\frac{1}{2}n}\int\dots\int\left(1+\frac{\mathbf{t}'\mathbf{t}}{n}\right)^{-\frac{1}{2}(n+p)}d\mathbf{t}, \tag{13}$$

taken over the domain defined by

$$\frac{1}{p}\,\mathbf{t}'\mathbf{t}\leqslant T_1^2,$$

and a spherical polar transformation in p dimensions reduces (13) to

$$\frac{\Gamma\frac{1}{2}(n+p)}{n^{\frac{1}{2}p}\Gamma\frac{1}{2}p\Gamma\frac{1}{2}n}\int(\chi^2)^{\frac{1}{2}(p-2)}\left(1+\frac{\chi^2}{n}\right)^{-\frac{1}{2}(n+p)}d(\chi^2),$$

taken over the region defined by

$$\frac{1}{p}\chi^2\leqslant T_1^2,$$

which is equivalent to the distribution of e^{2z} with $n_1=p$ and $n_2=n$. Fiducial probabilities associated with regions of the type defined by (12) are thus obtainable directly from the distribution of e^{2z} with parameters $n_1=p$ and $n_2=n$.

A comprehensive test which enables us to specify the aggregate of compound hypotheses contradicted by experimental or observational data at any assigned level of significance, is then obtained by computing the hyper-ellipsoid

$$(\boldsymbol{\xi}-\mathbf{x})'\,\mathbf{S}^{-1}(\boldsymbol{\xi}-\mathbf{x}) = pe^{2z}, \tag{14}$$

in which the value of e^{2z} corresponds to the chosen level of significance, P, when $n_1 = p$ and $n_2 = n$. The P% test of significance for a hypothesis specifying simultaneously the values of $\xi_1, \xi_2, \ldots, \xi_p$ will thus reject any hypothesis for which the value of the above quadratic expression exceeds the chosen value of pe^{2z} and will accept any hypothesis for which it is less than pe^{2z}. All acceptable values of the parameters constitute the interior and boundary of the above ellipsoid.

III. Illustrative Examples

(*a*) In a recent climatological investigation, the author had occasion to compute the regression of mean monthly rainfall on the altitude, latitude, and longitude of certain observing stations in South Australia, the original data comprising the monthly rainfall at each of 50 stations for the years 1888–1955 inclusive.

The following are the regression coefficients for the mean monthly rainfall of July on altitude, latitude, and longitude respectively,

$$\begin{aligned} b_1 &= 0\cdot237, \\ b_2 &= 0\cdot133, \\ b_3 &= -0\cdot253, \end{aligned}$$

the units being inches for rainfall, 10^2 feet for altitude, and 10^{-1} degree for both latitude and longitude, and the inverse of their variance–covariance matrix is

$$\frac{1}{0\cdot2042}\begin{bmatrix} 1138\cdot265 & -161\cdot151 & 215\cdot305 \\ -161\cdot151 & 534\cdot297 & -125\cdot495 \\ 215\cdot305 & -125\cdot495 & 199\cdot183 \end{bmatrix}$$

in which the estimated residual variance in the analysis of variance is $0\cdot2042$ on 46 degrees of freedom.

The 1% test of significance for a hypothesis specifying simultaneously the values of β_1, β_2, and β_3 will accept any hypothesis for which

$$\begin{aligned} &1138\cdot265(\beta_1-0\cdot237)^2+534\cdot297(\beta_2-0\cdot133)^2+199\cdot183(\beta_3+0\cdot253)^2 \\ &-2\times161\cdot151(\beta_1-0\cdot237)(\beta_2-0\cdot133)+2\times215\cdot305(\beta_1-0\cdot237)(\beta_3+0\cdot253) \\ &-2\times125\cdot495(\beta_2-0\cdot133)(\beta_3+0\cdot253)\leqslant 3\times0\cdot2042\times4\cdot24 = 2\cdot596, \end{aligned} \tag{15}$$

where $4\cdot24$ is the value of e^{2z} at the 1% point when $n_1 = 3$ and $n_2 = 46$. At this fiducial probability, all acceptable values of the parameters fall within, or on, the ellipsoid defined by (15) and centred at the point $(0\cdot237,\ 0\cdot133,\ -0\cdot253)$.

(*b*) The following data have been extracted from a comprehensive series of observations designed to determine the association of mean sea-level pressure patterns and temperature over the Australian continent south of 20°S. latitude

(L. G. Veitch, unpublished data). The mean pressures at stations 25 (lat. 25°S., long. 125°E.), 30 (lat. 30°S., long. 130°E.), 35 (lat. 35°S., long. 135°E.) and 39 (lat. 40°S., long. 140°E.) for the period January 22 to April 21 in the years 1950–1959 inclusive were, respectively, 1011·5, 1013·1, 1014·3, and 1011·7, the unit being millibars. The inverse of the variance–covariance matrix of the means is

$$\begin{bmatrix} 25\cdot044 & -24\cdot700 & 8\cdot552 & -1\cdot450 \\ -24\cdot700 & 42\cdot938 & -25\cdot991 & 9\cdot489 \\ 8\cdot552 & -25\cdot991 & 32\cdot041 & -17\cdot677 \\ -1\cdot450 & 9\cdot489 & -17\cdot677 & 12\cdot913 \end{bmatrix}$$

on 179 degrees of freedom, and the 1% test of significance for a hypothesis specifying simultaneously the true mean pressures will accept any hypothesis for which

$$\begin{aligned} &25\cdot044(\xi_1-1011\cdot5)^2+42\cdot938(\xi_2-1013\cdot1)^2+32\cdot041(\xi_3-1014\cdot3)^2 \\ &\quad+12\cdot913(\xi_4-1011\cdot7)^2-2\times24\cdot700(\xi_1-1011\cdot5)(\xi_2-1013\cdot1) \\ &\quad+2\times8\cdot552(\xi_1-1011\cdot5)(\xi_3-1014\cdot3)-2\times1\cdot450(\xi_1-1011\cdot5)(\xi_4-1011\cdot7) \\ &\quad-2\times25\cdot991(\xi_2-1013\cdot1)(\xi_3-1014\cdot3)+2\times9\cdot489(\xi_2-1013\cdot1)(\xi_4-1011\cdot7) \\ &\quad-2\times17\cdot677(\xi_3-1014\cdot3)(\xi_4-1011\cdot7)\leqslant4\times3\cdot427=13\cdot708, \end{aligned}$$

where 3·427 is the value of e^{2z} at the 1% point when $n_1 = 4$ and $n_2 = 179$.

(*c*) Observations on the shearing strength and density of North Queensland kauri have been recorded by Barnard and Ditchburne (1956). The mean shearing strength was 777·1 lb/in² and the density 24·9 lb/ft³, and the inverse of the variance–covariance matrix of the means was

$$\begin{bmatrix} 0\cdot0^2 6599194 & -0\cdot3399002 \\ -0\cdot3399002 & 21\cdot1158099 \end{bmatrix}$$

on 14 degrees of freedom. The 1% test of significance for a hypothesis specifying the true mean shearing strength and density, will accept any hypothesis for which

$$\begin{aligned} &0\cdot0^2 6599(\xi_1-777\cdot1)^2+21\cdot1158(\xi_2-24\cdot9)^2 \\ &\quad-0\cdot6798(\xi_1-777\cdot1)(\xi_2-24\cdot9)\leqslant2\times6\cdot51=13\cdot02, \end{aligned} \tag{16}$$

where 6·51 is the value of e^{2z} at the 1% point when $n_1 = 2$ and $n_2 = 14$.

IV. Comparison of the Fiducial and Confidence Regions

Figure 2 shows the ellipse comprising the 99% fiducial region (smaller ellipse) defined by (16) in example (*c*) of the previous section, contrasted with the corresponding confidence region *computed from Hotelling's distribution* in accordance with the procedure which has been advocated (larger ellipse). Thus, if p stands for the number of variates and n for the degrees of freedom on which the variance–covariance matrix is based,

$$\frac{n-p+1}{np}T^2 = \frac{n-p+1}{np}(\boldsymbol{\xi}-\mathbf{x})'\mathbf{S}^{-1}(\boldsymbol{\xi}-\mathbf{x})$$

is distributed as e^{2z} with $n_1 = p$ and $n_2 = n-p+1$, and the confidence region corresponding to a confidence coefficient $1-P$ is the interior and boundary of the ellipsoid

$$(\boldsymbol{\xi}-\mathbf{x})'\mathbf{S}^{-1}(\boldsymbol{\xi}-\mathbf{x}) = \frac{np}{n-p+1}\,e^{2z}, \tag{17}$$

where e^{2z} is the percentile at the probability P when $n_1 = p$ and $n_2 = n-p+1$ (*vide*, for example, Hotelling 1931; Anderson 1958).

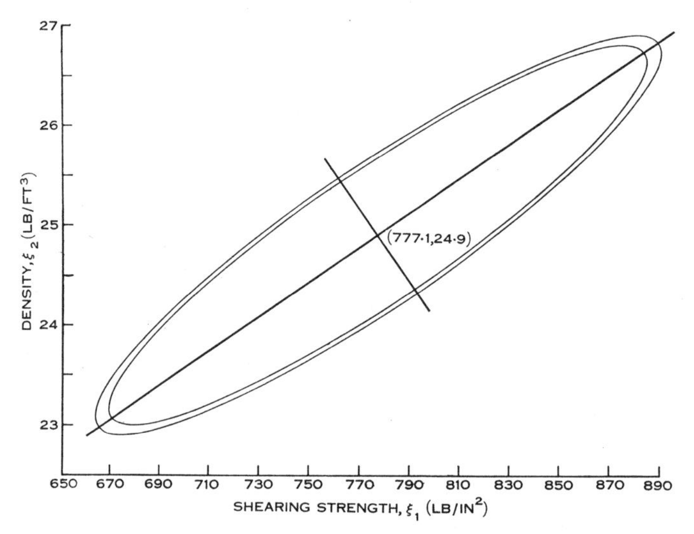

Fig. 2.—Ninety-nine per cent. fiducial and confidence regions for observations of example (*c*).

Excluding the conditions under which the estimated variance–covariance matrix is singular, comparison of (14) and (17) shows that the confidence region is the greater except for the special cases

(i) $p = 1$, when the two regions coincide,

(ii) $p-1 \geqslant n > 1$, when no confidence region is obtainable but a meaningful, though restricted, fiducial region exists.

For data of the type illustrated in example (*a*) above, the fiducial distribution of the parameters may be obtained in either of two ways, which yield identical results. Such observations may be regarded as a sample from the general multivariate normal distribution, and the fiducial distribution of the parameters may be derived in the manner outlined in preceding sections; alternatively, the problem may be examined by the direct procedure given by the author (Cornish 1960) which leads to the same result, thus confirming the general simultaneous fiducial distribution.

Since estimation is exhaustive for the class of data herein considered, it would be anticipated that the fiducial and confidence regions coincide; the discrepancy observed would seem to indicate that Hotelling's distribution is not appropriate for testing simultaneously the agreement of the observed quantities with the values proposed for them by hypothesis.

V. References

Anderson, T. W. (1958).—"An Introduction to Multivariate Statistical Analysis." (Wiley: New York.)

Barnard, M. M., and Ditchburne, N. (1956).—"Elementary Statistics for Use in Timber Research." (C.S.I.R.O. Aust.: Melbourne.)

Cornish, E. A. (1960).—*Aust. J. Statist.* **2**: 32–40.

Cramér, H. (1946).—"Mathematical Methods of Statistics." (Princeton Univ. Press.)

Erdélyi, A., Magnus, W., Oberhettinger, F., and Tricomi, F. G. (1954).—"Tables of Integral Transforms." Vol. 2. (McGraw-Hill: New York.)

Fisher, R. A. (1954).—*J. R. Statist. Soc.* B **16**: 212–13.

Hotelling, H. (1931).—*Ann. Math. Statist.* **2**: 360–78.

Kober, H. (1940).—*Quart. J. Math.* **11**: 193–211.

Printed by C.S.I.R.O., Melbourne

Commonwealth Scientific and Industrial Research Organization, Australia. Division of Mathematical Statistics Technical Paper No. 10, Melbourne, 1961.

INTER-STATION CORRELATIONS OF RAINFALL IN SOUTHERN AUSTRALIA

By E. A. Cornish,* G. W. Hill,† and Marilyn J. Evans*

[*Manuscript received June* 29, 1961]

Summary

Inter-station correlations of rainfall have been calculated for 55 stations in southern Australia, from records taken over the interval 1861–1955. A 6-day period was used as the subdivisional unit of the year, to provide accurate description and representation of the seasonal changes.

The investigation involved the calculation of 90,585 correlation coefficients, which, after transformation, furnished the basic material for determining the multiple regression relationships. The results completely substantiated those of a previous investigation.

Inter-station distances and bearings account for a substantial proportion of the variation observed among the transformed correlations. The relation with distance and bearing is simple, leading to a simple spatial distribution of the correlations, the axis of maximal correlation having an orientation approximating that of the fronts as they traverse the region under investigation. There is a well-defined seasonal change in this axis.

I. Introduction

The extent to which different parts of a country experience similar deviations from their average weather sequences is of paramount importance in several aspects of climatology. Firstly, a decision must frequently be made whether records of a meteorological element, e.g. rainfall, taken simultaneously at a number of observing stations for relatively short periods of time, can be used to replace observations taken at one station over a long time-interval. This question often arises in recently developed areas where reliable records are either unavailable or very rare, and the importance of any decision reached is due to the fact that its nature determines the manner in which forecasts of agricultural production from weather data are made, or the manner in which the characteristic climatic features of a given area are assessed. Forecasts are of both direct and indirect value in crop insurance, trade, and administration, and the more important climatic features are required for the purpose of developing agriculture and for use in problems of engineering and soil conservation. Thus, again taking rainfall as an illustration, suppose that the observations at four stations within a region are completely independent, then 25 years' data at each point will provide as much information on, say, the variability of rainfall as a record of 100 years at a single station. If, on the other hand, rainfalls at the four stations are correlated, the gain in information resulting from inclusion of additional stations will be reduced by an amount depending upon the

* Division of Mathematical Statistics, C.S.I.R.O., University of Adelaide.

† Division of Mathematical Statistics, C.S.I.R.O., Computation Laboratory, University of Melbourne.

magnitude of the correlation, and in the extreme case of perfect correlation, the inclusion of additional stations will provide no more information than that obtainable from the record of one station.

Secondly—and this applies particularly in regions where observing stations are sparsely distributed—it has become increasingly important to interpolate for meteorological variates. Various isopleths are required and, considering local demands only, those most urgently needed are isohyets. The accuracy with which rainfall, or any other meteorological variate, may be estimated, at a given point, from known observations at neighbouring points, must evidently depend upon the density of observing stations and their inter-correlations, and, knowing the latter, it is possible to state how densely located the stations should be to provide estimates with any assigned degree of accuracy.

Thirdly, the reliance which can be placed on all meteorological estimates or predictions is dependent upon the distribution of variance between localized and widespread fluctuations in the weather, and the only way of investigating this distribution is to determine empirically the manner in which the variance is partitioned.

Attention was first directed to monthly rainfall, and the problem of determining both the correlations and the distribution of variance was approached in two ways:

(i) by relating rainfall to the position and altitude of the observing stations within a relatively limited area,

(ii) by direct correlation of rainfall at stations dispersed over a much greater area, and determination of the relation to inter-station distance.

An investigation of type (i), with particular reference to the problem of interpolating monthly rainfall, has been reported (Coote and Cornish 1958) and the technique subsequently improved (Cornish and Evans, unpublished data). Research on problems of type (ii) was initiated as a subsidiary analysis in a more general investigation of the effects of rainfall on the yield of wheat in South Australia (Cornish 1950). Inter-station correlations of rainfall, which had been employed to describe, in part, the seasonal characteristics in the wheat belt, were found to be surprisingly high and demonstrated clearly the widespread nature of the major rain-bearing disturbances. This investigation embraced only winter rains in the broad subdivisions April–May, June–July–August, and September–October, so the decision was made to broaden the scope of these studies by extending them to rains throughout the year, using the calendar month as the unit of time (Stenhouse and Cornish 1958). The results completely substantiated those obtained previously and in fact extended them to all months of the year. Other important findings emerged from this work. There are four primary factors concerned with the formation of rains in the region studied, namely, frontal activity, latitudinal convergence, convection, and topography. The effects of topography, convection, and convergence were dominated by associations arising along the frontal lines, but at the same time they clearly modified the influence of frontal factors. The axis of maximal correlation was orientated approximately NW.–SE., but there were definite indications of an annual oscillation in the direction of this axis, and finally the simplicity of the changes in the correlations with varying distance and direction of displacement showed that relatively simple laws related these quantities over very considerable areas, laws which should provide sufficiently exact knowledge of the accuracy of monthly estimates based on a limited number of stations.

In the winter rainfall zone of southern Australia, the calendar month is too lengthy to enable us to follow accurately the rapid rate of change of rainfall with time during the periods autumn–early winter and spring–early summer. In order to

(i) provide the detail necessary for these purposes,

(ii) confirm the annual oscillation of the axis of maximal correlation,

(iii) determine the system of correlations existing in a more extensive area,

the scope of the investigation has been widened by increasing the number of stations to 55, by taking 6 days as the unit of subdivision, and by including eastern Eyre Peninsula from South Australia and a large tract of western Victoria. Further reference to the 6-day unit is made below in Section IV.

II. Data

The rainfall observations were extracted from the records of 55 stations, taken during 1861–1955 inclusive, the names and positions of the stations being given in Figure 11. The number of years for which simultaneous observations were obtainable was not constant for all pairs of stations, but differences in the weights of the correlations were small enough to be neglected in subsequent analyses. Rainfall records were made available by the Commonwealth Bureau of Meteorology, and the positions of the stations by the Department of Lands, South Australia. Inter-station distances and bearings were computed on WREDAC at the Weapons Research Establishment from the latitudes and longitudes of the stations.

III. Climatic and Topographic Features

Traversing the region from west to east, the principal topographic features are:

(i) the plains on the eastern boundary of Eyre Peninsula broken by minor elevated masses which vary in direction,

(ii) the meridional horsts of the Mount Lofty and Flinders Ranges,

(iii) the extensive plains of the Murray basin,

(iv) the plains of south-eastern South Australia which merge with the Western District plains of Victoria,

(v) the western highlands of Victoria containing the Great Dividing Range running west to east and lying between the Murray basin to the north and the Western District plains to the south.

The whole area lies in the winter rainfall zone of southern Australia, the season being marked by the strong winter incidence of its rains. Approximately 70–80% of the annual precipitation occurs within the period April–October inclusive, the summer months being characterized by hot, dry atmospheric conditions, low rainfall, and high evaporation. Milder summer conditions prevail in the western highlands and Western District plains of Victoria and the southern extremity of the south-eastern plains of South Australia.

The winter rainfall zone itself lies in the path of the migratory anticyclones, with their attendant cyclones, as they move eastward. The major component of the annual rainfall is due to activity on the associated cold fronts, the remainder being convectional rains occurring mainly in winter, and general rains, normally over large areas, due to convergence in the northerly or north-westerly stream ahead of the advancing fronts.

IV. Analysis

The main considerations leading to the choice of the 6-day interval have been enumerated above. This short interval, compared with the calendar month, could have introduced two undesirable features with serious consequences. A preliminary examination was made, therefore, to ensure that the extent to which these were present could have no important effect on the analysis. In the first place, if serial correlations of sufficient magnitude existed they would vitiate the findings, but

Table 1

MEAN TIMES IN DAYS TAKEN BY ANTICYCLONES TO TRAVERSE 10° LONGITUDE

Between	Jan.	Feb.	Mar.	Apr.	May	June	July	Aug.	Sept.	Oct.	Nov.	Dec.	Range
130–140°E.	1·29	1·33	1·48	1·47	1·79	1·62	1·56	1·27	1·38	1·11	1·22	1·32	0·68
140–150°E.	0·67	0·85	0·91	1·37	1·92	2·44	2·12	1·72	1·28	0·93	0·81	0·80	1·77

inspection of Kidson's (1925) data indicates that, for the 6-day interval, any such correlations present would be far too small to affect the analysis. Within the bounds 30–40°S. latitude and 110–170°E. longitude, the frequency distribution of anticyclonic life has a mode of 5 days, a mean of approximately 6 days, and a range of 22 days, so that, within the region under investigation (31°53′S.–38°24′S.; 135°52′E.–143°45′E.) the correlations must be negligible. This is, perhaps, more clearly shown by the mean speeds of the anticyclones, the figures given in Table 1 being the mean times (in days) to traverse 10° longitude from west to east. The second point, which concerns the form of distribution of rainfall for the 6-day period and which could only be checked after calculation of some of the correlations, is discussed below.

With respect to the solar year, the 6-day period is more variable than the calendar month, but, as will be shown, the comparatively slow seasonal changes in the correlation make correction for the error so introduced quite unnecessary; on the other hand, the 6-day intervals are of equal and constant duration whereas the calendar months are not.

The primary data were obtained by directly correlating the 6-day rainfall totals of all pairs of stations, yielding 1485 correlation coefficients (r) for each of the 61 periods of the year, a total of 90,585, which, for subsequent stages of analysis, were replaced by their inverse hyperbolic transformations $z = \tanh^{-1}r$. The advantages of using this transformation have been pointed out and emphasized by Fisher

(1958). When applied to correlations obtained from normally distributed data, the principal advantage is that the transform z is itself nearly normally distributed with a variance which, for all practical purposes, is dependent only on the size of the sample. The distribution of rainfall for the 6-day period, with its extreme positive skewness, is far from normal, so that it became vitally necessary to investigate the effect, on the distribution of z, of such departures from normality. Figure 1, which illustrates a typical distribution (period 30), shows that it is sufficiently near normality to validate the procedure which has been adopted.

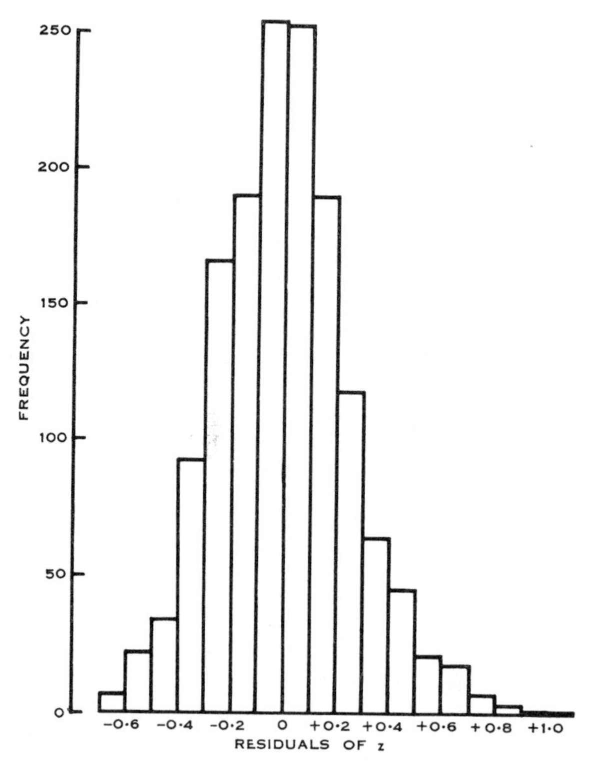

Fig. 1.—Frequency distribution of the residuals of z (period 30).

The crude sums of squares and products were obtained by progressive digiting, using a Powers-Samas sorter and tabulator, the sorter being fitted with a special device which made it possible to dispense with the very large amount of sorting otherwise necessary to obtain products of rainfall from stations recorded on different cards. At a later stage, ready access to WREDAC presented the opportunity to compute the correlations and their transforms and to complete other heavy stages of the computations.

As indicated by the previous analysis (Stenhouse and Cornish *ibid.*) the correlations were all positive, and this should hold even in the extended region. There are only 225 negative coefficients in the full set of 90,585 and these all represent

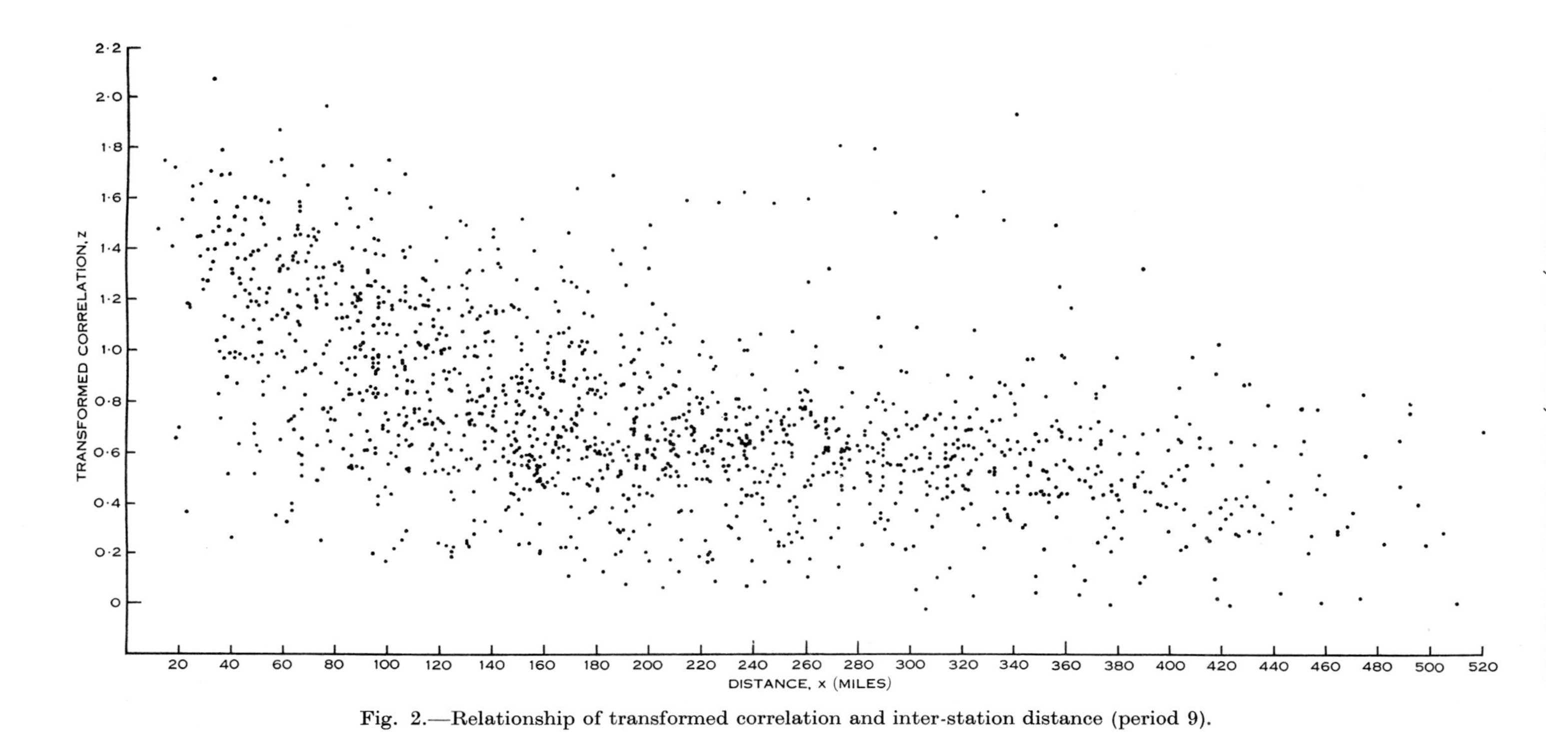

Fig. 2.—Relationship of transformed correlation and inter-station distance (period 9).

chance estimates of small positive correlations between very distant stations. The observations show clearly the inverse relationship with distance, and Figure 2 illustrates this relation for period 9. This is typical of all the data under examination.

With two stations close together, r approaches $+1$ and z tends to infinity, and for stations a great distance apart r and z both approach zero. The relation between z and distance is, therefore, very likely representable by a form in which z varies as some inverse power of distance, asymptoting to both axes. Over the extended range of distances at our disposal the relationship is well represented, except for the smallest distances, by the parabola

$$Z = a+b_1x+b_2x^2,$$

where x is the distance in miles. Convenience of fitting the regression formula has again dictated the use of this form.

As pointed out previously (Stenhouse and Cornish *ibid.*), the correlations in the neighbourhood of a given point did not decrease isotropically, and it was found that allowance for the effect of direction of displacement could be made effectively by modifying the regression formula to take the form

$$Z = k+b_1x+b_2x^2 +b_3(x \sin 2\theta) +b_4(x \cos 2\theta), \qquad (1)$$

where θ is the bearing of one station on another, measured counterclockwise from the north. With the data from this substantially greater region, there is no evidence to suggest that a more elaborate specification of direction is required. In the previous investigation inclusion of distance in the terms representing displacement improved the fit of the regression in all months of the year, and this has been confirmed in the present data by examination of the residuals from pairs of stations whose joins lie parallel to the axes of maximal and minimal correlation.

Complete tabulation of the results of this investigation has not been included since it is too extensive for reproduction herein.* Many of the derived quantities are presented in the diagrams, and these should give a very clear indication of such features of the data as, for example, variability. Figure 3 shows the values of z in the 61 periods, averaged over the 1485 pairs of stations and corrected for bias, and demonstrates the annual oscillation in the correlations occurring in this region. The seasonal changes are not marked, but the harmonic

$$Z = 0\cdot6932+0\cdot05429 \sin(\alpha+46°43'),$$

fitted to the data, represents a genuine cyclic change and is quite significant, thus confirming the observation and comment made previously, namely, that the data presented features of a graded sequence but that the calendar month was too long an interval to allow the trend to express itself because of interference by compensatory effects. Further reference to this seasonal change is made in the discussion which follows.

The linear terms on distance (b_1) are all negative and strongly significant ($P<0\cdot001$), while the quadratic terms (b_2) are all positive, except for period 2, and

* Copies have been filed and are available for inspection at the libraries of the C.S.I.R.O., 314 Albert Street, East Melbourne, Victoria, and of its Division of Mathematical Statistics, University of Adelaide, Adelaide.

all significant except for seven periods, including 2. The coefficients (b_3) of the displacement term involving the sine of the angle of bearing are all negative and all significant except for period 19. The significance is strongly marked, $P<0\cdot001$, except for periods 7 and 10, in which $P<0\cdot01$. There is no evidence of a seasonal

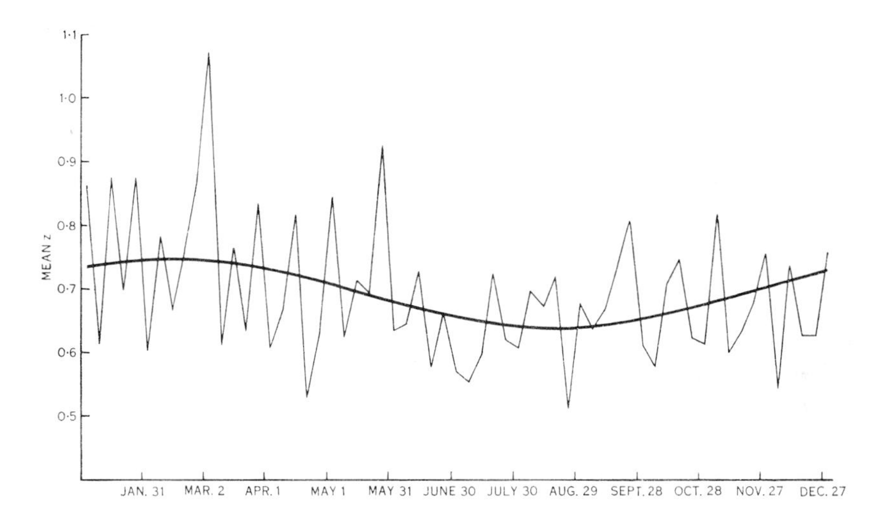

Fig. 3.—Seasonal trend of the means of transformed correlations (adjusted for bias).

change in any one of the three sets of coefficients, which are shown set out in temporal order in Figures 4, 5, and 6 respectively. It is remarkable that the measures, b_1 and b_2, of the reduction in the correlation should remain approximately constant, because on general grounds it might have been expected that they would have shown evidence

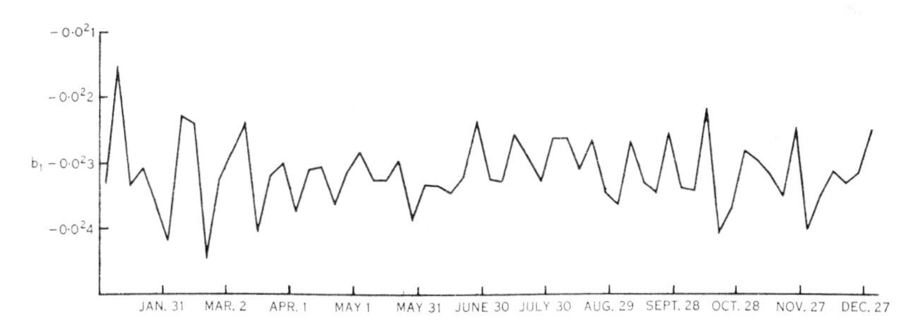

Fig. 4.—Coefficient of linear term on distance (b_1).

of a seasonal change. The extremely aberrant values of b_3 for periods 13 and 19 were suspect, but a complete check of the calculations confirmed them and consequently they have been retained.

On the other hand, the coefficient b_4 of the displacement term in the cosine of the angle of bearing exhibits a well-defined annual oscillation represented by

$$B_4 = -0\cdot0^{3}1378+0\cdot0^{3}2080(\sin\ \alpha+48°49'),$$

and illustrated in Figure 7.

The regressions on distance provide, by far, the more important contribution, both relatively and absolutely, to the total variation ascribable to the regression formula. The outstanding observation, however, is that up to 60% of the variance

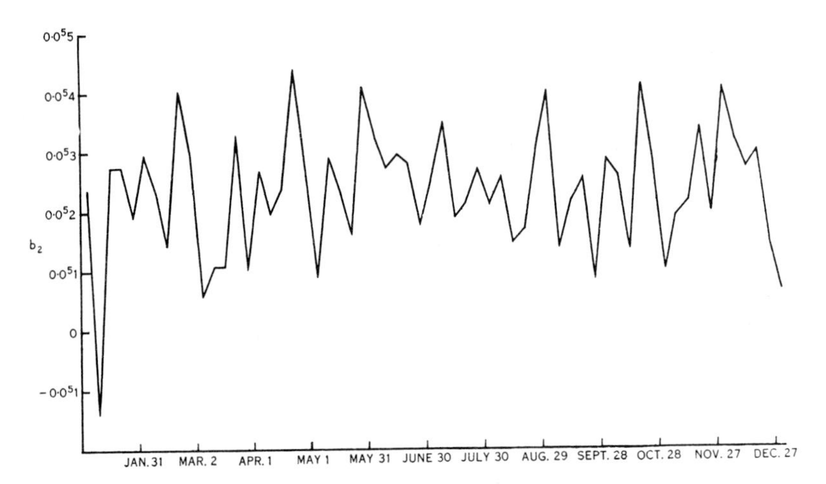

Fig. 5.—Coefficient of quadratic term on distance (b_2).

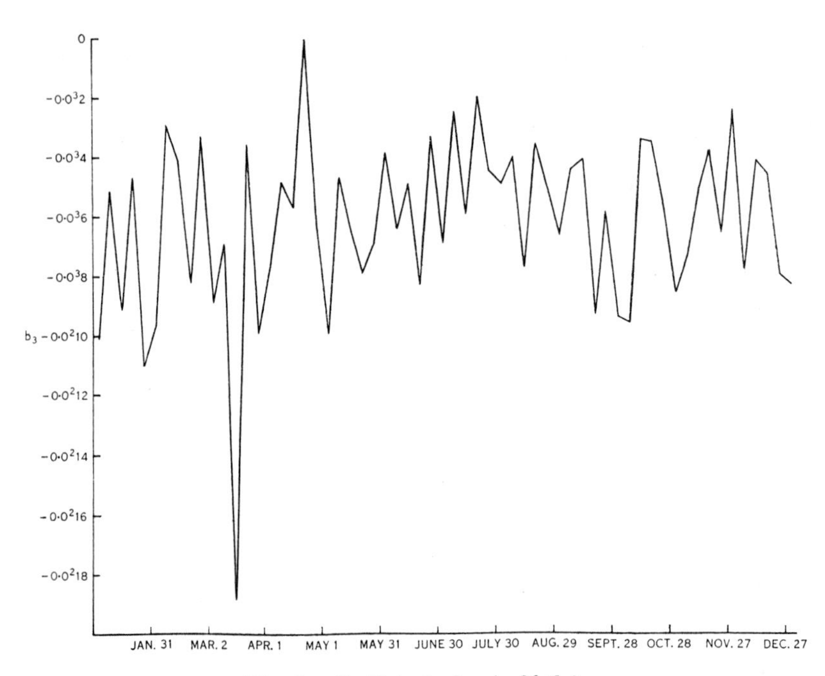

Fig. 6.—Coefficient of $x \sin 2\theta$ (b_3).

in the correlations, measured on the scale of the transform, is explicable in terms of the variates used, considering that the residual standard deviation includes, in addition to errors of observation, any systematic deviations from the formulae and irregularities arising from the erratic nature of the 6-day rainfall and from local topographical features or exposure of the stations. Both total variance and residual variance have a well-defined annual oscillation, but in each case the variability about

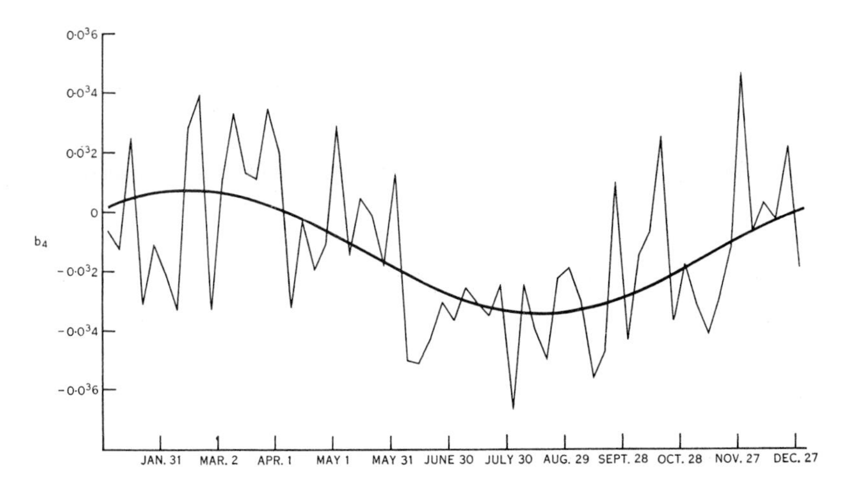

Fig. 7.—Seasonal trend of the coefficient of x cos 2θ (b_4).

the seasonal trend is not constant. To make this variability constant and independent of the variance itself, the transforms, log (total variance) and log (residual variance)

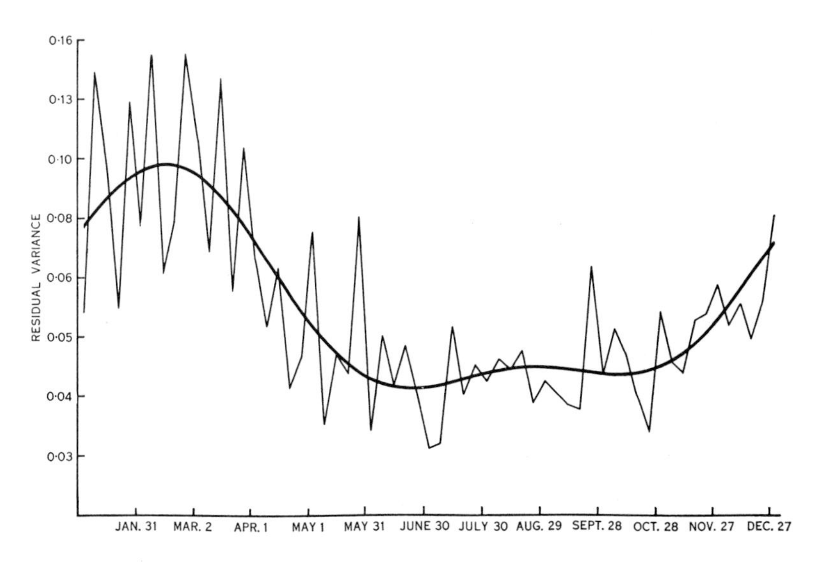

Fig. 8.—Seasonal trend of the residual variance (logarithmic scale).

have been taken and their seasonal trends examined; in this process differences between the larger values are reduced and those between the smaller values are exaggerated, thus giving a truer representation of the changes in progress. Figure 8 shows the trend for log (residual variance) which is represented by

$$-1\cdot 2525+0\cdot 1717\ \sin(\alpha+51°14')+0\cdot 07089\ \sin(2\alpha+2°21').$$

Log (total variance) follows a similar course.

The second harmonic of both curves is strongly significant ($P<0\cdot 01$). In each case its effect is to displace the minimum to a point in the latter part of June, thus inducing a rapid fall in autumn and simultaneously lengthening the rise in spring, which is, in fact, broken by an almost stationary period extending over August, September, and October. The period of high variability in the correlations coincides very closely with the period when the relative variability of the rainfall is greatest and the mean rainfall least, so that localized fluctuations in rainfall, which otherwise do not conform to any pattern in the spatial distribution of the rains, markedly influence the magnitude of the correlations, and this in turn induces the

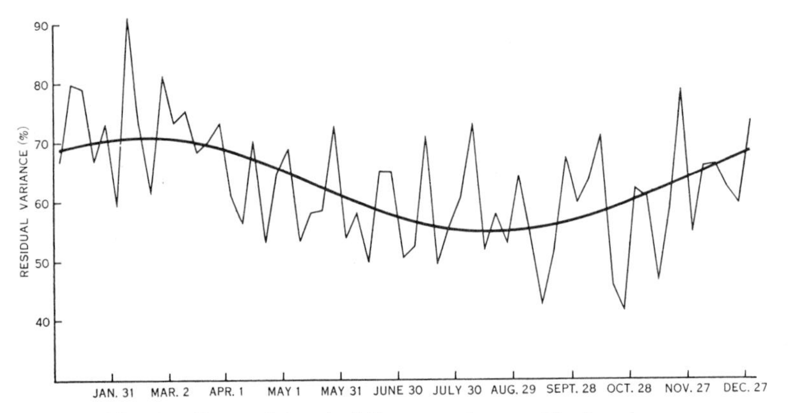

Fig. 9.—Seasonal trend of the percentage residual variance.

large variability observed among the correlations at that particular time. On the other hand, under the much more stable conditions which prevail after winter has definitely set in, the correlations of the region are much less variable. The seasonal change in residual variance of the correlations closely follows that of their total variance.

When considered relative to the mean, the residual standard deviation is slightly greater during January to March inclusive than during the remainder of the year, but the difference is so small that it may be regarded as constant for all practical purposes. On the other hand, the residual variance, relative to the total variance, displays a well-defined annual oscillation shown in Figure 9 and represented by

$$0\cdot 6275+0\cdot 07950\ \sin(\alpha+47°47').$$

Both absolutely and relatively, the residual variance is smaller during the winter months, and from the latter fact it is clear that the variables we have used in the regressions account for changes in the correlations more appreciably in the winter than in the summer. The unsatisfactory nature of the calendar month is brought out forcibly in this connexion because the previous analysis had indicated a small inverse effect.

In sampling from a bivariate normal distribution, the variance of the transform z may be regarded as depending only on the sample size, so that an accurate estimate of the average theoretical variance may be obtained from the harmonic mean of the numbers $(n_{ij}-3)$, n_{ij} being the number of pairs of observations for stations i and j. This average value for all periods is $0 \cdot 01805$ but in no case does the observed residual variance approach this theoretical value. Since the values of the transform z are sufficiently near normality, the discrepancies are due principally to the inadequacy of the regression formula we have used. No really definite evidence is yet available to indicate what other variables might be added to the regression formula; in addition to the length and direction of the vector joining a pair of stations, its position in the region under examination seems to possess significance, but, owing to complications

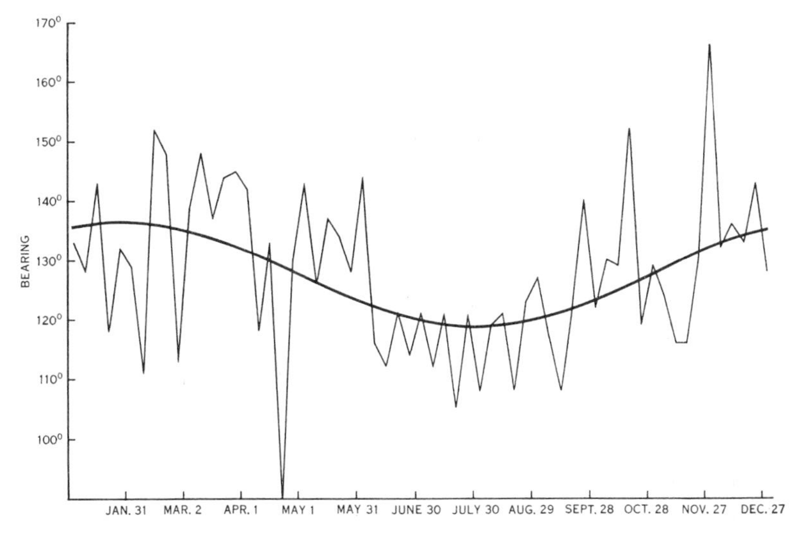

Fig. 10.—Seasonal trend of the axis of maximal correlation.

introduced by other unknown causes, its effects are not consistent. In all tests of significance the actual observed residual variances have been used. The residuals have been computed and are at present being examined.

Maximizing the right-hand side of the regression formula (1) for variations in θ yields the solution

$$\theta = \tfrac{1}{2} \arctan(b_3/b_4)$$

for the bearing of the axis of maximal correlation, and because of the manner in which the regression formula has been fitted, the axis of minimal correlation will be at right angles. The annual oscillation of the axis of maximal correlation is illustrated in Figure 10 and is represented by

$$\Theta = 127 \cdot 6144 + 8 \cdot 7808(\sin \alpha + 65°1').$$

Maps have been constructed from the regression formula (1) for all 61 periods by assigning a range of values of Z from $0 \cdot 40$ to $1 \cdot 25$ and a range of values of θ from 0° to 175° by steps of 5°, and solving the resultant quadratic equations for distance. The isohomeotropes were then found by interpolating in this basic framework. The 24 maps of Figure 11 are a representative selection from the full set of 61, chosen to illustrate the principal features of the annual sequence from the viewpoint of their spatial relationships. Peake has been taken as the reference point for constructing these diagrams because of its approximately central position, but it must be emphasized that any other point in the region could have been chosen, the effect of another choice being merely to translate the figure to a new origin. The axes of maximal and minimal correlation have been inserted in each map and, in addition, the N.–S. line through Peake. The three subdivisions represent the zones where $Z<0\cdot9$, $0\cdot9\leqslant Z\leqslant1\cdot2$, $Z\geqslant1\cdot2$, the values $0\cdot9$ and $1\cdot2$ of Z corresponding, respectively, to correlation coefficients of $0\cdot72$ and $0\cdot83$.

Depiction in this manner demonstrates very clearly the salient features of the annual sequence in the system of correlations in the region. These features are as follows:

(*a*) *The cyclic change in the axis of maximal correlation*

In Figure 10 the full significance of the swing in the axis of maximal correlation is not immediately obvious but, when shown as a change of orientation from NW. in summer to approximately WNW. in winter against the background of the map, it is quite marked. The annual oscillation in the eastward latitudinal paths of the migratory anticyclones as they pass over the region, which is very closely associated with the sequence of seasons during the year, leads to the oscillation in the orientation of the frontal lines over the area. In summer, the centres of the anticyclones have a mean track at a greater latitude than that of Peake giving a NW.–SE. orientation to the frontal line, but in winter the centres of the anticyclones have a mean track at a latitude less than that of Peake yielding a frontal line with an orientation of north of NW. Diagrammatic representation of the synoptic situation has been given in the paper cited (Stenhouse and Cornish *ibid.*, Fig. 7).

Two features of the movement of the pressure systems and their associated fronts produce effects which are superimposed on the oscillation of the frontal lines. First, a wave frequently forms on the frontal surface and travels southward along the front. This motion, in conjunction with the eastward movement of the front, gives to the centre of the wave a resultant course which has, in general, an orientation of west of NW. Pressure gradients are steepest at the centre of the wave, promoting greater uplift of the air mass and, therefore, generally more rain. Secondly, the two high pressure systems in advance of, and to the rear of the advancing front, extend into the atmosphere, but with increasing altitude lag behind, and at the same time are further northward than the centre at the surface. The stream of the trailing edge of the leading high is elevated by the front to yield rain, and its prevailing direction is west of NW. These effects occur almost entirely in the winter months and introduce a strong component with an axis of correlation in a direction west of NW.; and the correlations so introduced, when combined with those arising along the line of the front, yield the observed resultant axis of maximal correlation. Diagrammatic representation has been given in the paper cited (Stenhouse and Cornish *ibid.*, Figs. 8 and 9).

(b) The cyclic change in the magnitude of the correlations coupled with the effects of distance and direction of displacement

In a similar manner the change in the magnitude of the correlations and the effects of distance and direction of displacement are more readily comprehended against the background of the map. The generally higher correlations in the summer and their tendency to persist over greater distances, are brought out by the contrasts of the marked zones, and the approximately elliptical isohomeotropes emphasize the effects of direction of displacement. The former observation at first appears contrary to expectation, but it must be remembered that deviations below the mean rainfall contribute also to the correlations, and that fine weather prevails over extensive areas in summer, thus tending to increase the average value of the correlations. In addition, during winter the irregular local effects in the immediate vicinity of a rainfall station constitute an increased proportion of the total variance of rainfall, and the net effect of this is to reduce the average value of the correlations.

As pointed out earlier, a substantial proportion of the total variance of the correlations is explicable in terms of the variates employed in the regressions. It is of paramount importance that this high proportion can be represented by distance and direction of displacement in the simple manner indicated in the diagrams. Evidently simple laws describe the changes in the system of correlations over a very extensive area, despite the major topographical features of the region; the manner in which this information can be used to provide sufficiently exact knowledge of the accuracy of rainfall estimates, based on a limited number of stations, will be described elsewhere.

V. Acknowledgments

The authors are indebted to the Deputy Director of Meteorology, South Australia, for placing the records at their disposal, to the Department of Lands, South Australia, for information respecting the positions of the stations, and to Mr. M. C. Childs, of the Division of Biochemistry and General Nutrition, C.S.I.R.O., for preparation of the diagrams. Special acknowledgment is made to the Controller, Weapons Research Establishment, Department of Supply, Commonwealth of Australia, and officers of the Mathematical Services Division for their ready cooperation in making available the Department's computer, WREDAC.

VI. References

Coote, G. G., and Cornish, E. A. (1958).—C.S.I.R.O. Aust. Div. Math. Statist. Tech. Pap. No. 4.

Cornish, E. A. (1950).—*Aust. J. Sci. Res.* B **3**: 178–218.

Fisher, R. A. (1958).—"Statistical Methods for Research Workers." 13th Ed. (Oliver and Boyd: Edinburgh.)

Kidson, E. (1925).—Aust. Bur. Meteor. Bull. No. 17.

Stenhouse, N. S., and Cornish, E. A. (1958).—C.S.I.R.O. Aust. Div. Math. Statist. Tech. Pap. No. 5.

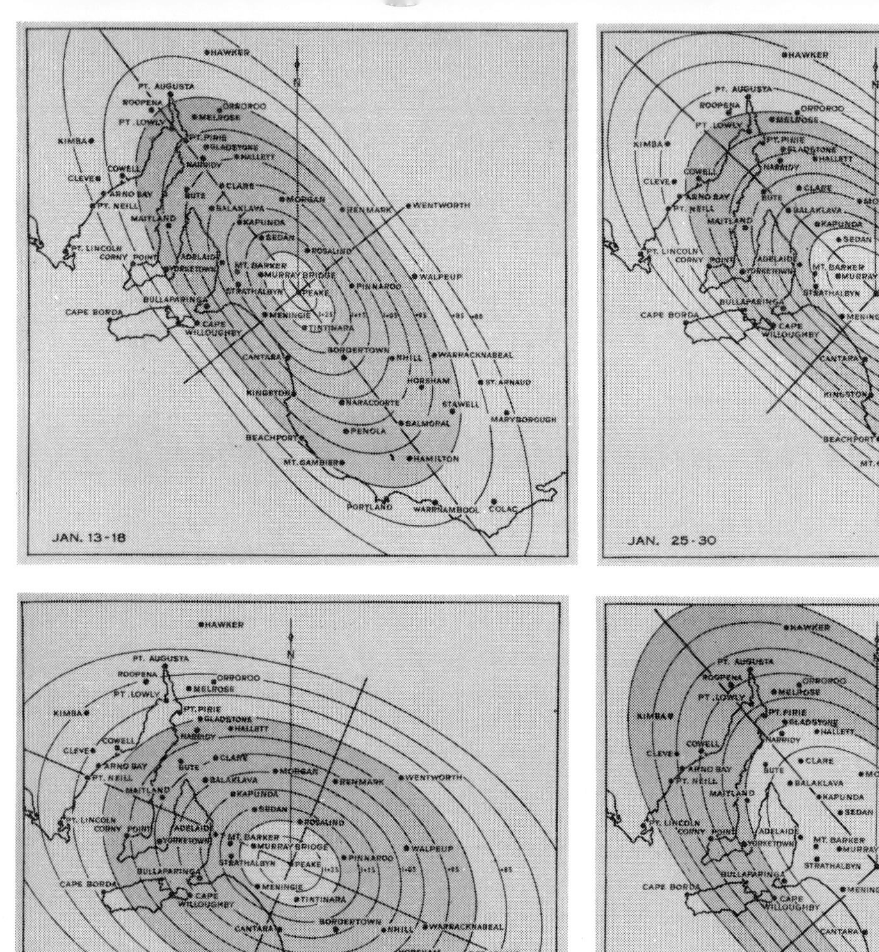
JAN. 13-18
JAN. 25-30
FEB. 24-MAR. 1
N
HAWKER
PT. AUGUSTA
ROOPENA
ORROROO
PT. LOWLY
MELROSE
PT. PIRIE
GLADSTONE
KIMBA
NARRIDY
HALLETT
COWELL
CLEVE
ARNO BAY
CLARE
BUTE
MORGAN
RENMARK
WENTWORTH
PT. NEILL
BALAKLAVA
MAITLAND
KAPUNDA
SEDAN
PT. LINCOLN
CORNY POINT
ADELAIDE
ROSALIND
YORKETOWN
MT. BARKER
MURRAY BRIDGE
WALPEUP
PINNAROO
STRATHALBYN
PEAKE
BULLAPARINGA
CAPE BORDA
MENINGIE
CAPE WILLOUGHBY
TINTINARA
BORDERTOWN
NHILL
WARRACKNABEAL
CANTARA
HORSHAM
ST. ARNAUD
KINGSTON
NARACOORTE
STAWELL
MARYBOROUGH
BALMORAL
PENOLA
BEACHPORT
HAMILTON
MT. GAMBIER
PORTLAND
WARRNAMBOOL
COLAC

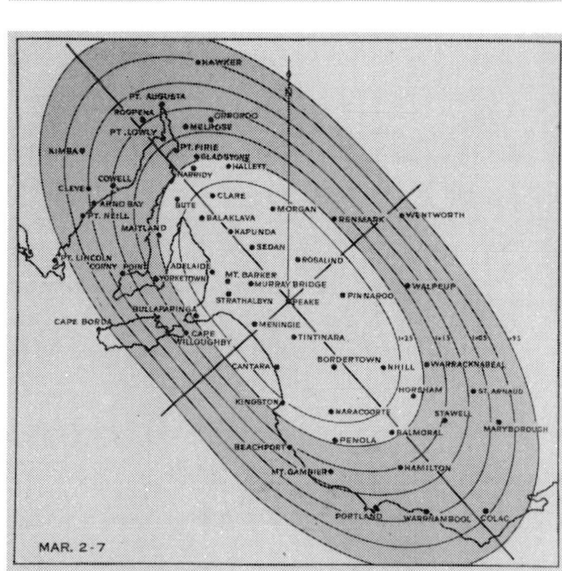
MAR. 2-7
N
HAWKER
PT. AUGUSTA
ROOPENA
ORROROO
PT. LOWLY
MELROSE
PT. PIRIE
GLADSTONE
KIMBA
NARRIDY
HALLETT
COWELL
CLEVE
ARNO BAY
CLARE
BUTE
MORGAN
RENMARK
WENTWORTH
PT. NEILL
BALAKLAVA
MAITLAND
KAPUNDA
SEDAN
PT. LINCOLN
CORNY POINT
ADELAIDE
ROSALIND
YORKETOWN
MT. BARKER
MURRAY BRIDGE
WALPEUP
PINNAROO
STRATHALBYN
PEAKE
BULLAPARINGA
CAPE BORDA
MENINGIE
CAPE WILLOUGHBY
TINTINARA
BORDERTOWN
NHILL
WARRACKNABEAL
CANTARA
HORSHAM
ST. ARNAUD
KINGSTON
NARACOORTE
STAWELL
MARYBOROUGH
BALMORAL
PENOLA
BEACHPORT
HAMILTON
MT. GAMBIER
PORTLAND
WARRNAMBOOL
COLAC

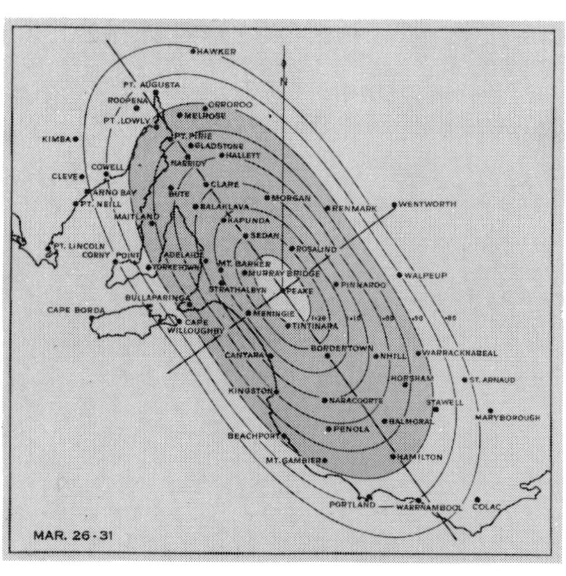
MAR. 26-31
N
HAWKER
PT. AUGUSTA
ROOPENA
ORROROO
PT. LOWLY
MELROSE
PT. PIRIE
GLADSTONE
KIMBA
NARRIDY
HALLETT
COWELL
CLEVE
ARNO BAY
CLARE
BUTE
MORGAN
RENMARK
WENTWORTH
PT. NEILL
BALAKLAVA
MAITLAND
KAPUNDA
SEDAN
PT. LINCOLN
CORNY POINT
ADELAIDE
ROSALIND
YORKETOWN
MT. BARKER
MURRAY BRIDGE
WALPEUP
PINNAROO
STRATHALBYN
PEAKE
BULLAPARINGA
CAPE BORDA
MENINGIE
CAPE WILLOUGHBY
TINTINARA
BORDERTOWN
NHILL
WARRACKNABEAL
CANTARA
HORSHAM
ST. ARNAUD
KINGSTON
NARACOORTE
STAWELL
MARYBOROUGH
BALMORAL
PENOLA
BEACHPORT
HAMILTON
MT. GAMBIER
PORTLAND
WARRNAMBOOL
COLAC

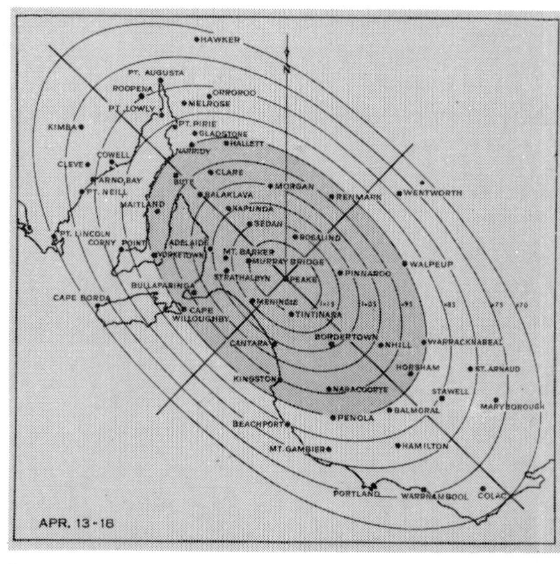
APR. 13-18
N
HAWKER
PT. AUGUSTA
ROOPENA
ORROROO
PT. LOWLY
MELROSE
PT. PIRIE
GLADSTONE
KIMBA
NARRIDY
HALLETT
COWELL
CLEVE
ARNO BAY
CLARE
BUTE
MORGAN
RENMARK
WENTWORTH
PT. NEILL
BALAKLAVA
MAITLAND
KAPUNDA
SEDAN
PT. LINCOLN
CORNY POINT
ADELAIDE
ROSALIND
YORKETOWN
MT. BARKER
MURRAY BRIDGE
WALPEUP
PINNAROO
STRATHALBYN
PEAKE
BULLAPARINGA
CAPE BORDA
MENINGIE
CAPE WILLOUGHBY
TINTINARA
BORDERTOWN
NHILL
WARRACKNABEAL
CANTARA
HORSHAM
ST. ARNAUD
KINGSTON
NARACOORTE
STAWELL
MARYBOROUGH
BALMORAL
PENOLA
BEACHPORT
HAMILTON
MT. GAMBIER
PORTLAND
WARRNAMBOOL
COLAC

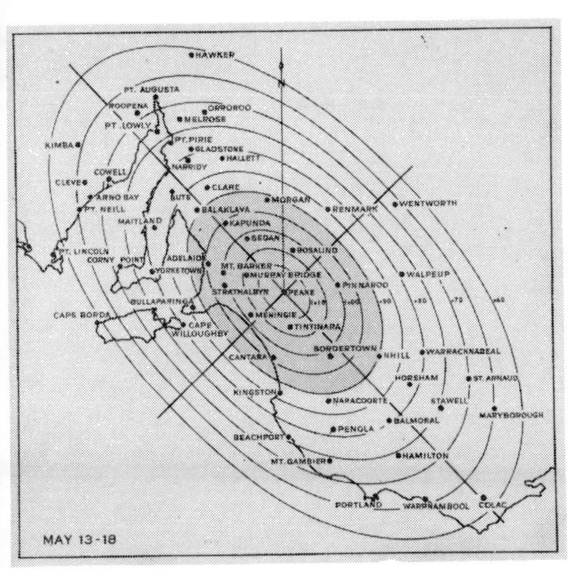
HAWKER
PT. AUGUSTA
ROOPENA
ORROROO
MELROSE
PT. LOWLY
KIMBA
PT. PIRIE
GLADSTONE
HALLETT
NARRIDY
COWELL
CLEVE
CLARE
ARNO BAY
BUTE
MORGAN
RENMARK
WENTWORTH
PT. NEILL
BALAKLAVA
MAITLAND
KAPUNDA
SEDAN
PT. LINCOLN
ROSALIND
CORNY POINT
ADELAIDE
YORKETOWN
MT. BARKER
MURRAY BRIDGE
WALPEUP
STRATHALBYN
PINNAROO
PEAKE
BULLAPARINGA
CAPE BORDA
MENINGIE
CAPE WILLOUGHBY
TINTINARA
BORDERTOWN
NHILL
WARRACKNABEAL
CANTARA
HORSHAM
ST. ARNAUD
KINGSTON
NARACOORTE
STAWELL
BALMORAL
MARYBOROUGH
PENOLA
BEACHPORT
MT. GAMBIER
HAMILTON
PORTLAND
WARRNAMBOOL
COLAC
N
MAY 13-18

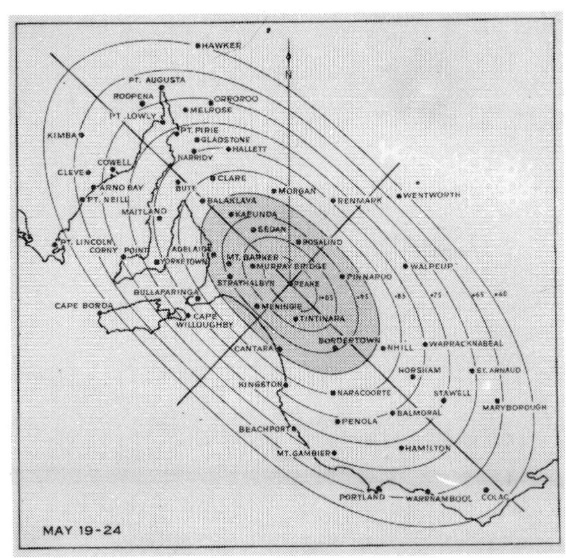
HAWKER
PT. AUGUSTA
ROOPENA
ORROROO
PT. LOWLY
MELROSE
KIMBA
PT. PIRIE
GLADSTONE
HALLETT
COWELL
NARRIDY
CLEVE
CLARE
ARNO BAY
BUTE
MORGAN
RENMARK
WENTWORTH
PT. NEILL
BALAKLAVA
MAITLAND
KAPUNDA
SEDAN
PT. LINCOLN
ROSALIND
CORNY POINT
ADELAIDE
YORKETOWN
MT. BARKER
MURRAY BRIDGE
WALPEUP
STRATHALBYN
PINNAROO
PEAKE
BULLAPARINGA
CAPE BORDA
MENINGIE
CAPE WILLOUGHBY
TINTINARA
BORDERTOWN
NHILL
WARRACKNABEAL
CANTARA
HORSHAM
ST. ARNAUD
KINGSTON
NARACOORTE
STAWELL
BALMORAL
MARYBOROUGH
PENOLA
BEACHPORT
MT. GAMBIER
HAMILTON
PORTLAND
WARRNAMBOOL
COLAC
N
MAY 19-24

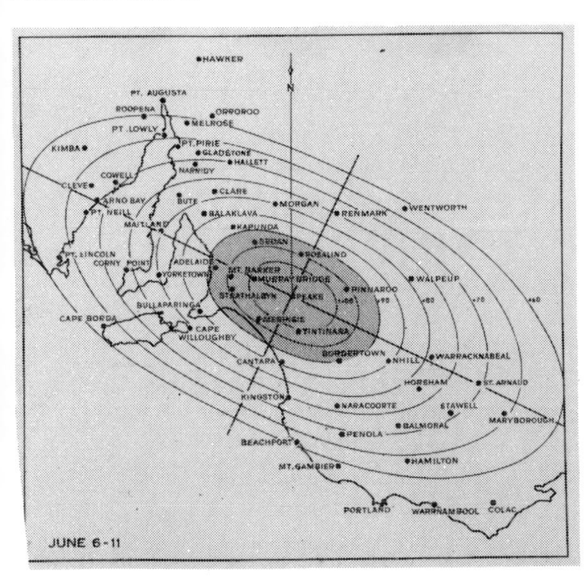
HAWKER
PT. AUGUSTA
ROOPENA
ORROROO
PT. LOWLY
MELROSE
KIMBA
PT. PIRIE
GLADSTONE
HALLETT
COWELL
NARRIDY
CLEVE
CLARE
ARNO BAY
BUTE
MORGAN
RENMARK
WENTWORTH
PT. NEILL
BALAKLAVA
MAITLAND
KAPUNDA
SEDAN
PT. LINCOLN
ROSALIND
CORNY POINT
ADELAIDE
YORKETOWN
MT. BARKER
MURRAY BRIDGE
PINNAROO
WALPEUP
STRATHALBYN
PEAKE
BULLAPARINGA
CAPE BORDA
MENINGIE
CAPE WILLOUGHBY
TINTINARA
BORDERTOWN
NHILL
WARRACKNABEAL
CANTARA
HORSHAM
ST. ARNAUD
KINGSTON
NARACOORTE
STAWELL
MARYBOROUGH
PENOLA
BALMORAL
BEACHPORT
MT. GAMBIER
HAMILTON
PORTLAND
WARRNAMBOOL
COLAC
N
JUNE 6-11

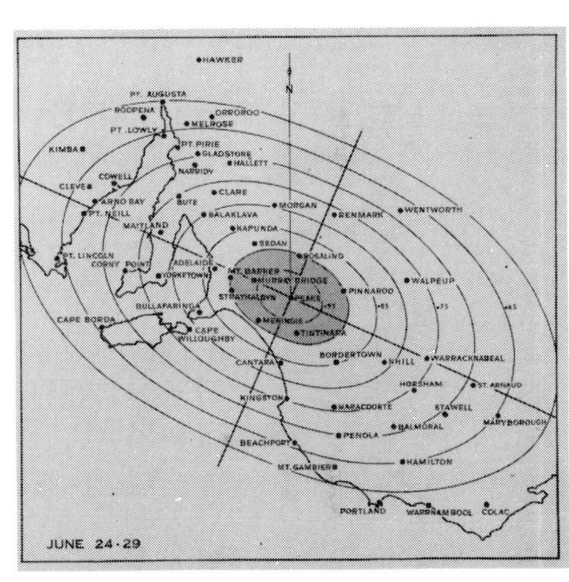
HAWKER
PT. AUGUSTA
ROOPENA
ORROROO
PT. LOWLY
MELROSE
KIMBA
PT. PIRIE
GLADSTONE
HALLETT
COWELL
NARRIDY
CLEVE
CLARE
ARNO BAY
BUTE
MORGAN
RENMARK
WENTWORTH
PT. NEILL
BALAKLAVA
MAITLAND
KAPUNDA
SEDAN
PT. LINCOLN
ROSALIND
CORNY POINT
ADELAIDE
YORKETOWN
MT. BARKER
MURRAY BRIDGE
PINNAROO
WALPEUP
STRATHALBYN
PEAKE
BULLAPARINGA
CAPE BORDA
MENINGIE
CAPE WILLOUGHBY
TINTINARA
BORDERTOWN
NHILL
WARRACKNABEAL
CANTARA
HORSHAM
ST. ARNAUD
KINGSTON
NARACOORTE
STAWELL
MARYBOROUGH
PENOLA
BALMORAL
BEACHPORT
HAMILTON
MT. GAMBIER
PORTLAND
WARRNAMBOOL
COLAC
N
JUNE 24-29

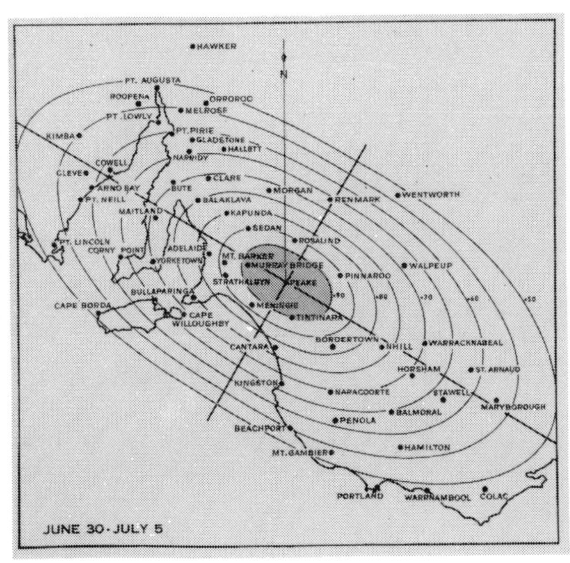
HAWKER
PT. AUGUSTA
ROOPENA
ORROROO
PT. LOWLY
MELROSE
KIMBA
PT. PIRIE
GLADSTONE
HALLETT
NARRIDY
COWELL
CLEVE
ARNO BAY
BUTE
CLARE
MORGAN
PT. NEILL
BALAKLAVA
RENMARK
WENTWORTH
MAITLAND
KAPUNDA
SEDAN
PT. LINCOLN
ROSALIND
CORNY POINT
ADELAIDE
YORKETOWN
MT. BARKER
MURRAY BRIDGE
WALPEUP
STRATHALBYN
PINNAROO
PEAKE
BULLAPARINGA
CAPE BORDA
MENINGIE
CAPE WILLOUGHBY
TINTINARA
BORDERTOWN
NHILL
WARRACKNABEAL
CANTARA
HORSHAM
ST. ARNAUD
KINGSTON
NARACOORTE
STAWELL
MARYBOROUGH
BEACHPORT
PENOLA
BALMORAL
MT. GAMBIER
HAMILTON
PORTLAND
WARRNAMBOOL
COLAC
N
JUNE 30-JULY 5

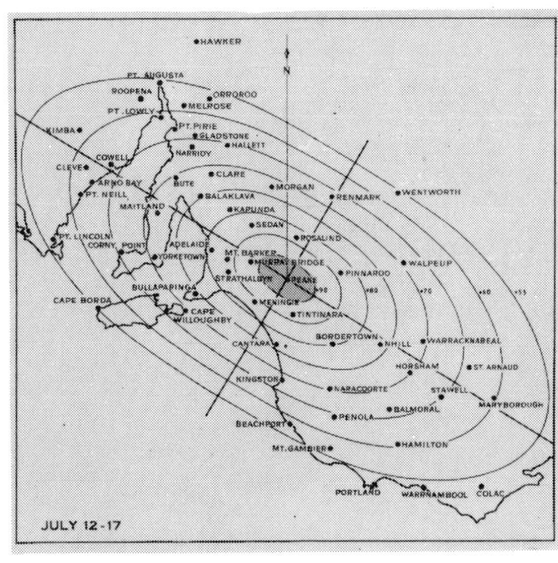
HAWKER
PT. AUGUSTA
ROOPENA
ORROROO
PT. LOWLY
MELROSE
KIMBA
PT. PIRIE
GLADSTONE
HALLETT
COWELL
NARRIDY
CLEVE
CLARE
ARNO BAY
BUTE
MORGAN
PT. NEILL
RENMARK
WENTWORTH
BALAKLAVA
MAITLAND
KAPUNDA
SEDAN
PT. LINCOLN
ROSALIND
CORNY POINT
ADELAIDE
YORKETOWN
MT. BARKER
MURRAY BRIDGE
PINNAROO
WALPEUP
STRATHALBYN
PEAKE
BULLAPARINGA
CAPE BORDA
MENINGIE
CAPE WILLOUGHBY
TINTINARA
BORDERTOWN
NHILL
WARRACKNABEAL
CANTARA
HORSHAM
ST. ARNAUD
KINGSTON
NARACOORTE
STAWELL
MARYBOROUGH
BALMORAL
PENOLA
BEACHPORT
MT. GAMBIER
HAMILTON
PORTLAND
WARRNAMBOOL
COLAC
N
JULY 12-17

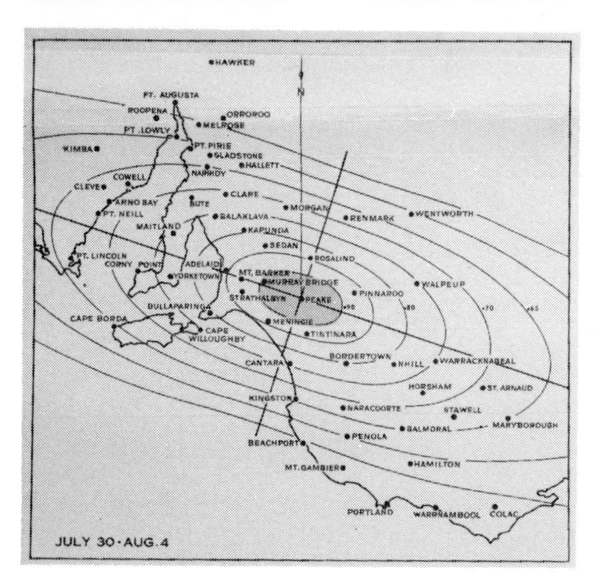

JULY 30-AUG. 4

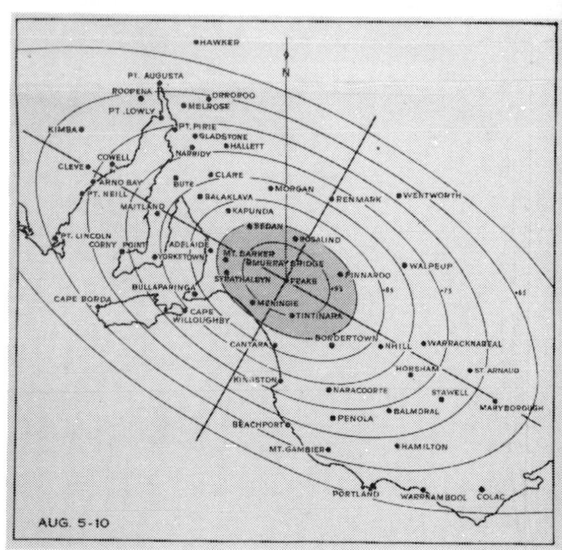

AUG. 5-10

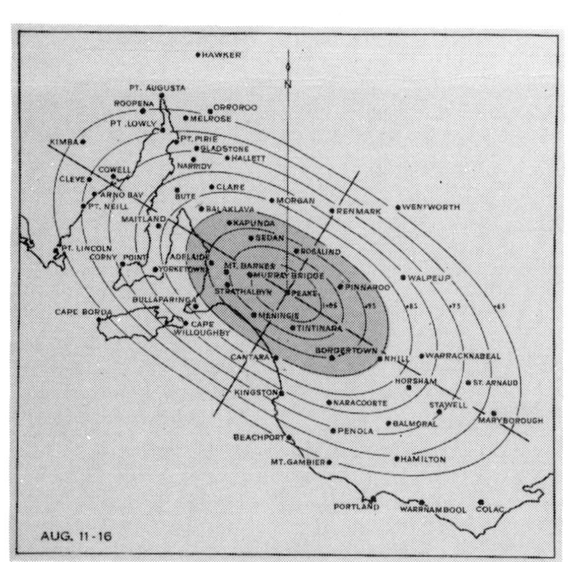

AUG. 11-16

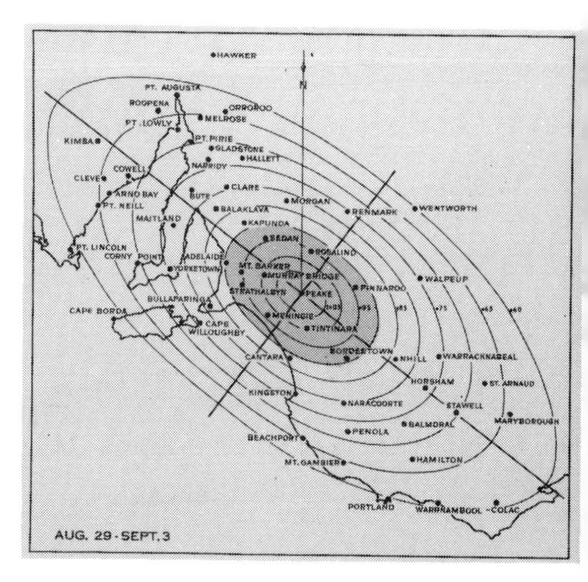

AUG. 29-SEPT. 3

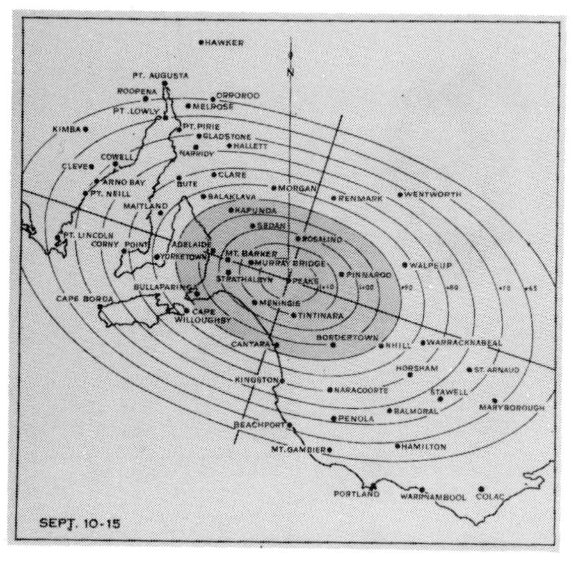

SEPT. 10-15

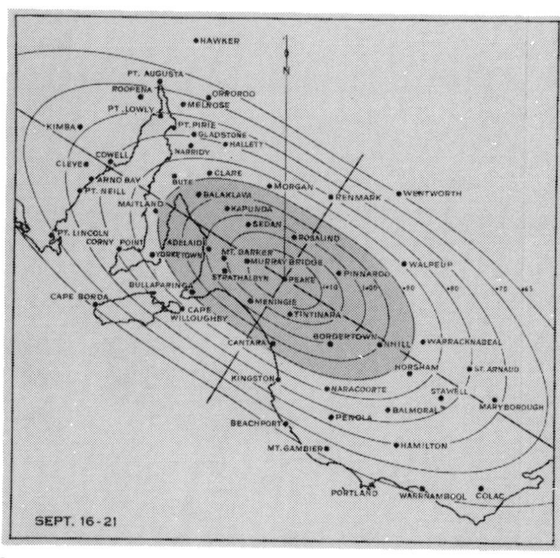

SEPT. 16-21

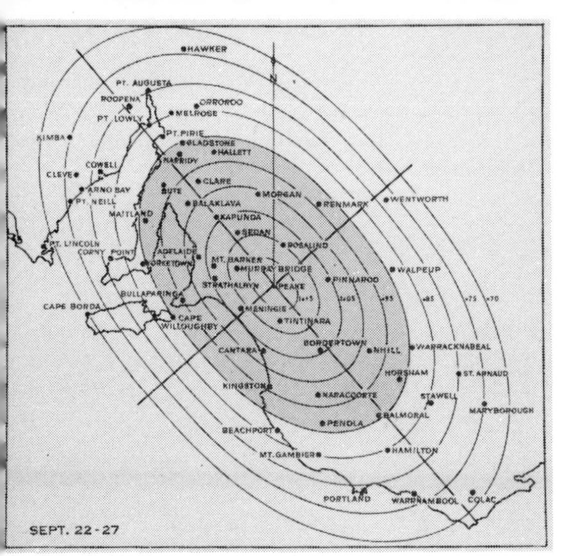
HAWKER
N
PT. AUGUSTA
ROOPENA
ORROROO
PT. LOWLY
MELROSE
KIMBA
PT. PIRIE
GLADSTONE
HALLETT
NARRIDY
COWELL
CLEVE
ARNO BAY
CLARE
BUTE
MORGAN
PT. NEILL
BALAKLAVA
RENMARK
WENTWORTH
MAITLAND
KAPUNDA
SEDAN
PT. LINCOLN
CORNY POINT
ROSALIND
ADELAIDE
YORKETOWN
MT. BARKER
MURRAY BRIDGE
PINNAROO
WALPEUP
STRATHALBYN
PEAKE
BULLAPARINGA
CAPE BORDA
MENINGIE
CAPE WILLOUGHBY
TINTINARA
BORDERTOWN
NHILL
WARRACKNABEAL
CANTARA
HORSHAM
ST. ARNAUD
KINGSTON
STAWELL
NARACOORTE
MARYBOROUGH
BALMORAL
PENOLA
BEACHPORT
MT. GAMBIER
HAMILTON
PORTLAND
WARRNAMBOOL
COLAC
SEPT. 22-27

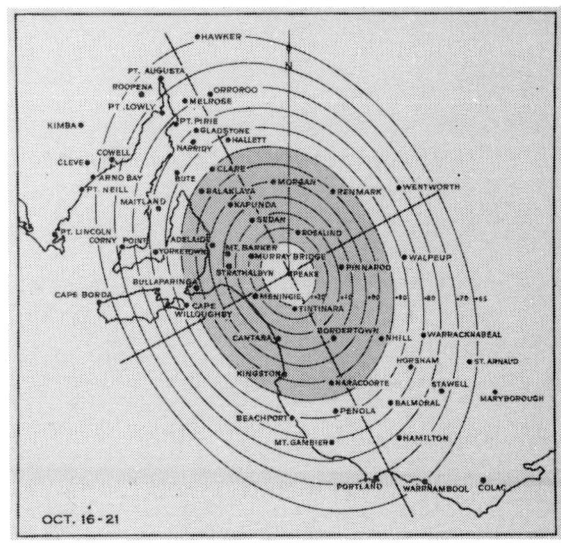
HAWKER
N
PT. AUGUSTA
ROOPENA
ORROROO
PT. LOWLY
MELROSE
KIMBA
PT. PIRIE
GLADSTONE
HALLETT
NARRIDY
COWELL
CLEVE
ARNO BAY
CLARE
BUTE
MORGAN
PT. NEILL
BALAKLAVA
RENMARK
WENTWORTH
MAITLAND
KAPUNDA
SEDAN
PT. LINCOLN
CORNY POINT
ROSALIND
ADELAIDE
YORKETOWN
MT. BARKER
MURRAY BRIDGE
PINNAROO
WALPEUP
STRATHALBYN
PEAKE
BULLAPARINGA
CAPE BORDA
MENINGIE
CAPE WILLOUGHBY
TINTINARA
BORDERTOWN
NHILL
WARRACKNABEAL
CANTARA
HORSHAM
ST. ARNAUD
KINGSTON
STAWELL
NARACOORTE
MARYBOROUGH
BALMORAL
PENOLA
BEACHPORT
MT. GAMBIER
HAMILTON
PORTLAND
WARRNAMBOOL
COLAC
OCT. 16-21

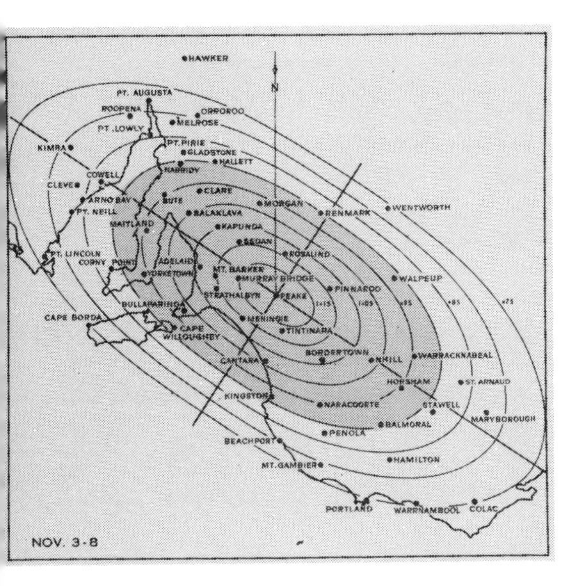
HAWKER
N
PT. AUGUSTA
ROOPENA
ORROROO
PT. LOWLY
MELROSE
KIMBA
PT. PIRIE
GLADSTONE
HALLETT
NARRIDY
COWELL
CLEVE
ARNO BAY
CLARE
BUTE
MORGAN
PT. NEILL
BALAKLAVA
RENMARK
WENTWORTH
MAITLAND
KAPUNDA
SEDAN
PT. LINCOLN
CORNY POINT
ROSALIND
ADELAIDE
YORKETOWN
MT. BARKER
MURRAY BRIDGE
PINNAROO
WALPEUP
STRATHALBYN
PEAKE
BULLAPARINGA
CAPE BORDA
MENINGIE
CAPE WILLOUGHBY
TINTINARA
BORDERTOWN
NHILL
WARRACKNABEAL
CANTARA
HORSHAM
ST. ARNAUD
KINGSTON
STAWELL
NARACOORTE
MARYBOROUGH
BALMORAL
PENOLA
BEACHPORT
MT. GAMBIER
HAMILTON
PORTLAND
WARRNAMBOOL
COLAC
NOV. 3-8

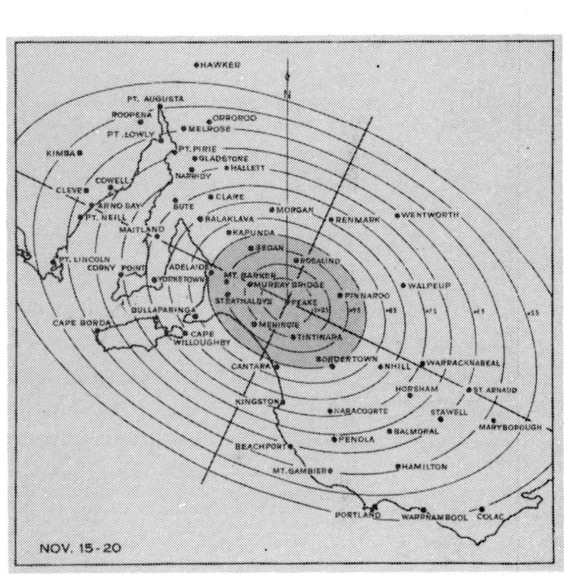
HAWKER
N
PT. AUGUSTA
ROOPENA
ORROROO
PT. LOWLY
MELROSE
KIMBA
PT. PIRIE
GLADSTONE
HALLETT
NARRIDY
COWELL
CLEVE
ARNO BAY
CLARE
BUTE
MORGAN
PT. NEILL
BALAKLAVA
RENMARK
WENTWORTH
MAITLAND
KAPUNDA
SEDAN
PT. LINCOLN
CORNY POINT
ROSALIND
ADELAIDE
YORKETOWN
MT. BARKER
MURRAY BRIDGE
PINNAROO
WALPEUP
STRATHALBYN
PEAKE
BULLAPARINGA
CAPE BORDA
MENINGIE
CAPE WILLOUGHBY
TINTINARA
BORDERTOWN
NHILL
WARRACKNABEAL
CANTARA
HORSHAM
ST. ARNAUD
KINGSTON
STAWELL
NARACOORTE
MARYBOROUGH
BALMORAL
PENOLA
BEACHPORT
MT. GAMBIER
HAMILTON
PORTLAND
WARRNAMBOOL
COLAC
NOV. 15-20

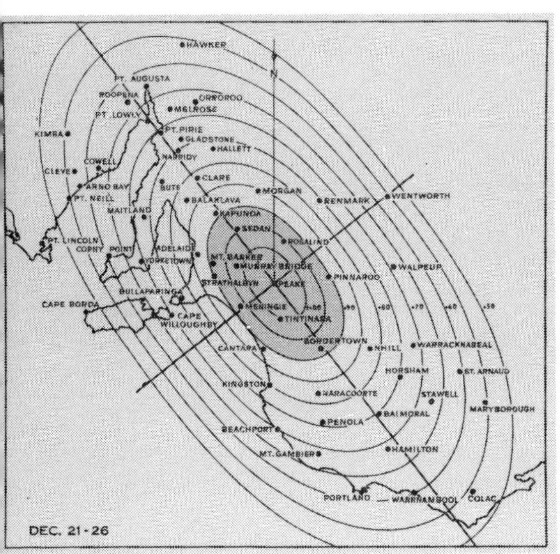
HAWKER
N
PT. AUGUSTA
ROOPENA
ORROROO
PT. LOWLY
MELROSE
KIMBA
PT. PIRIE
GLADSTONE
HALLETT
NARRIDY
COWELL
CLEVE
ARNO BAY
CLARE
BUTE
MORGAN
PT. NEILL
BALAKLAVA
RENMARK
WENTWORTH
MAITLAND
KAPUNDA
SEDAN
PT. LINCOLN
CORNY POINT
ROSALIND
ADELAIDE
YORKETOWN
MT. BARKER
MURRAY BRIDGE
PINNAROO
WALPEUP
STRATHALBYN
PEAKE
BULLAPARINGA
CAPE BORDA
MENINGIE
CAPE WILLOUGHBY
TINTINARA
BORDERTOWN
NHILL
WARRACKNABEAL
CANTARA
HORSHAM
ST. ARNAUD
KINGSTON
STAWELL
NARACOORTE
MARYBOROUGH
BALMORAL
PENOLA
BEACHPORT
MT. GAMBIER
HAMILTON
PORTLAND
WARRNAMBOOL
COLAC
DEC. 21-26

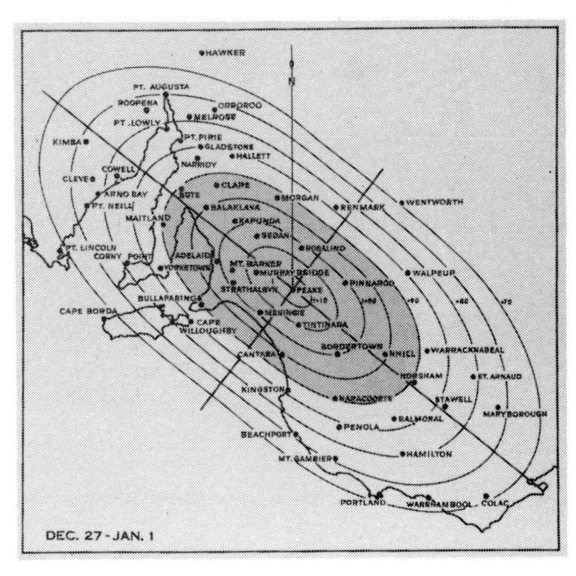
HAWKER
N
PT. AUGUSTA
ROOPENA
ORROROO
PT. LOWLY
MELROSE
KIMBA
PT. PIRIE
GLADSTONE
HALLETT
NARRIDY
COWELL
CLEVE
ARNO BAY
CLARE
BUTE
MORGAN
PT. NEILL
BALAKLAVA
RENMARK
WENTWORTH
MAITLAND
KAPUNDA
SEDAN
PT. LINCOLN
CORNY POINT
ROSALIND
ADELAIDE
YORKETOWN
MT. BARKER
MURRAY BRIDGE
PINNAROO
WALPEUP
STRATHALBYN
PEAKE
BULLAPARINGA
CAPE BORDA
MENINGIE
CAPE WILLOUGHBY
TINTINARA
BORDERTOWN
NHILL
WARRACKNABEAL
CANTARA
HORSHAM
ST. ARNAUD
KINGSTON
STAWELL
NARACOORTE
MARYBOROUGH
BALMORAL
PENOLA
BEACHPORT
MT. GAMBIER
HAMILTON
PORTLAND
WARRNAMBOOL
COLAC
DEC. 27-JAN. 1

Commonwealth Scientific and Industrial Research Organization, Australia. Division of Mathematical Statistics Technical Paper No. 11, Melbourne 1962.

FIDUCIAL REGIONS FOR PARAMETERS OF LOCATION IN COMPOUND HYPOTHESES

By E. A. CORNISH*

[*Manuscript received October* 30, 1961]

Summary

The applications of the simultaneous fiducial distribution of the location parameters from a multivariate normal distribution, in the practical testing of compound hypotheses respecting the values of these parameters, are illustrated by examples. Hitherto Hotelling's distribution has provided the basis for making such tests, but it is shown that, in the calculation of fiducial (or confidence) regions, this distribution is not as sensitive as the joint distribution of the parameters. This loss of sensitivity follows from the fact that Hotelling's T^2 statistic corresponds to the best linear discriminant calculated from the data, with a consequent loss of $p-1$ degrees of freedom (for p variates) from those available for estimating the variance-covariance matrix of the sample.

I. INTRODUCTION

In a series of papers, dating from 1930, Fisher has strongly advocated the fiducial argument as a tool of scientific inference. For the class of cases to which it is appropriate, this type of argument leads to rigorous probability statements about the parameters of the population, from which the observational record constitutes a random sample, without the assumption of any knowledge whatever, respecting their probability distributions *a priori*.

We may, for example, consider a set of N observations $x_1, x_2, \ldots, x_N$ taken from a normal distribution having a mean ξ, and variance σ^2, and calculate

$$\bar{x} = \frac{1}{N}\Sigma x,$$

$$s^2 = \frac{1}{N(N-1)}\Sigma(x-\bar{x})^2,$$

then Student's statistic t, defined by

$$t = (\bar{x}-\xi)/s,$$

is distributed in the well-known distribution

$$\frac{\Gamma\frac{1}{2}(n+1)}{(\pi n)^{\frac{1}{2}}\Gamma\frac{1}{2}n}\left(1+\frac{t^2}{n}\right)^{-\frac{1}{2}(n+1)} dt, \tag{1}$$

which depends only on the parameter $n = N-1$.

The distribution (1), and the relation

$$\xi = \bar{x}-st,$$

* Division of Mathematical Statistics, C.S.I.R.O., Adelaide.

together establish the fiducial distribution of ξ, based solely on the observed sample, in the form

$$\frac{\Gamma\frac{1}{2}(n+1)}{(\pi n)^{\frac{1}{2}}\Gamma\frac{1}{2}n}\left\{1+\frac{(\xi-\bar{x})^2}{ns^2}\right\}^{-\frac{1}{2}(n+1)}\frac{d\xi}{s}. \tag{2}$$

Fiducial probabilities which may be required in practice are obtainable by integrating (2) over appropriate ranges of ξ, but, clearly, are already available from tables of the probability integral of (1).

The statistic t may be regarded more generally as the ratio of any normally distributed quantity to an independent estimate of its standard deviation, and consequently, for appropriate observations, the distributions (1) and (2) provide the basis for testing hypotheses in the general problem of univariate tests respecting the values of a parameter of location.

The natural development is an extension to multivariate tests involving several parameters, and our immediate object is to consider tests which examine the possible variations of any hypothesis that specifies the values of the parameters simultaneously, while maintaining an assigned level of significance. Hypotheses such as this have been described as compound (Fisher 1959), and we shall continue to use the term.

For present purposes it is convenient to regard the extension to multivariate tests as following the two lines indicated below:

(i) A generalization of univariate analysis which is most readily treated by the notation and techniques of multivariate analysis. Reference is made, in particular, to an analysis of variance which has distinguished a number of components, independent or otherwise, due to different causes of variation, and we are required to test a compound hypothesis involving them. This type of problem has been considered in recent years by various workers including Scheffé (1953), Tukey (1953), Dunnett (1955), Fisher (1959), and Cornish (1960).

(ii) A direct extension to multivariate tests on the general multivariate normal distribution, in which we wish to examine the possible variations of a compound hypothesis respecting the elements of a vector of means. Hotelling (1931), Anderson (1958), Cornish (1961), and others have examined problems of this nature.

II. The Simultaneous Fiducial Distributions

The distribution (1) has been generalized (Cornish 1954; Dunnett and Sobel 1954) to give the simultaneous distribution of a number of t-statistics which are not only dependent through a common estimate of variance but may also be correlated because of the nature of the enquiry under consideration, and, in its turn, this multiple t-distribution has been used to establish the corresponding simultaneous fiducial distribution of the parameters, thus providing, in one sense, a generalization of (2).

Thus suppose $\mathbf{x}$ is a column vector with elements $x_1, x_2, \ldots, x_N$ which are normally and independently distributed with means $\xi_1, \xi_2, \ldots, \xi_N$ and common variance σ^2, and let

$$\mathbf{y} = \mathbf{H}\mathbf{x} = [h_{ij}]\mathbf{x} \qquad \begin{array}{l} i = 1, 2, \ldots, p \\ j = 1, 2, \ldots, N \end{array}$$

be any $p<N$, linearly independent, linear functions of the x_i, then the y_i are distributed in a multivariate normal distribution with a vector of means $\boldsymbol{\eta} = \mathbf{H}\boldsymbol{\xi}$ and variance-covariance matrix $\sigma^2\,\mathbf{HH}'$. If, now, s^2 is an estimate of σ^2, on n degrees of freedom, distributed independently of the x_i and, therefore, independently of the y_i, and

$$\mathbf{t} = \mathbf{Q}(\boldsymbol{\eta}-\mathbf{y})/s, \tag{3}$$

where $\mathbf{Q}$ is diagonal with its ith diagonal element equal to $(\mathbf{h}_i\mathbf{h}_i')^{\frac{1}{2}}$, $\mathbf{h}_i$ being the ith row vector of $\mathbf{H}$, the distribution of the t_i takes the form

$$\frac{\Gamma\frac{1}{2}(n+p)}{(\pi n)^{\frac{1}{2}p}\Gamma\frac{1}{2}n}\left(1+\frac{\mathbf{t}'\mathbf{R}^{-1}\mathbf{t}}{n}\right)^{-\frac{1}{2}(n+p)}\frac{d\mathbf{t}}{|\mathbf{R}|^{\frac{1}{2}}}, \tag{4}$$

where $\mathbf{R} = \mathbf{Q}(\mathbf{HH}')\mathbf{Q}$ is the non-singular, symmetric correlation matrix of the y_i, the correlation ρ_{ij} being

$$\frac{\mathbf{h}_i\mathbf{h}_j'}{(\mathbf{h}_i\mathbf{h}_i' \,.\, \mathbf{h}_j\mathbf{h}_j')^{\frac{1}{2}}}.$$

The elements of $\mathbf{H}$ are known constants satisfying the conditions

$$\sum_{j=1}^{N} h_{ij} = 0 \qquad \text{for } i = 1, 2, \ldots, p.$$

Using (3), and the relation

$$\mathbf{t}'\mathbf{R}^{-1}\mathbf{t} = (\boldsymbol{\eta}-\mathbf{y})'(s^2\mathbf{HH}')^{-1}(\boldsymbol{\eta}-\mathbf{y}), \tag{5}$$

the simultaneous fiducial distribution of the parameters is obtainable from (4) in the form

$$\frac{\Gamma\frac{1}{2}(n+p)}{(\pi n)^{\frac{1}{2}p}\Gamma\frac{1}{2}n}\left\{1+\frac{(\boldsymbol{\eta}-\mathbf{y})'(s^2\mathbf{HH}')^{-1}(\boldsymbol{\eta}-\mathbf{y})}{n}\right\}^{-\frac{1}{2}(n+p)}\frac{d\boldsymbol{\eta}}{|s^2\mathbf{HH}'|^{\frac{1}{2}}} \tag{6}$$

(Cornish 1960), from which any fiducial probability required may be calculated by integrating over appropriate ranges of the η_i.

Three important special cases may occur in practice:

(i) The rows of $\mathbf{H}$ are orthogonal, but not normalized, so that $\mathbf{HH}' = \mathbf{\Lambda}$, a diagonal matrix,

(ii) The rows of $\mathbf{H}$ are orthogonal, each having the same norm ν, so that $\mathbf{HH}' = \nu\mathbf{I}$,

(iii) The rows of $\mathbf{H}$ are orthogonal and normalized, so that $\mathbf{HH}' = \mathbf{I}$.

Fisher (1959, p. 206) has illustrated special case (i) and Bliss (1958) case (ii).

For the particular type of problem under consideration, the fiducial probability, corresponding to an assigned quantity T_1^2, is calculated by integrating (6) over the domain defined by

$$\frac{1}{p}(\boldsymbol{\eta}-\mathbf{y})'(s^2\mathbf{HH}')^{-1}(\boldsymbol{\eta}-\mathbf{y}) \leqslant T_1^2, \tag{7}$$

and it has been shown (Cornish 1960) that fiducial probabilities associated with regions such as those defined by (7), are obtainable directly from the distribution of e^{2z} with parameters $n_1 = p$ and $n_2 = n$. Equally, the result could have been derived by integration of (4) over the region defined by

$$\frac{1}{p}\mathbf{t}'\mathbf{R}^{-1}\mathbf{t} \leqslant T_1^2.$$

A comprehensive test which enables us to specify the aggregate of compound hypotheses contradicted by experimental or observational data, at any assigned level of significance, is then obtained by computing the hyperellipsoid

$$(\boldsymbol{\eta}-\mathbf{y})'(s^2\mathbf{HH}')^{-1}(\boldsymbol{\eta}-\mathbf{y}) = pe^{2z}, \tag{8}$$

in which the value of e^{2z} corresponds to the chosen level of significance when $n_1 = p$ and $n_2 = n$. All acceptable values of the parameters constitute the interior and boundary of the ellipsoid (8).

A more general distribution than (6), conjectured by Fisher (1954), has been established by the author (Cornish 1961). Thus, if $\mathbf{x}$ is a column vector whose elements are distributed in a multivariate normal distribution with a vector of means $\boldsymbol{\xi}' = [\xi_1\ \xi_2\ \ldots\ \xi_p]$, and variance-covariance matrix $\mathbf{\Sigma}$, of which there is available a non-singular estimate $\mathbf{S}$, on n degrees of freedom, and distributed independently of the x_i, the simultaneous distribution of the parameters is

$$\frac{\Gamma\frac{1}{2}(n+p)}{(\pi n)^{\frac{1}{2}p}\Gamma\frac{1}{2}n}\left\{1+\frac{(\boldsymbol{\xi}-\mathbf{x})'\mathbf{S}^{-1}(\boldsymbol{\xi}-\mathbf{x})}{n}\right\}^{-\frac{1}{2}(n+p)}\frac{d\boldsymbol{\xi}}{|\mathbf{S}|^{\frac{1}{2}}}. \tag{9}$$

In a manner similar to that above, to test a compound hypothesis respecting the values of the parameters, this is integrated over the region defined by

$$\frac{1}{p}(\boldsymbol{\xi}-\mathbf{x})'\mathbf{S}^{-1}(\boldsymbol{\xi}-\mathbf{x}) \leqslant T_1^2, \tag{10}$$

and, again, this fiducial probability is calculable from the distribution of e^{2z} with $n_1 = p$ and $n_2 = n$. The ellipsoid corresponding to (8) is then

$$(\xi - \mathbf{x})'\mathbf{S}^{-1}(\xi - \mathbf{x}) = pe^{2z}, \tag{11}$$

where the value of e^{2z} corresponds to the chosen fiducial probability when $n_1 = p$ and $n_2 = n$.

III. Desirable Features of the Inductive Procedure

There are fundamental differences in the methods of approach adopted by different groups of investigators in establishing tests of significance for compound hypotheses. Of those mentioned above, Fisher, Bliss, and Cornish have based their tests on Fisher's general theory of fiducial inference, whereas the remainder have followed the theory as advocated by Neyman, but apart from this, there are divergent views on what constitutes the appropriate distribution and the appropriate region of the parametric space to be used in the test of significance.

These are cardinal points in the construction of such a test. Since we are concerned with a test of the simultaneous agreement of a certain number of parameters with an equal number of observed quantities, it is natural, and reasonable, that the test should be based on the simultaneous distribution of an equal number of suitably chosen test quantities and, in addition, that the region of the parametric space, selected for making the inference, should have maximal average density of frequency for any assigned level of probability. By defining the region in this way, the simultaneous variation of the parameters is represented in accordance with the form of their simultaneous distribution (a similar remark applying to the test quantities), thus conferring on the test of significance a simple and natural interpretation. For example, the general cases represented by (6) and (9) have the ellipsoidal regions (8) and (11) respectively, while the special cases (i), (ii), and (iii), listed in Section II, have the ellipsoid

$$(\boldsymbol{\eta} - \mathbf{y})'(s^2\boldsymbol{\Lambda})^{-1}(\boldsymbol{\eta} - \mathbf{y}) = pe^{2z_1},$$

and the spheres

$$(\boldsymbol{\eta} - \mathbf{y})'(s^2\nu\mathbf{I})^{-1}(\boldsymbol{\eta} - \mathbf{y}) = pe^{2z_2},$$
$$(\boldsymbol{\eta} - \mathbf{y})'(s^2\mathbf{I})^{-1}(\boldsymbol{\eta} - \mathbf{y}) = pe^{2z_3},$$

respectively.

IV. Comparisons with Other Methods

The full implications of the two proposals made above become apparent on comparing properly chosen tests with those advocated by other workers. To illustrate first the difference in viewpoint regarding the choice of that region in the parametric space which shall be used for making the inference, we consider, for example, the method outlined by Dunnett (1955) for testing a compound hypothesis in the particular case where a number of treatments are compared with a control or standard.

A set of experimental observations has yielded $p+1$ means $\bar{x}_0$, $\bar{x}_1$, $\bar{x}_2$, . . . , $\bar{x}_p$ from a control and p other treatments respectively, together with an independent estimate of the variance, s^2, on n degrees of freedom. Dunnett's joint confidence region for the p differences $y_i = \bar{x}_i - \bar{x}_0$, $i = 1, 2, \ldots, p$, is obtained by a method which is equivalent to integrating (6) over the domain defined by the inequalities

$$y_i - ds_i \leqslant \eta_i \leqslant y_i + ds_i, \qquad i = 1, 2, \ldots, p,$$

where the s_i are the estimated standard deviations of the y_i (all based on s) and d is a multiple corresponding to the probability (or confidence coefficient) chosen (*vide* Dunnett's tables 1955).

Example 1

The following observations have been extracted from data on yield trials of new crossbred wheats at the Waite Agricultural Research Institute conducted during 1941, a year in which the season was characterized by a severe and widespread epidemic of wheat stem rust (Waite Agricultural Research Institute 1943). The mean yields of three varieties were

	Variety		
	Ranee 4H	Nabawa × Hope (37–76)	Gluwari
Mean yield (bush/ac)	16·03	40·31	26·06
Number of replicates	5	5	5

and the objective is to compare simultaneously the yields of the resistant varieties Nabawa × Hope and Gluwari with the standard non-resistant variety Ranee 4H.

The linear contrasts among the observations are, therefore,

$$40{\cdot}31 - 16{\cdot}03 = 24{\cdot}28,$$
$$26{\cdot}06 - 16{\cdot}03 = 10{\cdot}03,$$

and these are the elements of the vector $\mathbf{y}$ obtained from the original yield data by the matrix

$$\mathbf{H} = \begin{bmatrix} -1/5 & -1/5 & -1/5 & -1/5 & -1/5 & 1/5 & 1/5 & 1/5 & 1/5 & 1/5 & . & . & . & . & . \\ -1/5 & -1/5 & -1/5 & -1/5 & -1/5 & . & . & . & . & . & 1/5 & 1/5 & 1/5 & 1/5 & 1/5 \end{bmatrix}.$$

We then have

$$\mathbf{HH}' = \begin{bmatrix} 2/5 & 1/5 \\ 1/5 & 2/5 \end{bmatrix} \quad \text{and} \quad (\mathbf{HH}')^{-1} = \begin{bmatrix} 10/3 & -5/3 \\ -5/3 & 10/3 \end{bmatrix}.$$

The variance of a single observation is 11·33, on 12 degrees of freedom, so that the 1% test of significance for a hypothesis specifying the values of η_1 and η_2, will accept any hypothesis for which

$$(10/3)[(\eta_1 - 24{\cdot}28)^2 - (\eta_1 - 24{\cdot}28)(\eta_2 - 10{\cdot}03) + (\eta_2 - 10{\cdot}03)^2]$$
$$\leqslant 2 \times 11{\cdot}33 \times 6{\cdot}93 = 157{\cdot}06, \qquad (12)$$

where 6·93 is the 1% value of e^{2z} for $n_1 = 2$ and $n_2 = 12$. At this level of probability, all acceptable values of the parameters η_1 and η_2 fall within, or on, the boundary of the ellipse defined by (12), centred at the point (24·28, 10·03). This 99% fiducial region is the smaller ellipse of Figure 1, and in the first instance should be contrasted with the corresponding confidence region, the square in Figure 1, computed according to the procedure advocated by Dunnett. In each case the integral is 0·99, so that the contributions from the four sections A, B, C, and D, which should have been included, are exactly compensated by those from E, F, G,

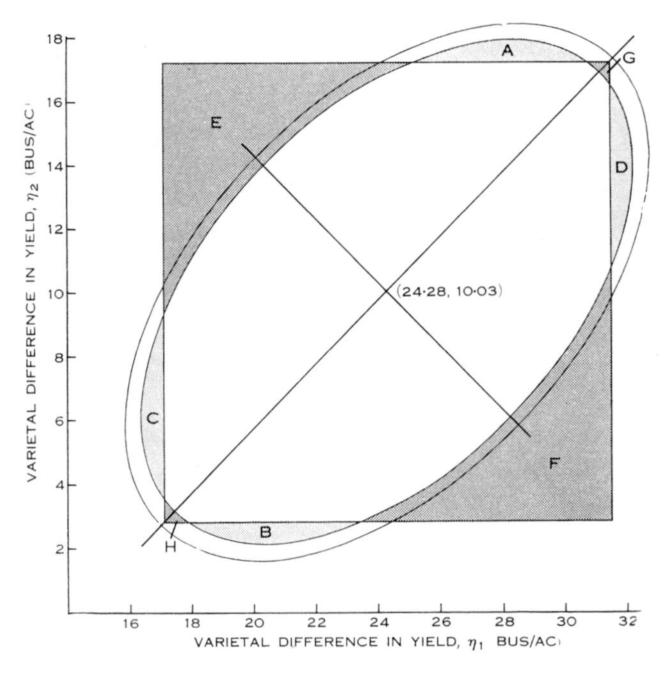

Fig. 1.—The 99% fiducial and confidence regions for observations of Example 1. *Smaller ellipse*: fiducial region calculated from the simultaneous fiducial distribution; *larger ellipse*: fiducial region calculated from Hotelling's distribution; *square*: Dunnett's confidence region.

and H. As already stated, the ellipse makes due allowance for the correlational properties of the distribution in testing a compound hypothesis and, in particular, emphasizes the fact that the limits for either parameter are conditional upon the value of the other parameter. At any assigned fiducial probability (or confidence coefficient), we may clearly take any fiducial (or confidence) region of arbitary form, but in a test of a compound hypothesis the only region really appropriate for our purposes is that which represents the simultaneous variation. Unconditional limits, such as those proposed by Dunnett, do not yield a common-sense answer to the question which has been posed.

Adverting now to consideration of the distribution which should be used in the test of significance, we first remark that the test based on the statistic T^2 was designed by Hotelling (1931) to provide a multivariate analogue of the test based on Student's statistic t. In particular, Hotelling's statistic has the important property of yielding exact probabilities independent of the unknown parameters. Moreover, as a test of the simultaneous deviations of a set of observed quantities from values proposed for them by hypothesis, it provides a means for testing a compound hypothesis in the sense in which we are using the term.

As, however, our object is to make a test of p quantities simultaneously, it seems pertinent to ask whether a more sensitive test would be obtained if it were based on p properly chosen test statistics rather than on just a single quantity, however suitable this might otherwise be, which itself provides a single measure of the simultaneous discrepancies. Attention has already been directed to the possibilities of following such a course (Cornish 1961).

Considering first the class of data and associated problem to which either the distribution (4) or the distribution (6) applies, the existence of a more sensitive test than that based on Hotelling's distribution is readily demonstrated. Two fiducial regions of the type indicated by (7) can be determined for this class of problem. The first is obtained from the fiducial distribution (6), or alternatively from the multiple t-distribution (4) using the relation (5), and the second could be derived from Hotelling's distribution using the fact that

$$\frac{n-p+1}{np}T^2 = \frac{n-p+1}{np}(\boldsymbol{\eta}-\mathbf{y})'(s^2\mathbf{HH}')^{-1}(\boldsymbol{\eta}-\mathbf{y})$$

is distributed as e^{2z} with $n_1 = p$ and $n_2 = n-p+1$, as first shown by Hotelling (1931). This second fiducial region is the ellipsoid

$$(\boldsymbol{\eta}-\mathbf{y})'(s^2\mathbf{HH}')^{-1}(\boldsymbol{\eta}-\mathbf{y}) = \frac{np}{n-p+1}e^{2z}, \tag{13}$$

where e^{2z} is the percentile at the chosen probability when $n_1 = p$ and $n_2 = n-p+1$. Comparison of (8) and (13) shows that they are homothetic ellipsoids, and that for an assigned probability, the former is always contained in the latter except for the special cases:

(i) $p = 1$, when the two regions coincide,

(ii) $p-1 \geqslant n > 1$, when the region based on Hotelling's distribution is unobtainable, but a meaningful, though restricted fiducial region, as defined by (8), is available.

For a given probability and given n, the discrepancy between the two regions increases as p increases, but for n large compared with p, whatever finite value is assigned the latter, the difference becomes smaller, tending to zero in the limit.

Example 2

Using the data from Example 1, the fiducial region defined by (13) is

$$\begin{aligned}(10/3)[(\eta_1-24\cdot28)^2-(\eta_1-24\cdot28)(\eta_2-10\cdot03)+(\eta_2-10\cdot03)^2]\\ = (2\times12/11)\times11\cdot33\times7\cdot20 = 178\cdot02,\end{aligned}$$

where $7\cdot20$ is the value of e^{2z} at the 1% point when $n_1 = 2$ and $n_2 = 11$. This is the larger ellipse in Figure 1.

It should be noted that advocates of the confidence theory could derive two confidence regions, the first from (4) and the second from Hotelling's distribution with p and $n-p+1$ degrees of freedom, coinciding respectively with the fiducial regions (8) and (13). The confidence region corresponding to (13) is the one which is currently advocated (*vide*, for example, Kullback 1959).

The situation respecting the general case of the multivariate normal distribution is more complex. In this connexion, it is of paramount importance that, on the basis of the fiducial theory, an inference can be made either in the form of a test of significance or as an induction of the fiducial type. The full force of this statement becomes apparent when we observe that in the theory of sampling from the general multivariate normal distribution there is, at present, no known distribution corresponding to (4) which could be used in a test of significance, but with the establishment of the fiducial distribution (9) an induction of the fiducial type is available. On the other hand, Hotelling's distribution is available for making either type of inference, and moreover has been used for computing a confidence region (Hotelling 1931), *but it should be noted that, at this juncture, advocates of the confidence theory are restricted to using the less sensitive procedure based on this distribution.*

Thus, for an assigned probability, the fiducial region calculable from (9) is the ellipsoid (11), whereas the fiducial region based on Hotelling's distribution is the ellipsoid

$$(\xi-\mathbf{x})'\mathbf{S}^{-1}(\xi-\mathbf{x}) = \frac{np}{n-p+1}e^{2z}, \tag{14}$$

in which e^{2z} corresponds to the probability chosen, when $n_1 = p$ and $n_2 = n-p+1$, and, as before, the ellipsoid (11) lies completely within the ellipsoid (14). The confidence region based on Hotelling's distribution coincides with (14).

Example 3

The following observations have been extracted from an extensive study of the variation of percentage sugar and percentage acid in maturing grapes and of the factors affecting them (B. C. Rankine, K. M. Cellier, and E. W. Boehm, unpublished data). The mean yield in lb/vine was 17·28, the mean sugar percentage was 23·83, and the inverse of the variance-covariance matrix of these means was

$$\begin{bmatrix} 30\cdot9013 & 3\cdot7453 \\ 3\cdot7453 & 0\cdot8867 \end{bmatrix}$$

on 15 degrees of freedom.

The test for a hypothesis, specifying the true mean yield and sugar percentage, based on the distribution (9) with a fiducial probability of 0·99, will accept any hypothesis for which

$$30\cdot90(\xi_1-23\cdot83)^2+7\cdot49(\xi_1-23\cdot83)(\xi_2-17\cdot28)+0\cdot89(\xi_2-17\cdot28)^2$$
$$\leqslant 2\times6\cdot36 = 12\cdot72,$$

where $6 \cdot 36$ is the value of e^{2z} at the 1% point when $n_1 = 2$ and $n_2 = 15$. This is the smaller ellipse in Figure 2.

The fiducial region computed on the basis of Hotelling's distribution is the ellipse

$$30 \cdot 90(\xi_1 - 23 \cdot 83)^2 + 7 \cdot 49(\xi_1 - 23 \cdot 83)(\xi_2 - 17 \cdot 28) + 0 \cdot 89(\xi_2 - 17 \cdot 28)^2 \leqslant (2 \times 15/14) \times 6 \cdot 51 = 13 \cdot 95,$$

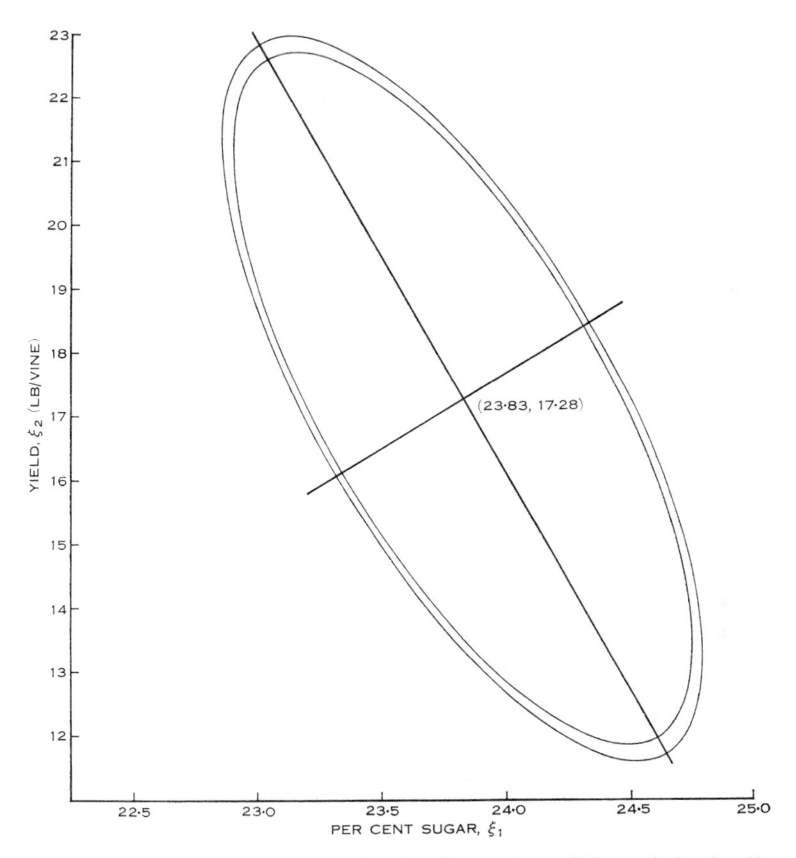

Fig. 2.—The 99% fiducial regions for the observations of Example 3. *Smaller ellipse*: Fiducial region calculated from the simultaneous fiducial distribution; *larger ellipse*: fiducial region calculated from Hotelling's distribution.

where $6 \cdot 51$ is the value of e^{2z} at the 1% point when $n_1 = 2$ and $n_2 = 14$. This is the larger ellipse in Figure 2, which also coincides with the confidence region based on Hotelling's distribution.

To facilitate diagrammatic representation we have taken $p = 2$ in the illustrations used above and, under this condition, the difference between the two fiducial regions or the difference between the fiducial and confidence regions is minimal. As an indication of the increasing discrepancy for larger values of p the example given below has been added.

Example 4

The following data have been extracted from a comprehensive series of observations designed to determine the association of mean sea-level pressure patterns and temperature over the Australian continent south of 20°S. latitude (L. G. Veitch, unpublished data).

Mean pressures at 14 stations have been determined for the period January 22 to April 21 during 1950–59 inclusive, the unit being 1 mb. The variance–covariance matrix of these means, on 179 degrees of freedom, has for its determinant the value 1,760,240,300, so that the hypervolume of the fiducial region (11), based on the fiducial distribution (9), is 624143×10^9 (e^{2z} at the 1% point $= 2 \cdot 1820$ when $n_1 = 14$ and $n_2 = 179$), which is only 57% of the volume of the fiducial region (14) based on Hotelling's distribution, namely, 1089×10^{12} (e^{2z} at the 1% point $= 2 \cdot 1910$ when $n_1 = 14$ and $n_2 = 166$).

V. Fiducial Limits for Linear Functions of the Components of the Mean Vector

Although not directly relevant, the situation which arises when inferences have to be made regarding arbitrary linear compounds of the mean vector has some connexion with the problem under consideration.

If $\boldsymbol{\lambda}$ is a non-null, column vector with p arbitrary, real, non-stochastic elements, the observable variate $\boldsymbol{\lambda}'\mathbf{x}$ is normally distributed with a mean $\boldsymbol{\lambda}'\boldsymbol{\xi}$ and variance $\boldsymbol{\lambda}'\boldsymbol{\Sigma}\boldsymbol{\lambda}$, and consequently

$$t = \boldsymbol{\lambda}'(\boldsymbol{\xi}-\mathbf{x})/(\boldsymbol{\lambda}'\mathbf{S}\boldsymbol{\lambda})^{\frac{1}{2}}$$

is distributed in Student's distribution with n degrees of freedom independently of $\boldsymbol{\lambda}'\boldsymbol{\xi}$ and $\boldsymbol{\lambda}'\boldsymbol{\Sigma}\boldsymbol{\lambda}$, and hence fiducial limits

$$\boldsymbol{\lambda}'\mathbf{x}-(\boldsymbol{\lambda}'\mathbf{S}\boldsymbol{\lambda})^{\frac{1}{2}}t_P \leqslant \boldsymbol{\lambda}'\boldsymbol{\xi} \leqslant \boldsymbol{\lambda}'\mathbf{x}+(\boldsymbol{\lambda}'\mathbf{S}\boldsymbol{\lambda})^{\frac{1}{2}}t_P \tag{15}$$

for the linear function, corresponding to a fiducial probability $1-P$, can be assigned.

Similar limits are also calculable from Hotelling's distribution. For a given positive constant k the statement that

$$\frac{|\boldsymbol{\lambda}'(\boldsymbol{\xi}-\mathbf{x})|}{(\boldsymbol{\lambda}'\mathbf{S}\boldsymbol{\lambda})^{\frac{1}{2}}} \leqslant k$$

is equivalent to

$$\frac{\boldsymbol{\lambda}'(\boldsymbol{\xi}-\mathbf{x})(\boldsymbol{\xi}-\mathbf{x})'\boldsymbol{\lambda}}{\boldsymbol{\lambda}'\mathbf{S}\boldsymbol{\lambda}} \leqslant k^2,$$

and, for given $\mathbf{x}$, $\boldsymbol{\xi}$, and $\mathbf{S}$, the maximal value of the first member of this inequality for variation of $\boldsymbol{\lambda}$ is

$$(\boldsymbol{\xi}-\mathbf{x})'\mathbf{S}^{-1}(\boldsymbol{\xi}-\mathbf{x}) = T^2,$$

which on the null hypothesis is distributed in Hotelling's distribution with p and $n-p+1$ degrees of freedom, independently of $\boldsymbol{\xi}$ and $\boldsymbol{\Sigma}$.

If

$$\Pr\left\{\frac{\boldsymbol{\lambda}'(\boldsymbol{\xi}-\mathbf{x})(\boldsymbol{\xi}-\mathbf{x})'\boldsymbol{\lambda}}{\boldsymbol{\lambda}'\mathbf{S}\boldsymbol{\lambda}} \leqslant k^2\right\} = P,$$

then $k^2 = T_P^2$, the percentile corresponding to the probability P in Hotelling's distribution with p and $n-p+1$ degrees of freedom. Thus the fiducial probability that

$$\boldsymbol{\lambda}'(\boldsymbol{\xi}-\mathbf{x})(\boldsymbol{\xi}-\mathbf{x})'\boldsymbol{\lambda} \leqslant T_P^2(\boldsymbol{\lambda}'\mathbf{S}\boldsymbol{\lambda})$$

is $1-P$ and this is also the fiducial probability that

$$\boldsymbol{\lambda}'\mathbf{x}-(T_P^2\boldsymbol{\lambda}'\mathbf{S}\boldsymbol{\lambda})^{\frac{1}{2}} \leqslant \boldsymbol{\lambda}'\boldsymbol{\xi} \leqslant \boldsymbol{\lambda}'\mathbf{x}+(T_P^2\boldsymbol{\lambda}'\mathbf{S}\boldsymbol{\lambda})^{\frac{1}{2}}, \tag{16}$$

giving fiducial limits for the linear compound for comparison with those in (15) (cf. Roy 1957).

In both cases the respective fiducial limits are perfectly valid as limits on the value of the compound itself, but it is difficult to appreciate why limits based on Hotelling's distribution should be preferred to the more precise and more easily calculated limits based on Student's t-statistic. The greater sensitivity of the limits in (15) is a direct consequence of the fact that the multivariate distribution (9) has been integrated over the domain defined by

$$\frac{\boldsymbol{\lambda}'(\boldsymbol{\xi}-\mathbf{x})}{(\boldsymbol{\lambda}'\mathbf{S}\boldsymbol{\lambda})^{\frac{1}{2}}} = t \leqslant t_P.$$

The inductive procedure should terminate at the stage where limits have been assigned to the linear function, and it is not valid to press refinement further to make inferences on the components of the mean vector by computing a joint fiducial region for them from the limits calculated for the linear function. Thus, taking for example the case based on Student's t, the relation

$$\boldsymbol{\lambda}'\boldsymbol{\xi} = \boldsymbol{\lambda}'\mathbf{x}+(\boldsymbol{\lambda}'\mathbf{S}\boldsymbol{\lambda})^{\frac{1}{2}}t_P$$

could be interpreted as a plane in the parametric space, and thus the two fiducial limits would yield two parallel planes in this space; these planes and the zone between them cannot, however, be validly regarded as a fiducial region of the type considered in previous sections, because the correlational properties have not been taken into account. The circumstances are even more extreme in this respect than in the computation of Dunnett's confidence region.

The same limits as those given in (15) and (16) could be obtained as the extremes of two confidence intervals, but the same criticism would apply to any attempt at extending the inference to the simultaneous occurrence of values of the individual components of the mean vector.

Example 5

The following observations for Stations 29 (lat. 35°S., long. 135°E.) and 30 (lat. 35°S., long. 145°E.) have been taken from the data of Example 4 for purposes of illustration. Let the vector of mean pressures be $\boldsymbol{\xi}' = [\xi_1\ \xi_2]$ and take $\boldsymbol{\lambda}' = [-1\ 1]$. With this value of $\boldsymbol{\lambda}'$ and the particular choice of stations, the linear function $\xi_2 - \xi_1$ has a well-defined physical meaning, namely, that, if positive, it represents a wind flow across the Adelaide region from north to south, and vice versa.

The observed mean pressures were $\bar{x}_1 = 1016 \cdot 28$ and $\bar{x}_2 = 1012 \cdot 97$, the unit as before being 1 mb, and their variance-covariance matrix was

$$\mathbf{S} = \begin{bmatrix} 0 \cdot 268897 & 0 \cdot 137244 \\ 0 \cdot 137244 & 0 \cdot 259408 \end{bmatrix}$$

on 119 degrees of freedom, of which the inverse is

$$\mathbf{S}^{-1} = \begin{bmatrix} 5 \cdot 094604 & -2 \cdot 695370 \\ -2 \cdot 695370 & 5 \cdot 280950 \end{bmatrix}.$$

Designating the observed mean vector by $\overline{\mathbf{x}}$,

$$\boldsymbol{\lambda}'\overline{\mathbf{x}} = -3 \cdot 31, \qquad \boldsymbol{\lambda}'\mathbf{S}\boldsymbol{\lambda} = 0 \cdot 253818, \qquad (\boldsymbol{\lambda}'\mathbf{S}\boldsymbol{\lambda})^{\frac{1}{2}} = 0 \cdot 504,$$

and taking $t = 2 \cdot 62$, the value at 1% for 119 degrees of freedom, the fiducial limits for the chosen compound based on Student's distribution are

$$-4 \cdot 63 \leqslant \boldsymbol{\lambda}'\boldsymbol{\xi} \leqslant -1 \cdot 99.$$

The value of e^{2z} at 1% for $n_1 = 2$ and $n_2 = 118$ is $4 \cdot 79$, so that the fiducial limits based on Hotelling's distribution are

$$-4 \cdot 88 \leqslant \boldsymbol{\lambda}'\boldsymbol{\xi} \leqslant -1 \cdot 74,$$

and the fiducial region given by (11) is the ellipse

$$5 \cdot 0946(\xi_1 - 1016 \cdot 28)^2 - 5 \cdot 3907(\xi_1 - 1016 \cdot 28)(\xi_2 - 1012 \cdot 97) + 5 \cdot 2810(\xi_2 - 1012 \cdot 97)^2 = 9 \cdot 576.$$

The three regions are represented in the two-dimensional space of the parameters in Figure 3; the dotted lines indicate, as near as can be judged, the upper and lower physical limits of the mean pressures.

VI. The Reason for the Loss of Sensitivity when using Hotelling's Distribution

The introduction of the linear discriminant function into multivariate analysis by Fisher (1936) presented a new method of deriving tests suitable for a multiplicity of variates. The inherent complexity of such problems was thus obviated and replaced by the very much simpler task of analysing a single variate obtained by linearly compounding the several variates, the compounding coefficients being chosen to maximize the value of a statistic suitable for a single variate. In the general case, these functions are such that they maximize the ratio of apportionment of two sums of squares, based on n_1 degrees of freedom for the selected category and

n_2 residual degrees of freedom, respectively, and in their simplest application, $n_1 = 1$, corresponding to the situation in which two groups are to be differentiated. Denoting the numbers of observations in the two groups by N_1 and N_2 and assuming there are p variates, the analysis of variance of the linear discriminant (or that

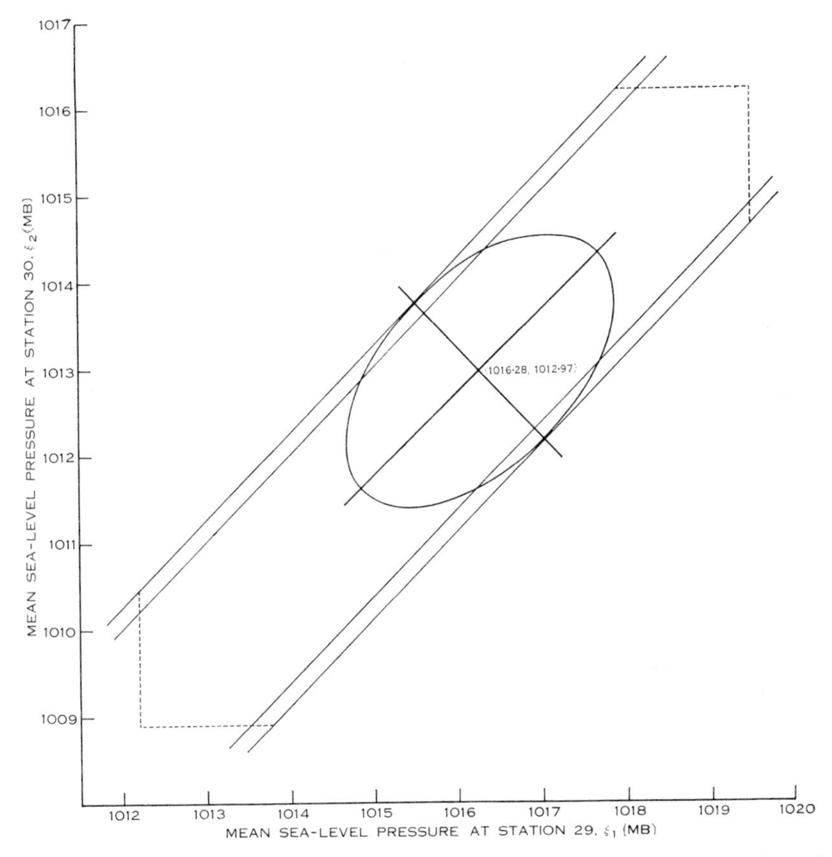

Fig. 3.—Correct 99% fiducial region and spurious fiducial regions from data of Example 5. *Ellipse*: correct fiducial region calculated from simultaneous fiducial distributions; *inner pair of parallel lines*: spurious fiducial region from the fiducial limits calculated using Student's distribution; *outer pair of parallel lines*: spurious fiducial region from the fiducial limits calculated using Hotelling's distribution.

of the multiple regression analogue) shows clearly the appropriate allocation of the degrees of freedom as follows:

Variation due to	Deg. Freedom
Between groups	p
Within groups	N_1+N_2-p-1
Total	N_1+N_2-1

in which $p-1$ degrees of freedom have been transferred from the residual to the component between groups, and the z test of significance from this analysis of variance is identical with the test of significance from Hotelling's distribution with degrees of freedom p and $n-p+1$ where $n = N_1+N_2-2$ (Fisher 1938).

The identity of the two tests of significance follows from the fact that the ratio of the two sums of squares in the above analysis is proportional to Hotelling's T^2. The allocation of the degrees of freedom for Hotelling's test is thus due to the correspondence of T^2 to a single variate, linearly compounded of the original variates and chosen to maximize the distinctness of the two groups. In order that the problem of comparing the two groups with respect to their multiple variates may be reduced to one of comparing these groups with respect to a specially chosen single variate, a premium must be paid, namely, the loss of $p-1$ degrees of freedom from those available for assessing the variation within the groups, with a consequent loss of sensitivity. The fact that the discrepancy between the tests outlined in previous sections increases with p for fixed n, conforms precisely with this view.

Similar remarks apply to other situations in which Hotelling's distribution has hitherto provided the test of significance; in all cases Hotelling's test is now superseded by the test based on the simultaneous fiducial distribution (9).

VII. References

Anderson, T. W. (1958).—"An Introduction to Multivariate Analysis." (Wiley: New York.)
Bliss, C. I. (1958).—Bull. Conn. Agric. Exp. Sta. No. 615.
Cornish, E. A. (1954).—*Aust. J. Phys.* **7**: 531–42.
Cornish, E. A. (1960).—*Aust. J. Statist.* **2**: 32–40.
Cornish, E. A. (1961).—C.S.I.R.O. Aust. Div. Math. Statist. Tech. Pap. No. 8.
Dunnett, C. W. (1955).—*J. Amer. Statist. Ass.* **50**: 1096–121.
Dunnett, C. W., and Sobel, M. (1954).—*Biometrika* **41**: 153–76.
Fisher, R. A. (1936).—*Ann. Eugen., Lond.* **7**: 179–88.
Fisher, R. A. (1938).—*Ann. Eugen., Lond.* **8**: 376–86.
Fisher, R. A. (1954).—*J. R. Statist. Soc.* B **16**: 212–3.
Fisher, R. A. (1959).—"The Design of Experiments." 7th Ed. (Oliver & Boyd: Edinburgh.)
Hotelling, H. (1931).—*Ann. Math. Statist.* **2**: 360–78.
Kullback, S. (1959).—"Information Theory and Statistics." (Wiley: New York.)
Roy, S. N. (1957).—"Some Aspects of Multivariate Analysis." (Wiley: New York.)
Scheffé, H. (1953).—*Biometrika* **40**: 87–104.
Tukey, J. W. (1953).—The problem of multiple comparisons. (Unpublished notes, Princeton University).
Waite Agricultural Research Institute (1943).—Report 1941–42.

Printed by C.S.I.R.O., Melbourne

Commonwealth Scientific and Industrial Research Organization, Australia. Division of Mathematical Statistics Technical Paper No. 12, Melbourne 1962.

TABLES FOR GRADUATING BY ORTHOGONAL POLYNOMIALS

By E. A. Cornish* and Marilyn J. Evans*

[*Manuscript received November 24, 1961*]

Summary

The multipliers which are required for calculating the coefficients of orthogonal polynomials up to the fifth degree from the successive differences of the observed series (containing up to 52 values) have been tabulated.

I. Introduction

If a polynomial regression line

$$Y = a+bx+cx^2+dx^3+\ldots$$

is to be fitted to a series of values y_i, $i = 1, 2, \ldots, N$ of a variate y which correspond to N successive values of a variate x taken at equal intervals, it is convenient (Fisher 1958) to use instead the equivalent regression

$$Y = A+B\xi_1+C\xi_2+\ldots,$$

in which $\xi_1, \xi_2, \ldots$ are orthogonal polynomials of degree 1, 2, ... in the variate x.

II. Methods of Fitting

There are two principal methods of determining the coefficients $A, B, C, \ldots$ and the polynomial values of Y corresponding to the N values of x,

(*a*) by successive addition (Fisher, ibid.),

(*b*) by multiplication using a table of values of the ξ or multiples of them $\xi' = \lambda\xi$ (Fisher and Yates 1957).

As indicated by these authors, various steps can be taken to reduce the arithmetic in both techniques, but the relative advantages of the two procedures depend also upon the facilities at the disposal of the computer. As a general rule, the second method is the more expeditious when desk calculating machines with easy and rapid facilities for multiplication are available, but for large values of N, and particularly where a number of different polynomials have to be fitted, successive addition using punched-card equipment is extremely rapid. For example, using a Powers Samas tabulator and summary card punch, all the quantities required to fit a polynomial of the 6th degree can be obtained for 6 sets of data, each with $N = 61$, in 15 minutes, excluding the time required to punch the variate values of y on the cards. With the increasing availability of electronic computers, investigators would, however, naturally have recourse to such equipment wherever possible.

* Division of Mathematical Statistics, C.S.I.R.O., Adelaide.

An alternate procedure, similar to (*b*) above, has been described by Fisher (ibid.). Consider a series of the variate y in which $N = 5$. The numerical values of the polynomials ξ_1', ξ_2', ξ_3', and ξ_4', from Table XXIII (Fisher and Yates ibid.) are

ξ_1'	ξ_2'	ξ_3'	ξ_4'
−2	2	−1	1
−1	−1	2	−4
0	−2	0	6
1	−1	−2	−4
2	2	1	1

and the successive differences of the series y_i are

	Δ	Δ^2	Δ^3	Δ^4
y_1				
	y_2-y_1			
y_2		$y_3-2y_2+y_1$		
	y_3-y_2		$y_4-3y_3+3y_2-y_1$	
y_3		$y_4-2y_3+y_2$		$y_5-4y_4+6y_3-4y_2+y_1$
	y_4-y_3		$y_5-3y_4+3y_3-y_2$	
y_4		$y_5-2y_4+y_3$		
	y_5-y_4			
y_5				

in which it will be noted that the coefficients of $y_1, y_2, \ldots, y_5$ respectively in Δ^4 are precisely the values of ξ_4', in their proper order, and hence

$$S(y\xi_4') = \Delta^4 y_1.$$

Inspection shows that

$$S(y\xi_1') = 2\Delta y_1 + 3\Delta y_2 + 3\Delta y_3 + 2\Delta y_4,$$

$$S(y\xi_2') = 2\Delta^2 y_1 + 3\Delta^2 y_2 + 2\Delta^2 y_3,$$

$$S(y\xi_3') = \Delta^3 y_1 + \Delta^3 y_2,$$

and this suggests an alternate method of computing the quantities $S(y\xi_i')$ which are needed to determine the coefficients of the fitted polynomial.

Similar results hold for any value of N, the coefficients of such expansions in differences being given by

$$1, \frac{(r+1)(N-r-1)}{1!(N-1)}, \frac{(r+1)(r+2)(N-r-1)(N-r-2)}{2!(N-1)(N-2)}, \ldots,$$

$$\frac{(r+1)(r+2)\ldots(r+j)(N-r-1)(N-r-2)\ldots(N-r-j)}{j!(N-1)(N-2)\ldots(N-j)},$$

the sequence terminating when $j = N-r-1$, since there are $N-r$ differences of order r. The quantities tabled herein are these coefficients, converted if necessary to

TABLE 1

INFANT MORTALITY RATES AND SUCCESSIVE DIFFERENCES OF SERIES

Year	y	Δ	Δ^2	Δ^3	Δ^4	Δ^5
1881	1165					
		192				
1882	1357		−327			
		−135		500		
1883	1222		173		−679	
		38		−179		811
1884	1260		−6		132	
		32		−47		−118
1885	1292		−53		14	
		−21		−33		212
1886	1271		−86		226	
		−107		193		−371
1887	1164		107		−145	
		0		48		−450
1888	1164		155		−595	
		155		−547		1844
1889	1319		−392		1249	
		−237		702		−2431
1890	1082		310		−1182	
		73		−480		2020
1891	1155		−170		838	
		−97		358		−1593
1892	1058		188		−755	
		91		−397		1460
1893	1149		−209		705	
		−118		308		−979
1894	1031		99		−274	
		−19		34		−85
1895	1012		133		−359	
		114		−325		1178
1896	1126		−192		819	
		−78		494		−1944
1897	1048		302		−1125	
		224		−631		2025
1898	1272		−329		900	
		−105		269		−909
1899	1167		−60		−9	
		−165		260		−452
1900	1002		200		−461	
		35		−201		663
1901	1037		−1		202	
		34		1		−517
1902	1071		0		−315	
		34		−314		1217
1903	1105		−314		902	
		−280		588		−1741
1904	825		274		−839	
		−6		−251		1028
1905	819		23		189	
		17		−62		−100
1906	836		−39		89	
		−22		27		−132
1907	814		−12		−43	
		−34		−16		182
1908	780		−28		139	
		−62		123		
1909	718		95			
		33				
1910	751					

integers by clearing fractions, the terminal value of ξ' for each combination of N and r supplying the appropriate factor.

When arranged in order, each set is symmetrical about the midpoint of the series, so that tabulation is necessary only to that point. The coefficients have been independently checked using the result that, if S is their sum for any particular combination of N and r, then

$$\frac{S^2(2\cdot 6)(6\cdot 10)\ldots\{(4r-2)(4r+2)\}}{N(N^2-1)(N^2-4)\ldots(N^2-r^2)}=\frac{\lambda^2(r!)^4(N+r)!}{(2r)!(2r+1)!(N-r-1)!},$$

where λ is the multiple given in Table XXIII (Fisher and Yates ibid.).

TABLE 2

SUMS OF PAIRS OF DIFFERENCES FROM TABLE 1

Δ	Δ^2	Δ^3	Δ^4	Δ^5
225				
−197	−232	623		
4	145	−195	−540	993
10	−18	−20	89	−250
−4	−92	−95	103	112
−113	−63	−58	415	657
−280	381	636	−984	−2191
189	−159	−861	307	3061
−203	−392	703	934	−2948
108	309	681	−980	2683
−262	30	618	377	−2045
−14	128	−128	−764	551
106	−538	−323	1605	1046
−97	401	528	−1399	−2029
114	−59	−325	460	1178

The advantages of this method are

(i) the coefficients are always positive,

(ii) the differences of the series are often smaller than the original observations, but these are rapidly offset by the additional labour involved in the computations for the higher values of N and r because, numerically, the coefficients are greater than the corresponding values of ξ'. When, however, the differences are required in other aspects of the investigation, or for exploratory purposes, the method is most expeditious even for the larger values of N and r.

The coefficients have been tabled only so far as $N = 52$, as from this point onward they become large.

III. Illustrative Example

The following example illustrates the procedure of fitting. The observed variate y is the infant mortality rate per 10^4 births in Australia during the period 1881–1910 inclusive (Knibbs 1911). Table 1 presents the mortality rates and their

Table 3

Analysis of variance. Infant mortality data

Variation due to	Deg. Freedom	Mean Square
Linear term	1	693021
Quadratic term	1	52175
Cubic term	1	11177
Quartic term	1	1183
Quintic term	1	48133
Residual	24	6114
Total	29	

successive differences up to the 5th, and Table 2 gives the sums of pairs of differences working to the mid-point of each series from the outer pair. From the values in

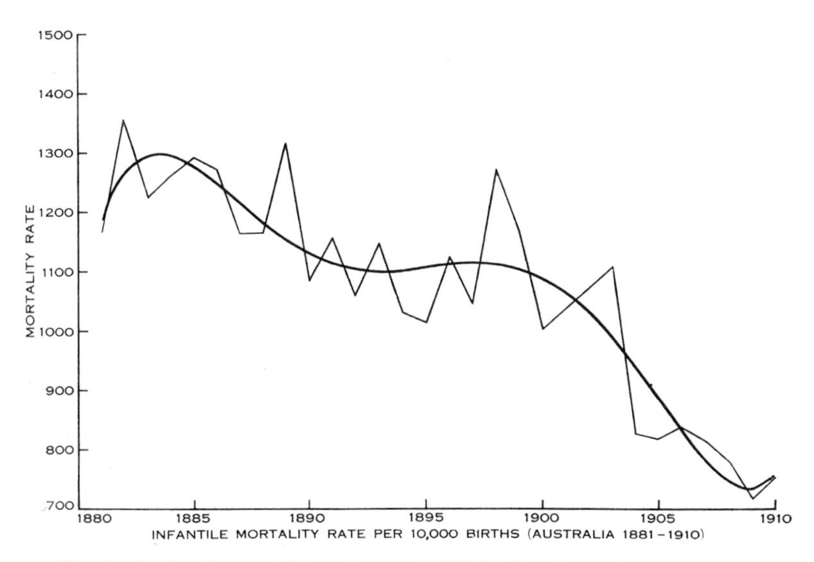

Fig. 1.—Infantile mortality rate per 10,000 births (Australia 1881–1910).

Table 2 and the multipliers (m_i corresponding to Δ^i) in Table 4 under $N = 30$, the following sums of products are obtained

$$Sm_1\Delta = -78932$$
$$Sm_2\Delta^2 = -125540$$
$$Sm_3\Delta^3 = -488606$$
$$Sm_4\Delta^4 = -2084078$$
$$Sm_5\Delta^5 = 10162688,$$

and, since $Sy = 32072$, the coefficients of the polynomial are

$$A = 1069\cdot07$$
$$B = -8\cdot77998$$
$$C = -0\cdot415607$$
$$D = -0\cdot0228746$$
$$E = -0\cdot0^3567623$$
$$F = 0\cdot0^2473623,$$

yielding the analysis of variance in Table 3.

Figure 1 illustrates the fitted curve.

IV. References

FISHER, R. A. (1958).—"Statistical Methods for Research Workers." 13th Ed. (Oliver & Boyd: Edinburgh.)

FISHER, R. A., and YATES, F. (1957).—"Statistical Tables for Biological, Agricultural and Medical Research." 5th Ed. (Oliver & Boyd: Edinburgh.)

KNIBBS, G. H. (1911).—*Trans. 9th Aust. Med. Congr.*: 670–9.

TABLE 4

MULTIPLIERS FOR CALCULATING COEFFICIENTS OF ORTHOGONAL POLYNOMIALS

$N = 3$

	Δ	Δ^2
1		
	1	
2		1

$N = 4$

	Δ	Δ^2	Δ^3
1			
	3		
2		1	
	4		1

$N = 5$

	Δ	Δ^2	Δ^3	Δ^4
1				
	2			
2		2		
	3		1	
3		3		1

$N = 6$

	Δ	Δ^2	Δ^3	Δ^4	Δ^5
1					
	5				
2		5			
	8		5		
3		9		1	
	9		8		1

$N = 7$

	Δ	Δ^2	Δ^3	Δ^4	Δ^5
1					
	3				
2		5			
	5		1		
3		10		3	
	6		2		1
4		12		5	

$N = 8$

	Δ	Δ^2	Δ^3	Δ^4	Δ^5
1					
	7				
2		7			
	12		7		
3		15		7	
	15		16		7
4		20		15	
	16		20		12

$N = 9$

	Δ	Δ^2	Δ^3	Δ^4	Δ^5
1					
	4				
2		28			
	7		14		
3		63		14	
	9		35		4
4		90		35	
	10		50		9
5		100		45	

$N = 10$

	Δ	Δ^2	Δ^3	Δ^4	Δ^5
1					
	9				
2		6			
	16		42		
3		14		18	
	21		112		6
4		21		50	
	24		175		16
5		25		75	
	25		200		21

$N = 11$

	Δ	Δ^2	Δ^3	Δ^4	Δ^5
1					
	5				
2		15			
	9		30		
3		36		6	
	12		84		3
4		56		18	
	14		140		9
5		70		30	
	15		175		14
6		75		35	

$N = 12$

	Δ	Δ^2	Δ^3	Δ^4	Δ^5
1					
	11				
2		55			
	20		33		
3		135		33	
	27		96		33
4		216		105	
	32		168		108
5		280		189	
	35		224		189
6		315		245	
	36		245		224

$N = 13$

	Δ	Δ^2	Δ^3	Δ^4	Δ^5
1					
	6				
2		22			
	11		11		
3		55		99	
	15		33		22
4		90		330	
	18		60		77
5		120		630	
	20		84		147
6		140		882	
	21		98		196
7		147		980	

$N = 14$

	Δ	Δ^2	Δ^3	Δ^4	Δ^5
1					
	13				
2		13			
	24		143		
3		33		143	
	33		440		143
4		55		495	
	40		825		528
5		75		990	
	45		1200		1078
6		90		1470	
	48		1470		1568
7		98		1764	
	49		1568		1764

$N = 15$

	Δ	Δ^2	Δ^3	Δ^4	Δ^5
1					
	7				
2		91			
	13		91		
3		234		1001	
	18		286		1001
4		396		3575	
	22		550		3861
5		550		7425	
	25		825		8316
6		675		11550	
	27		1050		12936
7		756		14700	
	28		1176		15876
8		784		15876	

$N = 16$

	Δ	Δ^2	Δ^3	Δ^4	Δ^5
1					
	15				
2		35			
	28		455		
3		91		273	
	39		1456		143
4		156		1001	
	48		2860		572
5		220		2145	
	55		4400		1287
6		275		3465	
	60		5775		2112
7		315		4620	
	63		6720		2772
8		336		5292	
	64		7056		3024

$N = 17$

	Δ	Δ^2	Δ^3	Δ^4	Δ^5
1					
	8				
2		40			
	15		28		
3		105		52	
	21		91		104
4		182		195	
	26		182		429
5		260		429	
	30		286		1001
6		330		715	
	33		385		1716
7		385		990	
	35		462		2376
8		420		1188	
	36		504		2772
9		432		1260	

TABLE 4 (*Continued*)

	$N = 18$				
	Δ	Δ^2	Δ^3	Δ^4	Δ^5
1					
	17				
2		68			
	32		68		
3		180		68	
	45		224		884
4		315		260	
	56		455		3744
5		455		585	
	65		728		9009
6		585		1001	
	72		1001		16016
7		693		1430	
	77		1232		23166
8		770		1782	
	80		1386		28512
9		810		1980	
	81		1440		30492

	$N = 19$				
	Δ	Δ^2	Δ^3	Δ^4	Δ^5
1					
	9				
2		51			
	17		204		
3		136		612	
	24		680		102
4		240		2380	
	30		1400		442
5		350		5460	
	35		2275		1092
6		455		9555	
	39		3185		2002
7		546		14014	
	42		4004		3003
8		616		18018	
	44		4620		3861
9		660		20790	
	45		4950		4356
10		675		21780	

	$N = 20$				
	Δ	Δ^2	Δ^3	Δ^4	Δ^5
1					
	19				
2		57			
	36		969		
3		153		1938	
	51		3264		1938
4		272		7650	
	64		6800		8568
5		400		17850	
	75		11200		21658
6		525		31850	
	84		15925		40768
7		637		47775	
	91		20384		63063
8		728		63063	
	96		24024		84084
9		792		75075	
	99		26400		99099
10		825		81675	
	100		27225		104544

	$N = 21$				
	Δ	Δ^2	Δ^3	Δ^4	Δ^5
1					
	10				
2		190			
	19		285		
3		513		969	
	27		969		3876
4		918		3876	
	34		2040		17442
5		1360		9180	
	40		3400		44982
6		1800		16660	
	45		4900		86632
7		2205		25480	
	49		6370		137592
8		2548		34398	
	52		7644		189189
9		2808		42042	
	54		8580		231231
10		2970		47190	
	55		9075		254826
11		3025		49005	

	$N = 22$				
	Δ	Δ^2	Δ^3	Δ^4	Δ^5
1					
	21				
2		35			
	40		133		
3		95		1197	
	57		456		2261
4		171		4845	
	72		969		10336
5		255		11628	
	85		1632		27132
6		340		21420	
	96		2380		53312
7		420		33320	
	105		3136		86632
8		490		45864	
	112		3822		122304
9		546		57330	
	117		4368		154154
10		585		66066	
	120		4719		176176
11		605		70785	
	121		4840		184041

	$N = 23$				
	Δ	Δ^2	Δ^3	Δ^4	Δ^5
1					
	11				
2		77			
	21		77		
3		210		1463	
	30		266		209
4		380		5985	
	38		570		969
5		570		14535	
	45		969		2584
6		765		27132	
	51		1428		5168
7		952		42840	
	56		1904		8568
8		1120		59976	
	60		2352		12376
9		1260		76440	
	63		2730		16016
10		1365		90090	
	65		3003		18876
11		1430		99099	
	66		3146		20449
12		1452		102245	

	$N = 24$				
	Δ	Δ^2	Δ^3	Δ^4	Δ^5
1					
	23				
2		253			
	44		1771		
3		693		253	
	63		6160		4807
4		1260		1045	
	80		13300		22572
5		1900		2565	
	95		22800		61047
6		2565		4845	
	108		33915		124032
7		3213		7752	
	119		45696		209304
8		3808		11016	
	128		57120		308448
9		4320		14280	
	135		67200		408408
10		4725		17160	
	140		75075		494208
11		5005		19305	
	143		80080		552123
12		5148		20449	
	144		81796		572572

	$N = 25$				
	Δ	Δ^2	Δ^3	Δ^4	Δ^5
1					
	12				
2		92			
	23		506		
3		253		1518	
	33		1771		1012
4		462		6325	
	42		3850		4807
5		700		15675	
	50		6650		13167
6		950		29925	
	57		9975		27132
7		1197		48450	
	63		13566		46512
8		1428		69768	
	68		17136		69768
9		1632		91800	
	72		20400		94248
10		1800		112200	
	75		23100		116688
11		1925		128700	
	77		25025		133848
12		2002		139425	
	78		26026		143143
13		2028		143143	

TABLE 4 (*Continued*)

$N = 26$	Δ	Δ^2	Δ^3	Δ^4	Δ^5
1					
	25				
2		50			
	48		1150		
3		138		2530	
	69		4048		2530
4		253		10626	
	88		8855		12144
5		385		26565	
	105		15400		33649
6		525		51205	
	120		23275		70224
7		665		83790	
	133		31920		122094
8		798		122094	
	144		40698		186048
9		918		162792	
	153		48960		255816
10		1020		201960	
	160		56100		323136
11		1100		235620	
	165		61600		379236
12		1155		260260	
	168		65065		416416
13		1183		273273	
	169		66248		429429

$N = 27$	Δ	Δ^2	Δ^3	Δ^4	Δ^5
1					
	13				
2		325			
	25		130		
3		900		2990	
	36		460		16445
4		1656		12650	
	46		1012		79695
5		2530		31878	
	55		1771		223146
6		3465		61985	
	63		2695		471086
7		4410		102410	
	70		3724		829521
8		5320		150822	
	76		4788		1281987
9		6156		203490	
	81		5814		1790712
10		6885		255816	
	85		6732		2302344
11		7480		302940	
	88		7480		2756754
12		7920		340340	
	90		8008		3097094
13		8190		364364	
	91		8281		3279276
14		8281		372645	

$N = 28$	Δ	Δ^2	Δ^3	Δ^4	Δ^5
1					
	27				
2		117			
	52		585		
3		325		1755	
	75		2080		13455
4		600		7475	
	96		4600		65780
5		920		18975	
	115		8096		185955
6		1265		37191	
	132		12397		396704
7		1617		61985	
	147		17248		706629
8		1960		92169	
	160		22344		1106028
9		2280		125685	
	171		27360		1566873
10		2565		159885	
	180		31977		2046528
11		2805		191862	
	187		35904		2494206
12		2992		218790	
	192		38896		2858856
13		3120		238238	
	195		40768		3097094
14		3185		248430	
	196		41405		3179904

	$N = 29$				
	Δ	Δ^2	Δ^3	Δ^4	Δ^5
1					
	14				
2		126			
	27		819		
3		351		4095	
	39		2925		8190
4		650		17550	
	50		6500		40365
5		1000		44850	
	60		11500		115115
6		1380		88550	
	69		17710		247940
7		1771		148764	
	77		24794		446292
8		2156		223146	
	84		32340		706629
9		2520		307230	
	90		39900		1013859
10		2850		395010	
	95		47025		1343034
11		3135		479655	
	99		53295		1662804
12		3366		554268	
	102		58344		1939938
13		3536		612612	
	104		61880		2144142
14		3640		649740	
	105		63700		2252432
15		3675		662480	

	$N = 30$				
	Δ	Δ^2	Δ^3	Δ^4	Δ^5
1					
	29				
2		203			
	56		1827		
3		567		23751	
	81		6552		16965
4		1053		102375	
	104		14625		84240
5		1625		263250	
	125		26000		242190
6		2250		523250	
	144		40250		526240
7		2898		885500	
	161		56672		956340
8		3542		1338876	
	176		74382		1530144
9		4158		1859550	
	189		92400		2220834
10		4725		2413950	
	200		109725		2979504
11		5225		2962575	
	209		125400		3741309
12		5643		3464175	
	216		138567		4434144
13		5967		3879876	
	221		148512		4988412
14		6188		4176900	
	224		154700		5346432
15		6300		4331600	
	225		156800		5470192

	$N = 31$				
	Δ	Δ^2	Δ^3	Δ^4	Δ^5
1					
	15				
2		145			
	29		1015		
3		406		783	
	42		3654		1131
4		756		3393	
	54		8190		5655
5		1170		8775	
	65		14625		16380
6		1625		17550	
	75		22750		35880
7		2100		29900	
	84		32200		65780
8		2576		45540	
	92		42504		106260
9		3036		63756	
	99		53130		155848
10		3465		83490	
	105		63525		211508
11		3850		103455	
	110		73150		268983
12		4180		122265	
	114		81510		323323
13		4446		138567	
	117		88179		369512
14		4641		151164	
	119		92820		403104
15		4760		159120	
	120		95200		420784
16		4800		161840	

TABLE 4 (*Continued*)

$N = 32$	Δ	Δ^2	Δ^3	Δ^4	Δ^5
1					
	31				
2		155			
	60		899		
3		435		899	
	87		3248		2697
4		812		3915	
	112		7308		13572
5		1260		10179	
	135		13104		39585
6		1755		20475	
	156		20475		87360
7		2275		35100	
	175		29120		161460
8		2800		53820	
	192		38640		263120
9		3312		75900	
	207		48576		389620
10		3795		100188	
	220		58443		534336
11		4235		125235	
	231		67760		687401
12		4620		149435	
	240		76076		836836
13		4940		171171	
	247		82992		969969
14		5187		188955	
	252		88179		1074944
15		5355		201552	
	255		91392		1142128
16		5440		208080	
	256		92480		1165248

$N = 33$	Δ	Δ^2	Δ^3	Δ^4	Δ^5
1					
	16				
2		496			
	31		248		
3		1395		7192	
	45		899		14384
4		2610		31465	
	58		2030		72819
5		4060		82215	
	70		3654		213759
6		5670		166257	
	81		5733		475020
7		7371		286650	
	91		8190		884520
8		9100		442260	
	100		10920		1453140
9		10800		627900	
	108		13800		2170740
10		12420		834900	
	115		16698		3005640
11		13915		1051974	
	121		19481		3907332
12		15246		1266265	
	126		22022		4811807
13		16380		1464463	
	130		24206		5648643
14		17290		1633905	
	133		25935		6348888
15		17955		1763580	
	135		27132		6852768
16		18360		1844976	
	136		27744		7116336
17		18496		1872720	

$N = 34$	Δ	Δ^2	Δ^3	Δ^4	Δ^5
1					
	33				
2		88			
	64		2728		
3		248		8184	
	93		9920		79112
4		465		35960	
	120		22475		402752
5		725		94395	
	145		40600		1189377
6		1015		191835	
	168		63945		2660112
7		1323		332514	
	189		91728		4987710
8		1638		515970	
	208		122850		8255520
9		1950		737100	
	225		156000		12432420
10		2250		986700	
	240		189750		17365920
11		2530		1252350	
	253		222640		22792770
12		2783		1519518	
	264		253253		28364336
13		3003		1772771	
	273		280280		33682649
14		3185		1996995	
	280		302575		38342304
15		3325		2178540	
	285		319200		41973204
16		3420		2306220	
	288		329460		44279424
17		3468		2372112	
	289		332928		45070128

	$N = 35$				
	Δ	Δ^2	Δ^3	Δ^4	Δ^5
1					
	17				
2		187			
	33		1496		
3		528		46376	
	48		5456		23188
4		992		204600	
	62		12400		118668
5		1550		539400	
	75		22475		352408
6		2175		1101275	
	87		35525		792918
7		2842		1918350	
	98		51156		1496313
8		3528		2992626	
	108		68796		2493855
9		4212		4299750	
	117		87750		3783780
10		4875		5791500	
	125		107250		5328180
11		5500		7400250	
	132		126500		7054905
12		6072		9044750	
	138		144716		8863855
13		6578		10636626	
	143		161161		10636626
14		7007		12087075	
	147		175175		12248236
15		7350		13313300	
	150		186200		13579566
16		7600		14244300	
	152		193800		14529186
17		7752		14825700	
	153		197676		15023376
18		7803		15023376	

	$N = 36$				
	Δ	Δ^2	Δ^3	Δ^4	Δ^5
1					
	35				
2		595			
	68		6545		
3		1683		5236	
	99		23936		162316
4		3168		23188	
	128		54560		834768
5		4960		61380	
	155		99200		2492028
6		6975		125860	
	180		157325		5638528
7		9135		220255	
	203		227360		10704393
8		11368		345303	
	224		306936		17955756
9		13608		498771	
	243		393120		27432405
10		15795		675675	
	260		482625		38918880
11		17875		868725	
	275		572000		51949755
12		19800		1068925	
	288		657800		65845780
13		21528		1266265	
	299		736736		79774695
14		23023		1450449	
	308		805805		92828736
15		24255		1611610	
	315		862400		104110006
16		25200		1740970	
	320		904400		112814856
17		25840		1831410	
	323		930240		118309086
18		26163		1877922	
	324		938961		120187008

	$N = 37$				
	Δ	Δ^2	Δ^3	Δ^4	Δ^5
1					
	18				
2		210			
	35		357		
3		595		11781	
	51		1309		4488
4		1122		52360	
	66		2992		23188
5		1760		139128	
	80		5456		69564
6		2480		286440	
	93		8680		158224
7		3255		503440	
	105		12586		302064
8		4060		792918	
	116		17052		509733
9		4872		1151010	
	126		21924		783783
10		5670		1567566	
	135		27027		1119690
11		6435		2027025	
	143		32175		1505790
12		7150		2509650	
	150		37180		1924065
13		7800		2992990	
	156		41860		2351635
14		8372		3453450	
	161		46046		2762760
15		8855		3867864	
	165		49588		3131128
16		9240		4214980	
	168		52360		3432198
17		9520		4476780	
	170		54264		3645378
18		9690		4639572	
	171		55233		3755844
19		9747		4694805	

TABLE 4 (*Continued*)

$N = 38$

	Δ	Δ^2	Δ^3	Δ^4	Δ^5
1					
	37				
2		111			
	72		777		
3		315		1887	
	105		2856		20757
4		595		8415	
	136		6545		107712
5		935		22440	
	165		11968		324632
6		1320		46376	
	192		19096		742016
7		1736		81840	
	217		27776		1424016
8		2170		129456	
	240		37758		2416512
9		2610		188790	
	261		48720		3738042
10		3045		258390	
	280		60291		5374512
11		3465		335907	
	297		72072		7277985
12		3861		418275	
	312		83655		9369360
13		4225		501930	
	325		94640		11544390
14		4550		583050	
	336		104650		13682240
15		4830		657800	
	345		113344		15655640
16		5060		722568	
	352		120428		17341632
17		5236		774180	
	357		125664		18631932
18		5355		810084	
	360		128877		19442016
19		5415		828495	
	361		129960		19718181

$N = 39$

	Δ	Δ^2	Δ^3	Δ^4	Δ^5
1					
	19				
2		703			
	37		2109		
3		1998		2109	
	54		7770		35853
4		3780		9435	
	70		17850		186813
5		5950		25245	
	85		32725		565488
6		8415		52360	
	99		52360		1298528
7		11088		92752	
	112		76384		2504304
8		13888		147312	
	124		104160		4272048
9		16740		215760	
	135		134850		6645408
10		19575		296670	
	145		167475		9612108
11		22330		387585	
	154		200970		13100373
12		24948		485199	
	162		234234		16981965
13		27378		585585	
	169		266175		21081060
14		29575		684450	
	175		295750		25187760
15		31500		777400	
	180		322000		29074760
16		33120		860200	
	184		344080		32515560
17		34408		929016	
	187		361284		35302608
18		35343		980628	
	189		373065		37263864
19		35910		1012605	
	190		379050		38276469
20		36100		1023435	

$N = 40$

	Δ	Δ^2	Δ^3	Δ^4	Δ^5
1					
	39				
2		247			
	76		9139		
3		703		82251	
	111		33744		9139
4		1332		369075	
	144		77700		47804
5		2100		990675	
	175		142800		145299
6		2975		2061675	
	204		229075		335104
7		3927		3665200	
	231		335104		649264
8		4928		5843376	
	256		458304		1113024
9		5952		8593200	
	279		595200		1740464
10		6975		11866800	
	300		741675		2531584
11		7975		15575175	
	319		893200		3471039
12		8932		19594575	
	336		1045044		4528524
13		9828		23774751	
	351		1192464		5660655
14		10647		27948375	
	364		1330875		6814080
15		11375		31941000	
	375		1456000		7929480
16		12000		35581000	
	384		1564000		8946080
17		12512		38709000	
	391		1651584		9806280
18		12903		41186376	
	396		1716099		10460032
19		13167		42902475	
	399		1755600		10868627
20		13300		43780275	
	400		1768900		11007612

$N = 41$

	Δ	Δ^2	Δ^3	Δ^4	Δ^5
1					
	20				
2		260			
	39		2470		
3		741		18278	
	57		9139		36556
4		1406		82251	
	74		21090		191919
5		2220		221445	
	90		38850		585599
6		3150		462315	
	105		62475		1356124
7		4165		824670	
	119		91630		2638944
8		5236		1319472	
	132		125664		4544848
9		6336		1947792	
	144		163680		7141904
10		7440		2700720	
	155		204600		10442784
11		8525		3560040	
	165		247225		14398384
12		9570		4499495	
	174		290290		18897879
13		10556		5486481	
	182		332514		23774751
14		11466		6484023	
	189		372645		28817880
15		12285		7452900	
	195		409500		33786480
16		13000		8353800	
	200		442000		38427480
17		13600		9149400	
	204		469200		42493880
18		14076		9806280	
	207		490314		45762640
19		14421		10296594	
	209		504735		48050772
20		14630		10599435	
	210		512050		49228487
21		14700		10701845	

$N = 42$

	Δ	Δ^2	Δ^3	Δ^4	Δ^5
1					
	41				
2		410			
	80		1066		
3		1170		20254	
	117		3952		749398
4		2223		91390	
	152		9139		3948048
5		3515		246753	
	185		16872		12090897
6		4995		516705	
	216		27195		28108752
7		6615		924630	
	245		39984		54923022
8		8330		1484406	
	272		54978		95001984
9		10098		2199120	
	297		71808		149979984
10		11880		3060816	
	320		90024		220378752
11		13640		4051080	
	341		109120		305451432
12		15345		5142280	
	360		128557		403154752
13		16965		6299293	
	377		147784		510242733
14		18473		7481565	
	392		166257		622466208
15		19845		8645364	
	405		183456		734855940
16		21060		9746100	
	416		198900		842063040
17		22100		10740600	
	425		212160		938728440
18		22950		11589240	
	432		222870		1019853120
19		23598		12257850	
	437		230736		1081142370
20		24035		12719322	
	440		235543		1119300336
21		24255		12954865	
	441		237160		1132255201

$N = 43$

	Δ	Δ^2	Δ^3	Δ^4	Δ^5
1					
	21				
2		287			
	41		574		
3		820		22386	
	60		2132		70889
4		1560		101270	
	78		4940		374699
5		2470		274170	
	95		9139		1151514
6		3515		575757	
	111		14763		2686866
7		4662		1033410	
	126		21756		5270391
8		5880		1664334	
	140		29988		9153837
9		7140		2474010	
	153		39270		14514192
10		8415		3455760	
	165		49368		21425712
11		9680		4591224	
	176		60016		29842956
12		10912		5851560	
	186		70928		39595556
13		12090		7199192	
	195		81809		50394344
14		13195		8589945	
	203		92365		61847604
15		14210		9975420	
	210		102312		73485594
16		15120		11305476	
	216		111384		84791070
17		15912		12530700	
	221		119340		95233320
18		16575		13604760	
	225		125970		104303160
19		17100		14486550	
	228		131100		111546435
20		17480		15142050	
	230		134596		116593785
21		17710		15545838	
	231		136367		119184758
22		17787		15682205	

TABLE 4 (*Continued*)

$N = 44$					
	Δ	Δ^2	Δ^3	Δ^4	Δ^5
1					
	43				
2		301			
	84		12341		
3		861		12341	
	123		45920		22919
4		1640		55965	
	160		106600		121524
5		2600		151905	
	195		197600		374699
6		3705		319865	
	228		319865		877344
7		4921		575757	
	259		472416		1727271
8		6216		930069	
	288		652680		3011652
9		7560		1386945	
	315		856800		4794867
10		8925		1943865	
	340		1079925		7108992
11		10285		2591820	
	363		1316480		9947652
12		11616		3315884	
	384		1560416		13263536
13		12896		4096092	
	403		1805440		16969524
14		14105		4908540	
	420		2045225		20943104
15		15225		5726630	
	435		2273600		25033554
16		16240		6522390	
	448		2484720		29071224
17		17136		7267806	
	459		2673216		32878170
18		17901		7936110	
	468		2834325		36279360
19		18525		8502975	
	475		2964000		39113685
20		19000		8947575	
	480		3059000		41244060
21		19320		9253475	
	483		3116960		42565985
22		19481		9409323	
	484		3136441		43014048

$N = 45$					
	Δ	Δ^2	Δ^3	Δ^4	Δ^5
1					
	22				
2		946			
	43		3311		
3		2709		19393	
	63		12341		38786
4		5166		88150	
	82		28700		206271
5		8200		239850	
	100		53300		638001
6		11700		506350	
	117		86450		1498796
7		15561		913900	
	133		127946		2961036
8		19684		1480518	
	148		177156		5181813
9		23976		2214450	
	162		233100		8282043
10		28350		3113550	
	175		294525		12329658
11		32725		4165425	
	187		359975		17328168
12		37026		5348200	
	198		427856		23211188
13		41184		6631768	
	208		496496		29842956
14		45136		7979400	
	217		564200		37024416
15		48825		9349600	
	225		629300		44504096
16		52200		10698100	
	232		690200		51992766
17		55216		11979900	
	238		745416		59180706
18		57834		13151268	
	243		793611		65756340
19		60021		14171625	
	247		833625		71424990
20		61750		15005250	
	250		864500		75926565
21		63000		15622750	
	252		885500		79051115
22		63756		16002250	
	253		896126		80651340
23		64009		16130268	

$N = 46$					
	Δ	Δ^2	Δ^3	Δ^4	Δ^5
1					
	45				
2		165			
	88		7095		
3		473		4257	
	129		26488		58179
4		903		19393	
	168		61705		310288
5		1435		52890	
	205		114800		962598
6		2050		111930	
	240		186550		2268448
7		2730		202540	
	273		276640		4496388
8		3458		329004	
	304		383838		7896096
9		4218		493506	
	333		506160		12666654
10		4995		695970	
	360		641025		18930384
11		5775		934065	
	385		785400		26714259
12		6545		1203345	
	408		935935		35939904
13		7293		1497496	
	429		1089088		46422376
14		8008		1808664	
	448		1241240		57877248
15		8680		2127840	
	465		1388800		69935008
16		9300		2445280	
	480		1528300		82161408
17		9860		2750940	
	493		1656480		94082148
18		10353		3034908	
	504		1770363		105210144
19		10773		3287817	
	513		1867320		115073595
20		11115		3501225	
	520		1945125		123243120
21		11375		3667950	
	525		2002000		129356370
22		11550		3782350	
	528		2036650		133138720
23		11638		3840540	
	529		2048288		134418900

$N = 47$	Δ	Δ^2	Δ^3	Δ^4	Δ^5
1					
	23				
2		345			
	45		759		
3		990		32637	
	66		2838		32637
4		1892		148995	
	86		6622		174537
5		3010		407253	
	105		12341		543004
6		4305		863870	
	123		20090		1283464
7		5740		1567020	
	140		29848		2552004
8		7280		2552004	
	156		41496		4496388
9		8892		3838380	
	171		54834		7238088
10		10545		5428566	
	185		69597		10857132
11		12210		7307685	
	198		85470		15380937
12		13860		9444435	
	210		102102		20777757
13		15470		11792781	
	221		119119		26954928
14		17017		14294280	
	231		136136		33761728
15		18480		16880864	
	240		152768		40996384
16		19840		19477920	
	248		168640		48416544
17		21080		22007520	
	255		183396		55752384
18		22185		24391668	
	261		196707		62721432
19		23142		26555445	
	266		208278		69044157
20		23940		28429947	
	270		217854		74459385
21		24570		29954925	
	273		225225		78738660
22		25025		31081050	
	275		230230		81698760
23		25300		31771740	
	276		232760		83211700
24		25392		32004500	

$N = 48$	Δ	Δ^2	Δ^3	Δ^4	Δ^5
1					
	47				
2		1081			
	92		3243		
3		3105		35673	
	135		12144		1533939
4		5940		163185	
	176		28380		8224524
5		9460		446985	
	215		52976		25656939
6		13545		950257	
	252		86387		60816448
7		18081		1727740	
	287		128576		121287348
8		22960		2820636	
	320		179088		214368336
9		28080		4253340	
	351		237120		346221876
10		33345		6031740	
	380		301587		521142336
11		38665		8142849	
	407		371184		740999259
12		43956		10555545	
	432		444444		1004887884
13		49140		13222209	
	455		519792		1308998691
14		54145		16081065	
	476		595595		1646701056
15		58905		19059040	
	495		670208		2008822816
16		63360		22074976	
	512		742016		2384097408
17		67456		25043040	
	527		809472		2759743008
18		71145		27876192	
	540		871131		3122133504
19		74385		30489585	
	551		925680		3457518939
20		77140		32803785	
	560		971964		3752753004
21		79380		34747713	
	567		1009008		3995986995
22		81081		36261225	
	572		1036035		4177293120
23		82225		37297260	
	575		1052480		4289184900
24		82800		37823500	
	576		1058000		4327008400

$N = 49$	Δ	Δ^2	Δ^3	Δ^4	Δ^5
1					
	24				
2		376			
	47		4324		
3		1081		38916	
	69		16215		95128
4		2070		178365	
	90		37950		511313
5		3300		489555	
	110		70950		1599213
6		4730		1042965	
	129		115885		3801028
7		6321		1900514	
	147		172774		7602056
8		8036		3109932	
	164		241080		13476372
9		9840		4701060	
	180		319800		21833812
10		11700		6683820	
	195		407550		32973512
11		13585		9047610	
	209		502645		47047572
12		15466		11761893	
	222		603174		64036973
13		17316		14777763	
	234		707070		83740657
14		19110		18030285	
	245		812175		105777672
15		20825		21441420	
	255		916300		129601472
16		22440		24923360	
	264		1017280		154524832
17		23936		28382112	
	272		1113024		179753376
18		25296		31721184	
	279		1201560		204425408
19		26505		34845240	
	285		1281075		227655568
20		27550		37663605	
	290		1349950		248579793
21		28420		40093515	
	294		1406790		266399133
22		29106		42063021	
	297		1450449		280420140
23		29601		43513470	
	299		1480050		290089800
24		29900		44401500	
	300		1495000		295023300
25		30000		44700500	

	Δ	Δ^2	Δ^3	Δ^4	Δ^5
1					
	49				
2		196			
	96		9212		
3		564		211876	
	141		34592		211876
4		1081		972900	
	184		81075		1141536
5		1725		2675475	
	225		151800		3579191
6		2475		5711475	
	264		248325		8529136
7		3311		10429650	
	301		370832		17104626
8		4214		17104626	
	336		518322		30408224
9		5166		25916100	
	369		688800		49413364
10		6150		36936900	
	400		879450		74858784
11		7150		50128650	
	429		1086800		107163914
12		8151		65343850	
	456		1306877		146370224
13		9139		82333251	
	481		1535352		192110919
14		10101		100757475	
	504		1767675		243609184
15		11025		120201900	
	525		1999200		299703404
16		11900		140193900	
	544		2225300		358896384
17		12716		160221600	
	561		2441472		419424544
18		13464		179753376	
	576		2643432		479342336
19		14136		198257400	
	589		2827200		536616696
20		14725		215220600	
	600		2989175		589226176
21		15225		230166475	
	609		3126200		635259471
22		15631		242671275	
	616		3235617		673008336
23		15939		252378126	
	621		3315312		701050350
24		16146		259008750	
	624		3363750		718317600
25		16250		262372500	
	625		3380000		724148100

	Δ	Δ^2	Δ^3	Δ^4	Δ^5
1					
	25				
2		1225			
	49		4900		
3		3528		46060	
	72		18424		75670
4		6768		211876	
	94		43240		408618
5		10810		583740	
	115		81075		1284228
6		15525		1248555	
	135		132825		3067878
7		20790		2284590	
	154		198660		6168393
8		26488		3754674	
	172		278124		10995831
9		32508		5701542	
	189		370230		17919132
10		38745		8145060	
	205		473550		27227772
11		45100		11081070	
	220		586300		39100347
12		51480		14481610	
	234		706420		53581957
13		57798		18296278	
	247		831649		70571358
14		63973		22454523	
	259		959595		89818092
15		69930		26868660	
	270		1087800		110929182
16		75600		31437420	
	280		1213800		133384482
17		80920		36049860	
	289		1335180		156559392
18		85833		40589472	
	297		1449624		179753376
19		90288		44938344	
	304		1554960		202222548
20		94240		48981240	
	310		1649200		223214508
21		97650		52609480	
	315		1730575		242003608
22		100485		55724515	
	319		1797565		257924898
23		102718		58241106	
	322		1848924		270405135
24		104328		60090030	
	324		1883700		278989425
25		105300		61220250	
	325		1901250		283362300
26		105625		61600500	

	Δ	Δ^2	Δ^3	Δ^4	Δ^5
1					
	51				
2		425			
	100		4165		
3		1225		3570	
	147		15680		55930
4		2352		16450	
	192		36848		302680
5		3760		45402	
	235		69184		953442
6		5405		97290	
	276		113505		2283072
7		7245		178365	
	315		170016		4601817
8		9240		293733	
	352		238392		8224524
9		11352		446985	
	387		317856		13439349
10		13545		639969	
	420		407253		20479008
11		15785		872685	
	451		505120		29496753
12		18040		1143285	
	480		609752		40548508
13		20280		1448161	
	507		719264		53581957
14		22477		1782105	
	532		831649		68432832
15		24605		2138526	
	555		944832		84828198
16		26640		2509710	
	576		1056720		102396168
17		28560		2887110	
	595		1165248		120681198
18		30345		3261654	
	612		1268421		139163904
19		31977		3624060	
	627		1364352		157284204
20		33440		3965148	
	640		1451296		174466512
21		34720		4276140	
	651		1527680		190145692
22		35805		4548940	
	660		1592129		203792512
23		36685		4776387	
	667		1643488		214937415
24		37352		4952475	
	672		1680840		223191540
25		37800		5072535	
	675		1703520		228264075
26		38025		5133375	
	676		1711125		229975200

Printed by C.S.I.R.O., Melbourne

Commonwealth Scientific and Industrial Research Organization, Australia. Division of Mathematical Statistics Technical Paper No. 13, Melbourne 1962

THE MULTIVARIATE t-DISTRIBUTION ASSOCIATED WITH THE GENERAL MULTIVARIATE NORMAL DISTRIBUTION

By E. A. Cornish*

[*Manuscript received July* 18, 1962]

Summary

The multivariate t-distribution, which is associated with the general multivariate normal distribution, is established, and its applications in various tests of significance are illustrated by examples.

I. Introduction

A multivariate generalization of Student's t-distribution has been considered by the author (Cornish 1954), and independently by Dunnett and Sobel (1954). Thus, if $\mathbf{x}$ is a column vector with elements $x_1, x_2, \ldots, x_N$ which are normally and independently distributed with means $\xi_1, \xi_2, \ldots, \xi_N$ and common variance σ^2 and

$$\mathbf{y} = \mathbf{H}\mathbf{x} = [h_{ij}]\mathbf{x} \qquad \begin{matrix} i = 1, 2, \ldots, p \\ j = 1, 2, \ldots, N \end{matrix}$$

are any $p < N$, linearly independent, linear functions of the x_i, the distribution of the y_i is well known to be

$$\{|\sigma^2\mathbf{HH}'|^{-\frac{1}{2}}/(2\pi)^{\frac{1}{2}p}\} \exp\{-\tfrac{1}{2}(\mathbf{y}-\boldsymbol{\alpha})'(\sigma^2\mathbf{HH}')^{-1}(\mathbf{y}-\boldsymbol{\alpha})\}d\mathbf{y},$$

with a vector of means

$$\boldsymbol{\alpha} = \mathbf{H}\boldsymbol{\xi}.$$

If s^2 is an estimate of σ^2 on n degrees of freedom, distributed independently of the x_i, and, therefore, independently of the y_i, and

$$\mathbf{t} = \mathbf{Q}(\mathbf{y}-\boldsymbol{\alpha})/s,$$

where $\mathbf{Q}$ is diagonal with its ith diagonal element equal to $(\mathbf{h}_i\mathbf{h}_i')^{-\frac{1}{2}}$, $\mathbf{h}_i$ being the ith row vector of $\mathbf{H}$, the distribution of the t_i takes the form

$$\frac{\Gamma\frac{1}{2}(n+p)}{(\pi n)^{\frac{1}{2}p}\Gamma\frac{1}{2}n}\left(1+\frac{\mathbf{t}'\mathbf{R}^{-1}\mathbf{t}}{n}\right)^{-\frac{1}{2}(n+p)}\frac{d\mathbf{t}}{|\mathbf{R}|^{\frac{1}{2}}}, \tag{1}$$

in which

$$\mathbf{R} = \mathbf{Q}(\mathbf{HH}')\mathbf{Q}$$

is the non-singular, symmetric correlation matrix of the y_i, the correlation ρ_{ij} being

$$\mathbf{h}_i\mathbf{h}_j'/(\mathbf{h}_i\mathbf{h}_i'.\,\mathbf{h}_j\mathbf{h}_j')^{\frac{1}{2}}.$$

* Division of Mathematical Statistics, C.S.I.R.O., University of Adelaide.

The elements of **H** are known constants satisfying the conditions

$$\sum_{j=1}^{N} h_{ij} = 0$$

for $i = 1, 2, \ldots, p$.

The principal properties of the distribution (1) have been studied (Cornish 1954), and the sampling distributions of statistics derived from it have been determined (Cornish 1955). The simultaneous fiducial distribution of the parameters α_i, corresponding to (1), has also been obtained and its practical applications have been illustrated (Cornish 1960). The distribution (1) extends to tests of significance involving multiple variates, the same advantages as those which attend tests of significance with Student's t-statistic in univariate analysis. Our object is to establish the counterpart of the distribution (1) for the general multivariate normal distribution, thus conferring the same advantages on tests of significance in multivariate analysis. The corresponding simultaneous fiducial distribution of the parameters of location, first conjectured by Fisher (1954), has recently been derived by direct proof (Cornish 1961).

II. The Multiple t-Distribution

Suppose $\mathbf{x}$ is a column vector whose elements $x_1, x_2, \ldots, x_p$ are distributed in a multivariate normal distribution with a vector of means $\boldsymbol{\xi}' = [\xi_1 \xi_2 \ldots \xi_p]$ and variance-covariance matrix $\boldsymbol{\Sigma}$, of which there is available a non-singular estimate $\mathbf{S} = [s_{ij}]$, $i, j = 1, 2, \ldots, p$, based on n degrees of freedom and distributed independently of the x_i. Corresponding to each of the quantities $x_i - \xi_i$ we may calculate a Student t-statistic

$$t_i = \frac{x_i - \xi_i}{s_{ii}^{\frac{1}{2}}} \qquad i = 1, 2, \ldots, p \tag{2}$$

and directly infer, from the simultaneous fiducial distribution of the parameters (Cornish 1961), that the simultaneous distribution of the t_i takes the form

$$\frac{\Gamma\frac{1}{2}(n+p)}{(\pi n)^{\frac{1}{2}p}\Gamma\frac{1}{2}n}\left(1+\frac{\mathbf{t}'\mathbf{R}^{-1}\mathbf{t}}{n}\right)^{-\frac{1}{2}(n+p)}\frac{\mathrm{d}\mathbf{t}}{|\mathbf{R}|^{\frac{1}{2}}}, \tag{3}$$

where **R** is the sample correlation matrix associated with **S**. As, however, such a derivation may not be universally accepted we shall establish this distribution by independent means.

The t_i will be distributed in some non-singular distribution which will exhibit the statistical dependence of the variates and must involve the elements of **R** as parameters. If, now, $\boldsymbol{\lambda}$ is a non-null column vector with p arbitrary, real, non-stochastic elements, the variate $\boldsymbol{\lambda}'\mathbf{x}$ is normally distributed with a mean $\boldsymbol{\lambda}'\boldsymbol{\xi}$ and variance $\boldsymbol{\lambda}'\boldsymbol{\Sigma}\boldsymbol{\lambda}$, and consequently

$$t = \boldsymbol{\lambda}'(\mathbf{x}-\boldsymbol{\xi})/(\boldsymbol{\lambda}'\mathbf{S}\boldsymbol{\lambda})^{\frac{1}{2}} \tag{4}$$

is distributed in Student's distribution with n degrees of freedom, independently of $\boldsymbol{\lambda}'\boldsymbol{\xi}$ and $\boldsymbol{\lambda}'\boldsymbol{\Sigma}\boldsymbol{\lambda}$. The elements of $\mathbf{S}$ are fixed by the observed values of the sample, but since $\boldsymbol{\lambda}$ is arbitrary, the statistic t, defined by (4), can be made an arbitrary linear function of the t_i as defined in (2).

No restriction is introduced by assuming that the frequency density is

$$f\{1+\phi(t_1, t_2, \ldots, t_p)\};$$

we now show that the *necessary* and *sufficient* condition for the function f to have the form (3), is that the multiple integral

$$\int \ldots \int f\{1+\phi(t_1, t_2, \ldots, t_p)\} dt_1 dt_2 \ldots dt_p,$$

when taken over the domain defined by

$$\boldsymbol{\mu}'\mathbf{t} = \tau \leqslant t_P$$

for *arbitrary* $\boldsymbol{\mu}$, is equal to

$$\frac{\Gamma\frac{1}{2}(n+1)}{(\pi n)^{\frac{1}{2}}\Gamma\frac{1}{2}n}\int_{-\infty}^{t_P} \frac{d\tau}{(1+\tau^2/n)^{\frac{1}{2}(n+1)}}. \tag{5}$$

If the distribution has the form given by (3), and the equations

$$\boldsymbol{\tau} = \mathbf{Ht}$$

define any $r \leqslant p$, linearly independent, linear functions of the t_i, the distribution of the τ_i is

$$\frac{\Gamma\frac{1}{2}(n+r)}{(\pi n)^{\frac{1}{2}r}\Gamma\frac{1}{2}n}\left\{1+\frac{\boldsymbol{\tau}'(\mathbf{HRH}')^{-1}\boldsymbol{\tau}}{n}\right\}^{-\frac{1}{2}(n+r)} \frac{d\boldsymbol{\tau}}{|\mathbf{HRH}'|^{\frac{1}{2}}}$$

(*vide infra* Section III(b)). Taking $r = 1$, it is clear that the condition is necessary.

From what has been stated above it is known that

$$\int \ldots \int f\{1+\phi(t_1, t_2, \ldots, t_p)\} dt_1 dt_2 \ldots dt_p, \tag{6}$$

when taken over the region defined by

$$\boldsymbol{\mu}'\mathbf{t} = \tau \leqslant t_P,$$

yields the result (5). The function f involves the elements of $\mathbf{R}$ as parameters, and to evaluate the integral (6) we may, since $\mathbf{R}$ is positive definite and symmetric, make the real, non-singular, linear transformation

$$\mathbf{t} = \mathbf{Cz} \tag{7}$$

with $\mathbf{C}$ chosen so that

$$\mathbf{C}'\mathbf{R}^{-1}\mathbf{C} = \mathbf{I}_p.$$

The jacobian of (7) is

$$|\mathbf{C}| = |\mathbf{R}|^{\frac{1}{2}};$$

thus (6) becomes

$$\int \ldots \int f\{1+\phi_1(z_1, z_2, \ldots, z_p)\}|\mathbf{R}|^{\frac{1}{2}} dz_1 dz_2 \ldots dz_p, \tag{8}$$

taken over the domain defined by

$$\boldsymbol{\mu}'\mathbf{Cz} = \tau \leqslant t_P,$$

where

$$\boldsymbol{\mu}' = \frac{1}{(\boldsymbol{\lambda}'\mathbf{S}\boldsymbol{\lambda})^{\frac{1}{2}}}[\lambda_1 s_{11}^{\frac{1}{2}} \lambda_2 s_{22}^{\frac{1}{2}} \ldots \lambda_p s_{pp}^{\frac{1}{2}}].$$

The matrix $\boldsymbol{\mu}'\mathbf{C}$ is of order $1 \times p$ and rank 1, and consequently the equation

$$\boldsymbol{\mu}'\mathbf{Cy} = 0 \tag{9}$$

has $p-1$ linearly independent solutions; in accordance with our needs we may choose an orthogonal and normalized set and take them as the elements of the rows of a matrix $\mathbf{Y}$ of order $(p-1) \times p$. With this choice of $\mathbf{Y}$

$$\mathbf{YY}' = \mathbf{I}_{p-1},$$

and from (9)

$$\boldsymbol{\mu}'\mathbf{CY}' = \mathbf{0}.$$

Moreover, taking the transpose

$$\mathbf{YC}'\boldsymbol{\mu} = \mathbf{0}.$$

Now let

$$\mathbf{v} = \mathbf{Yz},$$

and combine this with

$$\tau = \boldsymbol{\mu}'\mathbf{Cz}$$

to give the transformation

$$\begin{bmatrix} \tau \\ \mathbf{v} \end{bmatrix} = \begin{bmatrix} \boldsymbol{\mu}'\mathbf{C} \\ \mathbf{Y} \end{bmatrix} \mathbf{z} = \mathbf{Pz}.$$

Obviously $\mathbf{P}$ is non-singular, so that

$$\mathbf{z} = \mathbf{P}^{-1} \begin{bmatrix} \tau \\ \mathbf{v} \end{bmatrix}.$$

Furthermore,

$$\mathbf{PP}' = \begin{bmatrix}\boldsymbol{\mu}'\mathbf{C}\\ \mathbf{Y}\end{bmatrix}[\mathbf{C}'\boldsymbol{\mu}\ \mathbf{Y}']$$

$$= \begin{bmatrix}\boldsymbol{\mu}'\mathbf{CC}'\boldsymbol{\mu} & \cdot\\ \cdot & \mathbf{I}_{p-1}\end{bmatrix}$$

$$= \begin{bmatrix}\boldsymbol{\mu}'\mathbf{R}\boldsymbol{\mu} & \cdot\\ \cdot & \mathbf{I}_{p-1}\end{bmatrix}.$$

From the definition of $\boldsymbol{\mu}$, it follows that

$$\boldsymbol{\mu}'\mathbf{R}\boldsymbol{\mu} = 1,$$

thus $\mathbf{PP}' = \mathbf{I}_p$, and hence $\mathbf{P}$ is orthogonal.

The integral (8) then becomes

$$\int \cdots \int f\{1+\phi_2(\tau, v_1, v_2, \ldots, v_{p-1})\}|\mathbf{R}|^{\frac{1}{2}}d\tau dv_1 dv_2 \ldots dv_{p-1} \tag{10}$$

taken over the domain defined by

$$\tau \leqslant t_P.$$

We now have to trace, in reverse order, the steps by which the result (5) was obtained from the integral (10). By employing a similar argument to that used previously (Cornish 1961), the integral (10) must have been

$$\frac{\Gamma\frac{1}{2}(n+p)}{(\pi n)^{\frac{1}{2}p}\Gamma\frac{1}{2}n}\int_{-\infty}^{t_P}\int_{-\infty}^{\infty}\cdots\int_{-\infty}^{\infty}\left(1+\frac{\tau^2+\mathbf{v}'\mathbf{v}}{n}\right)^{-\frac{1}{2}(n+p)} d\tau dv_1 dv_2 \ldots dv_{p-1},$$

in which the integrand is *uniquely* determined. The function $\phi_1(z_1, z_2, \ldots, z_p)$ must, therefore, have been uniquely transformed to the real, positive definite, quadratic form

$$\frac{1}{n}(\tau^2+\mathbf{v}'\mathbf{v})$$

of rank p, by the real, non-singular, linear transformation

$$\mathbf{z} = \mathbf{P}^{-1}\begin{bmatrix}\tau\\ \mathbf{v}\end{bmatrix},$$

and consequently $\phi_1(z_1, z_2, \ldots, z_p)$ must be the real, positive definite, quadratic form

$$\frac{1}{n}\mathbf{z}'\mathbf{P}'\mathbf{P}\mathbf{z} = \frac{1}{n}\mathbf{z}'\mathbf{z}$$

of rank p. In addition, the function $\phi(t_1, t_2, \ldots, t_p)$ was transformed to the real, positive definite, quadratic form

$$\frac{1}{n}\mathbf{z}'\mathbf{z},$$

by the real, non-singular, linear transformation

$$\mathbf{t} = \mathbf{Cz}$$

and hence ϕ must be the real, positive definite, quadratic form

$$\frac{1}{n}\mathbf{t}'(\mathbf{C}^{-1})'\mathbf{C}^{-1}\mathbf{t} = \frac{1}{n}\mathbf{t}'\mathbf{R}^{-1}\mathbf{t}$$

of rank p. The expression (3) thus gives the distribution of the t_i, and hence the sufficiency of the condition is established.

The above proof is complete in its generality. It is equally applicable in establishing the form of the simultaneous fiducial distribution of the parameters, and supersedes the proof given previously (Cornish 1961).

III. Properties of the Distribution

(a) *Mean Values*

The vector of mean values $E(\mathbf{t})$ is null and consequently

$$E(t_it_j) = \frac{\Gamma\frac{1}{2}(n+p)}{(\pi n)^{\frac{1}{2}p}\Gamma\frac{1}{2}n}\int_{-\infty}^{\infty}\cdots\int_{-\infty}^{\infty} t_it_j\left(1+\frac{\mathbf{t}'\mathbf{R}^{-1}\mathbf{t}}{n}\right)^{-\frac{1}{2}(n+p)}\frac{d\mathbf{t}}{|\mathbf{R}|^{\frac{1}{2}}}.$$

The congruent transformation

$$\mathbf{t} = \mathbf{Pz}$$

with $\mathbf{P}'\mathbf{R}^{-1}\mathbf{P} = \mathbf{I}$ reduces this integral to

$$E(t_it_j) = \frac{\Gamma\frac{1}{2}(n+p)}{(\pi n)^{\frac{1}{2}p}\Gamma\frac{1}{2}n}\int_{-\infty}^{\infty}\cdots\int_{-\infty}^{\infty} \mathbf{p}_i'\mathbf{zz}'\mathbf{p}_j\left(1+\frac{\mathbf{z}'\mathbf{z}}{n}\right)^{-\frac{1}{2}(n+p)}d\mathbf{z},$$

where $\mathbf{p}_i'$ is the ith row of $\mathbf{P}$.

Since

$$\int_{-\infty}^{\infty}\cdots\int_{-\infty}^{\infty} z_lz_m\left(1+\frac{\mathbf{z}'\mathbf{z}}{n}\right)^{-\frac{1}{2}(n+p)}d\mathbf{z} \begin{cases} = 0 & \text{for } l \neq m, \\ = \dfrac{\pi^{\frac{1}{2}p}n^{\frac{1}{2}(p+2)}\Gamma\frac{1}{2}(n-2)}{2\Gamma\frac{1}{2}(n+p)} & \text{for } l = m, \end{cases}$$

$$E(t_it_j) = \frac{n}{n-2}r_{ij},$$

where r_{ij} is the (ij)th element of $\mathbf{R}$. The variance-covariance matrix of the t_i is thus

$$\frac{n}{n-2}\mathbf{R}.$$

(b) Distribution of Linear Functions

Suppose

$$\mathbf{x} = \mathbf{H}\mathbf{t}$$

are any $r \leqslant p$, linearly independent, linear functions of the t_i. In deriving the distribution of these functions we consider the two cases $r = p$ and $r < p$ separately.

(i) $r = p$

By definition, the distribution function of the x_i is

$$\frac{\Gamma\frac{1}{2}(n+p)}{(\pi n)^{\frac{1}{2}p}\Gamma\frac{1}{2}n}\int \ldots \int \left(1+\frac{\mathbf{t}'\mathbf{R}^{-1}\mathbf{t}}{n}\right)^{-\frac{1}{2}(n+p)} \frac{\mathbf{dt}}{|\mathbf{R}|^{\frac{1}{2}}}$$

taken over the domain defined by

$$\mathbf{h}_i'\mathbf{t} \leqslant X_i, \qquad i = 1, 2, \ldots, p,$$

where $\mathbf{h}_i'$ is the ith row of $\mathbf{H}$.

Since $\mathbf{H}$ is non-singular,

$$\mathbf{t} = \mathbf{H}^{-1}\mathbf{x},$$

the jacobian is $|\mathbf{H}|^{-1}$, and the distribution function becomes

$$\frac{\Gamma\frac{1}{2}(n+p)}{(\pi n)^{\frac{1}{2}p}\Gamma\frac{1}{2}n}\int \ldots \int \left\{1+\frac{\mathbf{x}'(\mathbf{HRH}')^{-1}\mathbf{x}}{n}\right\}^{-\frac{1}{2}(n+p)} \frac{\mathbf{dx}}{|\mathbf{HRH}'|^{\frac{1}{2}}}$$

over the region defined by

$$x_i \leqslant X_i, \qquad i = 1, 2, \ldots, p,$$

that is, a multivariate t-distribution of order p characterized by the matrix $(\mathbf{HRH}')^{-1}$.

(ii) $r < p$

The distribution function is now

$$\frac{\Gamma\frac{1}{2}(n+p)}{(\pi n)^{\frac{1}{2}p}\Gamma\frac{1}{2}n}\int \ldots \int \left(1+\frac{\mathbf{t}'\mathbf{R}^{-1}\mathbf{t}}{n}\right)^{-\frac{1}{2}(n+p)} \frac{\mathbf{dt}}{|\mathbf{R}|^{\frac{1}{2}}}$$

taken over the domain defined by

$$\mathbf{h}_i'\mathbf{t} \leqslant X_i, \qquad i = 1, 2, \ldots, r.$$

A congruent transformation $\mathbf{t} = \mathbf{Cz}$ reduces this to

$$\frac{\Gamma\frac{1}{2}(n+p)}{(\pi n)^{\frac{1}{2}p}\Gamma\frac{1}{2}n}\int \ldots \int \left(1+\frac{\mathbf{z}'\mathbf{z}}{n}\right)^{-\frac{1}{2}(n+p)} \mathbf{dz}$$

taken over the domain defined by

$$\mathbf{h}_i'\mathbf{Cz} \leqslant X_i, \qquad i = 1, 2, \ldots, r,$$

and then, following the procedure outlined in Section II, a second transformation

$$\begin{bmatrix}\mathbf{x}\\ \mathbf{v}\end{bmatrix} = \begin{bmatrix}\mathbf{HC}\\ \mathbf{Y}\end{bmatrix}\mathbf{z} = \mathbf{Pz},$$

for which $\mathbf{P}$ is non-singular and the jacobian is

$$|\mathbf{P}|^{-1} = |\mathbf{HRH}'|^{-\frac{1}{2}},$$

yields

$$\frac{\Gamma\frac{1}{2}(n+p)}{(\pi n)^{\frac{1}{2}p}\Gamma\frac{1}{2}n}\int\ldots\int\left\{1+\frac{\mathbf{x}'(\mathbf{HRH}')^{-1}\mathbf{x}+\mathbf{v}'\mathbf{v}}{n}\right\}^{-\frac{1}{2}(n+p)}\frac{d\mathbf{x}d\mathbf{v}}{|\mathbf{HRH}'|^{\frac{1}{2}}},$$

taken over the region defined by

$$x_i \leqslant X_i, \qquad i = 1, 2, \ldots, r.$$

Integrate with respect to $\mathbf{v}$, and the distribution function assumes the form

$$\frac{\Gamma\frac{1}{2}(n+r)}{(\pi n)^{\frac{1}{2}r}\Gamma\frac{1}{2}n}\int\ldots\int\left\{1+\frac{\mathbf{x}'(\mathbf{HRH}')^{-1}\mathbf{x}}{n}\right\}^{-\frac{1}{2}(n+r)}\frac{d\mathbf{x}}{|\mathbf{HRH}'|^{\frac{1}{2}}},$$

the integral being taken over the region where

$$x_i \leqslant X_i, \qquad i = 1, 2, \ldots, r,$$

that is, a multivariate t-distribution of order r characterized by the matrix $(\mathbf{HRH}')^{-1}$.

(c) *Marginal Distribution of* $t_1, t_2, \ldots, t_r$

The marginal distribution of $t_1, t_2, \ldots, t_r$, $r < p$, may be obtained by integrating for the remaining variates, but follows much more readily from the previous result by taking

$$\mathbf{H} = \begin{bmatrix}\mathbf{I}_r & \cdot\\ \cdot & \cdot\end{bmatrix},$$

and is thus a multivariate t-distribution of order r characterized by the matrix $\mathbf{R}_1^{-1}$, $\mathbf{R}_1$ being the leading submatrix of order $r \times r$ in $\mathbf{R}$. The variance-covariance matrix is $\{n/(n-2)\}\mathbf{R}_1$ and consequently the variances and covariances are identical with their values in the original distribution.

(d) *Conditional Distribution of* $t_1, t_2, \ldots, t_r$

Partition the matrix $\mathbf{R}^{-1}$ so that it takes the form

$$\begin{bmatrix}\mathbf{R}_1 & \mathbf{R}_3\\ \mathbf{R}_3' & \mathbf{R}_2\end{bmatrix},$$

where the submatrices $\mathbf{R}_1$, $\mathbf{R}_2$, and $\mathbf{R}_3$ are of order $r \times r$, $(p-r)\times(p-r)$, and $r \times (p-r)$ respectively, and $\mathbf{t}'$ so that

$$[t_1t_2 \ldots t_r|t_{r+1} \ldots t_p] = [\mathbf{t}_1'\mathbf{t}_2'].$$

Following the method outlined previously (Cornish 1954), and taking as the fixed set of values for the variables in $\mathbf{t}_2$, the elements of a vector $\mathbf{a}$, the conditional distribution function of $\mathbf{t}_1$ is

$$\frac{\Gamma\frac{1}{2}(n+p)|\mathbf{R}_1|^{\frac{1}{2}}}{(\pi n)^{\frac{1}{2}r}\Gamma\frac{1}{2}(n+p-r)}\left\{1+\frac{\mathbf{a}'(\mathbf{R}_2-\mathbf{R}_3'\mathbf{R}_1^{-1}\mathbf{R}_3)\mathbf{a}}{n}\right\}^{\frac{1}{2}(n+p-r)}$$

$$\times\int\ldots\int\left\{1+\frac{(\mathbf{t}_1+\mathbf{R}_1^{-1}\mathbf{R}_3\mathbf{a})'\mathbf{R}_1(\mathbf{t}_1+\mathbf{R}_1^{-1}\mathbf{R}_3\mathbf{a})+\mathbf{a}'(\mathbf{R}_2-\mathbf{R}_3'\mathbf{R}_1^{-1}\mathbf{R}_3)\mathbf{a}}{n}\right\}^{-\frac{1}{2}(n+p)}d\mathbf{t}_1,$$

the domain of integration being defined by

$$t_i \leqslant T_i, \qquad i = 1, 2, \ldots, r.$$

The mean value of $\mathbf{t}_1$ is $-\mathbf{R}_1^{-1}\mathbf{R}_3\mathbf{a}$ and the variance-covariance matrix is

$$\frac{n+\mathbf{a}'(\mathbf{R}_2-\mathbf{R}_3'\mathbf{R}_1^{-1}\mathbf{R}_3)\mathbf{a}}{n+p-(r+2)}\mathbf{R}_1^{-1}.$$

(e) *Distribution Function of* $\mathbf{t}'\mathbf{R}^{-1}\mathbf{t}$

The distribution function of the quadratic form $\frac{1}{p}\mathbf{t}'\mathbf{R}^{-1}\mathbf{t}$ is given by

$$\frac{\Gamma\frac{1}{2}(n+p)}{(\pi n)^{\frac{1}{2}p}\Gamma\frac{1}{2}n}\int\ldots\int\left(1+\frac{\mathbf{t}'\mathbf{R}^{-1}\mathbf{t}}{n}\right)^{-\frac{1}{2}(n+p)}\frac{d\mathbf{t}}{|\mathbf{R}|^{\frac{1}{2}}},$$

taken over the region defined by

$$\frac{1}{p}\mathbf{t}'\mathbf{R}^{-1}\mathbf{t} \leqslant T^2.$$

The congruent transformation $\mathbf{t} = \mathbf{Cz}$, reduces this to

$$\frac{\Gamma\frac{1}{2}(n+p)}{(\pi n)^{\frac{1}{2}p}\Gamma\frac{1}{2}n}\int\ldots\int\left(1+\frac{\mathbf{z}'\mathbf{z}}{n}\right)^{-\frac{1}{2}(n+p)}d\mathbf{z},$$

the integration now being taken over the region defined by

$$\frac{1}{p}\mathbf{z}'\mathbf{z} \leqslant T^2,$$

and a spherical polar transformation in p dimensions yields

$$\frac{\Gamma\frac{1}{2}(n+p)}{n^{\frac{1}{2}p}\Gamma\frac{1}{2}p\Gamma\frac{1}{2}n}\int(\chi^2)^{\frac{1}{2}(p-2)}\left(1+\frac{\chi^2}{n}\right)^{-\frac{1}{2}(n+p)}d(\chi^2) \tag{11}$$

taken over the region defined by

$$\frac{1}{p}\chi^2 \leqslant T^2,$$

and this is equivalent to the distribution of e^{2z} with $n_1 = p$ and $n_2 = n$.

IV. Information Matrix relative to the Parameters $\xi_1, \xi_2, \ldots, \xi_p$

The amount of information supplied by an observation x relative to the unknown parameter ξ (the mean) of a univariate normal distribution has been evaluated by Fisher (1959) as

$$(n+1)/(n+3)s^2,$$

where s^2 is a sufficient estimate of the unknown variance σ^2, based on n degrees of freedom. We now generalize this result to the multivariate normal distribution.

The probability that $t_1, t_2, \ldots, t_p$ shall lie in assigned ranges $dt_1, dt_2, \ldots, dt_p$ is given by (3) and this probability expressed in terms of the x_i and the ξ_i is

$$df = \frac{\Gamma\frac{1}{2}(n+p)}{(\pi n)^{\frac{1}{2}p}\Gamma\frac{1}{2}n}\left\{1+\frac{(\mathbf{x}-\boldsymbol{\xi})'\mathbf{S}^{-1}(\mathbf{x}-\boldsymbol{\xi})}{n}\right\}^{-\frac{1}{2}(n+p)}\frac{d\mathbf{x}}{|\mathbf{S}|^{\frac{1}{2}}},$$

in which $\mathbf{S}$ is the estimated variance-covariance matrix of the x_i on n degrees of freedom, whose elements constitute a jointly sufficient set of estimates of the population variances and covariances. Taking the derivative with respect to $\boldsymbol{\xi}$, squaring, dividing by df, and integrating the resultant expression over all values of the x_i yields

$$\frac{\Gamma\frac{1}{2}(n+p)}{(\pi n)^{\frac{1}{2}p}\Gamma\frac{1}{2}n}\frac{(n+p)^2}{n^2}\int_{-\infty}^{\infty}\cdots\int_{-\infty}^{\infty}\frac{\mathbf{S}^{-1}(\mathbf{x}-\boldsymbol{\xi})(\mathbf{x}-\boldsymbol{\xi})'\mathbf{S}^{-1}}{\{1+(\mathbf{x}-\boldsymbol{\xi})'\mathbf{S}^{-1}(\mathbf{x}-\boldsymbol{\xi})/n\}^{\frac{1}{2}(n+p+4)}}\frac{d\mathbf{x}}{|\mathbf{S}|^{\frac{1}{2}}}.$$

The congruent transformation

$$\mathbf{x}-\boldsymbol{\xi} = \mathbf{P}\mathbf{z}$$

with $\mathbf{P}'\mathbf{S}^{-1}\mathbf{P} = \mathbf{I}$ finally reduces this integral to

$$\frac{\Gamma\frac{1}{2}(n+p)}{(\pi n)^{\frac{1}{2}p}\Gamma\frac{1}{2}n}\frac{(n+p)^2}{n^2}\int_{-\infty}^{\infty}\cdots\int_{-\infty}^{\infty}\frac{(\mathbf{P}')^{-1}\mathbf{z}\mathbf{z}'\mathbf{P}^{-1}d\mathbf{z}}{(1+\mathbf{z}'\mathbf{z}/n)^{\frac{1}{2}(n+p+4)}},$$

$$= \frac{n+p}{n+p+2}\mathbf{S}^{-1}.$$

The scalar quantity $(n+1)/(n+3)s^2$ thus generalizes to an information matrix representing the information available in a single observation $x_1, x_2, \ldots, x_p$ on the parameters $\xi_1, \xi_2, \ldots, \xi_p$.

The corresponding result for the multiple t-distribution given by the author (Cornish 1954, 1960) is

$$\frac{n+p}{n+p+2}(s^2\mathbf{H}\mathbf{H}')^{-1},$$

in the notation of Section I.

V. Tests of Significance

The importance of the distribution (3) stems from the fact that a number of exact tests of significance can be based upon the property indicated by (11) which are, in all cases, more sensitive than those based on Hotelling's distribution as illustrated by Rao (1952), or Anderson (1958), or both. All tests enumerated and illustrated below have counterparts, as inductions of the fiducial type, which are dependent upon the fiducial distribution of the location parameters (Cornish 1962).

(a) *Tests with a Single Sample*

(i) *Test of the agreement of observed quantities with assigned values*

Suppose we have a sample of N observations from a multivariate normal distribution with a vector of means $\boldsymbol{\xi}$ and variance-covariance matrix $\boldsymbol{\Sigma}$, of which there is available a non-singular estimate $\mathbf{S}$, on n degrees of freedom. If $\bar{\mathbf{x}}$ is the vector of observed means, the hypothesis to be tested is that $\bar{\mathbf{x}}-\boldsymbol{\xi}$ is null.

The following observations have been extracted from data on the inheritance of resistance to bunt, *Tilletia foetida* (Wallr.) Liro, in certain varieties of wheat (Pugsley 1949). To illustrate this test the percentage infection produced by three collections of bunt on F_3 families (genotype $HhMm$) of Selection 1403 × White Federation 38 have been selected from Pugsley's Table 1, and their angular transformations (Fisher and Yates 1957) have been taken. The 32 families listed under this genotype yielded the following variance-covariance matrix for the means

		Bunt Collection		
		7	9	11
Bunt collection	7	0·7265	0·0219	−0·0422
	9	0·0219	1·3983	0·0252
	11	−0·0422	0·0252	1·5239,

with a corresponding correlation matrix $\mathbf{R}$

$$\begin{bmatrix} 1 & 0{\cdot}0217 & -0{\cdot}0401 \\ 0{\cdot}0217 & 1 & 0{\cdot}0173 \\ -0{\cdot}0401 & 0{\cdot}0173 & 1 \end{bmatrix},$$

of which the inverse is

$$\begin{bmatrix} 1{\cdot}0021 & -0{\cdot}0225 & 0{\cdot}0406 \\ -0{\cdot}0225 & 1{\cdot}0008 & -0{\cdot}0182 \\ 0{\cdot}0406 & -0{\cdot}0182 & 1{\cdot}0019 \end{bmatrix}.$$

The observed means and their standard deviations are

		Mean	S.D. of Mean
Bunt collection	7	14·75	0·852
	9	34·93	1·182
	11	32·42	1·234.

A test of the agreement of these observed means with assigned values 12, 32, and 30, respectively, yields

		Difference		t
Bunt collection	7	2·75	t_1	3·228
	9	2·93	t_2	2·479
	11	2·42	t_3	1·961,

from which

$$\frac{1}{p}\mathbf{t}'\mathbf{R}^{-1}\mathbf{t} = \frac{20\cdot423}{3} = 6\cdot81,$$

a value significant at 1% for $n_1 = 3$ and $n_2 = 31$ (required value of e^{2z} is 4·48).

Similarly, using the test based on Hotelling's distribution,

$$\frac{n-p+1}{np}T^2 = \frac{29}{3\times31}20\cdot423 = 6\cdot37,$$

which is significant at 1% for $n_1 = 3$ and $n_2 = 29$ (required value of e^{2z} is 4·54).

If, however, the proposed values for the means had been 12·5, 32·5, and 30·5, respectively, the differences with corresponding Student t-statistics would have been

		Difference		t
	7	2·25	t_1	2·641
Bunt collection	9	2·43	t_2	2·056
	11	1·92	t_3	1·556,

from which

$$\frac{1}{p}\mathbf{t}'\mathbf{R}^{-1}\mathbf{t} = \frac{13\cdot619}{3} = 4\cdot54,$$

which just reaches significance at the 1% point (4·48), but using the test based on Hotelling's distribution,

$$\frac{n-p+1}{np}T^2 = \frac{29}{3\times31}\times13\cdot619 = 4\cdot25,$$

which fails the test at 1%, but reaches significance at 5%, thus effectively illustrating the greater sensitivity of the test based on the multiple t-distribution.

With problems involving only a single sample, the introduction of p variates permits generalization in two ways, as indicated in (ii) and (iii) which follow.

(ii) *Test of the agreement of observed quantities with the hypothesis that all parameters are equal without specifying this parametric value*

Suppose x_{ij} ($i = 1, 2, \ldots, p$; $j = 1, 2, \ldots, N$) is a sample from a multivariate normal distribution with a vector of means $\boldsymbol{\xi}$ and variance-covariance matrix $\boldsymbol{\Sigma}$. Let $\mathbf{M}$ be any matrix of order $(p-1) \times p$ and rank $p-1$, such that

$$\mathbf{M}\boldsymbol{\varepsilon} = \mathbf{0},$$

where $\boldsymbol{\varepsilon}' = [1\ 1 \ldots 1]$. The linear functions

$$\mathbf{y} = \mathbf{M}\mathbf{x}$$

will be normally distributed with mean $\mathbf{M}\boldsymbol{\xi}$ and variance-covariance matrix $\mathbf{M}\boldsymbol{\Sigma}\mathbf{M}'$, and on the hypothesis that the ξ_i are all equal,

$$\mathbf{M}\boldsymbol{\xi} = \mathbf{0},$$

so that the test of this hypothesis reduces to a test of the departure of the observed $\mathbf{y}$ from the null vector, that is, a test similar to (i) above.

To conduct the test, the observations x_{ij} are replaced by the linear functions y_{kj}, the new column vector of means being

$$\bar{\mathbf{y}} = \frac{1}{N}\sum_{j=1}^{N}\mathbf{y}_j = \mathbf{M}\bar{\mathbf{x}},$$

and the estimated variance-covariance matrix

$$\mathbf{S} = \frac{1}{N-1}\sum_{j=1}^{N}(\mathbf{y}_j-\bar{\mathbf{y}})(\mathbf{y}_j-\bar{\mathbf{y}})'$$

$$= \frac{1}{N-1}\mathbf{M}\left\{\sum_{j=1}^{N}(\mathbf{x}_j-\bar{\mathbf{x}})(\mathbf{x}_j-\bar{\mathbf{x}})'\right\}\mathbf{M}',$$

with a corresponding correlation matrix $\mathbf{R}$.

If t_i is the Student t corresponding to the element $\bar{y}_i$ in $\bar{\mathbf{y}}$, the quadratic form

$$\frac{1}{p-1}\mathbf{t}'\mathbf{R}^{-1}\mathbf{t},$$

which has a known distribution on the null hypothesis, is invariant under any linear transformation in the $p-1$ dimensional space orthogonal to the vector $\boldsymbol{\varepsilon}$, and hence is independent of the matrix $\mathbf{M}$. $\mathbf{M}$ may thus be chosen arbitrarily provided that it is of rank $p-1$.

The test may be illustrated with further data on the inheritance of resistance to bunt (Pugsley 1950). The percentage infection, induced by each of three physiologic races of bunt, in 24 F_3 families of the cross Oro $\times$ Rapier have been extracted from Pugsley's Table 4 (beginning at family 38), and their angular transformations taken. If x_1, x_2, and x_3 represent the three variates, the contrasts chosen for making the test are

$$y_1 = x_1 - x_2,$$
$$y_2 = x_1 - x_3,$$

and the means, standard deviations of means, and t statistics relating to these contrasts are

	Mean	S.D. of Mean	t
Contrast y_1	$-3\cdot61$	2.277	$-1\cdot585$
Contrast y_2	$0\cdot03$	$2\cdot147$	$0\cdot014$.

The correlation matrix is

$$\begin{bmatrix} 1 & 0\cdot5960 \\ 0\cdot5960 & 1 \end{bmatrix},$$

and its inverse

$$\begin{bmatrix} 1\cdot5509 & -0\cdot9244 \\ -0\cdot9244 & 1\cdot5509 \end{bmatrix},$$

so that

$$\frac{1}{p}\mathbf{t}'\mathbf{R}^{-1}\mathbf{t} = \frac{3\cdot938}{2} = 1\cdot97,$$

which is not significant for $n_1 = 2$ and $n_2 = 23$. The data are, therefore, compatible with the view that this particular hybrid is equally susceptible to the three chosen races of bunt.

If the null hypothesis had been contradicted by the data, other considerations would have arisen, in particular, as pointed out by Rao (1952), it would have been of interest and importance to test the difference between an assigned contrast involving the p variates and the best contrast obtainable from the data, and to generalize this test to determine whether a set of k contrasts contains the best contrast. When the coefficients of a linear contrast are calculated from the data, there is necessarily a transfer of degrees of freedom from the n available for estimating the variance-covariance matrix, and the appropriate tests have been given by Rao (1952).

(iii) The second generalization provides a means for testing a $2p$-variate population of variates $y_1, y_2, \ldots, y_p, y_{p+1}, \ldots, y_{2p}$ to determine whether the means of y_i and y_{i+p} are equal for all values of i. The procedure is particularly useful, for example, in tests of morphological symmetry of organisms and in tests of the resemblance, with respect to a number of characters, of the members in families or groups of 2.

As an illustration we may use data on pearl shell of the oyster *Pinctada maxima* Jameson (Hynd, unpublished data), in a test for asymmetry of the 2 valves of the shell. The four characters are weight, heel depth, anteroposterior measurement, and dorsoventral measurement. Following Rao (1952), we take the difference between corresponding observations on the right and left valves and reduce the data to four variates. The test is then similar to (i) above.

The mean differences for the four characters (right valve—left valve), in the order given, were

$$-0\cdot64 \quad -0\cdot60 \quad -0\cdot0032 \quad -0\cdot0104,$$

giving corresponding values of t

$$-3\cdot368 \quad -4\cdot255 \quad -0\cdot041 \quad -0\cdot165,$$

each with 24 degrees of freedom.

The inverse of the correlation matrix is

$$\begin{bmatrix} 3\cdot0042 & -2\cdot5378 & -0\cdot9666 & 0\cdot6098 \\ -2\cdot5378 & 3\cdot5608 & 1\cdot0949 & -1\cdot3722 \\ -0\cdot9666 & 1\cdot0949 & 1\cdot6390 & -0\cdot9545 \\ 0\cdot6098 & -1\cdot3722 & -0\cdot9545 & 1\cdot9154 \end{bmatrix},$$

so that

$$\frac{1}{p}\mathbf{t}'\mathbf{R}^{-1}\mathbf{t} = 6\cdot18,$$

which is significant at 1% for $n_1 = 4$ and $n_2 = 24$, thus contradicting the hypothesis of symmetry.

(b) *Tests with Two Samples*

(i) *Test of the agreement of the means in two samples*

Suppose we have two samples of N_1 and N_2 observations, on each of p characters, giving mean vectors $\bar{\mathbf{x}}_1$ and $\bar{\mathbf{x}}_2$. The hypothesis to be tested is that the two samples have been drawn from the same multivariate normal distribution, in which case the test reduces to a test of the agreement of $\bar{\mathbf{x}}_1-\bar{\mathbf{x}}_2$ with the null vector.

To illustrate the test we shall take the two genotypes *Hhmm* and *hhMm* from Pugsley's (1949) Table 1, to represent the two samples, with $N_1 = N_2 = 17$. As before, the variates are the angular transformations of the percentage infections. The variances and covariances are obtained from the pooled sums of squares and products within samples, and the inverse of the correlation matrix is

$$\begin{bmatrix} 1\cdot2487 & -0\cdot5207 & -0\cdot1721 \\ -0\cdot5207 & 1\cdot2201 & 0\cdot0169 \\ -0\cdot1721 & 0\cdot0169 & 1\cdot0267 \end{bmatrix}.$$

The means are

		Genotype			
		Hhmm	*hhMm*	Difference	t
	7	31·99	28·36	3·63	1·517
Bunt collection	9	32·78	62·16	−29·38	−9·732
	11	69·33	28·75	40·58	17·517,

giving

$$\frac{1}{p}\mathbf{t}'\mathbf{R}^{-1}\mathbf{t} = \cancel{138\cdot24}\ 144\cdot64,$$

which is significant at 1% for $n_1 = 3$ and $n_2 = 32$.

(c) *Tests with Three or More Samples*

Suppose we have q samples, containing $N_1, N_2, \ldots, N_q$ observations from q populations each of p variates, with vectorial means $\xi_1, \xi_2, \ldots, \xi_q$ and common variance-covariance matrix $\mathbf{\Sigma}$ of which there is a non-singular estimate $\mathbf{S}$ based on n degrees of freedom. The hypothesis to be tested is

$$\sum_{i=1}^{q} a_i\xi_i = \xi,$$

where the a_i are given scalars and ξ is a given vector. On this basis, the expectation of the vector

$$\sum_{i=1}^{q} a_i\bar{\mathbf{x}}_i-\xi$$

is null, so that test reduces to that in (*a*) (i) above.

As an illustration we may consider the data presented by Clifford and Binet (1954) in a quantitative study of a presumed hybrid between *Eucalyptus elaeophora* F. Muell. and *E. goniocalyx* F. Muell. The three characters observed were fruit weight (W), fruit length (L), and length of fruiting umbel (P), and the hypothesis to be tested is that the hybrid has a gene structure such that any character assumes a value midway between the values of that character for the two species. Clifford and Binet transformed their data to metrics which equalized the variances and covariances within the three groups, and then determined the estimated variances and covariances from the pooled sums of squares and products within groups.

The transformed sample means are

	W	L	P
E. elaeophora (1)	122·10	91·20	99·78
Hybrid (2)	115·72	94·62	113·34
E. goniocalyx (3)	117·42	99·36	125·50,

each based on 50 observations, and the inverse of the correlation matrix is

$$\begin{bmatrix} 2\cdot2466 & -1\cdot6978 & 0\cdot0426 \\ -1\cdot6978 & 2\cdot7800 & -0\cdot8948 \\ 0\cdot0426 & -0\cdot8948 & 1\cdot4979 \end{bmatrix}.$$

The hypothesis specifies that the mean values of such quantities as

$$\bar{x}_{W_1}+\bar{x}_{W_3}-2\bar{x}_{W_2}$$

are 0, so that the corresponding t-statistic is

$$t_W = \frac{\bar{x}_{W_1}+\bar{x}_{W_3}-2\bar{x}_{W_2}}{\sqrt{(6s_W^2)}},$$

s_W^2 being the estimated variance of fruit weight, on 147 degrees of freedom. The three values are $t_W = 2\cdot131$, $t_L = 0\cdot704$, and $t_P = -0\cdot363$, so that

$$\frac{1}{p}\mathbf{t}'\mathbf{R}^{-1}\mathbf{t} = 2\cdot36,$$

a non-significant value for $n_1 = 3$ and $n_2 = 147$.

VI. References

Anderson, T. W. (1958).—"An Introduction to Multivariate Statistical Analysis." (Wiley: New York.)

Clifford, H. T., and Binet, F. E. (1954).—*Aust. J. Bot.* **2**: 325–36.

Cornish, E. A. (1954).—*Aust. J. Phys.* **7**: 531–42.

Cornish, E. A. (1955).—*Aust. J. Phys.* **8**: 193–9.

Cornish, E. A. (1960).—*Aust. J. Statist.* **2**: 32–40.

Cornish, E. A. (1961).—C.S.I.R.O. Aust. Div. Math. Statist. Tech. Pap. No. 8.

Cornish, E. A. (1962).—C.S.I.R.O. Aust. Div. Math. Statist. Tech. Pap. No. 11.

Dunnett, C. W., and Sobel, M. (1954).—*Biometrika* **41**: 153–76.

Fisher, R. A. (1954).—*J. R. Statist. Soc.* B **16**: 212–3.

Fisher, R. A. (1959).—"The Design of Experiments." 7th Ed. (Oliver and Boyd: Edinburgh.)

Fisher, R. A., and Yates, F. (1957).—"Statistical Tables for Biological, Agricultural and Medical Research." 5th Ed. (Oliver and Boyd: Edinburgh.)

Pugsley, A. T. (1949).—*J. Genet.* **49**: 177–82.

Pugsley, A. T. (1950).—*Aust. J. Agric. Res.* **1**: 391–400.

Rao, C. R. (1952).—"Advanced Statistical Methods in Biometric Research." (Wiley: New York.)

Reprinted from Biometrics
Vol. 19 No. 1 Pages 200–201 March 1963

The following contribution from Professor E. A. Cornish was received too late for inclusion in the December, 1962, *issue.*

SIR RONALD AYLMER FISHER, F.R.S.
ACTIVITIES IN AUSTRALIA, 1959–62

In March 1959, Sir Ronald Fisher came to Australia to take a post as Senior Research Fellow on the staff of the Division of Mathematical Statistics, C.S.I.R.O., his original intention being to stay for a period of six months, but in July he definitely decided to remain in the country, and conveyed this to me in a letter written in Sydney. The announcement that, after the expiration of his term of contract, his "few remaining years would be as well spent in Australia, in particular in Adelaide, as elsewhere" was given a tremendous reception by statisticians on the local scene.

Sir Ronald began his work with characteristic enthusiasm and energy, which remained unabated right up to this untimely death. Throughout the period, he travelled widely within Australia, visiting Perth, Melbourne, Canberra, Hobart, Sydney and Brisbane and also went abroad to New Zealand (1959), United States (1960), Japan (International Statistical Institute meeting, Tokyo 1960), Indian Statistical Institute, Calcutta (Dec. 1960–Feb. 1961 and Dec. 1961–Feb. 1962) and on a lengthy visit to England, France (International Statistical Institute meeting, Paris 1961), Italy (International Congress of Human Genetics, Rome 1961) and the United States during Apr.–Oct. 1961.

Despite his advancing years and long absences from Adelaide, it was a period of intense activity in which he completed the 7th edition of *The Design of Experiments*, the 2nd edition of *The Theory of Inbreeding* (with substantial additions), the 6th edition of *Statistical Tables for Biological, Agricultural and Medical Research* (in collaboration with F. Yates), made considerable progress with preparations for the 3rd edition of *Statistical Methods and Scientific Inference* and the 14th edition of *Statistical Methods for Research Workers*, and wrote fifteen papers, of which six represented major researches. At the same time he maintained a large international correspondence and was consulted on statistical and genetical problems by numerous people working in a wide range of research projects in Australia.

Throughout the three years, his personal interests were centred principally on further research into the fundamental problems of inductive inference, and in particular, during the first six months of 1962, he devoted considerable thought to the general use of pivotal quantities in forming a basis for the fiducial argument. Notwithstanding the criticisms levelled at the concept of fiducial probability, it will prove to be one of the greatest contributions to the advancement of statistical theory.

Ronald Fisher's associates, colleagues and friends throughout the world will be gratified to learn that contentment and happiness were the keynotes of his last three years. Unhampered by the responsibilities of administrative duties and, in the relative seclusion of Adelaide, sheltered from the irritations of scientific controversy, he could be himself, and the *real* Fisher was manifested in his supreme genius, depth of general knowledge, endearing manner, unbounded charm, generosity and intensely loyal friendship. Among those who care for quality and integrity, there will be an emptiness such as must always follow the loss of one who shows the way.

E. A. CORNISH

Commonwealth Scientific and Industrial Research Organization, Australia. Division of Mathematical Statistics Technical Paper No. 15, Melbourne 1963.

A METHOD OF FITTING STRAIGHT LINES AND PLANES TO OBSERVATIONS EXPRESSED AS TRILINEAR AND QUADRIPLANAR COORDINATES

By E. A. Cornish*

[*Manuscript received March* 1, 1963]

Summary

A method of fitting straight lines and planes to observations expressed as trilinear and quadriplanar coordinates is described and illustrated by a numerical example. Evidence is advanced to show that the residuals are distributed in such a manner that normal theory is applicable with sufficient accuracy. The general forms of the matrices, from which the asymptotic variances and covariances are obtainable by inversion, are given.

I. Introduction

The trilinear and quadriplanar coordinate systems, when referred respectively to an equilateral triangle and a regular tetrahedron, are frequently employed to provide graphical representation of percentages adding to 100%. When variable data are plotted in these two systems, the observed points may approximate a straight line in trilinear coordinates, or a plane in quadriplanar coordinates and, as in the fitting of lines and planes in ordinary regression problems, circumstances will arise which make the determination of the line or plane of best fit to the observations desirable or advantageous.

In this note we confine attention to straight lines and planes in the respective systems.

II. Procedure of Fitting

In devising an appropriate procedure of fitting, the points of difference between these new circumstances and the background to ordinary regression techniques must be considered. All variables are subject to variation, and there is not the distinction which exists between the dependent and independent variates of ordinary regression problems; on the contrary, the variates are more appropriately regarded as indistinguishable in this respect. Again, all variates are dimensionless quantities so that distance measured in any direction within the triangle or tetrahedron of reference has the same meaning. Finally, since a unique solution is required, ordinary methods are not feasible as they will not necessarily yield such a solution. The requirements are, however, met by the line or plane determined by minimizing the sum of the squares of the perpendiculars from the observed points. In either system of coordinates, the solution is unique and, moreover, is otherwise suitable, because it makes no distinction between the variates and, except for its sign, the direction of a perpendicular is immaterial.

* Division of Mathematical Statistics, C.S.I.R.O., University of Adelaide.

We assume that the perpendiculars are normally and independently distributed with constant variance σ^2. This is, of course, not strictly true, but, as will be shown, the distribution of the residuals is such that normal theory is applicable with sufficient accuracy.

The expression for a perpendicular in terms of either trilinear or quadriplanar coordinates is complex, thus rendering both systems unsuitable as starting points in the analysis, owing to the labour involved in the fitting, but this may be reduced considerably by using, instead, Cartesian coordinates referred either to a set of rectangular axes or to a set of oblique axes. A further labour-saving selection can be made by choosing oblique axes, as this avoids the transformation of all observed data to an orthogonal system.

(a) *Trilinear Coordinates*

Any vertex of the triangle may be chosen as origin, and the two sides meeting at this point as oblique axes. If x' and y' are the coordinates of a point referred to these oblique axes, the general equation of a straight line is

$$Ax'+By'+C = 0, \tag{1}$$

and the perpendicular from any point on to this line is

$$\frac{\sin\lambda(Ax'+By'+C)}{(A^2+B^2-2AB\cos\lambda)^{\frac{1}{2}}},$$

where λ is the angle between the axes, which, for an equilateral triangle, is $\frac{1}{3}\pi$.

The procedure may be simplified further by taking the coefficient of x', say, in (1) equal to unity, and rewriting (1) as

$$x'+by'+c = 0. \tag{2}$$

If, now, x and y represent the trilinear coordinates corresponding respectively to x' and y',

$$x' = 2x/\sqrt{3}, \qquad y' = 2y/\sqrt{3},$$

and the sum of the squares of the perpendiculars is

$$R = \frac{\Sigma\,(x+by+\frac{1}{2}c\sqrt{3})^2}{(b^2-b+1)},$$

since $\sin\frac{1}{3}\pi = \frac{1}{2}\sqrt{3}$ and $\cos\frac{1}{3}\pi = \frac{1}{2}$.

The logarithm of the likelihood (Fisher 1958) is then proportional to

$$L = -\frac{1}{2\sigma^2}\frac{\Sigma\,(x+by+\frac{1}{2}c\sqrt{3})^2}{(b^2-b+1)} - N\log\sigma,$$

N being the number of observations. The maximal likelihood estimates of b, c, and σ are obtained as solutions of the equations

$$\partial L/\partial b = \partial L/\partial c = \partial L/\partial\sigma = 0.$$

The second equation yields

$$\hat{c} = (-2/\surd 3)(\bar{x}+\hat{b}\bar{y}),$$

the bar over a symbol designating the arithmetic mean. Substitution of this value of $\hat{c}$ reduces the first equation to

$$-\{\Sigma\,(xy)+\tfrac{1}{2}\,\Sigma\,(y^2)\}\hat{b}^2+\{\Sigma\,(y^2)-\Sigma\,(x^2)\}\hat{b}+\Sigma\,(xy)+\tfrac{1}{2}\,\Sigma\,(x^2) = 0, \tag{3}$$

and finally using the solution for $\hat{c}$ and the appropriate value of $\hat{b}$ from (3), the residual standard deviation as obtained from the third equation is

$$\hat{\sigma} = \left\{\frac{\Sigma\,(x+\hat{b}y)^2}{N(\hat{b}^2-\hat{b}+1)}\right\}^{\frac{1}{2}}. \tag{4}$$

In (3) and (4) x *and* y *stand for deviations from the corresponding means.*

For a variety of purposes it is useful to express the equation for the line in the alternate form

$$\frac{x'}{\alpha}+\frac{y'}{\beta} = 1,$$

where α and β are the intercepts on the axes. It should be noted that, if the length of the perpendicular from any vertex of the triangle on to the opposite side is taken as 100 units, as is conventional, the length of the side is 115·5 units.

(b) *Quadriplanar Coordinates*

Any vertex of the tetrahedron may be chosen as the origin and the three edges meeting at this point as oblique axes. If x', y', and z' are the coordinates of a point referred to these oblique axes, the general equation of a plane is

$$Ax'+By'+Cz'+D = 0,$$

and the perpendicular from any point on to this plane is

$$\frac{k^{\frac{1}{2}}(Ax'+By'+Cz'+D)}{\{A^2\sin^2\lambda+B^2\sin^2\mu+C^2\sin^2\nu+2BC(\cos\mu\cos\nu-\cos\lambda)+2CA(\cos\nu\cos\lambda-\cos\mu)+2AB(\cos\lambda\cos\mu-\cos\nu)\}^{\frac{1}{2}}},$$

where λ, μ, and ν are the angles between the axes, and

$$k = 1-\cos^2\lambda-\cos^2\mu-\cos^2\nu+2\cos\lambda\cos\mu\cos\nu.$$

Make the coefficient A equal to 1, and rewrite the equation of the plane as

$$x'+by'+cz'+d = 0.$$

For a regular tetrahedron $\lambda = \mu = \nu = \frac{1}{3}\pi$, so that $k = \frac{1}{2}$; and if x, y, and z represent quadriplanar coordinates corresponding respectively to x', y', and z',

$$x' = x\surd(3/2), \qquad y' = y\surd(3/2), \qquad z' = z\surd(3/2),$$

since $\operatorname{cosec}\frac{1}{3}\pi\,.\,\operatorname{cosec}(\cos^{-1}\frac{1}{3}) = \surd(3/2)$, and the sum of the squares of the perpendiculars takes the form

$$R = \frac{3\,\Sigma\,(x+by+cz+\surd(2/3)d)^2}{\{3(1+b^2+c^2)-2bc-2b-2c\}}.$$

The logarithm of the likelihood is then proportional to

$$L = -\frac{1}{2\sigma^2}\frac{3\,\Sigma\,(x+by+cz+\surd(2/3)d)^2}{\{3(1+b^2+c^2)-2bc-2b-2c\}} - N\log\sigma,$$

and the equation $\partial L/\partial d = 0$ has the solution

$$\hat{d} = -\surd(3/2)(\bar{x}+\hat{b}\bar{y}+\hat{c}\bar{z}),$$

which when substituted in $\partial L/\partial b = 0$ and $\partial L/\partial c = 0$ reduces these equations to

$$\begin{aligned}
&-\{3\,\Sigma\,(xy)+\Sigma\,(y^2)\}\hat{b}^2+3\{\Sigma\,(y^2)-\Sigma\,(x^2)\}\hat{b}+3\,\Sigma\,(xy)+\Sigma\,(x^2)\\
&-\{3\,\Sigma\,(yz)+\Sigma\,(y^2)\}\hat{b}^2\hat{c}-2\{\Sigma\,(y^2)+3\,\Sigma\,(xz)\}\hat{b}\hat{c}+3\{\Sigma\,(y^2)-\Sigma\,(z^2)\}\hat{b}\hat{c}^2\\
&+\{3\,\Sigma\,(yz)+\Sigma\,(z^2)\}\hat{c}^3+\{3\,\Sigma\,(xy)-2\,\Sigma\,(yz)+2\,\Sigma\,(xz)+\Sigma\,(z^2)\}\hat{c}^2\\
&+\{3\,\Sigma\,(yz)-2\,\Sigma\,(xy)+2\,\Sigma\,(xz)+\Sigma\,(x^2)\}\hat{c} = 0,
\end{aligned} \tag{5}$$

$$\begin{aligned}
&-\{3\,\Sigma\,(xz)+\Sigma\,(z^2)\}\hat{c}^2+3\{\Sigma\,(z^2)-\Sigma\,(x^2)\}\hat{c}+3\,\Sigma\,(xz)+\Sigma\,(x^2)\\
&-\{3\,\Sigma\,(yz)+\Sigma\,(z^2)\}\hat{b}\hat{c}^2-2\{\Sigma\,(z^2)+3\,\Sigma\,(xy)\}\hat{b}\hat{c}+3\{\Sigma\,(z^2)-\Sigma\,(y^2)\}\hat{b}^2\hat{c}\\
&+\{3\,\Sigma\,(yz)+\Sigma\,(y^2)\}\hat{b}^3+\{3\,\Sigma\,(xz)-2\,\Sigma\,(yz)+2\,\Sigma\,(xy)+\Sigma\,(y^2)\}\hat{b}^2\\
&+\{3\,\Sigma\,(yz)-2\,\Sigma\,(xz)+2\,\Sigma\,(xy)+\Sigma\,(x^2)\}\hat{b} = 0.
\end{aligned}$$

The appropriate solution for $\hat{b}$ and $\hat{c}$ in (5), when substituted in $\partial L/\partial\sigma = 0$, then provides the equation

$$\hat{\sigma} = \left[\frac{3\,\Sigma\,(x+\hat{b}y+\hat{c}z)^2}{N\{3(1+\hat{b}^2+\hat{c}^2)-2(\hat{b}\hat{c}+\hat{b}+\hat{c})\}}\right]^{\frac{1}{2}} \tag{6}$$

for estimating the residual variance. In (5) and (6), x, y, *and* z *designate deviations from the corresponding arithmetic means.*

(c) *Illustrative Example*

A series of observations on the chemical analysis of feldspars for the constituents CaO (x), K_2O (y), Na_2O (z), and SiO_2 (w), as given by Dana (1892), Guppy (1931), Emmons (1953), Hewlett (1959), and Hess (1960), may be used for purposes of illustration. The requisite data for the example below were obtained by omitting other minor constituents from the analyses and recalculating the percentages of CaO, K_2O, Na_2O, and SiO_2 to bring their total to 100.

The necessary basic data are

$\Sigma\,(x) = 804\cdot65$, $\Sigma\,(y) = 930\cdot88$, $\Sigma\,(z) = 1060\cdot85$, $\Sigma\,(w) = 11803\cdot62$,

$\bar{x} = 5\cdot51$, $\bar{y} = 6\cdot38$, $\bar{z} = 7\cdot27$, $\bar{w} = 80\cdot85$,

$\Sigma\,(x-\bar{x})^2 = 9000$, $\Sigma\,(y-\bar{y})^2 = 7242$, $\Sigma\,(z-\bar{z})^2 = 2736$, $\Sigma\,(w-\bar{w})^2 = 2073$,

$\Sigma\,(x-\bar{x})(y-\bar{y}) = -4279$, $\Sigma\,(x-\bar{x})(z-\bar{z}) = -1079$, $\Sigma\,(x-\bar{x})(w-\bar{w}) = -3642$,

$\Sigma\,(y-\bar{y})(z-\bar{z}) = -3095$, $\Sigma\,(y-\bar{y})(w-\bar{w}) = 131$, $\Sigma\,(z-\bar{z})(w-\bar{w}) = 1438$,

$N = 146$.

Taking the origin at the vertex representing 100% SiO_2, and substituting in equations (5),

$$5595\hat{b}^2-5274\hat{b}-3837+2043\hat{b}^2\hat{c}-8010\hat{b}\hat{c}+13518\hat{b}\hat{c}^2-6549\hat{c}^3-6069\hat{c}^2+6115\hat{c}=0,$$
$$501\hat{c}^2-18792\hat{c}+5763+6549\hat{b}\hat{c}^2+20202\hat{b}\hat{c}-13518\hat{b}^2\hat{c}-2043\hat{b}^3+1637\hat{b}^2-6685\hat{b}=0.$$

Considerable labour may be involved in finding the appropriate solution of such equations. Approximate solutions must first be found by graphical or any other convenient means, and, provided these are sufficiently accurate, the solution which satisfies the condition of maximal likelihood is readily identified and, if necessary, may be improved by iteration.

In this example the approximate real solutions, calculated on the computer CSIRAC, are

$\hat{b}$	$\hat{c}$
$-0\cdot35$	$0\cdot30$
$1\cdot30$	$-0\cdot25$
$1\cdot55$	$2\cdot20$

of which the third is the one required. This was improved by Newton's method to give the values

$$\hat{b}=1\cdot5514, \qquad \hat{c}=2\cdot2245.$$

Since $\hat{d}=-38\cdot68$, the equation of the fitted plane in oblique coordinates is

$$x'+1\cdot5514y'+2\cdot2245z'-38\cdot68=0, \tag{7}$$

or, in terms of the intercepts,

$$\frac{x'}{38\cdot68}+\frac{y'}{24\cdot93}+\frac{z'}{17\cdot39}=1. \tag{8}$$

The corresponding alternate forms, omitting CaO, K_2O, and Na_2O in turn, derived by substituting

$$x'=100\sqrt{(3/2)}-y'-z'-w',$$

and so on in (8), are

$$\frac{y'}{-151\cdot97}+\frac{z'}{-68\cdot43}+\frac{w'}{83\cdot79}=1,$$

$$\frac{x'}{274\cdot44}+\frac{z'}{-224\cdot82}+\frac{w'}{97\cdot54}=1,$$

$$\frac{x'}{190\cdot91}+\frac{y'}{347\cdot30}+\frac{w'}{105\cdot09}=1.$$

Figure 1, in which the vertex representing 100% SiO_2 is taken as the origin, shows the plane in relation to the frame of reference.

It should be noted that, if the perpendicular from any vertex on to the opposite face is taken as 100 units, the length of an edge is 122·5 units.

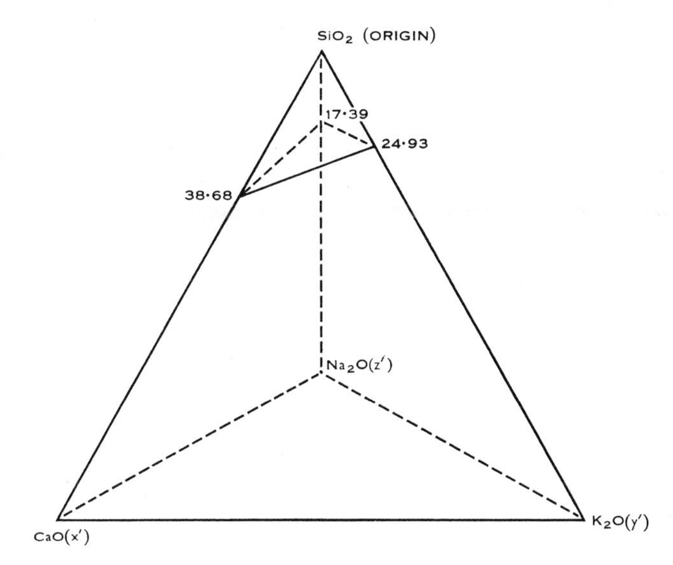

Fig. 1.—The plane $x'/38\cdot68+y'/24\cdot93+z'/17\cdot39 = 1$, fitted to the data on the chemical analyses of feldspars into the constituents SiO_2, CaO, K_2O, and Na_2O.

(d) *Errors of the Constants*

The asymptotic variance-covariance matrix of the constants is the inverse of the matrix whose elements are the negatives of the second partial derivatives of the log likelihood, in which the numerical values of the maximal likelihood estimates have been substituted. From (6), the residual sum of squares is 149·7145, and with $N = 146$, the residual variance is

$$s^2 = 1\cdot0254.$$

Numerically, the matrix of second derivatives is

$$\frac{1}{2\cdot050884}\begin{bmatrix} 7366\cdot0224 & 2102\cdot6777 & 429\cdot7465 & -0\cdot06962633 \\ 2102\cdot6777 & 5820\cdot6171 & 489\cdot7480 & 0\cdot07118153 \\ 429\cdot7465 & 489\cdot7480 & 55\cdot0333 & 0 \\ -0\cdot06962633 & 0\cdot07118153 & 0 & 584\cdot0000 \end{bmatrix},$$

having an inverse

$$\begin{bmatrix} 0\cdot0^2103418 & 0\cdot0^2121762 & -0\cdot01891148 & -0\cdot0^7251139 \\ 0\cdot0^2121762 & 0\cdot0^2283613 & -0\cdot03474724 & -0\cdot0^6200516 \\ -0\cdot01891148 & -0\cdot03474724 & 0\cdot49416254 & 0\cdot0^5198052 \\ -0\cdot0^7251139 & -0\cdot0^6200516 & 0\cdot0^5198052 & 0\cdot0^2351179 \end{bmatrix},$$

in which the order of the elements corresponds to the ordering $\hat{b}$, $\hat{c}$, $\hat{d}$, and $\hat{\sigma}$. The variances of the coefficients in the equation for the plane (7) are then

$$\mathrm{V}(\hat{b}) = 0\cdot0^2103418, \qquad \text{S.D.} = 0\cdot0322,$$
$$\mathrm{V}(\hat{c}) = 0\cdot0^2283613, \qquad \text{S.D.} = 0\cdot0533,$$
$$\mathrm{V}(\hat{d}) = 0\cdot494163, \qquad \text{S.D.} = 0\cdot703,$$

and the variances of the remaining intercepts, $-\hat{d}/\hat{b}$ and $-\hat{d}/\hat{c}$, in the form (8) are

$$\mathrm{V}(-\hat{d}/\hat{b}) = \frac{1}{\hat{b}^2}\mathrm{V}(\hat{d})+\frac{\hat{d}^2}{\hat{b}^4}\mathrm{V}(\hat{b})-\frac{2\hat{d}}{\hat{b}^3}\mathrm{Cov}(\hat{b}\hat{d}) = 0\cdot0806089, \qquad \text{S.D.} = 0\cdot284,$$

$$\mathrm{V}(-\hat{d}/\hat{c}) = \frac{1}{\hat{c}^2}\mathrm{V}(\hat{d})+\frac{\hat{d}^2}{\hat{c}^4}\mathrm{V}(\hat{c})-\frac{2\hat{d}}{\hat{c}^3}\mathrm{Cov}(\hat{c}\hat{d}) = 0\cdot0289547, \qquad \text{S.D.} = 0\cdot170.$$

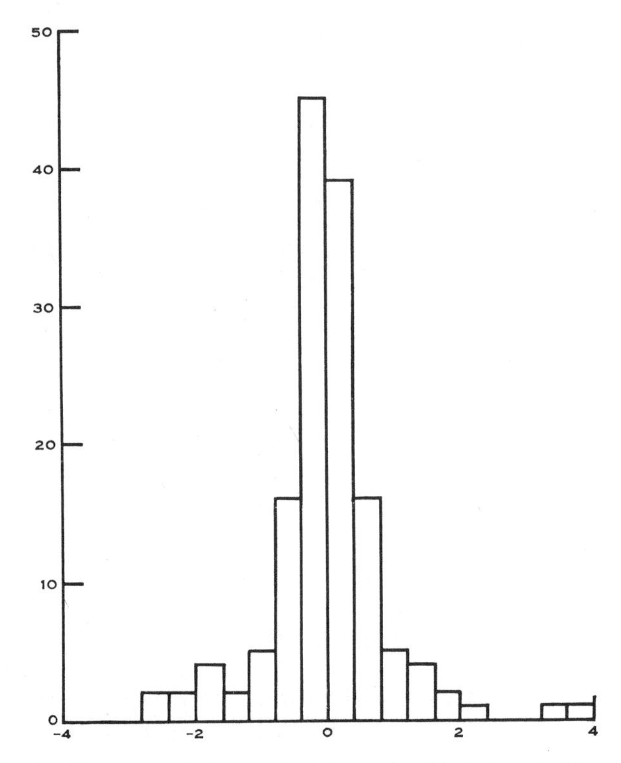

Fig. 2.—Histogram of the residuals from the fitted plane in Figure 1.

III. Examination of Residuals

Figures 2 and 3 illustrate histograms of the residuals obtained respectively from the chemical analyses of feldspars used above, and the mechanical analyses of volcanic ash soils from Mt. Gambier, South Australia, into the four constituents coarse sand, fine sand, silt, and clay (data from J. A. Prescott and J. T. Hutton). These are typical of the frequency distributions of residuals, and it may be safely

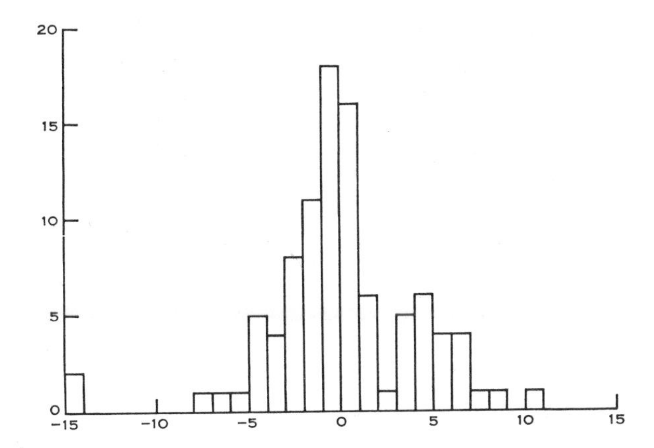

Fig. 3.—Histogram of the residuals from a plane fitted to data on the mechanical analyses of volcanic ash soils into the constituents coarse sand, fine sand, silt, and clay.

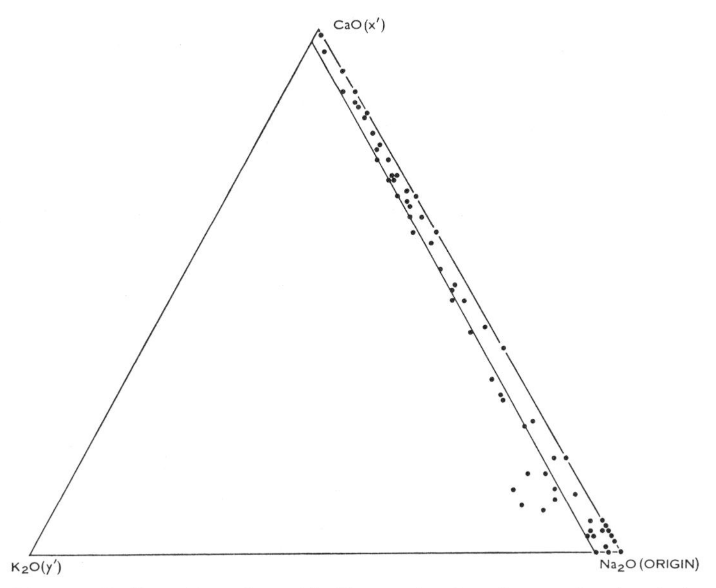

Fig. 4.—The line $x'/177 \cdot 27 + y'/3 \cdot 93 = 1$, fitted to data on the chemical analyses of feldspars into the constituents Na_2O, CaO, and K_2O.

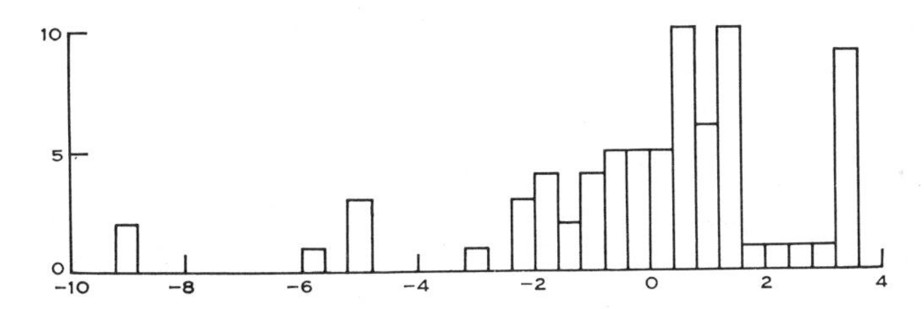

Fig. 5.—Histogram of the residuals from the fitted line in Figure 4.

assumed that such distributions will be represented with sufficient accuracy by the normal distribution, except under the extreme conditions imposed by close proximity in parallel to the bounding lines or planes of the frame of reference. Figures 4 and 5 illustrate such a case. The data were provided by chemical analyses of a group of feldspars for the three constituents CaO (x), K_2O (y), and Na_2O (z); the original observations and the line fitted to them are shown in Figure 4. Figure 5 gives the histogram of the residuals, which is not so very markedly different from what might occur with a sample of 72 from a normal distribution.

IV. Acknowledgments

I am indebted to Dr. J. B. Jones, of the Geology Department, University of Adelaide, for the references to the chemical analyses of the feldspars; to Professor J. A. Prescott, who first drew my attention to the problem and provided some of the analyses on the volcanic ash soils, to Mr. J. T. Hutton, Division of Soils, C.S.I.R.O., who supplied the remaining data on these soils, and to Messrs. T. Pearcey and R. P. Harris for computing the approximate solutions on CSIRAC.

V. References

Dana, E. S. (1892).—"System of Mineralogy of James Dwight Dana, 1837–1868: Descriptive Mineralogy." 6th Ed. (Wiley: New York.)

Emmons, R. C., Ed. (1953).—Geol. Soc. Amer. Memoir, No. 52.

Fisher, R. A. (1958).—"Statistical Methods for Research Workers." 13th Ed. (Oliver and Boyd: Edinburgh.)

Guppy, E. M. (1931).—Chemical analyses of igneous rocks, metamorphic rocks and minerals. Geological Survey of Great Britain, Memoir. (H.M. Stationery Office: London.)

Hess, H. H. (1960).—Geol. Soc. Amer. Memoir, No. 80.

Hewlett, C. G. (1959).—*Bull. Geol. Soc. Amer.* **70**: 511–38.

Appendix

The general forms of the matrices whose elements are the second partial derivatives of the likelihood (with changed sign) are as follows:

(a) *Trilinear Coordinates*

The ordering of the quantities which have been estimated by the method of maximal likelihood is $\hat{b}$, $\hat{c}$, and $\hat{\sigma}$, and writing

$$B = (2\hat{b}-1),$$
$$D = (\hat{b}^2-\hat{b}+1),$$
$$X = (x-\bar{x})+\hat{b}(y-\bar{y}),$$

and substituting for convenience

$$C = -\frac{2}{\sqrt{3}}(\bar{x}+\hat{b}\bar{y}),$$

the matrix is

$$\frac{1}{2\hat{\sigma}^2}\begin{bmatrix} \frac{2\,\Sigma\,(y^2)}{D}-\frac{4B\,\Sigma\,(yX)+2\,\Sigma\,(X^2)}{D^2}+\frac{2B^2\,\Sigma\,(X^2)}{D^3} & \frac{\sqrt{3}\,\Sigma\,(y)}{D} & -\frac{4\,\Sigma\,(yX)}{D\hat{\sigma}}+\frac{2B\,\Sigma\,(X^2)}{D^2\hat{\sigma}} \\ \frac{\sqrt{3}\,\Sigma\,(y)}{D} & \frac{3N}{2D} & 0 \\ -\frac{4\,\Sigma\,(yX)}{D\hat{\sigma}}+\frac{2B\,\Sigma\,(X^2)}{D^2\hat{\sigma}} & 0 & 4N \end{bmatrix}.$$

(b) *Quadriplanar Coordinates*

The ordering of the quantities which have been estimated is $\hat{b}$, $\hat{c}$, $\hat{d}$, and $\hat{\sigma}$ and writing

$$B = 3\hat{b}-\hat{c}-1,$$
$$C = 3\hat{c}-\hat{b}-1,$$
$$D = 3(1+\hat{b}^2+\hat{c}^2)-2\hat{b}\hat{c}-2\hat{b}-2\hat{c},$$
$$X = (x-\bar{x})+\hat{b}(y-\bar{y})+\hat{c}(z-\bar{z}),$$

and substituting for convenience

$$\hat{d} = -\sqrt{(3/2)}(\bar{x}+\hat{b}\bar{y}+\hat{c}\bar{z}),$$

the matrix is

$$\frac{1}{2\hat{\sigma}^2}\begin{bmatrix} \frac{6\,\Sigma\,(y^2)}{D}-\frac{\{24B\,\Sigma\,(yX)+18\,\Sigma\,(X^2)\}}{D^2}+\frac{24B^2\,\Sigma\,(X^2)}{D^3} & \frac{6\,\Sigma\,(yz)}{D}+\frac{6\,\Sigma\,(X^2)-12\{B\,\Sigma\,(zX)+C\,\Sigma\,(yX)\}}{D^2}+\frac{24BC\,\Sigma\,(X^2)}{D^3} & \frac{2\sqrt{6}\,\Sigma\,(y)}{D} & \frac{-12\,\Sigma\,(yX)}{D\hat{\sigma}}+\frac{12B\,\Sigma\,(X^2)}{D^2\hat{\sigma}} \\ \frac{6\,\Sigma\,(yz)}{D}+\frac{6\,\Sigma\,(X^2)-12\{B\,\Sigma\,(zX)+C\,\Sigma\,(yX)\}}{D^2}+\frac{24BC\,\Sigma\,(X^2)}{D^3} & \frac{6\,\Sigma\,(z^2)}{D}-\frac{\{24C\,\Sigma\,(zX)+18\,\Sigma\,(X^2)\}}{D^2}+\frac{24C^2\,\Sigma\,(X^2)}{D^3} & \frac{2\sqrt{6}\,\Sigma\,(z)}{D} & \frac{-12\,\Sigma\,(zX)}{D\hat{\sigma}}+\frac{12C\,\Sigma\,(X^2)}{D^2\hat{\sigma}} \\ \frac{2\sqrt{6}\,\Sigma\,(y)}{D} & \frac{2\sqrt{6}\,\Sigma\,(z)}{D} & \frac{4N}{D} & 0 \\ \frac{-12\,\Sigma\,(yX)}{D\hat{\sigma}}+\frac{12B\,\Sigma\,(X^2)}{D^2\hat{\sigma}} & \frac{-12\,\Sigma\,(zX)}{D\hat{\sigma}}+\frac{12C\,\Sigma\,(X^2)}{D^2\hat{\sigma}} & 0 & 4N \end{bmatrix}$$

The elements of the inverses of these matrices are the variances and covariances of the maximal likelihood estimates.

Reprinted from the Bulletin of the International Statistical Institute, Vol. 40, 1963. (34th Session Proc.).

A COMPARISON OF THE SIMULTANEOUS FIDUCIAL DISTRIBUTIONS DERIVED FROM THE MULTIVARIATE NORMAL DISTRIBUTION*

G. W. BENNETT and E. A. CORNISH
C.S.I.R.O., Division of Mathematical Statistics, Adelaide

1. INTRODUCTION

Extension of the theory of fiducial inference to more complex situations involving the determination of the simultaneous fiducial distribution of several parameters, in particular the multivariate normal distribution, has led, in recent years, to a number of different solutions for this problem. The principal reason for the diversity of the solutions and the differences of opinion regarding them is that the conditions under which a valid and unique solution exists are really not yet fully understood, with the result that implicit and explicit assumptions have been made, based either on analogous properties of sampling distributions or the results obtained from some simple inference problems.

It is clear that some revision of the initial conditions is necessary and, in view of the important central position occupied by the multivariate normal distribution in multivariate analysis, we propose to examine these conditions in relation to this distribution. Some results are available only for the bivariate normal distribution which is discussed in some detail.

2. MAXIMAL LIKELIHOOD ESTIMATION

The likelihood function and the maximal likelihood estimates of the parameters provide the basic material from which the fiducial distributions are to be constructed and consequently the form of the likelihood and the properties of the statistics are of prime importance in the fiducial argument.

The likelihood function for a sample of N from a bivariate normal distribution is

$$\mathfrak{L} \propto [\sigma_1\sigma_2\sqrt{(1-\rho^2)}]^{-N} \exp\left\{-\frac{1}{2(1-\rho^2)}\left[\frac{S(x_1-\xi_1)^2}{\sigma_1^2} - 2\rho\frac{S(x_1-\xi_1)(x_2-\xi_2)}{\sigma_1\sigma_2} + \frac{S(x_2-\xi_2)^2}{\sigma_2^2}\right]\right\}$$

so that

$$L = \log_e \mathfrak{L} = -\frac{N}{2}\log\sigma_1^2 - \frac{N}{2}\log\sigma_2^2 - \frac{N}{2}\log(1-\rho^2)$$

---

*Invited paper, 34th Session, I.S.I., Ottawa, Canada, 1963.

ERRATA

"A comparison of the simultaneous fiducial distributions derived from the multivariate normal distribution", by G.W. Bennett and E.A. Cornish

page	line	For	Read
904	12	dp^2	ds^2
	13	fabrication	factorization
905	3	r	τ
	8	(b)	(ii)
	14	(c)	(iii)
	16	(i)	(1)
906	end 7	x	r
	18	x_1 (bold type)	$\bar{x}$ (bold type)
	31	$\vec{x}^1$ (" ")	$\vec{x}'$ (" ")
907	16	Γ (" ")	Γ (ordinary type)
	30	$\left(\frac{1}{\upsilon_1+1}+\frac{1}{\upsilon_2+1}\right)\vert^{-\frac{1}{2}}$	$\left(\frac{1}{\upsilon_1+1}+\frac{1}{\upsilon_2+1}\right)\vert^{-\frac{1}{2}}$
908	2	$d\bar{x}d_2$	$d\bar{x}_2d$ (bold type x, each case)
	2	$S_2\upsilon^{-1}_1$	$S_2^{-1}\upsilon_1$ (" " S " ")
	8	λi	λ_i
	9	$\lambda+\lambda 2+\dots+\lambda p1$	$\lambda_1+\lambda_2+\dots+\lambda_p$
	21	Γ (bold type)	Γ (ordinary type)
	26	$\pi^{p\frac{1}{4}(p-1)}$	$\pi^{\frac{1}{4}p(p-1)}$
909	3	σ_2^2	σ_i^2
	4	(iii)$(\xi-\bar{x}_1)^1$	(iii)$(\xi-\bar{x}_1)'$ (bold type)
	5	$n_1=\upsilon_1-p+1$	$n_2=\upsilon_1-p+1$
911	3	$x=0$	$r=0$
	16	s_2^2	s_i^2
	17	$s_i^2\rho_i^2$	$s_i^2\rho^2$
912	30	t (bold type)	t (ordinary type)
	31	λ (ordinary ")	λ (bold ")
	34	(14)	(4)
913	1	$x=0$	$r=0$
	5	$1+t_1^2+t_1^2$	$1+t_1^2+t_2^2$
	16	$\bar{x}$ (ordinary type)	$\bar{x}$ (bold type)
	26	Σ (" ")	Σ (" ")
915	22	$+\partial F(t_1\,\theta)$	$\pm\partial F(t_1\,\theta)$
	43	$\{(x_1-\bar{x}_1)^2$	$\{S(x_1-\bar{x}_1)^2$
918	8	H^1SH (bold type)	$H'SH$ (bold type)

page	line	For	Read
905	17	(ii)	(2)
	20	(iii)	(3)
	21	in	on
	23	(iv)	(4)
	27	(a)	(i)

$$-\frac{1}{2(1-\rho^2)}\left[\frac{S(x_1-\xi_1)^2}{\sigma_1^2}-2\rho\frac{S(x_1-\xi_1)(x_2-\xi_2)}{\sigma_1\sigma_2}+\frac{S(x_2-\xi_2)^2}{\sigma_2^2}\right].$$

From this expression the five likelihood equations can be formed and solved simultaneously to yield the usual estimates of the parameters. For future reference some points about the solution of these equations are required. In the first place, the equations for ξ_1 and ξ_2 may be written as

$$\mathbf{A}S(\boldsymbol{x}-\boldsymbol{\xi}) = 0,$$

where

$$\mathbf{A}^{-1} = \begin{bmatrix} \sigma_1^2 & \rho\sigma_1\sigma_2 \\ \rho\sigma_1\sigma_2 & \sigma_2^2 \end{bmatrix},$$

and

$$S(\boldsymbol{x}-\boldsymbol{\xi}) = \begin{bmatrix} S(x_1-\xi_1) \\ S(x_2-\xi_2) \end{bmatrix},$$

and since $\mathbf{A}$ is non-singular, $\bar{x}_1$ and $\bar{x}_2$ are the estimates of ξ_1 and ξ_2, regardless of the knowledge available about σ_1, σ_2, and ρ; in particular, they are the estimates even if these latter parameters are known.

The same cannot be said for the equations $\partial L/\partial\sigma_1^2 = 0$, $\partial L/\partial\sigma_2^2 = 0$ and $\partial L/\partial\rho = 0$, for the solution evidently depends on ξ_1 and ξ_2. Moreover, the equations for σ_1^2 and σ_2^2 cannot be solved independently of that for ρ, for if this latter equation is ignored the estimates are

$$\bar{\sigma}_1^2 = \frac{N-1}{N}s_1^2\frac{(1-\rho r)}{(1-\rho^2)}, \qquad \bar{\sigma}_2^2 = \frac{N-1}{N}s_2^2\frac{(1-\rho r)}{(1-\rho^2)},$$

where, as usual, r denotes the sample correlation coefficient. Similarly $\partial L/\partial\rho = 0$ cannot be solved independently of the values of σ_1^2 and σ_2^2.

Thus the solution of the likelihood equations divides the parameters into two groups (ξ_1, ξ_2) and (σ_1, σ_2, ρ) while within each group there is no discrimination between members. Moreover, the five estimates are known to be jointly sufficient, ($\bar{x}_1$, $\bar{x}_2$) being jointly sufficient for (ξ_1, ξ_2), while (s_1, s_2, r) are sufficient only when the distribution of the observations is considered to be conditional on the calculated values of $\bar{x}_1$ and $\bar{x}_2$. No one of the estimates is individually sufficient.

3. CONDITIONS TO BE EXAMINED

The conditions to be examined in relation to various problems in inference are as follows:

A. Joint fiducial distributions may be constructed from marginal and conditional distributions which may be derived by inversion of the appropriate sampling distribution.

B. If this can be performed in several ways each will lead to the same result.

C. The formula $\pm\partial F(T, \theta)/\partial\theta$, where $F(T, \theta)$ is the cumulative distribution function of T, may be used to carry out the inversions, subject to restrictions of monotonicity and differentiability on F, considered as a function of θ, and provided

$$F(T, +\infty) = \begin{Bmatrix} 1 \\ 0 \end{Bmatrix} \qquad -\infty < T < \infty$$

$$F(T, -\infty) = \begin{Bmatrix} 0 \\ 1 \end{Bmatrix}$$

depending on whether F is an increasing, or decreasing, function of θ.

4. UNIVARIATE NORMAL DISTRIBUTION

We briefly summarize the known facts about this particular case.

(a) Sampling distribution

The sampling distribution of the mean and variance is

$$\frac{\sqrt{N}}{\sigma\sqrt{(2\pi)}} \exp[-N(\bar{x} - \xi)^2/2\sigma^2]d\bar{x} \frac{(N-1)^{\frac{1}{2}(N-1)}}{2^{\frac{1}{2}(N-1)}\Gamma\frac{1}{2}(N-1)\sigma^{N-1}} (s^2)^{\frac{1}{2}(N-3)} \exp[-(N-1)s^2/2\sigma^2]dp^2,$$

in which $\bar{x}$ and s^2 are distributed independently, and since the fabrication is complete, the marginal distributions are identical with the respective factors of the simultaneous distribution. Moreover, it should be noted that coupled with this factorization the statistic s^2, given $\bar{x}$, is sufficient for σ^2, and whether σ^2 is known or not, $\bar{x}$ alone is sufficient for ξ.

(b) Fiducial distribution

The simultaneous fiducial distribution is

$$\frac{\sqrt{N}}{\sigma\sqrt{(2\pi)}}\exp[-N(\xi-\bar{x})^2/2\sigma^2]d\xi \cdot \frac{(N-1)^{\frac{1}{2}(N-1)}(s^2)^{\frac{1}{2}(N-1)}}{2^{\frac{1}{2}(N-1)}\Gamma\frac{1}{2}(N-1)} \frac{\exp[-(N-1)s^2/2\sigma^2]}{(\sigma^2)^{\frac{1}{2}N}}d(\sigma^2)$$

in which (1) ξ and σ^2 are not independently distributed; (2) the marginal distribution of ξ is

$$\frac{\Gamma\frac{1}{2}N}{\Gamma\frac{1}{2}(N-1)\sqrt{[\pi(N-1)]}} \left\{1 + \frac{N(\xi - \bar{x})^2}{(N-1)s^2}\right\}^{-\frac{1}{2}N} \frac{\sqrt{N}}{s} d\xi$$

so that $(\xi - \bar{x})\sqrt{N}/s$ is distributed in the same manner as Student's t-statistic with $(N-1)$ degrees of freedom; and (3) the marginal distribution of σ^2 is identical with the corresponding factor of the simultaneous distribution, and if the distribution is transformed to that of $1/\sigma^2$, then $(N-1)s^2/\sigma^2$ is distributed in the same manner as χ^2 with $(N-1)$ degrees of freedom.

The simultaneous fiducial distribution may be derived in various ways; (i) using either of the two sets of pivotal quantities

$$r = (\bar{x} - \xi)\sqrt{N}/\sigma$$
$$\chi^2 = (N-1)s^2/\sigma^2$$

or

$$t = (\bar{x} - \xi)\sqrt{N}/s$$
$$\chi_N^2 = [(N-1)s^2 + N(\bar{x} - \xi)^2]/\sigma^2$$

the two sets being essentially equivalent (Fisher, 1941); (*b*) using the known simultaneous distribution of the two quantities

$$t = \frac{(\bar{x} - \bar{x}')\sqrt{[NN'(N + N' - 2)]}}{\sqrt{(N + N')}\sqrt{[(N-1)s^2 + (N'+1)s'^2]}},$$
$$z = \log(s/s'),$$

to obtain the simultaneous distribution of $\bar{x}'$ and s' in terms of $\bar{x}$ and s, and then taking the limiting form of the latter distribution as $N' \to \infty$ (Fisher, 1935); (*c*) using the fiducial argument in a step-by-step procedure to construct the simultaneous distribution of the parameters (Fisher, 1959), the principal steps being (i) $\bar{x}$ and s^2 are jointly sufficient for ξ and σ^2, and s^2, which is also sufficient for σ^2, given $\bar{x}$, is distributed independently of $\bar{x}$ and ξ; (ii) the marginal distribution of σ^2 is supplied by the pivotal quantity $(N-1)s^2/\sigma^2$, which has a known distribution independent of σ^2, or alternatively by application of the formula $-\partial F(s, \sigma)/\partial\sigma$; (iii) the marginal distribution of σ^2 is independent of $\bar{x}$, so that conditionally in any given σ^2, $\bar{x}$ is sufficient for ξ and further, since the pivotal quantity $(\xi - \bar{x})/\sigma$ has a known distribution independent of both parameters, it supplies the fiducial distribution of ξ for given σ^2; (iv) the product of the marginal distribution of σ^2 and the conditional distribution of ξ for given σ^2, then gives the bivariate distribution of ξ and σ^2.

In this simple case the method is essentially the same as using the first pair of pivotals under (*a*) above.

5. BIVARIATE NORMAL DISTRIBUTION

a. Sampling distribution

The sampling distribution of the five statistics may be written in the form

$$(1) \quad \frac{N}{2\pi\sigma_1\sigma_2(1-\rho^2)^{\frac{1}{2}}} \exp\left\{-\frac{N}{2(1-\rho^2)}\left[\frac{(\bar{x}_1 - \xi_1)^2}{\sigma_1^2} - \frac{2\rho(\bar{x}_1 - \xi_1)(\bar{x}_2 - \xi_2)}{\sigma_1\sigma_2} + \frac{(\bar{x}_2 - \xi_2)^2}{\sigma_2^2}\right]\right\} d\bar{x}_1 d\bar{x}_2 \times \frac{(N-1)^{N-1} s_1^{N-2} s_2^{N-2} (1-r^2)^{\frac{1}{2}(N-4)}}{\pi\Gamma(N-2)\sigma_1^{N-1}\sigma_2^{N-1}(1-\rho^2)^{\frac{1}{2}(N-1)}} \exp\left\{-\frac{N-1}{2(1-\rho^2)}\left[\frac{s_1^2}{\sigma_1^2} - \frac{2\rho r s_1 s_2}{\sigma_1\sigma_2} + \frac{s_2^2}{\sigma_2^2}\right]\right\} ds_1 ds_2 dr,$$

in which $\bar{x}_1$ and $\bar{x}_2$ are distributed independently of s_1, s_2 and r, so that the marginal distributions of these two groups of statistics are identical with the respective factors of the simultaneous distribution, but this latter cannot be factorized further whilst retaining these five statistics.

The marginal distributions of $\bar{x}_i$ $(i = 1, 2)$ and s_i^2 are identical with the corresponding marginal distributions of the univariate case, and similarly for the joint marginal distribution of $\bar{x}_i$ and s_i^2, while the marginal distribution of x may be written

$$\frac{N-2}{\pi}(1-\rho^2)^{\frac{1}{2}(N-1)}(1-r^2)^{\frac{1}{2}(N-4)}\int_0^{\infty}\frac{dz}{(\cosh z-\rho r)^{N-1}}\,dr, \tag{2}$$

since

$$\int_0^{\infty}(\cosh z+\cos\theta)^{-1}dz=\theta/\sin\theta.$$

b. Fiducial distribution

Attempts to determine the fiducial distribution corresponding to (1) and to generalize the results to the multivariate normal distribution have led to different solutions, thus raising difficulties and doubts about the application of the theory.

(1) *Segal's solution.* Segal (1938) used the two pivotal quantities

$$\mathbf{A}=\mathbf{\Sigma}^{-\frac{1}{2}}\mathbf{S}\mathbf{\Sigma}^{-\frac{1}{2}} \tag{3}$$

$$\mathbf{B}=\mathbf{\Sigma}^{-\frac{1}{2}}(\mathbf{x}-\boldsymbol{\xi})$$

and formally made the change of variables

$$(\mathbf{S},\bar{\mathbf{x}})\rightarrow(\mathbf{A},\mathbf{B})\rightarrow(\mathbf{\Sigma}^{-\frac{1}{2}},\boldsymbol{\xi});$$

for the bivariate distribution his result may be written

$$\frac{N}{2\pi\sigma_1\sigma_2(1-\rho^2)^{\frac{1}{2}}}\exp\left\{-\frac{N}{2(1-\rho^2)}\left[\frac{(\xi_1-\bar{x}_1)^2}{\sigma_1^2}-2\rho\frac{(\xi_1-\bar{x}_1)(\xi_2-\bar{x}_2)}{\sigma_1\sigma_2}+\frac{(\xi_2-\bar{x}_2)^2}{\sigma_2^2}\right]\right\}d\xi_1 d\xi_2\times\frac{(N-1)^{N-1}s_1^{N-1}s_2^{N-1}(1-r^2)^{\frac{1}{2}(N-1)}}{\pi\Gamma(N-2)\sigma_1^N\sigma_2^N(1-\rho^2)^{\frac{1}{2}(N+2)}}$$
$$\exp\left\{-\frac{N-1}{2(1-\rho^2)}\left[\frac{s_1^2}{\sigma_1^2}-\frac{2\rho r s_1 s_2}{\sigma_1\sigma_2}+\frac{s_2^2}{\sigma_2^2}\right]\right\}d\sigma_1 d\sigma_2 d\rho. \tag{4}$$

Without commenting on the validity of the derivation, we may note (i) ξ_1 and ξ_2 are not distributed independently of σ_1, σ_2, and ρ; (ii) the marginal distribution of ξ_i $(i = 1, 2)$ is

$$\frac{\Gamma\frac{1}{2}(N-1)}{\Gamma\frac{1}{2}(N-2)\sqrt{[\pi(N-2)]}}\left\{1+\frac{N(\xi_i-\bar{x}_i)^2}{(N-2)s_i^2}\right\}^{-\frac{1}{2}(N-1)}\frac{\sqrt{N}}{s_i}\,d\xi_i$$

so that $(\xi_i - \bar{x}_i)\sqrt{N}/s_i$ is distributed in the same manner as t with $(N-2)$ degrees of freedom; (iii) the marginal distribution of $(N-1)s_i^2/\sigma_i^2$ is that of χ^2 with $(N-2)$ degrees of freedom; (iv) $(\boldsymbol{\xi}-\bar{\mathbf{x}})'\mathbf{S}^{-1}(\boldsymbol{\xi}-\bar{\mathbf{x}})$ is distributed as Hotelling's T^2 with $n_1 = 2$ and $n_2 = N-2$ degrees of freedom; and (v) the marginal distribution of ρ is identical with that of r with ρ and r interchanged. Thus the conditions (A) and (B) are not satisfied and as will be shown, the distribution of ρ is not consistent with (C).

Segal's distribution may also be derived by a generalization of the limiting process used by Fisher (1935). As a preliminary two multivariate analogues of the quantities t and z and their sampling distributions are required. First, if $\mathbf{S}$ (of order $p \times p$) is an estimate of $\mathbf{\Sigma}$ on ν degrees of freedom, then

$$\mathbf{t} = \mathbf{S}^{-\frac{1}{2}}(\mathbf{x} - \boldsymbol{\xi}) \tag{5}$$

may be shown to have the distribution

$$\frac{\Gamma\frac{1}{2}(\nu + 1)}{(\pi\nu)^{\frac{1}{2}p}\Gamma\frac{1}{2}(\nu - p + 1)} \frac{dt}{|I + tt'/\nu|^{\frac{1}{2}(\nu+1)}}; \tag{6}$$

secondly, if $\mathbf{S}_1$ on ν_1 degrees of freedom and $\mathbf{S}_2$ on ν_2 degrees of freedom, from samples of $\nu_1 + 1$ and $\nu_2 + 1$ observations, are independent estimates of $\mathbf{\Sigma}$, the distribution of $\mathbf{W} = \mathbf{S}_2\mathbf{S}_1^{-1}$ is

$$\frac{\nu_1^{\frac{1}{2}\nu_1 p}\,\nu_2^{\frac{1}{2}\nu_2 p}}{\pi^{\frac{1}{4}p(p-1)}\beta_p(\frac{1}{2}\nu_1, \frac{1}{2}\nu_2)} \frac{|\mathbf{W}|^{\frac{1}{2}(\nu_2-p-1)}}{|\nu_1\mathbf{I} + \nu_2\mathbf{W}|^{\frac{1}{2}(\nu_1+\nu_2)}} d\mathbf{W} \tag{7}$$

where

$$\beta_p(\alpha, \beta) = \Gamma_p(\alpha)\Gamma_p(\beta)/\Gamma_p(\alpha + \beta)$$

and

$$\Gamma_p(\alpha) = \prod_{i=1}^{p} \Gamma(\alpha - i + 1).$$

Suppose now that a sample of $\nu_1 + 1$ observations gives the jointly sufficient statistics $\bar{\mathbf{x}}_1$ and $\mathbf{S}_1$, and the fiducial distribution of $\bar{\mathbf{x}}_2$ and $\mathbf{S}_2$, the corresponding estimates from an unmade sample of $\nu_2 + 1$ independent observations, is rerequired. It can be shown that

$$\mathbf{t} = \left[\frac{\nu_2\mathbf{S}_2 + \nu_1\mathbf{S}_1}{\nu_1 + \nu_2}\left(\frac{1}{\nu_1 + 1} + \frac{1}{\nu_2 + 1}\right)\right]^{-\frac{1}{2}}(\bar{\mathbf{x}}_2 - \bar{\mathbf{x}}_1)$$

and $\mathbf{W} = \mathbf{S}_1\mathbf{S}_2^{-1}$ are distributed independently in the forms (6) with $\nu_1 + \nu_2$ degrees of freedom and (7) with ν_1 and ν_2 degrees of freedom, respectively. Changing to the variables $\mathbf{S}_2^{-1}$ and $\bar{\mathbf{x}}_2$ yields the jacobian

$$|\mathbf{S}_1|^{\frac{1}{2}(p+1)} \left|\frac{\nu_2\mathbf{S}_2 + \nu_1\mathbf{S}_1}{\nu_1 + \nu_2}\left(\frac{1}{\nu_1 + 1} + \frac{1}{\nu_2 + 1}\right)\right|^{-\frac{1}{2}}$$

so that the distribution of $\bar{\mathbf{x}}_2$ and $\mathbf{S}_2^{-1}$ is

$$\frac{\nu_1^{\frac{1}{2}p\nu_1}\nu_2^{\frac{1}{2}p\nu_2}\Gamma\frac{1}{2}(\nu_1 + \nu_2 + 1)|\mathbf{S}_1|^{\frac{1}{2}(\nu_1-p-1)}|\mathbf{S}_2^{-1}|^{\frac{1}{2}(\nu_1-p-1)}|\mathbf{S}_1|^{\frac{1}{2}(p+1)}}{\pi^{\frac{1}{4}p(p-1)}\beta_p(\frac{1}{2}\nu_1, \frac{1}{2}\nu_2)\Gamma\frac{1}{2}(\nu_1 + \nu_2 - p + 1)[\pi(\nu_1 + \nu_2)]^{\frac{1}{2}p}|\nu_2\mathbf{I} + \nu_1\mathbf{S}_1\mathbf{S}_2^{-1}|^{\frac{1}{2}(\nu_1+\nu_2)}}$$

$$\left|\frac{\nu_2\mathbf{S}_2 + \nu_1\mathbf{S}_1}{\nu_1 + \nu_2}\left(\frac{1}{\nu_1 + 1} + \frac{1}{\nu_2 + 1}\right)\right|^{-\frac{1}{2}}\left|\mathbf{I} + \left[\frac{\nu_2\mathbf{S}_2 + \nu_1\mathbf{S}_1}{\nu_1 + \nu_2}\left(\frac{1}{\nu_1 + 1}\right.\right.\right.$$

$$\left.\left.\left. + \frac{1}{\nu_2 + 1}\right)\right]^{-1}\frac{(\bar{\mathbf{x}}_2 - \bar{\mathbf{x}}_1)(\bar{\mathbf{x}}_2 - \bar{\mathbf{x}}_1)'}{\nu_1 + \nu_2}\right|^{-\frac{1}{2}(\nu_1+\nu_2+1)} d\bar{\mathbf{x}}_2 d\mathbf{S}_2^{-1},$$

which may be simplified to

$$(8) \qquad \frac{\text{const.}|\mathbf{S}_1|^{\frac{1}{2}\nu_1}|\mathbf{S}_2^{-1}|^{\frac{1}{2}(\nu_1-p-1)}|\mathbf{S}_2^{-1}|^{\frac{1}{2}}d\bar{\mathbf{x}}d_2\mathbf{S}_2^{-1}}{\nu_2^{\frac{1}{2}p(\nu_1+\nu_2+1)}\left|\mathbf{I}+\dfrac{\mathbf{S}_2\,\nu_1^{-1}\mathbf{S}_1}{\nu_2}+\dfrac{1}{\nu_2}\dfrac{\mathbf{S}_2^{-1}(\bar{\mathbf{x}}_2-\bar{\mathbf{x}}_1)(\bar{\mathbf{x}}_2-\bar{\mathbf{x}}_1)'}{\dfrac{1}{\nu_1+1}+\dfrac{1}{\nu_2+1}}\right|^{\frac{1}{2}(\nu_1+\nu_2+1)}}.$$

Now let $\nu_2 \to \infty$; since $\mathbf{S}_2$ and $\bar{\mathbf{x}}_2$ are maximal likelihood estimates

$$\mathbf{S}_2^{-1} \to \mathbf{\Sigma}^{-1}$$

$$\bar{\mathbf{x}}_2 \to \xi.$$

Also

$$\lim_{m\to\infty}|\mathbf{I}+\mathbf{A}/m|^m$$

$$= \lim_{m\to\infty}\prod_{i=1}^{p}(1+\lambda i/m)^m$$

$$= e^{\lambda+\lambda_2+\cdots+\lambda p_1}$$

$$= \exp\,(\text{trace}\,\mathbf{A}),$$

where the λ_i are the latent roots of $\mathbf{A}$.

The expression (8) thus becomes

$$\lim_{\nu_2\to\infty}\frac{K}{\nu_2^{\frac{1}{2}p(\nu_1+\nu_2+1)}}|\mathbf{\Sigma}|^{-\frac{1}{2}}\exp\{\text{trace}-\tfrac{1}{2}(\nu_1+1)\mathbf{\Sigma}^{-1}(\xi-\bar{\mathbf{x}}_1)(\xi-\bar{\mathbf{x}}_1)'\}d\xi$$

$$|\mathbf{\Sigma}^{-1}|^{\frac{1}{2}(\nu_1-p-1)}|\mathbf{S}_1|^{\frac{1}{2}\nu_1}\exp\{\text{trace}-\tfrac{1}{2}\mathbf{\Sigma}^{-1}\nu_1\mathbf{S}_1\}d(\mathbf{\Sigma}^{-1}),$$

where

$$\frac{K}{\nu_2^{\frac{1}{2}p(\nu_1+\nu_2+1)}} = \frac{\nu_1^{\frac{1}{2}p\nu_1}\nu_2^{-\frac{1}{2}p(\nu_1+1)}\Gamma\frac{1}{2}(\nu_1+\nu_2+1)}{\pi^{\frac{1}{2}p}\pi^{\frac{1}{4}p(p-1)}\beta_p(\frac{1}{2}\nu_1,\frac{1}{2}\nu_2)\Gamma\frac{1}{2}(\nu_1+\nu_2-p+1)\left[\dfrac{1}{\nu_1+1}+\dfrac{1}{\nu_2+1}\right]^{\frac{1}{2}p}},$$

from which the factor

$$\frac{\nu_1^{\frac{1}{2}p\nu_1}(\nu_1+1)^{\frac{1}{2}p}}{\pi^{\frac{1}{2}p}\pi^{\frac{1}{4}p(p-1)}\Gamma_p(\frac{1}{2}\nu_1)}$$

may be separated.

Next, considering the factor $\nu_2^{-\frac{1}{2}\nu_1}$ with each of the p terms in

$$\Gamma_p\tfrac{1}{2}(\nu_1+\nu_2)/\Gamma_p(\tfrac{1}{2}\nu_2)$$

and using Stirling's formula yields $2^{-\frac{1}{2}p\nu_1}$ in the limit, and then combining $\nu_2^{-\frac{1}{2}p}$ with the other gamma functions gives the factor $2^{-\frac{1}{2}p}$, so that the distribution becomes

$$(9) \qquad \frac{(\nu_1+1)^{\frac{1}{2}p}|\mathbf{\Sigma}^{-1}|^{\frac{1}{2}}}{(2\pi)^{\frac{1}{2}p}}\exp\{\text{trace}-\tfrac{1}{2}(\nu_1+1)\mathbf{\Sigma}^{-1}(\xi-\bar{\mathbf{x}}_1)(\xi-\bar{\mathbf{x}}_1)'\}d\xi$$

$$\frac{\nu_1^{\frac{1}{2}p\nu_1}|\mathbf{S}_1|^{\frac{1}{2}\nu_1}|\mathbf{\Sigma}^{-1}|^{\frac{1}{2}(\nu_1-p-1)}}{2^{\frac{1}{2}p\nu_1}\pi^{p\frac{1}{4}(p-1)}\Gamma_p(\frac{1}{2}\nu_1)}\exp\{\text{trace}-\tfrac{1}{2}\mathbf{\Sigma}^{-1}\nu_1\mathbf{S}_1\}d(\mathbf{\Sigma}^{-1}),$$

which is Segal's general result.

From this distribution it may be shown that (i) the marginal distribution of ξ_i $(i = 1, 2, \ldots, p)$ is equivalent to the t distribution with $\nu_1 - p + 1$ degrees of freedom; (ii) the marginal distribution of σ_2^2 is equivalent to that of χ^2 with $\nu_1 - p + 1$ degrees of freedom; and (iii) $(\xi - \bar{\mathbf{x}}_1)^1 \mathbf{S}_1^{-1}(\xi - \bar{\mathbf{x}}_1)$ is distributed as Hotelling's T^2 with $n_1 = p$ and $n_1 = \nu_1 - p + 1$ degrees of freedom. The close connection existing between Hotelling's distribution and the above procedure is brought out by the transformation (5) leading to the result (6) from which Hotelling's form is immediately obtainable, and the relation of both to Segal's distribution arises from the connection between the pivotal quantities (3) and the limiting forms, $\boldsymbol{\Sigma}^{-1}$ and ξ, of $\mathbf{S}_2^{-1}$ and $\bar{\mathbf{x}}_2$ respectively.

(2) *Fisher's solution.* Fisher (1959) constructed his distribution using the fiducial argument in the following steps: (i) inversion of the marginal distribution of r using the formula $-\partial F(r, \rho)/\partial \rho$; (ii) inversion of the conditional distribution of s_1 and s_2 for given r and ρ; and (iii) inversion of the distribution of $\bar{x}_1$ and $\bar{x}_2$ for given σ_1, σ_2, and ρ, and obtained

$$(10)\qquad \frac{N-2}{\pi}(1-\rho^2)^{\frac{1}{2}(N-3)}(1-r^2)^{\frac{1}{2}(N-2)}\int_0^\infty \frac{\cosh z dz}{(\cosh z - \rho r)^{N-1}} d\rho$$

$$\frac{(N-1)^{N-1} s_1^{N-1} s_2^{N-1}}{(1-\rho^2)^{N-1}\sigma_1^N \sigma_2^N \Gamma(N-1)} \exp\left\{-\frac{N-1}{2(1-\rho^2)}\left[\frac{s_1^2}{\sigma_1^2} - \frac{2\rho r s_1 s_2}{\sigma_1\sigma_2} + \frac{s_2^2}{\sigma_2^2}\right]\right\}$$

$$d\sigma_1 d\sigma_2 \div \int_0^\infty \frac{dz}{(\cosh z - \rho r)^{N-1}} \times \frac{N}{2\pi\sigma_1\sigma_2(1-\rho^2)^{\frac{1}{2}}}$$

$$\exp\left\{-\frac{N}{2(1-\rho^2)}\left[\frac{(\xi_1-\bar{x}_1)^2}{\sigma_1^2} - 2\rho\frac{(\xi_1-\bar{x}_1)(\xi_2-\bar{x}_2)}{\sigma_1\sigma_2} + \frac{(\xi_2-\bar{x}_2)^2}{\sigma_2^2}\right]\right\} d\xi_1 d\xi_2 .$$

As pointed out in Section 2, $\bar{x}_1$ and $\bar{x}_2$ are the estimates of ξ_1 and ξ_2, respectively, even when $\boldsymbol{\Sigma}$ is assumed to be known, so that the conditional argument of the third step above is directly applicable. On the other hand, when ρ is known the estimates of variance are modified and the second step above takes no cognizance of this fact. It is clear from the second set of pivotals in Section 4 *b* (1) that Fisher recognized the necessity for modified estimates where appropriate.

Fisher did not give an explicit form for the marginal distribution of ρ, but this may be obtained in the following manner (Fraser communicated privately by Fisher). The Wishart distribution from the bivariate normal distribution may be factorized into the product of three independent distributions, namely, that of χ_{N-1}, χ_{N-2} and a standard normal variate z; thus

$$t_1 = \frac{s_1}{\sigma_1}\sqrt{(N-1)},\ t_2 = \frac{s_2\sqrt{(N-1)}\sqrt{(1-r^2)}}{\sigma_2\sqrt{(1-\rho^2)}},\ t_3 = \frac{\sqrt{(N-1)}}{\sqrt{(1-\rho^2)}}\left(\frac{rs_2}{\sigma_1} - \frac{\rho s_1}{\sigma_2}\right),$$

are distributed independently and respectively as χ_{N-1}, χ_{N-2} and z.

Moreover,

$$t_3 = \frac{r}{\sqrt{(1-r^2)}} t_2 - \frac{\rho}{\sqrt{(1-\rho^2)}} t_1$$

or

$$z = \frac{r}{\sqrt{(1-r^2)}} \chi_{N-2} - \frac{\rho}{\sqrt{(1-\rho^2)}} \chi_{N-1}. \tag{11}$$

Starting with the joint distribution of z, χ_{N-1} and χ_{N-2}, taking ρ as fixed and changing variables to r, χ_{N-1} and χ_{N-2} and integrating for the latter two variables, yields the marginal distribution of r (2). In a similar manner, taking r as fixed yields the marginal distribution of ρ as given in the first factor of (10) above.

We now show that the formula $-\partial F(r, \rho)/\partial \rho$ (Fisher, 1930) leads to the same density. Now

$$F(r, \rho) = \frac{N-2}{\pi}(1-\rho^2)^{\frac{1}{2}(N-1)} \int_{-1}^{r} (1-t^2)^{\frac{1}{2}(N-4)} \int_0^\infty \frac{dz}{(\cosh z - \rho t)^{N-1}} dt$$

so that

$$-\frac{\partial}{\partial\rho} F(r, \rho) = \frac{(N-1)(N-2)}{2\pi}(1-\rho^2)^{\frac{1}{2}(N-3)} 2\rho \int_{-1}^{r} (1-t^2)^{\frac{1}{2}(N-4)} \int_0^\infty \frac{dz}{(\cosh z - \rho t)^{N-1}} dt - \frac{N-2}{\pi}(1-\rho^2)^{\frac{1}{2}(N-1)} \int_{-1}^{r} (1-t^2)^{\frac{1}{2}(N-4)} \int_0^\infty \frac{(N-1) t\, dz}{(\cosh z - \rho t)^N} dt. \tag{12}$$

Taking the marginal distribution of ρ in the first factor of (10) and differentiating with respect to r gives

$$-\frac{(N-2)^2}{\pi}(1-\rho^2)^{\frac{1}{2}(N-3)}(1-r^2)^{\frac{1}{2}(N-4)} r \int_0^\infty \frac{\cosh z\, dz}{(\cosh z - \rho r)^{N-1}} + \frac{(N-1)(N-2)}{\pi}(1-\rho^2)^{\frac{1}{2}(N-3)}(1-r^2)^{\frac{1}{2}(N-2)} \rho \int_0^\infty \frac{\cosh z\, dz}{(\cosh z - \rho r)^N}, \tag{13}$$

and if the formula $-\partial F(r, \rho)/\partial\rho$ holds, then the derivative of (12), with respect to r, and (13) should be equal.

Equating these two expressions it follows that

$$(N-1)\rho \int_0^\infty \frac{dz}{(\cosh z - \rho r)^{N-1}} - (N-1)(1-\rho^2) r \int_0^\infty \frac{dz}{(\cosh z - \rho r)^N} = (N-1)(1-\rho^2)\rho \int_0^\infty \frac{\cosh z\, dz}{(\cosh z - \rho r)^N} - (N-2) r \int_0^\infty \frac{\cosh z\, dz}{(\cosh z - \rho r)^{N-1}}; \tag{14}$$

writing

$$I_\mu(\rho r) = \int_0^\infty \frac{dz}{(\cosh z - \rho r)^\mu},$$

we have

$$\int_0^\infty \frac{\cosh z\, dz}{(\cosh z - \rho r)^\mu} = I_{\mu-1}(\rho r) + \rho r I_\mu(\rho r),$$

so that (14) may be written

$$r[(N-2)I_{N-2}+(2N-3)\rho r I_{N-1}-(N-1)(1-\rho^2 r^2)I_N]=0, \tag{15}$$

and hence either $x = 0$ or the expression within the square bracket in (15) is equal to 0. If $I_\mu(\rho r)$ is evaluated by expanding the integrand in a power series of $1/\cosh z$ and integrating term by term, then

$$I_\mu(\rho r)=\sum_{\nu=0}^{\infty}\frac{\Gamma\frac{1}{2}(\mu+\nu)\Gamma\frac{1}{2}(\mu+\nu)}{\Gamma(\mu)\Gamma(\nu+1)}2^{\mu+\nu-2}(\rho r)^\nu \qquad |\rho r|<1$$

and it is readily verified that the recurrence relation holds. The pivotal quantity (11) and the expression $-\partial F(r,\rho)/\partial\rho$ thus lead to the same distribution for ρ.

The integrations necessary to obtain marginal distributions of individual means and variances from (10) are exceedingly difficult in the general case when $r \neq 0$, but for the particular value $r = 0$, they are readily performed and give the following results:

(*a*) the marginal distribution of σ_i has the frequency density

$$\frac{(N-2)(N-1)^{\frac{1}{2}(N-1)}\Gamma\frac{1}{2}(N-2)}{\pi^{\frac{1}{2}}\Gamma\frac{1}{2}(N-1)\Gamma\frac{1}{2}(N-1)2^{\frac{1}{2}(N-3)}}\frac{s_i^{N-1}}{\sigma_i^N}\int_0^1\frac{\exp\{-(N-1)s_i^2/2\sigma_i^2(1-\rho^2)\}}{(1-\rho^2)}d\rho$$

which does not simplify to the χ^2 density obtainable by inverting the distribution of s_2^2, for equating the two implies that

$$\int_0^1\frac{\exp\{-(N-1)s_i^2\rho_i^2/2\sigma_i^2(1-\rho^2)\}}{(1-\rho^2)}d\rho$$

is a function of N alone and this is evidently not so;

(*b*) writing

$$t_i=\frac{\sqrt{N}(\xi_i-\bar{x}_i)}{\sqrt{[S(x_i-\bar{x}_i)^2]}}, \qquad i=1,2$$

their distribution is

$$\frac{(\Gamma\frac{1}{2}N)^2}{\pi\{\Gamma\frac{1}{2}(N-1)\}^3(1+t_1^2)^{\frac{1}{2}N}(1+t_2^2)^{\frac{1}{2}N}} \tag{16}$$

$$\sum_{\mu=0}^{\infty}\frac{\Gamma(\mu+\frac{1}{2}N)\Gamma(\mu+\frac{1}{2}N)}{\Gamma[\mu+\frac{1}{2}(N+1)]\Gamma(\mu+1)}\left(\frac{t_1^2}{1+t_1^2}\cdot\frac{t_2^2}{1+t_2^2}\right)^\mu dt_1dt_2,$$

which for odd values of N may be expressed in terms of the complete elliptic integral

$$K(k)=\int_0^{\pi/2}(1-k^2\sin^2\theta)^{-\frac{1}{2}}d\theta. \qquad |k|<1.$$

Thus, although condition (C) is satisfied for the distribution of ρ, the other conditions are not satisfied for the means and variances. Moreover, the distribution for ρ does not agree with that derived from Segal's general distribution, but it should be noted that this is the only point of difference between the two solutions.

As far as the fiducial argument itself is concerned, the distribution of r in the bivariate case is the only one encountered, for which no simple pivotal quantity analogous to t or χ^2 is known to exist. On general grounds there is no reason to suppose that the formula $-\partial F(T, \theta)/\partial\theta$ will always be applicable, since this implies that $F(T, \theta)$ must be a cumulative distribution function with respect to θ, i.e., to say for all fixed T $(-\infty < T < \infty)$,

$F(T, \theta)$ must be monotonic in θ,

$$F(T, +\infty) = \left.\begin{matrix}1\\0\end{matrix}\right\} \quad F(T, -\infty) = \left.\begin{matrix}1\\0\end{matrix}\right\}$$

depending on whether F is an increasing or decreasing function of θ.

$F(r, \rho)$ is a monotonically decreasing function of ρ and it may be verified, for small numbers of degrees of freedom, that $F(r, -1) = 1$ while it is generally clear that $F(r, 1) = 0$, so that presumably the conditions will hold for arbitrary degrees of freedom. The conditions are necessary but may not be sufficient.

No definite evidence has, as yet, been adduced to establish the marginal distribution of r as the appropriate starting point for constructing the distribution of the parameters step-by-step. The mere facts that r is one of a set of sufficient statistics and that its distribution depends only on ρ may well not be the only conditions. In the univariate case s is one of two jointly sufficient statistics, having a marginal distribution dependent only on σ, but in addition this marginal distribution is identical with the factor involving s in the joint distribution of $\bar{x}$ and s.

(3) *Solution obtained from Fisher's marginal distribution of the means.* Fisher (1954) proposed that the marginal fiducial distribution of the means took the form

$$(17) \qquad \frac{\Gamma\frac{1}{2}(N+1)}{\pi(N-1)\Gamma\frac{1}{2}(N-1)}\left\{1 + \frac{(\boldsymbol{\xi} - \bar{\mathbf{x}})'\mathbf{S}^{-1}(\boldsymbol{\xi} - \bar{\mathbf{x}})}{N-1}\right\}^{-\frac{1}{2}(N+1)} \frac{d\xi_1 d\xi_2}{|\mathbf{S}|^{\frac{1}{2}}},$$

in which $\mathbf{S}$ stands for the 2×2 variance-covariance matrix of the observed means, and this has been generalized to p variates by Cornish (1961),

$$(18) \qquad \frac{\Gamma\frac{1}{2}(N+p-1)}{\{\pi(N-1)\}^{\frac{1}{2}p}\Gamma\frac{1}{2}(N-1)}\left\{1 + \frac{(\boldsymbol{\xi} - \bar{\mathbf{x}})'\mathbf{S}^{-1}(\boldsymbol{\xi} - \bar{\mathbf{x}})}{N-1}\right\}^{-\frac{1}{2}(N+p-1)} \frac{d\xi_1 d\xi_2 \ldots d\xi_p}{|\mathbf{S}|^{\frac{1}{2}}},$$

both results depending upon the fact that

$$\mathbf{t} = \frac{\boldsymbol{\lambda}'(\bar{\mathbf{x}} - \boldsymbol{\xi})}{(\boldsymbol{\lambda}'\mathbf{S}\boldsymbol{\lambda})^{\frac{1}{2}}},$$

for arbitrary λ, is distributed as Student's t statistic with $(N-1)$ degrees of freedom.

Fisher's result (17), with $r = 0$, and the corresponding distribution from Segal's expression (14), may be directly compared with (16) by expanding them in power series. Thus, after setting $x = 0$ in Segal's result and expressing it in terms of the t_i, the distribution appears as

$$\frac{\Gamma\frac{1}{2}N}{\pi\Gamma\frac{1}{2}(N-2)}\{1+t_1^2+t_2^2\}^{-\frac{1}{2}N}dt_1dt_2.$$

Since

$$1+t_1^2+t_1^2 = (1+t_1^2)(1+t_2^2)\left(1-\frac{t_1^2}{1+t_1^2}\cdot\frac{t_2^2}{1+t_2^2}\right),$$

expansion of the density in power series gives

$$(19)\qquad \frac{\Gamma\frac{1}{2}N}{\pi\Gamma\frac{1}{2}(N-2)}\cdot\frac{1}{(1+t_1^2)^{\frac{1}{2}N}}\cdot\frac{1}{(1+t_2^2)^{\frac{1}{2}N}}$$
$$\sum_{\mu=0}^{\infty}\frac{\Gamma(\mu+\frac{1}{2}N)}{\Gamma\frac{1}{2}N\Gamma(\mu+1)}\left(\frac{t_1^2}{1+t_1^2}\cdot\frac{t_2^2}{1+t_2^2}\right)^{\mu}dt_1dt_2.$$

In a similar manner (17) yields

$$(20)\qquad \frac{\Gamma\frac{1}{2}(N+1)}{\pi\Gamma\frac{1}{2}(N-1)}\cdot\frac{1}{(1+t_1^2)^{\frac{1}{2}(N+1)}}\cdot\frac{1}{(1+t_2^2)^{\frac{1}{2}(N+1)}}$$
$$\times\sum_{\mu=0}^{\infty}\frac{\Gamma[\mu+\frac{1}{2}(N+1)]}{\Gamma\frac{1}{2}(N+1)\Gamma(\mu+1)}\left(\frac{t_1^2}{1+t_1^2}\cdot\frac{t_2^2}{1+t_2^2}\right)^{\mu}dt_1dt_2.$$

Hence the three forms are not very different and, in fact, the power series terms tend to the same value very rapidly. At $t_1 = 0$, $t_2 = 0$ the densities, in increasing order of magnitude, are (19), (16), and (20).

Assuming conditions (A) and (B) hold and using the fact the fiducial distribution of the means, for known $\mathbf{\Sigma}$, is multivariate normal with a mean $\bar{x}$ and variance-covariance matrix $\mathbf{\Sigma}/N$, it may be shown that $\mathbf{\Sigma}^{-1}$ has a distribution of Wishart form, so that the joint distribution of all parameters is

$$(21)\qquad \frac{N^{\frac{1}{2}p}|\mathbf{\Sigma}^{-1}|^{\frac{1}{2}}}{(2\pi)^{\frac{1}{2}p}}\exp\{\text{trace } -\tfrac{1}{2}N\mathbf{\Sigma}^{-1}(\xi-\bar{\mathbf{x}})(\xi-\bar{\mathbf{x}})'\}d\xi$$
$$\frac{(N-1)^{\frac{1}{2}p(N+p-2)}|\mathbf{S}|^{\frac{1}{2}(N+p-2)}}{2^{\frac{1}{2}p(N+p-2)}\pi^{\frac{1}{4}p(p-1)}\Gamma_p(N+p-2)}$$
$$\exp\{\text{trace } -\tfrac{1}{2}(N-1)\mathbf{S}\mathbf{\Sigma}^{-1}\}|\mathbf{\Sigma}^{-1}|^{\frac{1}{2}(N-3)}d(\mathbf{\Sigma}^{-1}),$$

in which $\mathbf{S}$ now stands for the sample variance-covariance matrix, and it is to be noted that the "degrees of freedom" have been increased by $p-1$.

With this distribution all marginal distribution of means have the forms obtained by inversion of their corresponding marginal sampling distributions, and any leading submatrix ($k \times k$) with $k < p$ of Σ^{-1} has a marginal distribution of Wishart form with "degrees of freedom" increased by $k-1$.

For two variates it may be written

$$\frac{N}{2\pi\sigma_1\sigma_2(1-\rho^2)^{\frac{1}{2}}}\exp\left[-\frac{N}{2(1-\rho^2)}\left\{\frac{(\xi_1-\bar{x}_1)^2}{\sigma_1^2}-\frac{2\rho(\xi_1-\bar{x}_1)(\xi_2-\bar{x}_2)}{\sigma_1\sigma_2}\right.\right.$$

$$+ \frac{(\xi_2 - \bar{x}_2)^2}{\sigma_2^2}\Big\}\Big] d\xi_1 d\xi_2 \frac{(N-1)^N s_1^N s_2^N (1-r^2)^{\frac{1}{2}N}}{\pi\Gamma(N-1)\sigma_1^{N+1}\sigma_2^{N+1}(1-\rho^2)^{\frac{1}{2}(N+3)}}$$

$$\exp\left[-\frac{N-1}{2(1-\rho^2)}\left\{\frac{s_1^2}{\sigma_1^2} - \frac{2\rho r s_1 s_2}{\sigma_1\sigma_2} + \frac{s_2^2}{\sigma_2^2}\right\}\right] d\sigma_1 d\sigma_2 d\rho,$$

from which the marginal distribution of ρ is

$$\frac{N-1}{\pi}(1-r^2)^{\frac{1}{2}N}(1-\rho^2)^{\frac{1}{2}(N-3)} \int_0^\infty \frac{dz}{(\cosh z - \rho r)^N} d\rho,$$

thus differing from both Segal's and Fisher's forms.

One basic reason for the difference between Segal's general solution (9) and the above expression (21) is that essentially different types of t statistic have been used in the two cases.

For the bivariate case there are thus at least three expressions available for consideration as the joint distribution of the parameters, and all three conflict with one or more of the conditions listed. Further discussion of these initial conditions is dependent upon an examination of the problems arising in connection with pivotal quantities.

6. PIVOTAL QUANTITIES

Fisher suggested that an effective set of pivotal functions must comply with the following conditions:

(*A*) They must have a simultaneous distribution independent of all parameters (ignoring parameters of lower strata);

(*B*) Each must involve only one parameter of the stratum;

(*C*) Each shall vary monotonically with that parameter uniformly for variations fo the statistics;

(*D*) Jointly they involve a set of statistics exhaustive for the parameters;

and, as an alternative set:

(*A*) Each shall be monotonic in one "parameter" uniformly for possible values of the statistic;

(*B*) Their simultaneous distribution must be independent of all parameters;

(*C*) The statistics shall constitute an exhaustive set;

the term "parameter" of this second set being interpreted as a function of the original parameters which was to be regarded as a fiducial variate.

In the first set the notion of a stratum is introduced but it has never been defined. Fisher considered ξ and σ as occupying different strata (σ in the lower stratum) in the univariate case, and similarly (ξ_1, ξ_2) and (σ_1, σ_2, ρ) comprise two strata in the bivariate case, but the stratum (σ_1, σ_2, ρ) was partitioned further into (σ_1, σ_2) and (ρ), the latter being the lowest of the three strata.

The likelihood equations certainly support the grouping (ξ_1, ξ_2) and (σ_1, σ_2, ρ) but give no further information as regards partitioning the latter group. As remarked above, the marginal distribution of r depends only on ρ, but this alone is not sufficient justification for taking ρ as occupying the lowest stratum since the distribution of either estimate of variance depends only on the corresponding parameter.

The vital rôle of sufficient statistics has long been recognized, and in view of the properties of the maximal likelihood estimates it may be helpful to consider a system of stratification based on the solution of the likelihood equations, or factorization of the likelihood function itself. This would probably not cover all distributions but for some at least of those in common use such as classification has been useful.

The following modifications of the first and second conditions may be made:

(A) Joint fiducial distributions may be built up stratum by stratum from marginal distributions which can be obtained by inversion of the sampling distribution appropriate to the stratum as a whole;

(B) If this can be performed in various ways, each will lead to the same result.

In the cases so far examined these two conditions hold, thus, for example, Segal's original method involves the inversion first of the Wishart distribution of $\mathbf{S}$ for given $\mathbf{\Sigma}$, followed by the inversion of the distribution of $\bar{\mathbf{x}}$ for given $\mathbf{\Sigma}$.

The third condition is more troublesome, but provides some light on a limitation to the fiducial argument as it stands at the present time. An essential feature of a pivotal quantity is that it shall be monotonic. If $u = u(t, \theta)$ is the pivotal quantity, this requirement ensures that $u \leqslant u_1$ maps into the region $\theta \leqslant \theta_1$ or $\theta \geqslant \theta_1$ so that $+\partial F(t, \theta)/\partial\theta$ can be used; but if the pivotal quantity is not monotonic several regions may correspond to $u \leqslant u_1$. For a single variable with just one parameter all pivotals are essentially the same because they are all functions of the marginal cumulative distribution function which has, for all parameter values, a uniform distribution over [0, 1]. Thus, if this is not a monotonic function of θ with $F(t, \infty) = 0$ (or 1) and $F(t, -\infty) = 1$ (or 0) the fiducial argument in its present form cannot be applied. The conditions which must be imposed on $F(t, \theta)$ to ensure at least these properties then become conditions on the fiducial argument itself; the case of several variates and parameters is much more complex.

7. THE BIVARIATE NORMAL DISTRIBUTION WITH EQUAL VARIANCES

Cornish (1960) has considered this distribution and obtained an inversion by using Fisher's approach. The sampling distribution of r depends only on ρ and the derivatives $-\partial F(r, \rho)/\partial\rho$, and $-\partial F(s, \sigma|r, \rho)/\partial\sigma$ are easily evaluated. Multiplication of these two functions gives a density for σ and ρ to which is attached the normal conditional distribution of the means to obtain the final result. This problem may be approached in several different ways: (1) the method used by Cornish; (2) ordinary pivotal quantities; (3) matrix pivotal quantities; and (4) the limiting process. Method (2) is instructive because it brings out the care necessary in dealing with pivotal quantities. Using Cornish's notation,

$$s^2 = \{S(x_1 - \bar{x}_1)^2 + S(x_2 - \bar{x}_2)^2\}/2(N - 1)$$
$$r = 2S(x_1 - \bar{x}_1)(x_2 - \bar{x}_2)/\{(x_1 - \bar{x}_1)^2 + S(x_2 - \bar{x}_2)^2\},$$

it is not difficult to show that

$$(22) \qquad u = \frac{s\sqrt{(1+r)}}{\sigma\sqrt{(1+\rho)}}, \qquad v = \frac{s\sqrt{(1-r)}}{\sigma\sqrt{(1-\rho)}},$$

have essentially independent χ^2 distributions with $(N-1)$ degrees of freedom, thus

$$dF(u,v) = \frac{(N-1)^{N-1}(uv)^{N-2}}{2^{N-3}\{\Gamma\frac{1}{2}(N-1)\}^2}\exp\left\{-\frac{N-1}{2}(u^2+v^2)\right\}du dv.$$

Consider (Figure 1) the region in the (σ, ρ) plane defined by $0 \leqslant \sigma \leqslant \Sigma$, $-1 \leqslant \rho \leqslant P$ ($P > 0$ for definiteness).

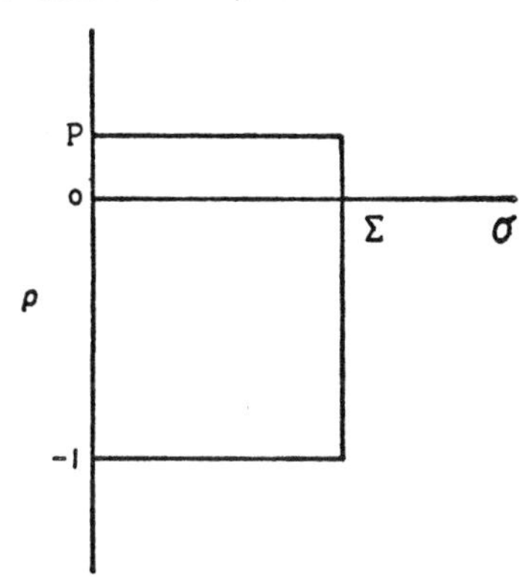

FIGURE 1.

Put

$$u = \frac{s\sqrt{(1+r)}}{\sigma\sqrt{(1+\rho)}}, \qquad v = \frac{s\sqrt{(1-r)}}{\sigma\sqrt{(1-\rho)}},$$

then

$$\frac{v}{u} = \sqrt{\frac{1-r}{1+r}}\sqrt{\frac{1+\rho}{1-\rho}},$$

which varies from 0 to

$$\sqrt{\frac{1-r}{1+r}}\sqrt{\frac{1+\rho}{1-\rho}},$$

and for given $\rho = \rho_1$,

$$\frac{s\sqrt{(1+r)}}{\Sigma\sqrt{(1+\rho_1)}} \leqslant u \leqslant \infty, \qquad \frac{s\sqrt{(1-r)}}{\Sigma\sqrt{(1-\rho_1)}} \leqslant v \leqslant \infty,$$

thus

$$\frac{s^2(1+r)}{u^2} + \frac{s^2(1-r)}{v^2} = 2\Sigma^2,$$

so that the corresponding region (Figure 2) in the (u, v) plane is R: Hence the double derivative

$$\frac{\partial^2}{\partial\sigma\partial\rho}F(u, v \mid \sigma, \rho)$$

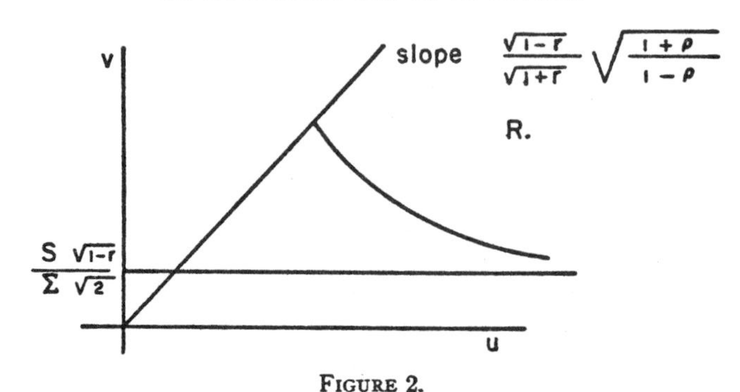

FIGURE 2.

would not give the correct frequency element since $F(u, v\,|\,\sigma, \rho)$ corresponds to an infinite rectangular region, whereas it is necessary to consider the above slice-shaped region. It is possible to consider

$$\Pr(u, v \in R),$$

and apply

$$\frac{\partial^2}{\partial\sigma\partial\rho}$$

after first changing to coordinates $w = v/u$ and

$$x = \frac{s^2(1+r)}{u^2} + \frac{s^2(1-r)}{v^2}.$$

The final result is

$$(23)\qquad \frac{(N-1)^{N-1}}{2^{N-3}[\Gamma\frac{1}{2}(N-1)]^2(1-\rho^2)^{\frac{1}{2}(N+1)}}\left(\frac{s}{\sigma}\right)^{2(N-1)}(1-r^2)^{\frac{1}{2}(N-1)}$$
$$\exp\left\{-\frac{(N-1)s^2(1-\rho r)}{\sigma^2(1-\rho^2)}\right\}\frac{d\sigma}{\sigma}\,d\rho,$$

as obtained by Cornish using Fisher's method.

Formally, the method of matrix pivotals is very simple; write

$$\mathbf{S} = \begin{bmatrix}(N-1)s^2 & (N-1)s^2r\\ (N-1)s^2r & (N-1)s^2\end{bmatrix}, \qquad \mathbf{\Sigma} = \begin{bmatrix}\sigma^2 & \sigma^2\rho\\ \sigma^2\rho & \sigma^2\end{bmatrix},$$

then $d\mathbf{S} = 2(N-1)^2s^3dsdr$ and the sampling distribution may be written as

$$\frac{|\mathbf{S}|^{\frac{1}{2}(N-3)}\exp\{\text{trace } -\frac{1}{2}\mathbf{\Sigma}^{-1}\mathbf{S}\}}{2^{N-2}[\Gamma\frac{1}{2}(N-1)]^2\,|\mathbf{\Sigma}|^{\frac{1}{2}(N-1)}}\,d\mathbf{S}.$$

In this matrix space, the jacobian of the transformation from $\mathbf{S}$ to $\mathbf{A}$ defined by

$$\mathbf{S} = \mathbf{\Sigma A}$$

is just $|\mathbf{\Sigma}|$, so that $\mathbf{A} = \mathbf{\Sigma}^{-1}\mathbf{S}$ has frequency element

$$\frac{|\mathbf{A}|^{\frac{1}{2}(N-3)}\exp\{\text{trace } -\frac{1}{2}\mathbf{A}\}}{2^{N-2}[\Gamma\frac{1}{2}(N-1)]^2}\,d\mathbf{A},$$

which yields a distribution of $\mathbf{\Sigma}^{-1}$

$$\frac{|\mathbf{S}|^{\frac{1}{2}(N-1)}\exp\{\text{trace } -\frac{1}{2}\mathbf{\Sigma}^{-1}\mathbf{S}\}}{2^{N-2}[\Gamma\frac{1}{2}(N-1)]^2\,|\mathbf{\Sigma}|^{\frac{1}{2}(N-3)}}\,d(\mathbf{\Sigma}^{-1}).$$

This is identical with the above result (23) since

$$d(\mathbf{\Sigma}^{-1}) = 2/[\sigma^5(1-\rho^2)^2]d\sigma d\rho.$$

However, these two methods are essentially the same for let $\mathbf{H}$ be the matrix

$$\begin{bmatrix} 1 & 1 \\ 1 & -1 \end{bmatrix},$$

then

$$\mathbf{H}^1\mathbf{S}\mathbf{H} = \begin{bmatrix} (N-1)s^2(1+r) & 0 \\ 0 & (N-1)s^2(1-r) \end{bmatrix},$$

so that the pivotals u and v (22) are essentially the elements of $\mathbf{\Sigma}^{-1}\mathbf{S}$ which in this case happens to be symmetric. The limiting process will undoubtedly yield this same result.

Since all these methods yield the same answer for this special case and fail to do so for the general bivariate Wishart distribution, it is interesting to compare the two with respect to various remarks made earlier about this latter expression. In the first place, although Cornish used $-\partial F(r, \rho)/\partial\rho$ to invert the distribution of r, it is, in fact, possible to find a simple pivotal quantity for r, viz.:

$$\alpha = \frac{\sqrt{(N-1)}}{\sqrt{(1-r^2)}} \cdot \frac{r-\rho}{\sqrt{(1-\rho^2)}}$$

which is distributed as Student's t with $(N-1)$ degrees of freedom. From this fact it is evident that a simple set of probability statements leading from r to ρ via α is available and these are analogous to the classical mean and variance fiducial arguments. Similarly the inversion of $F(s, \sigma | r, \rho)$ can be accomplished through

$$\beta = \frac{2(N-1)s^2(1-\rho r)}{\sigma^2(1-\rho^2)}$$

which has a χ^2 distribution with $2(N-1)$ degrees of freedom independent of α. The fact that these two quantities are increasing with respect to the appropriate statistics and decreasing when considered as functions of the corresponding parameters means that the formula $-\partial F(T, \theta)/\partial\theta$ can be properly applied.

Furthermore, the problem mentioned in connection with the combination of marginal and conditional distributions and the formulation of the latter does not arise because β, apart from the factor $2(N-1)$, is in fact the appropriate pivotal for σ when ρ is considered to be known.

SOMMAIRE

Extension de la théorie de l'induction fiduciaire aux situations plus complexes entraînant la détermination de la distribution fiduciaire simultanée de plusieurs paramètres, en particulier la distribution normale multivariée, a conduit, ces dernières années, à un nombre de différentes solutions à ce problème. La raison principale pour la diversité des solutions et les différences d'opinion les concernant est le fait que les conditions pour lesquelles une solution valide et unique existe ne sont réellement pas tout-à-fait entièrement comprises, d'où des suppositions implicites et explicites ont été faites, fondées, soit sur des propriétés analogues des distributions d'essai, soit sur les résultats obtenus d'un certain nombre de problèmes simples d'induction.

Les conditions suivantes on été examinées par rapport à divers problèmes d'inférence.

A. Des distributions fiduciaires communes peuvent être construites de distributions marginales et conditionnelles, qui peuvent être dérivées par inversion de la distribution d'échantillonage appropriée.

B. Si ceci peut être effectué de plusieurs manières chacune amènera au même résultat.

C. La formule $\pm\partial F(T,\theta)/\partial\theta$, où $F(T,\theta)$ est la fonction de distribution cumulative de T, peut être employée à accomplir les inversions, sujet à restrictions de monotonicité et de différenciabilité relatives à F, considérées comme une fonction de θ, et pourvu que

$$F(T, +\infty) = \begin{matrix}1\\0\end{matrix}\Bigg\}\qquad F(T, -\infty) = \begin{matrix}0\\1\end{matrix}\Bigg\}\qquad -\infty < T < \infty$$

suivant que F soit fonction croissante ou décroissante de θ, et il est montré que les diverses solutions disponibles pour la distribution commune des paramètres sont toutes en désaccord avec une ou plusieurs de ces conditions.

Un ensemble révisé de conditions basé sur la stratification des équations de vraisemblance ou sur la factorisation de la vraisemblance même est discuté brièvement.

La dérivation de la distribution sumiltanée des paramètres dans le cas spécial de la distribution bivariée normale avec variance égale est discutée en quelque détail.

REFERENCES

E. A. Cornish (1960), C.S.I.R.O., Aust. Div. Math. Statist., *Tech. Pap. 6.*

——— (1961), C.S.I.R.O., Aust. Div. Math. Statist., *Tech. Pap. 8.*

R. A. Fisher (1930), *Proc. Cambridge Philos. Soc. 26*, 528–35.

——— (1935), *Ann. Eugen. Lond. 6*, 391–8.

——— (1941), *Ann. Eugen. Lond. 11*, 141–72.

——— (1954), *J. R. Statist. Soc., B, 16*, 212–13.

——— (1950), *Statistical Methods and Scientific Inference* (2nd ed., Edinburgh: Oliver and Boyd).

I. E. Segal (1938), *Proc. Cambridge Philos. Soc., 34*, 41–7.

*Invited discussant.

Commonwealth Scientific and Industrial Research Organization, Australia. Division of Mathematical Statistics Technical Paper No. 17, Melbourne 1964.

AN ANALYSIS OF DAILY TEMPERATURES AT ADELAIDE, SOUTH AUSTRALIA

By E. A. CORNISH* and MARILYN J. EVANS*

[*Manuscript received August* 30, 1963]

Summary

A method of summarizing accurately and comprehensively the extensive mass of observations of a lengthy temperature record in terms of the daily maximum and minimum temperatures and their derived quantities, mean temperature and range, is discussed. The method is applied to the temperature record at Adelaide, South Australia, and the results are surveyed in terms of the variability of these four temperature variates and their interrelations. It is shown that a small fraction of the observed variability is attributable to secular changes of the temperature.

I. INTRODUCTION

The importance of the role of temperature in the effect of climatic factors on agriculture, in ecological studies of plants and animals, and on the geographical distribution of soils, needs no emphasis. References to such investigations abound, and it will suffice to mention the series of papers by Prescott (see Prescott, Collins, and Shirpurkar 1952; Prescott 1956 for a full set of references). Throughout studies of this nature, temperature data have usually been considered only in broad terms as, for example, annual or monthly means, and although this procedure is appropriate in the initial investigations needed to establish general principles, it is inevitable that greater detail will be required in subsequent work. For example, in southern Australia the calendar month is too lengthy to enable us to follow accurately the changes in temperature, especially during the autumn and spring, both of which are critical periods of the growing season in a zone of winter rainfall. In addition to establishing the seasonal changes of temperature in finer detail, we are concerned to illustrate a means for summarizing accurately and comprehensively, the extensive mass of temperature observations which have been accumulated in a lengthy record, so that the salient features of the particular region may be readily identified. Such information is always of value in its local applications, but its value may be enhanced considerably if similar information from other stations is available for comparison. At the same time the opportunity has been taken to examine the observations for secular changes in the temperature.

II. DATA

Temperature records at Adelaide have been compiled from thermometers exposed in both a Greenwich stand and a Stevenson screen. The Greenwich stand record was begun as early as 1856, but there were several changes of site involving considerable distances before the instruments were established at the Observatory

* Division of Mathematical Statistics, CSIRO, University of Adelaide.

in May 1860; this record was maintained until February 27, 1944, when the site was changed again to the present position of the Bureau of Meteorology, the record being continued here until July 31, 1947, when it ceased.

The Stevenson screen record began in January 1887 at the Observatory site and was maintained until February 27, 1944, but during this interval there were changes of screen in January 1898 and November 1925, the last change being to the standard screen. A new record with the standard screen was begun at the Bureau of Meteorology on July 10, 1943; this has been continued to the present day and now constitutes the official record.

Two circumstances are responsible for strongly marked discontinuities in the observations:

(i) There were differences in the respective immediate environs of the sites at the Observatory and the Meteorological Bureau although they were separated by only a short distance,

(ii) The methods of exposing the instruments in the Greenwich stand and the Stevenson screen are different, particularly with regard to reflected radiation from the ground.

The longest homogeneous and continuous record obtainable from the observations comprises those taken during the interval January 1, 1861, to December 31, 1943, with the Greenwich stand. The primary data were the daily maximum and minimum temperatures, and a 6-day unit was chosen for subdividing the year, a unit conforming with that used for the analysis of the Adelaide rainfall, and yet sufficiently fine to provide an accurate description of the annual temperature wave.

III. Method of Analysis

For each 6-day period the analysis of variance and covariance of the daily maximum temperature (M) and the daily minimum temperature (m) was constructed. Since the record extended over the interval 1861–1943 inclusive, this analysis assumed the following tabular form, in which the figures given are those for the 6-day period March 2–7 inclusive (period 11).

	Degrees of Freedom	Sums of Squares and Products			Mean Squares and Products		
		M^2	Mm	m^2	M^2	Mm	m^2
Between periods	82	24461	14640	12957	298·31	178·54	158·01
Within periods	415	28232	11871	15487	68·03	28·61	37·32
Total	497	52693	26511	28444			

The mean squares between periods and within periods (for both M and m) and the corresponding mean products, together with the correlations between periods and within periods, have been calculated; for example, for the eleventh period:

$$\text{Correlation of maximum and minimum (between periods)} = \frac{14640}{(24461 \times 12957)^{\frac{1}{2}}} = 0\cdot 8224,$$

$$\text{Correlation of maximum and minimum (within periods)} = \frac{11871}{(28232 \times 15487)^{\frac{1}{2}}} = 0\cdot 5677.$$

In all subsequent calculations relating to such correlations, their inverse hyperbolic transformations $z = \tanh^{-1} r$ have been used (Fisher 1958).

The original data were also examined from the viewpoint provided by the mean temperature (t) and the range (R) between the maximum and the minimum. Analyses of variance and covariance, similar to that given above, were constructed for each of the 61 periods of the year, but there was no necessity to make these calculations *ab initio*, as they were readily obtainable from the quantities evaluated for the maximum and minimum temperatures using the relations:

$$t = \tfrac{1}{2}(M+m),$$
$$R = M-m,$$

from which

$$\Sigma\,(R-\bar{R})^2 = \Sigma\,(M-\bar{M})^2 - 2\,\Sigma\,(M-\bar{M})(m-\bar{m}) + \Sigma\,(m-\bar{m})^2,$$
$$\Sigma\,(t-\bar{t})^2 = \tfrac{1}{4}\{\Sigma\,(M-\bar{M})^2 + 2\,\Sigma\,(M-\bar{M})(m-\bar{m}) + \Sigma\,(m-\bar{m})^2\},$$
$$\Sigma\,(R-\bar{R})(t-\bar{t}) = \tfrac{1}{2}\{\Sigma\,(M-\bar{M})^2 - \Sigma\,(m-\bar{m})^2\},$$

and so on.

IV. Survey of Results

(a) *The Significance of the Between-periods Mean Square and the Annual Sequence of the Intra-class Correlations*

Table 1 summarizes the mean squares between periods, within periods, and the corresponding mean products, maximum $\times$ minimum and mean $\times$ range, for each of the 61 subdivisions of the year. The principal features of this table are:

(i) The mean square between periods from year to year is always strongly significant in comparison with the mean square within periods for maximum, minimum, and their mean, but with the range it is significant up to period 39 and is non-significant, or significant only at $P = 0\cdot 05$, in all except five periods, from period 40 until the end of the year. Thus, there are real differences from year to year in the maximum and minimum temperatures of any subdivision and hence also in the mean temperature, but with the range this is true only of the subdivisions in the first half of the year.

(ii) The maximum temperature is nearly always more variable between periods from year to year than the minimum temperature, whereas for the range the overall significance of the between-period mean square is much less than in the other temperature variates. During the latter half of the year the range thus conforms with expectation, so that there is some interest in the fact that in the first half of the year the range shows significant variation between periods from year to year. We shall return to this point at a later stage (see below, Section IV(c)).

TABLE 1

MEAN SQUARES AND MEAN PRODUCTS OF THE TEMPERATURE VARIATES

Period*	Max., M		Min., m		Mm		Range, R		Mean, t		Rt	
	Between Periods	Within Periods	Between Periods	Within Periods	Between Periods	Within Periods	Between Periods	Within Periods	Between Periods	Within Periods	Between Periods	Within Periods
1	288·26	97·99	120·63	50·53	149·37	44·85	110·14	58·81	176·91	59·56	83·81	23·73
2	345·72	85·99	156·25	48·04	196·40	38·69	109·16	56·64	223·69	52·85	94·74	18·97
3	257·53	91·20	141·93	46·58	151·06	42·17	97·34	53·43	175·40	55·53	57·80	22·31
4	251·97	86·32	143·23	49·50	154·84	42·01	85·53	51·80	176·22	54·96	54·37	18·41
5	284·00	73·99	130·72	34·95	151·03	26·20	112·67	56·55	179·19	40·33	76·64	19·52
6	291·23	79·37	178·03	44·03	192·50	38·31	84·26	46·79	213·56	50·00	56·60	17·67
7	320·48	73·51	155·22	43·32	191·08	33·40	93·53	50·04	214·47	45·91	82·63	15·10
8	252·06	76·75	130·24	38·77	135·45	30·13	111·40	55·25	163·30	43·95	60·91	18·99
9	204·76	72·43	110·97	40·54	112·25	27·32	91·23	58·34	135·06	41·90	46·90	15·94
10	201·40	67·56	101·06	32·97	104·78	25·85	92·89	48·84	128·00	38·06	50·17	17·30
11	298·31	68·03	158·01	37·32	178·54	28·61	99·24	48·14	203·35	40·64	70·15	15·35
12	299·09	61·86	189·25	34·91	203·99	24·72	80·35	47·34	224·08	36·55	54·92	13·48
13	178·69	54·20	63·34	26·19	71·84	17·93	98·36	44·53	96·43	29·06	57·68	14·01
14	223·11	49·73	105·62	26·91	112·46	17·14	103·81	42·37	138·41	27·73	58·74	11·41
15	188·68	53·62	85·81	29·86	95·73	17·31	83·04	48·86	116·49	29·52	51·43	11·88
16	155·13	49·29	68·18	26·00	69·09	19·60	85·14	36·09	90·37	28·62	43·48	11·64
17	184·55	39·21	72·57	27·03	78·02	12·61	101·09	41·02	103·29	22·87	55·99	6·09
18	186·49	34·71	98·82	20·52	97·37	9·17	90·58	36·89	120·01	18·39	43·84	7·10
19	174·47	27·60	79·01	18·05	85·25	6·52	82·97	32·61	105·99	14·67	47·73	4·78
20	105·15	22·15	57·44	15·09	43·43	4·47	75·74	28·30	62·36	11·54	23·86	3·53
21	139·66	22·10	74·21	18·06	70·89	6·31	72·10	27·53	88·91	13·19	32·72	2·02
22	111·30	20·72	64·69	19·81	58·05	6·20	59·91	28·13	73·02	13·24	23·30	0·46
23	109·52	15·53	64·49	13·97	54·00	4·10	66·01	21·30	70·50	9·43	22·52	0·78
24	69·24	11·13	46·23	14·77	14·48	3·89	86·51	18·12	36·10	8·42	11·51	−1·82
25	70·41	10·73	56·45	14·51	30·79	3·95	65·27	17·34	47·11	8·29	6·98	−1·89
26	40·19	10·60	52·54	16·69	27·69	4·31	37·35	18·67	37·03	8·98	−6·18	−3·04
27	37·29	8·17	55·72	14·52	24·04	1·84	44·92	19·00	35·27	6·59	−9·21	−3·18
28	30·84	8·37	44·44	11·99	11·27	1·96	52·75	16·44	24·45	6·07	−6·80	−1·81
29	39·99	7·96	61·36	11·44	27·32	1·85	46·71	15·70	39·00	5·77	−10·68	−1·74
30	35·83	7·09	57·39	13·71	21·67	3·01	49·89	14·79	34·14	6·70	−10·78	−3·31
31	28·78	7·78	52·47	14·18	21·56	3·08	38·12	15·80	31·10	7·03	−11·84	−3·20
32	37·29	8·95	29·98	12·21	18·03	1·70	31·20	17·76	25·83	6·14	3·65	−1·63
33	33·20	9·49	50·20	13·30	27·10	1·90	29·19	18·99	34·40	6·65	−8·50	−1·90
34	26·66	6·90	56·32	11·58	22·20	1·45	38·56	15·58	31·85	5·34	−14·83	−2·34
35	30·22	8·60	49·36	13·40	22·51	2·39	34·56	17·21	31·15	6·70	−9·57	−2·40
36	30·59	9·15	37·90	13·18	13·54	1·90	41·41	18·52	23·89	6·53	−3·65	−2·01
37	54·57	11·39	48·63	12·94	26·94	2·52	49·33	19·30	39·27	7·34	2·97	−0·77
38	39·70	15·29	29·27	15·16	9·73	2·88	49·52	24·68	22·11	9·05	5·21	0·06
39	44·45	18·46	41·90	17·17	21·22	5·18	43·92	25·26	32·20	11·50	1·27	0·65
40	69·68	20·36	52·05	19·41	40·53	5·20	40·68	29·38	50·70	12·54	8·82	0·48
41	83·86	22·82	38·76	17·78	37·60	4·24	47·43	32·12	49·45	12·27	22·55	2·52
42	87·81	27·28	52·12	23·20	44·39	6·94	51·16	36·60	57·18	16·09	17·84	2·04
43	92·38	32·49	32·57	21·21	25·39	10·28	74·18	33·14	43·93	18·57	29·91	5·64
44	113·86	38·83	55·44	23·19	55·64	9·53	58·03	42·95	70·15	20·27	29·21	7·82
45	93·21	40·17	48·74	26·24	45·54	10·65	50·87	45·11	58·26	21·93	22·24	5·66
46	98·32	44·17	61·48	28·92	53·73	14·00	52·34	45·10	66·82	25·27	18·42	7·62
47	133·42	55·15	54·75	32·57	63·67	15·40	60·83	56·93	78·88	29·63	39·33	11·29
48	130·59	60·90	66·05	31·20	61·62	18·72	73·40	54·67	79·97	32·38	32·27	14·85
49	166·31	66·72	58·98	40·09	63·54	20·71	98·21	65·39	88·09	37·06	53·67	13·31
50	141·95	73·72	73·02	38·17	75·66	25·09	63·66	61·72	91·57	40·52	34·47	17·77
51	177·60	83·38	78·12	40·45	94·41	29·83	66·89	64·17	111·14	45·88	49·74	21·47
52	188·49	81·19	91·35	42·48	98·02	26·84	83·79	70·00	118·97	44·34	48·57	19·35
53	185·68	81·55	79·91	46·55	87·32	29·95	90·96	68·21	110·06	47·00	52·88	17·50
54	234·41	84·26	116·04	41·88	130·82	31·42	88·80	63·30	153·02	47·25	59·18	21·19
55	267·55	73·59	108·35	41·15	134·35	26·55	107·21	61·65	161·15	41·96	79·60	16·22
56	189·34	82·01	95·75	40·51	98·29	29·13	88·50	64·27	120·42	45·19	46·80	20·75
57	252·62	89·38	116·37	42·00	136·82	32·75	95·34	65·88	160·66	49·22	68·12	23·69
58	225·04	81·57	93·52	46·52	120·04	35·90	78·47	56·29	139·66	49·97	65·76	17·53
59	217·76	94·27	93·79	45·55	111·56	36·49	88·44	66·84	133·67	53·20	61·99	24·36
60	228·37	93·46	97·28	48·54	120·49	36·14	84·67	69·73	141·66	53·57	65·54	22·46
61	266·68	92·35	121·93	44·66	142·50	38·39	103·61	60·24	168·40	53·45	72·37	23·84

* For periods 11 to 61 inclusive the degrees of freedom are: Between periods 82, Within periods 415; and for periods 1 to 10 inclusive, where the record extended into 1944, they are 83 and 420 respectively.

These features, and the well-defined annual oscillation in the significance of the between-period mean square, are clearly demonstrated in the seasonal trends of the corresponding intra-class correlations displayed in Figures 1 and 2. The intra-class correlations (r) were calculated from the mean squares given in Table 1, and their inverse hyperbolic transforms (z) were derived from the relation

$$z = \tfrac{1}{2}\log\frac{1+(k-1)r}{1-r},$$

where k is the number of days in a period (Fisher 1958), in this case 6.

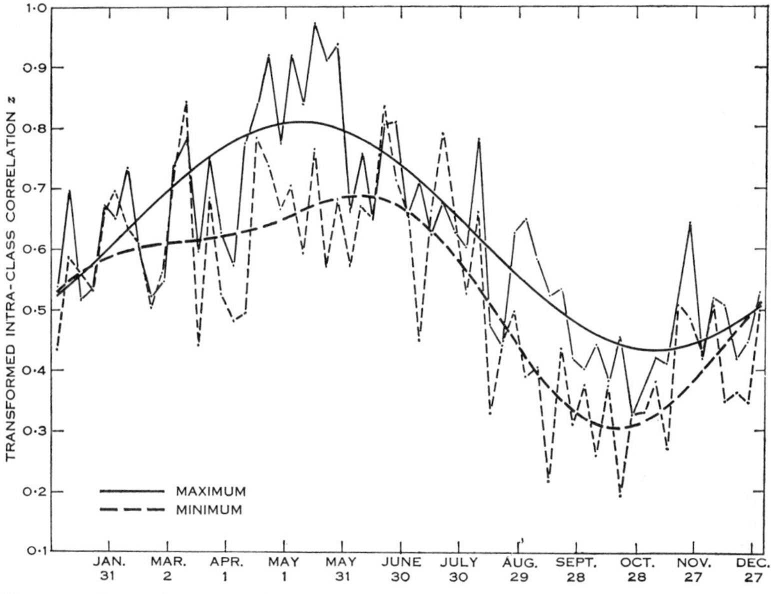

Fig. 1.—Intra-class correlations of the maximum and of the minimum temperatures.

The simple seasonal change in the course of the intra-class correlation of maximum temperatures is represented by

$$Z_M = 0\cdot62+0\cdot19\sin(\theta+327^\circ\ 51'),$$

while the rather more complicated trend of the intra-class correlation of the minimum temperatures is represented by

$$Z_m = 0\cdot53+0\cdot17\sin(\theta+337^\circ\ 48')+0\cdot06\sin(2\theta+81^\circ\ 9').$$

The diagram shows clearly that for an interval of about a month centred on the second week in May, maximum temperatures are more closely related from year to year than at any other stage of the season, whilst for a month centred on November 6, these correlations are weakest. The correlations of the minimum temperature increase very slowly to their greatest value, but thereafter follow a course similar in form to that of the maximum temperature, except that the lowest value is reached some 18 days earlier.

The seasonal changes in the correlations of the mean and the range, which are represented, respectively, by

$$Z_t = 0{\cdot}67+0{\cdot}18\sin(\theta+325^\circ\ 22'),$$
$$Z_R = 0{\cdot}31+0{\cdot}19\sin(\theta+323^\circ\ 34')+0{\cdot}05\sin(2\theta+115^\circ\ 24'),$$

are illustrated in Figure 2. The correlations of the mean temperature follow a course similar to that of the maximum temperature (the turning points being almost coincident on the axis of time), but throughout the year exceed the correlations of

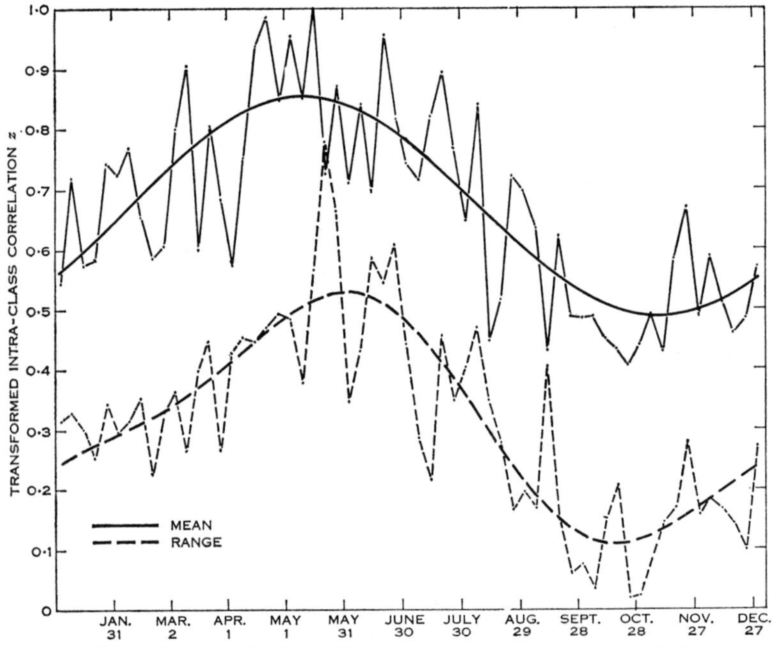

Fig. 2.—Intra-class correlations of the mean temperature and of the range.

the maximum temperature. On the other hand, the correlations of the range take a course strongly influenced by the minimum temperature, but are much lower throughout the year.

If v_p and v_w represent the components of variance between periods from year to year and within periods from day to day respectively, the intra-class correlations were calculated from

$$v_p/(v_p+v_w),$$

and the observed trends are directly attributable to the changes occurring in v_p and v_w as the season advances. We shall return to this point (see below, Section IV(c)) when discussing the variance within periods.

(b) Annual Sequence of the Means

Figure 3 shows the annual courses for each of the four temperature variates, represented by

Max. $72 \cdot 71 + 13 \cdot 58 \sin(\theta + 76° \, 54') + 1 \cdot 05 \sin(2\theta + 329° \, 7')$,
Min. $53 \cdot 14 + 8 \cdot 35 \sin(\theta + 67° \, 34') + 0 \cdot 51 \sin(2\theta + 345° \, 0')$,
Mean $62 \cdot 93 + 10 \cdot 93 \sin(\theta + 73° \, 20') + 0 \cdot 77 \sin(2\theta + 334° \, 15')$,
Range $19 \cdot 57 + 5 \cdot 51 \sin(\theta + 91° \, 8') + 0 \cdot 57 \sin(2\theta + 315° \, 19') + 0 \cdot 16 \sin(3\theta + 228° \, 27')$.

The maximum temperature falls more rapidly from its greatest value attained during period 5 (Jan. 25–30) than the minimum temperature falls from its greatest value, attained during period 6 (Jan. 31–Feb. 5). Similarly, the maximum temperature

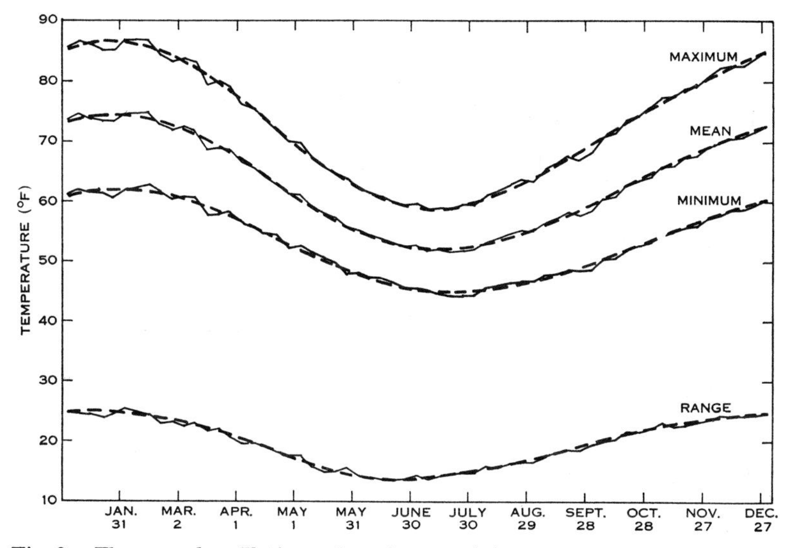

Fig. 3.—The annual oscillations of maximum, minimum, and mean temperatures and of the range.

increases more rapidly in the spring. The minimum temperature lags behind the maximum temperature throughout the year, and there is a difference of about 12 days in the dates of attainment of the two minima (maximum temperature, period 32, July 6–11; minimum temperature, period 34, July 18–23). As would be expected, the course of the mean temperature is intermediate between the maximum and minimum temperatures.

The range is high in summer, the greatest value being 25°, occurring at the end of period 2 (Jan. 12), and low in winter, the lowest value, which falls in period 29 (June 18–23), being just over half of the greatest value. The minimal value of the range occurs at such an early date as a consequence of the differential rates of change in the courses of the maximum and minimum temperatures and their difference in phase. In the summer a rise in day temperature (represented by the maximum) tends to be accompanied by a rise in night temperature (represented by the minimum), but, since the variances differ considerably, that for the maximum being the greater, the two deviations differ considerably thus yielding a large value for the range. In the winter the position is reversed. A fall in night temperature is associated with a fall in day temperature, but since the minimum temperature now has the larger variance, the two deviations still differ, though not nearly to the same extent as in

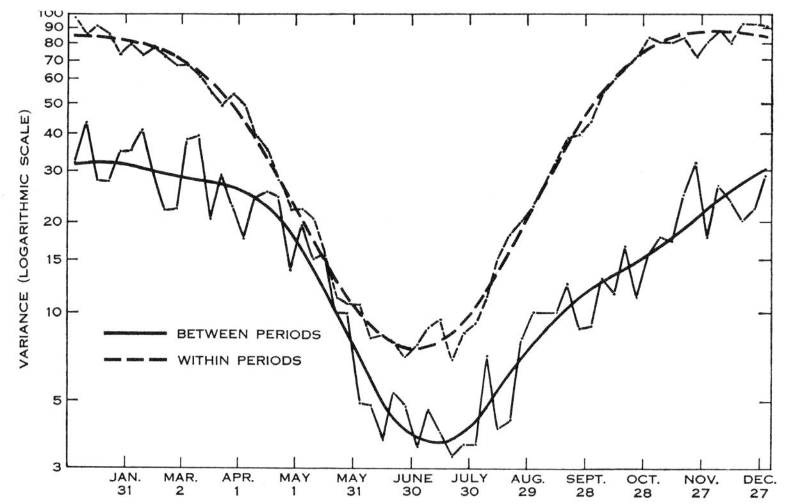

Fig. 4.—Components of variance of maximum temperature (*a*) between periods from year to year, (*b*) within periods from day to day.

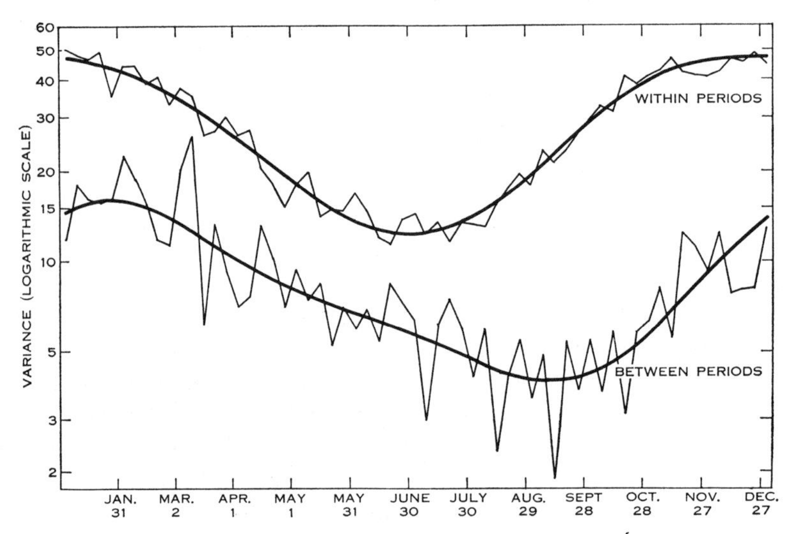

Fig. 5.—Components of variance of minimum temperature (*a*) between periods from year to year, (*b*) within periods from day to day.

summer, because the variances of both maximum and minimum temperature are only a fraction of their summer values. The net result is a small value for the range.

We shall demonstrate that these sequences are not permanent features of the climate, as they are being modified by secular changes (see below, Section IV(*h*)).

(c) *Variance from Day to Day within Periods*

Figures 4, 5, 6, and 7 show the seasonal trends of the variance from day to day within periods for the four temperature variates in the order maximum, minimum, mean, and range. All dependent variates have been measured on a logarithmic scale to make the variability about the trend constant and independent of the variance itself; at the same time this procedure gives a truer representation of the changes in progress. The fitted curves are as follows (in the order previously given)

$$1\cdot54+0\cdot53\sin(\theta+95^\circ\ 52')+0\cdot13\sin(2\theta+274^\circ\ 22'),$$
$$1\cdot41+0\cdot29\sin(\theta+98^\circ\ 1')+0\cdot03\sin(2\theta+261^\circ\ 7'),$$
$$1\cdot33+0\cdot48\sin(\theta+94^\circ\ 49')+0\cdot08\sin(2\theta+271^\circ\ 53'),$$
$$1\cdot57+0\cdot29\sin(\theta+102^\circ\ 57')+0\cdot09\sin(2\theta+265^\circ\ 28').$$

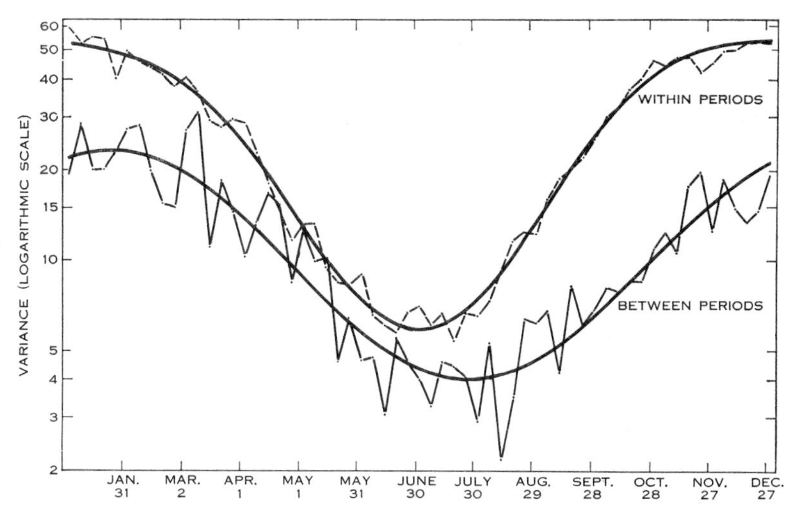

Fig. 6.—Components of variance of mean of daily maximum and minimum temperatures (*a*) between periods from year to year, (*b*) within periods from day to day.

The variability of the maximum temperature is greatest in summer, remaining practically constant from approximately November 1 to mid February, and then decreases rapidly to its smallest value, which occurs in period 30 (June 24–29). Similarly the variability of the minimum temperature is greatest in summer and least in winter, but for an interval almost centrally located in the calendar year (period 23 to period 39) it exceeds the variability of the maximum temperature.

The variability of the mean temperature is controlled by the high correlation existing between the maximum and minimum temperatures and their high variances in the two subdivisions of the year (i) from period 1 to period 14 (ii) from period 44 to period 61, thus giving it high values lying between the variances of the maximum and minimum. On the other hand, in the central section, periods 15 to 48, it is controlled by the low correlation characteristic of that interval and the variance of the maximum, particularly between periods 23 and 39, where, instead of falling between the variances of the maximum and minimum, it falls below that of the maximum.

The variability of the range is also greatest in summer and least in winter, being intermediate between the variabilities of the maximum and minimum temperatures for the intervals, periods 1–17 and periods 47–61, but for the central interval, periods 18–46, it exceeds the variability of both maximum and minimum, since at this stage their correlation is so weak (see below, Section IV(*d*)).

With all four variates the variability within periods contains a component due to the annual temperature wave, but, since the length of a period (6 days) is so small, this component must be negligible.

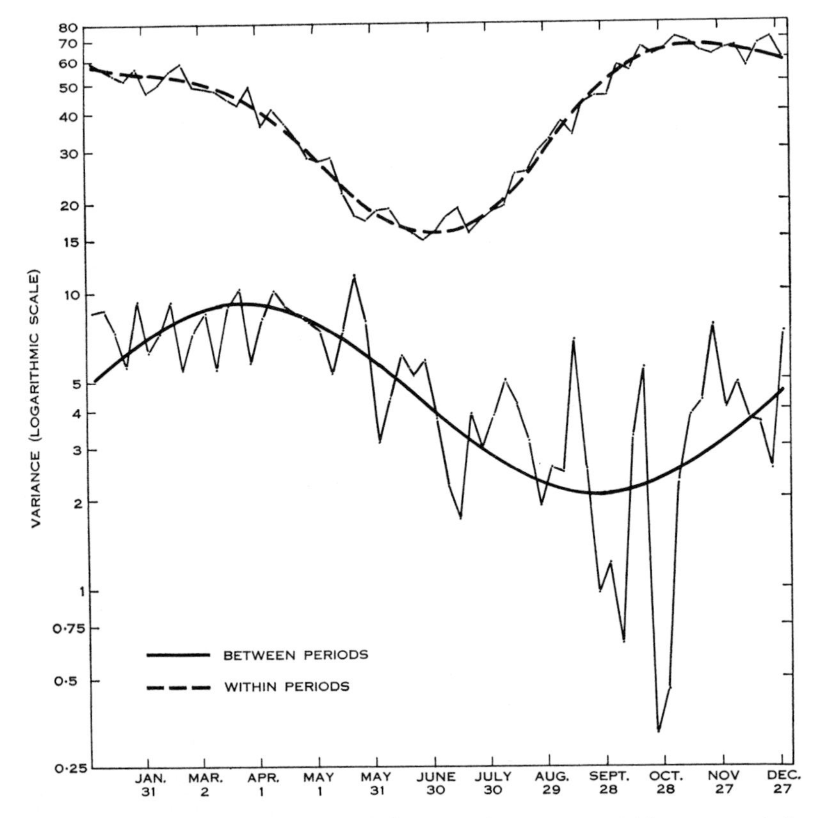

Fig. 7.—Components of variance of daily range of temperature (*a*) between periods from year to year, (*b*) within periods from day to day.

In Figures 4, 5, 6, and 7 the variances within periods, of the four temperature variates, are contrasted, each with its corresponding component of variance between periods from year to year (v_p in Section IV(*a*)), also on a logarithmic scale, the trends of the latter, in the order previously given, being represented by:

$1 \cdot 14 + 0 \cdot 43 \sin(\theta + 70° \; 20') + 0 \cdot 11 \sin(2\theta + 264° \; 42') + 0 \cdot 06 \sin(3\theta + 95° \; 20')$,

$0 \cdot 89 + 0 \cdot 27 \sin(\theta + 48° \; 25') + 0 \cdot 07 \sin(2\theta + 80° \; 14')$,

$0 \cdot 99 + 0 \cdot 38 \sin(\theta + 68° \; 13')$,

$0 \cdot 64 + 0 \cdot 32 \sin(\theta + 11° \; 10')$.

The origin of each feature of the trends given in Figures 1 and 2 can then be clearly seen by comparing corresponding curves in Figures 4, 5, 6, and 7. For example, the high correlation of maximum temperatures from year to year, for the month centred on the second week of May, follows from the fact that, at this time, the

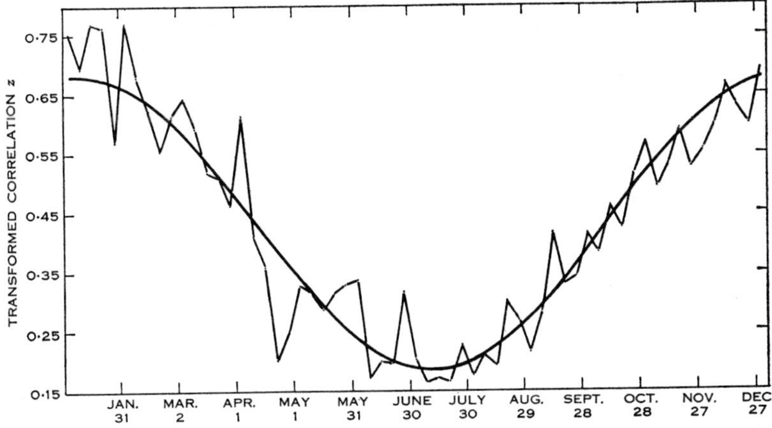

Fig. 8.—Correlation of maximum and minimum temperatures from day to day within periods.

component between periods, v_p, is nearly equal to the variance within periods v_w (Fig. 4); at all other stages their courses diverge; for the month centred on November 6, the variance within periods has increased to its greatest value relative to v_p, thus making the correlation take its lowest value.

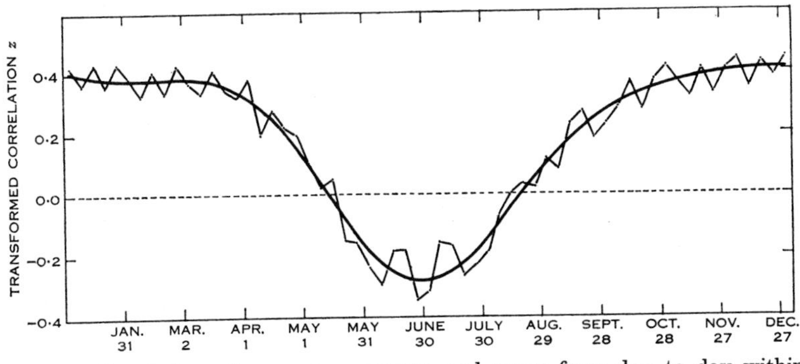

Fig. 9.—Correlation of mean temperature and range from day to day within periods.

Again, in Figure 5 these components of variance for minimum temperatures have courses which are slightly convergent from period 1 to period 28, thus yielding the slowly increasing correlation of the minimum temperatures during the first six months of the year.

Adverting to the point raised in Section IV(a)(ii) above, the significance of the variation in range between periods from year to year is directly traceable to the fact that the component v_p of maximum temperature maintained a consistently higher value during the first half of the year than during the second half (Fig. 4) coupled with the long protracted fall of the corresponding component for minimum temperature (Fig. 5). As a consequence, component v_p of the range is maintained at a consistently higher value in the first half of the year, thus leading to the significance of the range at that time, whereas, during the second half, this component is on the average lower and much more variable (Fig. 7). A longer record is required to definitely resolve this point.

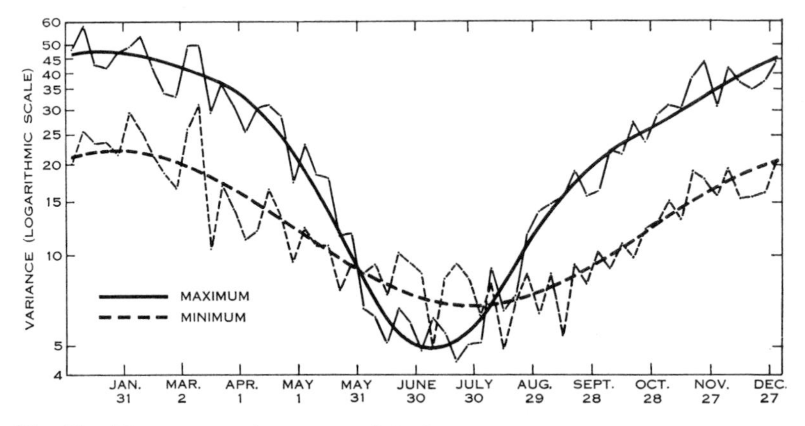

Fig. 10.—Mean squares between periods from year to year for maximum and minimum temperatures (on a per day basis).

(d) *Inter-class Correlation between Maximum and Minimum, Mean and Range, within Periods*

Figure 8 illustrates the course of the correlation between maximum and minimum temperatures calculated from the observed variation from day to day within periods and represented by

$$0\cdot43+0\cdot25 \sin(\theta+83^\circ\ 39').$$

This correlation is positive throughout the season, high in summer, and low in winter. During the summer, maximum and minimum temperatures are most variable but there is a definite tendency for a hot day to be accompanied by a hot night; as the season advances this tendency falls rapidly in the autumn, reaches its lowest point in midwinter and then rises again in the spring. In the winter, maximum temperatures are less variable than minimum temperatures, and the general tendency is for a cold night to have relatively little influence on the day temperature. Thus in the 83 years of record, the maximum temperature has been less than 50°F on only 17 occasions, whereas the minimum temperature has been less than 40°F as many as 1018 times. Furthermore, in the interval from period 25 to period 40 inclusive, 83% of maximum temperature readings have fallen in the range 55–65°F inclusive.

Figure 9 shows the correlation between mean temperature and range, the fitted curve being represented by

$$0\cdot18+0\cdot31 \sin(\theta+91^\circ\ 57')+0\cdot12 \sin(2\theta+280^\circ\ 17')+0\cdot03 \sin(3\theta+135^\circ\ 0').$$

This correlation attains its highest absolute values in summer, when it is positive; it falls to zero during May and August, and in the winter takes negative values, but the relations are weak. The negative correlations in the winter are closely related

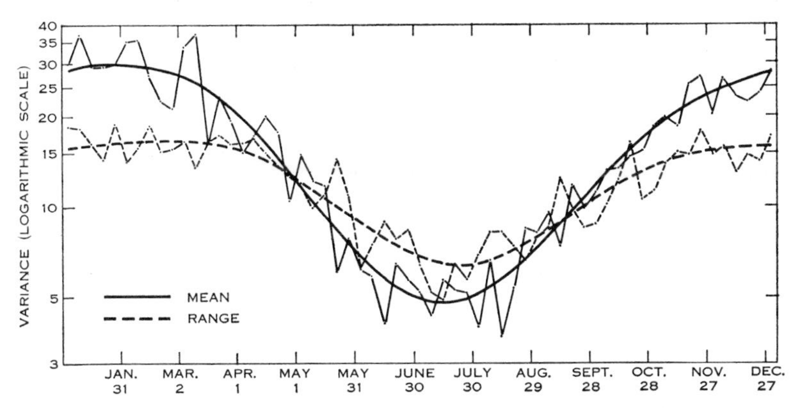

Fig. 11.—Mean squares between periods from year to year for the mean temperature and range (on a per day basis).

to the differential variability of the maximum and minimum temperatures and their low positive correlation. Although a fall in day temperature tends to accompany a fall in night temperature, thus contributing to the positive correlation of maximum and minimum and reducing the mean temperature, the drop in day temperature is

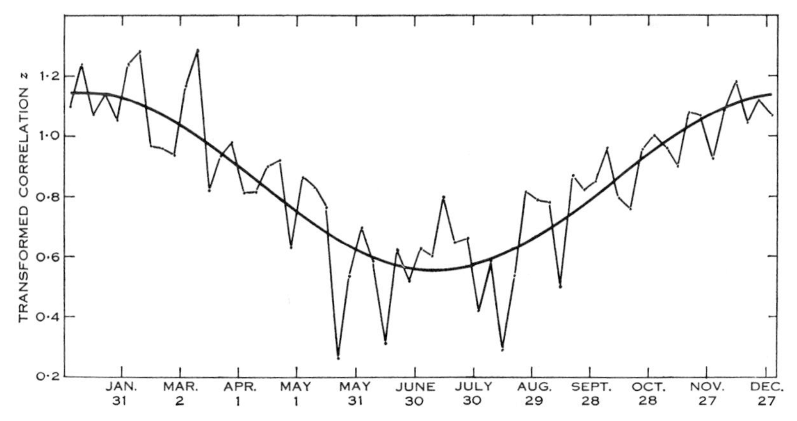

Fig. 12.—Correlation of maximum and minimum temperature between periods from year to year.

not nearly so great as the drop at night; this increases the range, which thus accompanies the reduction of the mean temperature. In summer the position is reversed, since now an increase in the mean is accompanied by a greater range. A hot day is associated with a hot night, thus contributing to a positive correlation and increasing the mean temperature, but the increase in night temperature is not as great as the increase of the day temperature, so increasing the range, which thus accompanies a rise in the mean temperature.

(e) *Variance between Periods from Year to Year*

The mean squares between periods from year to year in Table 1 were divided by 6 to reduce them to a per-day basis, and are illustrated in Figures 10 and 11, where they are set out on a logarithmic scale. Figure 10 shows the sequences for maximum and minimum temperatures, which are represented, respectively, by

$$1\cdot30+0\cdot45\sin(\theta+79^\circ\ 7')+0\cdot12\sin(2\theta+275^\circ\ 26')+0\cdot05\sin(3\theta+93^\circ\ 21'),$$
$$1\cdot09+0\cdot25\sin(\theta+67^\circ\ 52').$$

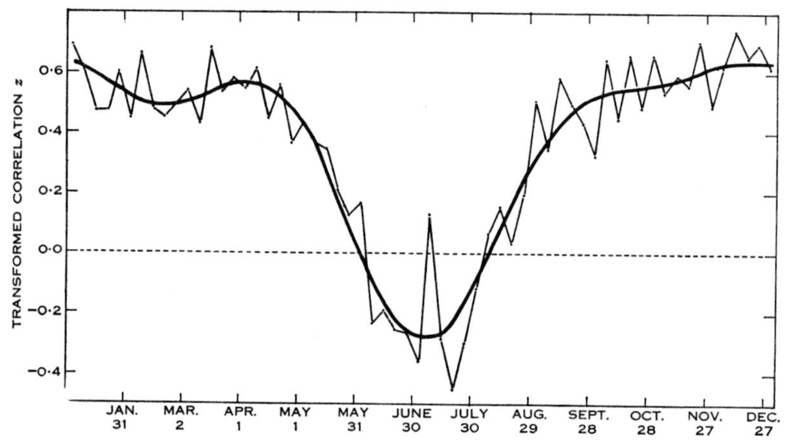

Fig. 13.—Correlation of mean temperature and range between periods from year to year.

The variability is greatest in summer and least in winter for both maximum and minimum, and thus for both measures there is not only a high variation from day to day, but also from year to year. Moreover, the variance of the maximum temperature is less than that of the minimum temperature in the winter from period 25 to period 36, and hence the observations made previously, namely, that day temperatures are not appreciably affected by night temperatures in this central section of the winter, holds in general for all winters. It is apparent, too, that the minimum lags behind the maximum.

Since maximum and minimum temperatures are fairly highly correlated in the summer, and in addition, both are very variable, the variability of the mean temperature (Fig. 11) takes an intermediate course, so that the mean may be regarded as providing a good approximation of the temperature at this time. On the other hand, the difference between a mild and a cold winter is more clearly shown by the minimum temperatures, but the variance of the mean is influenced by the more constant day temperatures.

The range follows a course similar in form to that of the mean but it is not as variable in the summer and is more variable in the winter. The fitted curves in

Figure 11 are represented by

$$1 \cdot 13 + 0 \cdot 39 \sin(\theta + 76° \, 44') + 0 \cdot 06 \sin(2\theta + 266° \, 32'),$$
$$1 \cdot 06 + 0 \cdot 20 \sin(\theta + 70° \, 6') + 0 \cdot 06 \sin(2\theta + 242° \, 15').$$

(*f*) *Inter-class Correlations of Maximum and Minimum, Mean and Range, between Periods from Year to Year*

Figure 12 illustrates the course of the correlation between maximum and minimum temperatures calculated from the observed variation between periods from year to year, the fitted harmonic being represented by

$$0 \cdot 85 + 0 \cdot 29 \sin(\theta + 84° \, 41').$$

This curve is similar in form to that of the corresponding correlation from day to day within periods, its amplitude and phase closely approximating their counterparts,

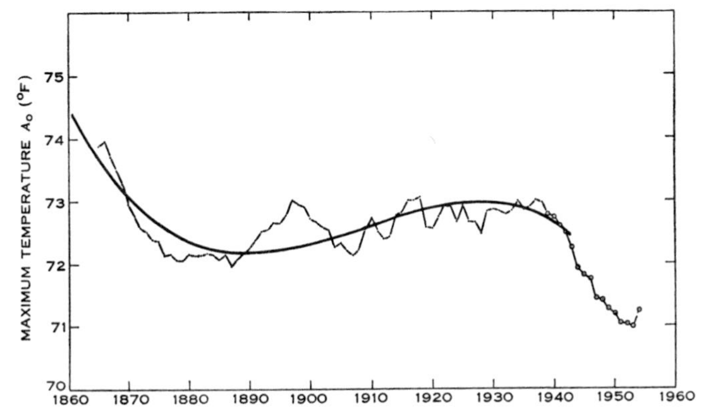

Fig. 14.—Ten-year means and polynomial curve of the mean annual maximum temperature (A_0).

but the mean is twice as great. As a consequence, the correlation is positive throughout the year, high in summer, low in winter, and at all stages greatly exceeds the correlation within periods. In the summer, a hot month or a cool month is thus reflected in both maximum and minimum temperatures, and this holds, in general, for all summers; moreover, this tendency persists to a lesser degree in the autumn and spring and even into the winter, where, although a cold night is not necessarily accompanied by a cold day, in say a cold June, the maximum temperature is none the less reduced, though not on the same scale as the minimum.

Figure 13 shows the correlation between mean temperature and range, also calculated from the observed variation between periods. The curve fitted to the observations is

$$0 \cdot 36 + 0 \cdot 36 \sin(\theta + 90° \, 35') + 0 \cdot 19 \sin(2\theta + 260° \, 1') + 0 \cdot 10 \sin(3\theta + 101° \, 31'),$$

in which the amplitudes and phase angles of the three harmonics agree reasonably well with those of the corresponding correlation within periods (Fig. 9), but the mean is twice as great. In consequence, the annual course of the correlations resembles that given in Figure 9. As before, the negative correlations of mean and range in the winter are closely related to the differential variability of the maximum and minimum temperatures and their lower positive correlation. The observation that during the winter, day temperatures are not markedly affected by night temperatures, holds in general for all winters, and hence although a fall in day temperature tends to accompany a fall in night temperature, thus contributing to the positive correlation of maximum and minimum and reducing the mean temperature, the drop in day temperature is not nearly so great as the drop at night; this increases the range, which thus accompanies the reduction of mean temperature, and these changes give rise to the negative correlations observed. In summer the position is reversed, since now an increase in the mean is accompanied by a greater range.

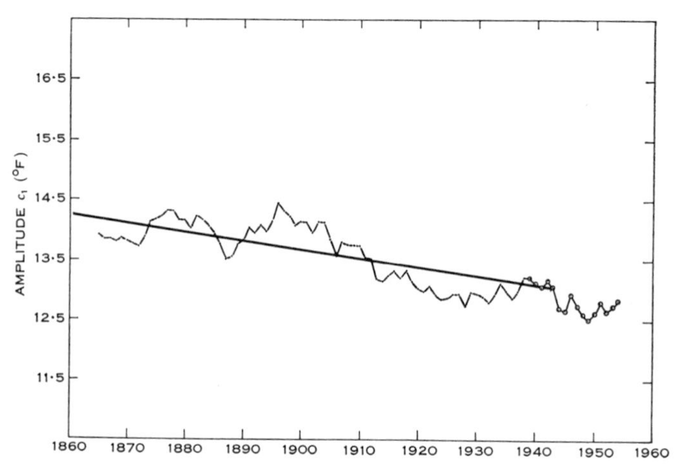

Fig. 15.—Ten-year means and linear regression for the amplitude of maximum temperature (c_1).

(*g*) *General Remarks*

The general pattern of the observed results follows directly from the differential variability of the temperature variates in the summer and winter seasons. The high variability characteristic of the summer is due to large changes of temperature which follow the alternation of movement of the air masses from the continent to the north and north-west, after intense heating, to an influx of cool air from the ocean to the west and south-west. During the winter there are not the same violent changes of temperature. The more northerly latitudinal track of the migratory anticyclones induces prevailing winds from north-west to south-west, bringing air principally from over the ocean. Moreover, with the northward movement of the Sun, the land mass to the north-west, from the coast to some considerable distance inland, is subject only to mild heating.

(*h*) *Secular Changes in Temperature*

The basic data for this aspect of the analysis comprise the 61 observations in each of the 83 years on each of the four temperature variates. Considering first maximum temperature, each annual sequence of 61 values was fitted with a harmonic series of the form

$$A_0+A_1 \cos \theta+B_1 \sin \theta+A_2 \cos 2\theta+B_2 \sin 2\theta,$$

thus furnishing five quantities with which the magnitude and seasonal changes of maximum temperature could be represented. For complete representation it was occasionally necessary to carry the series to the terms involving 3θ, but, as this was required in so few years and the contributions of such terms in the analysis of variance were so small, they have been omitted from the calculations.

To facilitate interpretation of the changes in progress, the coefficients A_1 and B_1 were replaced by the amplitude c_1 and the phase angle ϕ_1 of the first harmonic, and similarly A_2 and B_2 were replaced by the amplitude c_2 and the phase ϕ_2 of the second harmonic.

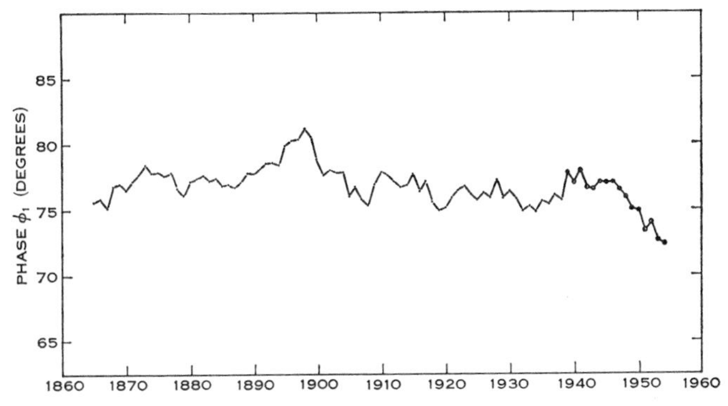

Fig. 16.—Ten-year means for the phase of maximum temperature (ϕ_1).

In Figures 14, 15, and 16, the courses of the changes in A_0, c_1, and ϕ_1 for maximum temperature have been depicted by plotting running 10-year means of each. In the first instance we confine the discussion to the period 1861–1943. Figure 14 shows that the mean maximum temperature fell from approximately 74°F about 1860 to just over 72° about 1890, and then gradually increased, reaching about 73° in 1930. Thereafter there is some indication that the mean began falling again.

On the other hand, the amplitude of the first harmonic in the annual wave of maximum temperature follows a much simpler course with a distinct and apparently uniform decrease throughout the 80 years of record (Fig. 15). The phase angle of the first harmonic shows no evidence of secular change, apparently fluctuating with a large standard deviation about its mean value (Fig. 16).

The five quantities were then examined in more detail by fitting each series of 83 values with an orthogonal polynomial function of time of sufficient degree to account for any changes observed. The procedure adopted at this stage may be illustrated by the calculations relating to the mean maximum temperature A_0. The following values were obtained, the unit being degrees F.

Mean	72·716627	x'_1	662·480	$(x'_1)^2$	438879·75
Coefficient of 1st degree	−0·048230	x'_2	−0·770	$(x'_2)^2$	0·59
Coefficient of 2nd degree	0·141850	x'_3	2·996	$(x'_3)^2$	8·98
Coefficient of 3rd degree	−0·112470	x'_4	−2·914	$(x'_4)^2$	8·49
Coefficient of 4th degree	0·041386	x'_5	1·276	$(x'_5)^2$	1·63.

The quantities x'_1, x'_2, . . . are orthogonal, normal linear functions of the 83 values of A_0, obtained from the corresponding quantities in the first column by multiplying the coefficient of the term of degree r by

$$\left\{\frac{(2r+1).\,83.84\ldots(83+r)}{82.81\ldots(83-r)}\right\}^{\frac{1}{2}},$$

and x'_2, x'_3, . . . represent the several components of change in the sequence of values of A_0. The total variation of the values of A_0 from their mean may be divided into two portions

(i) a sum of squares associated with regression on time, and due to a comparatively simple temporal trend predominating over the random fluctuations,

(ii) the remaining sum of squares which may be attributed to random annual variation,

and the first of these components may itself be partitioned further to give the individual contributions of the terms of the several degrees in time (the quantities $(x'_2)^2$, $(x'_3)^2$, . . . above) to the total for regression. The analysis of variance given in Table 2 establishes the strong significance ($P < 0{\cdot}01$) of the quadratic and cubic terms, confirming the feature observed in Figure 14.

TABLE 2

ANALYSIS OF VARIANCE OF MAXIMUM TEMPERATURE SEQUENCE (1861–1943)

Variation due to	Degrees of Freedom	Sum of Squares	Mean Square	Variance Ratio
Linear component	1	0·59	0·59	n.s.
Quadratic component	1	8·98	8·98	10·20
Cubic component	1	8·49	8·49	9·65
Quartic component	1	1·63	1·63	n.s.
Deviations from regression	78	68·62	0·88	
Total	82	88·31		

Table 3 provides a summary analysis of the five distribution constants for each temperature variate; as stated above, the fitting in each case has been carried only as far as the successive terms are significant.

The minimum temperature is affected by definite changes in the amplitude and phase of its first harmonic component, while the mean temperature follows closely the changes observed in maximum temperature. Both the mean and the amplitude of the first harmonic of the range follow rather more complex courses in which the quartic term is strongly significant. No significant changes have occurred in the amplitude or phase angle of the second harmonic component of any temperature variate.

TABLE 3

SUMMARY POLYNOMIAL ANALYSIS OF THE DISTRIBUTION CONSTANTS OF THE FOUR TEMPERATURE VARIATES (1861–1943)

Units: degF for A_0, c_1, c_2; degrees circular measure for ϕ_1 and ϕ_2.

Temperature Variate		Distribution Constant				
		A_0	c_1	ϕ_1	c_2	ϕ_2
Maximum temp.	Mean	72·717	13·646	76·918	1·617	239·25
	x'_2	−0·770	−3·142			
	x'_3	2·996	−0·841			
	x'_4	−2·914				
	x'_5	1·276				
	Stand. resid.	0·938	1·024	5·153	0·829	118·3
	t*		124·05(80)	134·94(82)	17·77(82)	18·40(82)
Minimum temp.	Mean	53·145	8·409	67·678	1·104	187·48
	x'_2		−1·185	1·573		
	x'_3		−1·865	−14·877		
	x'_4		1·474	0·565		
	Stand. resid.	0·718	0·808	5·953	0·536	123·4
	t		94·48(79)	104·12(79)	18·71(82)	13·89(82)
Mean temp.	Mean	62·929	10·989	73·356	1·300	205·50
	x'_2	−0·696	−2·102			
	x'_3	1·295	−1·394			
	x'_4	−2·026				
	x'_5	0·149				
	Stand. resid.	0·750	0·822	5·121	0·666	128·7
	t		122·10(80)	130·99(82)	17·81(82)	14·57(82)
Range	Mean	19·572	5·573	91·199	0·986	232·28
	x'_2	−0·158	−2·286			
	x'_3	3·383	1·289			
	x'_4	−1·778	0·692			
	x'_5	2·249	1·952			
	x'_6	0·592	1·183			
	Stand. resid.	0·725	0·792	9·231	0·433	115·5
	t		64·06(77)	90·30(82)	20·55(82)	18·29(82)

* The figures in parenthesis are the degrees of freedom of the t-statistics in tests of significance of the several means.

To illustrate the secular changes which have occurred in maximum temperature, the three years 1861, 1890, and 1928 have been selected from the period under review, and the annual waves of maximum temperature at these points in time have been reconstructed from

(i) the corresponding polynomial values of A_0, namely, 74·37, 72·17, and 72·97 (Fig. 14),

(ii) the polynomial values of c_1, namely, 14·24, 13·82, and 13·42 (Fig. 15),

(iii) the mean values of ϕ_1, c_2, and ϕ_2, 76° 55′, 1·62, and 239° 15′ respectively (Table 3).

The three courses are then represented by

$$74\cdot 37+14\cdot 24\sin(\theta+76^\circ\,55')+1\cdot 62\sin(2\theta+239^\circ\,15'),$$
$$72\cdot 17+13\cdot 82\sin(\theta+76^\circ\,55')+1\cdot 62\sin(2\theta+239^\circ\,15'),$$
$$72\cdot 97+13\cdot 42\sin(\theta+76^\circ\,55')+1\cdot 62\sin(2\theta+239^\circ\,15'),$$

and Figure 17 illustrates the effects of the differences of the mean and the amplitude of the first harmonic.

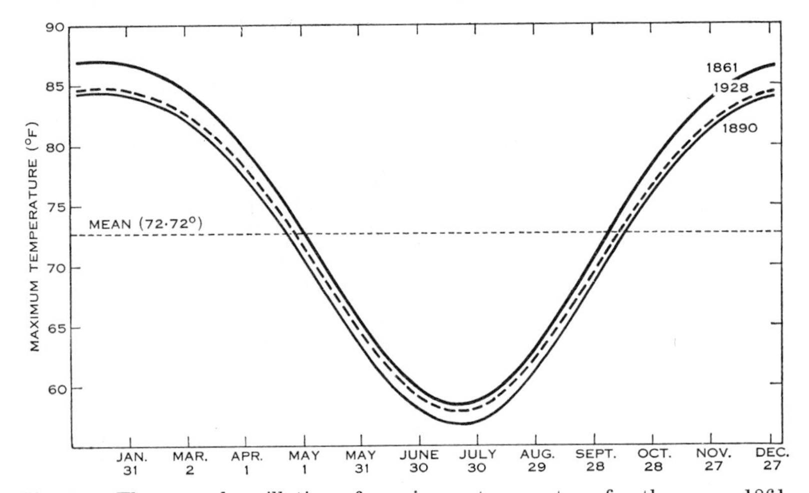

Fig. 17.—The annual oscillation of maximum temperature for the years 1861, 1890, and 1928, illustrating secular changes.

As indicated in Section II, temperature records from the Greenwich stand and the Stevenson screen were taken concurrently, during the period 1926–1947 and differences in site have been ignored since they are negligible in comparison with differences of the instruments. From these observations, the regression of Greenwich stand maximum on Stevenson screen maximum was taken, and the 61 regression coefficients so calculated were used to construct estimates of what the Greenwich stand observations might have been from July 31, 1947, to December 1959 from the observed Stevenson screen record of maximum temperature over the same period. With the observed Greenwich stand record from 1944 to July 1947, this supplied a set of 61 values for each of the years 1944–1959, which were treated in the same manner as the original observations had been. The additional 10-year running means are recorded in Figures 14, 15, and 16, and it is of very considerable moment that the members of each set continue the particular trend exhibited by their respective predecessors. In Figure 14, the fall in mean maximum temperature is extended from 1943 to 1959, reaching as low as 71°F. In Figure 15, the downward linear trend of the amplitude of the first harmonic is extended to 1959, and in Figure 16, the additional values agree with the earlier observations in showing no evidence of secular change.

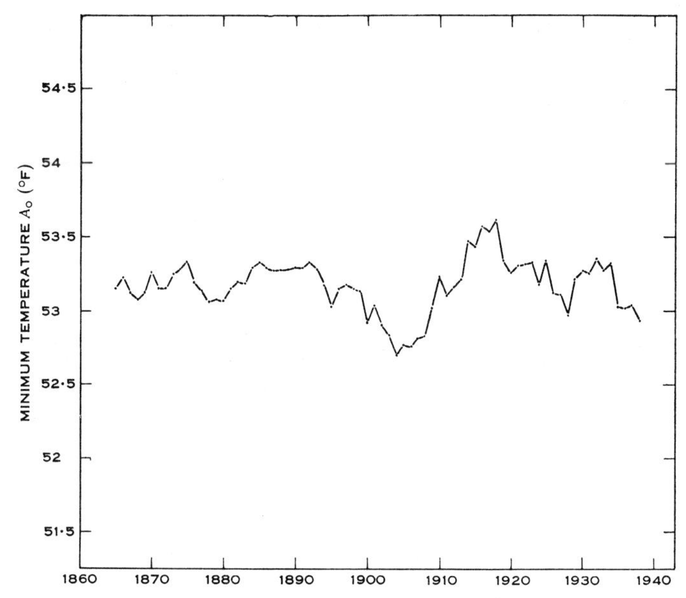

Fig. 18.—Ten-year means of the mean annual minimum temperature (A_0).

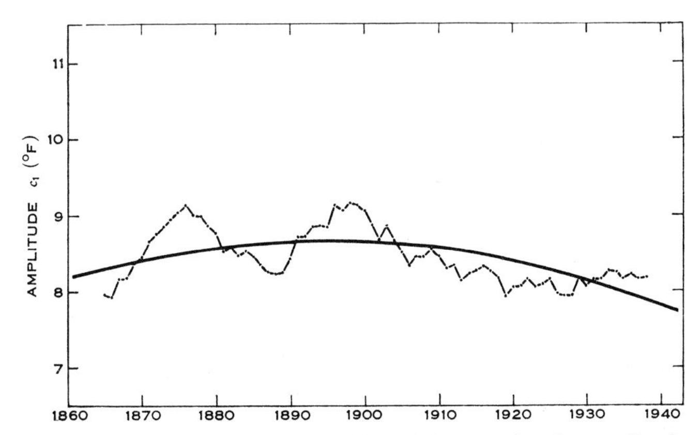

Fig. 19.—Ten-year means and polynomial curve for the amplitude of minimum temperature (c_1).

The fact that the additional observations on the amplitude of the first harmonic in the annual wave of maximum temperature conform with the previous trend provides strong confirmation of the conclusion that maximum temperature has undergone a genuine secular change. No attempt has been made to extend the polynomial fitting to include the additional estimates.

Owing to the difficulties encountered in the methods of recording the data, it has been impossible to obtain sufficiently reliable observations of minimum temperature to extend the analysis of this variate, and hence also the analysis of mean temperature and range.

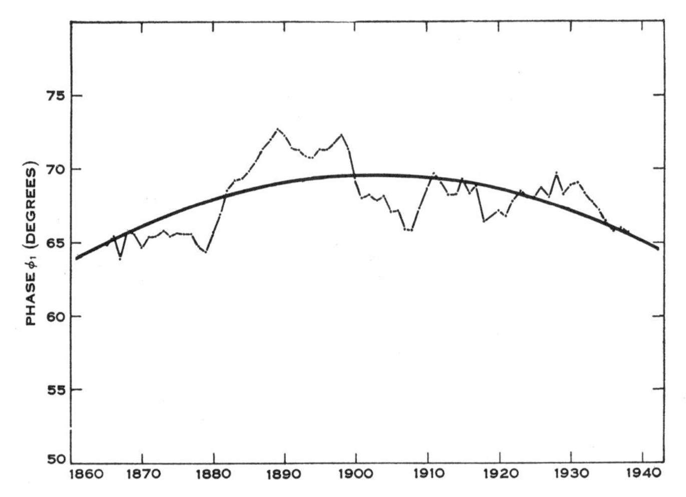

Fig. 20.—Ten-year means and polynomial curve for the phase of minimum temperature (ϕ_1).

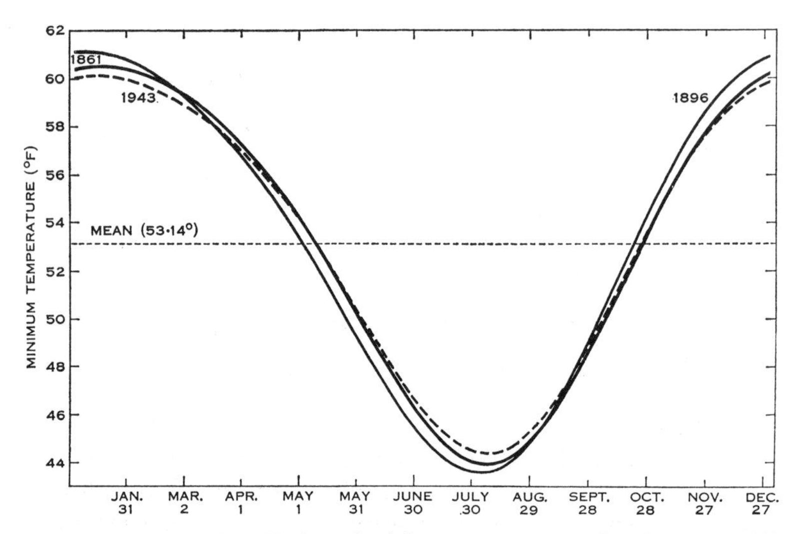

Fig. 21.—The annual oscillation of minimum temperature for the years 1861, 1896, and 1943, illustrating secular changes.

The fall in mean maximum temperature from 1930 onward agrees with the observation made by Deacon (1953) for the same period, but the rise in mean maximum from 1890 to 1930 is at variance with Deacon's result for the same years. Further investigation should explain the discrepancy, the explanation possibly depending on the location of the stations; Adelaide is on the coast some hundreds of miles west of the most westerly of Deacon's stations, which are all situated well inland.

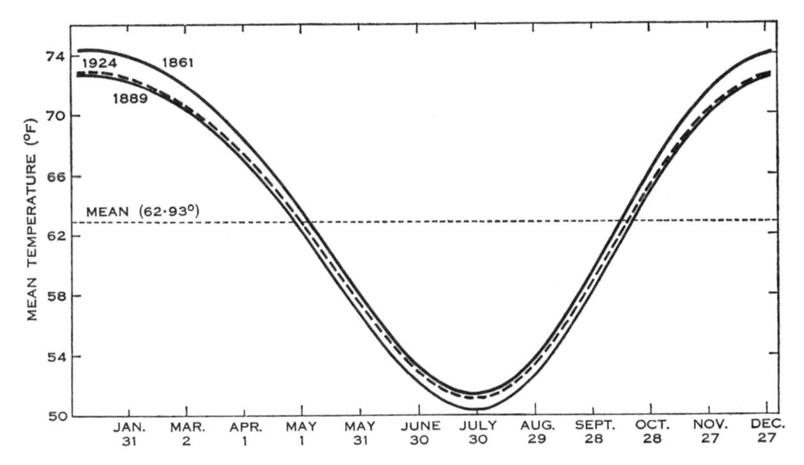

Fig. 22.—The annual oscillation of mean temperature for the years 1861, 1889, and 1924, illustrating secular changes.

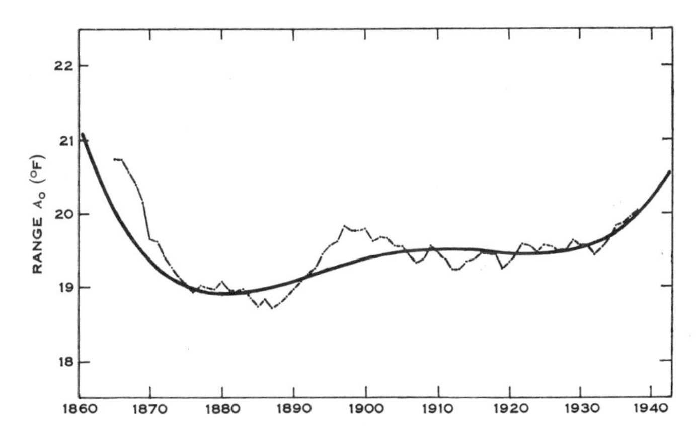

Fig. 23.—Ten-year means and polynomial curve of the mean annual range (A_0).

Figures 18, 19, and 20 give the 10-year means of the mean, and the amplitude and phase angle of the first harmonic of the minimum temperature, from which the years 1861, 1896, and 1943 have been selected. The annual courses of minimum temperature corresponding to these years are then

$$53\cdot14+8\cdot19\sin(\theta+63^\circ\,52')+1\cdot10\sin(2\theta+187^\circ\,29'),$$
$$53\cdot14+8\cdot66\sin(\theta+69^\circ\,21')+1\cdot10\sin(2\theta+187^\circ\,29'),$$
$$53\cdot14+7\cdot74\sin(\theta+64^\circ\,27')+1\cdot10\sin(2\theta+187^\circ\,29'),$$

from which Figure 21, illustrating the effects of the changes in the amplitude and phase angle of the first harmonic, has been constructed.

The secular changes of the mean temperature follow closely those of the maximum temperature, as given in Figures 14, 15, and 16, with appropriate adjustments to the several ordinates. Figure 22 illustrates the effects of the changes in the mean and the amplitude of the first harmonic, the selected years being 1861, 1889, and 1924, in which the annual sequences are represented by

$$63{\cdot}95+11{\cdot}38\sin(\theta+73^\circ\ 21')+1{\cdot}30\sin(2\theta+205^\circ\ 30'),$$
$$62{\cdot}64+11{\cdot}11\sin(\theta+73^\circ\ 21')+1{\cdot}30\sin(2\theta+205^\circ\ 30'),$$
$$63{\cdot}08+10{\cdot}78\sin(\theta+73^\circ\ 21')+1{\cdot}30\sin(2\theta+205^\circ\ 30').$$

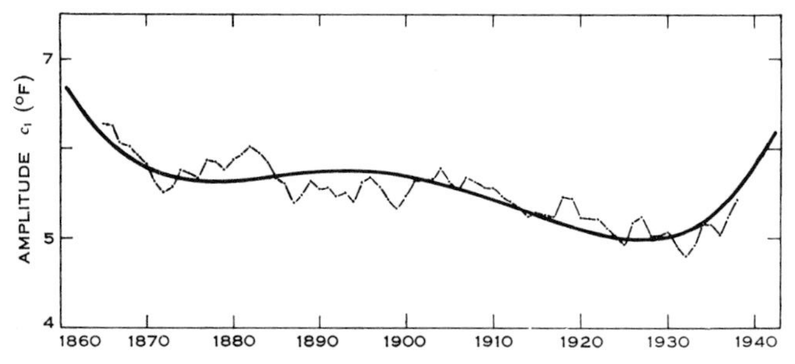

Fig. 24.—Ten-year means and polynomial curve for the amplitude of the range (c_1).

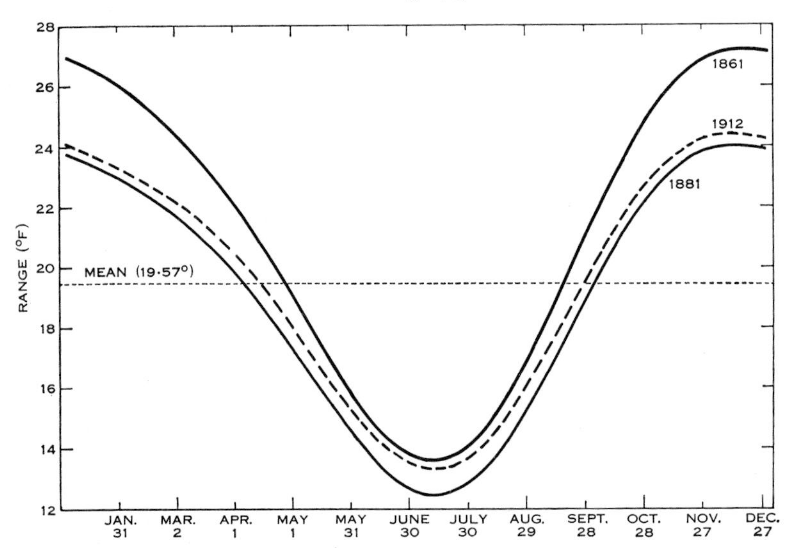

Fig. 25.—The annual oscillation of the range for the years 1861, 1881, and 1912, illustrating secular changes.

Figures 23 and 24 give the 10-year means of the mean, and the amplitude of the first harmonic of the range, from which the years 1861, 1881, and 1912 have been selected. The annual sequences corresponding to these years are

$$21{\cdot}08+6{\cdot}69\sin(\theta+91^\circ\ 12')+0{\cdot}99\sin(2\theta+232^\circ\ 17'),$$
$$18{\cdot}90+5{\cdot}65\sin(\theta+91^\circ\ 12')+0{\cdot}99\sin(2\theta+232^\circ\ 17'),$$
$$19{\cdot}51+5{\cdot}38\sin(\theta+91^\circ\ 12')+0{\cdot}99\sin(2\theta+232^\circ\ 17'),$$

from which Figure 25, illustrating the effects of the changes in the mean and the amplitude of the first harmonic, has been constructed.

(*i*) *Correlation Tables of Use in Comparative Climatology*

The correlation tables of the 61 subdivisions of the calendar year have been amalgamated to form 12 similar tables which correspond approximately with the calendar months, the 12th table including the odd 6-day period. These tables are presented as an appendix and provide a very convenient means for summarizing all the original observations.* It must be remembered that the variation exhibited in each table is not statistically homogeneous as it includes two components which can be separated as indicated (see above, Section III). It should be noted that any entry, such as 60°F, includes all original readings from 60·0 to 60·9.

V. Acknowledgments

The authors are indebted to the Deputy Director, Bureau of Meteorology, Adelaide, South Australia, for placing the temperature records at their disposal.

VI. References

Deacon, E. L. (1953).—*Aust. J. Phys.* **6**: 209–18.

Fisher, R. A. (1958).—"Statistical Methods for Research Workers." 13th Ed. (Oliver and Boyd: Edinburgh.)

Prescott, J. A. (1956).—7th Int. Grassld. Congr. N.Zld. 1956. Proc. pp. 24–31.

Prescott, J. A., Collins, Joyce A., and Shirpurkar, G. R. (1952).—*Geogr. Rev.* **62**: 118–33.

* Similar tables, with the 6-day periods in different combinations, will be made available on application to the Division of Mathematical Statistics, CSIRO, University of Adelaide, Adelaide, South Australia.

Appendix

Correlation Tables of Maximum and Minimum Temperatures

1861–1944, Periods 1–10

1861–1943, Periods 11–61

Periods 1–5

Minimum (°F)	Maximum (°F)																												
	61		63		65		67		69		71		73		75		77		79		81		83		85		87		89
45						1																							
46								1																					
47																1													
48													1	4	2	1		3							1				
49							1			1	1	1	2	1	2	2	2	1	2	1			1	1				1	
50					1			2	1		1		3	4	2	2	3	2	4	7	1		3		2				
51						1		1	1	1	4	5	4	4	3	6	5	3	1	4	3	2	1	1	3	4		1	
52							1		1	1	2	1	6	3	6	5	9	6	7	7	8	5	4	6	9	1	1	3	1
53	1					1	1	2	1	4	3	3	7	9	5	10	9	6	8	3	5	4	7	8	7	4	3	3	2
54	1			1	1		3	1	3	2	3	6	10	8	9	12	14	10	8	6	10	6	5	10	5	8	3	4	4
55			1				1	6	1	4	11	11	7	8	10	5	7	7	9	7	7	5	7	13	7	11	5	1	5
56	1								6	1	8	15	12	13	19	7	9	11	6	9	5	8	11	13	4	5	2	3	6
57					1	1	1		2	6	1	9	7	12	9	11	13	12	3	11	8	7	7	5	7	7	4	6	6
58							1		1	3	6	15	10	5	9	9	12	7	4	9	7	4	6	1	9	3	5	6	4
59					1				1	1	7	6	7	10	11	13	13	3	9	5	9	4	5	2	3	7	3	2	
60							1	1		1	3	4	5	6	12	6	7	9	2	8	2	3	4	1	5	6	3	5	1
61						1			2		1	2	5	2	4	5	5	4	3	4	1	4	4	3	4	4	2	5	2
62							1		1		1	3	1	2	5	3	5	5	6	3	1	1	4	2	2	1	5	3	1
63						1		1					3	1	7	6	3	3	3	4			1	5	2	3	1	3	2
64															3	2	1	2	4	2	1	2	1	1	1	4		3	2
65											1		2	1	3	1			2	1	2	1	2	3	1	1	1	2	1
66													1			2		2	1	3	1	4	2	1	1	1			4
67												1								2	1		1	2		2	1	1	
68										1	1				2	1	1	2	2					1		1	2	1	
69													1					1	1		1	1	1		1	3		1	
70														1		1		1		2		1	1		1		1	1	
71													1				2				1		2	1	1	1	2		1
72																1				2	1			1				1	
73																1		1				1	2		2		3		
74																								2			3		
75																				1				3			1		
76																			1						1				
77																						2							
78																													
79																											1		
80																													
81																													
82																													
83																													
84																													
85																													
86																													
87																													
88																													
89																													
90																													
Frequency	3	0	1	1	4	6	11	15	21	26	54	82	95	94	123	113	120	101	86	101	75	65	82	86	79	77	52	56	42

PERIODS 1–5

Maximum (°F)																												Frequency
	91		93		95		97		99		101		103		105		107		109		111		113		115		117	
																												1
																												1
																												1
																												12
																												20
																												38
1	2								1																			62
	1																											94
	1				2																							119
1			3	1	1	1																						160
3	4	1		2	1		2																					169
3	7	3	3	3	2		1			1	1																	198
3		6	2	3	2	6		1																				179
4	4			5	2	3	1	3	2	1	1	1	1															164
3	5	6	5	5	4	2	2	4	1																			159
7	2	2	2	1	3	4	3	4	1			2	1			1												128
1	7	5	3	5	1	2	1		3	3	1	1		2	1													103
1	1	5	6	2	2	6	2	4	3	2	1	2	1			1	1		1									97
4	2	2	5	4	3	3	5	3	2	2		1					1											86
4	3	4	4	1		1	3	3	2		5				2			2	1									64
2	4		1	2	4	7	2	4		3		3	2			1			1		1							62
2	4	3	2	4	3	4	2	3	3	5	5		1	2	2	1	1		1									71
	2			2	4	1		1	3	2	3	2	1	1			1					1						35
1	2	3	1	1	2	4	4	2	3	5		1	2	4	2	1	2	1										56
1	1	2	2	2	4	2	1	2	1	7	1	1	2	2	1	1	1	1	1			1						48
2		4		2		5	1	2	3	1	4		1		4	1			1	2		1						44
3	2	2	2		3	2	2	2		2	2		1	1	1		1			1	1							40
1		1	2	2	1			3	2	1	1	1	5	1		2	1			1							1	32
4		1		2	4	4	1	3	3	3	2	5	2	2		2	3						1	1				53
	1			1	1	1	3	2	2	2	2	1	6	3	1	1		1	1	1	1	1						37
				1					1	1	2	2	1	2	2	3	2				1		1					24
							1	1	1	5	1	2	3	1	3	1		3		1								25
	1	1			1	1			1	1	1		1		1	2	1	1		3	1		1	1				21
1				1		2		2		2	1	2	3	3	3	1			3	2		1	1			1		29
	1				1			1	2	2		1	1	2	2	1		1				1						17
1						1		1			1	3	1	1	2	1	1					1						14
	1										1	1	1			1	2	2	1	3	1							14
				1				2		1	1		1		1	1		3		2		1						14
		1								1		1	1		1	1								1				7
								1						1			1	3			1				1			8
													1						1	1		1						4
		1															1			1	1				1			5
																		1										1
																												0
									1			1							1									3
																1												1
53	58	53	43	53	51	62	37	54	41	53	37	34	40	28	29	25	20	19	13	18	8	9	4	3	2	1	1	2520

Periods 6–10

Minimum (°F)	Maximum (°F) 60	61	62	63	64	65	66	67	68	69	70	71	72	73	74	75	76	77	78	79	80	81	82	83	84	85	86	87
45															1													
46														1														
47												1		1	1	1		1										
48										1	1		2	1			2	3		1		1						
49						1		1		1	1	1			3		2	1	2									
50									2		2	3		1	8		2	2	2	4	1	1	3	3			1	1
51											1	2	3	1	4	3	4	3	10		2	2	4	4	2	1	1	3
52	1							1	4	1		1	2	3	4	5	9	6	1	7	4	6	3	3	1	1	1	
53					1				1	2	2	1	7	9	6	9	5	13	7	8	6	7	4	7	7	7	2	7
54							1		2	2	3	3	8	2	3	8	8	11	9	8	9	4	8	5	4	7	3	5
55				1	1	1			3	2	1	5	7	7	9	10	5	10	11	8	9	11	7	10	7	4	5	8
56							1		2	1	4	4	11	9	16	8	11	9	8	10	8	6	5	4	10	4	7	5
57					1			1	2	5	3	2	8	10	16	9	9	13	5	6	5	8	6	6	7	4	11	7
58							1	1	2	2	3	4	5	9	5	14	5	7	11	8	9	3	7	4	5	5	1	3
59									1	1		3	1	10	9	6	9	10	7	5	5	6	5	9	9	7	5	6
60										1	1	2	3	3	4	8	7	3	7	10	8	5	3	5	5	3	7	6
61						1					2	2	5	7	11	6	4	6	6	1	4	4	4	1	3	3	3	2
62								1	2		1	2	2		4	2	6	8	3	2	6	4	5	4	5	7	5	7
63								1		1		1	1		5		5	6	5	5	1	2	4	4	3	2	7	4
64												1		1	4	2	2	5	5		1	1	1	1		4		3
65									1			3	3	2	1		3	2	3	1	4	1	4	1	2	1	3	3
66								1				1		1			2	2	3	1		3	2	1	1	2	1	1
67												1			2		2	2	2		2		1	1	2	2	3	3
68														1	1				2	2			3	2	1	2	1	2
69																		1	1	1	1				1	1	1	1
70												1		1					2	1	1		1					
71																		1		2	2	1		1	1			
72															1							1	1	1	1	1	2	2
73																	1					1				1		2
74																		1						1				1
75																						1		1		1		
76																												
77																												
78																									1			
79																												
80																						1						
81																												
82																												
83																												1
84																												
85																												
86																												
87																												
88																												
89																												
90																												
Frequency	1	—	—	1	3	3	3	7	22	20	25	44	68	80	118	91	103	126	112	91	88	80	81	79	78	70	70	83

PERIODS 6–10

Maximum (°F)																										Frequency
88		90		92		94		96		98		100		102		104		106		108		110		112	113	
																										1
																										1
																										5
																										12
																										13
1																										37
																										50
1	1		1		1			1																		69
3	1		2			1																				125
1	5	2	1				1																			123
1	4	2	4	1	1																					155
5	1	2	6	6	4	3	2	2			1						1									176
9	4	6	6	2	4	2	2	3	1	2	1															186
6	8	3	4	2	1	3	2	2	3	1																149
4	3	5	3	6	4	5	3	2	2	3		1	1	1	1											158
6	5	2	3	6	7	2		3		1	3	2		1			1									133
6	5	5	5	5	1	4	4	7	5	3		1			1		1	1								129
2	2	1	3	3	6	3	4	3	2	2	1	1	2		1	2										114
2	1	5	4	3	4	5	2	6	5	3	3	2	1	2	1				1			1				108
3	1	6	1	4	5	3	1	2	1		4	3	1	2			2									70
3	1	3	1	3	2	4	3	5	3	4	1	4	3	1		2	1		1							83
2	3	3	1	4	5	4	1	3	6	4	2	1	1	1	2	2		1					1			69
1		2	4	1	4			1		3	3	4	2	3	2	1	1	2	1							58
	2	1	3	1	2	7	5	6	2	1	3	1	2		3	1	2	2		1				1		63
	3	3	2	2	4	1	4	2	5	3	2	3	3	2		1	1	2		1						52
2	2	2	3		1	5	1	2	2		1	2	4	3	2	2	1	2	1							45
	3	1	3		2	1	1	3	2	5	4		1	2	2	3		1			1					43
2		2	3	3	1	1	2	1	1	2	1	2	3	2		2	3	1	3		2					47
1	1	1		1		1	2	3	5	2	4	4	4	4	1	1	1	1	1				1			44
2			2			2	2	1	3	4	3		1	2	1	1	1		1	1		1				31
			1			2	1	2	3		3	2	2	1	1	6		1	2				1	1		32
	1		1		3		1	1		1	4	1	2	2	2		4	1			1		1			26
	1	1	1		1	1	1	3	1	2	1	1	2	1	2	2				1				1		23
				1	2	1		1			1	2			1	1		2		1	1	1	1			17
1			2		1				1		1	1	2	2	3		1	1		1		2	1	1		21
										1				1		3			2							8
													1		2	2	4	2	1							12
		1				1			1						1				3	3						10
						1								1	1	1				2	1					8
														1			1	1				1	1			5
								1											1	1						3
					1								1									1				3
																					1					1
														1												1
																									1	1
64	58	59	70	54	67	63	45	66	54	47	47	38	39	36	30	33	26	21	18	12	7	7	7	4	1	2520

Periods 11–15

Minimum (°F)	Maximum (°F)																										
	58		60		62		64		66		68		70		72		74		76		78		80		82		84
43															1												
44																											
45															1		1										
46												1	2		1	3	1										
47						1	1	1				1	1		3	2	2		2	1	1			1			
48	1						2				2	4	2	5	5	2	6	2			1	2	1				
49				1				1	3	4	1	4	7	5	3	4	5	7	3	2	1	1	2	2	2		
50							1	3			6	3	2	8	7	8	12	7	7	6	7	4	2	5	3		1
51							1	1	2	2	5	7	10	8	10	9	11	8	5	5	9	5	6	4	4	1	3
52									5	2	5	3	9	3	8	12	8	7	9	11	9	9	7	5	2	3	2
53						1	1	2	3	6	5	9	12	4	10	11	9	11	7	9	8	9	3	5	7	4	3
54								1	6	4	8	7	8	10	9	8	7	10	11	11	8	9	4	5	3	8	7
55					1	1	3	2	4	5	5	6	3	10	9	7	9	4	6	12	12	5	7	11	7	6	14
56								1	3	2	3	6	9	12	8	7	13	12	8	10	13	6	9	8	5	8	7
57									4	2	4	8	7	9	13	4	3	8	6	7	5	7	5	6	5	8	3
58								1	3	1	7	3	6	11	15	8	11	6	4	4	2	4	4	7	5	5	4
59					1		1		2	1	1	4	8	4	6	9	6	7	5	4	3	2	4	5	3	3	9
60								1		1	4	3	3	5	3	4	8	5	8	4	1	3	3	1	3	9	6
61						1	1		1		3	1	1	4	8	5	2	5	3		2	4	3	1	2	3	6
62									2		1		4	5		6	6	2	3	5	5	2	1	2	5	2	4
63														2		1	3	2	3	4		3	1	1	2	2	5
64											1			1	5	2	1	3	3	1	4	2	2	2	4	2	2
65														2	3	3	1	1	4	2	2		4		4	2	3
66									1			1		2	1	1	1	2	1	1		1		1	1	2	1
67												1	1					1	2				2	2			1
68													1		2		1	2			1	1	1	2	2		2
69																1			1		1			1			1
70															2	1		1	3		1	1	1	1	1		1
71														1								1			1		
72																					1	2				1	1
73																					1	1			1		
74																			1						1		
75																									1		
76																											
77																											
78																											
79																											
80																											
81																											
82																											
83																											
Frequency	1	—	—	1	2	4	11	14	39	30	61	72	96	111	133	118	127	113	105	99	98	84	72	78	74	69	86

Periods 11–15

Maximum (°F)																										Frequency
	86		88		90		92		94		96		98		100		102		104		106		108		110	
																										1
																										2
																										8
																										17
																										35
1			1																							60
	1	1																								94
2			2	1			1																			122
6	2	1		1			1																			130
2	3	3	1	1	2	1																				152
3	8	5	3	4	4	2					1	1		1												176
3	6	1	1	4	1	2	1					1														169
12	4	3	6	1	3	3	3	1	3		1															190
5	7	1	5	3	2	5		3	4			1														150
1	2	4	7	6	3	3	2		3	1	1	2	1	1												148
6	7	5	3	3	4	3	1	5			1		1	1	1		1									130
1	8	6	5	4	5	1	1	1	3	1	1		1	1		1										115
4	3	6	2	1	2	1	2	3			1		3	1	1											86
2	3	2	10	2	4	3	4	3	3	1	3	1	1		2											99
3	2	3	5	3	3	5	4		5		4	2			1											69
5	5	2	2	3	3	1	5	5	5	2	1	1	2		1	1		2								81
2	1	2		6	3	5		3	1	4	3	3		3	1	2	1	1								72
2	1	1	3	3	2	4	3	6	3		1	2	1	2	2		1	1		1						56
3	4	1	1		4	2	2	1	4	2	5	3	1	1	1				1							46
1		1		4		1	3	3		4	4	4	1	2	2											45
		1	4	2		3	1	4	4	3	1		1	2	1	1		1		1	1					36
1		1	1	2	1	3	1	2	1	2		1	2		1		2	1								35
				1		1	5			4	1		1	2	2	1		1								22
				1	1	1		2	4	4		3	3			1	2									27
	1	2							2			1		3	2	5		1							1	21
	1	2						1	1	2	2	2	4			2			2							21
			1			1	2	1	1	1					1	1		1								11
	1						1		1	1	1	1	1			1			1		2	2				13
									2							1	1	1	4				1			10
							1			1	1	1	2	2		1		1			1	1		1		13
	1								1		2		1	1			3		2							11
								1	1					1		1		1					2			7
																		1	1			1		1	1	5
																	1	1				1				3
																		1				1				2
65	71	54	63	56	47	51	44	45	52	33	35	30	27	24	19	19	12	15	11	2	4	6	3	2	2	2490

Periods 16–20

Minimum (°F)	Maximum (°F)																						
	52		54		56		58		60		62		64		66		68		70		72		74
40							1						1										
41									1				1			1			1				
42											1	1	1		1	2		1					
43							2		1	3			2	1	1	1		2	1				1
44									1	2	1	1	3	2	3	2	1	3	1	1	1		
45									3		3	1	4	2	4	7	2	5	2	5	4	1	2
46						1	1		3		3	4	7	5	2	2	5	4	2	2	5	4	4
47					1				1	5	2	5	5	12	3	12	5	6	17	5	9	6	2
48	1					1				4	5	7	4	11	6	4	8	11	11	10	16	6	5
49									3	1	5	7	8	10	7	10	12	5	8	10	9	10	10
50					1				1	6	4	5	10	10	8	10	9	6	13	12	10	8	8
51							1			3	4	4	10	9	10	12	12	8	7	16	6	5	14
52				1		2			1	3	5	4	11	12	10	14	9	4	9	13	8	6	13
53						1		1		1	4	9	12	14	15	20	15	7	7	16	10	7	8
54										1	2	8	14	8	14	15	19	10	13	4	8	7	6
55									1		1	5	6	14	9	16	13	15	11	8	10	4	3
56								1	1	1	1	2	3	7	15	9	12	10	9	7	14	5	6
57							1		2	1	1	2	4	6	8	9	8	11	15	8	9	4	2
58														2	1	11	6	12	7	4	7	3	8
59									1				6		5		6	3	9	5	6	6	2
60													1	2	2	2	2			5	3	5	2
61												1	1	1		1		2	2	2	5		3
62																1	1	3	3	2	3	1	5
63													2		1	1			1	2		3	1
64																				1		2	2
65																					3		1
66																	1				1		
67																			1				
68																		1					
69																				1	1		1
70																							
71																							
72																							
73																							
74																							
75																							
76																							
77																							
78																							
79																							
Frequency	1	—	—	1	2	5	6	2	20	31	42	66	116	128	125	162	146	129	150	139	148	93	109

PERIODS 16–20

Maximum (°F)																								Frequency
	76		78		80		82		84		86		88		90		92		94		96		98	
																								2
																								4
																								7
																								15
																								22
	1					1																		47
1	4																							59
4	2	1	1	1					1															106
1	4	2	3	2	1	1																		124
4	7	5	2	1	2		2	2	1															141
6	4	3	8	1	4	3																		150
7	10	10	4	3	1	3	3	2	1	1		1	1											168
8	2	7	4	6	6	6	6	1	3	3			1											178
6	6	5	4	9	2	5	5	4	2	2	2	1				1								201
5	9	6	4	4	5	3	1	3	3	1	3	1	1	1										179
7	8	1	2	3	7	4	5	5		6		2	3		1	1								171
12	4	6	3	2	3	1	8	7	5		3	2	3	1	3									166
4	5		2	3	2	4	3	2	4	5	6	3	2	2		2								140
6		3	3	1	2	5	5	5	4	2	3	2	3	2	3		3							113
4	3	3		3	4	1	5	4	6	3	3	2		6	2		1							99
3	4	3	1	2	5	1	2	5	4	1	2	2	3	4	1		1			1				69
1		2	1	2	3	5	1	1	2	3	3	4	3		1	1		1				1	1	54
2	3			3		3	1	4		3	3	2	3	2	1	3	1							53
4	3	2	2	1	4	5	1	3	1	6	1	3	2	4	4								1	58
1	2			1	2	2	2	3	1	3	2	1	1	7	1	1		2	1					38
1	1				2	1	2	1	3		3	1	4	4	2	1			1					31
1	3				1	1	1	1			4	2		1	1	1	1	1						21
		1	2	1	1		2		1	1	1	2		2	2			1		1				19
							1		1	2	2			1	1			1						10
			1	1			1					1		1		1		1		1				11
			1	1								1	1	1	2	1					1			9
	1				1					1	1	1	1	1		1	1		1					10
	1											1				2		1						5
				1						1				1					1		1			5
1										1														2
												2												2
																								—
																								—
																								—
																			1					1
89	87	60	48	52	58	55	57	53	43	45	42	37	32	41	25	16	8	8	5	3	2	1	2	2490

PERIODS 21–25

Minimum (°F)	Maximum (°F)																		
	52		54		56		58		60		62		64		66		68		70
36								1	1										
37					1	1		2			1								
38					1		1	2	1	1		1	1	1					
39					2	1	1	6	4	1	2	2							
40					2	4	2	2	3	3	4		1	1	1			1	
41						5	1	4	6	7	6	11	5	4	2	2	2		
42		1			1	1	1	8	6	8	10	2	12	8	5			1	
43					2	6	6	5	5	7	9	10	12	3	8	1	5		1
44			1		3	5	4	5	10	4	6	8	16	7	9	2	4	8	4
45			1	1		3	7	7	12	4	15	6	6	8	6	7	7	5	4
46	1		1	2	2	8	5	12	13	13	10	14	9	11	8	9	9	6	5
47				1	2	6	3	14	9	10	13	17	10	11	10	6	7	3	10
48		1			1	10	4	9	16	14	18	19	14	16	10	9	9	5	8
49		1	3	1	3	4	8	13	9	12	20	17	14	14	12	12	6	9	11
50					4	1	5	9	13	18	19	18	25	14	9	12	10	7	4
51					2		6	9	16	21	17	16	17	25	14	14	8	4	5
52				2	1	1	6	9	6	19	21	13	20	19	10	7	6	8	2
53							5	2	9	9	13	17	18	15	17	8	11	7	3
54						1			7	9	10	13	14	19	15	9	5	7	4
55					1		1	2		5	8	7	6	10	16	4	5	5	2
56									1	2	2	5	3	4	2	7	2		6
57								1			3	3	4	4	6	2	7	7	4
58										1		2	3	4	1	2	2	4	4
59											2	4		1			2	2	1
60									2			1		1			2		3
61											1						1	2	2
62																		1	
63												1		1	1				
64																			
65																			
66																			
67																			
68																			
69																			
Frequency	1	3	6	7	28	57	66	122	149	168	210	207	210	201	162	113	110	92	83

Periods 21–25

Maximum (°F)																			Frequency
	72		74		76		78		80		82		84		86		88	89	
																			2
																			5
																			9
																			19
																			24
																			55
	1																		65
1	1																		82
2	1				1														100
	1	1	1	1	1	1	1												106
9	9	4	2	1	1			1											165
6	2	3	4	2		1													150
5	8	4	4		3			2											189
7	4	2	3	2		1	1	2			1								192
5	4	4	5	2	2	3		1											194
5	6	8	6	1	1	5	2	2	2	1		1							214
3	5	2	6	2	2	2	2	1	1										176
5	4	10	5	5	4	2	2	2		1									174
5	6	2	5	6	4	2	1	3	1		2	1							151
1	7	2	5	3	1	3	5	2	2	3			1						107
3	6	4	1	4	6	2	6	1	2	2	1			1					73
6	2	5	2	6	1	1		2	2	1	2	1							72
2	3		2	2	4	3	1			1							1		42
1	3	4	3	3	2	1		3	1	2	1	1							37
2	3	1			1		3	1	2			1							23
4		1		2		2	2		1		3				1				22
	1		2		2				2			2							10
1					1	1						2	1	1		1			11
			1			1		1	1	2	1	1							8
			1		2									1					4
				1											2			1	4
				1						1									2
										1									1
					1												1		2
73	77	57	58	44	40	31	26	24	17	15	11	10	2	3	3	1	2	1	2490

PERIODS 26–30

Minimum (°F)	Maximum (°F)																													Frequency
	48		50		52		54		56		58		60		62		64		66		68		70		72		74		76	
32							1						1																	2
33																														
34							1	1	1	1																				4
35							1			3		2																		6
36					1	3	1	6	1	6	4	1	2	1																26
37						1	3	3	4	7	1	5	3	1	1	1	1													31
38			2			2	1		3	6	7	9	6	1	3	1	1	2												44
39				1		2	5	7	5	12	8	16	7	6	6	4		2												81
40					1	1	2	8	10	13	14	7	10	6	5	3	1	1	1											83
41		1			2	3	6	6	10	12	9	15	17	12	12	8	6	3	2	1	1									126
42	1			2		3	5	5	10	13	10	12	18	15	8	6	4	3	2				1							118
43	1	1	1	2	1	3	3	7	10	12	17	9	9	13	13	2	8	7	3	1		1	1							125
44			1			6	3	11	14	17	17	20	16	13	7	12	9	9	3	1	3		1	1						164
45					2	3	3	14	16	16	24	23	13	12	10	13	7	5	5	2		1	1							170
46		1			3	2	6	6	15	22	22	22	27	25	18	14	8	5	5	2	4			1	1					209
47			1		1	3	2	6	18	21	23	32	27	25	20	15	9	13	4	4	1	1	1	1	1					229
48					3	1	2	9	11	20	24	30	30	23	26	15	8	2	8	4		1		2						219
49					1	2	2	10	15	17	26	22	35	25	21	7	12	5	8	2	2	5	1							218
50						1	1	4	5	14	18	24	22	28	26	10	7	10	4	4	2	1	2			1				184
51						1	2	5	4	7	13	11	15	15	15	11	12	12	4	4	5	4		1						141
52							3		2	6	6	9	8	12	18	18	12	3	6	3	4	1		1						112
53								1	1	2	2	5	8	12	9	6	10	6	6	2	2	1	1		1	1				76
54									1	1	1		1	4	10	6	8	1	4	3	3				1					44
55								1	1	1	2		1	1	2	3	1	3	3	4	1	2		1						27
56												2	1	1		1	2	1	3	2					3	1			1	18
57												1	1		2	1		1	3			1	1							11
58														1		1	2	1				1		2						8
59																			1			1	1			2	1			6
60																							1							1
61																								2					1	3
62																							2				1			3
63																														
64																	1													1
Frequency	2	3	5	5	15	37	53	110	157	229	248	277	278	252	232	158	129	95	75	39	28	21	14	12	7	5	2	—	2	2490

Periods 31–35

Minimum (°F)	Maximum (°F)																												Frequency
	47		49		51		53		55		57		59		61		63		65		67		69		71		73	74	
32						1		1																					2
33									1		1																		2
34					1		1	2		2	2	2			1														11
35							3	2	4	5	2	1	1	1	1														20
36					1	2	4	6	8	6	9	3	6	4	5		1	1	1										57
37						4	3	7	8	15	12	13	5	5	5	1	4												82
38					3	4	4	4	12	13	13	10	8	10	3	2	2												88
39				1		4	5	12	12	14	19	16	10	6	7	8	3	1											118
40			1			6	3	6	9	18	17	17	15	5	9	6	4	3											119
41	1			1		4	4	8	12	22	25	17	19	10	14	6	7	1	3	2	1								157
42			2	1	1	5	8	8	12	20	19	23	17	11	11	6	6	1	3	1			1						156
43		1	1		1	6	9	15	15	28	26	34	22	19	12	10	8	2	8	3			1						221
44	1		1	1	1	4	6	9	16	28	20	32	29	15	10	7	5	5	3	3	1					1			198
45						4	5	14	19	22	25	28	30	23	18	16	5	9	2	5	1	2	1	1					230
46			1	1	1	5	3	6	10	24	32	27	28	30	18	10	8	7	7	3	5	1							227
47						2	2	7	8	23	17	27	27	29	21	19	8	9	8	2	2		2						213
48				1	1		4	5	4	13	23	23	28	25	15	6	7	9	5		2	3		1					175
49					1	2		3	5	10	11	16	15	21	20	16	8	7	2	4	4	1	1	1					148
50						1	1	1	1	6	9	10	16	8	6	11	6	3		4	1		1						85
51								3	2		7	11	8	10	9	7	3	4		2	3	1		1					71
52							1	1			1	2	8	6	7	4	2	4	2			2							40
53										1	1		1	1	4	6	5	4	1	2	2	1		1					30
54										2	1		2		2	3	3			2	4	1	1						21
55													1		2		1	1		1		1		1					8
56											1						1				1					1			4
57																			1							1	1		3
58																			1	1	1							1	4
Frequency	2	1	6	6	11	54	66	120	158	272	293	312	296	239	200	144	97	71	47	35	28	13	8	6	—	3	1	1	2490

PERIODS 36–40

Minimum (°F)	Maximum (°F)															
	49		51		53		55		57		59		61		63	
33								1								
34							1		2							
35			1			1	1	2	2	3		1	1	2		
36				1	1	1	3	7	6	3	4	3	2		1	
37						1	1	4	8	4	9	6	1	2	2	2
38					2	2	7	11	7	9	8	7	9	3	3	2
39			1		1	3	4	10	11	5	11	10	10	7	8	7
40				2	2	3	3	7	19	11	8	17	12	9	8	6
41			1	1	5	1	6	4	11	12	11	13	15	10	9	6
42		1		2	1	5	11	13	20	18	15	13	13	12	16	9
43	1	1			4	5	11	9	20	17	19	20	12	13	10	8
44			2	3	4	6	10	10	17	22	20	22	17	17	12	8
45					2	4	9	14	19	21	26	23	22	10	11	12
46				2	4	3	11	7	30	20	32	30	21	12	11	20
47	1		1	2	2	1	7	8	16	17	27	25	15	10	12	5
48	1				1	3	4	2	13	25	16	33	23	23	12	9
49				1	1	3	4	5	5	8	14	24	21	20	12	8
50					4	2		2	4	9	9	14	13	10	4	9
51					2	2		1	1	9	2	11	11	8	2	5
52					1		2		2	2	3	2	6	3	7	3
53						1		1	1	1	3	2		4	8	3
54						1	1	2	1	1		4	2	3	2	1
55								1			2		1	1	1	2
56												1	3		1	
57										1		1				1
58															1	
59													1			
60																
61																1
62																
63																
64																
65																
Frequency	3	2	6	14	37	48	96	121	215	218	239	282	231	179	153	127

Periods 36–40

Maximum (°F)																		Frequency
65		67		69		71		73		75		77		79		81	82	
																		1
																		3
																		14
		1																33
	1																	41
	3	1																74
1	1	1																91
5	2	1	3	2						1								121
8	3	4	3	1	1	2												127
3	8	6	5	1	2	2	1		1									178
10	8	4	3	3	2	1					1							182
7	7	9	3	2	4	1	2	1										206
6	5	8	7	6	1	3	1	2										212
4	10	6	5	4	3	2	1	1										239
7	8	6	5	4	5	2	1											187
4	4	11	8	1	4	2	1	1	3				1		1			206
10	7	4	8	2	3	1	1		2	1	1		1					167
5	6	5	4	4	2	4	2	1	3				1					117
5	7	3	2	7	3	4	3		1		2		1					92
4	2	2	4	4	3	3	1		1			1						56
4	1	2	2		3	2	2	1	2	1	1							45
2	3	3	1	1	3	2		2	1									36
1				1		1	2		1	1		1						16
	1		1	1	1		1			1								11
		2	1	1	2		1		1									11
1			1				1		1			1	2	2				10
					2				1									4
								1		1								2
				1														2
							1				1	1					1	4
									1									1
								1										1
87	87	79	66	46	44	32	22	11	19	6	6	4	6	2	1	—	1	2490

Periods 41–45

Minimum (°F)	Maximum (°F)																		
	50		52		54		56		58		60		62		64		66		68
34							1												
35									1	1									
36									1	3	1	1	3						
37					2	1			1	3	2	3		2			1		1
38				1		1	1	1	5	2	3	4	2	2		3	1		1
39				1	1		4	5	5	1	3	10	3	9	2	4	2	1	2
40	1				2	3	2	2	6	8	9	6	6	6	4	6	4	3	3
41				1		5	4	7	5	8	8	10	9	8	7	5	7	6	5
42		1	1		2	1	4	1	7	8	13	12	8	6	11	9	5	7	6
43					3	1	4	5	9	4	13	8	16	6	9	4	6	7	2
44					2	5	4	12	14	16	17	13	13	19	9	14	9	7	8
45				1		3	7	14	11	9	16	12	11	17	10	10	5	8	6
46						1	2	9	15	13	18	23	27	11	15	13	12	5	6
47				2	1	2	3	3	11	14	21	23	15	19	13	11	11	12	6
48				1		4	3	6	9	7	20	15	17	16	7	6	11	9	7
49					1		1	2	8	14	11	22	19	13	14	12	8	5	7
50				2	1	1		5	3	7	15	13	11	4	9	7	10	3	3
51				2	1	1	1	1	2	4	7	12	14	13	8	7	8	6	7
52				1	1	2	1		3	1	6	8	9	11	13	13	7	7	2
53									2	1	3	7	5	8	2	4	4	6	12
54						1					3		6	2	1	6	3	5	4
55								2		1	1	1	2	2	2	2		1	4
56									1				1	1	2	6		2	2
57											2		1	1	1	1		4	
58												1		2		1			2
59																	1	2	
60												2			1				
61														1					
62																		1	
63																			1
64																			
65																			
66																			
67																			
68																			
69																			
Frequency	1	1	1	12	17	32	42	75	119	125	192	206	198	179	140	144	115	107	97

PERIODS 41–45

Maximum (°F)																						Frequency
	70		72		74		76		78		80		82		84		86		88		90	
																						1
																						2
																						9
																						16
																						27
1	3	1										1										59
2	3		2				1															79
5	2	2	3	1	1																	109
6	5	4	1	2	3	1	1	1		1												127
2	3	2	6	5	2	2	2	1	1						1							124
6	2	2	4	4	3	3	2	1	2													191
7	7	7	5	2	3	2	2		1		3											179
5	4	5	2	4	4	5	2	2			1	1	1									206
1	3	3	4	6	4	3	3	7	2	2	2	1							1			209
4	5	10	1	5	4	3	3	2	3	1	2	2		1	1	1						186
4	9	5	8	8	5	5	3	3	2	1	2	1		1						1		195
5	4	7	1	3	5	5	3	4	2	2	3	2		3								143
1	3	2	3	5	3	1	3	3		3	1	5	1									128
6	1	6	4	3	4	3	1	3	4	2			2	1								125
5	2		2	1	4	3	2	5	3	4	2	1	1	1	1	3						94
6	3	5		3		5	2	1	1	1	1			1	2		1					63
3	2	1	3	6	2		1	1	3	2	1	2	1	1	1							48
1	4	2	2	4		4	3		4		3	2	1		1							46
1			1			2	3	1	4	3	1	2										28
1	1	1	1	1		2		1	1	2		1		1								19
	1			2			1			2		3				1						13
2	2	1		3	2	2	1	3	1	3							1				1	25
1	1		1		1	2					1											8
				1							2			2								6
					1		1				4				1							8
	1	1	1				1	1			1	1						1				8
		2		1				1			1			1				2				8
																						—
																						—
																						—
									1													1
75	71	69	55	70	51	53	41	41	35	29	31	25	7	13	8	5	2	3	1	1	1	2490

Periods 46–50

Minimum (°F)	Maximum (°F)																							
	54		56		58		60		62		64		66		68		70		72		74		76	
35					1																			
36		1											1											
37						1	2			1		1												
38						1	1		2		3	2	3											
39		1						3	2	2	4	1	1		1				1	1				1
40				1	2		1	3	1	6	4	1	3	1	1					1				
41		1			3	3	3	4	6	4	9	5	5	3	2		7	2	4	3		2	1	2
42		3	2		1	2	3	5	5	6	1	6	2	7	3	2	5	3	1	1	2		2	2
43		1	1		3		4	10	4	4	8	7	3	5	8	7	4	2	2	3	7	2	4	3
44			2		4	4	6		8	7	3	7	8	5	5	4	5	6	3	3	3	7	5	
45			1	2	2	5	6	7	12	4	12	9	11	7	10	8	6	11	4	11	6	2	5	3
46			1		4	9	5	11	13	13	11	11	7	10	9	5	8	1	9	6	7	4	4	1
47				2	1	2	6	8	11	9	14	10	12	9	4	8	4	6	7	4	5	7		7
48	1				4	5	7	7	5	14	19	9	12	11	10	12	4	1	4	3	6	4	11	1
49		1	1	1	2	5	5	10	6	10	18	14	12	9	9	8	4	3	9	4	2	8	2	1
50	1			1	1	1	2	10	9	9	8	18	12	10	8	9	4	7	5	6	8	3	2	3
51						3	3	5	4	12	16	13	11	17	7	5	6	8	8	3	6	3		2
52	1			1	2	1	2	3	2	7	12	10	8	11	11	4	8	6	5	4		4	5	1
53		1		1		2	2	4	4	5	6	8	12	6	7	8	2	2	4	4	2	6	3	4
54						2			6	3	5	5	10	3	8	9	5	6	3	3	3	3	2	1
55								2	1	3	1	4	5	7	4	5	4	2	2	6	1	1		2
56								1	1		3	3	1	6	4	5	1	4	1	6	1	3	3	1
57							2	1	1	1	1	3		3	1	2	5	4	1	3	5	3	2	2
58											1		2	2	6	2	2	5	3	1	2	1	1	3
59								1			1			3	1	1	2	2	1	1	1	2		2
60												2		1	3	1			1	1		2	1	1
61															3	1		1	1		2	2	4	1
62										1					2			1	3	1	1	1		
63												1					1		1		2	1		
64																	2	1			1	1		
65															1	2	1		1			1		1
66																	2							1
67															1		1			1				
68																						1		
69																					2			
70																								
71																								
72																								
73																								
74																								
75																								
76																								
77																								
Frequency	3	9	8	9	30	46	60	95	103	121	160	150	141	136	129	108	93	84	84	80	75	74	57	46

PERIODS 46–50

Maximum (°F) 78		80		82		84		86		88		90		92		94		96		98		100		102	Frequency
																									1
																									2
																									5
																									12
																									18
																									25
		2																							71
	2				1																				67
2			1																						95
5	4	1		1																					106
6	1	1	1			1	1	1			1														157
4	2	8	3	4	6		4		2						1										183
5	1	2	2		5	2		1	1		1														156
		7	3	2	4	1	1		1	1	1														171
1	6	2	5	3	3	2	2		1		1	1					1								172
1	6	2	3	2	2	2			3		3														161
2	1	7	5	4	3	2	3	1	2	2	1	1			1	1	1								169
3	3	3		2	4	2	2	2		1	1	2	2												135
4	4	1	2	2	5	4	2	1	1	1		3	2	1				1							127
4	4	3	3	3	1	1	4	2	2	1	1														106
1	1	1	3	1	1	2	2	2	1	2	2		1		2										72
1	7	3	1	2	2	1	2		2	2		1		1	1										70
3	5	1	3	1	4	1	5	2	2	1	1			1			1	1							72
2	4	5		3		3	2	1		2		1	4												58
1		2	1	1	2	3	2		3	1	3	1		2											40
2	1	1	2	4	1	2	3	2	3	2	2	3	2		1		1								45
	2		2	2	1	2		1	1			1		1					1						29
1	1	1	2		1		1	4		2	1	2	2			1						1			30
1		3	1	3	3	1	2		2	2	2	2	2	1		1									32
		1	1	4	2		1	1		4	1		1							1					22
			1	2	2			1	1	1		1	1			1									18
2		2		1		1	1	1	1	1					1				1						15
				1	2					2		1	1	1						1					12
				1			1		1		2	1	1	2		1									11
	1	1	1		2			1				1								1					10
						1									1							1			3
								1			1					1		1						1	5
																1									1
												1					1							1	3
1																	1								2
																									—
																									—
												1													1
52	56	60	46	49	57	34	41	25	30	28	25	24	19	10	8	7	6	3	2	3	—	2	—	2	2490

PERIODS 51–55

Minimum (°F)	Maximum (°F) 55	56	57	58	59	60	61	62	63	64	65	66	67	68	69	70	71	72	73	74	75	76	77	78	79	80	81	82	83
40						1						1	1																
41	1									1	1		1				1												
42								1			2	4		1	2	1		1				2		1					
43								1					3	1	3	3	1	3	1		1							1	
44			1	1	1	2	2	1	1	2	2	1	4	3	7	1	2	7	5	4	2	1		2	1	2	1	4	1
45							1	1		3	4	4	5	2	6	8	2	2	2	2	4	3		2	3	2		1	
46				1	1	1	1	2	4	2	2	4	4	4	7	6	6	5	10	6	5	2	6	5	5	4	3	1	2
47			1			3	1	1	3	3	5	3	5	5	6	4	9	5	9	2	8	2	4	3	5	1	4	2	5
48				1			4	3	3	7	7	8	11	11	8	5	5	10	12	5	3	5	3	3	4	3	2	3	6
49						1	1	3		5	3	9	10	7	10	11	5	4	6	5	5	7	6	9	6	4	4		3
50							2	2	2	10	10	7	8	14	8	7	7	2	4	6	5	3	5	2	8	4	9	4	5
51					1	1		2	6	4	8	7	15	18	16	11	9	11	6	5	6	3	1	6	3	5	5	2	3
52						1		4	3	5	5	10	8	12	15	11	7	14	11	7	5	7	5	3	4	1	4	7	1
53								1	3	3	7	10	5	15	13	7	8	7	7	7	8	1	2	4	7	1	2	2	4
54						1			2		5	10	5	14	11	14	11	9	7	6	8	5	5	2	6	2	3	2	6
55						1		1	1		5	3	4	11	10	14	6	6	12	7	8	3	2		4	2	3	2	
56										1	1	6	5	3	11	9	5	6	3	3		4	6	3	2	3	2	3	4
57						1		1	1			1	1	4	1	3	7	5	5	1	3	6	4	4	1	3	1	3	3
58							1				1	1	1	1	4	3	4	3	4	2	3		2	1	3	2		2	3
59								1			1	1	2	3	3		7	6	5	1	1	1	1	2	2	2		2	3
60								1	1				1	1	3	2	7		1	1	1		3	1	5	5	2	1	1
61												2	1	1		3	3	3	3	1	5	1	1	1		1		2	
62													1		2	3	2	2	3		2		2	2	3	2	1	4	1
63																1		3			3	1			2	1		2	2
64													1	2			1	1	1	1		1	1	1	1		1	4	4
65															1		2				1	1	1	1		1	1	1	1
66																1			1				1				1		
67																							1	2				1	
68																			1							1			
69																							1	2		1		1	
70																				1		1		2					
71																						1			1	1			
72																												1	
73																								1	1				
74																								1					1
75																												1	
76																													
77																													
78																													
79																													
80																													
81																													
82																													
Frequency	1	—	2	3	3	13	13	26	30	46	69	92	102	133	147	128	117	115	119	73	87	61	63	66	77	54	49	59	59

Periods 51–55

Maximum (°F)																														Frequency
	85		87		89		91		93		95		97		99		101		103		105		107		109		111		113	
																														3
																														5
																														15
																														18
2																														63
		1		1																										59
3			1																											103
1	1		1		1	1		1																						105
	4	1	2	3				1																						143
5	2	1	3	5	2	1	1	2																						146
6	4	4		2	3		2		5																					160
6	7		5	5	2	5	2	3	1	2																				192
1	4	2	6	2	2	5	3		1	2																				178
2	2	1	3	1	3	1	2		2	1	1				1															144
3	5	3	3		3	2	3	3	1		1	1				1														163
3	1	3	8	2	2	1	1		3	2	1	2																		134
1	2	5	4	6	4	1		3		1	1		1																	109
3	1	3	1	3	4	3	2	2	1	1	2		1	2	2															90
3		1		4	3	2		2		3	1	1			3	1						1								66
1			2	2		1	3		2		3	1	1	1																61
2	2	4	2	1	2			1	2	2		1			1	2														59
2	2	2	2	2	1	4	4	4	1	1	2		1	1				1												58
2		4	2	3	1	1	2	3	1	3	3	1	3	1							1									61
2	1	2	2	2	1	4	1		5		1	2	1	1	1	1														42
1	2	3	1	3		4	3	3	2	3		1	2	1	3	1														53
	1	1	2	1	2	1	2	2	3	3	2	3	1	3	2	1	1													42
	3	5	3	1	1	1	3	1	1	1	1	1	3	1	2	1		1	1	1										36
2	1	1	2	2	1	1	1	2	1	2	2	1	2		1															26
2		1		2		1	4		2	1		1	1		1		3													21
1	1		1	3		2		2	1	1		1	2	1	2	1	1		1		1									27
1	1		2							1	2	1	1	3	2	1		1			1									21
			1				1		1		4	2	2	1				1												16
			1		1	1	1	1	1			1					1		2		2									13
					2				1				2		1	2		1	1				1							13
											2				1		2	2												9
1			1		2	1			1		2	1				1	1	2	1											15
			1				1															1	1							4
						1		1						1		1	1		1										1	7
				1			2					1																		4
											1									1										2
																			1		1									2
																						1								1
																		1												1
56	47	48	62	57	43	45	44	37	39	30	32	23	24	17	23	14	10	10	8	2	6	3	2	—	—	—	—	—	1	2490

PERIODS 56–61

Minimum (°F)	Maximum (°F)																										
	61		63		65		67		69		71		73		75		77		79		81		83		85		87
43						1				1																	
44		1				1					1	1		1	1										1		
45			1			1		1	1	1	1	1	2	1													
46		1			1			1		1	2	1	1	1	2	1	1		1	2		1		1			
47							1		1		1	2	1	3	2	3	4		1	2	1		1	3			
48						1	2	2	1	3	6	2	3	7	1	7	4	3	1	3	3	1	3	1		1	1
49					1	3	2	2	2	3	2	3	5	4	8	5	2	9	7	4	3		1	3	2		3
50	1				1	1	6	1	4	6	4	8	10	10	3	7	5	2	6	9	5	6	4	3	1	6	2
51					1	2	2	4	6	8	11	10	7	10	9	4	10	9	6	8	4	9	4	5	1	2	4
52				1	2	4	7	9	11	8	13	11	11	11	4	9	13	12	9	9	9	8	8	8	9	5	7
53				1	5	2	5	6	10	13	14	19	10	17	16	9	12	5	4	6	7	6	8	6	7	8	4
54		1	1			3	5	8	9	13	11	14	14	9	13	12	8	5	10	4	3	7	5	5	3	5	7
55						2	4	5	9	14	14	16	10	9	12	6	12	7	11	4	6	3	3	5	5	4	8
56				1	1		2	3	7	11	22	9	11	14	13	16	11	5	5	6	1	8	6	4	4	11	4
57						1	1	7	5	11	2	11	9	9	10	8	12	3	7	5	9	6	4	3	2	6	4
58			1				1		3	2	9	10	13	5	11	8	4	6	3	1	5	7	3	7	4	4	5
59	1				1			1	1		4	3	6	10	9	9	6	5	5	3	2	5	2	3	5	4	5
60				1			1	1	1	1	1	3	4	9	10	1	2	3	3	3	2	1	1	4	2	1	2
61									3	3	3	3	3	3	8	3	6	5	13	2	7	1	4	1	1	3	1
62								2	1	1	1	3	2	5	1	4	3	3	2	3	2	1	2	1	3	1	2
63							1	1	1				3		3	4	2	1	1	1		2	5	1	2	4	5
64									1		1		2	3	2	2	1	3	1	4	3	1	2	1	3	1	1
65													1	1	1	1	3	4	1	3	1	2	1	4	3	4	2
66									1			2		1		1	1	1	1	2	1	2	3	2	2		1
67																2		1	1	1	3		1	2			
68														1	1		1		1	1	2	3		2	1	2	3
69														1		1	1		2	1	1	3		2		2	
70																1			1		1	1		2	1	1	
71														1				2	1	1		1	3	3	1	2	2
72																		1	1			1	2		1		
73																	1		1		1	1		1		1	1
74																						1					
75																		1								1	
76																											1
77																											
78																											
79																											1
80																											
81																											
82																											
83																											
84																											
85																											
86																											
87																											
88																											
Frequency	2	3	3	4	13	22	40	54	78	100	123	132	128	146	140	124	125	96	106	88	82	93	76	83	64	79	76

Periods 56–61

Maximum (°F)																											Frequency
	89		91		93		95		97		99		101		103		105		107		109		111		113	114	
																											2
																											7
																											10
																											18
	1																										27
																											56
1	1	2																									78
1								1																			113
3		1	2	1	1	1	1	1																			147
3	6	2	3	1	1	1	2	1																			218
3	5	4	2	3	1	2				3	1																224
1	4	3	1	1		1	1	2			2		1														192
2	5	3	7	2	3		6	1	2	1																	206
6	2	3	3	6	4	2	4	2			1	1	1	1													211
3	5	2	7	6	1	3	1	2	3	1	2	1	1														173
1	2	4	3	7	6	4	2	4	2	1	2		2						1								153
8	3	6	1	1	5	3	3	2	1	1		2	2			1			1								130
3	5	5	1	7	5	4	3	2	1	3	3	2	1		2												104
1	2		3	2	2	3	2	2	4	2	7	1		1	3			1									109
4	3	4	1	7	6	3	3	2	3	6	2	1	4	3		1	2										98
1	4	4		4	3		3	4	3	3	5	3	1	2			1	1	1								80
	7	2	3	1	5	2	1	1	3	3	1	2	1	1	5	1		1		1							73
2	7	2	2	3		1	4	3	3	2	4		5	2	1	1	1		1	1							77
2	1		3		4	3	1		2	1	1	7	2	1	1				1			1					52
3	2	2	6	1	3	2	4	1	2	2	1	1	1	3	4	4	1		1								55
1	1	2	2	2	2	4	1		4	1	4	3			3		1	1		1	1						52
1	1	1	3		2	2	2			3	2	1	3	1		1	1		2					1			41
		1	1	4	1	4	4	2	4	1	2	2	2	1	2	3		1	2		1	1					47
1		1	1	1		1	3	2	2		4	2	3	2	2			1			1					1	45
1		3		2	2		3	3		1	2	1	1	2			1										28
						1	1					2		2	1	2	1				1						18
1		1	1	1	2	1		2	1		1	2	2	3	1	2		1									23
2	1			1			1	2		4	1	1	1	1	3		2	2				3				1	28
		2		1			1	2			3	1	1	1	3	2	2	1				1					22
							1		2				1		1		1	2	2	1	1	1					13
	1					1	1	1		1		1				3	1		1								11
		1			1				1	1	2	1	1				1	1		2							13
	2						1				2	1			1		1	1	1	1	1			1			13
						1										1											2
														1			1	1	1			1		1			6
											1								1		1						3
																			1	1							2
																			1	1			1	1			4
																					1						1
																					1						1
																						1				1	2
55	71	61	56	65	60	50	60	45	43	41	56	39	37	28	33	22	18	15	18	9	9	9	1	4	0	3	2988

REPRINTED FROM
MULTIVARIATE ANALYSIS
AF33 (615)-3016

ACADEMIC PRESS INC., NEW YORK

A Multiple Behrens-Fisher Distribution

E. A. CORNISH
DIVISION OF MATHEMATICAL STATISTICS
C.S.I.R.O.
ADELAIDE, AUSTRALIA

1. RECAPITULATION OF THE UNIVARIATE CASE

If $\bar{x}_1$, with an estimated variance $s_1{}^2$ based on n_1 degrees of freedom is the mean of a sample of observations from a normal distribution whose mean is ξ, then

$$\xi = \bar{x}_1 + s_1 t_1,$$

where t_1 is distributed in Student's distribution with n_1 degrees of freedom. Similarly, for a sample of observations from a second population with the same mean ξ, the corresponding quantities are related by

$$\xi = \bar{x}_2 + s_2 t_2,$$

and t_2 is distributed in Student's distribution with n_2 degrees of freedom, independently of t_1.

Consequently, on the null hypothesis under consideration, $\bar{x}_1 - \bar{x}_2$ will be distributed as is

$$s_1 t_1 - s_2 t_2,$$

where s_1 and s_2 are independent and have values fixed by the observed data, and t_1 and t_2 have a known simultaneous distribution.

Since $d = \bar{x}_1 - \bar{x}_2$ is known, the null hypothesis may be tested by calculating the probability from the integral

$$\int\int \frac{\Gamma[(n_1+1)/2]}{(\pi n_1)^{1/2}\Gamma(n_1/2)} (1 + t_1{}^2/n_1)^{-(n_1+1)/2}$$
$$\times \frac{\Gamma[(n_2+1)/2]}{(\pi n_2)^{1/2}\Gamma(n_2/2)} (1 + t_2{}^2/n_2)^{-(n_2+1)/2} \, dt_1 \, dt_2$$

taken over the domain defined by

$$s_1t_1 - s_2t_2 > d$$

(Fisher [1, 2]).

2. A MULTIVARIATE DISTRIBUTION

The above result may be generalized readily to take account of multiple variates. If $\bar{x}_{11}, \bar{x}_{12}, \ldots, \bar{x}_{1p}$, each with estimated variance s_1^2 based on n_1 degrees of freedom, are independent means of samples from independent normal populations with means $\xi_1, \xi_2, \ldots, \xi_p$, respectively, then

$$\xi_j = \bar{x}_{1j} + s_1t_{1j} \qquad (j = 1, 2, \ldots, p),$$

where the t_{1j} are distributed in the multivariate Student's distribution

$$\frac{\Gamma[(n_1+p)/2]}{(\pi n_1)^{p/2}\Gamma(n_1/2)}(1 + \mathbf{t}_1'\mathbf{t}_1/n_1)^{-(n_1+p)/2}\, d\mathbf{t}_1$$

with n_1 degrees of freedom, in which $\mathbf{t}_1$ is the column vector with elements $t_{11}, t_{12}, \ldots, t_{1p}$.

Similarly, for samples of observations from a second set of normal populations with the same means ξ_j, the corresponding quantities are related by

$$\xi_j = \bar{x}_{2j} + s_2t_{2j}$$

and the t_{2j} are distributed independently of the t_{1j} in the multivariate Student distribution

$$\frac{\Gamma[(n_2+p)/2]}{(\pi n_2)^{p/2}\Gamma(n_2/2)}(1 + \mathbf{t}_2'\mathbf{t}_2/n_2)^{-(n_2+p)/2}\, d\mathbf{t}_2,$$

where $\mathbf{t}_2$ is the column vector with elements t_{2j}.

Consequently the $\bar{x}_{1j} - \bar{x}_{2j}$ will be simultaneously distributed, as are the

$$s_1t_{1j} - s_2t_{2j},$$

where s_1 and s_2 are known and fixed by the observations and the t_{1j} and t_{2j} have a known simultaneous distribution.

Since the vector $\mathbf{d} = \bar{\mathbf{x}}_1 - \bar{\mathbf{x}}_2$ is known, the null hypothesis may be tested by calculating the probability from the integral

$$\int \cdots \int \frac{\Gamma[(n_1+p)/2]}{(\pi n_1)^{p/2}\Gamma(n_1/2)}(1 + \mathbf{t}_1'\mathbf{t}_1/n_1)^{-(n_1+p)/2} \times \frac{\Gamma[(n_2+p)/2]}{(\pi n_2)^{p/2}\Gamma(n_2/2)}(1 + \mathbf{t}_2'\,\mathbf{t}_2/n_2)^{-(n_2+p)/2}\, d\mathbf{t}_1\, d\mathbf{t}_2$$

taken over the domain defined by

$$s_1\mathbf{t}_1 - s_2\mathbf{t}_2 > \mathbf{d}.$$

The generalization may be carried a step further to take account of multiple variates formed by taking linear functions of the original means. Suppose $\bar{\mathbf{x}}_1$ is a column vector with elements $\bar{x}_{11}, \bar{x}_{12}, \ldots, \bar{x}_{1p}$ and $\boldsymbol{\xi}$ is the column vector of population means $\xi_1, \xi_2, \ldots, \xi_p$.

Let

$$\begin{aligned}\mathbf{y}_1 &= \mathbf{H}\bar{\mathbf{x}}_1 \\ &= (h_{ij})\bar{\mathbf{x}}_1\end{aligned} \qquad (i = 1, 2, \ldots, q; \; j = 1, 2, \ldots, p)$$

and

$$\boldsymbol{\eta} = \mathbf{H}\boldsymbol{\xi}$$

be any $q < p$, linearly independent, linear functions of the $\bar{x}_{1j}$ and the ξ_j, respectively; then the elements of $\mathbf{y}_1$ are distributed in a multivariate normal distribution with a vector of means $\boldsymbol{\eta}$. The elements of $\mathbf{H}$ are known constants satisfying the conditions

$$\sum_{j=1}^{p} h_{ij} = 0 \qquad \text{for} \quad i = 1, 2, \ldots, q.$$

If, as before, $s_1{}^2$ is the estimated variance of the $\bar{x}_{1j}$ based on n_1 degrees of freedom, distributed independently of the $\bar{x}_{1j}$ and therefore independently of $\mathbf{y}_1$, then

$$\boldsymbol{\eta} = \mathbf{y}_1 + s_1\mathbf{Q}^{-1}\mathbf{t}_1$$

where $\mathbf{Q}$ is a diagonal matrix whose ith diagonal element is $(\mathbf{h}_i\mathbf{h}_i')^{-1/2}$, $\mathbf{h}_i$ being the ith row of $\mathbf{H}$, and the distribution of the t_{1j} takes the form

$$\frac{\Gamma[(n_1 + q)/2]}{(\pi n_1)^{q/2}\Gamma(n_1/2)}(1 + \mathbf{t}_1'\mathbf{R}^{-1}\mathbf{t}_1/n_1)^{-(n_1+q)/2}\frac{d\mathbf{t}_1}{|\mathbf{R}|^{1/2}},$$

in which $\mathbf{R} = \mathbf{Q}(\mathbf{H}\mathbf{H}')\mathbf{Q}$ is the nonsingular, symmetric correlation matrix of the elements of $\mathbf{y}_1$, the correlation ρ_{ij} being

$$\frac{\mathbf{h}_i\mathbf{h}_j'}{(\mathbf{h}_i\mathbf{h}_i' \,.\, \mathbf{h}_j\mathbf{h}_j')^{1/2}}.$$

Similarly, from the observations of the second set, assuming the original populations have the same means $\xi_1, \xi_2, \ldots, \xi_p$, the corresponding quantities are related by

$$\boldsymbol{\eta} = \mathbf{y}_2 + s_2\mathbf{Q}^{-1}\mathbf{t}_2$$

and the t_{2j} are distributed independently of the t_{1j} in the distribution

$$\frac{\Gamma[(n_2 + q)/2]}{(\pi n_2)^{q/2}\Gamma(n_2/2)}(1 + \mathbf{t}_2'\mathbf{R}^{-1}\mathbf{t}_2/n_2)^{-(n_2+q)/2}\frac{d\mathbf{t}_2}{|\mathbf{R}|^{1/2}}.$$

Consequently the elements of $\mathbf{y}_1 - \mathbf{y}_2$ will be distributed simultaneously, as are the elements of

$$s_1\mathbf{Q}^{-1}\mathbf{t}_1 - s_2\mathbf{Q}^{-1}\mathbf{t}_2,$$

in which s_1, s_2, and $\mathbf{Q}$ are known and $\mathbf{t}_1$ and $\mathbf{t}_2$ have a known simultaneous distribution.

Since the vector $\mathbf{d} = \mathbf{y}_1 - \mathbf{y}_2$ is known, the null hypothesis may be tested by calculating the probability from the integral

$$\int \cdots \int \frac{\Gamma[(n_1+q)/2]}{(\pi n_1)^{q/2}\Gamma(n_1/2)}(1+\mathbf{t}_1'\mathbf{R}^{-1}\mathbf{t}_1/n_1)^{-(n_1+q)/2} \times \frac{\Gamma[(n_2+q)/2]}{(\pi n_2)^{q/2}\Gamma(n_2/2)}(1+\mathbf{t}_2'\mathbf{R}^{-1}\mathbf{t}_2/n_2)^{-(n_2+q)/2}\frac{d\mathbf{t}_1\,d\mathbf{t}_2}{|\mathbf{R}|}$$

taken over the domain defined by

$$s_1\mathbf{Q}^{-1}\mathbf{t}_1 - s_2\mathbf{Q}^{-1}\mathbf{t}_2 > \mathbf{d}.$$

The transformation

$$\begin{bmatrix}\mathbf{t}_1\\ \mathbf{t}_2\end{bmatrix} = \begin{bmatrix}\mathbf{P} & \cdot\\ \cdot & \mathbf{P}\end{bmatrix}\begin{bmatrix}\boldsymbol{\tau}_1\\ \boldsymbol{\tau}_2\end{bmatrix},$$

with $\mathbf{P}'\mathbf{R}^{-1}\mathbf{P} = \mathbf{I}$, reduces this integral to

$$\int \cdots \int \frac{\Gamma[(n_1+q)/2]}{(\pi n_2)^{q/2}\Gamma(n_1/2)}(1+\boldsymbol{\tau}_1'\boldsymbol{\tau}_1/n_1)^{-(n_1+q)/2} \times \frac{\Gamma[(n_2+q)/2]}{(\pi n_2)^{q/2}\Gamma(n_2/2)}(1+\boldsymbol{\tau}_2'\boldsymbol{\tau}_2/n_2)^{-(n_2+q)/2}d\boldsymbol{\tau}_1\,d\boldsymbol{\tau}_2$$

taken over the domain defined by

$$s_1\mathbf{Q}^{-1}\mathbf{P}\boldsymbol{\tau}_1 - s_2\mathbf{Q}^{-1}\mathbf{P}\boldsymbol{\tau}_2 > \mathbf{d}.$$

At this juncture we may distinguish three important cases which may occur in practice:

(a) The rows of $\mathbf{H}$ are orthogonal, but not normalized, in which case $\mathbf{HH}' = \mathbf{Q}^{-2}$.

(b) The rows of $\mathbf{H}$ are orthogonal, each having the same norm v, in which case $\mathbf{HH}' = v\mathbf{I}$.

(c) The rows of $\mathbf{H}$ are orthogonal and normalized, in which case $\mathbf{HH}' = \mathbf{I}$.

Since $\mathbf{HH}'$ is positive-definite and symmetric, the general case may be reduced to any one of the three forms (a), (b), and (c), and equally either (a) or (b) may be reduced to (c). The course to be followed depends upon the nature of the enquiry; orthogonal comparisons, for example, may not provide just what is wanted.

3. DIFFERENT RESIDUAL VARIANCES IN THE SAME EXPERIMENT

It may happen that in the presence of one element of treatment the residual variance is much less than in its absence. Even in agricultural experiments it is possible, although it is not characteristic, that the experimental area is severely deficient in, for example, nitrogen, so that in portions of the experiment not receiving a nitrogenous manure, there may be much variation from plot to plot, owing to variation in the small amount naturally present, whereas under a nitrogenous dressing this cause of variation may have much less influence. We may imagine a set of varieties of a particular crop to be compared experimentally in respect to their reaction to nitrogen under the conditions mentioned above. In such cases the significance of the effect of nitrogen will be in little doubt. The comparison of the average yield of the dressed plots with that of the undressed plots, will, nonetheless, be essentially a comparison between the means of samples from normal populations with unequal variances, for which the univariate Behrens-Fisher distribution provides the test of significance. The average yield with and without nitrogen will not in such cases generally be of interest.

However, the significance of the interaction of nitrogen with varieties may not be obvious, and only the Behrens-Fisher test will be satisfactory for this. The components of this interaction may be tested individually or jointly, using, for the latter test, the multiple Behrens-Fisher distribution. Moreover, the main effects of varieties and any other factors, and their interactions among themselves, if not judged solely from the plots receiving nitrogenous dressings, may seem to have the error distribution of the weighted mean of two normal samples. Although tests of significance are available for reasonably large numbers of degrees of freedom (say 15 for each sample), the exact formulas are complicated [3] and the experimenter should give consideration to the possibility that his requirements will be best met by ignoring the more variable plots altogether (those without nitrogenous dressing in the illustration being considered).

The *unweighted* mean of the effects of varieties, with and without nitrogen, will, in these cases, have an error distribution given exactly by the Behrens-Fisher test. The use of the weighted mean is appropriate only if, as would not be usual in agricultural trials, the interaction was *known* to be zero.

REFERENCES

1. FISHER, R. A. (1935). The fiducial argument in statistical inference. *Ann. Eugenics* **6** 391–398.
2. FISHER, R. A. (1941). The asymptotic approach to Behren's integral, with further tables for the *d* test of significance. *Ann. Eugenics* **11** 141–172.
3. FISHER, R. A. (1961). The weighted mean of two normal samples with unknown variance ratio. *Sankhyā* **23** 103–114.

Commonwealth Scientific and Industrial Research Organization, Australia. Division of Mathematical Statistics Technical Paper No. 23, Melbourne, 1967.

THE SAMPLING AND FIDUCIAL DISTRIBUTIONS OF A STANDARD NORMAL DEVIATE

By E. A. Cornish*

[*Manuscript received May* 10, 1967]

Summary

The fiducial distribution of a standard normal deviate is derived and its form compared with the sampling distribution of a standard normal deviate. The fiducial distribution rapidly approaches normality and any desired percentile point is readily determined by using the Cornish–Fisher asymptotic expansion. Application of the fiducial distribution in assessing fiducial limits for normal probabilities is illustrated by examples.

I. Introduction

An important class of statistical problems arising in connection with the normal distribution is concerned with questions which involve a function of the two parameters, ξ the mean, and σ the standard deviation. Two types of problem belong to this class:

1. A probability is assigned, and hence the corresponding standard normal deviate α; fiducial limits are to be determined for the corresponding variate value, $\xi+\alpha\sigma$, when there is available a sample of N independent observations from which estimates of ξ and σ may be made.
2. A variate value of a particular normal distribution is assigned; fiducial limits are to be determined for the corresponding probability when there is available a sample of N independent observations, yielding estimates of ξ, σ, and α.

Our principal objective is to consider the second of these problems.

For problems of the first type, the fiducial distribution of the particular linear function, $\xi+\alpha\sigma$, is required and this is represented, in effect, by the non-central t-distribution or its equivalent, the sampling distribution of an estimated normal deviate, which was first given by Fisher (1931) as an application of tables of the Hermitian probability functions in Mathematical Tables, volume 1, of the British Association. Johnson and Welch (1939) have investigated this distribution and have given illustrations of its various applications, together with tables which greatly facilitate the several tests they describe. Resnikoff and Lieberman (1957) have also given examples illustrating applications of the distribution, and tables of the density function, the probability integral, and some percentile points. In illustrations of problems of the first type above, both groups of authors correctly apply the non-central t-distribution.

* Division of Mathematical Statistics, CSIRO, University of Adelaide.

For comparison with the fiducial distribution to be derived below, the sampling distribution of a normal deviate is stated in the form given by Fisher (1931). For a random sample of independent observations $x_1, x_2, \ldots, x_N$ from a normal distribution with a mean ξ and variance σ^2, Fisher used the maximal likelihood estimates

$$\bar{x} = N^{-1}S(x), \qquad s^2 = N^{-1}S(x-\bar{x})^2,$$

of the mean and variance respectively, and estimated the standard normal deviate

$$\alpha = (x-\xi)/\sigma$$

by

$$a = (x-\bar{x})/s,$$

where x is any observation. The distribution of a is then

$$\frac{\Gamma(N)}{2^{\frac{1}{2}(N-3)}\Gamma(\frac{1}{2}(N-1))}\exp\left(\frac{-\frac{1}{2}N\alpha^2}{1+a^2}\right)\frac{1}{(1+a^2)^{\frac{1}{2}N}}I_{N-1}\left(\frac{-\alpha a N^{\frac{1}{2}}}{(1+a^2)^{\frac{1}{2}}}\right)da, \tag{1}$$

where

$$I_m(x) = \{(2\pi)^{\frac{1}{2}}\Gamma(m+1)\}^{-1}\int_0^\infty y^m \exp\{-\tfrac{1}{2}(y+x)^2\}\,dy.$$

Note that this distribution is independent of the parameters ξ and σ and, when $\alpha = 0$, it reduces to

$$\{\pi^{-\frac{1}{2}}\Gamma(\tfrac{1}{2}N)/\Gamma(\tfrac{1}{2}(N-1))\}(1+a^2)^{-\frac{1}{2}N}\,da,$$

since

$$I_{N-1}(0) = 2^{-\frac{1}{2}(N+1)}/\Gamma(\tfrac{1}{2}(N+1)),$$

and consequently

$$a(N-1)^{\frac{1}{2}}$$

is distributed in Student's distribution with $N-1$ degrees of freedom.

Problems of the second type occur frequently and their solutions may often be of very considerable practical importance. Thus for several reasons, the most important of which is that on physical, chemical, biological, economic, engineering, or other grounds, variate values having definite significance in relation to the particular questions under consideration may be assigned, and we then have to estimate the proportion of individuals in the population with values exceeding these assigned amounts. For example a manufactured article will be defective, and therefore valueless, when a certain measurement, its length say, exceeds a specified amount. The manufacturer is vitally interested, for economic and other reasons, in estimating the proportion of defectives which the manufacturing process yields and, in particular, the accuracy of this estimate. Again, at a certain site we may calculate the calendar monthly rainfall which is needed to establish the break of the winter rainfall season, and it is of paramount importance to know how often the rainfall of the particular month in question can be expected to be greater than, or equal to, the specified quantity of rain, and how accurate this estimate is. The fiducial distribution of the standard normal deviate, which is derived below, correctly supplies the solution of this type of problem.

II. The Fiducial Distribution of α

Fisher (1941, 1959) has referred to the fiducial distribution of α but has not stated it explicitly. For values of $\bar{x}$ and s fixed by the sample observations, and a fixed by $\bar{x}$, s, and the variate value x, say, assigned in advance, $\alpha N^{\frac{1}{2}}$ is expressible as the sum of a standard normal variate and a multiple a of an independent χ-variate with $N-1$ degrees of freedom. Thus, if z is a standard normal variate,

$$\alpha N^{\frac{1}{2}} = z + a\chi \,, \tag{2}$$

and consequently, the distribution of α may be expressed in the form

$$\frac{N^{\frac{1}{2}}\,\Gamma(N-1)}{2^{\frac{1}{2}(N-3)}\,\Gamma\left(\frac{1}{2}(N-1)\right)} \exp\left(\frac{-\frac{1}{2}N\alpha^2}{1+a^2}\right)\left(\frac{1}{(1+a^2)^{\frac{1}{2}(N-1)}}\right) I_{N-2}\left(\frac{-\alpha a N^{\frac{1}{2}}}{(1+a^2)^{\frac{1}{2}}}\right) \mathrm{d}\alpha \,, \tag{3}$$

independently of the parameters ξ and σ. When $a = 0$, the distribution of α is

$$(2\pi)^{-\frac{1}{2}}\, N^{\frac{1}{2}} \exp(-\tfrac{1}{2}N\alpha^2)\,\mathrm{d}\alpha \,,$$

since $I_{N-2}(0) = 2^{-\frac{1}{2}N}/\Gamma(\frac{1}{2}N)$, so that α is normally distributed with zero mean and variance $1/N$ for all admissible values of N.

Since z and χ are independent and a is fixed, the cumulants of α are readily obtainable from those of z and χ, giving

$$\kappa_r(\alpha) = \kappa_r(z) + a^r\,\kappa_r(\chi)\,.$$

In particular,

$$\kappa_1(\alpha) = a\,\kappa_1(\chi)\,,$$

$$\kappa_2(\alpha) = 1 + a^2\,\kappa_2(\chi)\,,$$

and for $r > 2$

$$\kappa_r(\alpha) = a^r\,\kappa_r(\chi)\,.$$

All measures of departure from normality, $\gamma_r(\alpha)$, increase from 0 to $\gamma_r(\chi)$ as a increases from 0 to ∞ and are therefore always less than those of χ, which themselves are small enough to regard the distribution of χ as close to normal, even when the number of degrees of freedom is small. Under these conditions, percentile points of the fiducial distribution of α, for a given value of a, may be very accurately represented by the Cornish–Fisher asymptotic expansion, and this has been used herein as far as the sixth adjustment, based on the eighth cumulant (Fisher and Cornish 1960).

The last line of Table 1 gives the first four cumulants of the distribution of χ for 127 degrees of freedom and the values of γ_1 and γ_2 for comparison with the remaining entries, which give the first four cumulants, with corresponding values of γ_1 and γ_2, of the distribution of α with $N = 128$ for a range of values of a. The percentile points of the distribution of α with $N = 128$ are listed in Table 2 for the same values of a as used in Table 1. Table 3 gives the percentile points of the distribution of a with $N = 128$ for values of α equal to the values of a in the range of Table 2. Tables of the percentile points of the distributions of α and a, of form similar to that of Tables 2 and 3 and covering a comprehensive range of degrees of freedom, are at present under construction.

III. Illustrative Examples

The record for June rainfall at Adelaide may be employed to illustrate the application of the distribution of α. The rainfall distribution is positively skewed with an arithmetic mean 2·91 in., and we may assume that the square root of rainfall is sufficiently near normal for our purposes. With rainfall measured in points, the mean and standard deviation of the transformed metric calculated from a sample of 128 observations are respectively 16·4421 and 4·5669.

Table 1

CUMULANTS AND MEASURES OF DEPARTURE FROM NORMALITY OF THE DISTRIBUTION OF α FOR ASSIGNED VALUES OF a, WITH CUMULANTS AND MEASURES OF DEPARTURE FROM NORMALITY OF THE DISTRIBUTION OF χ

a	κ_1	κ_2	κ_3	κ_4	γ_1	γ_2
0	0	1·0	0	0	0	0
0·5	5·62363290	1·12475297	0·00277838	0·00000073	0·00232919	0·00000058
1·0	11·24726581	1·49901189	0·02222702	0·00001172	0·01211082	0·00000521
1·5	16·87089871	2·12277676	0·07501618	0·00005931	0·02425484	0·00001316
2·0	22·49453161	2·99604757	0·17781613	0·00018744	0·03428847	·00002088
2·5	28·11816452	4·11882433	0·34729714	0·00045762	0·04154716	0·00002698
3·0	33·74179742	5·49110703	0·60012945	0·00094893	0·04663963	0·00003147
3·5	39·36543032	7·11289568	0·95298335	0·00175801	0·05023607	0·00003475
4·0	44·98906322	8·98419028	1·42252908	0·00299909	0·05282539	0·00003716
χ*	11·24726581	0·49901189	0·02222702	0·00001172	0·06305432	0·00004705

* With 127 degrees of freedom.

From considerations based on Prescott's empirical formula, quoted, for example, by Prescott, Collins, and Shirpurkah (1952), the June rainfalls that are needed

(1) to establish a break in the winter rainfall season,

(2) to permit drainage through bare soil, and

(3) to balance evapotranspiration,

are respectively 75, 125, and 200 points. Taking the deviations of the square roots of these quantities from 16·4421 and standardizing with the standard deviation yields the estimated normal deviates −1·7039, −1·1521, and −0·5036, to which correspond, respectively, the estimated probabilities of exceeding these specified rainfalls (from the normal probability integral) 0·956, 0·875, and 0·693. To obtain, for example, the 0·95 and 0·99 fiducial limits for the true probabilities corresponding to these estimates, we first calculate the fiducial limits for α from the distribution (3), with $N = 128$ and the parameter a assigned the values −1·7039, −1·1521, and −0·5036, and then take the normal probabilities corresponding to the several values of α. Table 4 summarizes the results. Then, for example, the probability that June rainfall will be greater than, or equal to, 75 points is estimated to be 0·956, and the

TABLE 2

PERCENTILE POINTS OF THE DISTRIBUTION OF α FOR ASSIGNED VALUES OF a, WITH $N = 128$

The tabulated entries are the values of α_0 such that $P(\alpha > \alpha_0) = \epsilon$

a	Probability ϵ														
	0·999	0·995	0·990	0·975	0·950	0·900	0·750	0·500	0·250	0·100	0·050	0·025	0·010	0·005	0·001
0	−0·273	−0·228	−0·206	−0·173	−0·145	−0·113	−0·060	0·000	0·060	0·113	0·145	0·173	0·206	0·228	0·273
0·5	0·208	0·256	0·279	0·313	0·343	0·377	0·434	0·497	0·560	0·617	0·651	0·681	0·715	0·739	0·787
1·0	0·662	0·717	0·743	0·783	0·816	0·856	0·921	0·994	1·067	1·133	1·173	1·207	1·247	1·274	1·330
1·5	1·098	1·162	1·194	1·240	1·280	1·326	1·404	1·491	1·578	1·657	1·704	1·745	1·793	1·826	1·894
2·0	1·523	1·599	1·636	1·691	1·738	1·793	1·885	1·987	2·091	2·185	2·241	2·291	2·348	2·387	2·468
2·5	1·942	2·030	2·074	2·137	2·192	2·256	2·364	2·484	2·606	2·716	2·782	2·840	2·908	2·954	3·050
3·0	2·357	2·458	2·508	2·581	2·644	2·718	2·842	2·981	3·121	3·249	3·326	3·393	3·471	3·525	3·636
3·5	2·769	2·884	2·940	3·023	3·095	3·179	3·319	3·477	3·637	3·783	3·871	3·947	4·036	4·097	4·224
4·0	3·179	3·308	3·371	3·464	3·545	3·638	3·796	3·974	4·154	4·318	4·416	4·502	4·603	4·672	4·814

TABLE 3

PERCENTILE POINTS OF THE DISTRIBUTION OF a FOR ASSIGNED VALUES OF α, WITH $N = 128$

The tabulated entries are the values of a_0 such that $P(a > a_0) = \epsilon$

α	Probability ϵ														
	0·999	0·995	0·990	0·975	0·950	0·900	0·750	0·500	0·250	0·100	0·050	0·025	0·010	0·005	0·001
0	−0·280	−0·232	−0·209	−0·176	−0·147	−0·114	−0·060	0·000	0·060	0·114	0·147	0·176	0·209	0·232	0·280
0·5	0·225	0·270	0·292	0·324	0·352	0·385	0·440	0·503	0·568	0·628	0·664	0·697	0·736	0·762	0·819
1·0	0·699	0·746	0·770	0·805	0·836	0·872	0·934	1·006	1·082	1·153	1·197	1·236	1·283	1·316	1·387
1·5	1·152	1·206	1·233	1·273	1·309	1·351	1·424	1·509	1·600	1·686	1·739	1·787	1·845	1·886	1·973
2·0	1·593	1·656	1·687	1·734	1·776	1·825	1·911	2·013	2·120	2·223	2·288	2·346	2·416	2·465	2·570
2·5	2·027	2·100	2·136	2·191	2·239	2·297	2·397	2·516	2·643	2·764	2·840	2·908	2·991	3·049	3·174
3·0	2·457	2·540	2·582	2·645	2·700	2·767	2·882	3·019	3·166	3·306	3·394	3·474	3·570	3·637	3·782
3·5	2·884	2·978	3·026	3·097	3·160	3·235	3·367	3·523	3·689	3·849	3·950	4·041	4·150	4·227	4·393
4·0	3·310	3·415	3·468	3·548	3·619	3·703	3·851	4·026	4·213	4·394	4·507	4·609	4·732	4·819	5·006

fiducial probability that the true probability lies in the fiducial range (0·924, 0·976) is 0·95. The procedure is justifiable because the percentile points of the original distribution are invariant for any transformation for which the derivative with respect to the new variable is always positive.

IV. Remarks on the Argument used by Johnson and Welch

Johnson and Welch (1939) proposed that the non-central t-distribution should be used to solve problems of type 2 (Section I). In their notation z is a standard normal variate and χ is distributed independently of z with f degrees of freedom. Non-central t is defined by

$$t = (z+\delta)/\chi f^{-\frac{1}{2}},$$

where δ is a constant. The equivalences of the two systems of notation are

$$a = t/f^{\frac{1}{2}} \qquad \text{and} \qquad \alpha = \delta/(f+1)^{\frac{1}{2}}.$$

Table 4

FIDUCIAL LIMITS OF α AND CORRESPONDING FIDUCIAL LIMITS OF A NORMAL PROBABILITY

Rainfall (points)			0·95 Fiducial Limits for α		0·99 Fiducial Limits for α		0·95 Fiducial Limits for Probability		0·99 Fiducial Limits for Probability	
	a	P	α_1	α_2	α_1	α_2	P_1	P_2	P_1	P_2
75	—1·7039	0·956	—1·4308	—1·9742	—1·3473	—2·0614	0·924	0·976	0·911	0·980
125	—1·1521	0·875	—0·9270	—1·3744	—0·8576	—1·4456	0·823	0·915	0·804	0·926
200	—0·5036	0·693	—0·3188	—0·6867	—0·2011	—0·7440	0·625	0·754	0·603	0·772

The probability that $t > t_0$ is represented by $P(t > t_0 \mid f, \delta)$. In problems of type 2, t_0 has an assigned value and $\delta = \delta(f, t_0, \epsilon)$ has to be determined so that

$$P\{t > t_0 \mid f, \delta(f, t_0, \epsilon)\} = \epsilon.$$

This analytic problem cannot be approached directly so they use the fact that the inequality

$$t = (z+\delta)/\chi f^{-\frac{1}{2}} > t_0$$

is equivalent to the inequality

$$-z+(t_0/f^{\frac{1}{2}})\chi < \delta,$$

and hence that the statement

$$P\{t > t_0 \mid f, \delta\} = \epsilon$$

is equivalent to the statement

$$P\{Y = -z+(t_0/f^{\frac{1}{2}})\chi < \delta\} = \epsilon, \tag{4}$$

thus basing their solution on the variate Y, the percentile points of which supply the required values of δ. The variate Y also has the advantage of being nearly normally distributed with known cumulants, and an approximation based on the assumption that Y is normally distributed is presented as the third of a series of increasing accuracy. Finally, in Tables IV and V of their paper Johnson and Welch present exact values which have been calculated using the Cornish–Fisher asymptotic expansion of the percentile points of Y corresponding to chosen probabilities ϵ.

On the other hand, consider the problem of determining the fiducial distribution of α, given a particular value of a. Since the true deviation of any value x of the original variate in terms of the true standard deviation, and the apparent deviation in terms of the estimated standard deviation are related by

$$x = \bar{x}+as = \xi+\alpha\sigma,$$

the fiducial variate α is given by

$$\alpha N^{\frac{1}{2}} = z+a\chi,$$

where χ has $N-1$ degrees of freedom, and a percentile of $\alpha N^{\frac{1}{2}}$ (corresponding to a probability ϵ) is obtained by solving for $\alpha N^{\frac{1}{2}}$ the equation

$$P\{z+a\chi < \alpha N^{\frac{1}{2}}\} = \epsilon.$$

This equation is similar in form to (4) but the variate $z+a\chi$ now has a concrete interpretation as the fiducial variate $\alpha N^{\frac{1}{2}}$. The fortunate circumstance to which Johnson and Welch refer (1939, p. 388) is not the near normality of the variate Y but the similarity of this variate and the fiducial variate $\alpha N^{\frac{1}{2}}$, which follows from the close relation that exists between the definitions of non-central t and the fiducial variate.

V. Acknowledgment

The author is indebted to his colleague Dr. G. W. Hill for the programming and computation involved in the preparation of Tables 1, 2, 3, and 4.

VI. References

Fisher, R. A. (1931).—"British Association Mathematical Tables." Vol. 1, pp. XXVI–XXXV. (British Association: London.)

Fisher, R. A. (1941).—*Ann. Eugen.* **11**, 141–72.

Fisher, R. A. (1959).—"Statistical Methods and Scientific Inference." 2nd Ed. (Oliver & Boyd: Edinburgh.)

Fisher, R. A., and Cornish, E. A. (1960).—*Technometrics* **2**, 209–26.

Johnson, N. L., and Welch, B. L. (1939).—*Biometrika* **31**, 362–89.

Prescott, J. A., Collins, J. A., and Shirpurkah, G. R. (1952).—*Geogrl Rev.* **42**, 118–33.

Resnikoff, G. J., and Lieberman, G. J. (1957).—"Tables of the Non-central t-Distribution." (Stanford Univ. Press.)